MINING : CHALLENGES OF THE 21ST CENTURY

Proceedings of the International Conference on
Mining—Challenges of the 21st Century
New Delhi 25-27 November, 1999

MINING :
Challenges of the 21st Century

Editors
Prof. A.K. GHOSE
Prof. B.B. DHAR

A.P.H. PUBLISHING CORPORATION
7 ANSARI ROAD, DARYA GANJ
NEW DELHI-110 002

Published by
S.B. Nangia
A.P.H. Publishing Corporation
4435–36/7, Ansari Road, Darya Ganj,
New Delhi-110002
Phone: 011–23274050
e-mail: aphbooks@gmail.com

ISBN 81-7648-158-0

2026

Rs. 11995/-

Typesetting at
NEW APCON
25/2, Panchsheel Shopping Centre
New Delhi - 110 017
☎ 6490802
6494336

Printed in India at
Efficient Offset Printers
New Delhi-110 035

Preface

Mining Industry worldwide is under siege. Even as the new millennium beckons, a veritable war is being waged against the industry by environmental and social activist lobbies and increasingly strident criticism labelling the industry as an exploitive business and a despoiler of the environment is reverberating throughout the world. Such negative image, volatile commodity prices, legacy of a ravaged environment in some mining locations and lack of access to land for exploitation and mine development, have added to the woes of the mining industry.

That mining industry's survival is crucial to the sustainability of the spaceship earth is not disputed. But the burden of carrying a tarnished image and the emerging issues of environment and social responsibility, have exacerbated the problems of an already beleaguered industry, which has not been very pro-active thus far. The biggest challenge facing the global minerals sector is "issues management". Environmental responsibility, social and ethnic issues and sustainability *per se* pose serious threats to its long-term future.

The papers collated in this volume is a response of the global mining industry articulating the concerns, and mark an attempt at advertisement advocacy seeking to showcase the mining industry's innovative approaches in environmental, technological and management arenas, espousing the best practices for survival. Technology, will provide the piggyback for transitting into the new millennium and implementing a turnaround, protecting the bottomline of economic, social and environmental values. It is only through building of coalitions and global partnership and through new initiatives on an agreed agenda, can the mining industry reinvent itself and stave off the impending doom.

We thank the authors for their perceptive contributions. We must acknolwedge the cooperation received from Mr. S.B. Nangia of APH Publishing Corporation in printing this proceedings volume.

New Delhi
October, 1999

Prof. A.K. Ghose
Prof. B.B. Dhar
Editors

Contents

SECTION III : TECHNOLOGICAL CHALLENGES

SECTION IV : MINERAL EXPLORATION AND EXPLOITATION

SECTION VII : DEMAND & SUPPLY PERSPECTIVES AND FINANCING MINERAL SECTOR

SECTION VIII : R&D INITIATIVES

SECTION IX : INFORMATION TECHNOLOGY IN MINERAL INDUSTRY

POSTER SESSION PAPERS

Section I

GLOBAL MINERAL INDUSTRY—AN OVERVIEW

CHAPTER 1

Occupational health and safety in Mining: Some issues for the 21st Century

*Norman S. Jennings**

INTRODUCTION

Work-related accidents and diseases continue to be serious problems throughout the world. Workers suffer about 250 million accidents a year and at least 335,000 die as a result of accidents at work. Moreover, there are about 160 million cases of occupational diseases. Taking accidents and diseases together, the result is about 1 million work-related deaths each year. The consequential economic losses are enormous and the social damage to shattered families and communities is incalculable. About 4% of global GDP is spent on treatment, disability payments and survivors' benefits. This is more than the ODA to the world's developing countries. The situation is compounded in many developing countries due to material shortages and inadequate medical facilities. Unfortunately, mining takes more than its fair share of accidents and disease when the number of persons employed is taken into account.

There is a wide difference in occupational health and safety performance around the world, with fatality rates in highly industrialized countries about half those in Central and Eastern Europe, China and India. Some hazardous jobs, including in mining, can be 10-100 times riskier than the norm.

There has been a clear and steady decrease in the serious injury rate in industrialized countries as a result of changes in the nature of work and following sustained improvements in achieving a safer and healthier workplace — including the provision of first-aid and emergency care which can be crucial in saving lives after an accident. But the evolving nature of work is generating new occupational hazards — musculo-skeletal problems, stress and mental problems, asthmatic and other allergic reactions, and problems caused by exposure to hazardous and carcinogenic agents such as asbestos, chemicals and radiation.

* Senior Industrial Specialist, International Labour Office, Geneva, Switzerland.

An important starting point in improving safety performance in mining is the rejection of the notion that it is inherently dangerous and therefore work-related diseases and a certain number of mine accidents are inevitable, and that disasters are more "acts of God" which are bound to occur from time to time. Until this perception is put aside it will be difficult for those involved in mining to persevere with sustainable improvements in occupational safety and health.

A watershed in the international treatment of mine safety and health was reached in June 1995 with the adoption by the International Labour Conference of the Safety and Health in Mines Convention (No. 176) and its accompanying Recommendation (No. 183). This new standard is important for several reasons. *First*, because the special hazards faced by mineworkers are now dealt with in a specific instrument. *Second*, because the increasing importance of the mining industry in many countries is emphasizing the need to improve health and safety. *Third*, because earlier ILO standards on occupational safety and health, as well as existing legislation in many countries, are inadequate to deal with the specific needs of mining. The Convention has set the principle for national action on, and international monitoring of, the improvement of working conditions in the mining industry.

The Convention came into force in June 1998, one year after the first two ratifications. Eleven countries have so far ratified the Convention; several others are working towards it. A widely ratified and fully implemented Convention and Recommendation will make a material contribution to achieving significant and lasting improvements in safety and health in the mining industry. The ILO is ready to assist wherever it can in the attainment of this goal. But even before the Convention is ratified, its provisions can be included in regulations or collective agreements.

Factors that affect safe behaviour include: the design of systems, machinery, equipment; the working environment; and the organization and its relationship to the individual. Any programme to promote safe behaviour must address all three.

Experience has shown that significant improvements in occupational health and safety are not just a matter of technology; they are also a matter of culture. The progressive integration of safety principles and reflexes into the workplace is a fundamental prerequisite for any improvement in the incidence of occupational accidents, injuries and disease. Unfortunately, a safety culture and an awareness of the positive values of a safe and healthy working environment in terms of economic benefits and social justice are minimal or non-existent in many countries, particularly in smaller enterprises. The socio-cultural aspects of health and safety in mines will be to the forefront in the next millennium if improvements in performance are to be sustained.

AN INDIAN PERSPECTIVE

India is a major mining country, producing over 60 metals and minerals in tens of thousands of mines, large and small, with a workforce approaching 900,000. Ten years ago the workforce was about 1 million. DGMS data shows little change in the fatal accident and injury rates in metal mines over the last ten years. In fact, the number of deaths and serious accidents per year in metal mines has changed little since 1950. In coal mines, however, accident rates fell until about 1980 but have remained fairly static since then. The number of fatal accidents in coal mines has been 130-160 per year over a ten year period (leading to 140-240 deaths), whereas the number of serious accidents (and people seriously injured) per year has fallen from about 900 to 500 over the last ten years. Although the fatality rate per million tonnes of coal mined has fallen from 0.9 to 0.5, the rate per 10,000 people employed has remained static, at about 30. The number of mine inspections peaked in 1992, at about 10,000. In 1998 it was 7,090.

Has occupational safety in Indian mines reached a plateau? If so, what can be done to break through this ceiling? What is being done elsewhere that might be relevant in the Indian context? Privatization, foreign investment, new equipment, new work practices and a new work culture will all have an impact on safety and health in mining. What changes to legislation might be necessary? What will be the role of the regulator as the regulations and the mining industry change? How will any changes affect the duty and liability of the employer and of workers? What emphasis should there be on human rather than technical factors? On occupational health and surveillance? On risk assessment? On social dialogue? On training? These are the questions that will need to be answered throughout the mining industry — including by governments, employers and workers' organizations — if improvements in occupational health and safety are to be sustained.

TECHNICAL CHANGE

Technical and organizational changes in mining have been marked by greater mechanization and a trend towards automation, with larger concentrations of energy in plant and equipment. Greater mechanization has resulted in a move away from individual miners, each with his own job, to a complex interaction of closely coupled, tightly paced processes involving several people. Ore treatment is also more complex, with a wider use of process chemicals that incorporate hazardous substances. More parts of the production and processing function are being contracted out, changing the employment relationship and, sometimes, making responsibility less clear-cut. Rationalization of mining activities has resulted in a high turnover of mineworkers, particularly affecting older, more experienced workers, which has led to a loss of institutional memory — how similar incidents were dealt with in the past, for example. More intensive working of capital-intensive machinery, including extended shifts, has become a prerequisite for economic survival. Unfortunately, there is sometimes a gap between design standards and the requirements of legislation, especially as far as ergonomics is concerned. A 12-hour shift increases the exposure of workers to physical stressors (noise, dust, etc.) by 50% and the cumulative or synergistic effects of multiple stressors over an extended shift render fatigue proportionally greater. Unions are concerned about possible increases in fatigue levels and the impact on accident rates. But in some countries, workers have been enthusiastic about longer shifts, ignoring potential health effects, in view of the increase in free time that arises from such working arrangements. Some have even been able to get second jobs, the effects of which on the principal job are additional matters for study.

The impact of longer shifts underground on health and safety (exposure and fatigue) and on family life has not been widely studied and the ILO is funding a survey in Australia to obtain some base-line data that we hope will lead to a better understanding and to a wider assessment of this phenomenon. A related issue is the health and safety impact of "fly in-fly out" operations where intense periods of work in remote locations are followed by substantial periods of rest at "home". Such operations, similar to work on oil rigs, can have a significant impact on both working and personal lives. Changes in working time are being driven by the increasing competition in the mining industry, spurred by lower commodity prices. They are likely to be sustained.

Information technology (IT) is creating new opportunities for exchanging information, creating data bases, analysing practices and procedures and identifying where efforts should be focussed. IT can provide quick access to current legislation, codes of practice, bulletins, accident reports, etc. In other words, possible global solutions to local problems will be increasingly available, including at the workplace, so that workers will be better able to solve problems and thus increase their "ownership" of the processes concerned and the improvements made. Virtual reality is becoming a powerful training tool, with realistic computer programmes being developed for individual mines to cover a host of different scenarios, including accident prevention and analysis through 3-D simulation.

New technology has had a positive impact on the number and severity of mining accidents and diseases — an impact that has been developed and sustained over many years. But the effects have not been uniform. Some new technologies have been accompanied by new or intensified hazards — such as dust, noise, vibration, ergonomics-related problems, and electric current — that require special attention. For example, the steady increase in the size and operating speed of earth-moving equipment at surface mines has led to them being involved in the severest of accidents at some of these mines, despite an overall reduction in the number of accidents. Half the world's fleet of 10,000 mining trucks have a capacity exceeding 170 tonnes; 15% of loading equipment (but 30% of total capacity) exceeds 25 tonnes. In India, which is investing heavily in large opencast mining equipment, 75% of accidents at surface coal mines are caused by dump trucks and other transport machinery. Nearly 90% of these are due to negligent or unauthorized use. In view of the human factor, the importance of education and training and of instilling a safety culture is underscored.

HEALTH AS WELL AS SAFETY

It is fair to say that workers' health has often received less attention than their safety and that there is little awareness of the enormous cost of occupational health to much of the mining industry and to the general public. Regular health surveillance of mineworkers is by no means universal, and extreme care is needed in selecting, developing and implementing a surveillance programme. Occupational diseases (noise-induced hearing loss, vibration white finger, silicosis and pneumoconiosis) are becoming more pervasive with increased mechanization. Now-a-days, noise and vibration are major causes of sickness. The increasing use of VDUs is also having an impact on workers' health. These phenomena are by-products of more intensive working, particularly underground. Moreover, unlike silicosis or pneumoconiosis, they manifest themselves after relatively short exposure times. Stress-related illnesses are being increasingly recognized as being valid for compensation if they are proved to be work-induced. Increased work-shifts might have implications for occupational health. There have been several well-publicized court cases in which large amounts of compensation have been awarded to those suffering from the "new" diseases. An ILO code of practice on ambient factors at the workplace was published in 1999. It provides guidance on the roles and obligations of all those involved with regard to hazardous ambient factors. Moreover, infectious diseases, such as HIV, TB, cholera, malaria and typhoid, are still rife in many mining regions.

The three key areas of occupational health — testing, diagnosis and control — are being approached differently in different countries. At the national level, an approach that is fair to all is necessary; one that is practical, cost-effective and sustainable; and one that is philosophically, legally and morally acceptable and is effective in reducing exposure and mitigating the effects.

There is a shortage of trained occupational health professionals in the mining industry, particularly in developing countries. Even those that are available sometimes have problems in applying their knowledge — because of a lack of resources and/or motivation. Getting access to current occupational health information is still a problem in many areas, but there is growing international cooperation. One example is the ILO/WHO International Programme on the Global Elimination of Silicosis. International collaboration can contribute significantly to the development and strengthening of national capacities. Industry must play its part too, by collaborating with unions and other organizations, and in helping to set up and participating in occupational health and safety organizations that work to enhance occupational health and hygiene skills. As strengths are shared, weaknesses will be overcome and resources optimized. The ILO's International Programme on Chemical Safety develops, translates and disseminates clear and standardized information on the properties of chemical substances in the workplace. The recently published ILO *Encyclopaedia of Occupational Health and Safety* — in paper and CD-ROM versions — is a major source of reference for the mining and other industries.

Systems must be put in place that will identify the health problems due to, or made worse by, work activities, and establish priorities for addressing them. Relevant information should be collected and used as the basis for managing ill health. The workforce must be advised of the specific health risks they face and what to do about them. Staff should be trained to target ill health continually and apply their skills better to protect themselves. The effectiveness of actions taken should be monitored — feedback — and outputs and inputs measured. The commitment of those involved must be gained and held.

The importance of managing health risks has often been understated. In some cases the problem has been ignored because the relevant information that can be analysed is difficult to collect. This makes proper health management extremely hard. Ill health costs money, big money over many years, so there are plenty of economic, let alone humanitarian, incentives to do something about it now so that improvements in both health and safety in mines can proceed in step.

LEGISLATION AND THE ROLE OF THE REGULATOR

Legislation should provide an effective framework of standards and direction, and the regulatory authority charged with administering that legislation needs to do so in a professional, constructive and consistent manner. Effective administration by the regulator is best achieved using a combination of processes which create an interactive relationship with enterprises and their workforce, and with industry representative groups, so that enforcement sanctions rarely need to be invoked. But a *de facto* policy of non-prosecution can arise when the inspectorate is part of a ministry whose primary function is to assist companies to develop mineral resources. Moving the inspectorate to a safety body with a strong prosecution policy could ensure that offences under occupational health and safety regulations are dealt with as the regulators intended.

Prescriptive regulations are being reduced. New regulations are highlighting the responsibility and liability of mine managers and make it a regulatory requirement that they develop a variety of policies and procedures to manage health and safety and to provide a safe place to work. Voluntary initiatives on occupational health and safety are being developed in several industries and are quickly gaining acceptance.

Enforcement is not an end in itself; it must be matched by a continuing decline in preventable harm to miners. The long-term goal should be to ensure that, without exception, an individual can devote a lifetime to a mining career and emerge healthy and unharmed. As one safety slogan has it: "The most important thing to come out of a mine is the miner".

The regulator's basic objective is to ensure that the risks to the safety and health of the workforce from workplace activities are eliminated or effectively controlled. His task is to influence and advise on the achievement of minimum health and safety standards that are contained in legislation. The regulator's statutory powers mean that its advice should not be taken lightly. When a mine inspector calls, he should be regarded by the mine management not as merely an enforcer of regulations but as someone who is able to contribute to its thinking, able to interpret legislation and formal guidance sensibly and reasonably, in the particular situation and backed by scientific and technical resources that command respect. This will be increasingly important when foreign capital and highly experienced operators become increasingly used in a fast-globalizing mining industry. Thus, if the power of the inspectorate is to be sensibly deployed and effective, the quality of staff must be high and no compromises made. Regulators will continue to need a high level of technical and professional expertise but they will also need to be well equipped to deal with the increasingly important human and organizational factors of mining. The educational and advisory role of the regulator is growing as enterprises realize they can benefit from the regulator's overview of the industry and access to data. The regulator can also give advice on good or best practice.

The historical perception of the regulator managing safety at the mine has been put to rest with the recognition that the regulator's purpose is to be achieved largely through others — industry management, the workforce, and a range of intermediary groups such as industry associations, trade unions and suppliers.

CONVENTION NO. 176

Mining is of major economic importance in many countries, but it is often dangerous. Yet, before 1995, the existing ILO standards on occupational safety and health, as well as the legislation in a good number of countries, were inadequate to deal with the specific needs of mining for the prevention of occupational accidents and diseases. An important landmark in the ILO's continuing work on protecting workers against sickness, disease and injury arising out of their employment was, therefore, the adoption in June 1995 of the Safety and Health in Mines Convention (No. 176) and its accompanying Recommendation (No. 183).

The instruments cover both underground and surface sites where exploration and mineral extraction activities take place; they apply to all types of mining enterprises. Member States of the ILO which ratify the Convention undertake to adopt legislation for its implementation, including the provision for inspection services and the designation of the competent authority to monitor and regulate the various aspects of safety and health in mines. The Convention also sets out procedures for reporting and investigating disasters, accidents and dangerous occurrences related to mines, and for the compilation of the relevant statistics.

The Convention includes requirements for the provision of mine rescue, first-aid and other medical facilities. It makes it obligatory to provide self-rescue respiratory devices for workers in underground coal mines. It covers the manufacture, storage, transport and use of explosives and initiating devices. Furthermore, national legislation has to make provision for the safe storage and disposal of waste and for protective measures for abandoned workings.

Under the Convention, mineworkers have the right to report accidents, request inspections, obtain information and remove themselves from any location where there appears to be a serious danger to their safety or health. Workers' representatives have the right to consult the employer on their safety and participate in inspections and investigations undertaken at the workplace. A series of Articles requires employers to take all the necessary measures — in order of priority — to eliminate, control or minimize the risks to the safety and health of workers in mines.

The Convention provides a floor — the minimum safety requirement against which all changes to mine operations should be measured. The need for continued vigilance remains paramount during the substantial changes that are taking place in mining world-wide. Positive results should be examined and, where appropriate, replicated through the concerted efforts of all concerned parties. This implies a higher degree of information exchange and collaboration than has been the norm. The ILO could facilitate the collection and dissemination of examples of "best practice" in improving mine safety.

The ILO does not sit idly by after a Convention has been adopted. Under the terms of the ILO's Constitution, the government of each country that has ratified a Convention agrees to make a regular report to the ILO on the measures it has taken to give effect to the provisions of the Convention to which it is a party. Complaints concerning the application of the Convention can be made to the Governing Body of the ILO, in which case an investigation is undertaken in an attempt to resolve alleged shortcomings in the application of a ratified Convention.

The ILO's work to promote the ratification of Conventions and the application of its standards involves both its supervisory bodies and the provision of assistance directly to constituents. The ILO provides advice to help find solutions to difficulties in the application of standards, promotes the ratification of new standards, and it provides training in complying with the reporting procedures and assistance in the formulation and adoption of new labour legislation in conformity with international labour standards.

HUMAN FACTORS

By ensuring that the hazards in a physical working environment are well controlled, and having in place sound systems and procedures for the management of safety and health bring an organization only so far in protecting its workforce. In order to break through the ceiling, employees' behaviour must be the focus of attention. Behavioural safety is the key to effective accident reduction. The past focus on engineering and technical developments as a means to improve occupational health and safety was very successful. But such an approach is providing diminishing returns in many countries. There is, therefore, a need to provide officials, management and exposed workers with a better understanding of the different categories of human behaviour and performance-shaping factors that lie behind them. Without such an understanding, people will still be blamed for things they could not have done otherwise, and preventive measures will continue to be misdirected.

Effective safety management involves people. The interaction between the technical and social impact of systems needs to be handled through social dialogue at the enterprise level, as well as industry-wide, involving the workforce through its representatives. The workforce should also be involved in designing systems of work and in planning, organizing and carrying them out. All parties must act in good faith and not use occupational health and safety as a lever for other ends.

SAFETY CULTURE

Increasing awareness of the limitations of the culture of "blame" as far as health and safety are concerned has given rise to the promotion of a positive safety culture, including its incorporation in regulations. No matter how safety culture is defined, the best way to improve it is one step at a time, building on small successes and allowing a change in circumstances to create a change in attitude. In other words, identify circumstances that predispose unsafe behaviour and change them, rather than attempt to change behaviour directly.

Changing workplace culture sounds easy, but there are plenty of pitfalls. These include: looking for a "quick fix" — there is no such thing; setting too many goals or having objectives that workers cannot achieve; being inconsistent; having a lack of management commitment; and neglecting training, follow-up and perseverence.

Personal motivation is important if health and safety programmes are to succeed. A clear and credible answer to the question "What's in it for me?" must be provided. Everyone should know what the return on their investment in health and safety will be.

When an accident happens, the focus is on the person who triggered it. But designers, managers, supervisors, trainers and risk analysts are also part of human behaviour and affect the degree of risk at the workplace. Together their behaviour sets the agenda for safety culture and worker behaviour reflects it. Efforts to influence workers' risk behaviour must start by analysing safety culture and its effect on accidents.

Many companies, including in the mining industry, have safety "core beliefs" that have been developed and agreed through social dialogue. They form part of the contract of employment and influence behaviour from top to bottom of an enterprise. Examples of these core elements that are necessary for creating and maintaining a safety culture include:

- All fatalities, injuries and occupational diseases are preventable;
- No task is so important that it cannot be done safely;
- All hazards can be identified and their risks managed;
- Everyone has a personal responsibility for the safety and health of themselves and others;
- Safety and health performance can always be improved;
- Safety is a condition of employment;
- Training is an essential element of safe workplaces;
- Safety audits must be conducted to ensure compliance with standards;
- All injuries, unsafe practices and incidents with injury potential must be investigated and all deficiencies must be corrected promptly;
- Safety of the job is as important as safety on the job;
- It is good business to prevent injuries and illnesses;
- People are the most critical element in the success of a safety and health programme.

The achievement of a safety culture needs a genuine and visible commitment and leadership from the top, with health and safety being a line management responsibility. There should be an acceptance that improving health safety performance is a long-term goal that requires sustained effort and interest. To this end, there should be sound codes of practice and health and safety standards and a commitment to ensure that health and safety have adequate resources. Managers at all levels must regularly assess performance, and the factors that influence the behaviour of managers, supervisors and workers must be properly managed. The ILO proposes to develop guidelines on occupational safety and health management as an important plank in the development and promotion of a safety culture.

The long-term implementation of good safety, health and environmental practices — *i.e.* the continuous integration of a safety culture as an essential part of more general social cultures — is the only way to curtail the spiralling costs of health care delivery and environmental protection while increasing productivity. A safety culture can be built only through a long-term process of awareness-raising and education; consultation and consensus building among the social partners and other stakeholders involved in occupational health and safety and environment issues, and the participation of national economic and financial institutions.

The growth of a safety culture depends on the availability of reliable information as a tool for good decision making and the sound management of resources. Despite continuing efforts, the methods for collecting and interpreting occupational accident and diseases statistics are not well-developed, much less internationally harmonized and comparable. Data are essential for the establishment of international and national surveillance systems that are capable of providing early warnings of new occupational and environmental hazards. Data are also necessary for the definition of the indicators needed to measure the success or failure of occupational health, safety and environment programmes and to identify the areas where action may be needed. Sound data are also a prerequisite for targeting scarce resources and for developing best practices.

For comparisons to be made between different operations, a common definition of disabling injuries is needed, together with common indicators so that safety performance can be monitored and assessed. Safety is not competitive and so information should be shared, for example through an international safety network that would enable knowledge to be pooled, best practices to be identified, research to be carried out and information disseminated. While there is already collaboration among research agencies, the Internet provides a wide gateway for much broader comparison and collaboration.

Traditionally, prevention and protection actions have had relatively low priority in terms of investment and financing. This is partly due to the difficulty in measuring accurately the costs of accidents and disease. In purely money terms, the direct costs of a few deaths might matter little to a large company in the short-term. There could even be short-lived economic benefits in not devoting resources to preventing rare or catastrophic events, or even to preventing occupational diseases where, because of the delay between exposure and the onset of sickness, the discounted cost of investment in protection exceeds the eventual savings in reduced compensation. But such an approach is anathema to any industry with a long-term perspective. Nonetheless, safety and health need to be ensured through regulation and a safety culture.

SAFE WORK

In order to help meet the need for increased action to protect safety, health and the environment, the ILO is developing a global programme on "safe work for all". Targeting hazardous occupations, including mining, the objectives of the programme are, *inter alia,* to:

- Ensure that labour protection issues are given increased attention nationally and internationally;
- Ensure the ILO's constituents can develop preventive policies and programmes;
- Establish a global technical cooperation programme on occupational health, safety and the environment.

Through advocacy campaigns, statistical programmes, national programmes of action, training programmes and international exchanges of information, this new programme will create alliances and partnerships to promote government action and support national action to protect workers' health and welfare.

RISK MANAGEMENT AND RISK ASSESSMENT

Since human behaviour holds the key to further improvements, greater effort will need to be directed towards promoting an understanding of risk assessment, risk management and safety audits that are specifically based on the behavioural contribution to accidents. For example, through the development of management tools that allow a proactive approach to be taken to address behavioural aspects of safety performance. These include the identification of the potential for human error; examining attitudes to safety; training; equipment design and maintenance; and moving towards effective solutions.

Managing risks involves much more than undertaking a risk assessment, which itself does not improve safety and health performance. An overlying management system must be in place in order to undertake risk assessments effectively and, therefore, improve health and safety performance. In other words there must be a system that manages risk management.

The purpose of introducing risk management is to improve health and safety. The challenge is how to introduce risk assessment in a well thought out structured approach that is understood and practised by all, and is seen to be beneficial. The primary purpose of a systematic risk assessment of existing operations is the comprehensive identification of the hazards associated with a particular task or operation, the risks associated with those hazards and a list of appropriate risk control measures.

Improvements that are based on the introduction of risk assessment involve four issues:

- Ability to conduct comprehensive, systematic, high quality risk assessments;
- Creation of the infrastructure and having trained risk assessment teams;
- Creation of the process that leads to a widespread active involvement of staff and workers;
- Development of a management structure that ensures an effective speedy rational implementation of improved or extra controls that will lead to lower risk.

The accuracy of the decision as to whether the risks are adequately controlled is largely subjective, and is likely to remain so for a while. But where crucial safety decisions are being made on the basis of subjective judgements, it is essential that the decision incorporates all the necessary information, including consideration of the human factors.

It is inadequate to say that a risk is controlled by existing codes, rules and procedures if no one has checked whether they are practical; whether they are known and understood; and whether they are followed. Similarly, there is no defence against unsafe behaviour in the untested assumption that the training provided is adequate. Safety audits — used to back up and check risk assessment action — that simply check whether provision is made without addressing its effectiveness, are equally inadequate in promoting safe behaviour. The increasing practice of using accredited third party auditors who undertake structured audits of safety management systems and core risks, if carefully managed, can facilitate analysis and provide a sound basis for improved behaviour and performance.

EDUCATION AND TRAINING

The examination of the ability of staff to behave safely must be an integral part of risk assessment and other safety management tools such as audits. There is no point in advocating the importance of safety behaviour if it is not feasible to behave safely.

Education and training are the foundations for enabling improvement in occupational health and safety. Sound basic education for work, certified induction and refresher training and certified company-specific training as part of lifelong learning are crucial for the implementation of effective health and safety hazard prevention programmes, supplemented by specific seminars and refresher training and improvement programmes — for trainers and for supervisors and workers.

There is a clear link between poor or inadequate education and training and poor health and safety performance. Competence is increasingly being used as the basis for safe operations. Competence — a person's ability to perform at a satisfactory level in the workplace, including the person's ability to transfer skills and knowledge to new situations and achieve agreed outcomes — defines what people need to do and to what standard at every level in the industry or in an enterprise. Competency-based training has facilitated the introduction of health and safety competencies into various industry competency standards. This has proved to be a fresh approach to certification linked to career progression.

CONCLUSIONS

As health and safety regulations become less prescriptive and more emphasis is put on the human factors that affect health and safety performance, a key objective in the 21st century will be to instil a strong safety culture in all aspects of mining. An improved safety culture is partly a question of resources and technology. But above all, it requires better information, management and higher ethical standards in confronting the ever present, ever evolving changes at the workplace. Different strategies will be needed to meet different circumstances and degrees of economic and technological development, without compromising the final goal. Because the types of issues that predispose unsafe behaviour and their relative importance will vary from mine to mine, there is no single solution to the promotion of safe behaviour. The main strategies involve: compelling; facilitating; rewarding; training; informing; and participating. Also, it is clear that many issues arising from a more detailed, systematic evaluation of the factors influencing safe behaviour are unlikely to be immediately obvious. If they were, they would have been dealt with.

Careful data collection and analysis will enable risks to be identified and managed and information shared both nationally and internationally. The fostering of closer ties between manufacturers, regulators, employers, workers' representatives and the medical and safety professions, leading to meaningful social dialogue, will pave the way to break through the ceiling that has led to improvements in health and safety in mining becoming stalled in many mining operations.

Safety, health and environmental issues must be considered as an integral part of other mainstream social priorities, such as health, conditions of work, employment, training, industrial relations, social security, etc. The pressures on the world of work, be they driven by financial, technological, environmental or social factors of globalization, can only be dealt with by integrated management approaches at all levels.

REFERENCES

1. Jennings, Norman S.: "Improving safety and health in mines: The role of Convention 176". Paper presented to Mine Safe International 1996, Perth, Australia, 9-13 Sept. 1996.
2. *Journal of Occupational Health and Safety* (Sydney, CCH), various issues.
3. *Mining Magazine* (London, Mining Journal Ltd.), various issues.
4. Mine Safe International 1998: *Proceedings of Mine Safe International 1998*, Sun City, South Africa, 28 Sept. - 2 Oct. 1998.
5. Takala, J.: "Introductory report of the International Labour Office". Paper presented to 15th World Congress on Occupational Safety and Health, São Paulo, Brazil, 12 Apr. 1999.

CHAPTER 2

Issues for Global Mineral Industry : An Agenda of Problems and Opportunities in the New Millennium

*Ajoy K. Ghose**

ABSTRACT

The global mining industry is a major contributor to the material needs and economic well-being of the industrialized world with a contribution of over 6% of the global world product of some U.S.$ 30 trillion. On the eve of the 3rd millennium, the mineral industry worldwide is beleaguered by a wide array of problems, not the least of which are the environmental and social issues and the overall issue of sustainable development, as mineral resources are depletable and non-renewable. The paper seeks to present on a broad platter the emerging trends, vital issues and concerns that are likely to affect the global minerals sector in the coming millennium and the required responses of the mineral industry through technological and managerial innovations. In effect, this paper outlines the perceived issues of an essentially dynamic and changing world of tomorrow where new issues will be created and current issues altered over time and may even disappear and the mining industry has to evolve robust adaptive strategies for navigating through the challenges of change.

INTRODUCTION

On the eve of the new millennium, it is appropriate that we reflect on the future of the mining industry and its sustainability. Minerals and energy are factors of central significance in the sustainability of society and the world at large. We are operating today within the confluence of extraordinary change and the global mining industry has to navigate successfully through the turbulence of change in a hyper-competitive environment to be able to meet the burgeoning material needs of the society (Ghose & De, 1994). There have been strident voices raised in the past decade on the environmental damages wrought allegedly by mining; economic recession, depressed commodity prices and large-scale

---

* Chairman, Indian National Committee of World Mining Congress, and Vice Chairman, International Organizing Committee of World Mining Congress.

closures and lay-off in the mining industry worldwide have even prompted the doomsday predictors to dub mining as a "sunset industry". Mining is, was and shall continue to be the keystone of mankind's destiny, essential to the material well-being of the civilized society.

Mining is a global industry with a vast outreach; over 30 billion tonnes of minerals were extracted by the industry in 1999, including construction industry materials such as stone, sand, gravel and clay, to provide some 5 tonnes of materials per capita for a global population of 6 billion. Given the inertial force of demographic factors, under the inexorable pressure of population rise, and the soaring aspirations of the developing world for higher quality of life, the demand for mineral commodities is likely to rise many-folds and the mineral industry has to meet the demand in the face of depleting resources, shrinking exploration budget, and pervasive issues of land access for exploration and development, besides social, environmental and ethnic issues which are surfacing their hydra-heads in every mineral development project around the world. One of the omnipresent outputs of mining is energy and more than 92% of the world's primary energy production is based on mineral fuels .The mining industry's share of the global product of $ 30.00 trillion is estimated to be around 6%; according to a 1995 study by Western Economic Analysis Center in the United States, the American mining industries (metal mining, industrial minerals mining and coal mining) had a combined direct and indirect impact on the economy of the United States of $ 523.6 billion (Leaming, 1997) . Even the informal mining sector, small-scale mining, is an important phenomenon in many parts of the world with an employment of over 13 million with significant impact on national and regional economies (I.L.O, 1999) ; its contribution to the mining industry's poor image however is substantial. The leitmotif of the 18th World Mining Congress in the year 2000 –"Everything begins with Mining"- aptly reflects the ramifications of the mineral industry in society, underpinning the industrial base of the world.

It is in this context that we must examine the "issues", the emerging trends and concerns, that are likely to impact upon the future of global mining so that we can debate on these issues and their management for the long-term sustainability of the mineral industry. This presentation is essentially a personal worldview of how the global mining industry must address today's challenges and also prepare for tomorrow's opportunities in order to provide an appropriate backdrop for this International Conference.

A FISTFUL OF ISSUES

Issues that confront the global mining industry are many and varied and pose significant threats to mining industry's sustainability. In an increasingly interdependent world, with the emergence of a global mining village, there is a strong commonality between the issues being faced across the mining sites around the world, and which constitute the corporate dilemma for the minerals sector, if not a challenge. The principal issues are :

- Environmental degradation due to mining and its mitigation.
- Social and ethnic acceptability in a world of disparate groups.
- Augmenting the mineral resource base through massive investments in exploration efforts and high-technology.
- Innovations in technology for bulk mining, for maintaining the bottom line and for environmental compliance.

Some of the issues are the legacies of the 20th century, but will spill over into the new millennium with added momentum stirred by the growth in globalism and tribalism and the paradigm shift in expectations amongst communities close to mining sites. Mineral development has to be strategically planned to encompass all the "issues" of social acceptability and environmental accountability melded into the local needs and expectations. It is imperative that we highlight the fact that the "issues" which have surfaced or will be increasingly surfacing will need new prescriptions based on a new mindset as the future will not be a continuation of the past and discontinuities will be the rule rather than exception. The mining industry has to regenerate core strategies and rethink its fundamental assumptions on how best to contend with the emerging issues. The industry will have to create a meaningful vision of navigating in the turbulent times that the new millennium will offer.

ENVIRONMENTAL DIMENSIONS

The mining industry worldwide is being lambasted for its visible and not so visible impacts on the environment and this will be the key issue to be contended with and overcome in the coming decades. While some environmental wear is inevitable, and some of the impacts are transitory such as land use, responsible mining companies around the globe have demonstrated their commitment to upgrading the environment and attained improved performance standards through environmental audits, new technologies for reclamation, focus on occupational health and safety measures and embraced the basic tenets of sustainable development, both in theory and practice. The mining industry worldwide has an extraordinarily poor public relations image and has never been proactive in showcasing its environmental performance. Some 500 years ago, in his magnum opus, De re metallica, Georgius Agricola said :" Now a miner, before he begins to mine the veins, must consider seven things, namely: - the situation, the conditions, the water, the roads, the climate, the right of ownership, and the neighbours." (Agricola, G, 1556) which reads almost like the gist of an environmental impact statement and goes on to demonstrate that the mining industry had espoused environmental care since long.

It must be conceded however that increasing mineral production will exacerbate the adverse impacts on the environment, though not linearly. Increasing recourse to environment-friendly technologies and resort to high-technology in environmental clean-up will help in realising improved environmental standards. UNEP has done commendable work in brokering a new environmental agenda for the mining industry and its publications (Blinker, 1999) have promoted the concept of "benchmarking" whereby operators and regulatory authorities learn from, and adopt, the best practices of their peers.

It is important that we quantify, even if approximately, the dimensions of environmental challenges of the global mineral industry. For instance, the residuals from mining activities, especially of overburden, spoils, effluents and particulates are staggering. For mining and quarrying of some 30 billion tonnes of mineral commodities, the residuals aggregate to over 37 billion tonnes on a back of the thumb estimate, which need to be disposed of in an ecologically compatible manner. What is more of an intractable issue is the acid mine drainage, damage to and contamination of aquifers, and discharge of mercury to the environment in the case of informal gold mining. A study in the Upper Tapajos River Basin of the Brazilian Amazon showed that some 245 mining operations employing 30, 000 people and producing 35 tonnes of gold per year, were removing 67 million cubic metres of subsoil per year and releasing 12 tonnes of mercury to the air, soil and rivers (Anon, 1997). A number of ecological disasters have been caused in recent years by the failure of tailings dams, such as Harmony, South Africa (1994), Omai, Guyana (1995), Placer, Philippines (1995), Marcopper, Philippines (1996) and the recent failure of tailings dam at Les Frailes in Spain on April 24, 1998 which released an estimated 4-5 million cubic metres of acidic water besides tailings materials consisting sulphur, zinc, copper, iron and lead.

While corporate cooperation on environmental issues has emerged in the mining industry and increasing number of mining companies are joining the ISO 14001 club, there is significant increase in environmental compliance cost in many mining operations. Besides, there are increasing problems of land access for development and waste disposal, due to manifestation of the NIMBY syndrome which could put a serious road block to mineral development in the coming decades. One of the major environmental challenges in the new millennium will be the legacy of many closed mines and orphaned mine sites which pose serious environmental hazards. The cost of clean-up could be mind-boggling and initiatives are called for to set up a super-fund for rehabilitating such orphaned sites within a time schedule. There is no doubt whatsoever that the current trends indicate the evolution of worldwide legislation as a distinct possibility. Likewise, cleaner production is emerging as a desirable end by itself which could improve profitability in environmental management. With improved access to information, new environmental technologies and due diligence, a new order of environmental performance in the mining industry could be attained. Public debate and rhetoric on mining industry related environmental damages have to be proactively countered so that the industry's environmental stewardship is duly acknowledged. A new environmental agenda is evolving and more "prevention" oriented strategies based on understanding of social and environmental context will emerge.

THE SOCIAL AGENDA - CONTENTIOUS HIGH PRIORITY ISSUES

The last decade of the 20th century has witnessed an upsurge of public debate on the social responsibility of the mining industry besides resistance to development leading in some cases to violent social and ethnic conflicts. Such low credibility and social opposition appears to be a worldwide phenomenon, but more so in developing countries, and managing the socio-cultural issues could be the most serious challenge that the mining industry may have to grapple with in the coming millennium. One would not wish to see the recurrence of such events as the forced closure of operations which took place in Bougainville Copper in Papua New Guinea. Mining has often been viewed as a predator, an exploiter, which is creating wealth for some one far removed from the mining sites and the local communities receive only a token benefit and a derelict mine site, with tailings dumps, waste-rock piles and significant environmental wear. Unless the mining industry can develop a constructive agreement with the stakeholders in the community, the industry will have to face worldwide disorganised, imprecise, chaotic social demands, all of which will coalesce and congeal to build a veritable Maginot Line, a bulwark against mineral development and access. This is a serious issue in as much as mines have to operate and be located at places where a mineral deposit exists, and the problems are likely to be magnified in the 21st century with increasing discord as to what constitutes a fair distribution of the benefits of mining between the developer and the community. A number of communities in Peru and Bolivia, for example, oppose all new mining because of the visible contamination that encircle them due to a long legacy of mining and the transfer of wealth away from" their" land (Joyce and Thomson, 1998) .The mining industry's social acceptance will be assured and mining will be possible only if it can earn a social license through constructive engagement with the local community and other stakeholders, including the government. The fundamentals of long-term success for mining operations will be their ability to align their objectives with the communities in which they operate on the basis of mature and respectful partnerships. There is an urgent need for mineral-producing countries to develop agreements of the type of White Horse Initiative in Canada which support a vision of "environmentally, economically and s6cially sustainable mining industry, underpinned by community and political consensus" (McAllister and Alexander, 1997). This is the first successful cross-national, multi-stakeholder, mineral-sector initiative of its kind in the world, worthy of being emulated.

CHALLENGES FOR TECHNOLOGICAL INNOVATION

The realities at 2020, with a world populated by some 8 billion people whose basic needs for mineral commodities could be around 10 tonnes per capita , the sheer volume of demand for mineral commodities could rise to over 80 billion tonnes. The material expectations of *homo sapiens* in the 21st century can only be met by technological innovations across the board. The problem will be further compounded by depleting resources and the need to mine cleaner and in most cases deeper resources. As we approach the 21st century, the rate at which technology is evolving and impacting our world is more rapid and more profound than ever before and the ultimate resource of human intelligence and knowledge will have to provide the remarkable leap in the next 2 to 3 decades through innovative technologies to help meet the soaring demand for minerals and metals and at the same time help overcome the environmental problems through enabling technology of concurrent restoration in surface mining and in the case of underground mining, mineral processing at the extraction face and total return of the waste to the mining void. Yet another revolutionary concept to be actively pursued will be that of *in-situ* mining through chemical or biological processes. While the use of integrated GPS and GIS systems will become commonplace in surface mining, for real-time control through massive inputs of information technology, focussed R & D efforts will hopefully lead to the development of digital and intelligent mines and of tele-robotic mining for remote and inhospitable terrains in the Arctic or high up in the Himalayas. Continuing to innovate will be crucial for mankind's survival, and more so in the quest for new mineral deposits. In fact the challenge lies in leveraging technology to help balance the minerals, economy and ecology equation by developing and diffusing new, innovative, cost-effective and environmentally-compatible technologies with focus on continuous improvement in global best practices. New concepts in mineral exploration technology will spring from computing technology with the tools of virtual reality to create three dimensional models of deposits with "walk-through" capabilities; deep-seated deposits will be located through Mobile Metal Ion (MMI) process which holds the promise for enabling explorers to 'see' mineralisation with considerable accuracy at great depths beneath the surface (Birrell, 1996) .The use of low-orbit hyper-spectral imaging spectrometer, for instance, will allow of simultaneous exploration and environmental monitoring. The newly emerging arsenal of technology for bulk mining will enable the industry to move gargantuan proportions of materials, and using new beneficiation techniques sift them to produce the sought-after mineral commodities. Technological advances need to be pushed further through investment in R & D and developing strategic competence for implementation of ambitious long-term objectives. Developing an innovative culture will be the key to technological breakthroughs and one can predicate that the most impressive gains in mining technology will come from new and exciting developments in exploration technology and beneficiation using the oldest forms of life, namely bacteria. One could envision increasing application of bio-hydrometallurgy for metal extraction from effluents of mining operations. The mineral industry and mining processes will be re-engineered using the new tools of information technology for achieving what Susana Zhukoff very colourfully described as a striving to "automate, informate and transformate" the industry. Technology, and that too leading-edge technology, will be the backbone of the global mineral industry in the next millennium to expand the resource base, to economically produce from small deposits using high-tech and to limit environmental impacts. One of the major concerns in the future will be the need to attract and retain a diverse, talented, knowledge-based workforce in the mining industry, which is increasingly becoming a scarce commodity, what with the image of the mining industry. The attributes of mining engineers in the 21st century will include: learnability- the yen for lifelong learning, ability to work in a team, integrative skills, ability to employ information technology, commitment to sustainable development, adaptability and flexibility

THE MANAGEMENT DIMENSION

The global mineral industry will witness massive thrusts on acquisitions, mergers, consolidations and even strategic alliances for creating tomorrow's advantages. The perspective of the new millennium will be a vibrant industry composed of fewer companies, large in size and scope, operating in a cooperative global market and driven by the forces of technology. The concept of "bigger is better" in global market place has already become entrenched generating a wave of mergers and acquisitions and by the year 2020, the industry could be straddled by some 50 to 60 mega global mining corporations and widely dispersed small and medium-sized mining companies in the nations. The 21st century mining companies will be vision-directed, value and stakeholder-focussed, speed-oriented, innovative and entreprenurial learning organizations. We can expect a paradigm shift in management concepts and practices, with networked organizations, sharing knowledge and working together to develop more sustainable processes through collaboration across national and industry boundaries. The successful mining companies will be dynamic, integrated, effective and responsible.

CONCLUDING REMARKS

Mining industry has a vital role to play in ensuring a sustainable future for the spaceship earth, in terms of economic growth, protection of the environment and social progress. The challenge is both enormous and imperative as change for a new growth trajectory, meeting burgeoning material needs in a rapidly changing geopolitical landscape, will have to be achieved in a safe, responsible and affordable way. A vibrant array of new technologies, properly applied, will foster economic development, mitigate environmental impacts and move in the direction of minerals sustainability. The mineral industry will have to face a multitude of problems and challenges, but difficult situations also do offer opportunities to learn and to rise to the occasion. Like the Zen Buddhist, encountering unexpected challenges, the mining industry has to pin its faith by exclaiming "Oh, happy blessed opportunities" and cruise steadfastly ahead on the curve of change with wisdom and fortitude. Hopefully, the deliberations of this Conference will chart the pathways to the future and the global mining industry will be able to meet the rounded definition of sustainable development- " the development that our grandchildren will thank us for". It will call for willpower, imagination, and cooperation on a heroic scale of the global mining community to translate the vision into reality.

REFERENCES

Agricola, G.1556. *De re Metallica,* Book II. Translated from the Latin by Hoover, H.C. and Hoover, L.H.

Anon, 1997. Mining – facts and figures. UNEP Industry and Environment, October, pp.4-9.

Birrell, R. 1996. MMI geochemistry: mapping the depths. Mining Magazine, May, pp.306-307.

Blinker, L.R. 1999. UNEP's role and activities in mining and the environment. Proceedings of the International Congress "Mining for Tomorrow's World", Dusseldorf, pp.195-198.

Ghose, A.K. and P. De, 1994. Mining in the Changing World-Agenda for Action. Proceedings of International Workshop on "Mining in the Changing World", Calcutta. The Institution of Engineers (India) , pp.1-3.

International Labour Office, 1999. Social and labour issues in small-scale mines, Geneva. 99p.

Joyce, S.A. and I. Thomson, 1999. Earning a Social License. Mining Journal, June 11, pp. 441-443.

Leaming, G.F. 1997. Mining and the American Economy: Everything Begins with Mining, A Report for the National Mining Association, Section 9, pp.1-10.

McAllister, M. L. and C. J. Alexander, 1997. A Stake in the Future: Redefining the Canadian Mineral Industry.UPC Press, Vancouver, 240 p.

CHAPTER 3

World Best Longwall Mining Practices, What is Applicable to India?

*Walter von der Linden**

INTRODUCTION

"Today, longwall mining is the safest, most productive and cost-effective method of mining underground coal seams. It is common throughout all USA provinces" (McDonald, 1999). This statement would equally apply to underground coal mining in Australia, South Africa, Poland, Russia, China and Western Europe. In many cases longwall mining is likely to be the lowest cost deep mine production system but the right conditions must be in place to amortise the very high capital investment. These conditions comprise : (i) large coal reserves to facilitate production volumes required for amortisation of the investment, (ii) availability of capital to make such investments, and (iii) sufficient market potential. The reserve situation may be limited by either the volume of the mineable coal deposit and/or by geological faults which would not allow large enough longwall panels to be developed. The share of coal production by longwall method relative to total underground production is shown in Table No. 3.1 for some selected countries for 1997. Fig. 3.1 offers a view into a high production longwall face.

Table 3.1: Coal production by longwall method relative to total deep mine production

Country	*Total underground production (in million tons)*	*Tons per employee year (in millions)*	*Longwall production (in million tons)*	*Longwall % of total underground production*
PR China	1,240	180	-	> 70%
USA	378	7,710	146	39%
Australia	64.6	5,680	46.5	72%
South Africa	117	4,640	12.2	10%
India	72	220	6.5	9%

Source : Boyd, 1998

*Vice President, Deutsche Bergbau Technik., Germany.

Fig. 3.1 : Longwall face

These longwall production data do not include production from longwall entry development. This paper addresses higher production volume longwall mining. This mandates caving of the roof rather than back-filling methods. It is recognised that backfill to limit surface subsidence is mandated in some mining districts in India.

WHAT DEFINES WORLD BEST LONGWALL MINING PRACTICES?

There is broad consensus that world standard longwall mining practices are achieved by people, process and technology (Galvin, 1999). All elements are of equal importance, short falls of one aspect cannot be compensated by the other two and will have an adverse impact on overall results.

Results of coal mining practices are best demonstrated by employee productivity in tons per man year (Table No. 3.1) and in tons per production unit year. The large productivity differences can mainly be traced to differences in work and management practices, in work processes as well as in applied technology. Diversities in geological conditions may have an influence.

Naturally there is an interdependence between the three factors "people, process and technology". The people factor stands at the forefront, as efficient processes can only be maintained by trained, skilled and highly motivated employees. Complex technologies require increased skill and experience levels to sustain good work processes which may be described here by high standards of safety, consistency and availability.

BACKGROUND

DBT Deutsche Bergbau Technik GmbH (DBT) was established in 1995 through the merger of the three, at the time, well known companies Halbach & Braun, Hemscheidt and Westfalia Becorit, formerly Westfalia Lünen. These three companies had a long history in the field of longwall mining equipment. Many basic longwall mining technologies and machinery developments are linked to these companies, the most outstanding being the armoured face conveyor and coal plow by Westfalia, the centre chain concept by Halbach & Braun, innovative shield support developments by Hemscheidt and the PM-4 electronic shield control system again by Westfalia. A significant lift in low seam longwall productivity was attained by combining the PM-4 electronic control system with the plow coal cutting method. German mining technology had been exported to all coal mining regions world wide by these companies.

Today, DBT is a part of RAG (formerly Ruhrkohle AG), a global mining and technology company and owner of all underground coal mines in Germany. RAG has recently acquired major coal mine operations in USA, Australia and Venezuela and has diversified into other industries such as power generation, high tech manufacturing and petro-chemicals. DBT has established itself as the world market leader for longwall shield supports, armoured face conveyor (AFC) and plow mining systems (Fig. 3.2). The PM-4 control system is the outstanding product of DBT; it has contributed significantly to increases in production and productivity of longwall mining worldwide.

WHAT DRIVES GOOD PROCESSES?

Processes are performed by organisations and are determined by these key drivers: capacity, availability, utilisation, efficiency. The data from these drivers must be measured to gauge the performance of an organisation (DeMichiei, 1999). Capacity is maximum

Fig. 3.2 : 3D-drawing of thick seam 2-leg shield

possible output in tons per hour or tons per minute. Availability is measured by how much of the day, week or month the equipment is ready to produce. Utilisation is the percentage of time that equipment is actually used to produce coal. Efficiency is a measure of how much has been accomplished and what resources are required: tons per man-day, cost per ton, etc.

Continuous improvements

It is important to note that top performing operating companies have been achieving consistent improvements on a year by year basis. For example: Consolidation Coal Company in the USA, probably the world leader in underground coal mine performance, has raised the average productivity in all its mines from 9 t per 8 h man-day in 1978 to 36 t per a man-day in 1998. At the same time, production was increased from 54 million t to 69 million t while the number of operating coal mines was reduced from 54 in 1972 to only 25 operating mines in 1998. Today 95% of Consol's coal production comes from underground mines and 77% of this comes from longwall mines (Mc Donald, 1999).

Maximise longwall share

Improvements in technology in conjunction with new approaches in mine planning have made it possible to reduce less productive longwall development work in favour of a higher share of longwall mined coal. These process improvements have brought a substantial increase in mine productivity. In 1970 development tons for a 150 m wide by 1200 m long panel with 4 entry development were 20 to 25% of the total tons produced from this panel. Today's mines with 330 m wide by 3000 m long panels and 3 entry development extract less than 10% development coal and more than 90% of the production comes from the longwall operation. A number of external factors can interfere with the success of such measures. For example, geological faults can present a serious impediment. This highlights the need to incorporate geology and geotechnical engineering into the management process to avoid surprises and to build a data base of experience for dealing with geotechnical conditions. The revenue lost per day of downtime due to geological feature on a 3000 t per hours longwall is sufficient to employ a geologist fulltime for the order of 10 years (Galvin, 1999). In existing and older mines the laying out of large panels may no longer be feasible.

Predictive maintenance

Best practice longwalls are producing 3 shifts per day and 6 - 7 days per week to make optimal use of the invested capital. This procedure fully utilises the invested capital but it does not allow for scheduled, preventative maintenance and may therefore endanger by unexpected equipment breakdowns the gains sought from near full time operation. Predictive maintenance procedures provide an opportunity to perform repairs and/or replace components on an as needed basis. Predicting when a piece of equipment will fail, still allows the operator time to prepare for necessary repair arrangements, for procurement of parts and for having required specialists on hand to perform the maintenance job when required as quickly and efficiently as possible (McDonald, 1999). Elements of predictive maintenance are regular gear case oil and hydraulic fluid analysis to provide information about component wear in gear boxes and hydraulic circuits. The oil is analysed for water, dirt, chemical composition and viscosity. Vibration analysis is performed to locate and characterise impending bearing and gear failures. Deviations from the norm provide indications for impending problems and failures.

Monitoring of equipment performance data such as RPM, voltages, currents, temperatures and speeds are displayed, recorded and monitored to detect changes and trends. Remote fault diagnostics indicate which part has failed or no longer performs to specifications.

These predictive maintenance practices provide the opportunity to increase equipment utilisation. Unit production per year is increased and costs are reduced by more productive hours per day. This meets the key approach in mining: to reduce the number of production units. Predictive maintenance strategies require equipment and systems of high quality and reliability as well as trained, qualified, and committed personnel.

Partnership with equipment manufacturers

To successfully operate a longwall system is a highly complex undertaking. It truly involves people, processes and the application of technology. In more detail this would comprise data collection and planning, employee training and process design, procurement and logistics, management involvement and feed back, maintenance and repairs; all these efforts must be directed towards continuous improvement of the longwall operation. Increasingly, coal producers and manufacturers have formed long-term relationships, or strategic alliances. "To be truly successful, ... major equipment suppliers must be integrated into the management system. Alliances are a must for optimal performance." (DeMichiei, 1999). Some partnerships between coal mine operators and equipment manufacturers have developed and have significantly driven best practice goals. First of all, there is manufacturer's advice and training of the employees of the producer. Today's equipment is highly specialised and customised and requires many specific skills to operate at lowest possible cost over the longest possible time. The knowledge of the coal operator is often restricted to his own localities while the equipment manufacturer normally has a much broader perspective from his global activities. This has enabled the manufacturer to not only provide advice on equipment selection, but also on spare parts and repair requirements for optimum equipment performance and maximised equipment service life. Today's highly productive longwall systems require a reliable supply of spares, fast turnaround of rebuilt components. Major spare parts are supplied by the manufacturer under a long-term contract from a strategically located parts depots and thereby free the operator from maintaining large parts inventories and at the same time, through the manufacturer's supply guaranty, prevent potential idle times of the equipment due to parts shortages.

The strategic alliance process allows the supplier to voice problems he might see; this opens up a dialogue for possible solutions. Together, mutual long-term goals are being achieved and opportunities for continuous improvements are shared. An important factor in the success of the new operator – manufacturer relationship is believed to be open, forthright communication. "The manufacturer and coal producer that communicate the best -- and listening is the hardest part of communication – will be successful. We must learn more about each other, listen to each other and seek ways to create value for each other." (Chadwick, 1998).

A note of caution is appropriate at this point. The final objective of an operator – manufacturer partnership is mutual: to lower the cost per ton of coal mined. In the partnership approach this goal is attempted to be achieved by splitting the comprehensive task of mining coal into many sub-tasks and assigning them to the entity with the most expertise and capability. For example, within the expertise of the mine operator would be mine planning (possibly assisted by a consultant) and mine operation (could also be assumed by a mine contractor), while training for equipment operation and maintenance as well as parts supply, maintenance and repairs would be part of the manufacturer's expertise.

However, a reasonable activity and responsibility sharing should not be confused with risk shifting from the mine operator to the manufacturer. The assumption of the mining risk (geology, such as consistency of the coal seam, and operation) by the manufacturer would necessarily have to be reflected by risk premiums higher than what would have to be applied by the mine operator. This would then be counterproductive to the low production cost objective and would not serve either of the partners.

TECHNOLOGY

In the cost of coal production, personnel and capital make up the main components. In India, personnel costs are, by comparison, low while cost of capital is high. This requires appropriate consideration in the "best longwall mining practice for India" strategy. Also, social aspects to maintain employment may be a consideration. Short longwall faces of 100–150 m length appear to be the answer, as opposed to 300 – 400 m wide faces as is the current trend in the USA and Germany.

Previous longwall experience in India suggests that roof control and cutting of coal have frequently presented problems and have been impediments to economically successful longwall operations. Thick and strong roof layers found in many coal mining districts in India require Longwall shield supports of suitable design and technology. Hardness and strength of most coal seams place a high demand on longwall shearers to achieve satisfactory coal production rates and to ensure acceptable reliability and equipment service life. A discussion on longwall technology for India must, therefore, be focussed on roof control and coal cutting technology. The following observations are offered in this context (Campbell, 1999).

Shield Supports

Design: Finite Element Analysis (FEA) is used to study the stress effects which originate from a variety of load conditions which are imposed on the shields in the day-to-day operation (Fig. 3.3). The designer is then able to reinforce specific structure elements by alternate shapes and/or adding material to remain within safe stress limits (without to unduly increase weight and cost of components). This in turn will increase fatigue limits. Fig. 3.3 shows samples of FEA analysis on DBT shields.

The design failure mode and effects analysis (DFMEA) is a process that explores potential failure modes and places a rating to the severity of failure and probability of occurrence. The purpose is to prioritise design efforts and resources.

Welding technology is applied to facilitate ultrasonic testing of all major welds. It should be noted that in shield support structures, welds are more frequent on a per-ton-structure basis compared to most other structural applications. Stress levels in shield supports structures are chosen, by design, relatively high for the benefit of manageable unit weight and cost.

Prototype testing specifications, *i.e.*, test load conditions, load cycles and acceptance criteria have been developed. In general a prototype support is tested at 30,000 load cycles with a certain overload factor, depending on specific mine conditions.

A prototype unit must undergo extensive fitness for purpose testing to ensure compatibility and easy access for maintenance and repair.

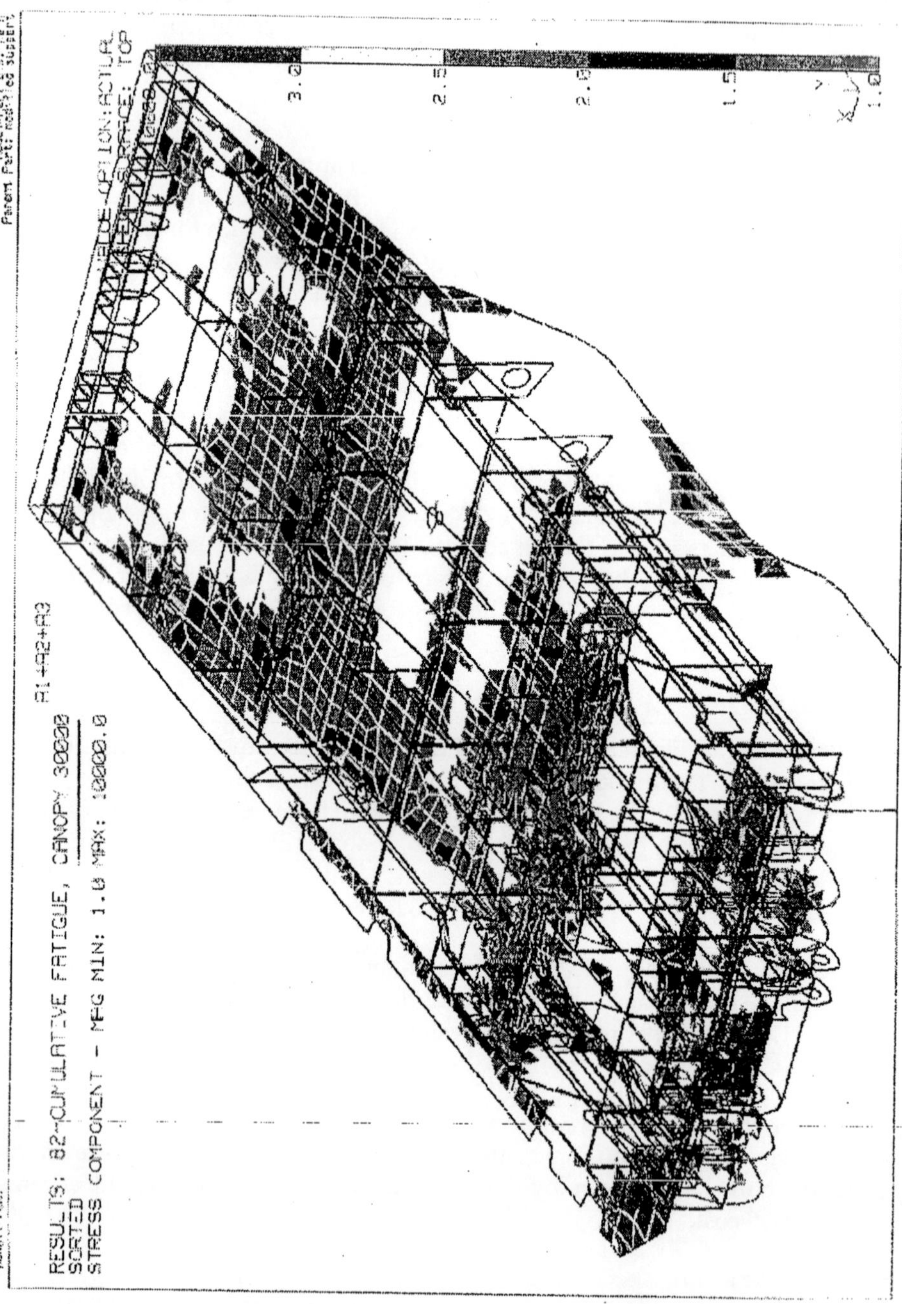

Fig. 3.3 : Results of finite element model of shield canopy

Materials: Criteria for materials used in the manufacture of structural components and props are being applied to ensure sufficient ductility to avoid premature material failure by fatigue.

Rapid acting yield or pressure relief valves: Specific roof conditions in some mining districts in the USA and with increasing importance in Australia, South Africa and India may cause, on certain occasions, extremely high forces being imposed on the shield supports. These high forces occur when massive immediate roof layers or main roof layers after having hung over a considerable distance suddenly break behind the supports. The rapid release of energy during this failure process requires the supports to rapidly yield in order to avoid destructive overload conditions. During the yielding a high volume of hydraulic fluid must be displaced from the legs to the atmosphere. In this instance only a limited internal pressure increase is permitted.

Today, rapid acting yield valves are manufactured by DBT with a displacement capacity of 3000 l/min at a pressure increase of 60% above nominal pressure. This is within the elastic limit of the prop tube and adjacent structure. 3000 l/min yield valves facilitate a yield rate of 500 mm/sec for 360 mm bore props which have a nominal load capacity of 4600 kN at 450 bar internal pressure. It should be noted that these high forces occur at the instant of a moment. Therefore the yield valves must react instantaneously within 5 -10 milliseconds.

Electronic controls, monitoring and automation: Nearly 100% of all longwall shield support installations installed since 1990 in Germany, USA, Australia and South Africa are fitted with electronic equipment to control and monitor practically all shield support functions and movements and interactions with the AFC and shearer system. Benefits provided to the user include increased production through consistent operation and a significant improvement in worker safety. The element of human error and neglect has been eliminated to a great extent, operating personnel is now far less exposed to dust and many functions can now be controlled from a remote location. Sensors and monitoring capability have raised reliability and availability of many of the heavy mechanical and hydraulic components employed in a state-of-the-art longwall mining system.

The success of electronic controls has been made possible by the high degree of component/system reliability, simple on face programming and interfacing capability. Electronic control system availability in excess of 99% is not uncommon, personnel can be trained for efficient operation within short periods of time.

Longwall Shearers

Shearer technology has been driven by the desire for increased capacity and availability for the longwall system. This has been achieved by increased total machine power, application of medium voltage (up to 4160 V) cutting and AC traction motors and high quality gearing and mechanical components (Fig. 3.4). The benefit of these developments to the user are an increase in production potential and equipment service life.

CASE HISTORIES

In 1997, shield supports by DBT have been supplied, one face in South Africa, three faces in Australia, in geotechnical environments and with a business strategy that may be applicable to India. In all cases the equipment operates under heavy massive roof, in shallow

Fig. 3.4 : Longwall shearer for high production face

depth and in short faces for the purpose to limit investments. The main data are listed in Table No. 3.2 (Campbell, 1999).

Table 3.2 : Short longwalls operating in shallow depth

Country	*South Africa*	*Australia*		
Mine	*Matla No. 2*	*Moonee*	*Metropolitan*	*West Wallsend*
Depth below surface	100m	120m	450m	200m
Face length	130m	90m	120m	135m
Extraction height	3.8 - 4m	3.1 to 3.6m	3.1 - 3.4m	4.8 - 5m
Support density	110 t/m^2	105 t/m^2	100 t/m^2	105 t/m^2
Support spacing	1.75m	1.75m	1.50m	1.75m
Prop dia	2 x 360mm	2 x 345mm	2 x 330mm	2 x 360mm

South Africa

A longwall system was started up at Matla Colliery (Fig. 3.5) in May 1997 and completed its first panel, 3.3 km long producing 2.0 million tons in 9 months for an average of over 200,000 t per month. In October 1997 longwall production was 345,000 t. Average monthly production in the second panel was over 250,000 t and over 300,000 in the third panel. Total production from the three panels mined so far was 6.4 million tons during 24 operating months during the calendar period May '97 to June '99. Each face move between panels was completed in approximately 4 weeks. 8 men per shift are employed in the face and headgate. Coal production is scheduled for 16 shifts per week. The coal has a compressive strength of 40 MPa and is extremely difficult to cut. Traming speed of the 1200 kW shearer is only 7m/min; this is a major factor to limit production.

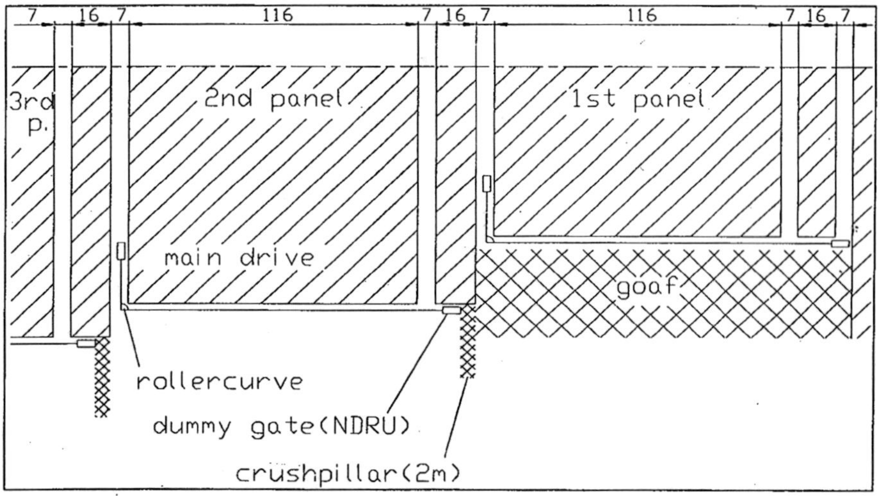

Fig. 3.5 : Panel layout, South Africa

The props are fitted with 3000 l/min rapid acting yield valves which facilitate a convergence velocity of the legs of 500 mm per second. In the first panel, the first significant caving of the roof occurred after approx. 60 m of face retreat. At that time the shields yielded by approx. 800 mm at a speed estimated to be 300 - 400 mm/sec. In the second panel, the first caving came also after approx. 60 m of advance. There were a few more but less severe yield occurrences after the first cavings. There was no damage to props, yield valves and shield structure after these heavy yield occurrences.

To date, after 3 panels and 6.4 million tons production, all yield valves and props as originally supplied are still in operation.

Optimal face length: The seam height of 4 m and high cost of capital have prompted the decision to operate a short longwall at Matla. By a combination of an economic analysis for lowest production cost and examination of 3 parameters (Fig. 6): (i) minimum face length to guarantee caving (80m), (ii) average planned achievable production in the development (35,000 t/month), and (iii) AFC without taildrive, the optimum face length was determined to be 132 - 135 m (Fig. 3.7). Meanwhile an average production of 52,000 t/month was achieved in the twin entry development with 17 m chain pillars and cross-cuts at 70 m spacing. This will support a longwall retreat of 620 m/month or an average production of approx. 470,000 t/month. This example shows that the planning and evaluation of longwall mining, especially when faces are short, must include entry development.

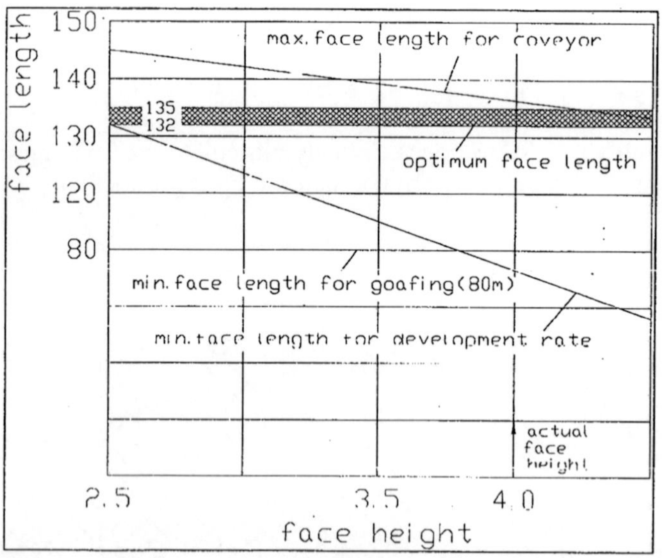

Fig. 3.6 : Criteria for optimal face length

Australia

In the last three years, DBT and its Australian daughter company have installed three "short" longwalls, these being: (Campbell, 1999).

(a) Metropolitan Colliery in 1995, 120 m face length.

(b) West Wallsend Colliery in 1997, 135 m in length.

(c) Moonee Colliery in 1997, 90 m face length.

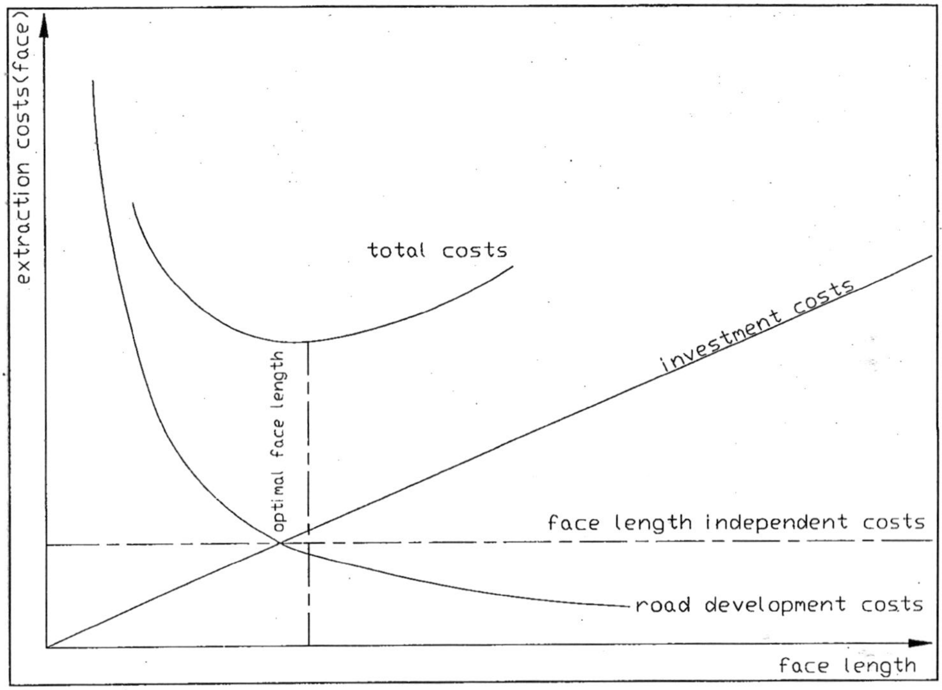

Fig. 3.7 : Development of production cost with face length

All these faces have exceeded customer expectations in particular in regard to productivity. Though the final equipment design varied slightly and the mining conditions were not identical, there were common customer ideas in each of the longwall installations.

Why a shortwall? The majority of recent longwall installations in Australia have been typically 250 m face length and designed for a production of 3,000 to 4,000 tph with a minimum target annual production of a 4 million tons. These large installations have principally been for new mines with a total initial capital cost of A$ 150 million to A$ 450 million. The mine must be able to produce at such a high rate in order to offset the capital cost of the initial mine infrastructure.

For older mines with existing infrastructure, a new longwall purchase must be designed to suit the existing belt conveying and preparation plant capacity and in most cases this limits the ability of the mine to install very high production faces (+ 3,000 tph). Instead these mines must choose lower production level equipment but at the same time ensure very high levels of availability with an overall high level in productivity are achieved in order to offset lower tons per hour. For Australia, a coal mine must have a productivity level of at least 10,000 tons per year per man employed, in order to be profitable. In many cases this needs to be more, such as low yields and steaming coal product mines.

All short longwalls installed to date in Australia have been in existing mines, constrained by existing markets and mine infrastructure. The peak production from these operations has varied from 900 tph (Metropolitan), 1500 tph (Moonee) to 2000 tph (West Wallsend). Because of these lower production rates it is also important to optimise capital expenditure, which certainly partly justifies the choice of shorter hence lower cost longwalls.

Periodic weighting and wind blasting problem: In some other Australian mines the occurrence of massive overburden strata had caused frequent face instability. The instability was manifested by fracturing of roof strata in front of the face and by heavy loading – periodic weighting of the supports. By shortening the longwall face it was deduced that the fracture planes would reach the coal seam level at or behind the shield supports. Subsequent practical experience has nearly eliminated the periodic weighting problem.

Narrowing the panel width introduced the risk of windblast by the ability of the strata to bridge the void created by extraction (Hamilton, 1999). The competent beams of the roof strata can increasingly cantilever back into the goof before failing. This failure can create a significant "piston effect" and displace large volumes of air. In circumstances where insufficient immediate roof has caved prior to large area competent beam failure, high velocity wind blast events can be generated in the face and maingate which can cause operational disruption and injuries to personnel. Wind blast management plans include personnel protection to prevent injuries, seismic monitoring for pre-warning of wind blast events to evacuate the face and maingate area (Hatherly, 1999). Recently hydro-fracturing of the strata has been employed by injecting water into 50 mm dia holes drilled from the face into the roof. The success of hydro-fracturing would depend on geological and strata specifics but appears to be promising. For conditions where hydro-fracturing can be developed as a reliable tool, wind blast events would no longer be unpredictable the magnitude of major roof caving and intensity of resulting wind blasts could be reduced.

CONCLUSION : WHAT IS APPLICABLE TO INDIA?

A review of "world best longwall mining practices" aspects reported in this paper should lead to relevant conclusions as to which of these practices may be applicable to India, respectively, what lessons can be learnt.

1. First of all, there is the need for continuous improvement. In today's competitive and global environment the need for continuous improvement must have top priority, irrespective of starting point, status, industry, or society. The achievements of Consolidation Coal Company and many other world class underground coal mine operators may provide a yardstick for improvement potentials and goals.

2. Learning! How to improve? A well accepted process today is to look over the fence. How are competitors performing and achieving their improvements? What processes and technologies are being used industry wide? Which strategies and tactics are being pursued? What skills have been acquired? The practice of "bench marking", for results and processes is now well accepted and has been around for some time.

3. Appropriate longwall technology is available, it can be adapted to suit specific conditions in India, not least with the objective to limit investments to what is tolerable by specific projects. In this context there is the simple optimisation possible by adjusting face length. A summarising statement may be that technology alone does not suffice. Processes designed to meet local conditions are of equal importance and are probably a dominating driver for continuous improvements.

4. Dialogue. Questions remain to probe the application of practices and technologies which have successfully been used in other countries. Such questions are valid because people must convince themselves to do or to change things. A constructive dialogue could assist in solving such questions and such dialogues should be conducted between appropriate stakeholders: operators, engineers, planners, administrators. The world mining industry has traditionally been open minded and communicative. Naturally, partners for a dialogue should be looked for in countries with mining conditions similar to India. Finally equipment suppliers have a keen reason to disseminate among operators the knowledge and information of what has worked well.

REFERENCES

1. Boyd, J.W. 1998. *Outlook for Worldwide Underground Coal Mining*. Proceedings of the Conference "Coaltrans 98", in Amsterdam, October 12-14, 1998. pp 347-370.

2. Campbell, R. and Richmond, A. 1999. *Short face longwall mining technology: experience from Australian and South Africa*. In: Underground Coal Mining Global Experiences, Lessons for Survival. Proceedings of a Conference jointly organised by the University of New South Wales and ACIRL Ltd, in Sydney, Australia, June 1999. pp 121-131.

3. Chadwick, John, 1998. *Partnerships*. In: MINING Magazine - July 1998. pp. 7-20.

4. DeMichiei, John. 1999. *State of technology in underground coal mining – What drives success at Cyprus Amax Coal?* In: Proceedings of the International Congress "Mining for Tomorrow's World", in Düsseldorf, June 8-10, 1999. pp 31-36.

5. Galvin, J. M. and Mc Carthy, P. J. 1999. *Sustainable underground mining – people, process and technology*. In: Underground Coal Mining Global Experiences, Lessons for Survival. Proceedings of a Conference jointly organised by the University of New South Wales and ACIRL Ltd, in Sydney, Australia, June 1999. pp 1-8.

6. Hamilton, Neville. 1999. *Single pass thick seam longwall experience at West Wallsend Colliery*. In: Underground Coal Mining Global Experiences, Lessons for Survival. Proceedings of a Conference jointly organised by the University of New South Wales and ACIRL Ltd, in Sydney, Australia, June 1999. pp 55-61.

7. Hatherly, Peter and Luo, Xun. 1999. *Microseismic monitoring – implications for longwall geomechanics and its role as an operational tool*. In: Underground Coal Mining Global Experiences, Lessons for Survival. Proceedings of a Conference jointly organised by the University of New South Wales and ACIRL Ltd, in Sydney, Australia, June 15-18, 1999. pp 65-73.

8. McDonald, C. Wesley and Dr.-Ing. Brune, Jürgen F. 1999. *High-performance longwalling and roadheading in hard coal mining*. In: Proceedings of the International Congress "Mining for Tomorrow's World", in Düsseldorf, June 8 – 10, 1999. pp 57-62.

CHAPTER 4

Vision 2020 for Mining Industry

*A. K. Dube**

ABSTRACT

Mining is one of the oldest professions of the world and it grew with the evolution of man and the civilization. We get the basic feed stock for the industrial growth and hence the mining occupies a paramount place in the advancement of a nation.

The Indian mineral industry is passing through an adverse phase. The profitability has gone down. Mining process has become difficult due to challenging geo-mining conditions. Cheaper mineral from abroad has changed the outlook and the mineral industry faces the prospects of losing its form. The environmental awareness and its compliance adds to the cost of mineral production and makes the mining venture still more costlier.

The next twenty years may see the privatization of the public sector. Gradual growth of private enterprise and the mutual competition may take place. The cheaper mineral offered by foreign countries may build up pressures on the indigenous mineral industry. Possibly, there may be growth of flow of imported minerals. It may be a blessing in disguise that our mineral wealth may be conserved and the environmental degradation may slow down. This may change the scenario of the industry and manpower requirements.

We may have to develop strategies for combating the challenges of the twenty first century. Development of consultancy services for mineral producing countries and even entering into joint venture for mining abroad may flourish. It will open up a new chapter in the annals of Indian Mineral Industry.

Currently, the surface mining provides bulk of our mineral production. Possibly, in the twenty first century there may be greater emphasis on underground mining. The industry has, therefore, to prepare itself for making underground mining viable.

The mining industry has to be dovetailed with other social developments. The rehabilitation of mined out areas can best be done by using such areas for some societal

*Central Mining Research Institute, (Council of Scientific & Industrial Research), Dhanbad.

purpose. Another opportunity for mining community is to help in development of underground space technology which is a much wider field of activity than the mining. The mining community will, therefore, has to look beyond mining.

INTRODUCTION

Mineral industry in India is poised for a turbulent period in the twenty first century. This situation calls for concerted efforts by the mining engineers to have an introspection and a firm resolve to get over the problems of the mining industry. Globalization of commerce and trade may throw numerous challenges and the mining being the basic industry may face special problems.

Mineral wealth is a wasting asset and its fast depletion with unscientific mining poses a serious problem to the mankind. The mineral getting process is very difficult one and, hence, there is an urge to make it safe, economical and environment friendly. The success of the industry will thus revolve around these aspects.

THE GLOBAL SCENARIO

The Agenda Twenty One of Rio Conference and WTO agreement on World Trade are two most powerful international tools to regulate entire commerce and trade of the world. The Green Agenda an off shoot of Agenda 21 means to a common man that we cannot compromise with the share of our future generations on the earth resources. The mineral wealth has, therefore, to be used sparingly, hence the bleak prospects for the mining. A compromise may be in the form of recycling the waste and rehabilitating and rejuvenating the mined out areas.

The environment cost of mineral exploitation is so high that most of the developed countries have stopped or in the process of stopping mineral production. These countries are importing cheaper minerals from the developing countries. They, therefore, are conserving their precious mineral wealth for the generations to come.

The WTO philosophy implicitly argues that the countries lacking proficiency in any kind of business activity should not try hand on it, instead they should depend on the proficient ones for the needs. This may have a greater effect on the economics of the poor and developing countries. The Indian mineral industry may also face similar predicament.

The global scenario is, therefore, heavily loaded against the developing countries and such countries will have to find out solutions to weather the storm in the offing. This situation needs a new strategy for the growth of the mineral industry. The overall economics of mineral getting may have to be gone into thoroughly and ways and means are to be found to mine the mineral at globally competitive prices. Failing which, the survival of Indian mineral industry will be in difficulties.

PROBLEMS OF MINERAL INDUSTRY

Mineral industry in India is beset by many problems. They are very well known to the industry and are listed briefly below :

1. Old and difficult mines.
2. Lacking infrastructure upgradation.

3. Higher cost of extraction.
4. Low productivity.
5. Excess labour.
6. Growing environmental cost.
7. Social concern and resistance to mining.

The above problems individually or collectively are making mining unfavourable and uneconomical. These factors are on rise and may adversely affect the mineral industry in the twenty first century. For the survival of the industry, serious efforts are to be made to mitigate them. A strategy is, therefore, needed to be made.

VISION FOR 2020 AND BEYOND

Visionary is a crystal ball gazer. He looks into a distant future and foretells about the events to come. It is an extremely difficult task and, therefore, not an easy one to accomplish. The author is fortunate to have some insight into the global scenario after economic liberalisation. This has a set agenda and due to this it is a bit helpful to make a guess about the events to follow and its consequences to be perceptible. The following points are worth mentioning :

1. Agenda 21 and its aftermath.
2. Increasing cost of mining due to environmental imperatives.
3. Public Sector loses its place.
4. Prominence of private sector.
5. Shrinking job market for mining engineers.
6. Recycling of waste.
7. Mining of low grade minerals.

Agenda 21 and Its Aftermath

To a common man, the agenda means the exploitation of minerals without compromise of the interest of the generations to come. Mining is a wasting asset and hence cannot be exploited without compromising the interest of future generations. There is thus a need for a via media. Possibly this may get a revised definition in the 21st century.

Environmental Imperatives Discourage Mining

The environmental management costs lot of money. This pushes up the cost of mineral exploitation. The advanced countries have taken a course of progressively not exploiting their resources. This may be an environmental gesture but the truth is otherwise. They are getting cheap minerals from developing countries, hence, they are willy-nilly concerning their mineral resources for the generations to come.

PUBLIC SECTOR LOSES ITS PLACE

In India, majority of mining business is held by public sector companies. Due to various reasons well known to the mining community, the sector is not doing well and it is losing its prominence. It may be on the down slide in the twenty first century and may need revival package or privatization.

PROMINENCE TO PRIVATE SECTOR

The motives for privatization are as follows :

Reduced role for government – Government has to govern and other activities can be left to the professionals to do the needful.

Increased efficiency – The experience has shown that the privatization is more efficient, hence, its economic parameters are better. These had been observed all over the world.

Customer satisfaction - The customer feels satisfied as he gets worth his expectations. Being shareholders, they feel interested in well being of the enterprise. This further boosts the moral of the industry and it grows with accelerated pace.

Enriches coffers of government – Government gets various types of taxes which can be used for social good. In the developed countries of the west, the government can afford to invest in infrastructure, social security, education, health care and even in speculative market driven activities. The taxes can be reduced and borrowings, if any, can be liquidated.

Prominence of Private Sector - The twenty first century will, therefore, see the door of privatization and this process may largely be completed by the end of the second decade of the twenty first century.

Shrinking Job Market For Mining Engineers - Progressively reducing mining activity may have an adverse impact on job prospects. There are reports that mining companies are not recruiting mining engineers. These engineers are seeking greener pastures elsewhere. This will have an impact in the years to come and we may face a great shortage of capable personnel. Closure of mines may render many active engineers jobless.

Recycling of Waste - Metallic waste may turn to be a valuable material. The recycling industry may flourish. What shape it will take in the twenty first century is difficult to say at this stage.

Mining Low Grade Minerals – Coal and other metalliferous minerals are of low grade in India. The impurities can be removed by mineral dressing process which inflates the cost of production. Rich mineral resources are depleting fast. In our country iron ore is rejected if it has less than 50% of Fe content. In the developing countries, specially in Russia ores as low as 20% in Fe content are used in steel plants. It is these plants which have contributed to the troubles of our steel industry. Mining engineers will have to give a careful thought to this important aspect as it may soon become a serious problem.

WHAT TO DO

Mining engineer is a jack of all trades but master of mining. He is capable of working out a solution for the industry and possibly as a Britisher could find solutions to coal mining industry in their country.

The scenario which may emerge in the next few years will mostly be the nationalization of mining industry and the curtailment of mining. We may, therefore, have to look into these two aspects.

NATIONALISATION

This seems most certain in view of the global changes, specially after the break up of USSR. The Russians and CIS countries managed their problems in a faster manner and possibly leaving many irritants. In the Indian context, the nationalization can take up "with human face". United Kingdom (U.K.) has experienced a full cycle of nationalization and denationalization of their core industrial sector. During the second world war and prior to that most of the industries were private owned and were under the patronage of the British Crown. Soon after the end of the great war, the labour government took the decision to go in for massive nationalization. The British rail, aviation, mining shipping, gas, etc., were nationalized. This went well up to the seventh decade and during the later part of the decade, public sector went into a non-performing mode. In early eighties, a move was made to gradually privatize public sector industries. In the following paragraphs, some cases are being discussed and suggestions are made for adoption in India.

BRITISH RAIL

In the U.K., the railways had been denationalized. The rail network, platforms, signalling systems etc. remain the state property. The trains are run by private companies. They pay rental for the use of infrastructure. It seems, the system is working well. There are reports that there is customer satisfaction and the competition amongst the companies may reduce the rail fare.

We can adopt this model in the mining sector. The infrastructure presently available in the mines can remain the property of the state. The mining companies can be facilitators and the private companies may run the production apparatus. They will also handle marketing and sales and transportation within the mining areas or up to the road or rail sidings. A certain via media exists in our mines and it can be formalised in an appropriate way.

The owner facilitator may get revenue as per an agreed formula. We can look into the details of the practice adopted by British rail and translate the same in such a way as to suite our requirements. The issue may have to be debated further to work out the modality and the mechanism.

BRITISH ELECTRICITY INDUSTRY

The electricity industry in Britain was privatized to reduce government role, give benefits to customers, to increase efficiency through competition and to increase revenue to the exchequer. There was an urge to regulate the industry so as to serve the nation in a better manner.

The process began by separating generation from transmission. This already exists in India, but still in the hands of public sector. New generating companies as independent power projects are coming in India, but the progress is tardy. In Britain, the private power generation companies brought in competition which resulted in benefiting the end user, the consumer.

The transmission and distribution network bifurcation to private companies may not be beneficial as it will result in duplication or triplication of lines network, hence, this remains intact, possibly the number of bulk consumers increased and the distribution passed into the hands of bulk consumers who further gave benefits to small consumers.

The generation by thermal, hydel and nuclear means were separated as they have different operational characteristics. The larger generating companies were broken into convenient parts and sold to others. Thus new entrants came into market giving boost to competition, which helped in lowering the prices and winning the confidence of the customer.

We can adopt this model also for privatization of coal mining. Here, we may consider breaking up large companies into smaller parts. These parts can be sold to different companies which can compete with each other. The sales and marketing be handled by different companies which may deliver coal up to main dispatch centre. These companies may handle the entire mining operations and own them also.

The competition amongst producing companies may help in customer satisfaction and lowering the product prices.

In India, we can evolve our own mechanism of privatization suiting to our requirements. Foreign capital, know-how and selling public sector to foreign companies may have to be looked in to carefully. In the initial stage, it appears to be lucrative, but it may have far reaching consequences and they may not be good for the Indian economy.

MANPOWER

The mining industry employs about 6 million persons. It is believed that manpower is surplus in most of the public sector ventures. The closure of mines which may be unviable and privatization may render many qualified persons jobless. These problems have been got over by Britishers in many ways. Some of them are listed below :

1. Entering into consultancy business abroad.
2. Involving in overseas mining ventures.
3. Educational activities.
4. Developing underground space technology.

The British mining engineers opened up a number of consultancy companies or attached themselves with established companies. These companies have many assignments to their credit going on in the developing countries. We may not have to look beyond India to find a model. RITES a reputed company was started by some railway experts who had a vision. Today, RITES has national and international business. They have stakes in railways, highways, civil engineering, transportation and energy business. The company phenomenally grew in last 20 years and is a name to reckon with. The mining engineers may also think on these lines and can float a multi-faceted company. The IRCON is another major construction company manned by railway engineers who take contracts for rail and other construction activities nationally and internationally.

Mining engineers can involve themselves in mining activities overseas. It may be modelled on the pattern of ONGC. The mining companies can take up lease in foreign countries and run the business opening up another field of activity for surplus mining engineers. It can be done under the patronage of Government of India or any consortium of mining companies.

Education is another money charming activity. The mushroom growth of foreign based institutions offering courses in India can teach us many lessons in this field. Leading mining educational institutions, in collaboration with foreign institutions can offer education in the developing countries. This needs to be further studied by the educational institutions.

Underground space technology is a well developed subject in the developed countries. It is no different from mining engineering and we can definitely develop this expertise for us as well as for helping the third world countries. A beginning can be made by introducing courses on underground space technology.

The mining expertise available so far in India is adequate for the planning, design and construction of underground facilities. We can have civic facilities like tunnel traffic by pass pedestrian crossing, storage, service tunnels, car parking and shopping malls. For the industrial use, underground storage for oil, chemicals and water can be considered. Defence underground infrastructure, nuclear wastes disposal facilities and civil defence shelters are also very important from the point of view of national defence. The subject is very vast and needs detailed deliberations for making a strategy for the twenty first century.

CONCLUSION

The above given narration is nothing new to us, we are poised to embark upon a new course of action in the twenty first century to face the challenges ahead. The message is clear and loud. Besides, excelling in the field of mining we must look beyond mining for the survival of the mining community in the twenty first century.

CHAPTER 5

Mining Challenges of the 21st Century

*Olle Östensson**

INTRODUCTION

This paper will offer some notes on the mining industry from the perspective of an outside observer. It will review the situation of the international mining industry at the end of the twentieth century and try to identify the challenges that it is facing as a new century begins. First, the prospects for continued increases in demand for metals and minerals will be assessed, against the background of globalization and trade liberalization. Second, supply side developments, particularly as regards the recent and potential changes in industry structure, will be addressed. Finally, the importance of environmental management will be discussed.

DEMAND

Consumption of minerals and metals has continued to increase in recent decades, spurred on by relatively stable world economic growth. Figure 5.1 shows world consumption of major nonferrous metals during the period 1969 to 1998. The figure is in logarithmic scale to show the rate of growth more clearly. As seen from the figures, although consumption growth slowed down in connexion with the two oil crises, it has been relatively steady over the past decade and a half, growing at a rate about equal to or slightly lower than that of world GDP.

The increase in demand has in recent years been supported mainly by a rapid consumption increase in the fast growing economies of Asia. The expansion of consumption in these countries conforms with the "intensity of use" hypothesis. Intensity of use is defined as consumption of a material in relation to some measure of economic development, such as GDP, GDP per capita or industrial production. It has been observed that, historically, in developed countries, intensity of use, however measured, grew rapidly from an initial low

---

* United Nations Conference on Trade and Development.**

** The views expressed in this paper are those of the author and do not necessarily represent the views of the UNCTAD Secretariat.

level. Subsequently, the rate of growth slowed down and eventually, intensity of use declined. Thus, intensity of use, when graphed, described an inverted U shape.[1] It has been assumed that all or most countries would exhibit the same development of intensity of use over time, and this assumption appears to have been borne out in the case of the Asian countries. It has also been noted that intensity of use tends to be higher in fast growing economies, presumably due to the high metals intensity of investment compared to consumption (Tilton, J., and Lohani, P.R., 1993). Again, the development of metals consumption in the Asian economies appears to confirm this hypothesis.

Fig. 5.1 : World metal consumption, 1969-1998, thousand metric tons

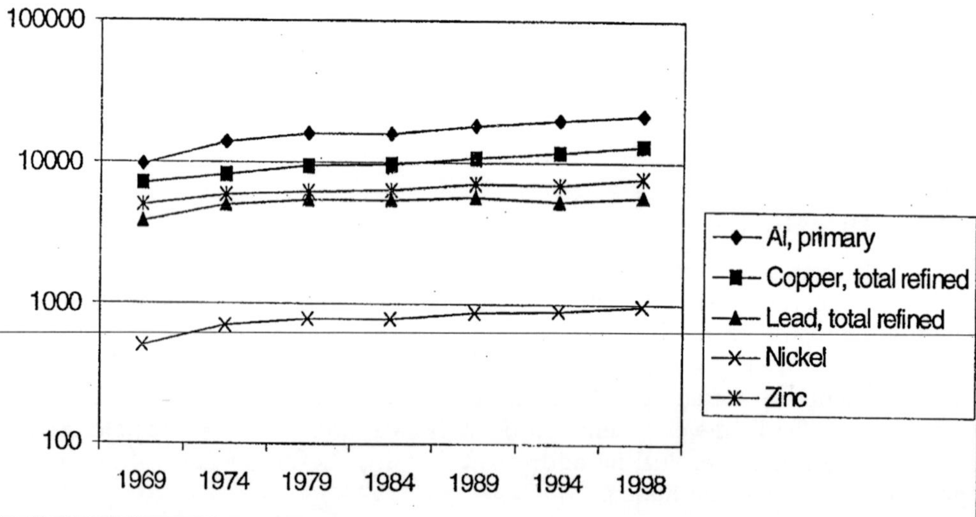

Source : World Bureau of Metal Statistics.

Metals consumption in individual countries is conventionally defined as consumption of unwrought metal in semi-manufactured products. However, the consumption of semi-manufactured products used in the production of finished products provides a measure that is preferable from many points of view, particularly since it would be expected to be more closely related to the level of economic development.[2] There is significant international trade in semifabricated metal products, and this trade has grown fast in recent years. World exports of semifabricated products of both aluminum and copper increased at an average annual rate of between 6.5 and 7 per cent from 1990 to 1996 (UNCTAD, 1997a, 1997b), partly as a result of reductions in tariffs on these products, but more importantly, owing to increased exports of finished products from countries with insufficient semifabricating facilities, including Asian countries. Although these countries have increased their semi-

1. See Malenbaum (1975) for the first formulation of the intensity of use hypothesis. Further discussion of the hypothesis is found in Tilton (1990). For a longer perspective on intensity of use for metals, see Crowson (1998).

2. Empirical tests of the intensity of use hypothesis have sometimes yielded ambiguous results, probably at least partly because consumption has not been measured as use of semi-fabricated products.

fabricating capacity dramatically and have even become important exporters of these products, their net imports have increased steadily. Imports of semimanufactured aluminum and copper products into the group of rapidly growing Asian economies (defined here as China, Hong Kong, Indonesia, Malaysia, Philippines, the Republic of Korea, Singapore, Taiwan province of China and Thailand) increased at average annual rates of 15.6 and 20.2 per cent respectively during the same period. Table 5.1 shows the dramatic increase in the share of these countries in world imports of unwrought and semifabricated copper and aluminium and in their share of world exports of semifabricated products of these metals. The relatively low share for unwrought copper imports is of course a consequence of the large copper production in several of the countries concerned. Thus, a focus only on consumption of crude metals significantly understates the importance of this group of countries to world metals consumption. If any confirmation of their importance was needed, it was amply provided by the dramatic impact on metals markets of the financial crisis that occurred in several of these countries in 1997.

Table 5.1 : Share of fast growing Asian economies in world metals trade

Product	*Share of world trade, per cent*	
	1990	*1996*
Unwrought aluminium imports	11.2	20.7
Unwrought copper imports	4.5	7.5
Semifabricated aluminum imports	9.3	18.8
Semifabricated copper imports	15.8	30.7
Semifabricated aluminium exports	3.6	5.0
Semifabricated copper exports	9.1	16.0

Sources: UNCTAD 1997a, 1997b.

While economic growth will continue to be a positive influence on metals demand, other factors will also be important. The composition of products using metals is influenced by technological change and by the prices of metals relative to other materials. These factors affect the amount of metals used per unit of output, and in the past their effects have been mainly negative on metals demand. Advances in materials technology are likely to continue and production costs for "new" materials may decline, making them stronger competitors with minerals and metals. Moreover, the rate of recycling, which has increased significantly over the last couple of decades, is likely to increase further, partly as a result of regulatory changes motivated by environmental concerns, and will have a negative effect on demand for primary metals.[3]

Competitive pressures, both from other materials and within the industry, has kept the price trend for minerals and metals on a declining long term trend in real terms. Even

3. The increase in the rate of recycling has been most striking for lead, where the proportion of world refined production coming from scrap increased from 30 per cent in 1976 to 48 per cent in 1997, and for aluminium, where the proportion of recycled material in production rose from under 18 per cent to 25 per cent over the same period (World Bureau of Metal Statistics).

in current US dollar terms, the trend has been pointing downwards in the 1990s. Figure 5.2 shows the development of the UNCTAD price index for minerals, ores and metals during the period 1985 to 1998. The index is calculated in current US dollar terms.

As already noted, trade liberalization has provided a powerful stimulus to metals consumption. It has allowed production of manufactured goods to take place in countries with lower costs, thus lowering the price of products and making them accessible to larger numbers of consumers. Without the decline in finished product prices brought on by trade liberalization, consumption of such goods, and consequently also of metals, would have been significantly lower. The continuation of the process of trade liberalization is, therefore, of paramount importance to the international mining community and offers scope for further expansion of demand for metals and minerals. Only increasing demand for finished products using metals can counter the effects of substitution and of recycling.

Fig. 5.2 : World metal consumption, 1969-1998, thousand metric tons

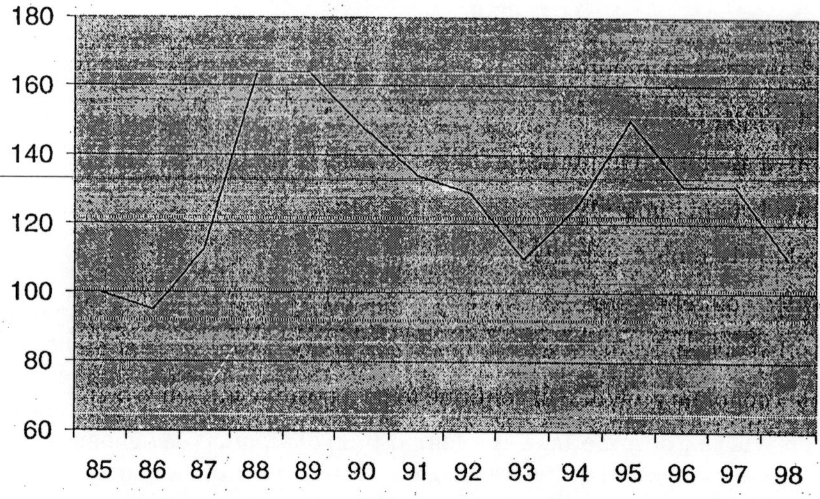

Source: UNCTAD monthly commodity price bulletin

As seen from Figure 5.2, fluctuations in metals prices have been dramatic over he past decade. It should be noted that the figure does not capture variations in prices of individual metals or short term variations, which have been considerable. In addition to changes in supply and demand fundamentals, ever larger amounts of speculative funds and increasingly closer links between financial and metals markets have aggravated the instability of the latter, particularly in recent turbulent periods in the world economy.

The relationship between financial and metals markets works in several ways. One of the more obvious is that the liquidity of metals futures markets, combined with their large fluctuations, may at times make them attractive to investors who are drawn by the potentially large gains and not deterred by the correspondingly large risks of losses. Thus, the growth in speculative funds over the past decade or so has contributed to increasing instability of metals prices, although it does not change market fundamentals. "In periods of heightened uncertainty about the future movements of commodity prices, or exchange rates, or inflation rates, very large amounts of funds can be switched, either for hedging or for speculation, from other asset markets into commodities, or vice versa" (Maizels, 1994, p. 1692). "The incursion of commodity funds as significant investors means that commodity prices increasingly anticipate, rather than reflect, changes in industrial production" (Financial Times, 1999) or in other aspects of the real economy that concern products with established futures markets.

Secondly, since profits of metals companies tend to fluctuate in tandem with, but more than, metals prices, investment in the shares of these companies, mainly mining and metals companies, will also attract investors who are looking for large potential gains. While investors would in theory be expected to base their valuations of shares on long-term profit expectations, their behaviour seems to indicate that increasing weight is accorded to short-term profit prospects. Shares in gold mining companies have traditionally been the preferred objective of such investment, but the shares of other mining companies have also been intensively traded at times of price fluctuations. It has been noted that the share prices of mining companies tend to anticipate price movements in metals. This has certainly been the case in the latest crisis, when metals price increases in the spring of 1999 were preceded by large increases in the shares of mining companies. For example, initially reasonable rises in copper and mining shares, anticipating certain closures (coupled by speculative rises in prices), had gone so far in late May 1999 that it was feared they would "render less likely the very closures on which they were predicated" (Financial Times, 1999). Thus, an important effect of such rises in share prices is that, although temporary and speculative, they affect decisions concerning the real economy by reinforcing the cyclical behaviour of the companies concerned.

SUPPLY

The world mining and metals industry has responded to the declining real price trend by efforts to reduce costs and raise productivity. The identification and development of high quality deposits which can be exploited at low cost has been an important part of such efforts. One of the most significant trends in minerals and metals production in the 1990s has been the rapid increase in developing countries' (except Africa's) share of world production, mainly as a result of an expansion of foreign investment, particularly in Latin America. From 1990 to 1995, the share of developing countries in the value of total nonfuel mineral production at the mining stage increased from 46.6 to 55.2 per cent (UNCTAD, 1997a and 1997b). The increase was particularly rapid for metals, including gold. It can be expected that the increase will continue, since exploration, which is a good indicator of future mineral investment and production, has shifted geographically from developed to developing countries. While in 1991, three developed countries (Australia, Canada and the United States) accounted for 61 per cent of exploration spending by large mining companies and Africa, Latin America and developing countries in the Southeast Asia/ Pacific region for 35 per cent, the proportions had changed by 1998 to 37 and 56 per cent (according to annual press releases by the Metals Economics Group, Halifax, Canada). Developing countries have also been able to increase their share of world production of refined metals, which has grown steadily for several decades and is approaching their share for unprocessed minerals. Table 5.2 shows the evolution of the share of developing countries in world production of some refined metals. Due to increased processing in developing countries, both of ores to metals and of metals into semi-fabricated products, the share of developing countries of world exports of unwrought metals has increased less than that of world production.

Table 5.2 : Developing countries' share of world production of refined metals

Metal	*Share of world production, per cent*	
	1975	*1988*
Aluminium	9.7	31.9
Copper	23.8	41.1
Lead	15.4	30.2
Nickel	19.4	24.7
Zinc	14.7	38.0

Source: World Bureau of Metal Statistics.

Efforts to improve profitability through cost reductions and productivity increases appear to take place in spurts and to be triggered mainly by crises, whether structural or cyclical. In such circumstances, high-cost labour intensive capacity is shut down (Tilton, 1999). Restructuring of this type took place during the 1980s and early 1990s, mainly in response to overcapacity. Labour productivity in metal mining in Canada increased at an annual rate of 12 per cent during the period 1986 to 1996, while that in nonferrous metal smelting increased at an even higher rate of 13 per cent per year. In the United States, labour productivity in nonferrous metal smelting increased at an annual rate of slightly over 6 per cent during the same period (Natural Resources and Energy Canada, 1998). Given the technological homogeneity of the mining and metals industry and the weight of these two countries in world production, the figures are likely to be broadly representative of productivity gains made in the industry as a whole, particularly since the early globalization of minerals and metals markets has exerted a strong equalizing pressure on costs across countries.

Much of the productivity increase is the result of an increase in the size of mining operations. Figures 5.3 and 5.4 show how the size composition of surface and open pit mines have changed over the past 30 years. The increase in the number of mines with an annual production of more than 3 million tons is evident, particularly for surface mines. Over roughly the same period, from 1970 to 1996, the average annual ore production of metal mines in the United States increased from 540,000 tons to 5.5 million tons (US Bureau of Mines/US Geological Survey). Thus, economies of scale and the increased mechanization possible in larger mines have been the preferred response to falling prices and have made possible the rise in labour productivity mentioned above. Technological developments, such as the introduction of SX-EW for copper, have of course also contributed to the improvement in productivity.

The attempt to counter declining prices with the scaling up of operations and replacement of labour with capital follows partly from the industry tradition of focusing on cash operating costs when making investment decisions. A common decision making rule has been to go ahead with an investment if cash operating costs are projected to be in the lowest industry quartile. The reasoning is that this will allow the mine or smelter to survive during times of low prices, while periods of high prices are counted on to provide enough cash flow to pay off loans and offer shareholders a good return on capital. However, the application of this and similar rules across the industry has resulted in a flattening of the industry operating cost curve for most metals. Fixed costs are a high proportion of total costs, and exit barriers, in the form of high capital charges, high decommissioning costs due to environmental regulations and long-term contracts for inputs such as electric power, are important. As producers tend to go on producing as long as they can cover their relatively low variable costs, prices have to fall far before closures occur. In this industry with a homogeneous cost structure, this means that continuing production in spite of declining prices is a general phenomenon. Unwillingness to undertake production cuts while demand was still relatively strong although prices were falling contributed to the severity of the recent price falls.

Fig. 5.3 : Mine sizes, surface mines, 1967-1997

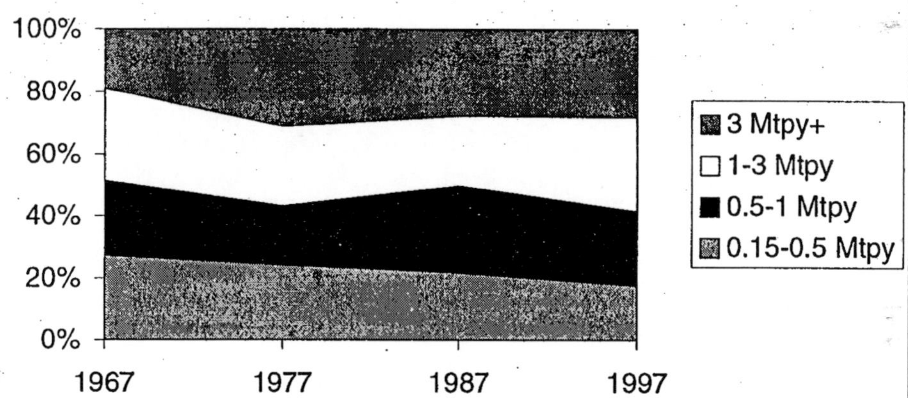

Source: Mining Magazine

Fig. 5.4 : Mine sizes, underground mines, 1967-1997

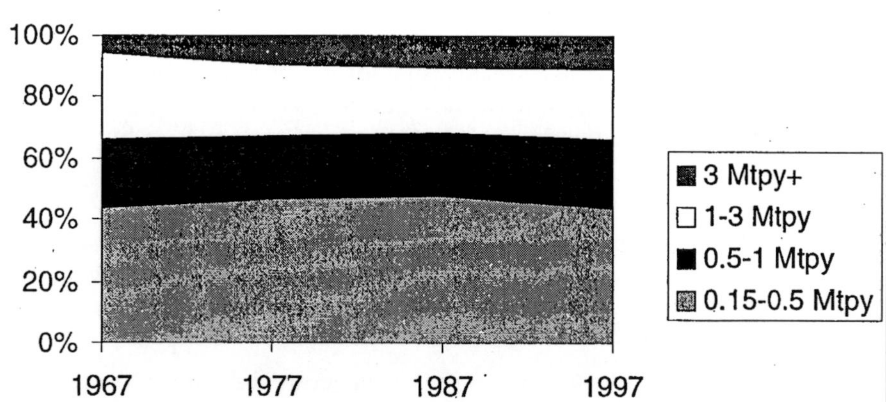

Source: Mining Magazine

Increased operational flexibility has also made it somewhat easier for mining and metals companies to respond quickly to increased demand. In most mines, production can be increased relatively quickly, particularly in large open pit mines which account for an increasing proportion of production. Metallurgical plants find it somewhat more difficult to raise production quickly, but as a legacy of the latest round of restructuring, most major metals companies have reserve high-cost capacity which has been "semi-retired" but which can be brought back into production in a matter of months in order to profit from a temporary price peak. For instance, over 10 per cent of the world's aluminium smelter capacity was closed more or less temporarily at the end of both 1997 and 1998 (U.S. Geological Survey, 1999). A large portion of this production capacity could be brought back into production relatively fast.

As a consequence of the reluctance to cut output when prices are low and the facility with which output is increased in response to price increases, the price troughs risk becoming longer and deeper, while the duration of the price peaks may become shorter. The links between financial and metals markets discussed in the preceding section may reinforce this tendency. Moreover, investment in the mining and metals industry tends to be procyclical, that is, it increases when prices are high (Linnell, 1999, Spilimbergo, 1999). Since delays in projects are costly, the investment activity is usually continued even when prices fall. Consequently, the additional capacity usually comes into operation when prices have turned down, thus reinforcing the downward pressure on prices. In contrast, projects on which construction has not yet started tend to be postponed when prices fall.

The tendency to delay cutbacks in times of low prices and to try to recuperate losses in times of high prices has had a negative impact on the profitability of the mining and metals industry.[4] Recent major mergers in the copper and aluminium industries may be seen as attempts to create groups that are large enough to react to market events in a way that takes longer term objectives into account and as first indications of further mergers and takeovers. It may also be significant that among the arguments presented in favour of the merger between Alcan, Alusuisse and Pechiney was the creation of a strong market presence in packaging. Vertical integration, including into finished products, may thus be seen as a complementary strategy that will enable mining and metals companies to exercise a degree of control over prices and escape from the unstable environment of commodity markets. The challenge to the industry is to market its products in competition with other materials producers and to focus increasingly on consumer needs and customer services.

THE ENVIRONMENT

Over the last two decades, possibly no issue has absorbed more of the attention of the mining and metals industry than the natural environment. Although many other activities have much more serious and wide-ranging effects on human health and ecosystems, and although environmental disturbances from mining are local and very insignificant in terms of the size of the area affected[5], it has a dramatic and highly visual

4. An example is provided by the Australian mining industry, where the return on average assets employed decreased over the last ten years from 8.9 to 1 per cent, while the return on average shareholders' funds fell from 18 to 1.8 per cent during the same period (Minerals Council of Australia, 1998, *see also* Linnell, 1999).

5. Over the period 1930-1980, only 0.25 per cent of the total land area of the United States was used for surface mining, disposal of wastes from surface and underground mines, and disposal of wastes from mineral beneficiation and further processing. Some 47 per cent of the land affected by mining and waste disposal had been reclaimed by the end of that period (Johnson, Paone, 1982).

impact on the environment. Accordingly, it became an early target of environmental criticism. Moreover, since mines often stay in operation for decades and even centuries, the newest technology and environmental management systems co-exist with mine designs that were determined in the early years of industrialization and that are difficult to change. For this reason, the environmental performance of mining companies has sometimes been assessed on the basis of capital investments of a much older vintage than for enterprises in other sectors.

It is clear, however, that major improvements have taken place in the way that the industry handles environmental problems. Moreover, it has been possible to undertake the improvements at relatively low costs.

Estimates of the costs as a proportion of total production costs of course vary from negligible to crippling. Those who are trying to demonstrate that the costs are low point to examples of mines where investments made for environmental reasons have led to higher productivity and lower costs, while those who aim to show that the costs threaten the industry's existence quote the full cost of investments, including elements that have little if anything to do with environmental requirements. It is easy to find examples supporting either version of the facts, partly because it is very difficult to identify the exact portion of the investment that is due to environmental objectives and partly because circumstances differ from one operation to another depending on geology, climate and other factors, including, most importantly, whether the investment is made as an addition to an existing facility or in a new one. In 1992, a reputable consulting firm predicted that "environmental upgrading" would cost the mining and mineral processing industry some US$ 6,000 million and that this would have "an enormous effect on exploration and new smelters as industry cuts back due to the environmental downside" (Metal Bulletin, 1992). While the figure may indeed appear enormous and is likely to be exaggerated, it deserves to be pointed out that it corresponds to less than 5 per cent of the value of world non-fuel mineral production in 1990 (UNCTAD, 1997a). It appears reasonable to assume that the costs of environmental protection are indeed positive and significant, but that they do not constitute an important constraint on the expansion of production.

The industry has demonstrated an impressive capacity to adjust to regulatory changes (usually after first having complained bitterly about the costs), and has achieved results when it comes to reducing environmental impacts that would hardly have been believed two decades ago. This is due partly to the development of new technologies, but above all to the introduction of management and planning methods that integrate environmental impacts in the project planning process from the outset of a project and that do not see environmental mitigation as something to be added as an afterthought once the project design has been decided. Clearly, the industry has realized that the costs of project delays and possibly closures due to the need to take mitigating measures that were not foreseen but which impose themselves as a result, for instance, of changed regulations are likely to be higher than the cost of doing it right from the outset. Moreover, pressure from shareholders, lenders and consumers, as well as the risk of being held liable for damages have provided powerful incentives for companies to improve their environmental management. The change in the way mining companies present and describe themselves and their environmental performance to the rest of the world is clear from the stated goals of the companies contained in their environmental policies and similar documents.[6]

6. One of the most radical formulations is found in the environmental policy of the Noranda Group, according to which the Group companies and operations "...commit to the principle of sustainable development, which means that our economic decisions will not take priority over considerations of health, safety and the natural environment." (www.noranda.com).

In addition to the various factors just described, and which may be assumed to determine the "enlightened self-interest", other factors may contribute to the integration of sustainable development type objectives into the goal structure of mining companies. These factors do not influence the economic situation of the company but rather the behaviour of its employees and the type of technical and economic solutions that they are likely to prefer.

First, employees of mining companies at all levels, up to and including managing directors, are human beings like the rest of us and do not function in a social vacuum. In a situation where environmental responsibility has become a social norm, at least among developed country professionals, decision makers in the mining industry are likely to lean towards the more environmentally friendly solution, other things being equal (and possibly sometimes when they are not).

Second, environmental management has become an important part of the curriculum in all schools for mining engineers, and it is generally accepted that the solutions to engineering problems should take into account environmental factors. Thus, ability to manage environmental impacts is one of the parameters that define the professional competence of a large portion of the people influencing technical solutions in the industry and environmental consciousness has become part of the professional culture of mining company employees. This factor provides a powerful bias in favour of environmentally friendly solutions.

A third and final factor is the phenomenon of standardization of processes and equipment. While miners like to say that every mine is unique, they tend to buy their equipment from the same sources and use the same processes, often because it is easier to use a solution "off the shelf" than to develop something new. Since equipment makers respond to the needs of their customers, and since a large portion of their best customers require environment friendly equipment for all the reasons already given, it may be simpler to supply the same equipment even to those who do not specifically ask for it rather than to design environmentally "unfriendly" equipment. As a consequence, even the performance of companies with very limited environmental consciousness is likely to improve over time.

This does not mean that all is well and that the environmental problems associated with mining can be considered as having been solved. There is still have a long way to go to the situation where a mine can be closed down and forgotten, with no need for monitoring or precautionary measures. Moreover, new and more ambitious targets will continue to be set as regards the reduction and mitigation of environmental impacts. Mining companies will have to be alert to new developments and prepared to take action to deal with new requirements.

One of the largest environmental challenges facing the international mining community today is posed by the many closed mine sites in developing countries and countries in transition which constitute major hazards for the environment and where it is not possible to use proceeds from production to finance a clean-up. Many of the sites may give rise to considerable environmental damage if left unattended. While the need to rehabilitate these "orphan sites" is widely recognized, it is also clear that the governments of the countries concerned - unlike governments in developed countries, where this problem is being dealt with - lack the financial means to do so and that no other realistic solutions to the financing problem have been found. In fact, the dimensions of the financing problem are not even well known, possibly because it is feared that any effort to arrive at a cost estimate might lead to resignation rather than action. So far, progress with the clean-up has been piecemeal and improvised, and there is no assurance that the most serious situations in terms of human health, ecological impact or economic importance have been addressed on a priority basis. The problem clearly has to be addressed in the context of abandoned hazardous sites in general, and it is to be hoped that institutional and financial solutions can be found that permit an elimination of these continuing dangers to the lives and health of local populations. Governments and industry have to work together to determine the size of the problem and identify realistic rehabilitation methods, financing solutions and a time schedule for rehabilitation. The time schedule can be long - what is important is to begin to approach the problem in a serious manner. While the industry can not be held collectively responsible for the actions of some individual companies, it does have the knowledge about the methods that could be used to deal with the "orphan sites" and it should be concerned about the public impact of the problem. Therefore, the industry may consider it to be in its own interest to be a part of the solution to the problem of orphan sites, rather than fostering the impression that it is only a part of the problem.

What has just been stated with respect to the need for the industry to demonstrate that it is prepared to contribute constructively to society goes beyond the issue of abandoned minesites. Mining does not have a positive image in the minds of the public, and while this may be partly undeserved, more could be done to demonstrate the willingness of the industry to solve problems. The following quotation from a book published by a Filipino NGO is representative of some of the opinions expressed about mining:

> "Globalization of mining has legitimized imperialist plunder and seized the sovereignty of the peoples of the world. It has dislocated workers, ravaged the environment, dispossessed indigenous peoples, displaced small-scale miners, and shattered dreams of national industrialization on a global scale." (Tujan, Guzman, 1998)

While these opinions are not shared by most NGOs or the public in general, they do express a concern about the impact of mining and demonstrates the need for the mining industry to interact more constructively with the people who are most directly affected by its activities. The establishment of large areas, particularly in developed countries, where mineral development is prohibited is the result of public perceptions of mining as an inherently and unavoidably damaging activity for the natural environment. As already mentioned, many major mining companies have improved the manner in which they interact with local communities and the general public, aware of the fact that information about events in isolated areas may quickly attract world wide attention. However, action by individual companies may not be sufficient. It is in this context that recent discussions about the need for global guidelines on environmental management and interaction with society, subscribed to by a large majority of mining companies, assume their importance. Such guidelines would go a long way towards assuring the public that the industry will behave responsibly and strive to maximize its positive contribution to sustainable economic and social development.

REFERENCES

Crowson, P., 1998, *Inside Mining: the Economics of the Supply and Demand of minerals and Metals*, Mining Journal books, London.

Financial Times, 1999, *Analysts see recovery in base metals*, 20 May.

Johnson, J., and Paone, J., 1982, *Land Utilization and Reclamation in the Mining Industry, 1930-1980*, Bureau of Mines Circular 8862, Washington, D.C.

Linnell, Richard, 1999, Untitled paper prepared for the UNCTAD Expert Meeting on the Impact of Changing Supply and Demand Market Structures on Commodity Prices and Exports of Major Interest to Developing Countries, Geneva, 7-9 July.

Maizels, A., 1994, *The Continuing Commodity Crisis of Developing Countries*, World Development, Vol. 22, No. 11.

Malenbaum, W., 1975, *Law of Demand for Minerals*, Proceedings of the Council of Economics, 104th Annual Meeting of the American Institute of Mining, Metallurgical and Petroleum Engineers.

Metal Bulletin, 1992, *Base metal victims of environmental assault*, London, 5 November.

Minerals Council of Australia, 1998, *98 Survey Report*.

Mining Magazine, *Annual Survey of planned projects*, various issues.

Natural Resources and Energy Canada, 1998, *From Mineral Resources to Manufactured Products: Toward a Value-Added Mineral and Metal Strategy for Canada*.

Spilimbergo, A., 1999, *Copper and the Chilean Economy, 1960-98*, IMF Working Paper No. 99/57.

Tilton, J., (ed.), 1990, *World Metal Demand: Trends and Prospects*, Resources for the Future, Washington, D.C.

Tilton, J., and Lohani, P.R., 1993, *A cross section analysis of metal intensity of use in the less developed countries*, Resources Policy, June.

Tilton, J., 1999, *Labour Productivity, Costs, and Mine Survival During a Recession*, paper presented at the Eighth Annual Meeting of the Mineral Economics and Management Society, Ottawa, 15-17 April.

Tujan, A.A. Jr., and Guzman, R-B, 1998, *Globalizing Philippine Mining*, IBON Books.

UNCTAD, 1997a: *Handbook of World Mineral Trade Statistics 1990-1995* (UNCTAD/ITCD/COM/2), New York and Geneva.

UNCTAD, 1997b: *Handbook of World Mineral Trade Statistics 1991-1996* (UNCTAD/ITCD/COM/9), New York and Geneva.

US Bureau of Mines/US Geological Survey, *Mineral Yearbook*, various issues

U.S. Geological Survey, 1999, *Mineral Commodity Summaries*, Washington, D.C.

World Bureau of Metal Statistics, *World Metal Statistics*, various issues.

CHAPTER 6

Challenges for the Indian Mining in 21st Century

*A.N. Bose**

ABSTRACT

India is the largest democracy in the world with proven federal system of more than five decades, sustaining the second largest population of the world over 975 million. Industrial revolution has got momentum in fifties' and today, the Indian economy is the fifth largest in the world and second largest amongst the developing countries based on purchasing power, parity and poised to become a 'developed' nation by the year 2025.

Obviously, to meet the ever increasing demand of minerals, which is the principal raw material for any industry, Indian mining industry has registered a substantial growth both in terms of quantity and value of minerals produced from US $58 million in 1947 to over US $8.83 billion in 1998-99. India has already come a long way from the small manual quarrying operations to mega-sized operations like Kudremukh iron ore and Neyveli lignite mines. The Indian mining industry is maintaining an average growth rate of about 6% per annum providing daily employment to nearly 0.7 million persons.

In 1991, India initiated a wide ranging programme of economic reforms. These reforms have transformed the business environment and created an atmosphere for development of energy, mining and infrastructural projects. Integration of Indian economy with that of the world has started showing encouraging results and more than 52 proposals aggregating to about US $0.73 billion have already been approved by the Foreign Investment Promotion Board. However, in a group of 3 largest mineral producing nations in Asia, *viz.* China, India and Russia, which have only recently opened their doors to international mineral development, India's ranking for 'development potential' and in terms of 'level of foreign investment' is not upto the desired level. This calls for major initiatives in technology upgradation, development of infrastructural facilities and investor-friendly climate.

---

*Controller General, Indian Bureau of Mines, Nagpur.

The paper deals with the evolution of the Indian mining industry since independence and its projections for the initial years of the next millenium and an attempt has been made to analyse the changing mining scenario and some of the challenges that are likely to be faced by the Indian mining industry.

INTRODUCTION

The history of mineral development is as old as civilisation itself. Through the ages, the earth has proved itself as a marvellous treasure of minerals which have now intrinsically woven into the fabric of human society. In India, development of minerals dates back to ancient times. Remnants of old workings can still be seen in some parts of the country. Mining of lead and zinc at Zawar, copper at Khetri, gold at Kolar are nothing but the continuance of mining activities which started in ancient times.

India is richly endowed with mineral resources. These comprise fossil fuels, ferrous and non-ferrous ores, fertilisers, refractory and other industrial minerals.

Immediately after independence and adoption of Industrial Policy Resolution, the search for minerals particularly for those essential to industrial development was intensified. The Geological Survey of India which was the only organisation set up to undertake the investigations of mineral deposits prior to independence was strengthened and Indian Bureau of Mines was established assigning the pivotal role for carrying out development of country's mineral resources with due regard to conservation of minerals. Atomic Energy Commission and National Laboratories particularly those connected directly with mineral development, such as Central Fuel Research Institute, National Metallurgical Laboratory, Central Glass & Ceramic Research Institute, etc., were set up to ensure development of mineral industry in the country on scientific lines and to provide facilities for fundamental and applied research. In the subsequent years, many more R&D organisations, mineral exploration and development agencies, regulatory authorities were established. All these efforts have given very encouraging results.

If we look towards mining industry, India is having a well developed mining sector, which has vast geological potential with over 20,000 known mineral deposits. As many as, 87 minerals comprising of 4 fuel, 11 metallic, 50 non-metallic and 22 minor minerals are being exploited aggregating about 500 million tonnes of production per annum.

INDIAN MINERAL INDUSTRY IN RETROSPECT

Keeping with spirit of Industrial Policy for higher targets for mineral production, the expansion and augmentation of mineral based industry in the country was quite obvious. Therefore, ambitious programmes were launched to increase the production of minerals to meet the growing demand of the core industries like steel, cement, power, non-ferrous metals, fertilisers, etc., and also in view of higher exports for much needed foreign exchange.

Mineral Resources

Concerted exploration carried out in the post-independence period both in known as well as virgin areas, has established large resources of minerals. Indian mineral industry has a well established exploration infrastructure. Geological Survey of India and the Mineral Exploration Corporation Ltd. are the main exploration agencies supplemented by various State Directorate of Mining and Geology have carried out extensive exploration. As a result, the mineral base has increased from 34 to 87 minerals. Table 6.1 shows the relative availability of various minerals. It can be seen that country has to rely heavily on imports of many minerals categorised as deficit and scarce. Table 6.2 shows the recoverable reserves of important mineral from 1970 to 1995, and life index of minerals beyond 1-4-1999.

Table 6.1 : Relative availability of various minerals

Sl.No.	*Mineral Grouping*	*Abundant*	*Adequate*	*Deficit*	*Scarce*
1.	Fuel	Coal	Lignite	Coal (Coking)	Petroleum crude
2.	Metallic Minerals				
	(i) Ferrous	Iron Ore	Chromite (Metallurgical grade) Manganese Ore	Chromite (Refractory grade)	Nickel, Tungsten, Cobalt, Molybdenum, Vanadium
	(ii) Non-ferrous	Bauxite (Metallurgical grade)	Zinc	Bauxite (Chem. & Refractory grade) Copper, Lead	Antimony, Gold, Platinum group metals, Silver, Tin
3.	Industrial Minerals	Barytes, Ball-, clay, Bentonite, Dolomite (excluding low silica grade) felspar, fire clay, Fuller's earth/ Garnet, Gypsum, Ilmenite & Rutile, Kaolin, Limestone, Magnesite, Mica, Pyrophyllite, Pyrite, Ochre, Quartz & Silica Sand, Quartzite, Beach sand, Sillimanite, Zircon, Steatite	Corundum, Graphite Rocksalt, Vermiculite, Wollastonite	Apatite & Rock-phosphate, Asbestos Fluorite, Kyanite	Andalusite, Borax, Diaspore, Diatomite, Native, Sulphur, Potash
4.	Precious Stone				Diamond Emerald

Table 4.2 : Recoverable reserves of major minerals in India 1970 to 1995

Minerals	*Unit*	*1970*	*1980*	*1990*	*1995*	*Life Index beyond 1-4-99 (in years)*
Fuel Minerals						
Coal	Million tonnes	94000	118505	186044.24	203000	NA
Lignite	-do-	2268	2100	5687	26900	NA
Metallic Minerals						
Bauxite	Million tonnes	233*	2489*	2525	2462	214
Chromite	-do-	9	111	88	88	35
Copper Ore	-do-	244	456	325	4.3	71
Metal	-do-	3.5	5.7	3.2	417	NA
Iron Ore :	-do-	8244	11470	9602	11057	132
Haematite						
Magnetite	-do-	2025	6096	3143	3408	132
Lead-Zinc Ore	-do-	107	351	186	179	45
Metal Lead	-do-	3.1	4.6	2.0	2.3	-
Metal Zinc	-do-	3.8	12.2	7.8	10	-
Manganese Ore	-do-	108	117	176	167	52
Gold Ore	-do-	5	8.7	20	18	NA
Metal Gold	Tonnes	47.4	64.9	97	67.9	-
Non-Metallic Minerals						
Barytes	Million tonnes	1.4	73	70	87	157
Dolomite	-do-	1152	4946	4967	4387	413
Diamond	'000 Carats	668	579	1196	981	85
Limestone	Million tonnes	40361	73199	76446	75659	533
Rock Phosphate	-do-	61	130	115	145	NA

*Bauxite & Diaspore.

NA—Not available

Mining

At the time of independence, the mining in the country was conventional and by and large labour intensive. A change from manual to mechanised methods of mines was brought about particularly in iron ore and limestone. Similarly, there is tremendous growth in the number of reporting mines. At the time of independence, there were 1977 working mines (excluding petroleum, atomic and minor minerals) which have gone up maximum 4746 in the year 1987. During the year 1998-99, there were about 3150 working mines in the country including about 550 for coal and lignite. Out of these, 2602 mines were in metalliferrous sector covering a lease area of 6,92,491 hects. comprising of 9244 leases. In the metalliferrous sector, about 350 mines are of 'A' category and 282 were operated by Public Sector Undertakings[1&2]. Till 1950s, almost all mines even those of coal, iron ore and limestone depended heavily on manual labour. Unskilled workers dominated the mining industry. This scenario underwent a complete change with the mechanisation of mines with the increased participation of the Public Sector in the mining industry. The percentage share of Public Sector in the value of mineral production in the year 1970 was just 30%. However, with the nationalisation of coal mines and new mega project came up in public sector, its share has increased upto 90%. There has been growing emphasis over the recent years on mechanisation of mining operations, specially in iron ore, limestone and coal and lignite. India has already come a long way from the small manual quarrying operations to mega-sized operations like Kudremukh iron ore and Neyveli lignite mines. Over the past decade, the Indian mineral industry has been maintaining an average growth rate of about 6% per annum[3]. Fig. 6.1 shows the annual growth rate in mining industry.

Mineral Production and its income

The Indian mining sector has registered a rapid growth since independence. During the last five decades, the production of all the principal minerals except gold and mica increased enormously. Though, the increase was almost continuous, it is most discernible in late seventies and eighties when the index recorded a faster growth in mineral production which has been made possible due to significant development in the mining industry in general and fuel minerals in particular. Table 6.3 indicates the production of important minerals in India from 1950 to 1998-99. The total value of mineral production has consistently increased over the past five decades and has risen from US $ 58 million in 1947 to over US $ 8.83 billion in 1998-99 (excluding atomic & minor minerals) and thus contributing a range of 2 to 3.5% of its GDP[2]. Although, the contribution of mining industry to GDP is not significant, India being a developing nation cannot afford to deprioritise the endeavours for mineral search and their cost effective exploitation due to absolute dependence of the core industry and the consumer world as minerals and mineral based products. The performance of the Indian mineral industry towards the National Income since independence is given in Table 6.4.

PRESENT MINING - ON THE THRESHOLD

In 1991, India initiated a wide ranging programme of economic reforms. In tune with this, a new era was ushered in when the new National Mineral Policy was announced in March, 1993 and the mining sector was thrown open for private initiative and investment. The policy envisages foreign technology and equity participation in exploration and exploitation of scarce and high value minerals. Foreign equity participation in joint ventures in mining promoted by Indian companies is also encouraged. Processing units which wish to develop captive mines to secure assured supplies of raw materials are allowed equity

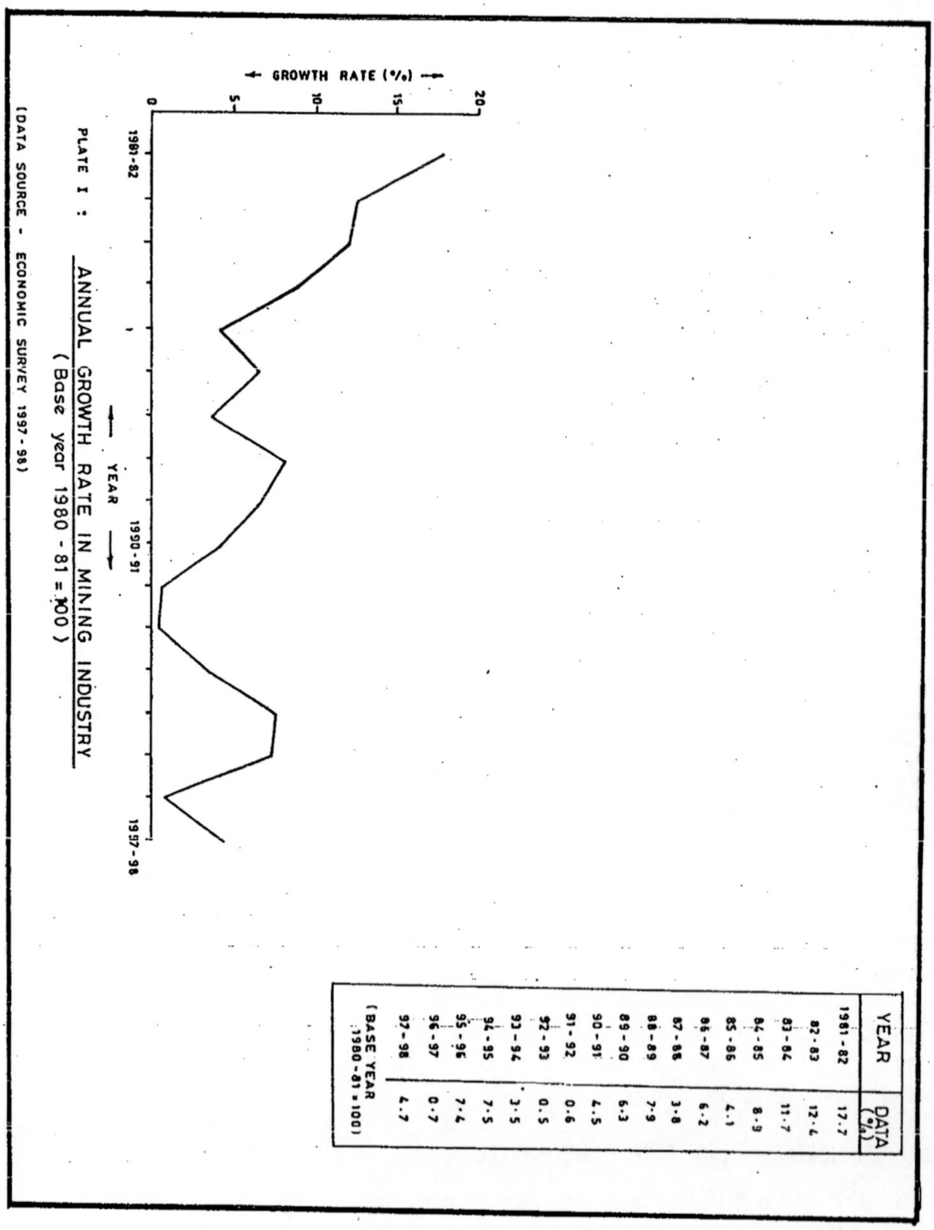

Fig. 6.1 : Annual Growth Rate in Mining Industry (Base year 1980-81=100)

Table 6.3 : Production of Important Minerals in India

Minerals	*Unit of quanity*	*1950*	*1960*	*1970*	*1980-81*	*1990-91*	*1991-92*	*1992-93*	*1993-94*	*1994-95*	*1995-96*	*1996-97*	*1997-98*	*1998-99*
Fuel Minerals														
Coal	'000 t	32805	52593	73698	114012	211616	229354	241998	249000	258000	273000	286080	297699	293565
Lignite	-do	21	47	3545	5104	14073	15811	16618	18000	19000	22000	22540	23047	23009
Metallic Minerals														
Bauxite	'000 t	65	387	1374	1936	4984	5013	5145	5535	4913	5565	6036	5984	6452
Chromite	-do-	17	107	274	327	940	1082	1071	1065	1132	1700	1456	1564	1404
Copper Ore	-do-	3636	448	518	2010	5255	5207	5210	5009	4767	4737	3896	4499	4253
Iron Ore	-do-	3144	16609	31366	42167	55591	58534	57495	59645	64506	67418	68173	73445	70683
Lead (Conc.)	Tonnes	2008	6245	3880	18647	44237	53255	60648	54000	53000	62000	60271	60972	63856
Zinc (Conc.)	-do-	666	9787	15888	49741	136958	252540	301283	290000	269000	289000	277000	292498	349934
Manganese Ore	'000 t	935	1452	1702	1634	1492	1640	1903	1696	1681	1837	1871	1606	1526
Gold	Kg.	6125	4995	3241	2404	2207	2041	1850	2075	2373	2036	2710	2636	2463
Non-Metallic Minerals														
Dolomite	'000 t	54	650	1148	2065	2648	2949	3232	3443	3419	3718	3400	2800	2908
Diamond	Carats	2679	1159	20325	14397	17976	18213	18017	19222	25518	29931	31836	31018	34579
Limestone	'000 t	4942	12935	23843	30202	70125	77185	80206	83159	93207	96832	104029	107568	109835
Apatite & Rock Phosphate	-do-	3	15	172	532	684	603	707	1048	1108	1319	1350	1192	1623

participation in a manner and to the extent applicable to such processing units. It is also intended to promote small-scale mining of small and isolated deposits of minerals while safeguarding vital environment and ecological imperatives. Emphasis in the new policy is on the export of minerals in value-added form and recycling of metallic scrap and mineral waste. In line with the objectives of the new National Mineral Policy, the Mines and Minerals (Regulation & Development) Act, 1957 was also amended in 1994 to encourage in-flow of foreign investment and technology into mineral sector. To further remove the bottlenecks and also to provide transparency, the Central Government issued a set of guidelines in 1996 to facilitate grant of large areas for prospecting, which may go upto 5000 sq. kms. for a single prospecting licence, and 10,000 sq. kms. for a party for the whole country.

Table 6.4 : National Income from Mining Sector (Excluding atomic minerals) (At current prices)

(Rupees in crores)

Year	*National income from mining sector*	*% contribution to National Income*
1948-49	56	0.65
1950-51	64	0.75
1960-61	134	0.95
1970-71	331	0.91
1980-81	1557	1.41
1989-90	11203	3.23
1992-93	14589	2.3
1993-94	16818	2.3
1994-95	18879	2.1
1995-96	20554	2.0
1996-97	21285	1.8
1997-98	22682	1.8(e)
1998-99	22222	1.8(e)

e - estimated

(*Source* – IBM 1992 & IBM Compilation).

These reforms have transformed the business environment and created an atmosphere for development of energy, mining, infrastructural projects etc. Integration of Indian economy with that of the world has resulted in the flow of investment in India. The measures taken in new National Mineral Policy, 1993, have started yielding encouraging results and more than 52 proposal aggregating to about US $ 0.73 billion have already approved by the Foreign Investment Promotion Board in mining sector. Further, new proposals are in the pipeline. Some more legislative changes contemplated which will attempt to decentralise powers over mineral reserves by handing over key decisions to State Government will further liberalise the mineral sector and make States major players in the development of their mineral resources.

CHALLENGES AHEAD

Increasing Population

Any discussion on the mineral sector invariably turns to the population growth rate in India. While it took 300 years for the population to double by the beginning of this century, it doubled again within 70 years and is likely to overtake that of China by the year 2030. Therefore, it is likely that brunt of this population growth will be more serious concern on all our natural resources such as land, water, air, etc., including that of on mineral commodities. Table 6.5 gives the important socio-economic parameters in India and its projections.

Table 6.5 : Important socio-economic parameters in India

	1970	*1980*	*1990*	*1993*	*2000**	*2010**
Population (million)	541	679	837	890	1000	1140
GDP (US $ billion)	110	149	256	296	420	700
Rate of growth %	3.5	3.1	5.6	5.0	5.1	5.2
GDP per capita (US $)	203	219	306	333	420	631
Rate of growth %	1.0	1.0	3.4	3.2	3.2	4.2
Urban population %	19.1	23.3	25.7	27.8	30.5	36.6
Electricity Generation (TWh)	61	119	289	355	540	986
Rate of growth %	11.6	6.9	9.3	7.1	6.2	6.2
Per capita KWh	113	175	346	400	540	865

* Projected

(*Source* : Sethna et al 1995)

Technology development

In the sphere of technology, domestic minerals sector represents a unique blend of small scale and large scale mining operations. While the country has successfully developed high capacity surface mines, in regard to metalliferrous sector, there seems to be some technology gap in regard to productivity, process efficiency and by-product recovery. The fact remains that the country is yet to fully tap its potential for utilising a large number of existing low grade deposits for substituting the export of unprocessed raw materials by value addition. Recovery of residual mineral in tailings is also an objective worth pursuing on both economic and environmental considerations. Considering a wide gap between the projected demand and supply, exploration of important minerals like base metals, gold, diamonds, platinum group of metals and fertiliser minerals needs to be emphasised and given priority. At the same time, a proper strategy would have to be evolved giving primacy to technological upgradation and optimum utilisation of metal and manpower resources. In metalliferrous sector, in spite of large-scale operations such as Kudremukh, Malanjkhand,

Panchapatmalli, Khetri, Zawar, Bailadila, etc. India is essentially a country of small-scale mining since, as much as 87% operations can be defined as small-scale. In respect of coal sector, though we have giant operations like Neyveli, Piparawar, old coal mining areas such as Jharia Coalfields are characterised by fire fissures and subsidence on account of spontaneous combustion.

As far as mining machinery development and deployment is concerned, India is lagging behind as compared to advanced countries. While 300 tonnes capacity dumpers are used in other countries, in India 170 tonner dumpers are the latest ones used in coal mines. Front end loader of 25.2m^3 bucket capacity, excavator with 42 m^3 bucket for oil sand mining, draglines of 168 m^3 bucket capacity on a 94.4 m boom are in vogue in advanced countries, in India draglines ranging from 24-39 m^3 capacity have been deployed [(4)]. Therefore, there is need to improve this position by development of indigenous technology for the higher capacity mining machinery to attain the desired level of output.

The areas where there is scope of technology development in Indian mining industry can be summarised below :

- Technology for development of thick coal seams.
- Concept for mobile/semimobile in pit crushing.
- To explore the slurry transportation option in order to reduce cost of material handling.
- Development of technology for conversion of underground mining into opencast mining and vice-versa.
- Effective technology to combat the problems of spontaneous combustion in coal mining area.
- Development of higher capacity mining machineries and deployment of modern machinery such as surface miners etc.
- Effective system for roof support and to combat problems of rock-burst in deep underground mining.
- Wider applications of information technology in mining industry.
- Development of eco-friendly area specific mining technology.

Ocean mining

Given that the present known reserves of many important minerals are depleting fast, the country requires much more sustained and intensive efforts to augment the resource base including that the mineral wealth hidden beneath the oceans. Considering the fact that India is deficient or barely self-sufficient in many basic and strategic minerals and Indian ocean holds a sizeable percentage of the precious polymetallic nodules, there is a need to acquire knowledge and experience in the field of ocean mining and to accelerate the preliminary work presently undertaken in the field of ocean mining.

Foreign participation

With the change of perception that has taken place in the last few years, India has become an excellent international target for transnational investment in mining. In comparison to other target countries, India has several advantages such as availability of basic geological and exploration data, institutional framework, highly qualified scientists, sound legal framework, well established markets to absorb the mineral based products. In spite of this, in a group of 3 largest mineral producing nations in Asia, *viz.* China, India and Russia, which have only recently opened their doors to international mineral development, India's ranking for 'development potential' and in terms of 'level of foreign investment' is not upto the desired level [5]. This calls for investor friendly legislative changes, in order to effectively harness the financial and technological muscle of the foreign mining sector, as also the change of attitude at the ground level.

Environmental concern

There is a growing awareness of pollution problem from mines, specially the problems of waste disposal in surface mines, disposal of tailings in metalliferrous mining, air and water pollution in mining belts, and destruction of vegetation and forests. Public concern and opposition is growing to mining projects, such as limestone mining in the foothills of the Himalayas which has leds to barren hills and land slides because of reckless mining. In India, being a developing nation, such environmental mishaps are common while striking a balance in developing activities and environmental requirements. This is a stupendous task, more so in case of India where land area is only 2.5 percent of the globe and population is as high as 16 percent of the world's total. Therefore, there is a great challenge to implement the various legal measures, blending with environmental awareness campaigns without jeopardising the interests of developing activities.

Human Resource Development

India's manpower strength is a part of its wealth. With a large scientific and technological manpower, the ability to absorb new technology and adopt them to local circumstances is not likely to be a constraint. However, in order to improve the competitive edge of the national mining industry, emphasis is being laid on mechanisation, computerisation and automation of the existing and new mining units. The manpower development programme need to suitably reoriented for this purpose. The existing facilities for basic and specialised training also needs to be constantly reviewed and upgraded from time to time to ensure that adequate trained-manpower at all levels is available for the development of mines and mineral industries.

Infrastructural development

As India's most of the mineral deposits are located in backward, isolated and undeveloped areas and in order to tap the mineral resources optimally, it is essential to develop the necessary infrastructural facilities around the mining areas. In mineral bearing areas, strengthening of railway, road handling, development of port facilities and availability of power, information facilities like telecom, fax, etc., requires immediate attention.

Research and Development

Although, for promoting exploration and exploitation of mineral resources, India has established many R&D institutes, the fact cannot be denied that it would be impossible to survive in this era of high competition created due to opening up of economy, liberalisation of policies, free import of technology alongwith domestic crunch of research fund. Therefore, keeping in view the changed national and international scenario, our R&D institutes has to redefine as to provide scientific and technical inputs to mineral and associated industries for the optimisation of mining technologies for better safety, economy, conservation and environmental management. Such efforts will necessarily call for mutual co-operation between India and many other countries.

VISION FOR THE 21st CENTURY

The future scenario is not easy to visualise. In the present world of scientific and technological explosion, picture changes drastically. The constraints under which the industry has to operate are many ranging from low capital availability to a large population and growing unemployment. As India is poised for a 'developed nation' lot of industrial and infrastructural growth such as construction of express highways, etc., are expected. Therefore, there is likely increase in production of important minerals is foreseen like coal for power sector, iron ore for steel sector, limestone for cement for housing and infrastructural development. Table 6.6 shows the domestic demand projections of mineral commodities in the initial years of the next millenium. As we are entering the next millenium, the mega trends for the minerals sector could well be in the areas of new strategies for mineral exploration, boom in mineral sector through foreign direct investment, major structural change in industry, increasing consumption of minerals, new challenges posed by environmental concerns, rise of small scale mining sector and massive R&D investment for multi-purpose utilisation of low grade ores/wastes. Sustainable mineral development in the face of rising expectations for better quality of life would call for a massive addition to our mineral resource endowment.

Table 6.6 : Domestic demand projections of Mineral commodities in the initial years of the next millenium (in '000tonnes)

Sl.No.	*Mineral*	*1999-00*	*2000-01*	*2001-02*	*2006-07*	*2011-12*
1	*2*	*3*	*4*	*5*	*6*	*7*
REFRACTORY MINERALS						
1.	Mangesite	305	326	349	489	686
2.	Kyanite	49	53	56.6	77	108
3.	Sillimanite	16	17	18.5	26	37
4.	Fire clay	525	561	601	842	1182
5.	Bauxite (RG)	389	410	439	616	863
6.	Chromite (RG)	74	80	85	119	166
7.	Dolomite	5707	6106	6534	9164	12853
8.	Graphite (B)	15	16	17	24	33
9.	Zircon	21.9	23.5	25	35	49
10.	Silica Mineral	1889	2021	2162	3031	4251
11.	Kaolin	496	531	569	797	1118
12.	Ball clay	190	203	218	306	429
13.	Feldspar	134	143	154	215	300
14.	Wollastonite	48	52	56	78	109
FLUX & CONSTRUCTION MINERALS						
1.	Limestone & other calcareous minerals	137004	147964	159801	234000	344998

1	2	3	4	5	6	7
2.	Gypsum	4032	4314	4616	6475	9081
3.	Asbestos	106	114	122	171	239
4.	Ochre	97	104	111	156	219
FERTILISER & CHEMICAL MINERALS						
1.	Apatite & rock phosphate	3604	3856	4127	5788	8118
2.	Potash	789	844	904	1268	1778
3.	Sulphur	1618	1731	1852	2598	3644
4.	Fluorspar	49	53	57	79	112
FILLER, ABRASIVE & OTHER MINERALS						
1.	Steatite	415	444	475	666	935
2.	Pyrophyllite	3.2	3.4	3.7	5.1	7.2
3.	Calcite	40	42	45	63	89
4.	Corundum	0.36	0.39	0.42	0.59	0.82
5.	Garnet (Ab)	0.15	0.16	0.17	0.23	0.33
6.	Vermiculite	1.2	1.3	1.4	1.88	2.63
EXPORT POTENTIAL MINERALS						
1.	Mica					
	(i) Blocks/Fil Splitings	1.76	1.90	2.06	3.02	4.42
	(ii) Mica products	0.30	0.33	0.35	0.52	0.76
	(iii) Mica scrap	10.50	11.34	12.24	17.98	26.41
2.	Diamond (in '000 carats)	57204	61780	66723	95122	139766
3.	Barytes	165	178	192	282	413
4.	Bentonite	231	249	269	395	579
5.	Fuller's earth	19.19	10.90	11.80	17.31	25.41
6.	Granite	With substantial reserves in the country, the production of these minerals may be increased as per domestic & export demands				
7.	Marble					
8.	Slate					

(RG) Refractory Grade; (B) Beneficiated; (Ab) Abrasive

Source: Report of the Working Group in Mineral Development (Other than Coal & Lignite), IX Five Year Plan.

CONCLUSIONS

Indian mining has come up a long way from prehistoric mining with prized flints of the stone-age miner to the atomic minerals of the nuclear scientists and has made vital contributions to the engineering and growth of civilisations. The technological status of Indian mining today is a mixed fare ranging from small-scale primitive mining to sophisticated technology. Presently, Indian mining is on the threshold of major structural reforms to attract foreign participation both in technologically as well as financially. The country has also to balance between various options, such as achievement of self-sufficiency through maximum mineral production; adoption of further industrial diversification and pursuit of growth through increased processing of minerals; exporting surpluses to earn valuable foreign exchange; and conservation of indigenous resources. However, while entering into the new millenium the mining sector have to overcome the challenges such as increasing population; technology development; know-how for ocean mining; conductive atmosphere for foreign participation; environmental concerns; human resource development; infrastructural development in mining areas and so on. Therefore, the time has come to unleash the creative potential of our engineers and scientists and innovators at grass root level. Only then we can make India truly self-reliant and a leader in sustainable technologies.

The views expressed in this paper are author's own and not necessarily of the organisation to which he belongs.

REFERENCES

1. Growth of Indian Mineral Industry since independence (1947-1991) 1992 Indian Bureau of Mines, pp. 68, 69.
2. Monthly Statistics of Mineral Production, March, 1999, Indian Bureau of Mines, pp. xiv, 40.
3. Editorial – Global economy and the Indian minerals sector, Journal of Mines, Metals & Fuels, June-July 1998, p. 213.
4. Ghosh D.K. & Kale Y.G., 1999, Scenario of Mining Machinery in India with particular reference to non-coal mines. Paper prepared for National Seminar on Mining Machinery to be held on November 26-27, 1999 at Dhanbad (under publication).
5. Clark Dr. Allen L. & Naito Koh, 1997, Structural reform of the mining industry in Asia and the Pacific region, 1997, Asia/Pacific Mining Yearbook. Asian Journal of Mining, pp. 36-37.
6. Sethna H.N., Manaktala S.P., Chandramouli R, 1995. Energy for a Billion People Technorama Spring Issue, June, 1996, The Institution of Engineers (India), pp. 6-11.
7. Economic Survey 1997-98, Ministry of Finance, Government of India, pp. 96.
8. Report of the Working Group in mineral development (other than coal and lignite), IX Five Year Plan.

CHAPTER 7

Mining Environment Scenario Beyond 2001

*Bharat B. Dhar**

ABSTRACT

The environmental changing scenario is not only confined to India, but cuts across the globe because of its socio-economic impact and more because of the technological advancements in mining sector and expectations for better quality of life. Today the world is facing the challenge of increasing globalization in most of its economic activities, and mining industry is not an exception. The core sector mining industry typically in the developing world has to face these environmental challenges not only within the country but also from the developed world.

The standards of environmental care of natural resource industries have to increase greatly in the 21st Century and the increased high levels of per capita consumption of energy and goods has to be kept in pace. Thus there is a continuous pressure on the existing resource base and the recognition of the environmental issues has become a matter of great concern in the next millennium.

To safeguard the environment in the coming millennium the stress should be on statutes and regulations world wide in order to improve the quality of overall environment and ecology, and accountability of land and water resources, assessment and combating of natural hazards to ensure development with least destruction of the nature. Although developed countries have been able to take care and make mining eco-friendly and reduced mining impacts, but same can not be the case with developing world where economic development is dependent on country's natural wealth to a great extent. The challenges, therefore, before the environmentalists particularly in developing countries is to make sustainable development a process of change thus, the need is to identify the major environmental issues for sustainable development of natural resource industries in their respective countries and to find ways and means to implement best mining practices.

*Director (Research), Association of Indian University, New Delhi, India.

Therefore, an attempt has been made here to focus on the changing mining and the environmental scenario for 21st Century together with important issues like **land resource conservation and management; hydrology and water management; preservation of biological diversity; control of environmental hazards; waste disposal and management; environmental management of mined out areas; utilisation of fly ash** - a problem to thermal power plant; **limitations of environmental regulation; mining in ecologically fragile areas; alternate source of energy; carrying capacity approach for regional planning; transfer of environmental research to commercial practices; and information technology** and **implementation of clean or green technology.**

INTRODUCTION

Sustainable development of Coal and Mineral sector industries has become, a global concern facing all the nations particularly the developing countries, where population and development is rapidly outpacing the resource base and affecting the carrying capacity of the regions. According to an estimate, over eighty million people are added to world's population every year. World population is expected to reach between eight and twelve billion by the middle of the 21st century. Very recently, a report of the United Nations Agency projected the size of world population to be between 6 to 18 billion by 2100 AD and around 24 billion by the middle of the next millennium. The population growth coupled with the rapidly increasing economic activities has exerted pressure on the resource base, resulting into new environmental problems. Poverty and income has also influenced the way in which the population affects the ecology and the environment (Dhar, 1994). However, for any country development and prosperity is a prime objective, but it must be linked to the eco-friendly environment, so that sustainable development is ensured (Ponayotou, 1993).

Mining in India, is one of the major core sector industries which plays a crucial role in the progress of country's economic development with of course some unavoidable environmental impacts (**Fig. 7.1**). While working out the strategy for development, a composite environmental view has to be taken right from exploration to mining, to processing and metal extraction. Since reserves for bauxite, chromite, and iron ores ranks considerably high in the global scene, production wise, India ranks high in a few minerals such as mica, barytes, lignite, coal, bauxite iron & manganese ore, but to keep pace with the emerging world trends and the onset of 21st century, new grounds are to be broken and new mineral deposits are to be explored fast to move ahead and remain in tune with the world production & consumption. Mining in developed and developing countries has substantially drawn the attention of international organisations and NGO's who are septic towards safe and eco-friendly mining. The problem perhaps is not of economic concern, but whether proper techniques, management policies and practices are utilised to reduce the impact together with optimising country's mineral production, and to enhance sustainable development of the region. While sustainable development is the key issue, the countries have to ensure that their policies are build with a link between development and the environment to achieve a positive impact with minimum degradation of the environment. In short a balance has to be established between development, environment and mining.

The major global environmental problems and issues that the technologist in the coming millennium must answer for sustainable mining and to meet the rising expectations for better quality of life are :

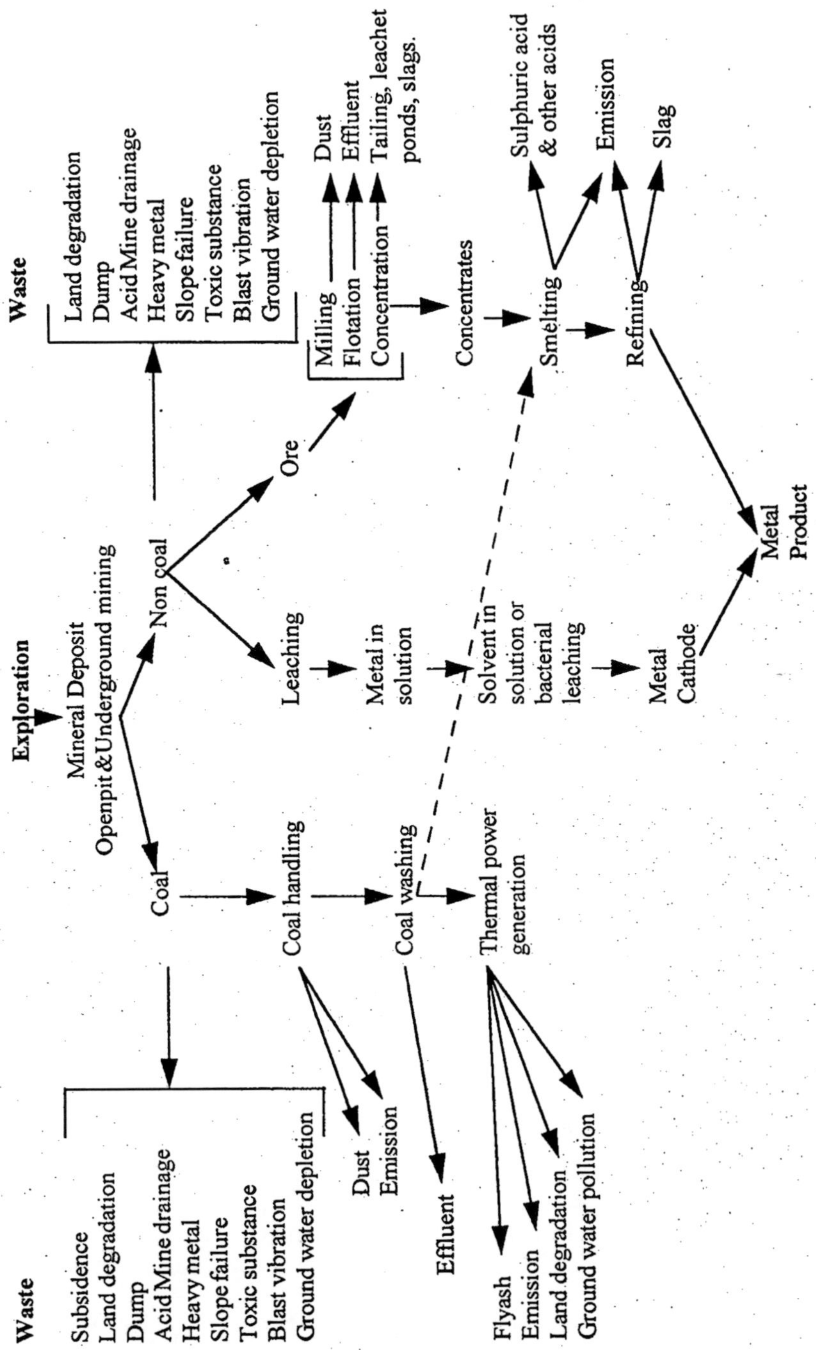

Fig. 7.1 : Environmental Degradation due to Mining and Beneficiation

- Resource conservation and management;
- preservation of biological diversity;
- Solid waste disposal and management;
- Environmental management of mined out areas;
- Flyash Utilisation - A problem to thermal power plant;
- Limitations of environmental regulation;
- Mining in ecologically fragile areas;
- Alternative source of energy;
- Carrying Capacity approach for Regional Planning; and
- Transfer of Environmental Research to Commercial practices.

RESOURCE CONSERVATION AND MANAGEMENT

There has been a massive consumption of energy in recent years the world over. Oil reserves are depleting and no new oil fields have been located. The world wide production of petroleum is projected to rise from 653 billion barrels in 1993 to as high as 1700 billion barrels by the year 2020. One half of the conventional oil resources will have been extracted by the year 2020. For each unit of energy extracted, coal emits more carbon dioxide than either petroleum or natural gas. Therefore, a systematic resource consumption in the coming years is required to be assessed.

According to Indian Bureau of Mines, India is self sufficient in 35 minerals that constitute the raw materials of basic industries, but our resources in petroleum and natural gas, ores of copper, lead, zinc, tungsten, etc., are inadequate. India also lacks in special grade minerals for very specific requirements of certain industries. Thus the policy in regard to mineral resources conservation should be, to exercise utmost restraint in the use of those minerals/fuels which are imported, of course keeping the export of minerals for which we have excess reserves. However, care needs to be taken in terms of best quality of ores / minerals that the country needs and therefore not to be exported.

In the process of conservation, equally important responsibility of the mines management is to ensure minimum damage to the environment caused during the mining operations. Since the recovery of minerals requires removal of vegetation cover with underlying soil mantle and excavating the overburden (the volume of which commonly exceeds that of materials sought), as such preservation and proper management is essential. Mining often results in reshaping of the topography, generation of large volumes of waste and disruption of surface and groundwater. The network of roads for servicing the mining operations further aggravates land degradation. All these activities, particularly during opencast mining tend to cause excessive sheet and gully erosion, landslides, pollution of surface and ground waters, etc. However, much depends on the ingenuity of the engineers to minimise these hazards. The conservation of mineral resources must be based on :

— scientific and rational utilisation with minimum waste in extraction;

— judicious blending of submarginal and high-grade ores; and

— beneficiation of low-grade ores so that optimum use is made of the mineral resource.

Simultaneously the mineral conservation policy should include :

- discovery of new deposits;
- finding substitutes for present minerals; and
- recycling of used metals and scraps.

HYDROGEOLOGY AND WATER MANAGEMENT

India is endowed with large resources of both surface and ground water, but parts of the country suffer from acute shortage of water. There is a greater use of water in industries and irrigation therefore the demand for water, particularly groundwater, has gone up substantially within the last 30 years (**Fig. 7.2**). Management of water resource is difficult in tropical countries like India, because of the seasonal nature of rainfall, thus effective arrangements for shortage of both surface and ground water for utilisation in dry season has to be made. Unfortunately, in recent years, creation of large reservoirs with high banks are generally considered ecologically unsound due to displacement of large number of people from the reservoir areas, especially in developing country where pressure of population is tantamount. Thus, increasing attention is being given to micro watershed management and developing large groundwater storage. For this purpose, water harvesting arrangements have to be made to see that a greater part of the excess rainfall during the monsoons seeps underground particularly in the upper catchment areas of the river systems. This is, therefore, a priority environmental issue.

Another issue to be considered is the quality of the surface and ground water - some quality problems are of natural origin e.g. problem of salinity of groundwater, iodine deficiency and excessive fluoride, arsenic and iron content in both surface and ground water. At present little in known about these problems, except to demarcate these areas, and to carry out detailed quantitative assessment, on the basis of which measures for iron removal, defluoridization and reiodization have to be taken. The other issue is quality problems due to anthropogenic pressure. Wastes from industries, fertilisers and pesticide residues are pollution not only the surface water but in places also ground water. An Environmental Engineer has a special responsibility in these respect, viz. to map out the areas now affected by ground water pollution and the potential areas for such pollution, and through his knowledge of ground water flow characteristics and the nature and geometry of the aquifers, he should try to control such pollution. Artificial recharging of aquifers may be considered in selected areas but only with great caution so as not to pollute the aquifers.

This particular case is an example of how an environmental geologist must help frame policy decisions on such emerging problems in the coming millennium.

PRESERVATION OF BIOLOGICAL DIVERSITY

The mining on forest ecosystem is of particular concern in India as well as world, as there is a considerable overlap between tropical forests and mineral reserves. This is evident in the Ghat region, Chhotanagpur plateau, the north east regions. Since the forest area covered by non-coal mining leases in the state of Bihar alone is estimated to be about 7200 ha. which accounts for about 8.4% of the total leased area of 86,000 ha. In Goa, 90% of the

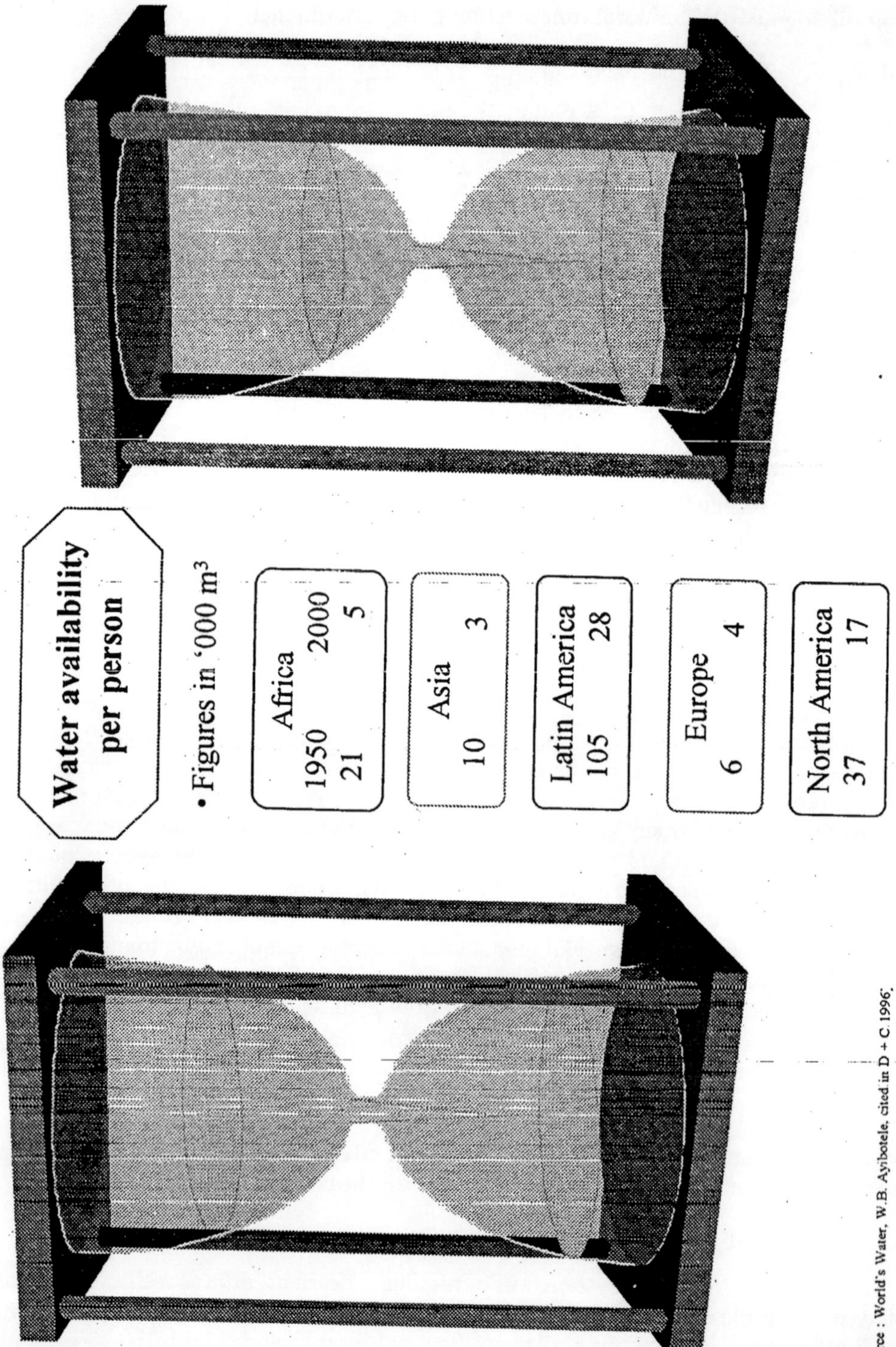

Fig. 7.2 : Declining Water Resources

(Source : World's Water, W.B. Ayibotele, cited in D + C 1996)

mining leases are in forest regions (Dhar, 1996). Forest loss leads to habitat loss. The earth has around 14 million different species of which only about 1.7 million have been identified and described. An estimated 1 percent to 11 percent of the world's species have become or will become extinct per decade between 1975 and 2015. A study has revealed that the average annual deforestation rates fluctuated according to average annual population growth rates and consequently utilisation of resource.

Densely populated Bangladesh and Hongkong have already lost more than 90% of their native wildlife habitat, while Philippines and Vietnam have lost about 80% (Ponayotou, 1993).

The approaches for preservation of biodiversity can be described in terms of : (a) in-site conservation, (b) ex-site conservation and (c) forestry approaches.

The combination of the above three approaches based on site specific consideration can produce the final revegetation strategy. The combination of the three approach is the 'Ecological Approach' which establish the biological processes which as nitrogen fixation, decomposition, nutrient cycling and retention, and important biotic interactions. It is these that indicate proper ecosystem functioning, which is as important as the careful selection of plant species in providing the primary vegetation structure.

SOLID WASTE DISPOSAL AND MANAGEMENT

Man is a producer of colossal amounts of solid wastes and pollutants through different activities in agriculture and industry and through the "Cumeristic" life-style in urban centres particularly in developed countries. The technologists and design engineers are therefore required to determine the factors influencing suitability of sites for disposal of wastes (e.g. radio-active wastes) and pollutants, and also to identify the nature and causes of pollution.

Mining is one such industry which perhaps generates maximum amount of solid wastes particularly in open pit operations and thus a direct impact on the surrounding area. It results in the coverage of extra land, soil erosion and water pollution. To illustrate an example the operations at Dimension Stone Mines, Rajasthan, India, have posed a serious threat to the region over the years due to large quantities of wastes. Kota stone is being used for flooring and wall cladding, 12-75 mm thickness, and is best suited for extraction. At present, the naturally splittable layer (thickness 15 m) mined manually with 47% recovery. A 9 m thick massive layer above this load is rejected as interburden, although it has the same physical and chemical properties, is an example of its type in India.

Lack of clean mining technology, practices of primitive extraction techniques, reluctance to adoption of scientifically planned operation, and disregard towards environment have led to wasteful mining, poor mineral recovery, generation of huge mine waste, seasonal scarcity of ground water and multifarious serious environmental treats, like drastic change in landscape, disorder in natural drainage pattern, loss and lock-up of large area of fertile land under waste dumps, and perhaps reduction in rainfall rate in the mining belt, especially since the boom in this industry from early 1980's. Presently around 2 million tonnes of mine waste is generated every year. Government of India has constituted a committee for the environmental management of Kota-stone mining. On the basis of recommendations of high level committee and R&D efforts, a technique has been developed whereby mining is proceeded with pre-splitting, resulting into reduction of waste generation at the pit head.

Environmental auditing has also shown that if these rock layers are mined with mechanical pre-splitting, the waste generation will be reduced by 40% at the pit head (Dhar, 1996) and cleaner material can be produced with minimum waste.

ENVIRONMENTAL MAMANGEMENT OF MINED OUT AREAS

Abandoned and inactive mines are the focus of significant environmental concern, many active mining operations have developed innovative technologies to address comparable problems. Runoff from abandoned and inactive mines is particularly a destructive and problematic aspect of the environmental problems. Methods to control contaminated runoff, implement optimum waste pile controls, minimise human health and ecosystem exposure to risk, and plan for full reclamation while working with a multitude of jurisdictional authorities and public interest groups. Such operations are also characteristic of responsible and effective environmental management in active mining operations. The research hypothesis is that techniques and approaches in active mining operations may offer insight into the management of non-effective and efficient management of pollution from mines, both active and abandoned, will be the availability of accurate, timely and objective information.

Best practices of biological reclamation would largely depend on the selection of appropriate plant species. It, therefore, requires utmost care (**Fig. 7.3**) and the issues that need to be taken care of are :

(a) The plant species should preferably be local and native to the soil;

(b) Nitrogen fixing leguminous species of plants should be included;

(c) The species diversity should be in conformity with the natural forest of the particular area of mining;

(d) The species should have greenery and aesthetic value to the land;

(e) Requirement of local people should be considered while selecting the species; and

(f) The species should be selected considering the climatic conditions and quality of the dumping materials.

Central Mining Research Institute (CSIR) had shown an excellent rehabilitation result on old coal mine waste dumps lying abandoned for over 15 to 30 years. The project was sponsored by the Ministry of Coal, Govt. of India, under its R&D programme.

Prior to the field study a pilot study was carried out in the laboratory for five different phyto-geographical regions of the country. The study showed that the native tree species performed better both in pot and field trail experiments, and that leguminous pulses can also be grown. Grass species were grown on the slopes of dumps and were found to be very suitable for the area. Amendment of bio-fertilizers and mycorrhizae enhanced shoot and root development, ultimately leading to better plant growth and survival on the mine dumps (**Fig. 7.4**).

UTILISATION OF FLY ASH — A PROBLEM ASSOCIATED WITH THERMAL POWER PLANTS

In India coal contributes about 72% of the overall power generation. Presently, 72 coal fired thermal power plants are generating around 75 million tonnes of flyash, per

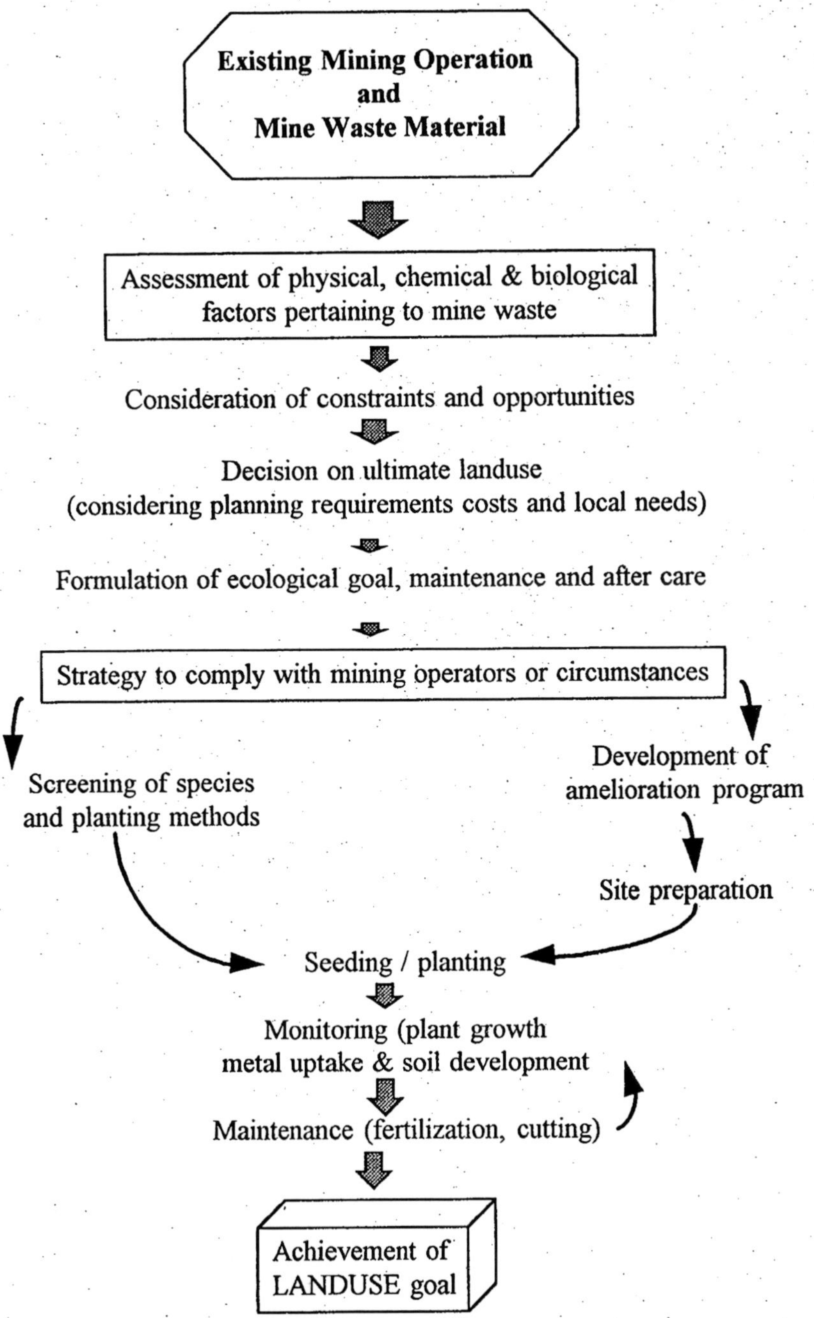

Fig. 7.3 : Flow chart of the stages of achieving successful rehabilitation

year. So, generation of such a huge quantity of fly ash, contaminates the atmosphere, the land surface, and surface and ground water bodies. The major issues before the technologists and engineers world over is how best to utilise the fly ash for different purposes. Some such practices developed over the time period are :

(a) Preparation of fly ash bricks;

(b) Use of fly ash for underground goal filling (hydraulic stowing) and exhausted open pits;

(c) Use of fly ash as a amendment for biological reclamation work; and

(d) Use in construction of roads and dams etc.

Asia's biggest fly ash brick plant has been constructed at Tribeni, West Bengal in 1994. This has created a new era in the conversion of a waste into wealth for an essential and the commonest requirement of human habitat, namely, the building bricks for the houses. The plant has the capacity to produce 24 million bricks per annum. These bricks are made with 80 percent of dry fly ash.

In 1993, Department of Science and Technology, Govt. of India had constituted a task force and identified mine areas for it's utilisation. The "filling of underground voids of mining Areas" was one of the identified tasks of which the author was the chairman. This economic utilisation of flyash is expected to be environmental friendly as it is likely to reduce (a) mine subsidence, (b) dumping areas, (c) air and water pollution, and (d) control of mine fires - a national problem in certain coalfields in the country. Underground utilization as substitute to sand has been successfully conducted in Singareni coal mines by CMRI (CSIR) together with NTPC (National Thermal Power Corporation) and the mining company.

LIMITATIONS OF ENVIRONMENTAL REGULATION

Regulatory frameworks to safeguard the quality of land, water, and air as a result of mining and mineral processing activities are growing in number and complexity. The norm in environmental regulation is that governments set maximum permissible discharge levels or minimum levels of acceptable environmental quality. Such 'command and control' mechanisms include : (i) best available technology standards, (ii) clean water and air acts, (iii) super funds for clean-up and liability determination, and (iv) a range of site-specific permitting procedures which tend to be the responsibility of local governments within nationally approved regulatory regimes. Three issues are relevant in industrialised country's (N. America, Australia, Japan and Europe) environmental regulations to reduce environmental degradation and improve environmental management practices :

1. There is a trend away from a 'Pollutee suffers' to 'Polluter pays' principle. This highlights the tendency of such environmental regulations to deal with the symptoms of environmental mismanagement (pollution) rather than its causes. (economic constraints, technical constraints or information about better environmental management practices).

2. Best Available Technology (BAT) standards may be appropriate at plant start-up, but their specified effluent and emission levels are not necessarily achievable throughout the life of the plant.

3. BAT standards and environmental regulations of the command and control type tend to presume a static technology a best technology at any one time.

Environmental regulations designed specifically for mining and mineral processing have, until recently, been uncommon in developing countries, although most countries now have in place basic standards for water quality and, less commonly, air quality. A few developing countries (Chile and Brazil) have recently adopted extensive regulatory frame works. This growing concern about environmental degradation is occurring during a period of rapid liberalisation in developing countries. In India, an approach for sustainable closure of mine has been depicted (**Fig. 7.4**) which finds expression in new policies to promote foreign investment, privatisation schemes, and the availability of loan capital. These conditions also influence the regulatory regime of developing countries.

Should the developing country pose less onerous environmental burdens on the potential investor to improve the terms of the investment by implying lower compliance costs or a greater assumption by the state of the environmental costs associated with mineral development projects? This is an important decision that a country has to take. The other question equally important to considered are :

1. Should agreements be signed which release new investors from any liability for environmental damage caused by previous mine owners under less-restrictive regulatory regimes?
2. Or will a clear and strict regulatory regime be more likely to facilitate credit flows from increasingly more environmentally conscious lending agencies?

Environmental regulations tend to be of the blanket-type which specify maximum levels of emitted substances, minimum levels of environmental quality and best available technology standards. They do not tend to reflect the propensity of a particular operation to pollute, which in part depends on local site-specific conditions (geology, geography, and climate) as well as economic, infrastructural and technology-related constraints. Since many developing country regulations are often copied from the statute books of the industrialised countries (Chile and Brazil) whose regulations are adapted to suit their circumstances, they may not be appropriate for the site-specific characteristics of mines in either tropical regions or deserts. They may result in unnecessary and costly adaptations on the one hand, and the lack of necessary control, on the other hand.

Another recently discussed example of potentially contradictory policy issues revolves around the **International Basal Convention** which restricts the inter-country transportation of toxic wastes. Since certain scrap metals fit theoretically into this category on account of their heavy metal content, this would undoubtedly restrict trade in scrap metal and metal recycling. This is considered to contradict many of the new intentions of European and American governments to encourage the recycling of metal-containing materials at the expense of new primary production.

MININING IN ECOLOGICALLY FRAGILE AREA

Ecosystem which is prone to be easily disturbed may be considered as fragile area. Some of the ecosystems which are fragile and more susceptible to disturbance than others are listed below :

- Ecosystems with unique properties;

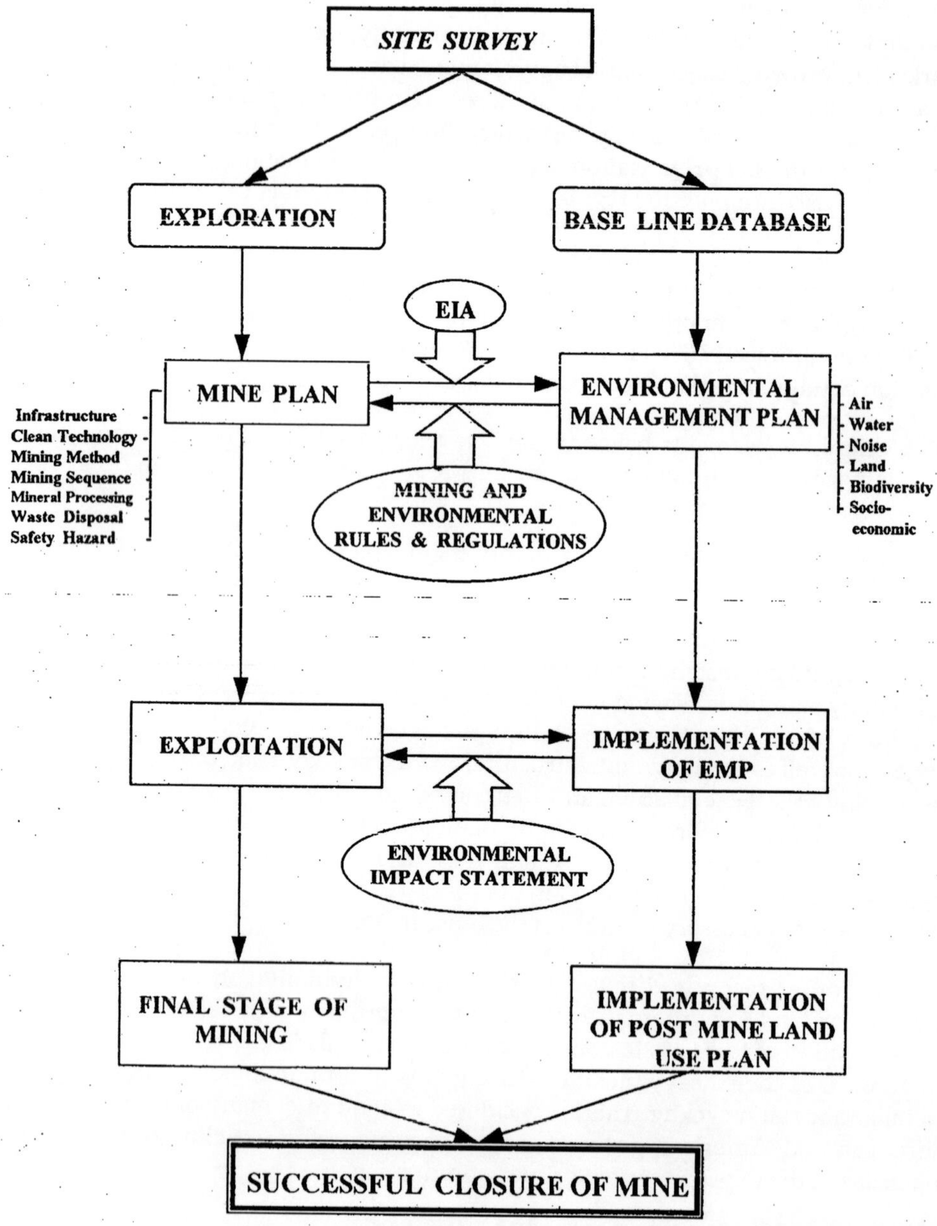

Fig. 7.4 : An approach for Sustainable Closure of Mine

- Ecosystems with intrinsically low resilience;
- Ecosystems with high species richness and bio-diversity;
- Ecosystems susceptible to species loss;
- Ecosystems linking two or more protected ecosystems;
- Ecosystems with aquifers and water recharge areas of mountain springs; and
- Areas with active geological faults and seismic hazards.

India has a diverse variety of land forms, like, snowy peaks of Himalayas with fragile foot hills, desert and arid and semi-arid areas, vast coastlines, plains and rich forests enjoying a multitude of bio-diversity. Consequently problems of mining are varied and may, therefore, need site specific solutions. An integrated mining and environmental management plans including the new technology for the exploration, exploitation and processing of mineral is essential for eco-friendly mining and sustainable development of ecological fragile areas if the economic progress for the country has to be maintained/sustained.

Unfortunately in India, the past reckless and unscientific mining practices have brought an imbalance in some pockets of environmentally sensitive and fragile areas of Himalayan range. This malady today has been realised, and, now with a reasonable and clean scientific temper, for such sensitive operations, it is possible to exploit mineral wealth in an eco-friendly manner with well designed and advanced scientific planning for exploitation. However, it has to be kept in mind that these techniques are costly and also all the deposits may not tend to these techniques because of other constraints such as locations, etc.

Considering all the problems associated with these small scale mines in this area, Central Mining Research Institute, Dhanbad, India, had taken a National Project for this area and an integrated mining and environmental management plan suggested for sustainable development and exploitation of minerals in the region. Similar studies are suggested for similar other mining regions in the country and in other parts of the world where cluster of mines are in operation. In short an well designed integrated approach is the only way out, if mining has to continue in fragile eco-systems in particular and other regions in general. Each time an approach has to be site specific and not one common approach/technique for all.

ALTERNATE SOURCES OF ENERGY

Population growth has led to massive energy consumption. Energy consumption increased by 144 percent in Africa and by 185 percent in Asia during 1973-93. Coal and oil reserves are depleting. This growing understanding on the fossil fuels which is going to be exhausted in near feature (particularly oil and natural gas) and the increased use of such fuels will lead to global warming (because of CO_2 emission), concerted efforts are being made to develop alternative viable sources of energy. Hydroelectricity, though a good supplement the production of which does not cause pollution, has its limitation in India because of seasonal rainfall. In India, there is an additional problem of firewood, char-coal and cow-dung being a major energy source (estimated at about 14% of the total energy demand) in rural and remote areas - is an important cause for heavy depletion of the forest wealth. This issue is of concern and demands a long lasting solution, if an environmental and ecological balance has to be maintained.

The ideal solution of this problem would be to replace the fossil fuel as well as fire-wood by non-conventional energy sources, which do not generate carbon dioxide nor produce air pollution. The Ministry of Non-conventional Energy Sources, in India has estimated an enormous annual potential of such energy sources, as can be judged from Table 7.1.

Table 7.1 : Annual Potential Of Non-Conventional Energy Sources

Energy Source	*Potential*
Wind energy	30,000 - 50,000 MW
Small & mini-micro hydel	5,000 - 10,000 MW
Bio-mass	17,000 - 60,000 MW
Solar energy	5,70,000 MW

In particular, solar energy has maximum potential. The technologists and engineers with their first hand knowledge of terrain characteristics could play a major role in the selection of appropriate sites for producing and harvesting non-conventional energy sources economically on large scale.

Though nuclear energy is considered pollution free and on the whole economic but because of several disastrous accidents in nuclear power plants within the last 15 years it has not gained popular acceptance. However, if proper and stringent precautions are taken risk of accidents is said to be negligible. Unfortunately, India's uranium reserves are low and in view of the capital-intensive nature, nuclear power plants are not likely to make much headway in near future. Besides these are strong and active NGO groups world over not in favour of nuclear energy from several angles including environment and dumping of nuclear wastes.

CARRYING CAPACITY APPROACH FOR REGIONAL PLANNING — A MUST FOR FUTURE DEVELOPMENT

The regional imbalance between supply and demand may increase in course of time since population demand is growing fast. Water available for human use is gradually shrinking because of pollution from agriculture, industry and other human activities. Underground water deposit are being pumped out faster than they are being replenished by surface water. Therefore, carrying capacity approach is essential.

Planning for development within the limits of carrying capacity recognises that mankind is dependent on the productive capacity of ecosystem and that some minimal level of ecosystem degradation is, therefore, essential to human survival (Fig. 7.5).

To identify the environmental impact of any project in isolation proves futile because of its serious limitations. More over the planning process mainly depends on the population growth structure and therefore without recognising the biosphere capability for supporting them. No long term planning is going to be sustainable. Thus to remove this ambiguity, the concept has changed and the approach is gradually getting modified by introducing large regional carrying capacity studies. The broad aims of such a study are as follows :

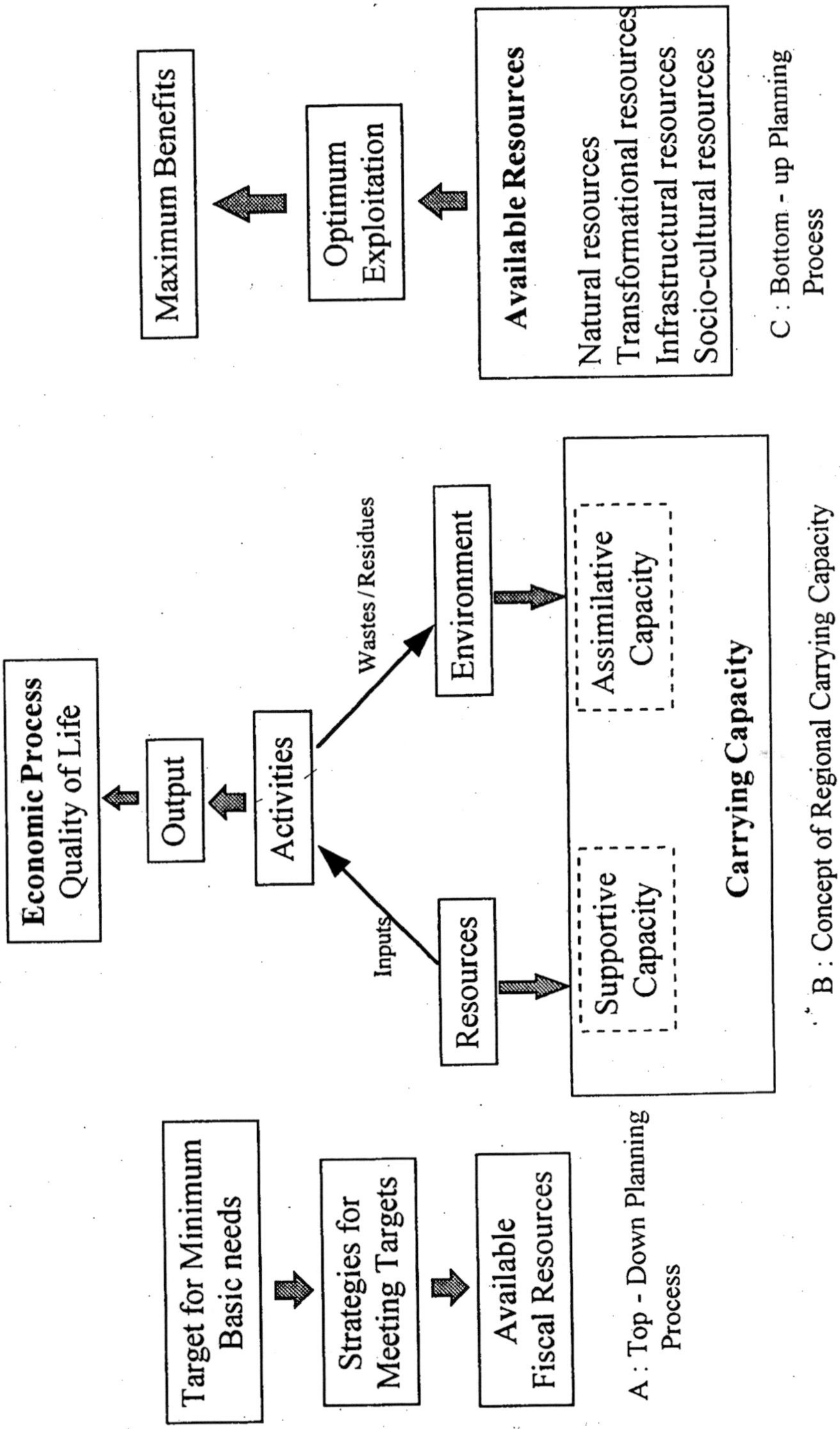

Fig. 7.5 : The Planning Process, Traditional and Proposed

(a) Shifting of population base to resource base planning.

(b) Evaluating the environmental impact of the project in terms of assimilation and supportive capacity of the area.

(c) Allocation of resources to various socio-economic activities for maximising the quality of life.

(d) Delineation of optimum resource allocation strategy.

(e) Preparation of short and long term development plan and choosing the best alternative and preferred development

The Government of India, Ministry of Environment and Forest needs to be complemented who conceived this idea of regional planning for the protection of country's environment and increasing population base approached premier scientific organisations like NEERI and CMRI to take up national mission projects to carry out Carrying Capacity based studies in known highly polluted areas and come up with new and modern approaches for national development. Some such studies completed and ongoing have been listed below for ready reference :

— Carrying Capacity of National Capital Region (New Delhi and surrounding areas);

— Carrying Capacity of Doon Valley (Himalayan range of rich limestone);

— Carrying Capacity of Damodar River Basin (rich coal mining, thermal power plants, steel industry region); and

— Carrying Capacity of Tapashwari Basin (rich in petroleum).

The developing countries must therefore bear in mind this concept for their mineral resource rich areas which will help sustainable development, where population plays an important role.

TRANSFER OF TECHNOLOGY TO COMMERCIAL PRACTICE

The practising engineers and designers of major projects must examine the techniques of transferring the environmental research knowledge to practical applications. The analysis should be based atleast on the following two observations :

1. The management, the financial and the engineering skills required to convert fundamental knowledge to practical application which is normally not found within the university or research organisation has to be encouraged and practice established.

2. Today there is a need of possible commercial uses of the environmental research and to choose the time to initiate the process of transfer so that findings can reach to the grassroots level.

This, however, is a difficult question but nevertheless needs an attention if mitigation measures have to be adopted for better and sustained environment.

CONCLUSIONS

The environmental impacts give way to that issues which are not only confined to extraction and processing stage alone, but serious problems at times are associated with the transportation and with storage. Moreover, the use of natural resources as inputs into specific industrial processes may generate hazardous waste residues and result in industrial pollution. Finally, the end-use of products manufactured from natural resources may result in further environmental problem and there is, therefore, need of environmentally sound product disposal technologies or for recycling programmes the sustainable future. Clean technologies are the processes of future thinking now.

For non-renewable and renewable resource such a minerals, forests and fisheries, problems of depletion exhaustion and unsustainability may arise. In addition to environmental impacts, the natural resource industries often have a profound socio-cultural and economic impacts.

The policy statement strategies to achieve the environmental objectives, some of the important aspects are :

(i) The total mine planning and design should be environmentally acceptable.

(ii) The reclamation of mined out areas should aim to achieve a post mining condition consistent with the proposed final land use.

(iii) The progressive reclamation must be carried out as soon as the mined out areas become dormant.

(iv) The base line environmental survey should be carried out in new mines to serve as a basis for comparison later.

(v) The funds and equipment allotted for environmental work must not be diverted to any other work.

(vi) Environmental consideration should be the criteria for evaluating the overall performance of the company and individual project officials.

(vii) Each company should have a well designed environmental policy.

(viii) Resource full persons should take charge of environmental protection and safeguard and not discourage officers of the company.

(ix) It is high time in India to have corporate culture in EMP and EMD with a board level executive to head the latter.

(x) Training in the beginning and awareness programmes in general must form a part of the company culture.

(xi) Environmental costs should not be seen in isolation.

The future, therefore, can be safe only if the problems are fought locally, while thinking globally together with an outlook of mutual trust and co-operation. To achieve sustainable development particularly in developing countries, it is essential to ensure environmental awareness of all levels of education both formal and informal (Dhar & Badrinath, 1995). The developed world has to lend their support to the developing countries if an international sustainable development has to be maintained and practised.

Indian Scientists/Technologists and Engineers together with international experts can play a vital role in safeguarding the natural ecosystem and helping in achieving the best mining practices.

REFERENCES

1. Choudhuri, S.K. (1998); Status of Environment in Mining Indusrty of India. Jr. of Mines, Metals and Fuels. pp. 399-404.

2. Dhar, B.B. (1994); Sustainable development in South Asia and Pacific region vis-a-vis Indian contribution. Int. Seminar on minerals & Mineral based Industries In Escape region : Trade & Technology Co-operation. 6-9 Dec. Kathmandu.

3. Dhar, B.B. (1994); Environmental Impact Assessment for Mining : The Indian Scenario. Proc. of the seminar on Capacity building for environmental management in Asia / Pacific mining. UNCTAD 6-8 Sept. Jakarta.

4. Dhar, B.B. and Badrinath, S.D. (1995); Environmental Education for sustainable development. pp 757-770. Mining Environment, WOMEC '95, Oxford & IBH publishing, New Delhi.

5. Dhar, B.B. and Thakur, D.N. (1995); (ed.) Proc. of the First World Mining Environment congress, 11-14 December, Oxford & IBH publishing, New Delhi.

6. Dhar, B.B. (1996); Major environmental issues for mineral sector : Problems and challenges; Proc. of the National Workshop EMICS, India.

7. Dhar, B.B. (1996); Environmental Auditing. Case studies illustrating Environmental Practices in mining and metallurgical processes. UNEPA & ICME. U.N. Publications.

8. Dhar, B.B. (1996); Environmental management system for closure and best practices in Indian Mining Industry. Sixth workshop of the Mining and Environmental Research Network. 4-9 August. Harare.

9. Jha, A.K. (1999); Population and Environment. Employment News Vol. XXIV No. 12. 1-2.

10. Panday, P. (1998); Global and Indian scenarios of coal mining in 21st century. Jr. of Mines, Metals and Fuels. pp 376-382.

11. Ponayotou, T. (1993); The environment in South East Asia : Problem and Policies. En. Sci. Tech Vol. 27 No. 12, 2270-74.

Section II

COUNTRY MINING PERSPECTIVES

CHAPTER 8

India's Mining and Mineral Industry : Some Issues for Next Millennium

B.B. Dhar and A.K. Soni***

The 21st century could easily see a terminal environmental and ecological crisis for humanity. The issues of resources and population are already threatening. We probably have little more than 25 years in which to get a grip on things. It is our capability, how effectively we mobilise all those elements of society that reduces the adverse effect of environmental crisis on human beings. Since the dimensions of crisis will grow with an accelerated pace in the coming days, formal government can only become a part for managing it. Parliaments, popular assemblies, NGO's, businesses and professional associations all have a crucial part to play. Environmental governance embraces the whole exciting patchwork of institutions, processes and people which together make society.

(Commission on Global Governance, 1994: 3, After Hamphel, 1996)

India is a mineral rich country and its mining/mineral industry is well developed. In terms of manpower deployment and capital investment mining industry in India, rank among first five major industries. In south-east Asia, India continued to be the major mineral power, feeding its huge industrial base and meeting external industrial demands.

Nearly, 84 minerals of both major and minor categories are produced in India, of which 4 are fuel minerals, 11 metallic and 69 non-metallic minerals of industrial categories. The country, however, is self sufficient in only 36 minerals resulting in huge deficit in several others. The import continues to be higher than export by more than Rs. 9000 crores worth (Jha, 1997). The industry is significantly important from the viewpoint of export earning and GDP contribution (3.2% approximately) to national income. After liberalisation and stress on free market economy in the last decade of twentieth century, the industry has witnessed significant changes in terms of–

* Director (Research), Association of Indian Universities, New Delhi.

** Scientist, Central Mining Research Institute, Regional Centre, Roorkee (U.P.)

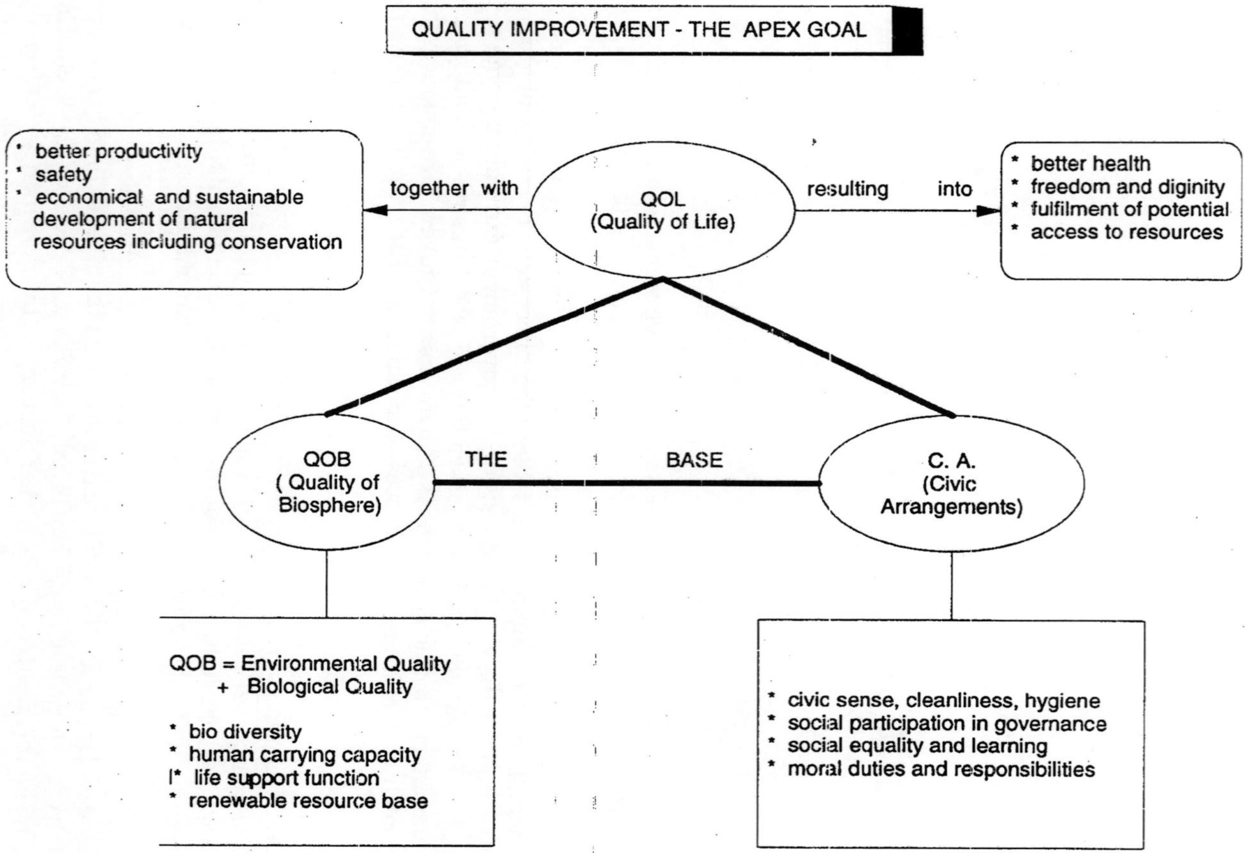

Fig. 8.1 : The Sustainability Traingle (modified from Hamphel, 1996)

- policy changes, liberalisation, de-licensing, taxation, liabilities etc.
- open and wider investment opportunities.

In this paper an attempt has been made to focus issues/discussions central to the conference theme i.e. mining challenges of the 21st century. It is evident that the topic chosen for description is such that it covers a host of issues from planning, policy to human resource use etc. and can't be described in limited number of pages. Therefore, some designed principles required for policy formulation with particular reference to mining industry have been formulated, which could be used suitably according to the requirement. This discussion paper also raises some potential issues, to be pondered for next century taking into account the overall scenario of mining industry.

CHANGING PARADIGM

As the next century approaches, it will be essential that governments (both federal and state) and the public at large find a balance between social, economic and environmental issues. This integration is the only way in which sustainable development in mining can become a reality. There has been concerned efforts in the Indian minerals industry to match pace with the current global trends (Soni, 1998) as the present decade was the decade of globalisation. In order to break new grounds, we must take a serious view to link the future technological options with the national mineral resource base. A tabular summary of changing paradigms relating to mining and mineral industry in India has been given in Table 8.1. The table summarises the issues where the industry lacks and what needs to be done in future.

Policies declared till date, for the mining and mineral industry recognise the need for sustainable, environmental friendly and cost effective management at nearly every stage of the mining project from exploration to exploitation. Sustainable communities in the mining are made possible by the basic model of sustainability triangle in a balanced and mutually enhancing fashion (Fig. 8.1).

Table 8.1 : Changing paradigms for mining industry in India

Past/present	*Future*
1	2
* Mining techniques are a mix of traditional and modern.	* Eco-friendly technologies, clean technology options.
* Low to very low standard of living in mining areas besides large influx of money.	* Raising of quality of life/ standard of living in general
* Age old, orthodox and typed culture in mining.	* Cultural changes
* Poor civic utilities/infrastructure/ transportation/energy and power utilities/information links etc.	* Improvement with modern amenities and information technology
* Comparatively less strong quality human resource base in both technical and non technical cadres.	* Plan policies to attract strong human resource base.

1	2
* Neglected responsibilities.	* Responsibility to be ensured and implementation of *polluter pays principle* in total mining industry irrespective of scale and size of mine (s) and operation.
* Lack of innovation	* Quality inputs required.
* Dull image of industry.	* Better image projection.

The concept of control and command of the industry by the government through statutes, rules, regulations and bye-laws is essentially required in the mining industry because of the risky nature of industry. Therefore, the need is that in future the existing gap(s) needs to be plugged for better efficiency and productivity. Some policy statements/ principles which have been designed only for mining/mineral industry can be helpful in the future policy making. These are based on the Agenda 21 of the Rio Declaration on Environment and development (UNCED, 1992).

Principle 1 : Human beings are at the centre of concerns for sustainable development. They are entitled to a healthy and productive life in harmony with nature. The guiding principle of next century should be based on the fact that–No developmental plans will be expected to bear fruit unless efforts are made to limit population growth.

Principle 2 : Central or state governments which have sovereign right to exploit minerals or other resources pursuant to their own environmental developmental policies and the responsibility should ensure that activities within their jurisdiction or control do not cause damage to the environment of other states or of areas beyond the limits of national jurisdiction.

Principle 3 : The right to development must be fulfilled so as to equitably meet out developmental and environmental needs of present and future generations.

Principle 4 : In order to achieve sustainable development, environmental protection shall constitute an integral part of the mining processes and *Integrated Approach* must be adopted for managing environment and conservation of mineral resources.

- Implementation of *pollution prevention approach*. Pollution prevention approach uses processes, practises, material, products or energy that avoid or minimise the creation of pollutants and waste without creating new risks to human health and environment.
- Enforcement of *polluter pays principle.*
- Environmental issues are best handled with the participation of all concerned citizens at various relevant levels. Organisations shall facilitate and encourage public awareness and participation by making information widely available. Those mining companies involved in handling of hazardous minerals or materials shall not hide the information (if, any) which concerns to common peoples.
- Indigenous peoples and local communities have a vital role in environmental management and development shall be recognised because of their knowledge about

traditional practises. Their identity, culture and interest to enable effective participation in the achievement of sustainable development shall be duly supported.

- To protect the environment, the precautionary approach shall be widely adopted. Mining operations shall not be ordinarily taken up in ecologically fragile/sensitive and biologically rich areas.
- Environmental impact assessment as a national instrument shall be undertaken for the implementation purpose in both proposed and ongoing projects which are likely to have a significant adverse impact on the environment and are subject to the decisions of a competent authority.
- Mining belts are potential areas for land degradation thereby generating *wasteland*. *Post Mining Land Use* should be promoted as far as possible and mine closure policies should be formulated separately.
- Organisations and states shall co-operate in good faith and in spirit of partnership in fulfilment of requirement of an approved mining and Environmental Management Plan (EMP)

Principle 5 : Based on enacted environmental and mining laws/regulations, quality standards, the management objectives and priorities should be lined up. National authorities should endeavour to promote the internalisation of costs and make use of economic instruments taking into account the "polluter pays principle" without distorting the nation's company's image.

Principle 6 : To achieve higher *quality of life* for all peoples associated and all organisations involved in mining trade steps should be taken to reduce and eliminate unsustainable patterns of production and consumption and promote appropriate policies guided towards eradicating poverty and improve the standard of living.

Mining units are generally located in interior areas and has direct links with rural economy. Mining companies should be made responsive and play active role in national mission programmes such as poverty eradication, population containment and literacy campaigns. This should be done without affecting the environment, natural resources of rural people. Rural people under oppression, domination and occupation shall be protected.

Principle 7 : States possessing the mineral rights should encourage, motivate and be mentally prepared to implement and strengthen indigenous technologies which suits to Indian geo-mining conditions. The principle that enhancement, adaptation, diffusion and transfer of new and innovative indigenous technologies not only promotes industry but helps a lot in indigenous capacity building must be adopted in to practice.

Principle 8 : Mineral resource utilisation policy of the next millennium should be sensible enough to promote, invest and enhance in value additive, foreign exchange earning and self reliant mining industry which is competitive in international mineral trade too.

Principle 9 : Safety of man and materials by modern instruments and tools must be ensured and implemented to make the mining operations safe and less hazardous.

Principle 10 : Human resource development through quality education/awareness schemes and appropriate utilisation of developed skills must be ensured and implemented to make the mining industry attractive, leading and lucrative for professionals. Principle of timely updating of technical/non technical skills through various means must be adopted in practice to match pace with competitive world.

ISSUES FOR NEXT MILLENNIUM

In the next millennium the importance of industry will continue to grow as far as India is concerned but it is certain that industry has to face the changes and challenges of time. The study and experience shows that the following points need quick attention (if, not immediate) for healthier growth of industry as a whole.

- Need base updating of policy issues on labour, infrastructure, privatisation and foreign investment, mining taxation and human resource management.
- Rapid introduction of computer and information technologies at the production units i.e. at the mines for various unit operations.

  (It may be noted that compared to other industries computerisation trend has not picked up that fast in mining industry for various obvious reasons of remote locations of production units, labour intensive operations etc.)
- Techno-financial management of mining and mineral based organisations for production optimisation.
- Automation of mineral benefaction techniques procedures and systems including application of robots and remove control operations of hazardous processes in mining.
- New, improved and environmentally friendly mining and mineral evaluation technologies (e.g. geo-statistics, research on techno-economic viability of transition from surface mining to underground mining etc.) for assessment and exploitation of lower grade ores, low value dimensional stone and high ash content fuel minerals.
- Improved mine safety technologies including rescue techniques.
- Quality standards and its implementation at various stages.
- Rural-urban interface for sustainable development.
- Use of information technology (IT) for environmental management of mining operation including data base creation.
- Introduction of E-commerce in mining to enhance marketability of different minerals and setting international minerals market by adopting modern management techniques.
- Application based R&D in relatively virgin areas such as mineral economics, mining taxation, clean technologies etc. on long term perspective basis.
- Modernisation and updating of mining engineering education in Indian universities.
- International collaboration in different areas of the mining and mineral engineering with an objective to make Indian mining industry globally compatible in terms of standards and mineral market value.
- Research to combat *geo-politics* in a globally competitive mineral market for strategic minerals such as tungsten, cobalt, vanadium, mercury, silver, atomic energy minerals etc.

Without ending the above mentioned long list of issues which requires debate, author(s) would like to keep these issues alive before the gathering of eminent mining engineers so that the *challenges* which are ahead for the Indian mining industry can be faced or tackled easily and conveniently.

SOME CRITICS

In the light of the above mentioned policy principles and broader issues there exist a considerable scope for improvement in the state of art of mining on modern lines. Both *micro* and *macro* areas must be attended to and rationalised in tune with national interest and goal. It has also been made evident from the designed principles that mining and environment are inseparable components and therefore complimentary to each other for sustenance. Several guidelines developed by international agencies (BOX-1) lays emphasis on similar principles.

It has been observed that norms, rules and principles usually develop interactively. Policy makers frequently downplay normative analysis in their rush for politically advantageous rules. Moral values and principles may be loudly proclaimed but economic self interest and coalition politics usually dictate the choice of rules. Can we keep a distance from it in the coming days of next millennium?

BOX-1

SOME NOTED INTERNATIONAL GUIDELINES ON MINING AND ENVIRONMENT

1. **UNDP's Environmental Management Guidelines.**
2. **Berlin's Guidelines on Mining and Environment, 1991**
3. **UNDP-UNRFNRE Guidelines**
4. **UNEP Technical guidelines on Environmental aspects of Mining of Non-ferrous Metal Ores**
5. **UNDCD Draft Environmental Principles and Guidelines on Mining in Developing Countries.**

EPILOGUE

Though, this paper won't have the necessity of conclusion but as a customary practise we would like to sum up as follows:

Can the Indian mining industry afford the human and ecological price of satisfying its mineral need? If yes, let us join hands and work together for improving the quality of life of the mining associated communities who are looking towards you for making the life easier and comfortable in the next millennium.

REFERENCES

Dhar B.B. (1998), "Mining in Public and Private Sector in 21st Century–A Trend" Proceeding of Platinum, Jubilee Workshop on Mining in Public and Private Sectors in 21st Century, Department of Mining Engineering, IT BHU, Varanasi, pp. 40-48.

Ghose A.K. (1993), Policy Issues related to Mining Environment–A Global Overview, Proceedings of National Seminar on Environmental policy Issues in Mineral industry, Calcutta, Mining Geological and Metallurgical Institute of India, pp. 25-34.

Hamphel L.C. (1996), Environmental Governnacc - The Global Challenge, Affiliated East-West Press Private Ltd., New Delhi, p. 291.

Jha, C.S. (1997), Major Issues in Techno-Financial Management in Mineral Industry, Technorama, Vol. 47, No. 1-2, Institution of Engineers (India), pp. 17-25.

Natrajan R. (1998), The Energy, Environment and Ecology-Dimensions of Sustainable Development, Technorama, Vol. 48 No. 1, Institution of Engineers (India), , pp. 4-14.

Soni A.K. (1998), Mining in Public and Private Sector in India, Technological Developments and Currents Global Trend, Saket Mineral Industry Handbook, Ahmedabad, pp. 383-395.

UNCED (1992), Earth Summit-Agenda 21 of the Rio Declaration, United Nations Conference on Environment and Development pp. 11-13.

WOMEC (1995), Souvenir, World Mining Environment Congress (WOMEC), New Delhi.

CHAPTER 9

Coal Industry of Russia in the Early 21st Century

*Yu. N. Malyshev**

At the beginning of the restructuring of the Russian economy, in the late 1980s, the coal industry of the country had an outdated and worn production potential. It was a sector that could not compete in market economy. Only 15% of all underground mines met modern mining standards. Every year the Government allocated substantial subsidies for the sector from the central budget. Therefore, in 1993, the comprehensive restructuring of the Russian coal industry was initiated for the development of coal mines and companies, which could form a sector able to efficiently operate in market economy.

The improvement of the competitiveness of Russian coals in the national and foreign energy markets through radical upgrading of the structure of the sector reserves and resources, comprehensive restructuring of its production potential, introduction of novel technologies and equipment rather than the quantitative expansion of coal output is a top-priority objective of this process.

To improve the efficiency of the coal sector operation the balance (proven) coal reserves have been re-assessed, with the rejection of coal-winning sections and the closure of mines operating in unfavourable geological conditions (thin, steep, disturbed coal seams and those liable to sudden outbursts of coal and gas), as well as those mines, which extract low-quality coals (low BTU, extremely high ash and sulphur content, etc.). As a result, of the total 210 billion ton of balance reserves nearly 40% were rejected. The rest form a reliable coal reserve basis, which will be sufficient for the sector operation for several centuries.

At the same time, coal production has been concentrated in coal-winning sections and mine takes with the so-called "technologically feasible" reserves, which are mineable with an application of the most efficient high-performance machinery, with the utmost use

*Prof. Dr. Sc. (Eng.) President, Russian Union of Coal Producers, Russian Academy of Mining Sciences, Moscow, Russia.

of its potential (flat non-disturbed seams of a medium thickness, non-gassy sections and mine takes or those with low methane content, etc.).

In 1998, Russia produced 232.2 million metric tons of coal (in 1997 the country's coal output was 244.3 million tons). However, with a nearly two-fold decline of gross coal output during one decade (in 1988, the country's coal output amounted to 425 million tons) the sector operation efficiency has been substantially improved. For instance, the structure of coal production has been substantially improved due to the top-priority development of the most efficient surface mining method. Its share in the total coal output of the country has increased from 55 % to 65 %, with a clear trend towards its further growth to 70 – 75 % in the future. At the same time, over a five-year period of the sector restructuring coal production has been ceased at 150 extremely loss-making mines of a 42 million ton aggregate annual capacity.

At the operating promising mines of the sector work is in progress on the concentration of production and technological refurbishing aimed at the expansion of the share of completely mechanized coal mining to 95 % and daily coal output per face to at least 1,000 and to 5,000 – 10,000 tons in more favourable geological conditions.

As a result, over a 5-year period labour productivity in the coal sector has increased by 35 %, while coal production costs have substantially decreased, and the quality of coal supplied to main consumers has been improved to a certain extent.

The development of a number of new highly-profitable coalfields, first of all, the Yerunak Coalfield of the Kuznetsk Basin (Western Siberia), has been initiated. Underground and opencast mines of a new engineering level are being constructed there. By technical and economic indices they will be competitive with their foreign analogies, including those operating in the USA and Australia.

By present, several highly efficient mines of a new engineering and technological level have been designed, and now the construction is in progress (Table 9.1). These projects are the synthesis of national and foreign advanced findings and developments in mining and successful practices, including all the elements of the technological circuit of a mine. They are aimed at the exclusively efficient development of favorable coal reserves of the most profitable coalfields of Russia (mainly those of the Kuznetsk Coal Basin and Western Siberia).

For coal-face operations high-performance heavy-duty mechanized coal-winning systems of the national and foreign design are envisaged there. New engineering approaches for main and auxiliary processes provide for the following:

- application of the most reasonable "mine-longwall" technology;
- highest concentration of mining processes – up to 10,000 ton daily coal output per face;
- implicit reliability of all the elements of the mine technological circuit;
- application of modern management systems meeting the conditions of the concentration of mining processes.

Table 9.1 : Planned Economic and Performance Indices of the Mines of a New Engineering Level in the Russian Federation

Index	*Mine*			
	No. 7	*Kotinskaya*	*Taldinskaya-Yuzhnaya*	*Obukhov-skaya No 1*
Annual production capacity, '000 t	2,800	1,500	1,500	1,800
Construction period, year	5	3	3.5	4
Average number of operating coal faces	1	1	1	2
Average daily output per face, t	10,000	5,000	5,000	3,000
Number of man on books	334	285	286	745
Monthly coal output per man, t	791*	498*	496*	232*
Capital expenditures for mine construction, million $ US	121	75	76	235.5
Coal production costs, $ per ton	8	8	8	15
Profitability, %	47	25	45	15
Average capital investment pay-off-period, year	5	8	5.6	6.6

* At these mines monthly coal output per man will exceed that of the US coal mines and it will be 3-4-times higher than that of the modern mines of England, Germany and France.

Such advanced engineering and technological approaches in combination with the most favorable geological conditions provide for the achievement of high economic and performance indices: monthly coal output per man will reach 500 – 800 t, coal production costs will not exceed $8 US per metric ton, capital investments for an underground mine of a 1.5 – 1.8 million t annual productive capacity will be $ 75 – 120 million US accordingly.

Over recent years for the above mentioned and other promising mines the defense-related machinery manufacturers have designed a number of relatively cheap mechanized coal-winning systems of a new generation. Now these systems are commercially available. They demonstrate high performance operating in nearly all geological conditions in flat and inclined seams.

All these measures contribute to the refurbishing and upgrading of the Russian coal industry with the advent of the 21st century.

For underground coal mining in Russia a special Program for the manufacture of modern mining equipment, such as coal-winning systems, roadheaders, mine transport facilities, auxiliary equipment, etc., has been elaborated.

This machinery, equipment and relating technologies will dominate for long.

At the same time, in parallel with the above, some other technologies, such as hydro-abrasive cutting technology and other methods of coal winning and coal-bed methane extraction, are being developed.

In Russia bore-hole methods are used for commercial utilization of coal-bed methane. Its resources are rather plentiful and comparable to those of natural gas (Table 9.2).

Table 9.2 : Prognostic Coal-Bed Methane Resources of the Principal Coal Basins and Coalfields of Russia (billion m³)

Coal Basins Coalfields	*Total*	*Including*			
		takes of operating mines	*explored promising sections and areas being prospected*	*upper level of coal basins, to a 1,200 m depth (to a 900m abs.)*	*lower level of coal basins, to a 1,200-1,800 m depth (-900-1,500 m abs.)*
Kuznetsk Basin	13,085	211.5	12,873.5	7,448	5,637
Pechora Basin	1,942	26.4	1,915.6	1,260	682
Donets Basin incl.	1.178	495.0	683.0	-	-
Eastern Donets Basin	97	2.2	94.5	-	-
Bureya Basin	312	-	-	223	89
Alsat Basin	55	55	-	55	-
Sakhalin Basin	48	4.2	43.8	22	26
Partizansk Basin	22	6.8	15.2	-	-
Total:	16,642 (199%)	798.9 (4.9 %)	15,531.1 (95.1 %)	9,008 (58.5 %)	6,434 (41.7%)
Tungus Basin*	20,000	-	-	-	-
Lena Basin*	6,000	-	-	-	-
Taimyr Basin*	5,500	-	-	-	-
Zaryansky Basin*	1,234	-	-	-	-
Total	32,734				

* Due to insufficient geological data on these basins only rough estimates of coal-bed methane reserves of coal mines are available.

At the same time, coal mines of the sector annually emit to the air basin over 1.5 billion m^3 of methane, while methane utilization is effected at a rather narrow scale. However, recently a new utilization technology for methane-air mixtures with a varying methane concentration has been developed in Russia. By present, work has been completed on the development of the utilization technology and facilities for catalytic oxidation of methane brought to the surface by mine fans, with eventual heat and power generation.

In 1996, a comprehensive intersectoral project on advance coal-bed methane extraction was elaborated and included in the Federal Program "Mineral Wealth of Russia".

To provide the prospects for the development of surface coal mining the following engineering approaches are applied:

- machinery, equipment and technologies for coal and rock excavation with complete elimination of blasting processes;
- methods and facilities providing for more precise selection and lower share of stone in mined coal in the development of coal deposits of a complicated structure;
- broad-scale application of continuous and cyclic-continuous technologies with the use of conveyor and combined transport systems.

Rather promising is the technology envisaging the application of wide-web excavating machinery for layer-by-layer cutting in coal winning and stripping operations. This technology allows to completely eliminate blasting operations in surface mining. A pilot KSM-2000P surface miner, known as the Russian KSM Project, has been developed by Russian engineers in close co-operation with the Krupp-Fordertechnik, Germany. Now it is successfully operated at the Taldinsky opencast mine in the Kuznetsk Basin. A two-year commercial practice has proved an opportunity for the co-manufacture of such systems, which will be able to cut the strata of 100 – 120 mPa compression strength.

In conditions of the Taldinsky opencast mine the guaranteed monthly coal output per man operating such systems exceeds 1,000 metric tons (it is four times higher than the level already achieved there with the conventional machinery), while coal production costs decline at least by 30%. Metal consumption for the entire system intended for stripping and coal winning operations is by far lower than that of conventional systems.

Complete elimination of blasting operations due to this novel technology will also result in the radical improvement of the ecological situation of an opencast mine and higher quality of coal produced there. Essentially, it is the technology of the 21st century, which is ecologically clean, efficient, it improves the quality of mined coal and completely eliminates blasting processes in surface mining.

It is expected that in the nearest future a broad-scale program will be initiated. It is aimed at the re-equipment of surface coal mines with such new systems. Opencast mines of the Central Kuznetsk Basin, Transbaikal Region, South Yakutia and some other principal coal-producing areas of Russia will be included in this program.

The Program for the development of novel surface mining equipment and machinery is oriented towards the expansion of the annual surface coal output to 250 – 300 million metric tons that is expected for the nearest decades.

In general, new coal mining technologies for both underground and surface mining are aimed at the multiple growth of coal-winning machinery performance and coal output per time unit. Coal quality improvement problem is solved but partially in this case.

Actually, this objective is attained to a certain extent through conventional and novel coal preparation methods, as well as comprehensive thermal conversion technologies. If this trend is actually realized, even in the early 21st century the supply of raw high-ash steam coals, as well as coking coals with a relatively high sulphur content will be ceased.

In the area of coal preparation the creation of efficient methods for the processing of coals, particularly those with high sulphur content, aimed at the maximum possible removal of the latter is the most acute and complicated problem. Work is in progress on the material analysis of the organic matter of high-sulphur coals and development of the relevant coal processing technologies. To radically cut ash content of coals new types of jigs and some other equipment are being introduced.

In the area of comprehensive coal conversion in Russia experimental research is in progress. It is aimed at a more complete utilization of the energy-chemical potential of various coal grades, ranging from brown coals to anthracites. Namely, pilot technologies have been developed for hot-cure chemical systems for the briquetting of brown coals from the Kansk-Achinsk Basin, as well as hard coal (anthracite) slurries and fines. Work is in progress on the improvement of various technologies for the synthesis of liquid and high-BTU solid fuels. Namely, opportunities for the implementation of the Russian-American Project based on the so-called "mild pyrolysis" (LFC) technology for the synthesis of coal-derived liquid hydrocarbons and relatively clean solid fuels are being investigated. "Young" coals of the Kuznetsk Basin (Western Siberia) of a relatively low calorific value ranging from 5,000 – 5,800 kcal/kg form a potential feedstock for this technology of the 21st century. In Russia, research is in progress into the alternative methods for the production of coal non-fuel derivatives.

All these research directions, findings and developments serve a basis for actual upgrading of coal resources including coal-derived products with new consumer characteristics. This is the basis, which will open an actual opportunity for further expansion of the application of the Russian coal and its derivatives in the 21st century.

Technologies for preparation, conveyance and combustion of coal-water slurries, a new type of coal-derived organic fuel, completely meet the requirements to coal, as a widely-used energy source, in the 21st century. In this area Russia has accumulated a substantially high research and production potential. It has served a basis for the practical realization of the project of a pilot coal-water slurry pipeline connecting the town of Belovo with the Heat and Power Plant in Novosibirsk. The 264-km pipeline annual capacity is estimated at 3 million. ton on a dry coal basis. As compared with the railway transportation costs, operating and capital expenses of the pipeline are 1.5 – 2 –times lower. This technology allows to produce and transport such a coal-water fuel, which is actually an ecologically clean energy source. Coal slurry combustion results in lower ash and reduced So_x and No_x emission. The 21st century will be the time of a broad-scale application of coal-water fuels, and presumably it will require the construction of several main coal-water slurry pipelines from the Kuznetsk Basin through Urals to the lands along the Volga River. Realization of this program is of paramount importance for the expansion of coal application in the 21st century from the viewpoint of economic and ecological situation.

Improvement of the current ecological situation in general has acquired an ever-growing importance for the Russian coal industry. The share of the coal sector in the total harmful environmental effect in Russia is estimated as follows: air pollution with harmful agents – 4%; waste and contaminated water effluents – from 9 to 11 %; area of disturbed lands – 9%; solid wastes – 40 to 45%. With the expected top-priority development of surface mining in the 21st century special attention to the reclamation of disturbed lands in coal-mining regions of Russia must become an essential long-term trend. Moreover, it is proposed to include the strategy of land reclamation in the Law "On the Restructuring of the Coal Sector of Russia". The Draft Law has been prepared for the eventual hearings and approval by the Lower Chamber of the Russian Parliament (Duma).

Particularly important ecological problem will be faced by the sector in the early 21st century. This problem concerns both the coal industry and the power engineering. The current distribution of coal-fired power plants over the vast territory of Russia has allowed so far to avoid the effect of excessive concentrations of solid particles and acids emitted to the environment. However, with the growth of coal consumption by power plants the problem of clean coal combustion processes will acquire top-priority and strategic importance for ecological safety.

The advent of the 21st century is closely associated with another important ecological problem concerning global climate change. Some experts believe that coal combustion is one of the main factors contributing to the so-called greenhouse effect. Dealing with this problem one should take into account that the level of heat emission per unit of area in Russia is minimal. At the same time the country possesses 20% of the planet's forests, which are "the lungs of the Earth". Within the limits of equal unit norms and quotas it allows both to develop the energy sector and to neutralize the excessive emissions of industrially developed countries (of course for a fair compensation). For this purpose it seems feasible to establish an information system of industrial ecological monitoring for operating enterprises. It will help to radically improve a multiform and painstaking work aimed at the reduction of anthropogenic effect of mining enterprises.

Such a system is being elaborated in Russia by the coal sector experts in close co-operation with the Space Research Institute of the Russian Academy of Sciences.

In general, paramount importance is being attached to the solution of the ecological problems in Russia on the eve of the 21st century. Comprehensive approaches will be used for the purpose, and top-priority will be given to the issue.

To our view, for the 21st century the most probable could be the general trend towards the development and broad-scale application of ecologically clean highly efficient technologies for the complete cycle of coal production and supply to main consumers. With the realization of the available potential and depending on coal market demand in the nearest future, a trend may appear towards stabilization and presumably towards certain gradual growth of coal output in Russia to 250 - 300 million t per year. Even today (according to the data of 1998) 218 million t of the Russian coals are supplied to main consumers including 100 million t for thermal power plants, 32 million t for coke by-product plants and 22 million t for exports.

It is quite evident that in a more distant future the coal output of Russia will be gradually expanded. This trend is determined by vast coal reserves of the country. It will give an opportunity to turn the Russian power generation sector to a broad-scale application of coal, as a basic strategic resource, for long years.

Actual implementation of this trend will depend upon three main conditions (factors):

- competitiveness of steam coals as compared with natural gas;
- improvement of the quality of saleable coal and its derivatives including those with new consumer characteristics, such as the products of thermochemical coal conversion, namely synthetic gas and liquid motor fuel, etc.;
- adequate ecological safety of production and especially utilization of coal and its derivatives.

To meet the above conditions the relevant special-purpose investments are required. Such investments could be also provided by those countries, citizens and companies which are willing to become the consumers of the Russian coals.

According to our forecasts these main objectives will be completely attained even in the first decades of the 21st century. By the 2005 the sector labour productivity will compete with that of the industrially developed countries of the Western Europe. Russian coal industry will enter upon the early 21st century as a completely upgraded profitable sector of the national economy providing for 300 million ton annual supply of competitive coal. Predominantly, large promising mines with qualitatively more perfect and ecologically clean equipment and technologies will be in operation. Moreover, coal production and processing wastes will be used as a fuel for mini-TPPs.

CHAPTER 10

Slovak Coal Today and in Future

Jan Fabian, Vladmir Misut & Cicmanec

ABSTRACT

Utilisation of own mineral resources in the Slovak energy industry is directed by the strategy objects accepted in the document "Energy Conception of the Slovak Republic till 2005". This paper deals with a present position in 3 coal centres (Hornonitrianske bane–Prievidza, Bana Dolina–Vel'ký krtiš and Bana Záhorie–Holíc.) Decisive producer of brown coal in Slovakia is Hornonitrianske bane in Prievidza. We are analysing current problems in the power energy of Slovakia, concerning solid fuels and other sources of electrical energy.

INTRODUCTION

The Slovak coal mining is based on the extraction of brown coal. It features a rich history dating to the last century, when the extraction of brown coal started in the mine Handlová. Later on, in the 1940 's, the extraction started in the mine Nováky, in 1952, in the mine Dolina, in the 1960 's in the mine Cigel', and in the 1990 's in the mine Záhorie–the coal and lignite mines.

The period till 1988 may be characterised as a period favourable for the coal industry. The extraction in Slovakia was subsidised by redistributing the profit of the Federal Ministry of Fuel and Power Industry. This period left some unfavourable elements of inertia in assessing a certain superiority of the person employed in the coal extraction, and, in addition to it, caused that the citizens and employees of other industries began to consider the mining activities as non-necessary for our economy, and as seriously ecologically harming the environment.

Since 1997, the intensive efforts had been made to make the brown coal extraction more effective, so the Ministry of Fuels focused its conceptual objectives to liquidation of the most cost demanding operations. These objectives resulted in structural changes and the damping programme in the coal mining industry for the year 1990 through 1992. The management of mines in Slovakia passed on under the control of the Slovak Republic and within it under the Ministry of Economy in Bratislava. Our Government, without any proper detailed analysis of the condition and needs of power sources, had taken over the damping ideas about the coal mining industry, and in March 1991, issued the Resolution No. 125, which launched the concrete liquidation of coal extraction in Slovakia. The suspension of extraction till 30.12.1992 had to represent its 9% reduction as well as the reduction in the number of employees by 1,900 person.

The suspension of capacities caused a decline in extraction. Since 1988, when the extraction reached 5821.8 th. tons, the later was gradually decreasing until 1992 by 37.4 % down to 3,522 th. tons, and the number of employees was reduced by 3,532 persons, i.e. from 15,508 down to 11,974 (see Fig. 10.1)

The most important change in extraction occurred in 1992. The number of employees continued to fall following the process of rationalisation, and at present amounts to approx. ten thousand persons employed in the coal mining industry in Slovakia.

This extremely difficult period from 1988 to 1992 was overcome for our coal mining industry, and the mining units were classified on the basis of the so-called deposit principle, which meant one enterprise are deposit.

In the course of the entire period of 1990-1992 as well as in 1993 we had arranged numerous negotiations concerning the need of Slovakia to develop and accept its own conception of power industry. The Slovak Republic belongs to the poorest states world-wide in assessing its own primary sources (PEZ). We have to import up to 90% of them.

We argued with the need to safe for Slovakia at least one more important source of power (except of hydro-potential), which we have got, namely the brown coal extraction. The SR PEZ structure is shown in Fig. 10.2

In August 1993, on the basis of the SR Government's Resolution No. 562, the Conception of Power for Slovakia until 2005 was accepted. The document was passed as open, which is requested to be modified, made more precise and confronted with the succession of economic reforms and restructuring of the national economy. The document is an expression of the need to extract brown coal in the SR at the level of 3,6-4,0 mil. tons per year.

This volume was confirmed also in the upgraded Conception of Power for the SR until 2005, with the outlook till 2010. The Conception designate the brown coal extraction in Slovakia as an activity having the strategic importance.

The Fig. 10.3 points out to a gradual decline in the brown coal consumption in the Slovak Republic. But the most decisive is the fact that already in 2010 the consumption, expected by prognosticators, is higher than our production birth in the power industry and sorted kinds of fuels. The difference was and will be covered by import most probably from the Czech Republic, because the consumers in Slovakia were constructed for this kind of coal.

The fact is that the position of solid fuels in the power balance of our state, regardless of whether it concerns the sources imported or own, is most important.

Fig. 10.4 points out to the full comparability with the OECD countries.

International conferences and leading experts emphasise the need to continuously clear any issues of the future relation of coal to other PEZ, also despite the global objective to limit the growth of carbon dioxide in the atmosphere and to decelerate the process of warming our planet.

The advantages of national sources of solid fuels lay mostly in the fact that they are:

- competitive against other carriers of power;
- independent on foreign-political influences;
- representing rather smaller risks against the risks of large-technical alternatives and technologies;
- and that there exists a possibility of a long-term prospective computability of such national carriers of power, without any most significant influence of the growth of world prices; and
- and that such a national source may be immediately applied in the ecologically pure technologies with the completed development.

Mostly the new power-station technique opens better chances. With new power-stations, equipped with the optimised technique of the installed facilities and complex purification of waste gases in the internal recuperation of heat up to the combined brown coal power-stations with the para-gaseous cycle, where the advance turbine is driven with gas made of brown coal, a rather higher efficiency may be expected, with a positive reflection both in the economics and ecology.

The concrete situation in the development of solid fuels in our country is influenced by market mechanisms (the price of coal were gradually liberalised), and the availability of coal from abroad.

Thus, the solution of efficiency of all three mining coal organisations in our country (the mines Prievidza and Dolina in the Horná Nitra region, and the mine Záhorie–the joint-stock companies) remains only in their practical own skills and capabilities, using the racionalisation elements in the production process and active price policy, what allows to establish themselves in the market and to compete with imported Czech coal. This task is extremely demanding and difficult, mainly for the mentioned competition of imported coal. The Result of the sizuation is an insufficient volume of investment construction. The reserves of coal in the existing mining organisations, suitable for their opening and preparation, are relatively sufficient, but our access to financial sources is extremely difficult. The banking sphere remains unfavourable to the mining industry. Though we have evaluated in details the conditions and possibilities of opening the next prospective deposits of brown coal (Beladice with the balance reserves of 93,0 mil. tons) and of lignite (Kúty and Štefanov with the balance reserves of 216 mil. tons), under thc present conditions the idea of taking the deposits under the own administration and of their opening is just a courageous vision.

Based on the real situation and expected possibilities, our strategic objectives, processed in each organisation, may be summarised in an elementary axiom: to extract rationally the reserves of coal at the existing five locations, which dispose the balance reserves of 154,289 th. tons in total, out of 45,615 th. tons of extractable reserves. The extraction will be gradually decreasing when applying the mentioned rational manner of mining. The extraction in 2000 is expected to reach 3, 950 th. tons, in 2005 already 3,340 th.

tons, and in 2010 only 2,600 th. tons of coal. The mining activity is expected to by terminated in the mine Dolina in 2006, and in the mines Cigel' and Handlová, the branch-plants of Hornonitrianske Bane Prievidza, a.s., in the second decade of the next century.

At present, the technology of mining in the mines does not allow any more expressive changes in the assortment in favour of the sorted kinds, thus the producers must rely on their traditional consumers–the power station /ENO/ Zemianske Kostol'any with the annual consumption of 2.8 mil. tons, and partially the heating plant Zvolen and Chemko Strázske. In the oncoming years ENO Zemianske Kostol'any notifies a certain reduction in consumption, reasoned by leaving in operation the block 3 and 4 of ENO be until their definite wear-out. The main our objective is to reach the operation in the stated restructured blocks of ENO Zemianske Kostol'any as they consume our coal.

CONCLUSION

During the last years, the extraction of coal in Slovakia has undergone expressive changes. It features the lack of funds, warm out mining machines and devices, a strong pressure on the economics of mining, processing and use of the national sources of brown coal, deeply extracted.

REFERENCES

1. Cicmanec, P. : Actual problems of the Slovak coal mining. The conference: 45, Years of BERG in Košice, Košice, 1997.

2. Fabián, J.: Perspective development trends in the mining of ores and coal, globally and in Slovakia. The conference: 50Years of the Technical Museum Košice, May 1997.

3. Mišút, V.: Statistic mining annual 1995, Bratislava.

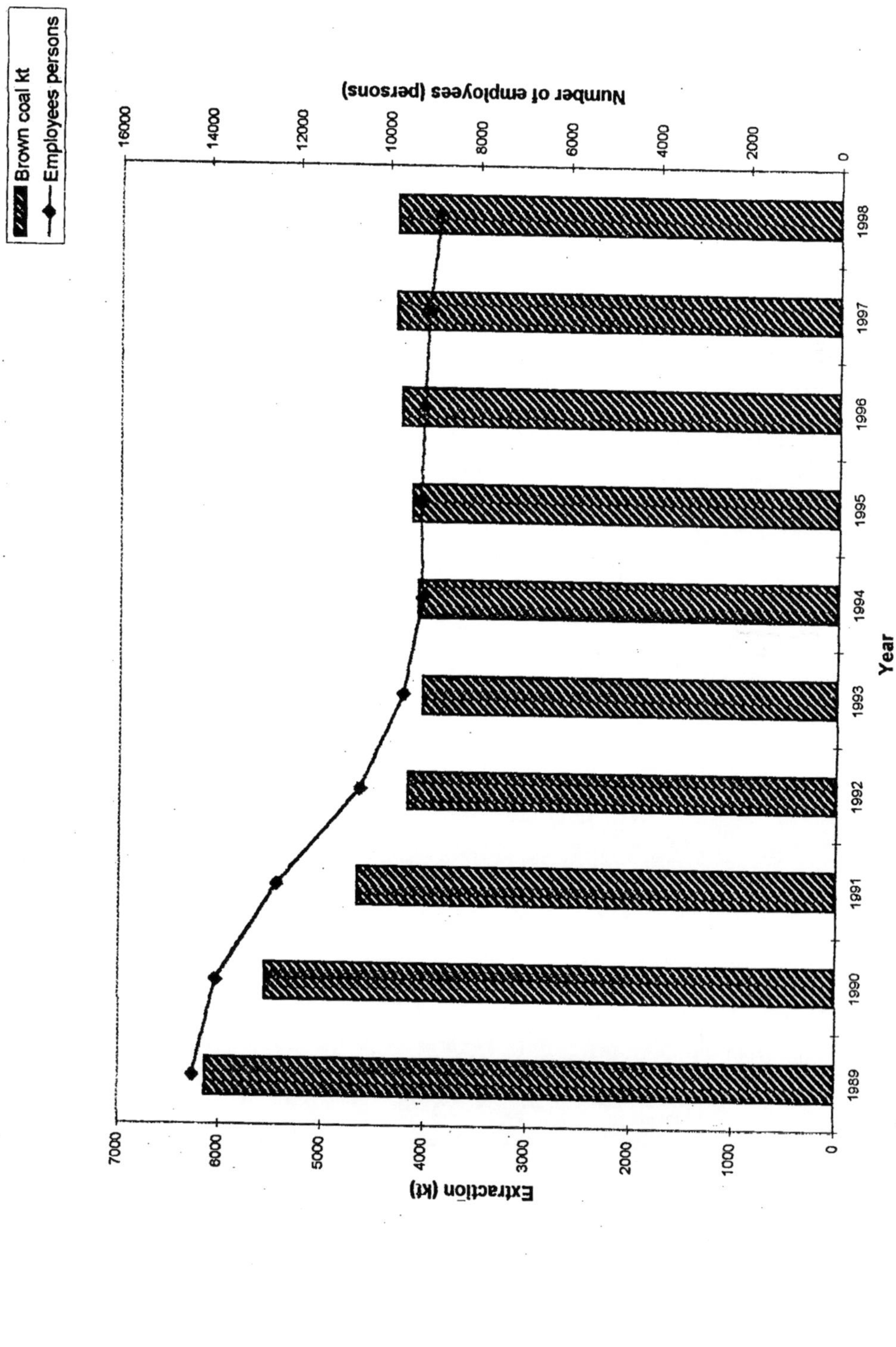

Fig. 10.1 :Development of the brown coal extraction in the SR and of the number of employees

Fig. 10.2 : Expected structure of the primary sources of power in the SR National sources and import in 2000

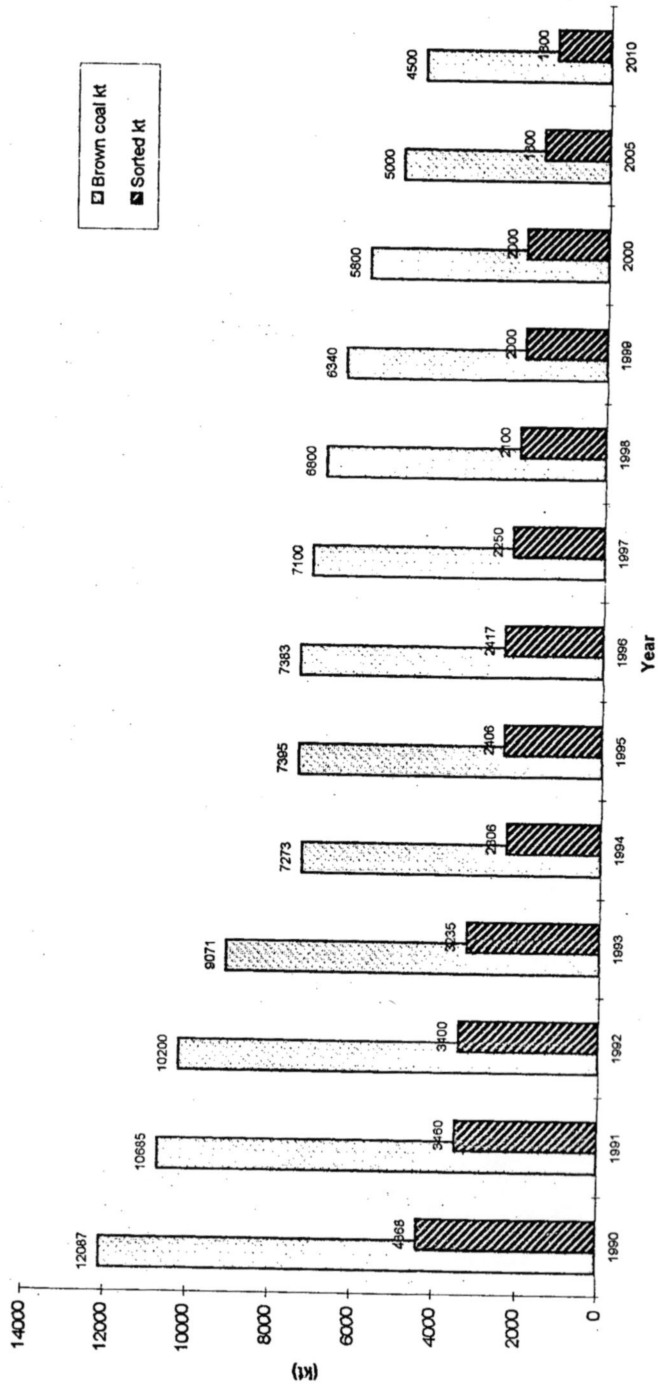

Fig. 10.3 : Consumption of brown coal in the SR

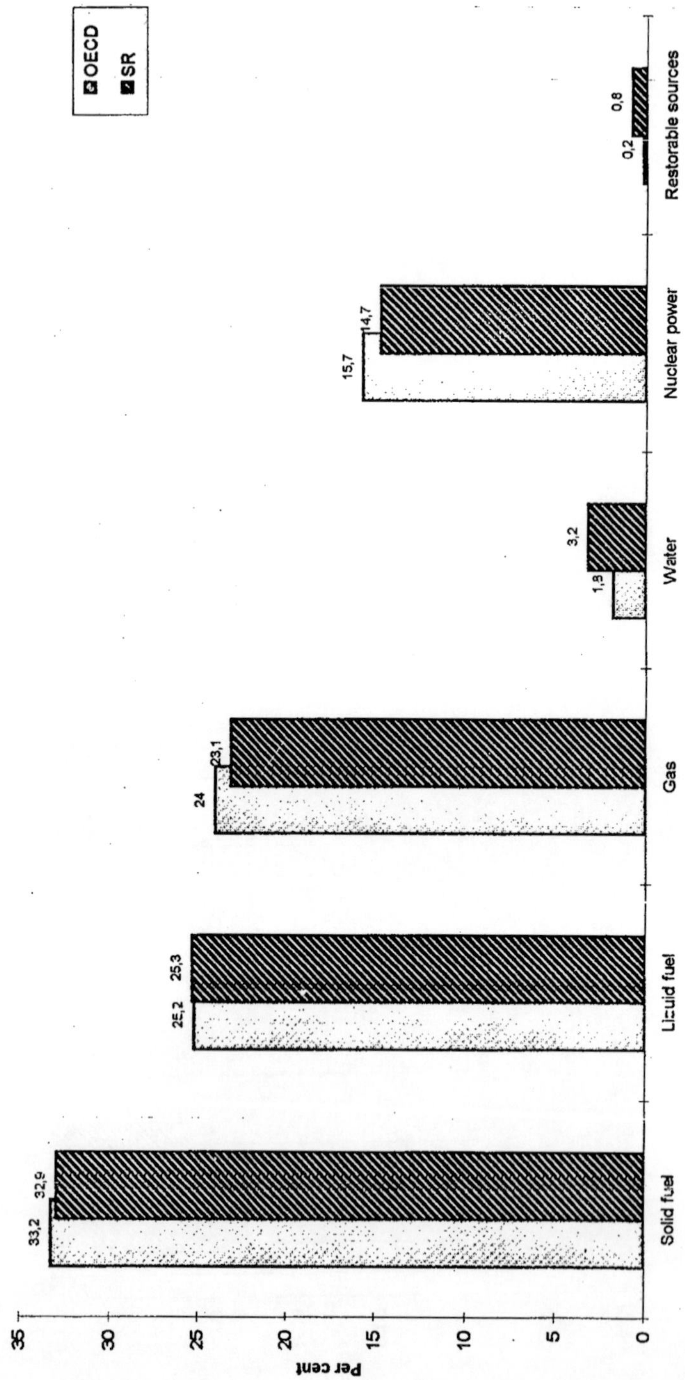

Fig. 10.4 : Primary Power Sources 1996

CHAPTER 11

Coal Mining in Romania at the Beginning of the 21st Century

Dumitru Fodor[1], *Gavril Baican*[2], *Constantin Pasarin*[3], *Gheorghe Bonci*[4]

INTRODUCTION

The territory of Romania is characterized by a very complex geological structure and a large variety of useful minerals in underground. Among these minerals a special attention should be given to coal deposits as, besides other utilizations, coal provides also 40% of the Romanian electric power.

The Romanian power system has an installed power of 20,000 MW of which 39.5% are produced in power stations operating with coal, 30.1% in hydro power plants and 27.5% in power stations running with hydrocarbon and 3.4% in atomic-electric plants, Fig. 11.1.

The coal deposits of Romania are located and grouped in four zones, as per Fig. 11.2.

Zone I, mainly located in the Southern Carpathian Mountains, include all the high grade coal deposits such as anthracite, brown coal and pit coal deposits from Petrosani, Anina and Tebea Brad basins.

Zone II, located within the Pre-Carpathian creep, between Olt river and Valea Buzaului river includes the lignite deposits of Campulung, Sotanga, Filipestii de Padure and Ceptura. The coal basins of the Eastern Carpathian are also included within this zone: Baraolt-Varghis - lignite and Comanesti Bacau - brown coal.

---

1. Head of "Resources" Commission of the Romanian, National College for Research and Development
2. General Manager of CNL "Oltenia"
3. Scientific Secretary of S.C. ICSITPML S.A. Craiova
4. Technical Manager of S.C. ICSITPML S.A. Craiova

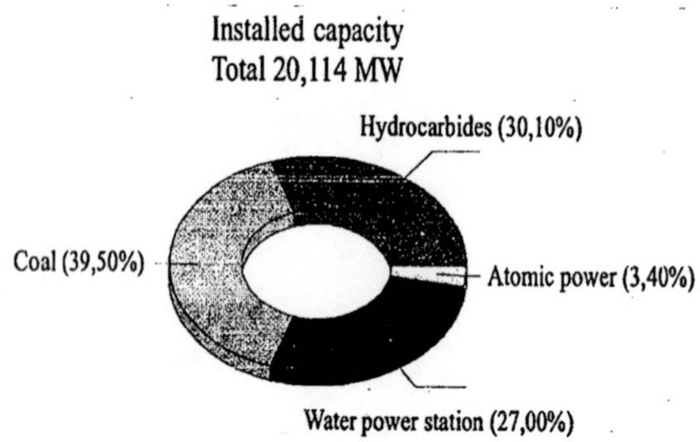

Fig. 11.1 : Contribution of raw materials at the production of electric power in Romania

Zone III, is located in the Sub-Carpathian creep of the Getic Plateau, between Olt river and the Danube and this zone includes the lignite deposits of Rovinari, Motru, Jilt, Berbesti-Alunu and Mehedinti.

Zone IV, is located within the Panonian creep in the North-West part of Transylvania and includes the brown coal and lignite deposits of Sarmasag, Voevozi, Surduc and Borod.

The total coal reserves of Romania amount to approximately 1.0 billion t of pit coal and 3.0 billion t of brown coal and lignite.

More than 90% of the Romanian coal reserves are located within the second zone, namely in the mining basins of Oltenia Region. More than 80% of the Romanian lignite reserves are profitably mineable in open cast, while the remaining 20% is mineable in underground mines.

The Romanian coal deposits are low quality deposits as the lignite has only 1,600-1,800 Kcal/kg and the pit coal provides about 3,000-4,000 Kcal/kg. The average humidity varies from 42% in the case of lignite to 10% for the pit coal. The sulphur content is 0.8 + 1.2% for lignite and 2% for the pit coal.

The trend of coal production in Romania trough the years is given in Table 11.1.

PIT COAL MINING

The most important pit coal reserves of Romania are located within the mining basin of Valea Jiului, situated in Petrosani mountaineering creep and the deposit has a triunghiular asymmetrical syncline shape with a length of 48 km and a width of 2 km to the western side and 9 km to the eastern side and covering a total surface area of 155.5 sq.km, Fig. 11.3.

The balance sheet reserves are distributed in 13 mining perimeters wherefrom the power and coke pit coal is extracted, but the pit coal quality varies from one perimeter to another, from one layer to another.

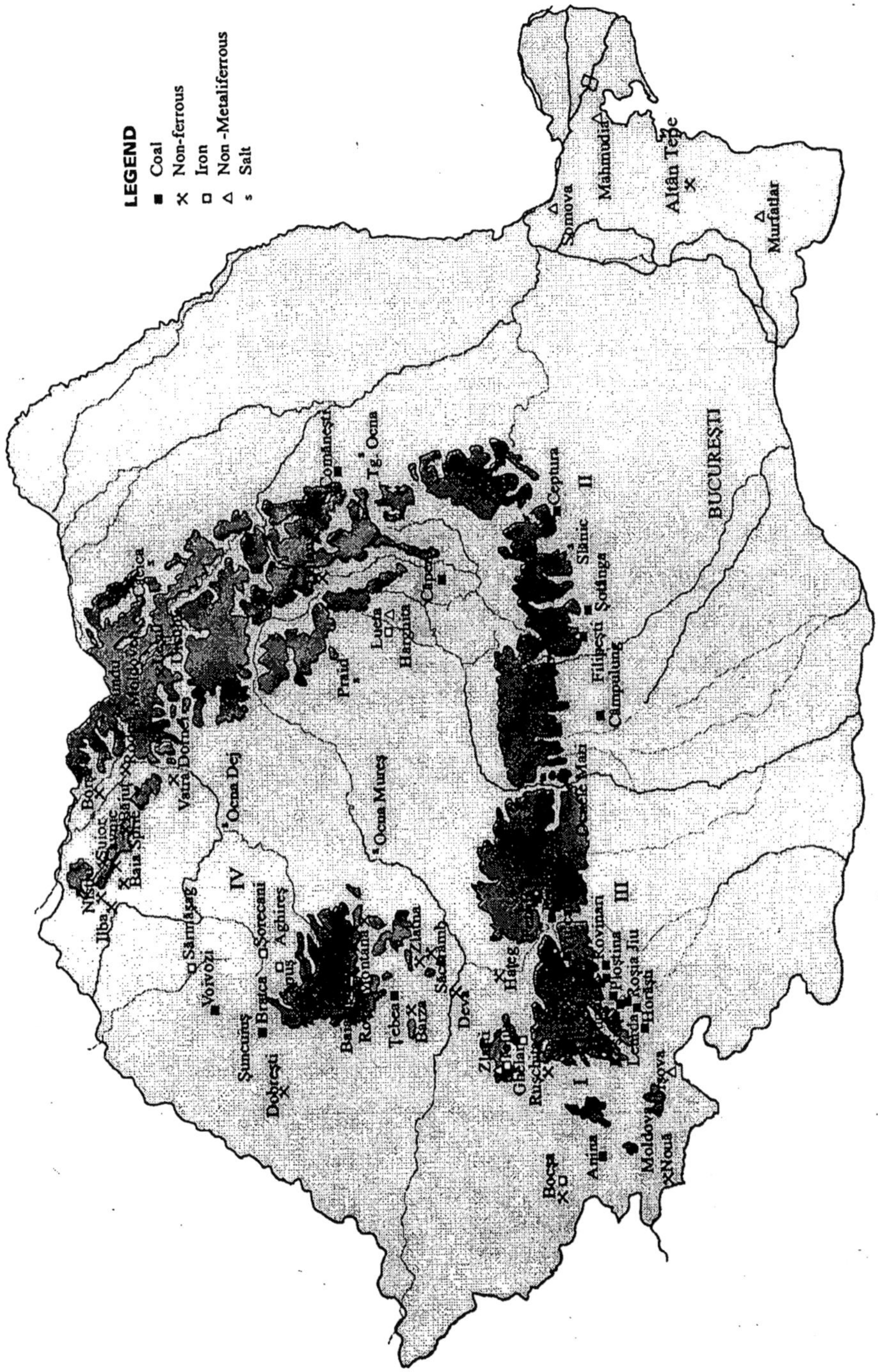

Fig. 11.2 : Regional distribution of coal deposits in Romania

Table 11.1 : Trend of physical production of coal over the last 10 years

	Extracted in year										*Program*
	1989	*1990*	*1991*	*1992*	*1993*	*1994*	*1995*	*1996*	*1997*	*1998*	*1999*
Net total coal	615	376	324	384	397	405	411	418	334	2623	2770
—CNL Oltenia		336	285	341	353	355	323	327	258	2040	222
—SNC Ploiesti		(*)	(*)	(*)	(*)	(*)	37	37	32	26	24
—RAH Petrosani		40	39	43	44	48	51	54	44	325	313
Lignit and brown coal	532	337	286	343	355	357	362	365	291	2302	246
—CNL Oltenia		336	285	341	533	355	323	327	258	20415	222
—SNC Ploiesti		(*)	(*)	(*)	(*)	(*)	37	37	32	26	241
—RAH Petrosani		1	1	2	2	2	2	1	1	5	6
Net pit coal (RAH)	83	39	38	41	42	48	49	53	43	32	3075
—power coal	51	26	29	31	37	44	45	50	40	30	26
—steel plant coal	32	13	9	10	5	4	4	3	3	2	475
Crude pit coal extracted	112	59	54	56	57	63	64	70	56	50	480

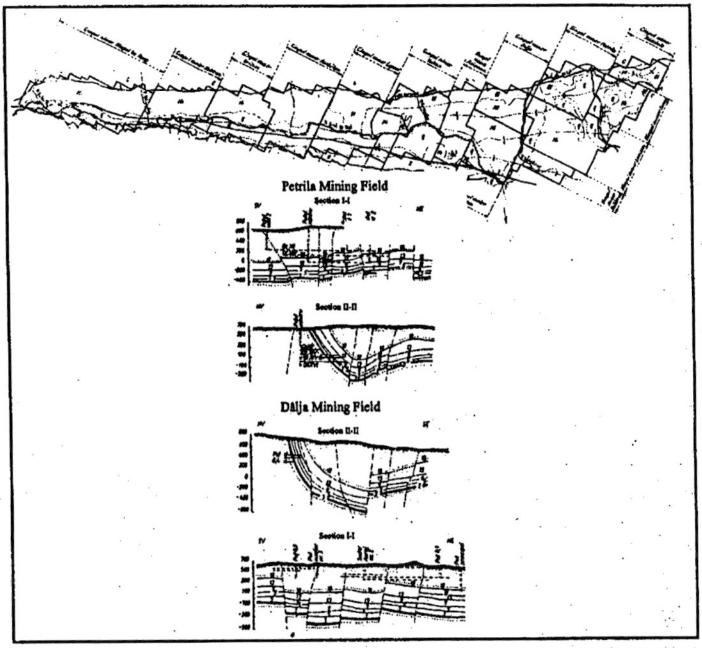

Fig. 11.3 : Configuration and extension of Valea Jului basin

The pit coal deposit of Valea Jiului is characterized by complex tectonics as the deposit is divided in numerous cross and long slices and in smaller tectonic blocks which sizes vary between 200 and 600 m.

There were identified 20 coal layers within the Oligocene formation. Some of these layers are very thin and they cannot be mined out, while others are only sporadically mined. Taking into consideration the economic efficiency of these layers, they were classified as follows:

- Layer 3 - it is the main layer of the basin and it contains about 48% of the deposit reserves;
- Layer 5 - this is the second important layer where approximately 16% of the deposit reserve is located;
- Layer 13 - it is constantly developed in most of the perimeters and represents about 10% of the reserves;
- Layer 4, 6, 7, 8, 9, 12, 15, 17 and 18 - are thinner and they are discontinuous in the mining perimeter, but anyway, each of them contributes with 1-3% of the basin reserves;
- Layers 1, 2, 10, 11, 19 and 20 - are very thin and they are not mined out;
- Layers 14 and 16 - are very thin, but sporadically they have been mined.

Considering the coal layers disposal, their geometry and their vertical development the opening of the mining perimeters was carried out using vertical shafts drilled from the surface, adits and vertical shafts drilled in the underground and crosscut drilled within the floor of the coal layers.

The mining operations are performed under very difficult conditions due to several reasons:

- complex tectonics of the deposit;
- presence of the methane gas and coal dust which is explosive;
- the self-ignition properties of the pit coal which can cause underground fires.

The diversity of the deposit conditions existing in Valea Jiului mining bassin led to the application of different mining methods and technologies for the extraction and beneficiation of coal.

Keeping the pace with the evolution of the mining technologies new working methods have been introduced and the older ones have been improved.

During the last decades the long face stops with individual support and mechanized cutting has been extended together with the mechanized complexes stopes.

Today, the following mining methods are being used in Valea Jiului:

- long face method using long directional or inclined walls for mining the thick horizontal slices extracting the coal either in slices or through mechanized stopes;
- mining the thick layers with inclined slices and the extraction of coal in face stopes, Fig. 11.4, 11.5 and 11.6;
- mining of thick coal layers with inclined slices and extraction of coal in front stopes.

The utilization of the complex mechanized methods, mentioned above provided the increase of the mining field production by 30-50% and of the work efficiency by 15-25% reducing the power and material consumption and the labor.

Throughout the last years the long directional pillar sub-caving method has been used in Valea Jiului as per Fig. 11.7 and 11.8, and the economic results have been better than those obtained using the old methods.

The trend of coal production in Romania troughs the years is given in Table 11.1.

MINING OF THE LIGNITE DEPOSITS

The mineable coal layers of Romania are 1.0-8.0 m thick and they occur as packs with waste intercalations and in their neighborhood there are aquifers formations both in the floor and the roof, free or captive causing much mining difficulty and additional costs, both for the opening works and the mining ones as there are required some de watering operations as per Fig. 11.9.

The lignite deposits of Romania are located within areas where the younger geological formations are prevailing – Dacian, Levantine and Quaternary – consisting of soft rocks (marns, clay, sand etc.).

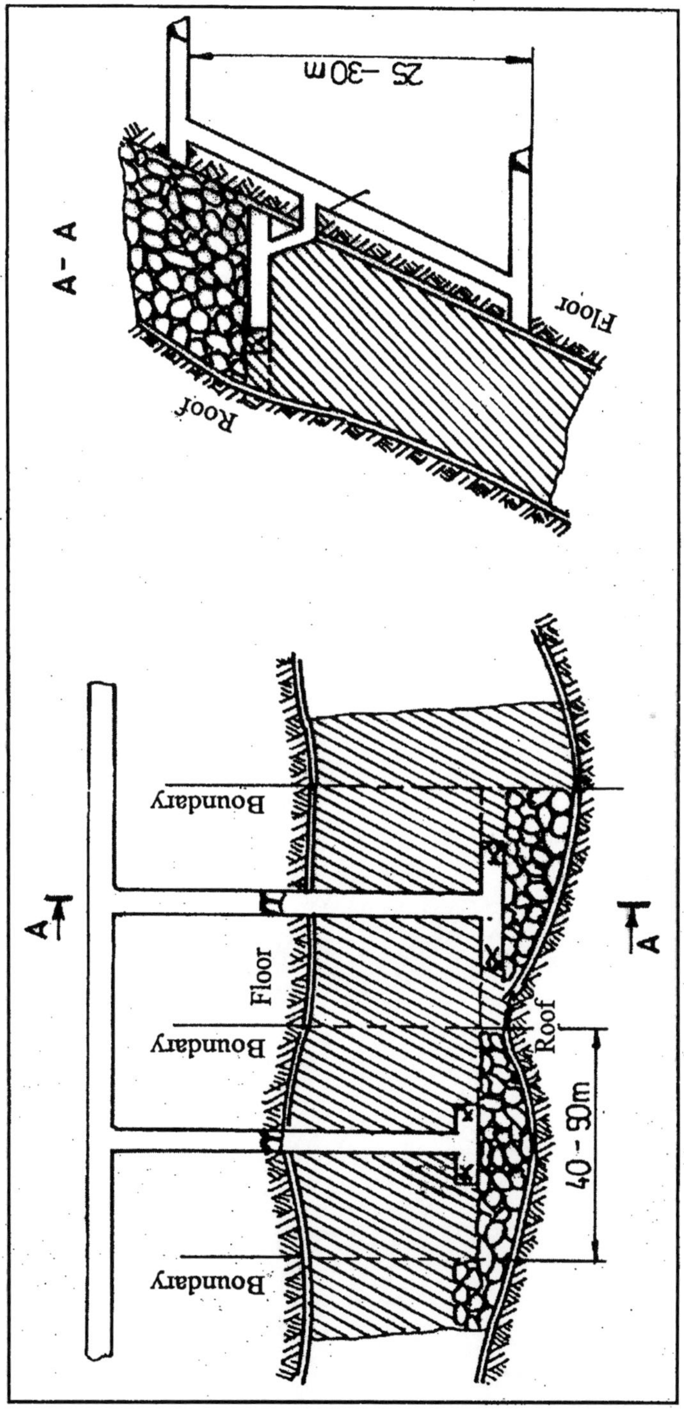

Fig. 11.4 : Horizontal slice method (extraction of horizontal slices on directional bands)

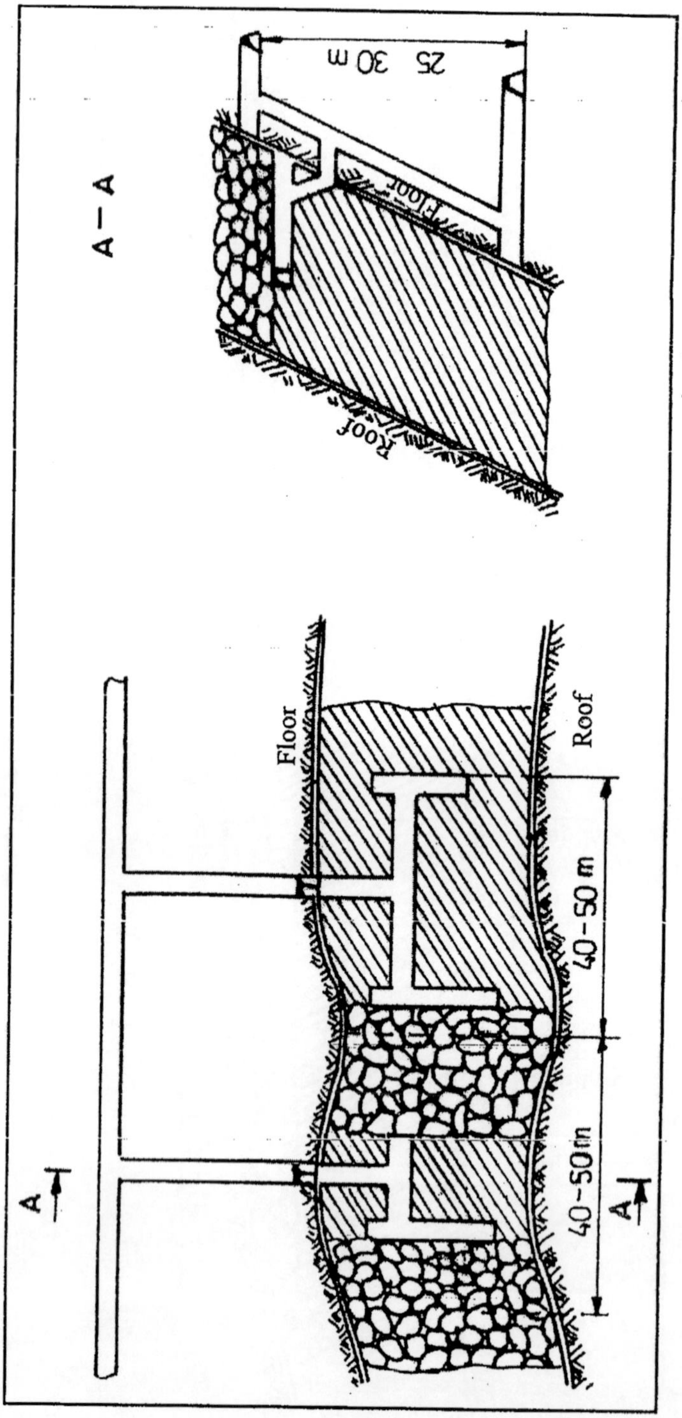

Fig. 11.5 : Horizontal slice method (horizontal slice extraction on crossing bands)

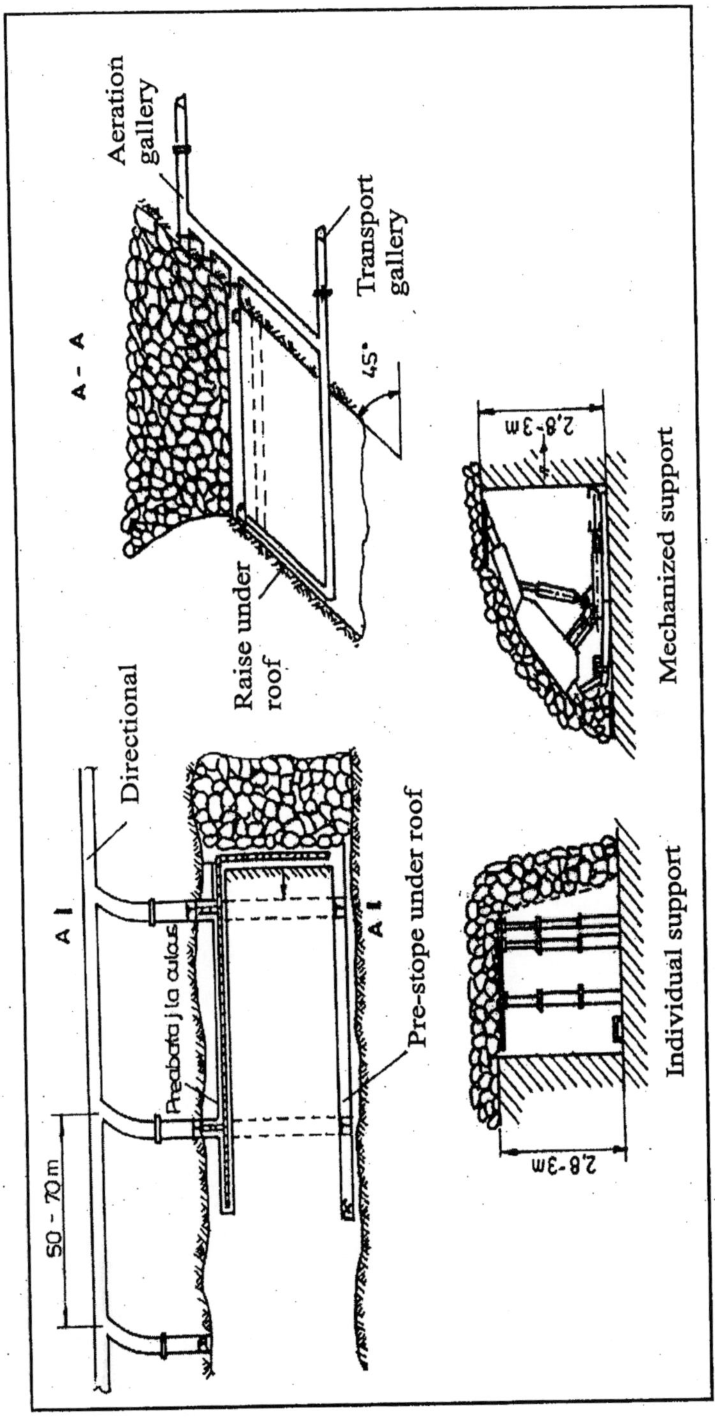

Fig. 11.6 : Horizontal slice methods with directional long face

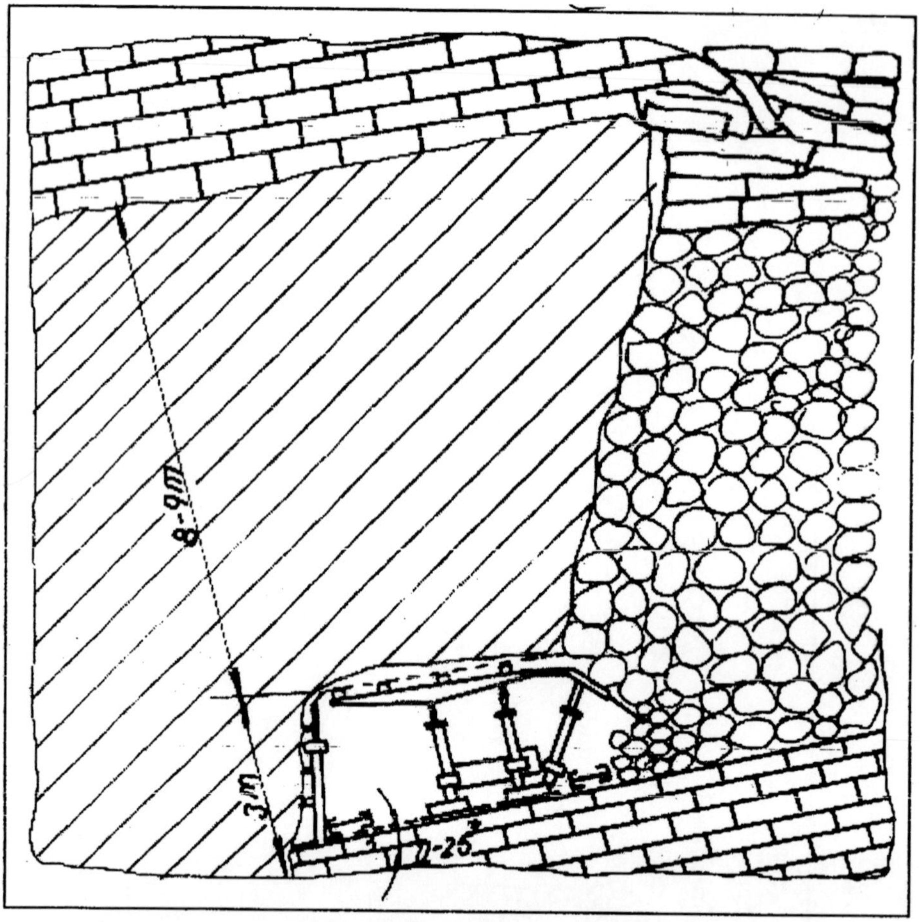

Fig. 11.7 : Long wall-caving over the whole depth of the layer

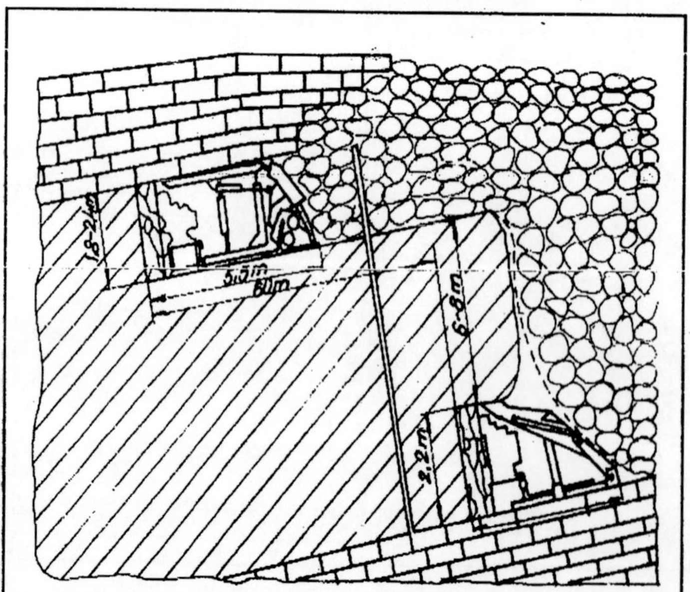

Fig. 11.8 : Long wall slice sub-caving

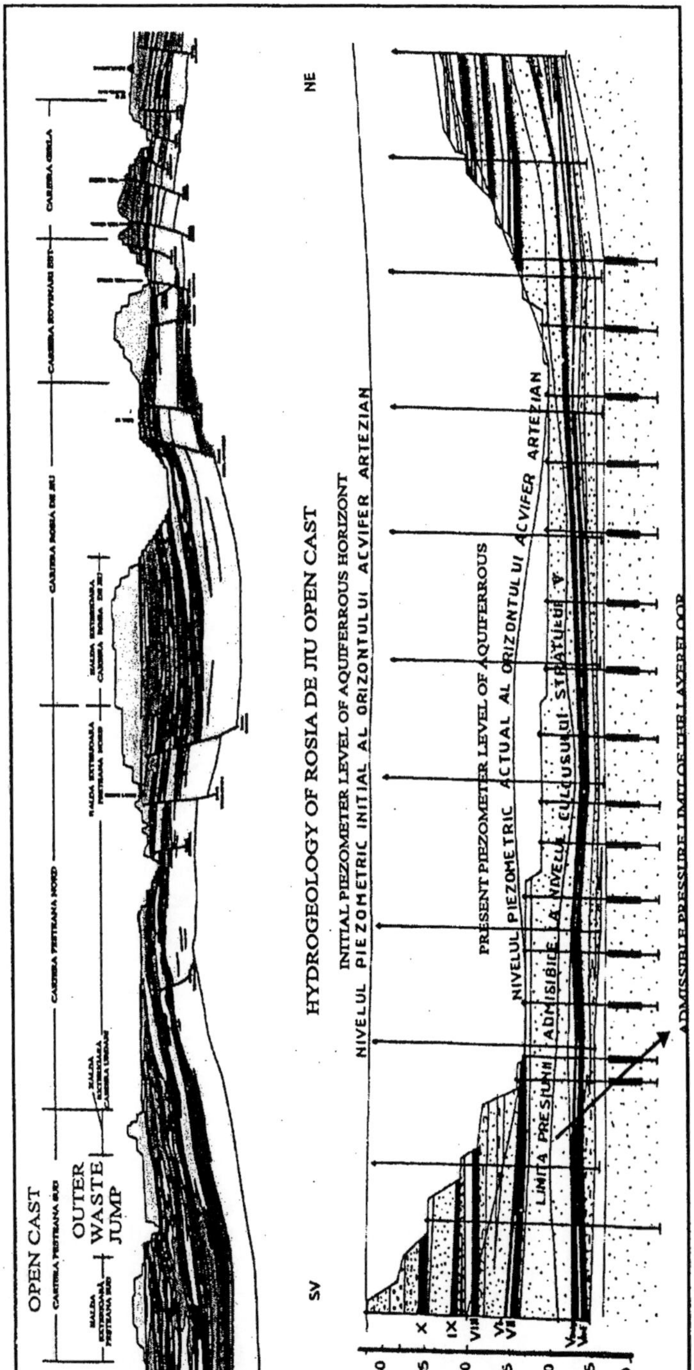

Fig. 11.9 : Geological and hydrogelogy of Rovinari coal basin

The lignite extraction is carried out both in open cast and underground mines, while the brown coal is mined only in underground mines.

Underground Mining

The opening of the lignite mines, is carried out either by adits, inclined platforms or mine shafts depending on the land configuration and the depth of the reserve location.

The most commonly used mining method is the inclined ramp method (Fig. 11.10) with circular section the radius of the free section ranging between 1,300 mm and 1,500 mm with concrete support using also mining metal frames or prefabricated block mansonry. Recently, there have been tested the supporting method using big prefabricated blocks and the results have been satisfactory.

The preparation of the stope panels (Fig. 11.10) is performed by drilling two parallel galleries which trace the coal layer direction. The free section radius of the galleries is 1,300 mm or 1,500 mm, and for the working supports there have been used metal SG frames located in variable fields ranging between 0.6-0.7 m, and wrapped with wire net.

Depending on the local specific conditions the opening index varies between 0.7 and 2.2 m/1,000 tonnes, and the preparation work index vary between 7.0 and 15.0 m/ 1000 tonnes.

The underground mining of the lignite and brown coal deposits is carried out, in almost all the mines, using long-wall stopes with mechanized cutting and the stops support with mechanized combines, Fig. 11.11.

This mining method was preferred as it ensures a high concentration of the production, an efficient survey and control of the production and a significant increase of the work productivity and lower operational costs.

The length of the stops panel is 600-800 m, depending on the extraction capacities and the maintenance of the preparation galleries and on the requirements of mechanized stops combines in order to obtain the highest economic efficiency parameters.

About 85% of the opening and preparation works are performed using complexe combines and most of the underground transport is carried out by belt conveyors or scraper conveyers.

Presently, the mechanized combines, the stops equipments, the combines for the mining workings drilling are manufactured in Romania as well as the scraper and belt conveyers.

Despite the results which are lower than those obtained in other countries where lignite deposits are mined out, it is worth mentioning here that the stopes advance is more than 250 m/month and the production amounted to 70,000 t/month/stope and the work efficiency is 50 t's/job and even greater.

The lignite underground productions had been increasing until 1980 then there was noticed a slow diminution of these figures until 1989, and between 1990-1991 the production dramatically decreased.

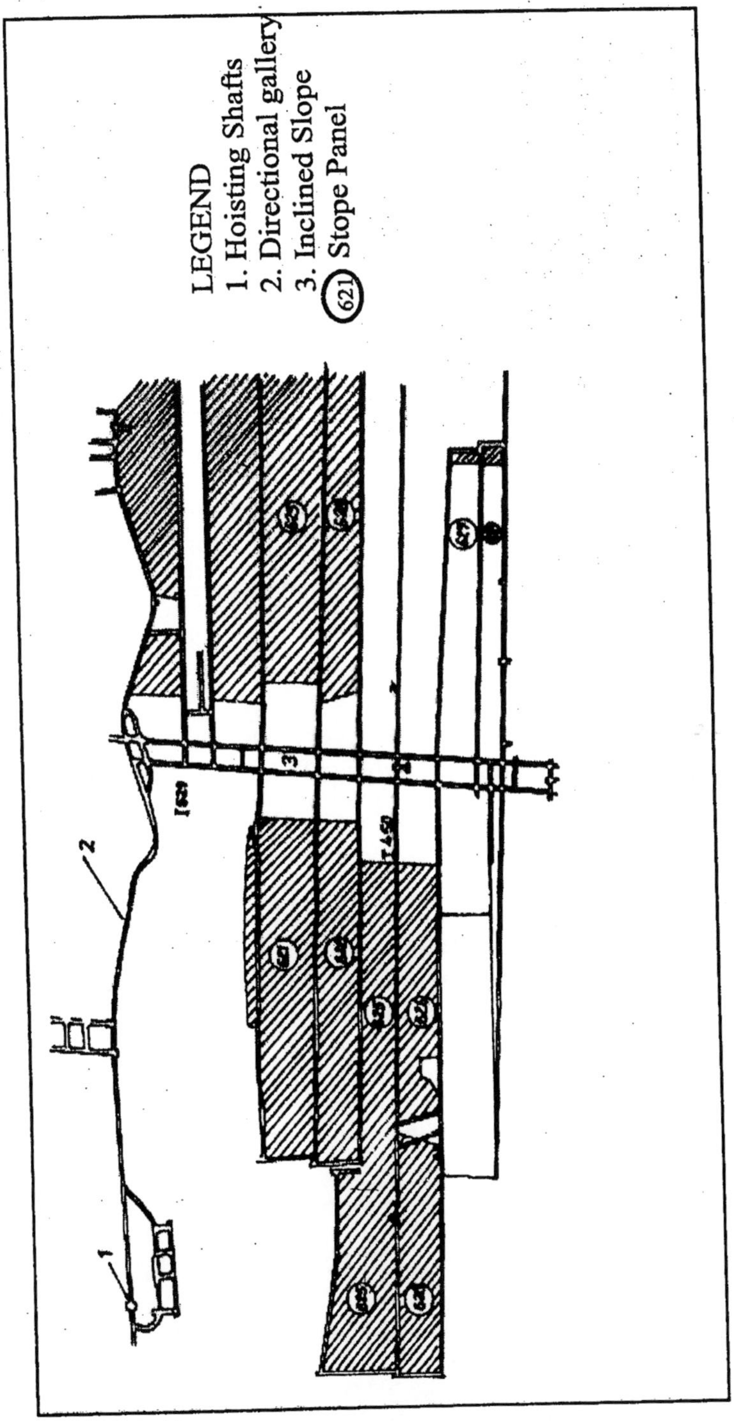

Fig. 11.10 : Opening, preparation and mining flow-chart applied at the lignit mines of Romania

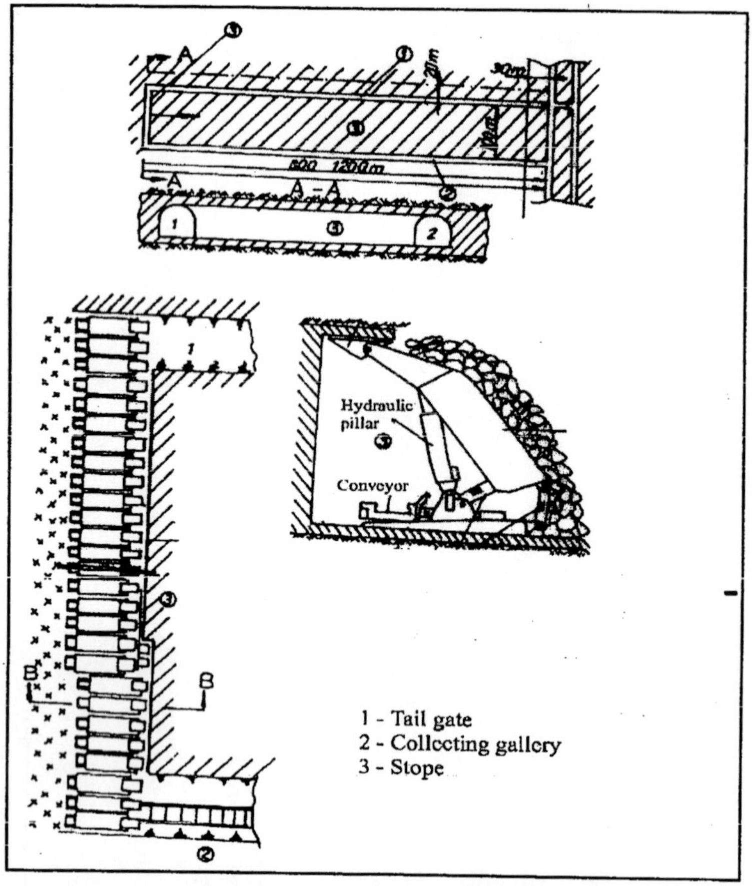

Fig. 11.11 : Longwall Stope with mechanized support

In comparison with the whole lignite production of Romania, there has been noticed a slight decrease of the underground production since 1992.

The strategy of the National companies which activity implies the mining and beneficiation of the lignite deposits for the next years, provides the maintenance of a constant production level for the underground mines provided that the work efficiency increases and the operations run profitably. Beginning with 1999, the lignite production will be concentrated within areas with most favorable geological and mining conditions thus ensuring an increased work efficiency and cost of less than 15 $/t's of lignite.

The mines where the costs are higher than 15 $/t are included in a program for mine activity restriction and closing.

Open Cast Operations

The mining operations are developed in open casts where the lignite deposits are located at small depth and where the land configuration ensures the suitable conditions for such operations. The open casts were designed for production capacities ranging between one and eight million tonnes per year.

Depending on the land configuration and the elevation, the depth of the open casts varies between 40-110 m within plain regions and up to 180 m's in the hilly ones; the stripping ratio for the entire mining field is between 2.5 and eight c.m. of waste/t of mined lignite.

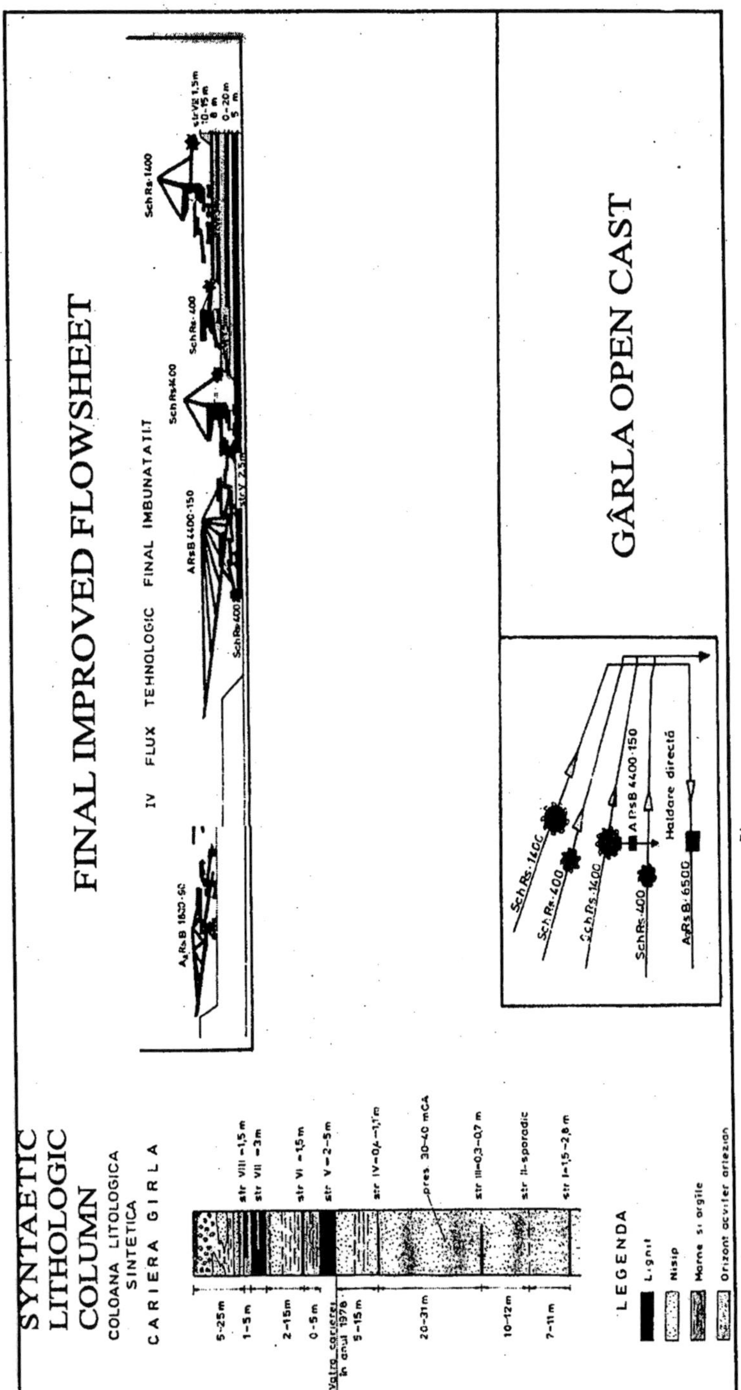

Fig. 11.12 : Garla open case

In the beginning the open casts were equipped with discontinous flow sheet technologies consisting of bucket shovels and truck transportation both for the waste and for the coal. Later on, in 1960, there were designed and completed the continued flow sheet open casts (Fig. 11.12) equipped with bucket wheel excavators CBS trucks belt conveyers which widths are between 1,200 mm and 2,200 mm, dumping equipments for the waste with capacities between 2,500 and 12,500 c.m. per hour and stokers and stocking and excavating machines for the stockpiles.

The Romanian open casts are equipped presently, with: SRS-470, SRS-1,300, Sch RS-1,400 and SRS-2,000 rotor cutting bucket excavators.

The classical technologies are mostly used in smaller open casts and for the preliminary excavations within hilly regions of the large open casts provided with continuing flow sheet and traveling platforms for the rotor excavators.

The modern flow sheet may be easier or more complicated than the classical one, depending on the land morphology, the thickness of the overburden, the number and thickness of the coal mine able layers.

In several open casts where the coal layers are clearly separate from the waste, the technological flow sheet is easier and each shovel operates either in the coal or in the waste on benches forming separate technological lines.

There are also some open casts where the coal layers are separated by waste bands on several benches and this makes the extraction process difficult and influences the quality of the extracted coal.

In some open casts the mining operations are performed on separate benches and the selective-alternative excavation on the same bench are performed.

When the last coal layer of a coal field had been mined there was used the combined mining method which implies the dumping of a part of the waste on waste dumps and the remaining waste is dumped on inner dumps using the 90, 120 and 170 arm dumpers.

The working benches in the open casts are 15-25 m high, the slope angle is 60°-65° and the lengths vary between 1,000 and 2,000 m. The general slope angles of the open casts vary between 9 and 120 depending on the rock type and the working technologies used.

The waste resulted at the opening trenches works were transported and dumped on the outer waste dumps located at variable distances from the mining perimeters depending on the land configuration from the neighbouring areas.

Within the plain regions the waste dumps were located in the neighborhood of the open cast final slope and vertical benches of maximum 15 m were provided.

Depending on the characteristics of the dumped material (humidity, consistence, loosening factor, grain size, etc.) as well as the designed heights of the outer waste dumps, the general slope angle ranged between 6 and 90°.

Within the hilly regions the outer waste dumps were built by filling the valleys from the open cast neighbourhood and the waste was transported to the dumps, in some cases, over more than 10 km.

Today, in Romania, the lignite extraction in open casts is performed in 20 open casts provided with discontinuous flow sheet and 19 large continuous flow sheets and their total designed capacity is 40 mil. tonnes/year.

The 19 open casts mentioned above are equipped with the following equipments:

- 97 rotor and cutting bucket shovels;
- 50 dumpers;
- 400 km of belt conveyors with widths between 1,400 and 2,200 mm;
- 38 pilling and extraction machines;
- 58 belt track carriages;
- 187 power conversions' stations of 20/6 KV and an installed power of 1,200 MVA.

There have been excavated 2,405,387,112 t's of material since the open cast mining start up, of which 367,407,000 t's of coal.

On the basis of the economic assessment of the 19 continuous flow sheets open casts, it was concluded that only 8 open casts have been running profitably, obtaining operational costs bellow 11 $/t of coal. The other open casts are in an opening stage and they have not reached the designed capacity yet.

REORGANIZATION AND MODERNIZATION OF OPERATIONS

Since 1997, the Romanian economy has been submitted to a complex process of changes and reorganization on the basis of the free market economy principles.

At the beginning of 1997, the electric and thermal power price was modified according to the free market rules and during the second part of that year the restructuring of the mining sector was started.

Thus, the volume of subsidies was significantly reduced and several mines which were running with losses were closed.

In September 1997, the number of the employees of the mining sector was dramatically reduced and to achieve this, three methods were used:

- the auxiliary activities got separated from the main activity, and these activities were organized in commercial ventures;
- early retirement;
- unemployed granting compensating salaries.

The diminuation of the employees numbers from the coal industry in Romania is presented in Table 11.2.

In 1998, the volume of the subsidies granted for the mining products was much reduced and once the prices were adjusted to the free market principles, some products such as the lignite were no more subsidized.

Table 11.2 : Number of the employees of the National companies and commercial ventures for coal between 1995/1998 in Romania

Unit	1995	1996	1997	1998
CNH Petrosani	42933	44754	37808	23265
CNLO Tg. Jiu	70503	51115	32976	22768
SNC Ploiesti	17443	16709	14641	9648
Total	**130879**	**112578**	**95425**	**55671**

In 1999, and in the next years the main customers of the coal industry assessed their requirements of coal at approximately 27 mil. t's of coal.

The total required production of coal can be divided as follows: 23.8 mil. t's of lignite and brown coal and 3.2 mil. t's of pit coal, of which 3.0 mil. t's of power pit coal and 0,2 mil. t pits coal for steel plants.

The existing operational capacities of Romania could ensure the supply of about 44 mil. t's of lignite and brown coal and of 8 mil. t's per year of pit coal; thus, the domestic requirements of coal of the Romanian market are met.

The Restructuring Program of the Mining Sector issued and approved by the Romanian Government stipulates the closing of 64 mining projects where either the reserves are depleted or there are very difficult mining-geological conditions and thus, the operational costs are high.

In order to improve the economic-financial situation of the pit coal mining exploitations, it was decided to close the mines where the reserve tonnage is low and the operational costs are high; also it was decided that Lupeni, Livezeni and Paroseni mines would be modernized. New technologies will be used in some other pit coal and brown coal mines in order to improve the product quality by 10%.

The studies elaborated for the lignite mines indicated that the production activity should be concentrated in the areas where the reserve tonnages are higher so that to obtain work efficiencies of 600 t/man and year in the underground mines and 1,200 t/man and year in the open pit.

The underground operations will be developed in 4-5 mines which will provide approximately 4,6 mil. t per year with maximum economic efficiency while the remaining 25 mil. t's of coal will be obtained in 6 large open casts.

In order to render the open cast operations profitable, all the capacities should run at a utilization rate of 22-24% and this rate will be achieved by the rehabilitation of the technological lines.

The mechanical and electrical parts of the belt conveyers from Rosia de Jiu, Pinoasa, Jilt Sud, Lupoaia, Rosiuta and Husnicioara open casts have been rehabilitated so far.

The modernization of a first lot of 6 SchRS – 1400 30/7 shovels started in 1998. The modernization operations are carried out by KRUPP company from Germany and S.C. ROMINEX S.A. Timisoara, Romania.

The modernization of the 6 shovels will be performed in three stages scheduled more than two years.

By concentrating the production activity in only some large open casts and working with a stripping ratio of maximum 8 c.m. of waste/tonne of coal and by modernizing all the main equipments and machineries (shovels, belt conveyers, dumpers) the utilization factor for the technological lines will reach 22-24%, even more, and thus all the operational open casts will run with profits and the lignite could be an alternative power source.

CONCLUSIONS

In conclusion it can be said that Romania has significant high quality coal and lignite reserves which beneficiation extends over many years. Thus, under the Romanian economy circumstances the power pit coal and the lignite remain the primary source of power over a long term (50-60 years). In the next year new technical and technological solutions of mining and beneficiation should be applied so that these products become a lasting energy source for the Romanian economy.

The processing and the supply technologies should meet the European standards regarding the environment safety.

CHAPTER 12

Coal Mining Prospects in the Changing Global Scenario

*T.N. Singh**

ABSTRACT

Fast changing global economic scenario is reflected in all sectors including coal mining in India. Coal in India being inferior quality with high ash, particularly non-coking coal accounting for the 85 per cent reserve is to sustain the main burnt of the power sector. The surface mining has to improve its level of performance to make this inferior grade coal techno-economically viable even in the remote power plants. The underground mining suffering from past legacy has to undertake drastic revision policy in terms of sinking of shaft or development of incline opening of new deposits and introduction of new technology at the earliest to meet the rising demand of quality coal.

The plateau of surface mining expected by the first decade of the next millennium has to be supplemented by the underground mining which will be under deeper cover and may need strong dose of introducing new technology and heavy capital investment. The performance of the underground mining at present is deplorably bad and going the worse mainly because of surplus manpower, low technology input and lack of vision and initiative. The steel, cement and other sectors looking for quality coal have to support the coal sector to make it strong enough to meet their rising demand under open economy. The support of the global fraternity in this endeavour on business terms has to be encouraged to make best use of the national resources.

COAL MINING PROSPECTS IN THE CHANGING GLOBAL SCENARIO

Coal in Indian sub continent are mainly of Permo-Carboniferous period formed along with that of the Australian and South African deposits. These deposits coincided with the projected Hercynian activities when the process of sedimentation was extremely disturbed due to geotectonic activities. The deposit invariably from the drifted wood material along

*Former Director, Central Mining Research Institute, (Council of Scientific & Industrial Research), Dhanbad, India

with the weathered base soil resulted in thick seams and with high inclusion of inert material resulting in high ash content. The coal reserve of the country is estimated as 206.239 billion tons up to 1200 m depth cover, inclusive of 205.352 billion tons of Gondwana stage. Nearly 69.3% of the total reserve is in depth cover within 300m and another 16% in between 300-600m depth cover. The GSI has identified 57 individual coalfields of Permian stage over total 14550 sq km area. The coking coal reserve of different category up to 1200m depth cover is estimated as 14.6% of the total coal is as follows:

Prime coking coal	5313.26	million tons
Medium coking coal	23211.49	million tons
Semi coking coal	1607.88	million tons
Total	**30132.63**	**million tons**

The coking coal of Indian subcontinent in general has high ash content, classified as steel I and II grade with ash content below 15% and in between 15-18%, Washery I to IV with maximum ash below 21%, 24%, 28% and 36% respectively. Except Jharia coalfield, there is no source of S-I and II grade coal in India and Washery Grade I and IV coals are primarily produced in Jharia coalfield, small amount in Bokaro, Karanpura and Chanch Victoria area close to Raniganj coalfield. The production of coking coals of different grades in Jharia coalfield was nearly 22 million tons while another 16 million tons came from Bokaro and Karapura coalfield.

Non coking coal reserve of India has high ash content; about 15% estimated to be wish ash below 25% and rest with ash content as high as 45%. The non coking coal is graded from A to G with A category having highest useful heating value. Nearly 48% of the total non coking coal reserve comes under A to C category while the rest falls under D to F category. The coal produced in India is of very inferior grade with heat value within 2400-3360 Kcal/kg accounting of SECL-59% MCL-93% and ECL-30% in the year 1995-96 with ash content as high as 40%. Fortunately, the sulphur content in this grade of coal is again very low in the range of 0.5-0.8%.

MINING OPTIONS

About 30% of the Indian coal reserve is available under suitable stripping ratio, amenable to surface mining. The share of this process remained nil till 1947; mainly because of the quality consideration. With the First Five Year Plan, the surface mining of coal was initiated and the share of the underground mining was 80% in 1970-71 before the nationalisation and 55% in the year 1985-86. The percentage has come down around 26% in 1996-97. The trend in favour of surface mining has continued in spite of large investment made in underground mines to achieve 60:40 ratio for the underground and surface production. Nearly 80% of the total workforce was responsible for the 26% production from underground while only 20% work force produced rest 74% coal from the surface mines in 1996-97. With the rising wages of the miners and nearly static productivity, major share of the investment has gone towards subsiding towards wages in case of underground mining, which alone accounted for 88% of the total budget in 1997.

Future mining scenario is heavily tilted towards surface mining of power grade coal from shallow deposits within 150 m depth cover. The areas with huge reserve of the desired coal like North Karanpura with 8156mt, Rajmahal-8933mt. Singrauli-6350 mt, Korba 7668 mt, Mand-Raigarh - 8551mt, Wardha-3420mt, Godavari valley-5429mt, Ib River coalfield-13339 mt and Talcher-23355 mt may sustain power sector for next 100 years, provided proper blending of technology, value addition and resource utilisation was realised.

Surface mines operating with giant machine drills as well as heavy blasting, remote handling, loading and transport of coal have quality dilution due to mixing of burden, bands and intrusions as the heavy machines can not differentiate between the coal and the non coal matter. This has been responsible in lowering the heat value, affecting the efficiency of the plants and the market cost to the producing unit. The power sector however has to improve the combustion process to match the supply of inferior grade coal. The average heat value of power grade coal has come down from 5900 Kcal/kg in 1960-61 to 4200 in 1980 and 4100 in 1992-93.

COAL FOR POWER SECTOR

Largest market for the Indian coal is the power sector with demand of 215 million tons, followed by steel sector 40.5 million tons and cement industry 17.5 million tons in the year 1996-97. This demand by 2009-10 is expected to be 690 million tons total inclusive of 500 million tons for the power sector alone. This demand is based on 4.6% annual projected growth in energy demand, targeting 14500·PJ (500 Mtce) by 2010 and nearly 70% power generation from coal as the main fuel. With large amount of coal amenable to surface mining and of desired quality, the demand has been fulfilled so far though short fall of nearly 35 million tons was expected by 1997 and nearly 200 million tons by the year 2010. Global partnership in power generation was planned in 1991 as a part of economic liberalisation and 240 projects proposals for 90 GW power generation were submitted to the government at an estimated cost of US $100 billion. As a confidence building measure, the government agreed for counter guarantee to 8 fast track projects for 3780 MW including 5 coal based projects at a cost of Rs. 160 billion. The 1000MW Manglore Power Project was one of such coal based scheme with Australian coal. In the beginning of VIII Plan, schemes worth over 80000 MW were in channel with 50% based on coal fuel. Over 69 new mining blocks were open to private entrepreneurs to bring down the deficit in coal demand to these power projects. By the terminal year of tenth Five year Project, demand of 500 million tons of coal was guestimated for generating 1.1 lakh MW. The NTPC alone under the program planned to add 13000 MW thermal power capacity by 2007, involving an investment of nearly Rs. 50,000 crores.

MINING FOR POWER SECTOR

The power grade coal of D to F grade is in abundance in Central Indian coalfields-Orissa, Madhya Pradesh, Maharastra under shallow depth cover amenable to surface mining. The production from the surface mines of the Coal India has increased from 16.4 million tons in 1973-74 to 159.7 million tons in 1993-94 recording annual growth of 12%. The CIL had 449 mines including 168 surface mines. The price of the D, E, F and G grade coal as per 1997-98 latest revision stands Rs. 694, 538, 428 and 327 only. Mass production of inferior grade coal at low price is possible only by the surface mining, deploying giant coaling, loading and transporting machines.

Surface mining has undergone tremendous change with the industrial revolution, development in machine design and growth in information technology. The heavy earth moving machines are growing size every day as there is no restriction on dimension. The machines used for the purpose are continuous and cyclic in nature.

Continuous surface miners

Bucket wheel Miner

Cutting Drum Miner

Cutting Boom Miner

Cyclic Heavy Earth Moving Machine

Heavy drilling machines and blasting cycle

Shovel-Dipper shovel or rope shovel

Transport System-Conveyors, dumpers, rails

Inpit crushing, conveying system

The surface mining segment of the Indian coal industry can claim to be leader with many success stories in recent years. Of the aggregate coal production of 283 million tons in 1996-97, nearly 212 million tons came from the surface mines, a share nearly 74% of the total production. Some of the technological milestones are as follows:

1990	Trial with inpit crushing and conveying at Padampur project of WCL
1993-94	Commissioning of 25m^3 shovel and mobile crusher at Piparwar mine
1994-95	Inpit crushing and conveying at Ramgundam opencast Phase II
1998-99	Surface miner (Proposed for MCL mine)

The introduction of walking draglines with greater reach, better manoeuvrability, better handling capacity and under given condition is the lowest cost O/B handling machine. The bucket capacity of BE 2570-WS dragline is 122.3 cum, boom length 109.7 to 128m and maximum digging and dumping height of 60m. The cost of the machine at the same time is very high and demands high productivity to make it economically viable. There are 40 such units under operation in CIL Mines with fair improvement in productivity. The productivity has been improving in the surface mines since mid 1970 with output ranging 0.76 t/man shift in 1974-75 to 4.01 in 1993/94 (Table 12.1). The productivity of surface mines of West Bokaro under TISCO was reported to be 10.58 in the year 1996-97 compared to average 5.12 for CIL. The productivity of MCL for the year was 12.82. Compared to this, the average productivity in the US surface mines has been 37.5 tons/man shift and as high as 265 tons/man shift in Powder River Basin, Wyoming, U.S.A.

Table 12.1: Opencast mining growth in CIL

Year	*Opencast production-mt*	*Average stripping ratio*	*OMS*
1973-74	16.40	0.80	0.76
1978-79	28.92	1.43	1.13
1984-85	70.31	1.89	2.10
1989-90	119.91	2.21	2.88
1993-94	159.50	2.04	4.01
1996-97	212.50	-	5.12

There appears to be ample scope for improvement in the surface mining performance as the average production per man year from the mines in India is around 3220 tons in comparison to 15000 tons in Australian and US mines. The sense of complacency because of comparative low productivity in underground mining has to be cared about and economics improved further by better utilisation of the resources to make it globally competitive. This is all the more important because of the inferior grade coal reserve available for the sector.

MINING OF QUALITY COAL

The operation of underground mining is influenced by the geo-mining domain, productivity and efficiency of men and machinery and the managerial skill. *Cost of coal production under the identical condition remains the same irrespective of the quality.* On thee other hand coal produced from the same depth cover fetches 5-8 times market value with 3 to 4 times higher heat value. The industry is as such again on the verge of another wave of slaughter mining when the better quality underground coal will be mined in any fashion, even at the cost of losing the inferior grade coal for ever.

Quality of coal improves with the depth of the deposit in general and the underground mining was adopted immediately after the discovery of coal in Raniganj coalfield. The scenario is different in Jharia coalfield where the top most seams were prime quality coking in nature. The underground mining with manual or with miniature equipment always catered the need of the quality coal; delineating the shale of stone bands. Underground mining is invariably by conventional drilling, blasting handling and transport, using miniaturised tools and machines and from a close distance. Human touch, quick and real time response to the quality variation and scope of beneficiation offer quality improvement with the underground mining. Sectional mining of thick seams leaving bands in roof or floor or selective mining was the possible options resulting in improvement in quality. The type of coals are in demand for steel and cement sector, conventional power plants and other industries, many of which are looking towards imported quality coal from China, Australia and Indonesia. Major share of coking coal the most valuable input to the steel making process is coming from the underground mines of Jharia coalfield containing nearly 5313 mt of prime coking coal, inclusive of 4043mt over 600m depth cover. The operating mines of the coalfield-mainly underground are extremely uneconomical and in view of past legacy, present chequered mining and surface constraints offer very limited scope for surface mining of prime coking coal. The coking coal import starting in early eighty's has now reached to 10 million tons mark; endangering the very existence of Jharia coalfield. The indigenous availability of metallurgical coal is stagnating around 10 million tons in 1997-98 to 50 million tons in 2006-07, demanding nearly 36 million tons of coking coal. The projected shortfall 2006-07 is guestimated to be 23.09 million tons as per production target even inclusive of the private sector. Improvement in mining technology, conservation, level of recovery and productivity in case of underground mines and diversion of more land for the surface mining are the only means to meet the demand and save the viability of the coalfield.

The cement industry required quality coal with ash content within 24-27% and consumed 220 kg of coal for each ton of cement. The current coal requirement for the sector was estimated as 14-16 million tons but with the growth trend of the sector, is expected 30 million tons by 2007. The quality of coal supplied to the cement sector is deteriorating with heat value from 20GJ/ton in 1976-77 to 15 GJ/ton in 1996-97 reflecting lower efficiency of the plants and hence search of alternative coal resource of better quality. The average ash content of the coal supplied to the cement plants over the year has increased to 35%. The captive mining prospect of good quality coal in some of the ideal location, improvement in the quality of supply from the mines may contain any attempt of coal import by the coastal cement units in days to come. Underground mining in selected central Indian coalfields by the sector as captive deposit was in the best interest of all concerned.

UNDERGROUND MINING OF COAL

The technology of underground mining was influenced by the strata control practices, space constrain, environmental problems, managerial skill and efficiency and size of machinery. The mining technology remained rudimentary for 200 years with scanty introduction of coal cutting machines and basket loading of loosened coal. The introduction of delay detonators and safe explosive encouraged off the solid blasting and even the CCM lost the ground, in favour of drilling machines in early 1960's. The continuous miners with shuttle car loaders were introduced in selected 4 mines with the first five years ambitious plan. The SDL / LHD entered in the coal mines to minimise dredgery of basket loading and nearly 650 SDL and 54 LHD were operating in CIL mines in 1996-97. The longwall with friction or hydraulic props entered the Indian mines along with shearer. The powered supported longwall faces were started in early 1980 and today 23 sets are installed including 14 in CIL and 9 in SCC in 15 mines. The contribution of these faces remained 1.58 million tons for CIL and 2.29 million tons for SCCL for the year 1996-97. It is interesting to note that 5-6 longwall systems in CIL and 2 in SCCL have been lying idle as they were not found suitable for the sites they were earmarked. In spite of substantial investment (Nearly Rs. 7000 crores) in underground mines by Coal India since its formation there has been no improvement in production and productivity remaining 70 million tons and 0.55 tons/ man shift. Nearly 70% of the total output came from age-old conventional board and pillar mining with basket loading and nearly 28% from SDL/LHD faces. The production from the development work has been 60% contrary to normal ratio of 36:60 even including longwall mining.

ECONOMICS OF UNDERGROUND MINING

The underground mining started with the board and pillar system in early 19th century on line to that of the British mining practice, with cheap labour and primitive mechanisation. Most of the mines continued to work on age old pattern with basket loading of coal in tubs and the drills and the haulages were the only machines used underground. The mines operated at low OMS dropped from 0.66 in 1975-76 to 0.53 in year the year 1995-96. The wage and allowances below 10% of the coal price in 1960, increased to 20-30% in 1970's and is nearly 80% with recent wage revision. The earning per man shift was Rs. 14 in 1973 and increased to Rs. 186.39 per man shift in 1993-94. According to an estimate, the employees per capita emolument has increased from Rs. 5920 in 1971-72 to Rs. 113,003 in 1996-97 while the productivity has gone down over these years. The underground production over these years dropped from 65.2 million tons in 1975-76 to 55.2 million tons in the year 1996-97 in CIL mines. As on date major component of the production cost in the coal mines are wages and allowance, material supply, power, drilling and explosive cost, capital and specially maintenance and depreciation in case of mechanised mines. In case of a manual board and pillar mine, the cost of wages and allowances alone accounted for 50-60% with average district OMS of 1.6. The population of the loaders the main working force responsible for coal evacuation was decreasing every day and was below 20% of the total work force attached with the underground mining.

The mechanised loading with LHD/SDL on conveyor or in tubs has become popular in SECL, minimising the dredgery of basket loading and improving the OMS to 3-4 and the mine OMS over 1.1. Taking the depreciation, interest and power cost in to account, there was 20-25% reduction in the production cost of the coal with this simple mechanisation. Availability of coal, location of conveyor and suitable support may improve the performance with the mechanised loading. The bolts, cables or rope dowels with quick setting cement were found to change the economics of the underground mining.

The longwall introduced in 1960's with individual props and early 1980 with powered support have failed to make any dint in underground production. The production from about 12 operating mechanised longwall faces over India was just 1.15 million tons in 1993-94, mainly because of poor utilisation factor of man and machine within 26-35%. The major down time has been breakdown in out bye transport, shearer, AFC and powered support itself amounting nearly 32.2% (Jain 1994). According to the study. Jhajhra face operated for 29% utilisation hours, Moonidih 26.9% while Singareni faces—VK7, GDK9 operated at utilisation hour around 34.4%. Only. The rated capacity of the faces have been around 470 to 1030 tpd at Monodih, 3000 tpd for Jhanjhra 4th set and 3540 tpd for the Kottadih face. Even this achievement has not been possible in any of the faces and all the sets have been written off, refurbished or overhauled with total production within 0.45 million tons (Dhemomain) to 2.42 million tons (Monidih). The OMS for the Indian longwall faces is 1 ton/man shift in comparison to 4 tons in UK and Poland and 16 tons in the USA. This has been possible by deployment of heavy duty machines including powered supports at exorbitant high cost. According to an estimate the investment for longwall faces in the US was equivalent to Rs. 70000 crores for producing 12% share of the coal. A longwall project upgradation in Australia was estimated to cost US $ 550 million to produce 4 million tons of coal per year.

TECHNOLOGY FOR SURVIVAL

The productivity of underground mining is suffering due to surplus employment, over 1.5 lakh added over night during nationalisation, poor R&D input, primitive technology, vision and motivation. Surplus non productive workers; poor training for adopting even intermediate level of mechanisation and half hearted mechanisation due to financial crunch has reflected badly over the health of the coal companies already suffering from the legacy of pre independence mining. Hard decisions are required by the policy makers in terms of new project planning for underground mining. It may of interest of note that China produces nearly 95% of its total coal, USA 41.5%, South Africa 70% and UK 75% from the underground mines at globally competitive rate. China the neighbouring giant has planned to introduce heavy mechanisation in state mines including 60% mechanised loading in drifts, 61% mines fully mechanised loading and 30 fully mechanised crew to produce a million tons of coal per year per face in 1989. The state planned 23 mines in East China in 1985, 16 mines in 1986 to 1989 and 19 mines during 1991-95 for annual production of 40 million tons. The country invited global partnership in these ventures.

The other neighbouring country—Australia had problem identical to that of India and had nearly 9 million tons less production in 1989 in comparison to that of 1987 mainly because of labour unrest and low productivity in most of the mines. More than 17 mines were closed during the period in New South Wales, when 3500 people lost their job because of uneconomic production cost. The nation decided to go for heavy mechanisation in underground mines, primarily for long wall to over come this phase, utilisation of equipment was improved by adopting 9 hours shift and 6 day week concept resulting in better production. Bulga mine of Australia was producing over 24000 tons of coal per man over years and the average mines produced 13000—15000 tons. South Africa had identical trend again because of the labour unrest in 1987 when about 40000 miners lost their job. The reason for this was identified as 20% rise in wages, 70% rise in rail freight and rise in warfage, loading and ocean freight. Planning for mechanised mining, transport and loading and improvement in men and machinery utilisation factor resulted overall improvement in the health of the industry.

India with large market and demand of inferior to prime quality coal has to learn from the global experience and introduce corrective measures in terms of the following:

- Better utilisation of men, better training to the manpower and working environment
- Optimum utilisation of the existing machinery and resources.
- Planning of new mines; mechanised for high productivity, reduced down time and high reliability.
- Redeployment of workforce in new ventures after suitable training.
- Involvement of private participation in mining, particularly for the power projects.
- Involvement of the private participation in otherwise uneconomical underground mines.
- Formation of Underground Mining Authority with fiscal support for developmental activities.
- Technological renovation for underground mining.
- Quality improvement for high ash coal to make it competitive even along the coast.

Coal mining prospects in India are bright in spite of all odds because of the scarcity of the other energy sources. The surface mining activities have still large scope of improvement right now because of favourable OB ratio. Most of the coalfields will operate below 150 m depth cover by next decade and if the planned depth of 300 m is to be realised, the OB ratio may increase up to 10. The environmental awareness, particularly social consciousness may restrict this option and a plateau for surface mining may come much earlier than envisaged. The industry has to post its resources for the development and underground mining of deep reserve in terms of access—shaft or inclines; technology—longwall or other alternatives for thick coal seams and value addition to the inferior grade coal.

For its survival in the fast changing present scenario, the industry has to change its attitude and be more responsive to the customers in terms of quality, time schedule and reliability. Need is there to save the industry from onslaught to avoid repetition of history.

REFERENCES

Jain D K and Roy U (1994), Utilisation of longwall mining equipment and analysis, National Seminar on Mine Productivity and Technology, July 8/9th Calcutta

Sarkar SK and dhar, BB (1994), A stratgy to improve the productivity in coal mining industry in the changing technology scenario July 8/9, Calcutta.

Mishra M P (1998): Longwall extraction in India—Why and how to go about, Indian Mining and Ennginееring Journal, March, 1988.

Ghose A. K. (1997): Indian coal industry—search of a strategic direction, Journal of Mines Metals and Fuels, Vol. XLV, Special Issue.

Rajsekhar, T.R. (1998) : Surface miners-Future Prospective of opencast coal mining. The Indian Mining and Engineering Journal, Special Issue on Coal, June.

Singh T.N. (1998): Indian coal mining industry—Emerging Trend in the Light of Global Trade, the Indian Mining and Engineering Journal, Special Issue on IMME, Sept.

Singh T.N. (1998): Environment Issues and Mines Safety—Underground vs surface coal mining, Indian Mining Summit, Oct.

CHAPTER 13

Supply-Demand of Mineral Resources in Taiwan

*Wu, Jung-Hua**

1. INTRODUCTION

Taiwan is a densely populated island and is endowed with limited mineral resources, especially energy and metallic mineral resources. In the past few decades, Taiwan has achieved an amazingly high economic growth rate. This high growth rate has led to incessant increasing demand for mineral resources.

However, the depletion of mineral resources together with the increasing stringent restrictions posed on mining activities made Taiwan rely heavily on imported mineral resources. Practically, most of the sizeable demand of energy and metallic minerals by the industry depends on imports thus the Taiwan economy is highly vulnerable to mineral supply instability.

This paper attempts to provide briefings on the following aspects namely, overview of the role of mineral resources in the Taiwan economy, the supply-demand pattern as well as future prospects of mineral resources of Taiwan, and salient features of mineral resources policy.

2. THE ROLE OF MINERAL RESOURCES IN THE TAIWAN ECONOMY

In the past two decades, Taiwan experienced a high growth rate of economic development, which is considered as a miracle in the developing countries. Per capita gross national product(GNP) increased from US$ 1,301 in 1977 to US$ 13,303 in 1997. However, Taiwan is relatively inadequate in indigenous mineral resources to support its dynamic economic growth. Hence, over the past two decades, GNP gained steady growth. On the contrary, the output of the domestic mining sectors exhibited declining trend in general (Figure 13.1).

---

*Department of Resources Engineering, National Cheng-Kung University, Tainan, Taiwan Republic of China

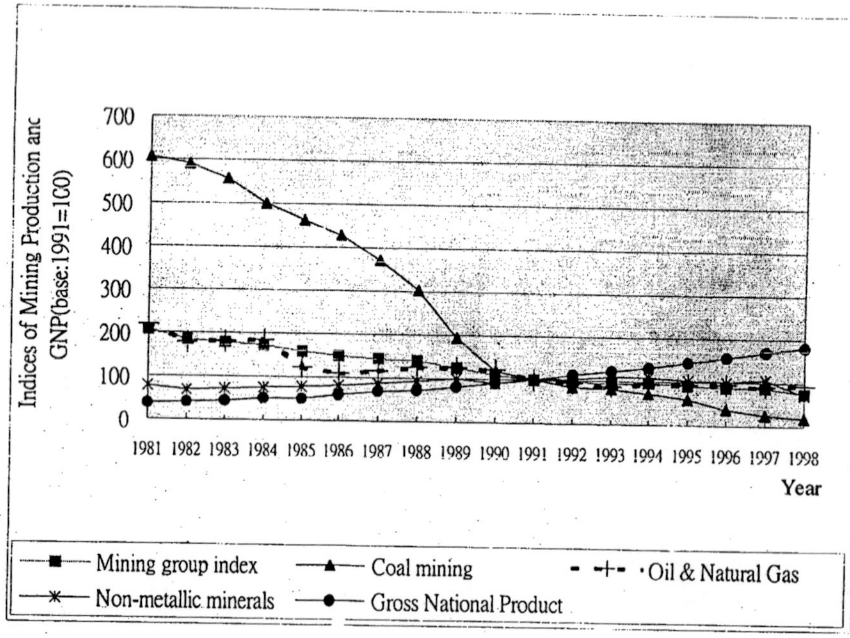

Figure 13.1 : Indices of Mining Production and GNP in Taiwan

Source: Department of Mine, Taiwan, R.O.C. (1998)

Figure 13.2 shows the role of mineral resources in the Taiwan economy in 1997. Basically, the output of domestic mining sector in Taiwan is limited. However, the contribution of mineral-related industries to the economy is quite significant. The production value of domestic minerals totaled NT$ 43.2 billion in 1997, representing only 0.57% of GNP. However, the aggregate outputs of industries which are highly related to mineral industry, such as oil industry, electricity, cement, glass ... etc. amounted to NT$ 1,248 billion, which constituted 13.4% of GNP. Imports of mineral totaled NT$ 309.4 billion, most for the payment of crude oil. The import value of scrap and processed mineral products was NT$ 20.4 billion and 458.1 billion, respectively. Exports of mineral and their processed products amounted to 441.5 billion.

3. MINERAL RESOURCES AND PRODUCTION IN TAIWAN

There are various kinds of mineral resource in Taiwan, however, only some of them are considered to be of economic value. They could be classified into four categories as follows:

A. energy minerals: coal, crude oil, natural gas, and geothermal;

B. metallic minerals: gold, silver, copper, iron, and zircon;

C. industrial minerals: sulfur, marble, limestone, serpentine, dolomite, talc, asbestos; mica, china clay, feldspar, fire clay, gem stone, quartz, and gypsum; and

D. sand and gravel.

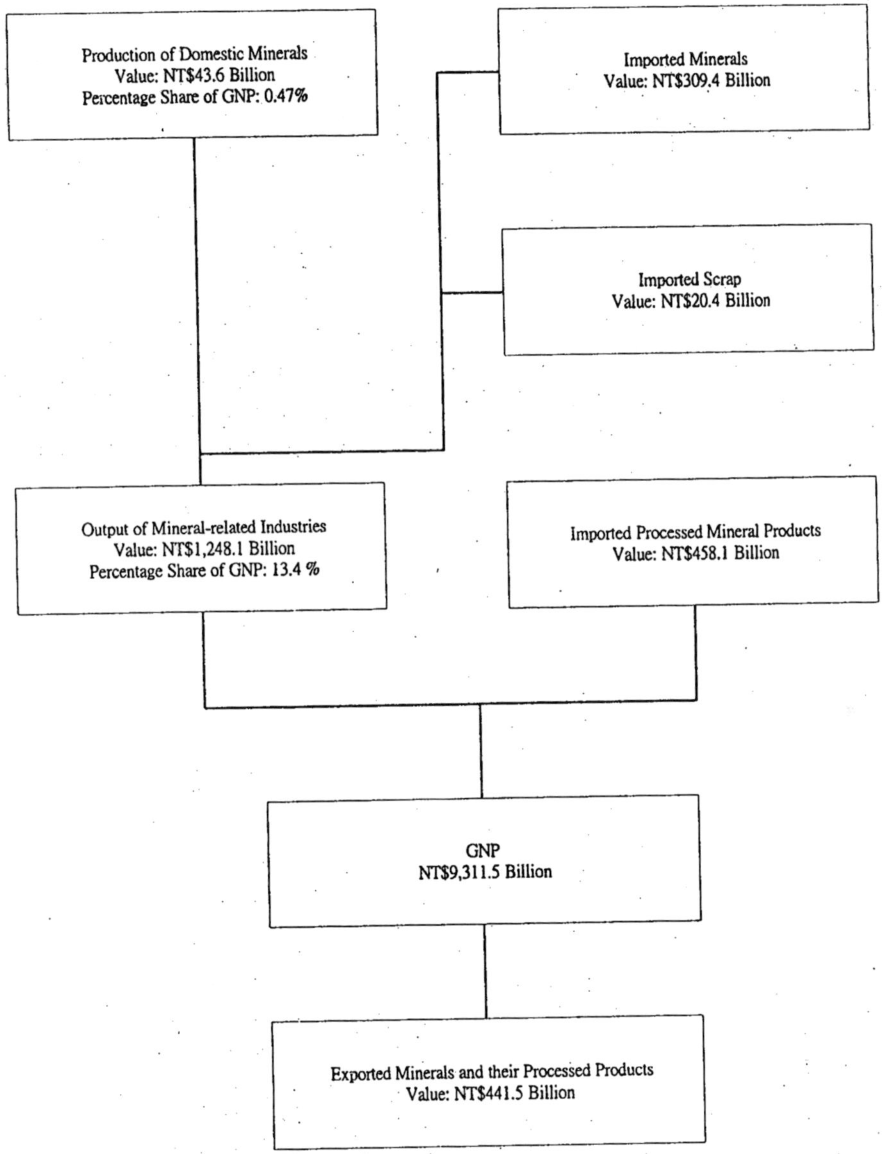

Source: Department of Mine, Taiwan, R.O.C. (1998)

Fig. 13.2 : The Role of Mineral Resources in the Taiwan Economy (1997)

Coal mining in the northern region of Taiwan used to be quite active in the 1960s, with production level above 5 million metric tons per year. However, depletion of reserves and competition from the imported coal has resulted in a sharp drop in production. The production of was less than 0.1 million metric tons in 1997(Table 13.1).

No commercial oil field has been developed so far. From 1980 to 1990, production level of natural gas maintained at the level of 1 billion cubic meters per year and gradually declined thereafter. Exploration of offshore oil deposits is still ongoing.

The Chinkuashih gold mine in the northern Taiwan enjoyed a brilliant production record in the 1940s, however, the mining operations came to a complete stop in 1988. To evaluate the possibility of revitalizing this mine, the government conducted the assessmeed on updated geological information. However, further information would be required to reach the final decision.

As to other minerals, in descending order of quantity significance, the top five are: sand and gravel, limestone and marble, serpentine, salt and dolomite. The eastern part of Taiwan is endowed with abundant reserves of limestone, marble, and serpentine. These resources are mainly used as industrial raw materials. They can also be used as construction, furniture and handicraft due to the fine grains and attractive colors.

In terms of total production value, sand and gravel accounts for 70%. Energy minerals and industrial minerals share the remaining 30%.

Table 13.1. Production of Minerals in Taiwan(1997)

Commodity	*Quantity*		*Value*	
	Unit		*NT$ 10,000*	*%*
1	2	3	4	5
Energy Minerals			619,842	14.21
Natural Gas	Km^3	849,282	574,469	13.17
Coal	MT	99,089	27,715	0.64
Petroleum	KL	50,670	17,658	0.40
Metallic Minerals				
Gold	Hg	90	198	0.01
Industrial Minerals			657,139	15.06
Marble	MT	18,071,024	154,041	3.53
Limestone	MT	15,446,875	250,147	5.73
Serpentine	MT	284,406	15,120	0.35
Salt	MT	270,548	160,683	3.68
Dolomite	MT	200,174	13,969	0.32
Sulfur	MT	163,964	18,168	0.42

1	2	3	4	5
Marble(DS)	m^3	87,347	18,034	0.41
Serpentine(DS)	m^3	56,836	21,286	0.49
Clay	MT	27,491	1,608	0.04
Mica	MT	7,806	2,342	0.05
Gem Stone	KG	5,108	126	*
Fire Clay	MT	4,725	257	0.01
Gypsum	MT	2,317	1,131	0.03
Talc	MT	1,331	133	*
Limestone(DS)	m^3	206	38	*
Quartz	MT	64	56	*
Sand and Gravel	KMT	201,630	3,084,939	70.72
Total			**4,362,118**	**100.00**

Source: Department of Mine, Taiwan, R.O.C.(1998)

Note: (1) DS (Dimension Stone), CS (Crushed Stone)

(2)* represents less than 0.01%

4. DEMAND/SUPPLY OF MINERAL RESOURCES IN TAIWAN

Energy Demand

Energy is an indispensable factor input in fuelling economic growth. Energy consumption in Taiwan increased from 20.8 million kiloliters of oil equivalent (KLOE) in 1977 to 78.2 million KLOE in 1997(Figure 13.3). The annual average growth rate during this period was 6.8%, while that for the GDP was 7.7%, hence the energy demand elasticity was 0.98.

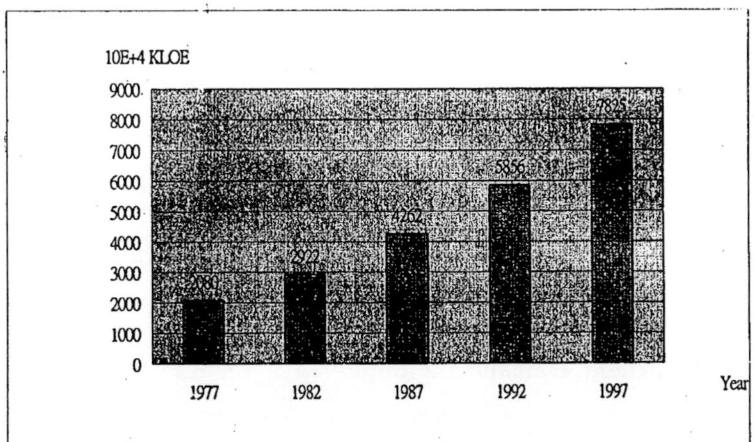

Figure 13.3 : Energy Consumption m Taiwar

Source: Energy Commission, Taiwan, R.O.C. (1998)

The structure of energy consumption by major sectors experienced significant change over this period. In 1977, industry consumed 64% of the total energy but in 1997 its consumption declined to 57%. Agriculture decreased from 4% to 2% and other sectors dropped from 8% to 6%. On the other hand, transportation increased from 10% to 17%. The residential sector increased from 10% to 12% and the commercial sector climbed from 2% to 5%.

As to the structure of energy consumption by energy forms, the consumption of coal increased from 9% in 1977 to 15% in 1997, oil decreased from 49% to 41%, natural gas dropped from 8% to 4%, while electricity increased from 34%to 40% (Figure 13.4).

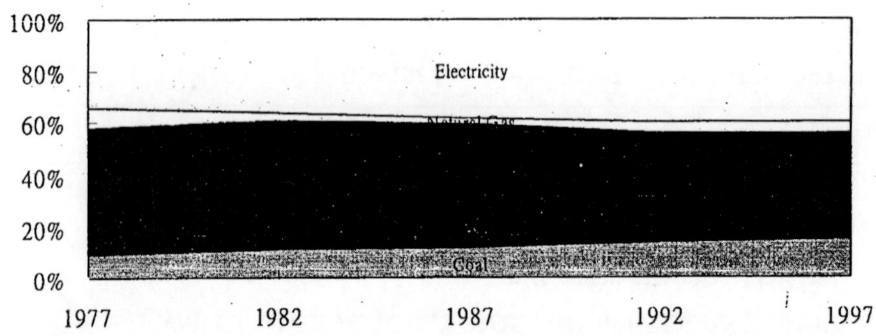

Figure 13.4 : Structure of Energy Consumption in Taiwan (by energy forms)

Source: Energy Commission, Taiwan, R.O.C. (1998)

Energy Supply

In respect to the energy supply, it totalled 24.7million KLOE in 1977 of which 21% was indigenous energy and the remaining 79% was imported (Figure 13.5). Due to lack of abundant indigenous energy and the soaring demand arising from rapid economic growth, in 1997, energy supply increased to 87.6 million KLOE of which 4% came from indigenous resources while the remaining 96% was imported.

The structure of energy supply in Taiwan also changed drastically over the past two decades. Coal, which contributed 11% of the total energy supply in 1977, increased to 30% in 1997. On the contrary, oil dropped from 77% in 1977 to 51% in 1997. Nuclear power, first introduced in 1977 held a share of 10% in 1997 mainly due to the promotion of diversification of energy sources after energy crises. Hydro-power dropped from 4% to 3%. Natural gas decreased from 8% to 6% (Figure 13.6).

Non-Energy Mineral Demand and Supply

As to the non-energy minerals, the demand has been increasing relative to the growth of the manufacturing and construction activities. On the supply side, we rely heavily on importation of a majority of required minerals except for selected non-metallic minerals.

In view of the very broad scope of minerals in the entire fabric of Taiwan industrial activities, four mineral-related sectors are selected and described to give a general picture of non-energy mineral requirements in Taiwan. They are metals, ceramic, refractory and construction minerals. Table 13.2 displays the import of major minerals to Taiwan in 1997.

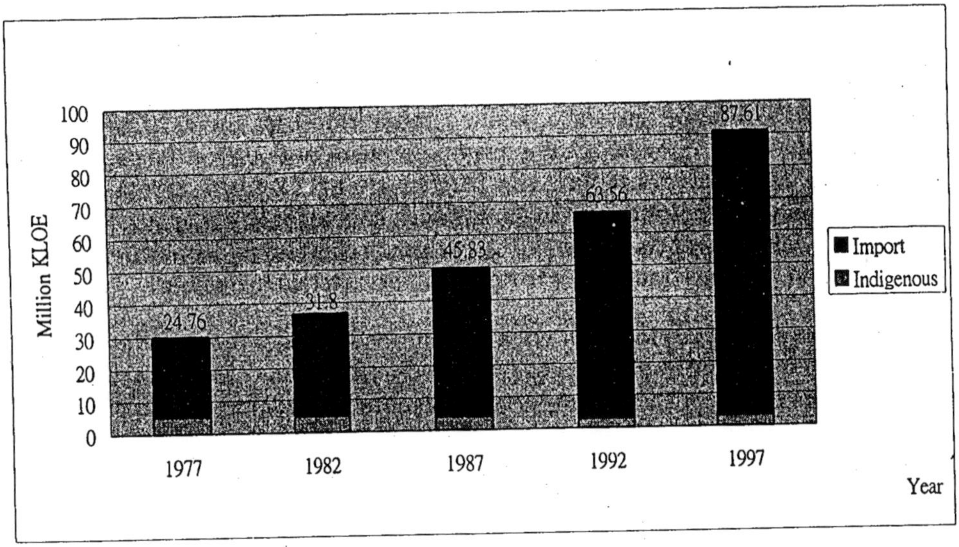

Figure 13.5. Energy Supply in Taiwan

Source: Energy Commission, Taiwan, R.O.C. (1998)

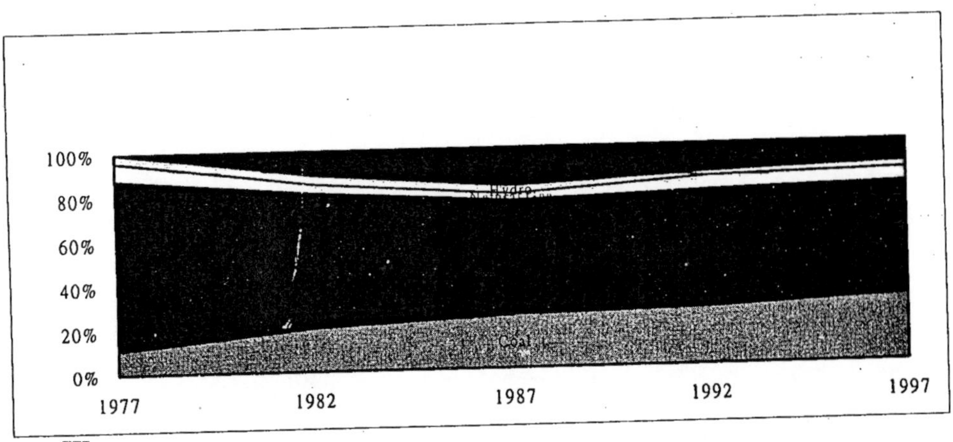

Figure 13.6. Structure of Energy Supply in Taiwan (by energy form s)

Source: Energy Commission, Taiwan, R.O.C. (1998)

Output of primary metal is largely limited to iron and steel, all of which required imported raw ore materials. China Steel Corp. (CSC) is the only integrated iron and steel facility in Taiwan. The overall steel production in 1998 was around 10 million metric tons. Most of the required minerals for iron and steel making depend on imports. For example, CSC imported 14 million metric tons of iron ores in 1997. Aluminum smelting activity was shut down since October of 1982 because of high energy cost. However, the fabrication plant was taken over by CSC. The required aluminum ingot totally relied on import.

The conventional ceramic industry of Taiwan was developed into a major industry in the 1980s. There are about six hundred ceramic factories employing more than fifty thousand workers, the total value the production being estimated to be NT$17 billion during 1987 (O'Driscoll, 1995). Despite Taiwan has viable ceramic industries, trends in the economic structure of Taiwan seems to project a somewhat pessimistic outlook for the conventional ceramic industries as a result of building construction slump. Domestic ceramic factories are dependent highly on import of raw materials such as feldspars, ball clays, and kaolin for making high quality products (Table 13.2). Low-grade clay and feldspars are, however, sufficient in Taiwan. The importation for ceramic minerals tends to decline.

The refractory materials industry currently imports most of their raw materials in need. It is obvious that as the industry steps the production to meet the demands of steel mills and foundries the proportion of imported raw materials will increase in proportion, pending the developments of our local resources. The industry will also be totally dependent on importation with respect to the following items: chromium ores, aluminum ores, manganese ores, natural graphite, high quality ball clays...etc.

As to the present, the local cement industry has a production capacity of more than 25 million metric tons per year. Gypsum is imported and used to meet most of the cement production. High quality marble and granite materials are being imported for the construction of modem office buildings, monumental and decorative uses.

Table 13.2. Statistics of Minerals Imported in Taiwan(1997)

Commodity	*Quantity*		*Value*	
	Unit		*NT$ 10,000*	*%*
1	*2*	*3*	*4*	*5*
Energy Minerals				91.56
Coal	MT	36,717,661	47,343,141	20.49
Petroleum	MT	34,477,161	145,895,190	63.15
LNG	MT	3,250,227	18,299,488	7.92
Metallic Minerals				5.53
Iron Ores	MT	14,042,962	10,939,300	4.73
Slag & Ash	MT	1,657,703	684,243	0.30
Containing Metals				
Titanium Ores	MT	117,431	1,522,940	0.66

1	2	3	4	5
Aluminum Ores	MT	79,063	175,847	0.08
Manganese Ores	MT	65,971	176,782	0.08
Zirconium Ores	MT	8,187	101,844	0.04
Chromium Ores	MT	10	434	*
Cinnabar Blende	MT	9	1,659	*
Industrial Minerals				2.91
Kaolin	MT	2,145,565	1,105,069	0.48
Limestone	MT	1,558,025	623,179	0.27
Salt	MT	991,491	865,805	0.37
Gypsum	MT	806,484	689,924	0.30
Feldspar	MT	755,071	804,786	0.35
Silica Sand	MT	503,782	455,927	0.20
Phosphate Rock	MT	265,234	482,699	0.21
Natural Sand/Gravel	MT	123,956	169,227	0.07
Clay	MT	96,262	225,913	0.10
Mica	MT	30,502	42,043	0.02
Cryolite, Borate	MT	14,184	214,985	0.09
Graphite	MT	8,744	104,506	0.05
Barite	MT	5,538	23,360	0.01
Dolomite	MT	5,474	30,438	0.01
Dimension Stone	MT	2,577	10,322	*
Emery, Abrasive	MT	2,217	14,489	*
Quartz	MT	1,114	27,198	0.01
Magnesium Carbonate	MT	703	4,254	*
Sulfur	MT	114	309	*
Total			**231,035,301**	**100.00**

Note: 1. Ores include concentrates.
2. Statistics exclude cement products and refined petroleum products.
3. *denotes less than 0.01

5. OUTLOOK AND POLICY

Energy

On the basis of the GNP annual average growth rate of 5.5% for the next 14 years, the total energy demand is estimated to increase from 78.2 million KLOE in 1997 to 136.8 million KLOE in the year 2010, with an annual average growth rate of 3.7%.

In the structure of total final energy requirements, electric power share will increase from 40% in 1997 to 44% in the year 2010, while oil products will decrease from 41% to 39%. Electric power consumption will constitute the major portion of future energy demand. The coal demand will decrease from 15% to 14% in the same period (Figure 13.7).

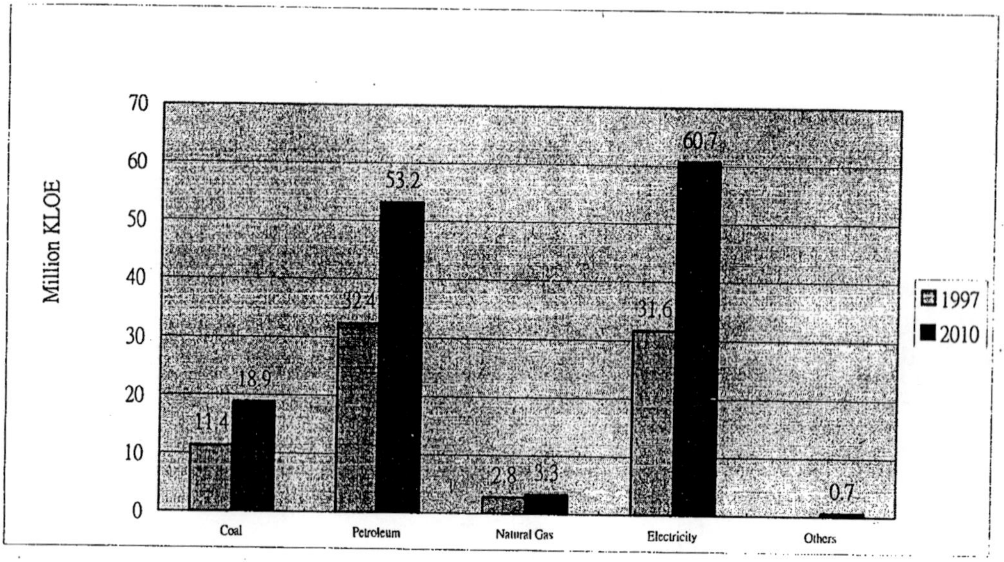

Figure **13.7. Forecast of Energy Consumption in Taiwar**

Source: Energy Commission, Taiwan, R.O.C. (1998)

Due to the development of hydropower and new energy, it is estimated that the share of indigenous energy supply will remain at 4% to the year 2010, while imported energy will stand at 96% in the same period.

During the same period, the oil share in total energy supply will decrease from 51 % in 1997 to 46% in the year 2010. Coal supply will decrease from 30% to 28%, natural gas will climb from 6% to 1%, nuclear will decrease from 10% to 9%, hydropower will decrease from 3% to 2%, and the other energy will share 4% in the year 2010 (Figure 13.8).

As Taiwan lacks of indigenous energy, to maintain a stable export-oriented market economy, its energy policy has emphasized on the following considerations:

- To ensure the stability of energy supply.
- To promote the efficiency of energy utilization.
- To deregulate the energy sector.
- To prevent energy-related environmental pollution.
- To enhance energy research and development.
- To promote energy education and promulgation.

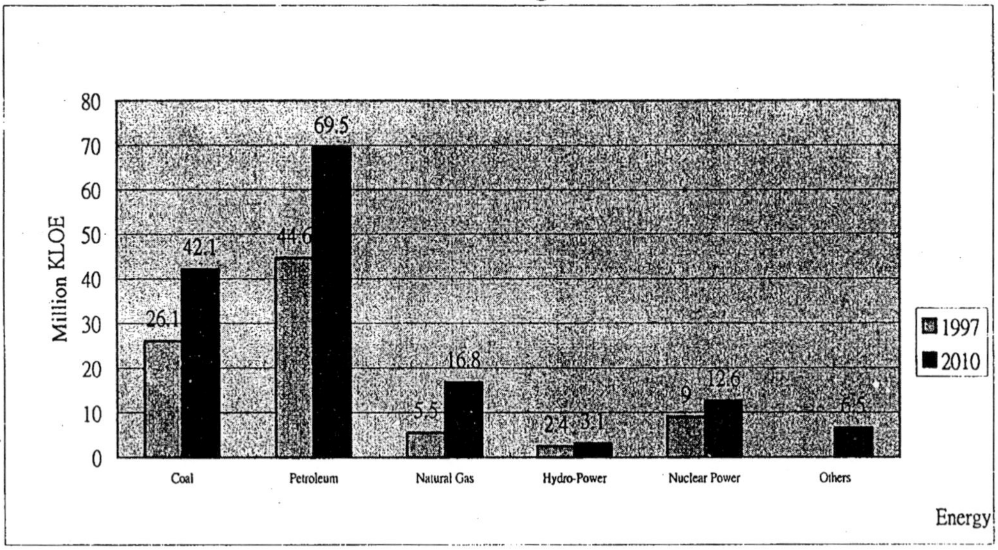

Figure 13.8. Forecast of Energy Supply in Taiwan

Source: Energy Commission, Taiwan, R.O.C. (1998)

Mineral

In parallel with rapid economic development, the consumption of mineral resources in Taiwan also increased accordingly. Since a stable supply of mineral commodity is of vital importance for Taiwan, how to ensure a continuously steady supply is the main concern of mineral policy. Some of the salient features of mineral policy might be noted as follows:

- Encouraging investment in mining and mineral processing industries.
- Providing technical supports to mining industry.
- Encouraging private mining companies to have joint venture in mining projects outside the country.
- Promoting and soliciting bilateral agreements on mining cooperation with interested countries.
- Diversifying supply sources to minimize the impact of possible supply shortage.

REFERENCES

Department of Mine (1998), Mineral Statistic Yearbook, Taiwan Area, the Republic of China 1997.

Energy Commission (1998), The Energy Situation in Taiwan, Republic of China.

Energy and Mining Research/Service Organization, Industrial Technology Research Institute (1988), An Overview of the Mineral Supply and Demand Status of Taiwan, R.O.C..

O' Driscoll, M. (1995), "Taiwan Tiger Still Burning Bright," Industrial Minerals August 1995, pp. 17-37.

Wu, R.H., F.S. Yen and H.S. Liu (1996), "Overview of Ceramic Raw Material Requirements and Ceramic Industries of Taiwan," Proc. of East Asian Ceramic: Minerals and Markets Outlook, Kuala Lumpur, Malaysia.

CHAPTER 14

Indian Mining in Global Perspective

*B. K. Kejriwal**

Nature has blessed India with rich mineral resources that can be matched by very few countries in the world. It has huge reserves of iron-ore, manganese, bauxite, limestone, magnesite, titanium, zinc, mica, barytes, chromite, non-coking coal, lignite and rare earths, besides smaller reserves of many other minerals. The known reserves of these minerals have been established using traditional methods of exploration. Given the country's geographical area and the favourable geological environment, there is a great potential for discovery of many more rich mineral deposits with the help of modern techniques of exploration. Hitherto, lack of adequate capital investment in exploration, restrictive laws and the closed economic regime had hindered application of sophisticated technology for exploration on a large scale. Now, with economic liberalisation and amendments of the archaic regulations, the field is open for large scale exploration for minerals, both on land and in off-shore areas.

Today India produces 84 minerals which include 4 fuel minerals, 11 metallic, 49 non-metallic and 20 minor minerals. The total value of the annual production is over 9 billion US dollars which constitutes about 3.5% of the GDP and 11% of the country's total industrial production. India exports iron-ore, manganese-ore, chromite, mica, granite and cut and polished diamonds. The total value of the annual exports is about 4.5 billion US dollars. On the other hand, it imports petroleum, coking coal, gold, copper, lead, nickel, tin, asbestos, potash, etc. at a cost of about 7 billion US dollars; the imports exceeding the exports by about 2.5 billion US dollars. The country has well-established mining companies, both in the private and public sectors and with the government now encouraging new entrants and foreign investments, there is great scope of substantial increase in mineral production. India has the potential to become a leading producer and exporter of minerals and mineral products in the world.

Another strength of India's mineral sector is the availability of well-trained mining engineers, geologists, technicians and skilled workers. A large number of Indian universities

*Head, Department of Mining Engineering, Indian School of Mines, Dhanbad - 826 004 (India)

offer specialised courses in Mining Engineering and Geology and the present annual outturn of mining engineers and geologists in the country is a little more than its internal requirements. Similarly, a large number of diploma institutions turn out high caliber technicians. Moreover, the statutory vocational training of all workers employed in mines has created a large work force of skilled miners. With relatively low wage-levels in the country, this human capital can be optimally used not only for expansion of the mineral industry within the country, but also for taking up mineral exploration and exploitation activities in other developing countries. Moreover, the country can offer attractive packages for the education and training of mining personnel from other developing countries.

Excellent facilities have been built up for research and development in mining. Besides the well established Central Mining Research Institute, several other research laboratories are working in areas related to mining and mineral beneficiation. Many university departments also have good facilities for research and development work in problems of the mining industry. With better management of the resources and creation of conducive work environment, results of R & D can take the industry to great heights.

India has a fairly well-developed industrial base for manufacture of machines and equipment for mining and processing of minerals. With some more capital investment the capacities can be expanded substantially to meet not only the bulk of its internal requirements but also for exports.

Apparently, India has everything that is needed to make it a mining giant in the world, yet it is a laggard both technologically and economically. What is holding it back ? What is preventing it from taking off ? One thing and one thing alone : the prevailing work culture. A hard look at the current scenario will tell us what has led to this lackadaisical attitude to work.

PUBLIC SECTOR

The concept of a large public sector to develop basic industries and essential services was indeed excellent to start with. Many such industries could not have otherwise started for want of the large investments required. However, in course of time, because of poor management on one hand and political and bureaucratic interference on the other, public sector has proved to be an utopian dream.

One of the aims of such undertakings was to create increasing opportunities for gainful employment and improve living standards. Unfortunately, under political pressure, many of the undertakings recruited people much in excess of their actual requirements. They built large townships, schools, hospitals and other welfare facilities and provided various types of services at huge costs. In their urge to become model employers and discharge their social responsibilities, they totally neglected their economic responsibilities. So long as they operated in a protected market and enjoyed government patronage, they managed to survive; but when exposed to an open economy, they failed to stand upto the competition of the market place.

Over-employment is one of the most effective ways of killing an organisation. People who are not fully occupied indulge in politics and create imaginary dissatisfaction among workers leading to all sorts of troubles. Coal India Ltd. is a glaring example of the ills of over-employment: with 20% less employees it would produce more coal at less cost.

It is not the Government's job to run businesses; its job is to uphold the law and to create an environment for business to flourish. All public sector undertakings must be privatised through disinvestment in a phased manner within a period of 5 to 10 years. And, so long as the public sector is there, each enterprise must run on business lines. A decisive transformation is required from a philanthropic outlook to a hard-nosed professional one. Strict economic discipline must be applied and all unnecessary expenditure must be cut down. If an enterprise needs 5000 persons and it has 7000, it must get rid of the additional 2000 to ensure that the 5000 work efficiently.

TRADE UNIONS AND THE LABOUR SITUATION

Every political party in India has a labour wing of its own and the trade unions are generally controlled by strong politicians whose primary concern is acquiring money and power and not the welfare of workers. The local union leaders in different industrial units also follow in the footsteps of their leaders and raise money from workers by giving them all sorts of promises on behalf of the leader. Soon they themselves become centres of power and begin to aspire for entering politics. They have greater loyalty to their political parties than to the workers. In their pursuit of power, they foment unrest in the workers and convert labour disputes into political issues. In the process they completely destroy the work ethic and turn the worker away from the pride of his work. Feelings of antagonism and mutual suspicion are created between the management and labour.

In order that industry may prosper, it is essential that labour unions work in close cooperation with management. Each side must understand that its future is linked with the success of the other. The management must understand that labour is the root source of quality and productivity and it should try to create the right environment to foster social and industrial discipline. On the other hand, the workers must shed the "we/they" attitude; they must feel an essential part of the organisation and must take pride in their work. Trade unions should organise short training programmes to create this awareness and bring about attitudinal changes in workers.

At present employers do not have the freedom to close down uneconomic industrial units. Government clearance is required for closures and this is usually not granted on grounds of public interest. This is having a devastating effect on uneconomic units. Government's anxiety to protect employment is understandable but there is no justification in continuing to operate a loss making unit which has no potential for a turn around. Mrs. Thatcher's government had set a glowing example by courageously facing the ordeal of coal miner's strike in 1984 which lasted for over a year in order to establish the basic right of an employer to close down an uneconomic unit.

Our labour laws guarantee almost 100% job security. Once a person has been appointed in any large organisation, particularly in the government sector, he is assured of a life-time employment. The laws give him so much protection that it is almost impossible to dismiss him even when he doesn't do anything or is disgruntled to the extent of being a drag on the organisation. This leads to low motivation of staff and reduced feeling of responsibility.

We need to follow the American model. Every employee, right from the highest executive to the lowest category of worker, must either perform or be ready to be sacked. It is necessary to create an environment for perfomance in which all individuals make their best contribution towards organisational goals.

In the matter of pay, perks and promotions, managements in the organised sector have time and again given in to the demands of the trade unions with the result that we have created islands of prosperity amidst an ocean of poverty. In a country where 40% of the population lives below the poverty line and where the vast majority of rural workers earn an average of less than Rs. 30 a day, the industrial worker has an EMS of over Rs.300. This disparity is nothing but a social injustice against the unorganised rural worker .

Moreover, the system of annual increments in pay and time-bound promotions are prevalent in many organisations. These work as disincentive to efficiency and performance. Guaranteed periodical wage hikes, unrelated to productivity, increase the cost of production and cause inflationary pressure on the economy in the long run. The advantage of low labour cost enjoyed by Indian products till a few years back has now been lost in several industries. The basic principle governing pay hikes and promotions should be performance. The only way to boost productivity and ensure a reasonable return on investment is to reward merit and good performance and to penalise poor performance.

The system of linking wage increase in the form of dearness allowance, etc., to the price index should be discontinued. Increased dearness allowance itself causes inflation and the rise in prices entitles workers to a further dose of dearness allowance. This vicious circle must be stopped.

In fact, the mining industry should insist on a wage freeze till a reasonable level of productivity is reached and thereafter all wage increases should be related to increased productivity. This is the only way to make the Indian mining industry globally competitive.

MISMANAGEMENT OF TALENT

Some three to five decades back, the best engineering students in India used to opt for the mining course and proudly took up jobs in the mining industry because the pay, perks and status enjoyed by mining engineers were the envy of others. However, since nationalisation of mines, the situation has changed totally and although the mining engineers continue to shoulder all the statutory responsibilities, they get the same pay and perks as their colleagues in other disciplines. Naturally enough, the best engineering talents shy away from the mining profession and the result is that mining engineering is among the least preferred branches in engineering colleges. Even the students who have to join the mining course due to their low ranks in the entrance examinations begin to think of alternatives from day one. In fact, they treat the degree as a stepping stone for more attractive careers in fields such as business management and the civil services. In recent years, almost 75% of mining graduates from the three best colleges in the country have moved over to higher paying management jobs or the more prestigious civil services.

The mining industry must take steps to attract and retain the best talents. For this it will be necesary to give a better deal to the young engineers which must include not only higher pay and perks but also better career prospects and job satisfaction. Instead of giving them a boring, routine type of work, they should be given assignments that make demands on their talent and imagination. The work must stimulate creative thinking. The industry has so long been working on borrowed technology. Our young engineers must be encouraged to come up with innovative ideas which will lead to the development of indigenous technology.

Our workers must be encouraged to develop skills. For this it will be necessary to completely re-structure our policy on wages. There must be only three categories of workers: skilled, semi-skilled and unskilled; and there must be a substantial difference in the wages of these three categories. Semi-skilled workers must get much more than unskilled workers and similarly, skilled workers must get much more than semi-skilled workers. If the wage difference is small, there is no incentive for development of skills. Moreover, special attention must be paid to the development of the right attitude towards work. The workers' thought process has to be suitably modified so that they enjoy doing the work and take pride in their contribution to the organisation.

CONCLUSION

It is now universally recognised that a nation's wealth does not depend only on natural resources. The decisive factor is the quality of its people: their abilities, discipline, sense of duty and national dedication. It is time for us to take hard decisions and instil in the people a sense of discipline, duty and dedication. Our political leaders have to show the way by taking principled decisions even if these may mean risking their popularity.

CHAPTER 15

Exploitation of India's Coal Resources — Strategies for Next Century

*L. K. Bose**

ABSTRACT

Demand of coal in India is likely to go up in future at a faster rate than the present, matching with future economic development of the country as coal based power generation has been identified as the main source of energy growth.

Presently, only 36% of India's coal reserves come under the category of proved reserves. India possesses only 6% of global proven coal reserves as against about 16% of world's population. There is need for intensification of coal exploration for taking up adequate coal projects in due time.

It has been observed that recovery of coal reserves by existing system of underground mining is only 15 to 35% of the net geological reserves and it is 50 to 70% in the case of opencast mining. Optimisation in production technology and adoption of efficient extraction techniques are essential to ensure sustainable growth in coal production in future years. This paper highlights the various strategies to be adopted for exploration and exploitation to achieve required growth in coal production in future.

As coal is recognised as essential energy input, there is urgent need to remove all impediments for development of coal industry. Rules, regulations and approval system needs to be simplified to attract more investment by private entrepreneurs so that production from private mines pick up. Government should take necessary steps to provide all facilities for fast development of country's coal industry failing which projected GDP growth of the economy will not take place.

INTRODUCTION

Coal, being the most important and proven energy form available, is considered as the backbone of Indian economy. Truly in the words of late Mohan Kumarmangalam, the founder of nationalised coal industry, **"Coal is the heart of India"**.

---

*Former Executive Director, Coal India.

Contributing nearly **two-thirds of commercial energy requirement,** coal is presently the principal source of primary commercial energy. Dominant role of coal in energy scenario has emerged since the oil crisis in 1970s. Since then Government has decided to make coal-based thermal power generation, the backbone of India's energy scenario.

Dependence on indigenous coal is likely to continue for meeting bulk of India's energy needs in foreseeable future as well. This is mainly because of availability of adequate coal reserves as compared to limited life of other sources of commercial energy and on the basis of overall economy.

The per capita consumption of primary energy in India is only 0.24 toe (tonne of oil equivalent) as against world average of 1 toe. The consumption of energy has registered an average annual growth rate of 5.8% during the last two decades. Matching with projected economic growth of the country, the consumption of energy resources, in the year 2011-12 is projected as 468 million toe which means a per capita consumption of 0.40 toe at the projected population level of 1196 million.

India's coal reserves of 206 Billion Tonnes (as on 1.1.1998) are considered to be **adequate to sustain the present level of production** for more than 200 years. However, India possesses only 6% of global proven coal reserves as against about 16% of world's population. Presently' only 36% of India's coal reserves comes under the category of proved reserves. For meeting the increased demand of coal there is **need for intensification of coal exploration** for taking up adequate projects in due time.

In view of our dependence on coal for future energy scenario, it is necessary to take care of **conservation aspects** of coal reserves. Since majority of our coal reserves are in seams thicker than 4.5 metre, recent study shows that recoverable reserves by present system of underground mining is only 15 to 35% of the net Geological reserves and it is 50 to 70% in the case of opencast mining.

Optimisation in production technology and adoptation of efficient extraction techniques are essential to ensure sustainable growth in coal production in future years. This paper deals with various strategies to be adopted for exploitation to achieve required growth in coal production in future.

DEMAND

Overall All India coal consumption at the end of VIII Plan (1996-97) was 295.9 Mt as against 235 Mt consumed in 1991-92. Thus the coal consumption increased by 60 Mt during the plan period. During 1996-97, consumption by different sectors comprised of Power 199 Mt (67.25%), Steel 35.00 Mt (11.83%), Cement 11.34 Mt (3.83%) and others 50.57 Mt (17.09%).

Due to recession in industrial sector and delay in materialisation of new power stations the coal consumption has remained more or less stagnant during the last two years.

The coal demand by the end of the Ninth Plan i.e. 2001-02 is estimated by Planning Commission as 412 Mt. However, this may be scaled down to around 360 Mt in view of the present trend. Coal demand is likely to increase at the rate of 100 - 120 Mt during the next two plan periods. Thus by 2011-12 the total coal demand may be of the order of 550 - 600 Mt.

As per the demand projections for coal, three major coal consuming sectors, viz. electricity generation, steel and cement require 90% of the total coal demand (power 70%, steel 15%, cement 5%). The power sector will continue to dominate the future coal demand scenario. Therefore, the future coal demand will depend mainly on materialisation of new power stations.

COAL SUPPLY SCENARIO

The indigenous coal production has gone up from a level of 77.9 Mt in 1973-74, which was the first year of nationalisation of coal industry, to 299.2 Mt in 1997-98. During this period, production of Coal India (CIL) has gone up from 69.8 Mt (comprising 89.6%) to 260.6 Mt (i.e. 87.1%), and that of Singareni Coal Co. (SCCL) from 5.3 Mt (i.e. 6.8%) to 28.9 Mt (i.e. 9.7%), whereas production from other sources including TISCO/IISCO Collieries have increased from 2.76 Mt to 9.7 Mt.

There has been gradual increase in import of coal during the last decade. With liberalisation policy adopted by the Government, the import duty on coking coal has been gradually reduced to 3% from earlier 35% and that non-coking coal to 10% from earlier 85%. Import of coal has been permitted under Open General Lincence (OGL) since 1993-94. The import of coking and non-coking coal has increased from a level of 5.3 Mt and 0.03 Mt respectively in 1991-92 to 11.7 Mt and 4.7 Mt respectively in 1997-98. Initially only steel sector was importing coking coal. Import is now resorted to by some of the power and cement plants located in Southern and Eastern India as imported coal has become more attractive after recent Government decision to reduce customs duty.

As per recent estimate, availability from the existing sources may be around 326 Mt during 2001-02. In addition there will be supply of about 2 Mt form captive mining sources. Since estimate of coal demand is also being scaled down, the demand-supply projection for 2001-02 will more or less match taking into consideration the present level of import of coal.

During X and XI Plans the existing sources alone will not be able to fully meet the increased demand of 100 - 120 Mt during next two plan periods. Expected growth from existing sources may be around 60 - 70 Mt during the next two 5-years periods. Thus balance requirement of about 40 - 50 Mt for 5 year periods has to come from captive mining sources as efforts need to be made to restrict import of coal considering foreign exchange implications.

COAL RESERVE STATUS

Global Status

The known coal reserves of the world are spread over 100 countries. It is estimated that at the current level of production these reserves will last for more than 200 years. Coal reserves vis-a-vis production status in some of the selected countries, which together account for 90% of coal reserves, are given in Table 15.1.

It may be noted from the statement given at Table 15.1, that nearly **80%** of total proved coal and lignite resources in the world are contained in only **six countries including India.**

Table 15.1 : Reserves of Coal and Lignite in Selected Countries [Figures in Billion tonnes (Bt)]

	Countries	*Total Proved Coal & Lignite Reserves (1995) (Bt)*	*Share of Total Reserves (%)*	*Share of World Production (%)*	*Reserve/ Production Ratio (years)*
1.	USA	240.6	23.3	24.6	258
2.	Former Soviet Union	214.0	23.4	9.2	>500
3.	China	114.5	11.1	29.5	88
4.	Australia	90.9	8.8	5.7	375
5.	**INDIA**	**69.9**	**6.8**	**5.9**	**245**
6.	Germany	67.3	6.5	3.4	273
7.	South Africa	53.3	5.4	4.8	272
8.	Poland	42.1	4.1	3.9	212
9.	Canada	8.6	0.8	1.8	115
10.	United Kingdom	2.5	0.2	1.4	48
TOTAL WORLD		**1031.6**			**228**

(Source : B. P. Statistical Review of World Energy, 1996)

Indian Scenario

The **reserves of coal** (as on 1.1.98) considering seams of 0.9 metre and above thickness upto a depth of 1200 metres as estimated by Geological Survey of India (GSI) are given in Table 15.2.

Table 15.2 : Indian Coal Reserves (as on 1.1.98) [Figures in Billion tonnes (Bt)]

		Type	*Coal Reserves (Bt)*	*Share of Total Reserves (%)*
	1.1	Prime Coking	5.31	2.57
	1.2	Medium Coking	23.21	11.26
	1.3	Semi-Coking	**1.61**	**0.78**
1.0	Sub-Total Coking		30.13	14.61
2.0	Non-Coking & Ungraded Tertiary Coal		176.11	85.39
	TOTAL		**206.24**	**100.00**

SALIENT FEATURES :

The salient features of Indian coal reserves are as follows :

- The **proved reserves** constitute only **36.4%** of the total reserves.
- Majority of the coal reserves i.e. **62.7% occur within 300 m depth** and 30.1% occur in the depth range 300 – 600 m. Only about **7.2 % occur beyond 600 m depth** in deep coal basins like Jharia and Raniganj Coalfields.
- **Coking coal** reserves are only **14.6%** of the total coal reserves confined to coalfields located in three States viz. Bihar, West Bengal and Madhya Pradesh. Prime Coking coals are available only in Jharia Coalfields in Bihar.
- Reserves of **non-coking coal** amount to about **85.4%** of the total reserves of which only 12% are of superior grades (i.e. A, B & C grades), majority of which are in Raniganj Coalfields.
- Bulk of the non-coking coal reserves (proved and indicated category) of 108.2 Bt or **81.1% are of power/inferior grade** (i.e. E, F & G) occurring in thick inter-banded seams and located in coalfields like Singrauli, Talcher, Ib Valley, North Karanpura, Rajmahal and Korba. Reserves of Power Grade coals of E, F & G grades constitute 84.9 Bt., i.e. 63.7% of non-coking coal reserves.
- Over **90%** of the proved coal reserves are mainly under the **command areas of Coal India.**
- Out of the proved reserves of 47 Bt. of non-coking coal under Coal India's command area, only 10.8 Bt. i.e. **23% are of superior grade** (i.e. A, B & C) and the balance 36.2 Bt., i.e. **77% are of power/inferior grade** coal (i.e. D, E, F & G).
- Nearly **90%** of coal reserves are concentrated in only four States viz. **Bihar (33.1%), Orissa (23.4%), Madhya Pradesh (20.3%), West Bengal** (12.6%), all located in the **eastern part of country.**
- Majority of the **virgin coal reserves lie below forest cover.**

Recoverable Reserve

It has been observed that extractable/recoverable reserve by present underground mining system normally varies from 15 to 35% only of the net Geological reserves. Since majority of our coal reserves are in seams greater than 4 metre thickness, considerable coal is lost in panel barriers, in partings between working sections, in protective pillars below surface features, in depillaring operations, etc. Even in present opencast workings the extractable/recoverable reserves are of the order of 50 to 70% of the overall reserves.

Based on analysis of the proved reserves done by CMPDI recently, the following observations are made :-

- Out of 75 Bt. of proved reserves available upto a depth of 600 m, **8.5 Bt.** have been consumed by **past mining activities** and a balance of 66.5 Bt. are available for assessment of recoverable reserves.
- Out of balance 66.5 Bt. of available reserves, 20.2 Bt. are locked up in lower seams which cannot be mined unless upper seams are exploited and include reserves left

below surface features. Another 5.2 Bt. reserves have doubtful economics. Thus only about 41.1' Bt. of proved reserves are available for exploitation.

- Out of this 41.1 Bt. of proved reserves, 17.2 Bt. are covered in existing mines and projects, 7.3 Bt. are earmarked for captive mining and 3.2 Bt. are identified for 9th Plan projects. Thus about **13.45 Bt.** only are available **for planning for future projects** in CIL/SCCL Command area in Xth Plan and beyond.
- Considering average recovery of around 40%, the available reserve of 13.4 Bt. can sustain additional production of about **200 Mt per annum** for **25 years** in **CIL/SCCL** Command area. Similarly, 7.3 Bt. of reserves earmarked for **captive mining** can sustain a production level of around **100 Mt.** per annum for **25 years.**
- A tentative assessment shows that out of the recoverable reserves about **70% is available for opencast production** and **30% is exploitable by underground mining** under present economic scenario.

STRATEGIES FOR COAL EXPLORATION

The analysis of proved coal reserves available for projectisation as explained in previous para indicates that reserves available for planning of projects for contribution beyond IX Plan, i.e. 2001-02 onwards, are not adequate. There is need for intensification of detailed exploration so that about 150 to 200 Mty of production capacity could be developed during each of the X Plan (upto 2006-07) and IX Plan (upto 2011-12). This also calls for increase in Regional and/Promotional exploration so that more reserves are available in "Indicated: Category to be taken up for detailed exploration.

Guide criteria and strategies for Regional and Detailed exploration (including those identified by the Working Group) in coming years are listed below :-

- In the known coalfields the uncovered areas are to be examined and advanced technologies should be deployed for identification and exploration of potential areas/ blocks.
- Detailed "Master Plan" for prospective coalfields are to be prepared to fix priorities for exploration.
- Due importance should be given to areas/blocks nearer to areas having infrastructural facilities. Areas away from infrastructural facilities may be given lower priority unless dictated by region-wise quality-wise demand pattern.
- Preference should continue to be given for location for opencastable coal to sustain present production mix of opencast/underground production.
- Greater thrust is to be given to prove power grade coal reserves, viz. Singrauli, Parts of Raniganj, North Karanpura, Rajmahal Group, Ib Valley, Talcher, Korba, Godavari, Hasndeo, etc. The reserves beyond present day depth cutoffs should be proved well in advance.
- Efforts may be made to locate coking coal reserves outside the known areas of Damodar Valley Coalfields.
- Exploration has to be extended to deeper levels for opencastable prospects in coalfields like Talcher, Ib, Mand-Raigarh, Korba, Singrauli, North Karanpura, Rajmahal, etc.

- Detailed exploration should be supplemented by 100% geophysical logging coverage, and intensification of surface geophysic, hydrogeological investigation, coal characterisation studies, etc.
- Basic data for assessing the Coal Bed Methane potential should be generated.

STRATEGIES FOR RESERVE EXPLOITATION AND PRODUCTION

Major share of increased coal demand in future years will be from power sector as noted earlier. During the last decade increased production of power grade coal came mostly from **Singrauli, Korba, Ib, Talcher, Rajmahal, North Karanpura, Wardha Valley and Singareni.**

As Wardha Valley and Singareni have reached a plateau of production growth, **increased production will depend on other six coalfields** as note above and new coalfields like **Mand-Raigarh, Hansdeo-Anand, Rajmahal etc.**

All the above coalfields are endowed with **thick interbanded coal seams at shallow depths amenable to opencast mining.** More **high capacity opencast mines have to be developed** in these coalfields to ensure quick growth in coal production matching with **fast increase in demand of power sector.**

Preference of opencast mining will continue because of **large capacity, competitive economics, less capital investment compared** to underground mining. **Short gestation period, more assured production, better safety** and **conservation** are added advantages which favour opencast mining.

Opencast mines have to take care of **better coal quality assurance, more environmental safeguards, more depth in operation, worsen OB : Coal ratios.**

Higher capacity excavators and **transport equipment** as per global trend are to be introduced to ensure higher productivity, **lesser fleet in large mines, working in greater depths** with more **adverse stripping ratios.**

Manufacturers of Heavy earth moving equipment are to gear up their production capability and also take up manufacture of larger size equipment to match with the demand.

With stringent quality assurance parameters incorporated in the agreements with IPP's, there has to be more concentrated efforts towards **coal quality aspects** in opencast mining. **Selective mining, controlled blasting, identification of stone bands in advance, separate removal of partings** etc are to followed as per the necessity.

As per notification of the Ministry of Environment & Forests, the following power stations are required to use beneficiated coal having not more than 34% ash with effect from 01.06.2001 : (a) located more than 1000 Km from the pit-head, (b) located in urban/sensitive/critically polluted area. **Beneficiation of non-coking coal** have to be taken up in some selected mines to ensure **consistent quality of coal supply, provided consumers are agreeable to bear the cost of beneficiation.** Possibility of blending with low ash imported coal or superior grade indigenous is also to be studied to meet the above stipulation.

Government has to take steps to develop **infrastructure in virgin coalfields** like Mand-Raigarh, Hansdeo-Anand, Rajmahal group (Chupervita, Panchwara, Brahmi Blocks) to enable development of these coalfields by private entrepreneurs.

Due care is to be given to **COAL CONSERVATION** aspects as our reserve position in some of the coalfields is **not comfortable. Technology to ensure higher percentage recovery even in opencast mines** need to be introduced.

Design of opencast mines upto a **depth of 300 m** is to be taken up for exploitation of **deeper deposits** of thick seams for higher percentage of recovery.

In underground mines **new technology** is to be introduced to ensure **high level of production with high productivity** and **better percentage of extraction in selected mines.** Indigenous manufacturer should take up development of such new equipment to ensure supply of adequate spares.

Our **geological conditions do not permit** large-scale introduction of **longwall technology**. In four mines of **SECL longwall equipment** have been introduced. If found suitable, more mines could be identified for introduction of such equipment.

Problem of full extraction by underground mining of thick means seams as available in most of the coalfields, is yet to be resolved. Old technology with sand stowing has limitation of low productivity, poor production rate and low percentage of extraction apart from constraints due to non-availability of sand. **Majority of coal reserves** being in seams **greater than 4.5 m thickness**, this problem needs to be solved.

Mechanisation with LHD and SDL'S has attained success in some of the coalfields specially in development work. Such **mechanisation in depillaring** operations need to be adopted suitably with development of high-speed supporting system.

Bulk of underground production still comes from **old underground mines** having limited production capacity and low productivity which results in huge loss. **Mines with limited resources** need to be **weeded** out to reduce loss.

Many old underground mines suffer from presence of **fire, water or roof collapse in upper seams.** Application of appropriate technology is to be developed for such mines.

CONCLUSION

Demand of coal is likely to go up in future at a faster rate matching with economic development of the country as **coal based power generation** has been identified as main source of energy growth.

As coal is recognised as essential energy input, there is urgent need to **remove all impediments for development of coal industry.** There is need for **capital fund for development, prompt approval for coal projects, simplification of rules for clearance of forest land, environmental clearance, land acquisition** etc.

Rules, regulations and approval system need to be simplified to **attract more investment** by private entrepreneurs so that production from private mines pick up.

Government should take necessary steps to provide **all facilities for fast development of country's coal industry** failing which projected GDP growth of the economy will not take place.

REFERENCES

1. Annual Performance Report, 1997-98, Coal India Ltd.

2. Annual Sales & Marketing Report, 1997-98, Coal India Ltd.

3. Kumar, U, 1999 – Coal Marketability through quality improvement. Proceedings of the National Seminar on Marketing of Indian Coals and its challenges ahead. MGMI, Bilaspur Chapter.

4. Tripathy, Y. P. 1999 – Coal Reserves Mineable upto the year 2000. Proceedings of the National Seminar on Marketing of Indian Coals and its challenges ahead. MGMI, Bilaspur Chapter.

CHAPTER 16

Coal Mining Industry in India–Past, Present and Future

*N.K. Sharma**

EARLY EVIDENCE

The recorded evidence of coal mining in the country was obtained through the Memorial by M/S. Sumner and Heatley in 1774. It records receipt of a proposal by the Council of Revenue from Summer and Heatley of M/s. East India Company for working coal mines and selling of coal as a consequence of their discoveries of a number of coal mines in the Ranigunj Coalfields. However, as described by M/s. Sumner and Heatley, iron was already being manufactured in that area and perhaps outcrops of coal seams were being mined for the purpose of smelting.

PRE-INDEPENDENCE DEVELOPMENT

The abolition of the monopoly of M/s. East India Company in 1813 paved the way for rapid development of private commercial enterprises. The first joint stock company registered as M/s. Bengal Coal Co. Ltd. came into being in 1843. From a level of output of about 400 tonnes per annum during 1815-1823, it reached a level of 91,000 tonnes during 1846. By this time, steam engines were already introduced and the demand for coal continued to pick up.

From as early as 1920, a number of committees, commissions and consultants made observations on the question of conservation and scientific exploitation of coal, working conditions of mines and mines safety etc. and thereby bringing in regulation and control of coal industry in India in some form or other. The indications of all of them were directed towards state ownership of the coal mines. However, most of these suggestions, recommendations were not considered for quite some time. The distinction of the first government owned coal company in India goes to the Singareni Collieries Co. Ltd. because in 1945, the government of Nizam of Hyderabad purchased all the shares of the company

*Director (Technical), Coal India Ltd.

and the mines were brought under the ownership of the Indian Trust Fund of the ex-Nizam's Government.

POST INDEPENDENCE DEVELOPMENT

In 1956, National Coal Development Corporation came into existence in the public sector with the coal mines of the State Railways as nucleus. During sixties, the coal industry passed through a period of lull principally owing to cheap availability of oil. Situation, however, took a dramatic turn with the advent of the seventies when coal demand accelerated as the oil prices started spiralling up. Coupled with this, there were reports of unethical labour practices, unsafe and unscientific mining practices and reluctance to go for require capital investment by the erstwhile private mine owners. All these came under serious considerations of the Central Government before it took the decision to bring the coal mines under control in phases. By this time (1970-71) coal production in the country reached a level of 73 million tonnes per annum. The coking coal mines were taken over by the Central Government in 1971 with the objectives of re-organising and reconstructing the coking coal mines, as well as coke even plants, for the purpose of protecting, consuming and promoting scientific development of the coking coal resources to meet the growing requirement of the iron and steel industry. The mines were nationalised in 1972 and Bharat Coking Coal Limited (BCCL) came into being. In the second phase of nationalisation, the Coal Mines (Taking over of Management) Ordinance, 1973 came into force in January, 1973 and the mines were nationalised finally in May, 1973 thus bringing the non-coking coal mines within the purview of the Public Sector Company "Coal Mines Authority Ltd" (CMAL). BCCL and CMAL merged later on to give birth of the holding company Coal India Ltd(CIL) in 1975 with five subsidiary companies viz. Eastern Coalfields Limited, (ECL) with headquarters at Sanctoria (West Bengal), Bharat Coking Coal Limited (BCCL) with headquarters at Dhanbad (Bihar), Central Coalfields Limited(CCL) with headquarters at Ranchi (Bihar), Western Coalfields Limited(WCL) with headquarters at Nagpur (Maharashtra) and Central Mine Planning & Design Institute Ltd.,(CMPDIL) with headquarters at Ranchi (Bihar). CMPDIL became the planning and design subsidiary of CIL. A few more subsidiary companies were formed by carving out of these subsidiaries for better management and administrative control and thus Coal India Ltd had finally seven producing subsidiaries and one planning and design subsidiary. The three additional producing subsidiaries were Northern Coalfields Ltd(NCL) with headquarters at Singrauli (Madhya Pradesh), South Eastern Coalfields Limited (SECL) with headquarters at Bilaspur (Madhya Pradesh) and Mahanadi Coalfields Limited (MCL) with headquarters at Sambalpur (Orissa).

COMMERCIAL ENERGY SCENE

The dominance of coal continues in the commercial energy spectrum of the country as can be seen from the Table 16.1 below showing the production of various energy resources during 1998-99. It can be seen from the Table 16.1 that the share of coal is 67%.

Table 16.1

	Energy Production (in Mtoe)	*%age*
1	2	3
Resource		
Coal	150.1	67.1
Lignite	11.3	5.0

1	2	3
Crude Oil	33.3	14.9
Natural Gas	21.4	9.6
Hydro	6.8	3.0
Nuclear	1.0	0.4
Total	**223.9**	**100.0**

*Mtoe represents million tonne oil equivalent.

PER CAPITA ENERGY CONSUMPTION

Even the per capita energy consumption in India is much lower as compared to the developed nations as can be seen from the Table 16.2.

Table 16.2

Countries	*In toe*
USA	7.94
Japan	3.93
European Nation	3.62
India	0.25

* toe represents tonne oil equivalent.

All these underscore the potential for growth of the energy resource industries and especially coal which has a fairly large resource potential as compared to any other energy forms in the country.

INVENTORY OF GEOLOGICAL RESERVES

With a total inventory of geological reserves of 208.8 billion tonnes as on 01.01.1999 the coal industry will continue to remain in a position to meet the growing energy demands of the country. State wise reserves of coal in the country upto a depth of 1200 metres is as below.

Table 16.3

State	*Geological Reserves (In Billion Tonnes) As on 01.01.1999*	
	Absolute	*%age*
West Bengal	25.9	12.4
Bihar	68.7	32.9
Madhya Pradesh	42.8	20.5
Uttar Pradesh	1.1	0.5
Maharashtra	6.9	3.3
Orissa	49.1	23.5
Andhra Pradesh	13.3	6.4
North Eastern States	1.0	0.5
Total	**208.8**	**100.0**

Out of the total reserves of 208.8 Billion Tonnes the proved reserves are estimated to be 79.2 Billion tonnes (38%). The share of coal reserves in the states of Bihar, Orissa and MP being high, these states will continue to dominate the coal production scenario in the country both in current term as well as in the long term.

COAL CONSUMPTION

The coal consumption which was at a meagre level of about 71.7 million tonnes just prior to the take over of the coal mines by the Central Government (1970-71), has reached a level of about 301.2 million tonnes in 1998-99, with growth of about 229.5 million tonnes over a span of twenty eight (28) years or about 8.2 million tonnes per annum. Major sector wise distribution is as under:

Table 16.4

(In Million Tonnes)

Sector	*1970-71*	*1998-99*	*Growth*
Power	13.20	204.89	15.5 times
Steel	13.50	32.72	2.4 times
Cement	3.50	13.56	3.9 times
Others	41.50	49.99	1.2 times
Total	**71.70**	**301.16**	**4.2 times**

The Table 16.4 shows that whereas the overall consumption has grown 4.2 times during the time span of twenty eight years from 1970-71 to 1998-99, the consumption of power sector has grown by 15.5 times, Steel Sector 2.4 times and Cement Sector 3.9 times. The aggregate of other sectors has grown 1.2 times only principally due to decline in consumption of Railways Sector during this time span.

COAL PRODUCTION

The growth in coal consumption was backed up by commensurate growth in coal production. From a level of All India coal production of about 73 million tonnes during 1971-72, the level achieved during 1998-99 is 292 million tonnes. India to-day ranks third in coal production coming after China and USA. The Table 16.5 summarises the trend in coal production during the various National Five Year Plan Periods.

Table 16.5

Terminal Year	*Five Year Plan Period*	*Production(Mt)*
1973-74	IV Plan	78.17
1978-79	V Plan	101.95
1984-85	VI Plan	147.41
1989-90	VII Plan	200.89
1996-97	VIII Plan	285.63

CIL has the major share in the coal production portfolio, followed by SCCL and other captive mines both in public and private sector. The contribution from the different coal companies for the year 1998-99 is shown in Table 16.6.

Table 16.6

Company	*Coal Production (1998-99)*	
	(Mt)	*Share(%)*
CIL	256.620	87.9
SCCL	27.020	9.2
TISCO./IISCO/DVC	6.342	2.2
Sub Total (Public Sector)	**289.982**	**99.3**
Captive Sector (Private)	2.061	0.7
Total All India	**292.043**	**100.0**

TECHNOLOGY

Such a major boost in coal production has been made possible through exploitation of a large number of large sized opencast mines. Out of the two broad modes of mining i.e. opencast and underground, the emphasis has been on the former due to its favourable economics, safety, large coal recovery, greater potential for large sized production units and lower gestation period. The Table 16.7 below shows the trend of growth of coal production of CIL only in terms of the two broad technologies:

Table 16.7

(In Million tonnes.)

Year	*Opencast*	*Underground*	*Total*
1974-75	20.79 (26%)	58.22 (74%)	78.99 (100%)
1984-85	70.31 (54%)	60.50 (46%)	130.81 (100%)
1994-95	167.46 (75%)	55.60 (25%)	223.06 (100%)
1996-97	195.93 (78%)	55.20 (22%)	250.63 (100%)
1998-99	203.21 (79%)	53.31 (21%)	256.52 (100%)

Thus, the share of opencast production of CIL went up from about 26% just after the nationalisation of the coal industry to about 79% in 1998-99. A large share of opencast was essential to match the rapidly increasing coal demand of the power sector industries in the country. In near future also, with advancement of technology and availability of large size of equipment, opencast may continue to greater depth. Application of in-pit shiftable crusher conveyor has already been made in both CIL and SCCL. A number of limestone mines in the country have been working with Surface Miners successfully. It is being introduced in a few opencast mines of CIL also. In fact surface miners may provide an answer to selective mining in opencast mines.

But at the same time, we can not lose sight of the fact that open cast resources are not unlimited and therefore a breakthrough in underground production is necessary. So far in our country, Longwall Powered Support, a widely proven underground technology around the world has not been successful. The technology failed due to varied reasons both social and technological. These are to be overcome as longwall suits to meet the requirement for mass production underground technology to effect economies of scale for meeting the accelerated demand for coal. Three sets of longwall powered support face equipment from China are being commissioned at three mines of SECL. Apart from longwall, mechanised Room & Pillar/Bord & Pillar practiced in a few coal producing countries which also offer mass production are also being envisaged on risk-gain sharing basis with foreign association.

A new technology Coal Bed Methane (CBM) has also been introduced in the country through foreign collaboration, Methane gas is extracted directly from the coal seams *in situ* through bore holes. The technology has been widely practiced in USA where about 900 wells are engaged in Coal Bed Methane extraction. The output of CBM there rose from 0.5 giga cum (1988) to more than 9 giga cum (1991). In India initially three companies evinced interest in CBM and out of them, one company has already drilled pilot hole in Ranigunj Coalfield. Methane extracted can find use in Power Plant, as Natural gas, as fuel for burning etc.

SAFETY

Side by side, due regard has been given to improve the status of safety in the coal mines. Safety in the coal mines of the country is enforced broadly by the provisions of Mines Act, 1952 and the Rules, Regulations and Bye laws made thereunder. Director General of Mines Safety (DGMS) under the Ministry of Labour is the statutory authority to oversee the improvement of the provisions of the Mines Act and Coal Mines Regulation. However, the basic responsibility of maintaining safety in mines rests with the Owner, the Agents and the Manager of the mine. Safety performance measured generally in terms of fatality rate per million tonnes of coal produced in the country and also 300,000 man shifts engaged shows that there has been a steady improvement in the safety status since the coal mines were brought under the state control.

Table 16.8

Year	*Fatality Rate(All India)*	
	Per Mt.	*Per 300,000 manshifts*
1973	2.05	0.37
1980	1.35	0.28
1990	0.75	0.28
1996	0.43	0.24

Mines Rescue Services have been completely re-organised and each subsidiary company is further revamping its rescue services to attain the requisite standard.

COAL QUALITY

Quality is accorded its due recognition by the coal industry. In case of coking coal, reduction in %ash in coking coal as demanded by the Steel Plants, is being tried at the cost of yield of washeries. In non-coking coal, two new beneficiation plants viz. Piparwar and Bina Deshaling have already been producing washed non-coking coal and the coal is being supplied to the Power Plants. Three more coking coal washeries have also been converted to non-coaking coal washeries. It is principally the reluctance of the consumers, which has been responsible for the tardy growth in the non-coking coal washing capacity in the country. Participation of outside parties is now being sought to build, own and operate non-coking coal beneficiation plants. Monitoring of quality is also being done through Third Party Sampling, Joint Sampling and Check samples at loading end and also at consumers' end as per the agreement with the consumers. Blending of coal of different grades to produce coal with ash below 34% is also being envisaged at suitable sites in the country. Coal India is also going for legally enforceable Fuel Supply Agreement with the consumers.

LIBERALISATION OF THE INDUSTRY

In consonance with the Government decision to liberalise the economy, a number of steps have been taken in the coal industry as well. Coal Mines Nationalisation Act was amended in 1993 to allow captive mining of Power, Steel and other industries. 10 blocks are offered to 17 parties from Steel, Power, Cement and Sponge Iron Sectors.

Government Budgetary Support to Coal India has been withdrawn and for the Ninth Plan period (1997-2002), the capital outlay of CIL is being financed by internal resources and extra budgetary resources like borrowing from both domestic as well as international resources. An International funding of the order of $ 1.06 billion has been organised which is being contributed by World Bank - ($ 530 million) and co-financed by EXIM Bank of Japan-($ 530 million).

To ensure that the nationalised coal industry faces competition, import duty of coking coal has been kept at 9.72% and that of non-coking coal at 21.6% which include Basic Duty, Special Duty and Special additional duty as applicable in the respective categories of coal.

PLANS AHEAD

Against the All India raw coal demand estimated by Planning Commission at 325.38 Mt. during 1998-99, the actual off take has been 301.16 Mt. The presently estimated raw coal demand for 1999-2000 is 311.83 Mt. Demand projected for the terminal year (2001-02) of IX Plan is 370.64 Mt which is rather unlikely to materialise. The perspective All India demand for the terminal year (2006-07) of X Plan is 410 Mt. as per current projection of CIL.

In consonance with the growth in the consumers' demand, the production programme of coal has been projected as below for the country:

Table 16.9

Year	*Production in Mt.*
1998-99(Actual)	292.04
1999-2000	300.45
2001-02	347.86
2006-07	401.60

A number of large sized open cast mines with higher capacity HEMM are being envisaged to offer economies of scale. Greater application on In-pit crusher-conveyor system and Surface Miner may be favoured in future application. In underground also, there will be greater thrust to substitute the manual loaders by mechanised system. SDL & LHD would continue to be deployed for facilitating the loading operations in underground mines keeping in view the need to improve their productivity. If the longwall faces now being worked prove successful, few more such faces may be taken up at appropriate geo-mining setting. As earlier stated, continuous miners with shuttle cars and roof-bolting are also being envisaged for future application in CIL mines. Problem facing coal industry now is to mobilise the vast sums of money required to pay for the new investment being envisaged in the mines. Under the liberalised regime, government is no longer in a position to offer budgetary support to the coal companies. For this, coal sector may be opened up further to private investment both domestic and foreign. The Central Government has already deregulated superior grade (A to D) thermal coal. This will be extended further to cover the inferior grade (E to G) thermal coal also by 2000. A team of consultants has also been commissioned to review the regulatory frameworks in the coal industry of the country covering mining environmental and labour laws, regulation and control system and suggest improvement so that contemporary technologies practiced around the world and private participation both indigenous as well as foreign are encouraged. Quantum of import of coal will be higher so also the beneficiation of non coking coal. In the emerging competitive environment, coal industry in India will have greater focus on quality of its products. Environmental Management will be part and parcel in every mining operation. Land reclamation, tree plantation will be a must in the future mines. Simultaneously, safeguard against water pollution, air pollution and noise pollution will be required to be provided in the new mines. Rehabilitation of Project Affected People will also be provided as far as practicable. The new beneficiation plants will work in close-circuit thereby eliminating the discharge of effluents to natural waterways.

From environmental angle, clear and more efficient coal combustion technologies like pressurised fluidised bed combustion and integrated coal gassification combined cycle would have to be given a serious consideration by the Power Sector for future power generation plants of the country. Computers and Telecommunications will make further in roads in various functions of the industry in future so as to expedite decision-making process, enhance safety, better control on production performance and improved utilisation of men, machine and materials.

Coal industry in India is passing through a phase of transition. Greater participation of private sector and greater flow of imported coal may create a new competitive environment for the industry. Within such an environment, the nationalised coal industry will became more efficient, cost-effective and adaptive to the interests of all its stakeholders.

CHAPTER 17

Dimensional Stone Industry from Romania in the New Millennium

Adrian Florea & Dumitru Fodor***

ABSTRACT

Romania is a country with many ornamental rock deposits (more over 300) and an old tradition in quarrying activity and utiiisation of these types of rocks. For example, the mining activity in the Ruschita marble deposit started more than one hundred years ago, in 1883. Here it's mined out very beautiful white, yellow and pink marble.

The paper presents the now-a-days situation in the dimensional stone industry both, in Romania and around the world and in correlation with general tendency for the Romanian mining industry. We also present the most important ornamental rock deposits from Romania, located in various regions of the country and mainly in the Alpine - Carpathion and Hercinic units, and the mining methods and techniques applied in quarrying activity.

INTRODUCTION

The geological research operations undertaken on the Romanian territory of 237500 km^2 have made a multitude of igneous, metamorphic and sedimentary rock deposits with various and important uses in different industrial branches.

To this effect, it can be shown that the mining of useful rocks and especially of ornamental rocks in Romania represents an important activity that has been pursued since ancient times. The monuments, constructions and fortifications on the country's territory as well as the abandoned quarries for marble deposits in the Southern Carpathians, the

*Adrian Florea - Lecturer Engineer - Faculty of Mining Engineering

**Dumitru Fodor - Prof. Ph.D. - Faculty of Mining Engineering

Mining Technologies and Geology Department - University of Petrosani - Romania Str. Universitatii 20; 2675 Petrosani; Jud. Hunedoara
Fax : 00 (40) 54-54.57.51; E-mail : adflorea@dec.integorj.ro

limestone deposits in Dobrogea, and the igneous and metamorphic rock deposits in Banat stand proof to it.

GENETIC TYPES AND SPREADING

The ornamental rock deposits in Romania are different according to the rock genesis type, located in the various regions of the country and mainly along the Alpine - Carpathion and Hercinic units (Fig. 17.1).

On the Romanian territory there are the next genetic types of deposits:

- igneous deposits of granite, granodiorite, syenite, diorite, gabbrow, riolite, porphyries, dacite, andesite, basalt;
- sedimentary deposits of aragonite, limestone, travertine, volcanic tuffs, silicolite, gypsum;
- metamorphic deposits of marble, amphibole, serpentine, green schist, grits stones (sandstones).

Table 17.1 contain the main characteristics of these rocks (colour, structure, texture, volumetric weight, compressive strength) and the distribution of deposits in Romania.

Table 17.2 contains the physical - mechanical characteristics of several ornamental rocks from different Romanian deposits.

Many of the known deposits are of considerable dimensions, comprising appreciable reserves while others are smaller, nevertheless enabling an economical mining and achieving a finite production of good and very good quality.

TENDENCY OF STONE INDUSTRY

The statistics show that the global dimensional stone yearly production exceeded 50 million tons after 1996 and the tendency is to increasing with few percents every year. China and Italy are the leaders, with a yearly production over 10 million tons. They are followed by India and Spain with a yearly production around 5 million tons. Another group of major producers with a yearly production among 1 and 5 million tons is Brazil, Greece, France, Portugal and South Korea. The other producers have the yearly production under 1 million tons.

After December 1989, the mining industry from Romania recorded a continuous decrease of production levels for almost all types of products (iron, cooper, coal, etc.). The stone industry is one exception from this rule. The Romanian dimensional stone yearly production was around 160 thousand tons in 1996, exceeded 200 thousand tons in 1998 and the prevision is to reach 300 thousand tons in 2003.

From technological point of view, in our country is possible to find a wide range of technologies. In case of smaller or isolated deposits and also for dividing large size blocks in some quarry for purpose to obtaining commercial blocks is still used the classical technology of breaking off the rock mass with wood and/or steel wedges.

The bigger marble quarry from Romania is Ruschita (photo 1). In this quarry are still being used, in the southern zone of the deposit, concave milling cutter machines (SM-

Table 17.1 : The main characteristics of Ornamental Rocks, Their distribution in Romania

The Rock	*The Colour*	*The Structure*	*The Texture*	*The volumetric Weight (g/cm³)*	*The Compressive Strength (dN/cm²)*	*The Distribution of Rocks in Romania*
1	2	3	4	5	6	7
Granites	White: Yellowish-white; Green; Red; Blucah	Grained Holocrystalline	Massive	2,3 - 2,7	1200-288	Greci-Macin, Iacondeal-Tulceaa; Ogreadeana-MH; Pietroasa-Beius, Ronda, Cladova Arad
Granodiorites	White-grey	Microcrystalline	Massive	2,4 - 2,6	1200-2800	Soimus-Arad; Maidan-Oravita; Nadrag-Caransebes
Syenites	Yellow-grey; white	Grained porphyry	Massive	2,7 - 2,9	1100-2600	Ditrau-Harghitsa; Malaia
Diorites	Grey; Dark Green	Holocrystalline	Massive	2,7 - 3	2000-2700	Birzava, Paulis-Arad; Forotic-Caras Severin
Gabbrows	Black; Black-greenish	Holocrystalline	Massive	2,8 - 3,2	1800-2500	Iuti-Orsova; Ghihan-Sibiu
Riolites	White; Yellowish-white; greenish	Porphyry Hypocrystalline	Massive Fluidal	2,0 - 2,7	1200-2000	Bala Sprie-Baia Mare; Rosia Montana-Alba
Porphyrys	Red : Green	Porphyry	Massive Fluidal	2,1 - 2,5	900-2400	Camena, Movila Goalm, Iglicioara-Tulcea; Dalul lui Manole-Turcomia; Selski-Orsova
Dacites	Grey;Blackish-brown	Porphyry	Massive	2,3 - 2,5	1300-1900	Poieni-Huedin; Singeorz Bai-Bistrita Nasaud; Danesti Maramures; Lunca-Turda
Andesites	Grey : Brick-red; Blackish; Green	Porphyry Grained	Massive fluidal	2,6 - 2,8	700-2500	Malnes-Covasna; Pietroasa-Deva; Cicirlau Maramurer Carpitus-Baile Tusnad
Basalts	Black	Porphyry	Massive Fluidal	2,5 - 3,2	2500-3500	Racosul de Jos, Bogata, Comana-Rupea; Branisca Hunedoara; Sanovita, Lucaret-Timis
Aragonites	White; Grey; Greenish-grey; Violet	Radiar Fibrous	Massive Concretional	2,9	800-7100	Corund-Hargita; Luncani-Timisoara; Suncuius-Bihor
Limestone	White; Yellowish; Grey; Brownish; Reddish	Microcristalline Criptocrystalline	Compact Organogenous	1,4-2,7	500-1500	Albesti, Mateias-Cimpulung Muscel; Moneasa, Pojoga Arad; Ighiu-Alba; Dognecea-Resita; Vascau-Adam Clisi; Bucova-Caransebes; Podeni-Aiud; Luncani-Lugoj; Bran Brasov; Caprioara-Savirain
Travertines	White-Grey;Yellowish; Brown-yellowish	Amorphous Criptocristalline	Organogenous	1,4 - 2,5	200-500	Borsec-Hargita; Carpinis, Banpotoc, Geoagiu Bai-Hunedoara
Volcanic Tuffs	White-greenish	Paammitic Aleuropelitic	Stratified Parallel	2,2 - 2.5	130-1270	Persani Rupea-Brasov; Valea Morii-Hunedoara; Reteag-Dej; Ocna Dejului, Tioc, Apahida-Cluj

1	2	3	4	5	6	7
Silicolites (radiolarites, japs)	Red; Green; Grey	Criptocrystalline	Compact Organogenous	1,8 - 2,1	700-1800	Adamclisi-Constants; Donoseni, Filia, Herculian-Covasna; Gornet, Patirlagele-Buzau; Brad-Hunedoara
Gypsum (Alabaster) Anhidrites	White	Grained Fibrous	Compact	2,3	150-650	Aghireau, Jebuc-Cluj; Copaceni-Cluj; Soveja Vrancea; Maneci-Prahovs
Marbles	White; Red; Rose: Yellow; Brown; Green; Azure	Granoblastic porphyryblastic	Compact Banded	2,6 - 2,8	600-1000	Ruschita-Caransebes; Alun-Luncani-Hunedoara; Porumbacu-Sibiu; Aries-Nasaud
Amphibols	Dark-green; Blackish	Granoblastic	Massive Shistous Banded	3 - 3,2	1300-2000	Lotrioara, Valea Marului, Valea Boului - Sibiu
Serpentines	Green; Greenish-Grey; Greenish-Brown	Lepidoblastic Nematoblastic	Schistous Banded	2,5 - 2,7	700-1300	Orasul lui Petru-Mehedinti; Izvorul Fecioarei-Orsova; Virghis - Sf. Gheorghe
Green Schists	Green; Blackish-greenish	Lepidoblastic	Schistous Banded	1,6 - 1,7	300-1500	Casincea, Cogialac, Dealul Morii, Sibioara-Constanta
Grits Stones	White; Red; Green; Blackish	Macrogranular Microgranular	Compact Banded	1,9- 2,7	300-2500	Caralita, Cotumbita-Moinesti; Margines Pietrosului tg. Secuisac; Grumazup Focsani

Table 17.2. The Physical-Mechanical characteristics of some ornamental rocks in Romania*

The physical-mechanical characteristics	*Limestone from Bapotoc-Carpinis*	*Polychrome limestone from Vascau*	*Marble from Alun*	*Marble from Caprioara*	*Marble from Moneasa (red)*	*Marble from Ruschita*	*Marble from Varanic Guri Vali*	*Travertine from Borsec*	*Travertine from Geoagiu*	*Andesite from Pietroasa Deva*
Density, kg/dm^3	2,66	2,71	2,74	2,74	2,73	2,72	2,72	2,69	2,70	2,75
Apparent density, kg/dm^3	2,55	2,70	2,72	2,72	2,70	2,70	2,70	2,25	2,41	2,67
Compactness, %	95,10	99,70	99,74	99,74	99,08	99,50	-	85	89,3	97,10
Total porosity, %	4,90	0,30	0,53	0,53	0,92	0,50	-	16	10,7	2,9
Water absorption at normal pressure and temperature, %	1,45	0,27	0,15	0,15	0,13	0,12	0,30	3,66	1,8	0,71
Water absorption by boiling, %	2,02	-	0,20	0,20	0,18	0,15	-	6,97	-	-
Fracture strength to compression in dry state, daN/cm^2	950	1150	1000	1150	1000	900	1200	325	510	1707
Fracture strength to compression in saturated state with water at normal pressure and temperature, daN/cm^2	850	990	950	950	1100	850	1100	295	380	-
Fracture strength to compression after 25 cycles of freezing - thawing, daN/cm^2	775	980	900	900	1000	800	900	220	260	1691
Softening factor after saturation with water at normal pressure and temperature, %	12	-	7	7	2,75	5	8,33	9	23,4	-
Softening factor after 25 cycles of freezin-thawing, %	18	11,9	13	13	9,40	11	25	42	48,2	-
Wearing resistance by frection (Bohme) to 440 revolutions g/cm^2	0,36	0,35	0,35	0,35	0,28	0,58	0,25	0,44	0,70	0,05
Strength under mechanical shock daN/cm^3	15	-	20	20	21	27	56	10	6,6	94

***- average (values)**

177A, SM-428) equipped with one, two or more concave milling cutters with cemented carbide type that cut simultaneous or successive in the rock mass (photo 2). This technology is in use from more than 30 years in the marble quarry Ruschita and limestone quarry Moneasa.

In some limestone and travertine quarries (Geoagiu, Moneasa) is still in use the conventional wiresaw but the productivity is very dependent by the physical-mechanical characteristics of rock, by the mineralogical composition of rock and by the quality of abrasive material that are used for making the cut.

In the new economic condition that appeared in Romania after the year 1989 it was necessary to improve the mining and processing technologies of ornamental rocks and replaced the older wear, moral and physic, equipment's with high productivity equipment's available on the market.

Hereby it was introduced the state-of-the-art technology based on chain cutter reinforced with synthetic diamonds and the diamond wiresaw. This technology applied in the marble, limestone and travertine ensured high productivity and perfect surface of the blocks.

For increasing the economical efficiency in the ornamental stone mining. some quarries have started to use in the last years the blasting technology with detonating cord.

After the technical-economical analysis the conclusion was that the mostly efficient quarrying technologies it's the combination between the above mentioned technologies: cutting chain and diamond wire, cutting chain and detonating cord, diamond wire and detonating cord in the specific conditions of each rock deposit.

At this time in Romania it is interest for developing the water jet cutting technology in the ornamental stone quarries. This technology has more advantages in case of quarrying hard rock deposits or rocks with high abrasivity where the tools used in the mechanical cutting technologies are a very short work life.

Due to the great development of the quarrying equipment's in the last years it was possible to open in 1996 first underground activity in the marble deposit Bucova (photo 3) based on state-of-the-art underground chain cutter made by Fantini. Company - Italy.

The people involved in processing activity are very skilled and are able to liven the stone making different objects from marble, limestone, andesit, etc. (photo 4).

REFERENCES

1. FODOR , D. - Exploatarea si valorificarea rocilor omamentale din Romania - Revista minelor - 4, 6, 7/1995.
2. FODOR D..FLOREA A. - Ornamental rock deposits from Romania. Genetic types, spreading and mining - 2nd Balcanic conference on ornamental rocks -Sofia 1994
3. FODOR D..FLOREA A. - Exploatarea zacamintelor de roci omamentale din Romania - Simpozion stiintific - Universitatea Tehnica Petrosani - 1995
4. FODOR D..FLOREA A. - Aspects of mining the ornamental deposits from Romania - 3rd International Scientific Opencast Mining Conference OMC'96 Belgrade24-28. Sept.1996
5. ****, Monografia rocilor omamentale din Romania - Min. Geologiei Bucuresti, 1985.

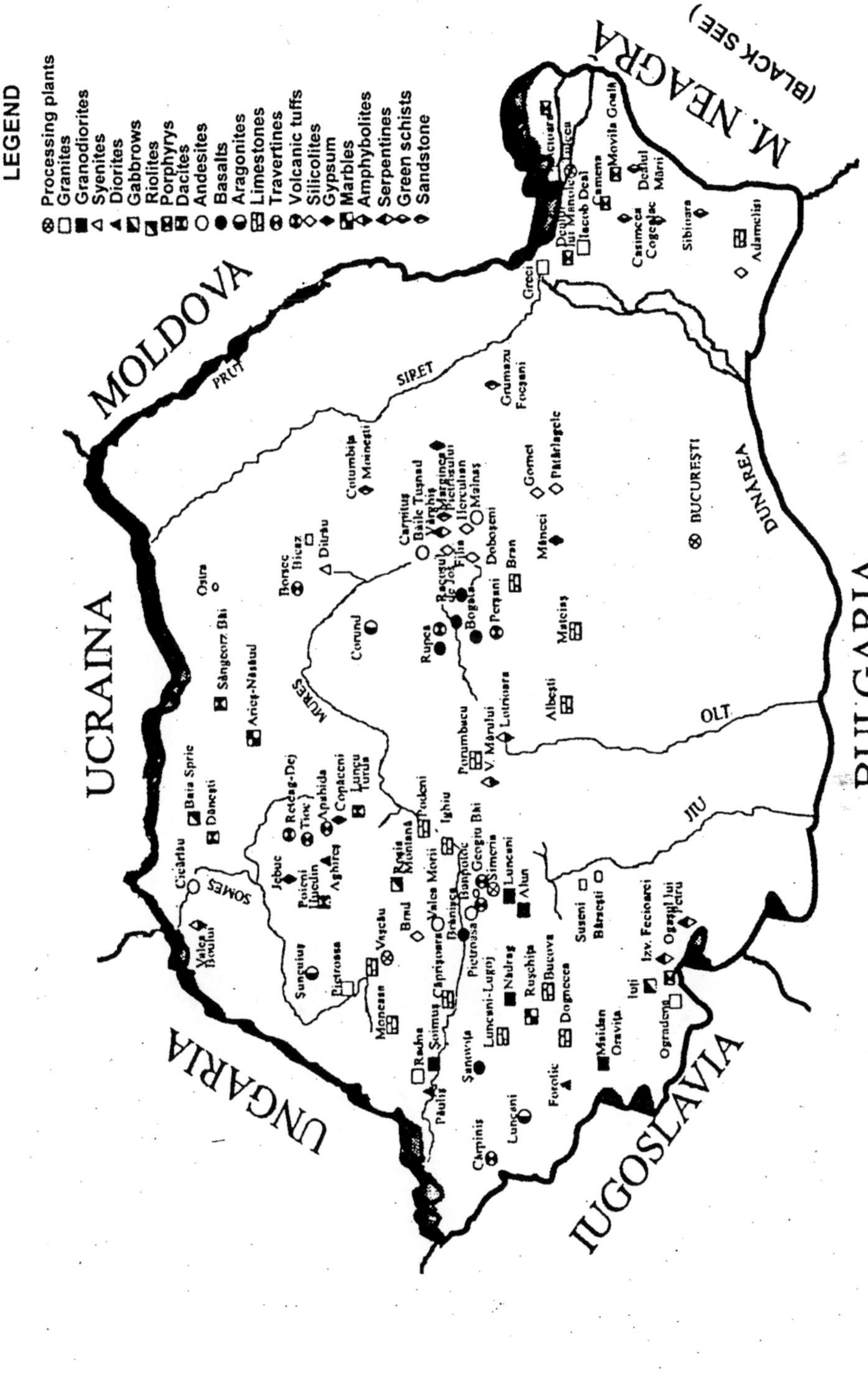

Fig. 1 : The distribution of the main ornamental rock deposits on the Romania territory

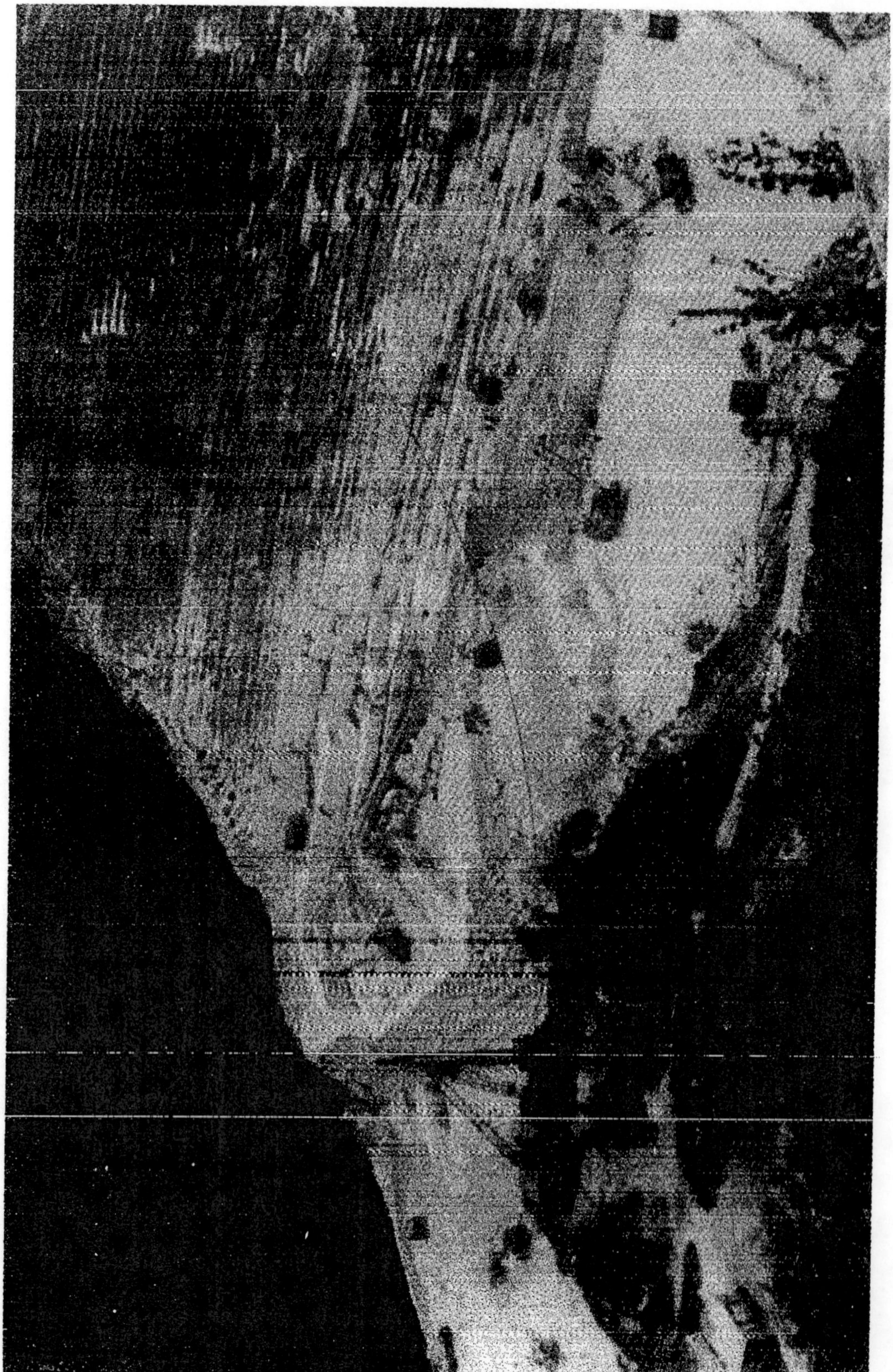

Photo 1

Photo 2

Photo 3

Photo 4

CHAPTER 18

Copper Mining Industry and Investment Potential in India

*Lokeshwar Prasad**

1. INTRODUCTION

1.1 The average price of Copper during 1995-1996 was 2844 US Dollars and in 1998-99, the average price was 1599 US Dollars, a fall of 1245 US Dollars over a period of three years. In spite of this, investment in the Copper Industry is growing at a fast pace. In the year 1998 the investment in World Copper Industry is projected at 16.35 billion US Dollars, an increase of 30% compared to the investment in 1997. This compares very favourably with the projected investment of 5.34 and 2.58 billion US Dollars for Aluminium and Lead-Zinc Industry and an increase in investment of 1.3 and 2.5% respectively compared to the investment in 1997.

1.2 In the Copper mining industry, most of the investment is in surface mining. Between the period 1996 - 2000 the world mining capacity is projected to increase from around 11.85 million tons to 14.08 million tons. 96% of this projected increase is being accounted for by surface mining. The cost of production of surface mines is said to be 1/3 of the cost of production for under ground mines.

1.3 The Solvent Extraction - Electro Winning (SX-EW) process of extraction of metal from ore is growing at a fast pace. Between 1996 - 2000, in SX - EW the world capacity is projected to grow from 1.6 million tons to around 3 million tonne of Copper registering a growth of 87%. During the same period, the capacity through Pyrometallurgical route is projected to grow a mere 5% from 12.9 million tons to 13.5 million tons per annum. SX - EW process offers a low cost technology for direct extraction of metal from ore. It is eco-friendly and Sulphur is recovered in elemental form through this process. The cost of production of Copper metal through SX-EW route is said to be only 60% of the cost of extraction through Pyrometallurgical process.

*Former Director General of Mines Safety, Ministry of Labour, Govt. of India.

1.4 All the new major mining project have been designed to produce upward of 25 million tons of ore per year. It has been estimated that a ten-fold increase in volume of production would reduce the mining and beneficiation cost by four fold.

1.5 The investment of the Copper industry is increasing in America and Latin American countries viz. Chile, Argentina, Peru and Brazil. Over 10 billion US Dollars investment is planned in these countries between the year 1996 - 2000. It is estimated that 76% of the total world Copper mining capacity increase will be in Chile and Argentina alone during this period.

1.6 This fast pace of expansion is said to be the direct outcome of the change in regulatory environment favouring investment. Chile offers no restriction on foreign investment and tax stability for 20 years on investment in excess of 50 million US Dollars. In Argentina no government approval is needed for mining projects, capital and dividends can be repatriated freely. Similarly Peru and Brazil also have removed restrictions on investment. Free repatriation of funds and tax stability for 10 to 20 years is also being offered. United States offers tax exemption of 5 to 22% on the minerals mined in the form of depletion allowance and royalty free exploitation of mineral deposits on Federal land.

2. PRODUCTION AND CONSUMPTION OF COPPER IN INDIA

India is witnessing a fast increase in consumption of Copper metal. The annual consumption has increased at an average of 8% during the last five years (Annexure -1). The total consumption of Copper for the year 1998-99 is estimated at 399 K tons. Against this consumption, the total Copper metal production from Indian mines was only around 37 K tons. Thus only 10% of the indigenous Copper consumption is met through production by the Indian mining industry. Further no new investment in mining industry is envisaged in the near future.

3. ORE RESERVES IN INDIA

3.1 The known ore reserves position in the country is placed at Annexure 2. India has known *In-Situ* Copper ore reserves of 588 million tonnes containing 8.1 million tonnes of metal. In addition, it has 674 million tonnes of conditional resources containing 1.9 million tonnes of Copper metal. Prospective resources stand at 770 million tonnes. The total potential of the known ore deposits is 2032 million tonnes.

3.2 The Copper ore reserves in proved and probable category stand at 42 million tonnes which is only 22% of the total potential of the known Copper ore reserves.

3.3 The States of Madhya Pradesh, Bihar and Rajasthan have concentration of these known Copper deposits. 93% of the total *In-Situ* reserves of India occur in these States with Madhya Pradesh, Bihar and Rajasthan accounting for 43%, 28% and 22% respectively of the total In-Situ reserves in the country (Annexure - 3).

4. STEPPING UP EXPLORATION

4.1 In India we have 203 known Copper deposits. These deposits are comparatively small in size and have not been fully explored. We need to intensify exploration for discovery of economically viable Copper deposits. The Government on its part, has initiated number of steps to attract foreign capital and 'State of Art' Technology in the field of Exploration and Mining.

4.2 The present regulatory environment is conducive for investment in the field of Exploration and Mining. The new policy envisages the following :

(a) 13 minerals reserved for Public Sector exploitation opened to Private Sector participation. List includes Iron, Manganese, Chrome, Sulphur, Gold, Diamond, Platinum group of metals, Copper, Lead, Zinc, Molybdenum, Tungsten and Nickel.

(b) Size of single prospecting licence increased to 5,000 sq. km. while the limit for multiple prospecting licence for a single company set at 10,000 sq. km.

(c) Aerial surveys and mapping allowed for foreign and private investors.

(d) Automatic approval for equity participation in mining ventures by foreign investors upto 50 per cent.

(e) Automatic approval for equity participation by foreign investors for mineral processing upto 74 per cent.

(f) Case by case approval for higher equity participation by foreign investors.

(g) Devolution of powers to State Governments for single window clearance of mining leases.

(h) Ad valorem rate of royalty for 17(non-coal) minerals.

4.3 With these changes in policy, the process of exploration for discovery of new Copper deposits has been stepped up and the Govt. has already approved prospecting licences in 23 cases covering an area of 30,000 sq. km.

4.4 Now, we need to expeditiously address problems of industry related to Trade Unions Act, 1926 and Industrial Disputes Act, 1947 in keeping with the changing business environment.

5. LIKELY GROWTH SCENARIO

One interesting feature of the Indian Copper deposits is that all deposits are comparatively small in size. This prohibits setting up of Smelters at mine site. In the present state the mines can only supply concentrates to the existing Smelters in India. But, if viable, the Copper ore can be subjected to cheap and eco-friendly Hydro-metallurgical route to produce Copper cathodes at the mine site itself. Simultaneous with this development, the search for world class deposits is expected to be speeded up.

6. CONCLUSION

Availability of Copper metal is directly linked to industrial growth of a Nation and therefore cannot be ignored. The Copper metal demand in India is projected to grow at an annual rate of minimum 8% during the next decade. With the present state of the mining industry, the Indian industry will remain dependent on import of Copper concentrate/ metal for its requirement. There is thus an urgent need for stepping up Copper production from existing viable mines in the country and go for large scale exploration to open new mines. To meet this objective, we need to make the regulatory environment conducive to investment. We firmly believe that Copper industry has a bright future in India.

ANNEXURE - 1 : COPPER CONSUMPTION IN INDIA

(UNIT –TONNES)

Year	*Consumption*	*% increase over last year*
1994 - 1995	287330	38
1995 – 1996	321080	12
1996 – 1997	347980	8
1997 - 1998	376660	8
1998 - 1999	399260	6
		AVERAGE OF LAST 5 YEARS- 8%

ANNEXURE –2 : COPPER ORE IN INDIA

(UNIT IN MT)

		Proved	*Probable*	*Possible*	*Total*
A	INSITU ORE RESERVES				
	ORE	231	212	145	588
	METAL	3.1	2.9	2.1	8.1
B	CONDITIONAL RESOURCES				
	ORE	72	88	514	674
	METAL	0.3	0.6	1.0	1.9
C	PROSPECTIVE RESOURCES	#	#	#	770

ANNEXURE –3 : DISTRIBUTION OF IN-SITU COPPER ORE RESERVES IN INDIA

(UNIT IN MT)

	Madhya Pradesh	*Bihar*	*Rajasthan*	*Other*	*Total*
ORE RESERVES	250	168	127	42	588
METAL	3.6	2.2	1.7	0.6	8.1
% DISTRIBUTION					
ORE	43	28	22	7	100
METAL	45	27	21	7	100

Section III

TECHNOLOGICAL CHALLENGES

CHAPTER 19

Automation in Mining - Where it is Proceeding in 21st Century

Vladimir Strakoš and Boleslaw Firganek***

SUMMARY

Predicting the development of mining control in the 21st century seems to be almost impossible task. Nevertheless there are some hints, which may be used to try to discuss some probable possibilities. Authors assume that mining will be performed by means of intelligent, fully automated machines-robots, equipped with sensors and control programmes, making possible the autonomous work with full recognition of the internal state and external environment. Introduction of such machines would change the technology of excavation and transport, eliminating a man from the technological process and real time control. Authors discuss research works to be done and results which are likely to be achieved in the fields of real time control, modeling and simulation and application of virtual reality methods.

INTRODUCTION

It is perhaps not possible to predict where shall the automation of the mining processes come at the end of the 21st century, especially when we consider of what progress was achieved in the technology in the ending one. Despite of this difficult or even impossible task, the authors, who spent their lives dealing with mining automation, decided to discuss some possibilities which seems to them the most probable.

One assumption is without a doubt - mining will be performed by fully automated devices! Whether we shall call them machines or robots it is rather philosofical question. It is also certain that a man will almost never enter the underground areas, devoting his mind, thinking, knowledge and experience to the control technologies, especially in the field of middle-term and long-term control.

---

*Technical University of Ostrava, Czech Republic, Vicechairman of ICAMC

**MICON Ltd., Katowice, Poland, Chairman of ICAMC

ANALYSIS OF THE PRODUCTION PROCESS

The process of mining is very complicated because it contains a whole complex of activities from the phase of geological survey, designing, development works and mining, to the phases of closing and removing the consequences of mining activities. The environment defenders are continuously in conflict with miners only making the situation more difficult, by wanting mining works to be removed to other countries or to other parts of the world. It is, however, necessary to keep in mind that what a society needs for its life and development must be extracted. When it is not here, then it should be somewhere else.

Since mining involves many complicated problems, we shall omit certain phases of a "mine life", limiting our discussion only to the period of its full development, when a mine is neither being opened nor closed and we shall neglect also maintenance works. Talking about mining processes we shall have in mind only the mining of raw materials.

Each mining process has a simple scheme: cutting, loading, transport, treatment and shipment. In deep mines the protecting of the working area is added. One can notice that making tunnels or energetic resorvoirs in rock are of the same character and belong to the mining activities.

It is also possible to look at technological processes in another way, which is needed for control and automation, dividing them on the product and technical supply with electrical energy, water, air, materials and other media. The subsystem of the technical provision of production also has almost unified structure and consists of source, network and consumers.

In order to consider automation, we must begin with thinking about the development of mechanization. An interesting problem is in cutting the rock formation itself. At present we are using picks for continuous mining and they require a proportionally great force. In our opinion there exists other possibility in the future for continuous cutting work. Through small dosages of explosives we can continuously cut the rock formation, which calls for a change in a principle of construction of mining machines. A change in mining pronciple will cause a change in principle of loading and transport which most probably will be performed by pipelines. A change in principle of mining complexes will also force a change in mining technology, especially in deep mining, which again will cause change in principle of development works. There are perhaps many other "changes of principle" but if we succeed in removing the picks from the mining devices such changes will take place - must take place.

Such a viewpoint on future mechanization and on mining methods is now a basis for thinking about control and monitoring of the production processes.

How will it be with the technical assistance subsystem? The main principle of energy and materials distribution from the source to the consumers will rather not be changed. But what may change is that the converters of energy may be located at the end of the main distribution frame of energy. What we have in mind now is almost unthinkable assumption, that one day the energy contained in coal matter will be easily relieved and transfered to perform mechanical or electrical work. It will be possible then to produce "energetic tins", which can be brought about by robots and used for long time of their activity.

AUTOMATIC ACQUISITION OF INFORMATION

Having done the assumptions about future development of mechanization, we may now try to think about the automatic processes themselves. They are composed of three basic processes, which are: measuring, processing of information and control. Necessary information for control of the movable as well as immovable machines can be consider as information about their internal and external states.

Information about internal states corresponds to the function of sensing units in the human body, which measure temperature, pressure, reciprocal positions, tension in materials (in human muscles) and oscilations. Is there anything at all to do to improve the sensors of the interior state of the machines? Even now they are mostly small units integrated with transducers, so a considerable change resulted from further development should rather not be expected. However the great progress may occur in transfer of information from the sensors to the central unit via computer network. It means that all sensors will be connected only to a single transfer route, which might be wireless and which will ensure the transmision of signals in form of information. It will not be a change in principle but will allow to built the "intelligent interior measuring system". It differs from the concept of an intelligent sensor, because one sensor in practice cannot be very intelligent, but a group of sensors can. Developing a perfect analysis of the information content of several sensors, which are reciprocally connected to the construction of a machine, it is an unresearched area at this time and it could last another decade, when we shall approach the capability to perceive the interior state of the machine similary to a live organism.

Information about external states should deal with the environment and reciprocal positions of mechanisms towards the surrounding objects, which means that the analysis of the scene should be performed. At present it is much less worked out than the analysis of the internal state. This also obviously deals with similarities in human senses.

Sight is the most significant sense of perceiving the surroundings and it is quite perfectly worked out for a human. A human has sensors of excellent properties, however the technology will be able to "see" in a much wider spectrum than the human eye. Even now there is a possibility of seeing countries from satellites. We also can suitably combine pictures acquired from the sensors measuring various wave lenghs. Human eyes are less perfect from that point of view, but the technical system of processing pictures cannot work from a distance as quickly as the human brain and technical systems mostly do not know how to process disassembled pictures as well as humans or animals. Further development will continue in this direction and it again will take a long time.

Hearing is another important sense for perceiving the surroundings and a human has much more perfect capability as opposed to technical systems. A human dismantles sound by means of two groups of sensing units (ears) and through an inconceivable method it separates the received frequencies into sounds belonging to different sources, such as for example human speech, music, background and the steps of other people. A technical system receives sound by several microphones, in which different sounds are added up and we still not know how they are to be divided. We do not know how to install for example three loudspeakers in such a way that each one could reproduce independently, for example the spoken word, music and the footsteps of another person. It is an area, which is also prepared for further demanding development work. The range of "hearing" of a technical system is again much wider than for a human. At present a human can perceive sound in acoustic band, which means up to 20kHz, while microphones and vibration sensors can receive sound and frequency of tens to hundreds MHz.

Smell and taste is equivalent to the chemical analysis of gas or liquid in technical systems. The equipment here is far less perfect, as opposed to living organisms, also in the field of measuring sensors. At present, the sensing units used for automatic chemical analysis are big, energy consuming, slow and unsensitive. Research and development will be long and expensive, but the results may be very important.

Touch is the last of the senses, with which a human perceives the surrounding world. There have been many attempts over a number of years to construct "artificial skin", through which a robot could perceive the shape of an object also in that field the technology has not progressed even the smallest little step and everything is still ahead of us. Some results my be achieved by means of proximity sensors which to a certain extent can replace part of touching (feeling) abilities.

We are expecting a long and demanding development with complex evaluations of all perceptions and in the end we will be able to built a machine which will quite safety orient itself in space, which is very important for machines in mining. It will take dozens of years, but we will get there eventually.

CONTROL

Control, which means automatic or remonte control, is perhaps presently the best worked out area of automatic control and so there perhaps it is not a big problem to estimate its future development. One should rather not expect any important changes, perhaps only in a small proportional valves controlled by serial digital code, or in the relatively small servo-mechanisms, which even in our days are already used for toys and models.

DATA PROCESSING

Processing of the information and working over informational content for control signals, such that it fulfills the goal of control, is a fundamental basis of control. Control is, therefore, basicly work with information. In this area computer equipment and microelectronics have made such a big step, that it is practically impossible to estimate its development in the future. It is however some limit of technology which seems to equalize the human capability. We hope that the technology will never achieve such a high level because it would mean the end of life on earth. Machine-robots would replace humans and then men would be useless. It seems as it is from a fantasy film, but we are convinced to be right. Nevertheless, the high technology, whether we want or not, as a goal imitates the highest capability of humans.

When we consider that in the area of sight, hearing and the mechanical strength the technical systems approach human capability, it is clear that such a mechanisms and machines of the future will significantly expand the possibilities of the technological processes. Let us look at the possibilities, which offers us development in area of processing information.

GPS - Global Positioning System has presently been very perfectly worked out and will be of great significance for automatic mining work. It is certain that to achieve and accuracy of position of 1 cm is already feasible and that is why we can easily imagine that in the short future cars and excavators in open cast mines will be mostly controlled automatically. In the first phase a human will set where each car should go, but later it would be enough to determine the quantity of mineral to be excavated in certain sections and next only the short-term mining plans.

In the future the system of sight will analyze the scene in bands of ultrared and infra-red radiation, radar waves and of course in the band of the visible spectrum. It will be not difficult to observe the scene with two sensing units, similary to two eyes, but evaluating these signals, finding a suitable and necessary information content and judging the significance of what is seen, for further control activities in real time, it is a question of a distant future. Nevertheless, development in this direction must take place. Presently, a big barrier are the relatively small speed of computers, the small memory capacity and generally insufficient knowledge on the whole area of pattern recognition.

Hearing is another area of obtaining knowledge about scenes. What the eye or camera senses is the appearance of a scene, but the sound is always generated by the movement of some element. It is different kind of information and it complements the sight observation. Neglecting the differences between sound and vibration, it is clear that from the point of view of the modern sensing technology it is possible to measure the vibrations in a very wide band, greater that a human is able to do, and certainly greater if the vibrations are transferred by air (ultrasounds). Similarly as for seeing, the development will be directed to the better and quicker processing of information from the vibrating sensing units. It is relatively easy, but much more complicated is the analysis of the informational content of signals and especially its use for machine control in real time.

Solving the problem of utilization of these two sources of information will have fundamental influence on the automation of mining activities and generally on mining. Let us imagine that we have a machine, which can see and hear and at the same time in the space in which it works there does not have to be light, there can be dust in the air, the air does not have to contain oxygen, harmful gases do not matter and the person who controls the machine is in a safe space and only gives basic orders for the machine work. Such a solution will be advantegous not only for open cast mining, for example at night, but especially for deep mining.

The chemical analysis of materials (smell, taste) is another source of information about the surroundings of a human or machine. These kind of analysis has not essential value for the autonomous work of mining machines. Of significantly greater importance for analyzers is however the treatment of raw materials.

For the treatment of minerals and for the transport control very important will be development of analyzers of solid materials. For transport control it means almost perfect selection of mined materials. For treatment it means reducing losses and changes in the construction of machinery as well as methods of control. In the case of coal mining it means also better shippment. The development of quick and reliable operational analyzers of solid matter is ahead of us. Perhaps there will not be any essential change in known principles, but perfect automation of taking and treatment of samples, as well as improvement in signal analysis, can significiantly influence the present state and efficiency of the final treatment of mines products before their sale.

In conclusion of this chapter we would like to repeat that the essential development in future will lead to obtaining a great amount of information from measured signals through efficient analysis and then by looking for interconnections among information obtained from various souces. For automation in mining it means the possibility of remote, semi-automatic or automatic control of moving mining machinary, which as a final consequence will lead to automatic mining without the presence of humans.

MODELING AND SIMULATION

Modeling and simulation are important areas of processing information which are developing very quickly at the present time. Here the development is the quickest and from its results it is possible to expect the most significant changes in automatic control. It is necessary to point out, that the problems of automation discussed above have an essential influence on the real time control of machines and mechanisms. Modelling and simulation deal with operational and middle-term control. It means that it is solving problems for technicians controlling mining processes or technical assistance subsystems.

From a detailed analysis of technological processes in mining, but also in other industry, follows that thc basis of each production process is a parametric or non-parametric transport network. Models are available for transport control in networks, but presently they are still not perfect and do not contain all properties necessary for simulation. Development to the extent of 10 to 20 years in this area will deal with improvement the models of the network and then the models of technological processes.

For automatic control in mining we must however solve one serious problem and it is the training of engineers, who will be able to use such models for control. This is not an easy thing. The model of the technological process it is not a simple formula into which we can substitute data and gain results. The model acts itself and will act even better as a real "live" system and we must understand such a system from the start. A model is a complicated instrument with many changeable variables and input/output data. We must understand each variable in order to understand interconnections and the principles of intervening into the network. There is a problem to use models and simulation for control and it is not possible to learn that in a few hours. The training of engineers using models for control will last dozens years, because presently at all universities in the world, which we have visited, very little time has been devoted to it.

VIRTUAL REALITY

Virtual reality is newer, interesting and perhaps even entertaining technology, which begins to expand quickly. Just now our mining processes are almost predetermined to use virtual reality in control activitiy. Using virtual reality for control of the technological process must be divided into two or more time levels. On the first level the virtual reality is used for control in real time, where we can control the movement of machines in the virtual world by creating a combination of the real world and virtual world on the basis of modelling and simulation. On the second level the control of mining deposits will be in in a virtual world created by geological survey and the modelling of geological areas for middle-term and long-term control.

It is certain that it is a new field of research works for the creation of the virtual world as well as for the principles of simulation of movement in the virtual world, for dealing with operational situations and for updating virtual reality on the basis of automaticaly gethered information. Another field it is the behavior of the virtual world of geological areas disturbed by virtual mining activities on the basis of information models from the range of geotechnics.

In open cast mines there is again the creation of virtual world according to information obtained during excavation, decision making and simulation in virtual world for transport the overlay rock, building the waste dumps, re-cultivation and other activities.

If we summarize the above mentioned problems, then virtual reality should be consider as a new area, which strongly influences the use of models for the control of mining processes and perhaps we are not far from the truth estimating that mining engineers will use these possibilites in control not earlier then after several tens of years and it is time during which research and development works must deal intensively with these problems.

CONTROL OF TECHNOLOGICAL PROCESSES

In the proceeding chapters we have gone through practically all problems concerning automation, modeling, simulation and virtual reality, nevertheless we would like to point out once more that the control of technological processes is being gradually implemented in each mining plant.

Control in real time deals with the automation of machines and mechanisms and development is finally leading to fully-automated mechanisms-robots.

Operation control (shift and day control) is among the second level of control, which means mining control itself, tratment and shipment control, of the technical provision of production and safety control, as far as people will be present in the production process. Automation in these areas of control are exclusively using models and basically means a control in the parametric and non-parametric networks. For example, for control of mining there is a plan how many and from which section should be excavated for each shift, to which places the bulk should be transported at which times, what transport means should be used for shipment when they are leaving the plant, and so on.

Middle-term control already uses different types of models because it means decision making about mining deposits. Here, similary like in long-term control, in the close future it will be unavoidable using the three dimensional drawings of deposits. In the further future there will be virtual reality and economical models of mining, which are rather behind the field of automation and control of technological processes.

SAFETY CONTROL

It seems that the most dangerous hazards now-a-days like gas and fire will not exist in the future deep mines, because of low concentration of oxygen in air. It means also substantional reduction, or even lack of ventilation, smaller cross-sections of the galleries and other consequences. The only real danger may be coused by roof falls and bumps, but due to the modelling and simualtion of strata behaviour they should be predicted and avoided.

CONCLUSION

In conclusion we would like to underline that the development of automation, as it was described, will have a significant influence on all mining. New machines shall be constructed, new mining methods worked out and mining without the presence of humans developed. Control models and virtual reality will be implemented and various situations and decisions will be tetested by simulation with good three-dimensional models. We will be moving in virtual world of mining workings several years earlier then real mining will begin. It will result in much better use of deposits and increase of recoverable resources. This again means reduction of mining damages of the surface and better environment pretection. By means of modeling of the landscape of the re-cultivating dumps it will be possible to create the environment suitable for human life or to buit underground locations for commercial or prodaction use. Full automation in mining will establish an ideal possibility to create underground transport routes with a minimal influence on the surface, which could remove part of the transported materials from the surface, reducing the load on highways. Shortly speaking, everything which will be possible and where a human will not have to be present will be placed bellow the surface. Here are the consequences of the full automation of mining activity at the end of the next century.

CHAPTER 20

Technological Challenges for Mineral Beneficiation in the 21st Century

*T. C. Rao**

ABSTRACT

Mineral processing is one of ancient professions of the mankind. This paper briefly describes the challenges facing the mineral processing industry in the next century. Some of the emerging advanced mineral processing technologies in the area of enhanced gravity concentration and column flotation are discussed. The problems of technology assimilation in mineral processing industry are examined and a strategy for smooth transition of new technology to industrial application is suggested through close interaction among technology marketers, industries and R & D institutions.

INTRODUCTION

Radical changes in the thinking and functioning of mineral industry in the next century are inevitable for survival. High processing costs due to depleting high-grade resources coupled with the falling metal prices have pushed the mineral processing industry to economically vulnerable stage. Developments in synthetic composite materials have resulted in replacing metals with light, high strength materials leading to energy conservation. Stringent product specifications and quality control are essential to face the competition from composite materials. The lowgrade ores need to be ground to finer sizes to achieve reasonable liberation. However, economically viable fines/ultrafines processing techniques are yet to be established and the associated problems in dewatering/filtration of fines still remain unsolved. The final blow comes from the environmental concerns related to the mining and tailing disposal.

However, these challenges also bring new opportunities and wisdom to the one of history's oldest profession. The mineral industry needs to synergise the technical, economical and environmental issues to bring out the best performance. Long-term planning and vision would be crucial for a mineral industry to survive in the next century. Introduction of

*Director, Regional Research Laboratory, Bhopal – 462026. India **Email :** tcrao@bom6.vsnl.net.in

efficient advanced processing technologies is essential which means that well calculated risks have to be taken, as this would require significant capital investment.

This paper identifies some of the promising beneficiation technologies, which will play crucial role in processing the low grade resources. Couple of case studies wherein the advanced technologies provided solutions to some long-term problems, are also briefly discussed. Further, the strategies required to creatively use the advanced technologies to maximise the economical gains of the mineral processing industry are outlined.

EMERGING TECHNOLOGIES

Unlike the other engineering fields, the development of new technologies has been slow in the field of mineral processing. The need to develop not only technologically superior but also cost-effective technology is not a easy task. In a classical paper, Napier-Munn (1997) describes the problems of the inventors and emphasises the need for long-term vision, commitment and obsession to develop a new technology. Typically, a mineral processing invention takes 10-15 years to find its first application in the industry and subsequent penetration of such technology into the existing circuits is also very slow. Despite these problems, there have been significant developments in new processing technologies as given in the following sections.

Enhanced Gravity Concentration

Gravity concentration is the oldest beneficiation technique adopted by the mankind. Gravity concentration is simple, cheap and efficient to treat coarse high-grade ores. However, the low-grade ores need to be reduced to fine sizes for better liberation and traditional gravity concentration techniques are not capable of treating fines. After the invention of froth flotation in the beginning of 20th century, gravity concentration was pushed to the back seat. But the gravity concentration made a comeback by the end of 20th century due to:

- Availability of new advanced techniques capable of treating fines and ultra-fines.
- Rising costs of reagents for flotation operation.
- Environmental concerns related to the flotation process.

The advanced gravity concentration techniques essentially use centrifugal force to provide enhanced gravity to efficiently treat fines. Some of the successful Enhanced Gravity Concentration technologies that found application in the mineral processing industry are Falcon Concentrator (Clifford, 1999), Kelsey Centrifugal Jig (Beniuk et al, 1994, Riley et al 1995, Honaker, et al 1996, Clifford, 1999), Knelson Concentrator (Holland-Batt, 1998) and Mozley Multi-Gravity Separator (Hall, 1991, Mozley, 1993, Patil, et al 1999).

Falcon Concentrator is a rotating bowl with the smooth inclined wall. The feed slurry enters the bottom of the bowl and gets stratified along the bowl wall. The heavier material settles on the wall while the lighter particles are pushed to the lip of the bowl and discharged as the overflow. The heavier fraction settled on the bowl wall is continuously removed through a nozzle located at the top of the bowl. A schematic diagram of the Falcon Concentrator is shown in Fig. 20.1.

Kelsey centrifugal jig utilises the principles of jigging in a centrifugal field. Figure 20.2 shows the schematic diagram of Kelsey Jig. The feed slurry is introduced to jig through

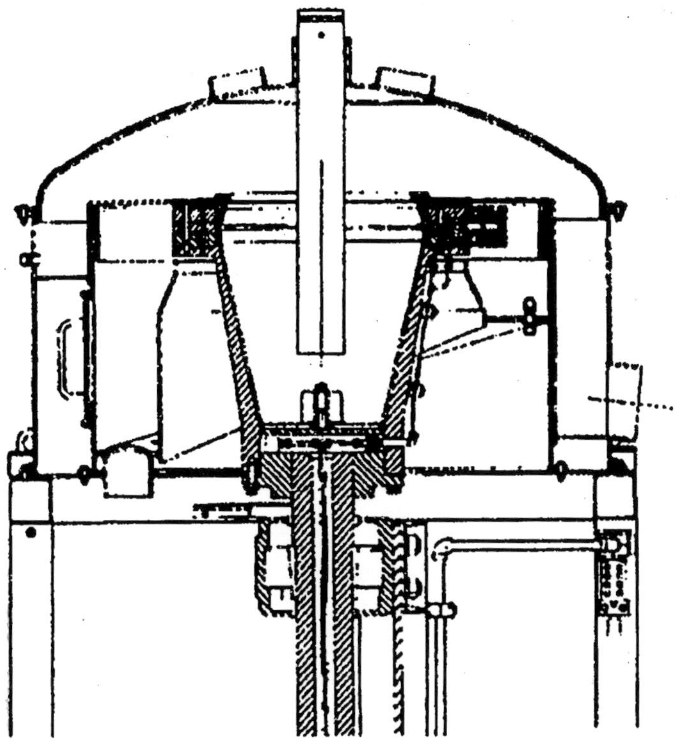

Fig. 20.1 : Falcon Concentrator (coutsey Falcon Concentrators)

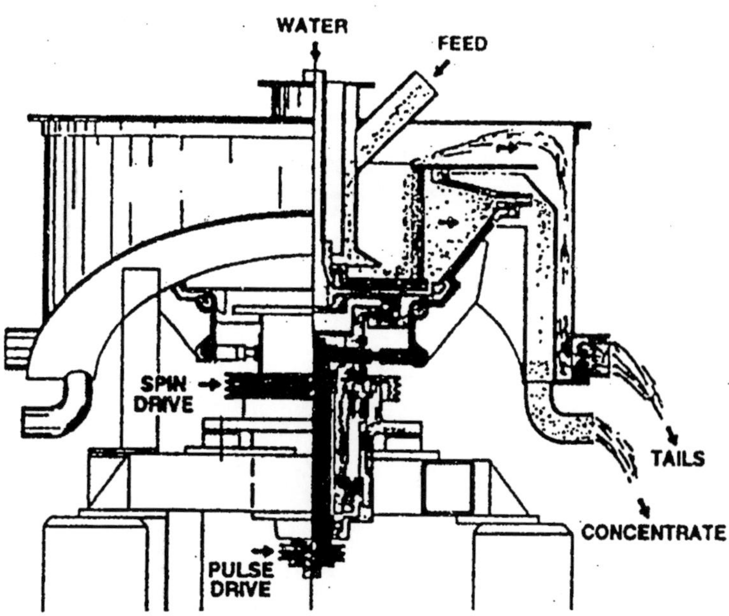

Fig. 20.2. Kelsey Centrifugal Jig (courtesy Geo-Logics)

central feed pipe and spreads uniformly over the ragging bed due to spinning action of bowl. Inward pulse action coupled with centrifugal forces causes the particles to separate according to their specific gravity. The particles affected by centrifugal forces pass through the ragging material and the screen against the current of water, whereas, the particles unaffected by g-force are displaced by tailing stream water and incoming feed.

A schematic diagram of the Knelson Concentrator is shown in Fig. 20.3. It has gone through several modifications, since its introduction as a semi-batch unit in 1982 to a continuous mode of operation (Knelson and Jones, 1994). The equipment essentially has a 'V' shaped cone with a series of parallel riffles and many small holes inside the cone. The cone rotates at a speed of 1500 RPM, producing a centrifugal force of 60 g. Water is fluidised through the small holes and the lighter particles are carried away by the water to tailings stream.

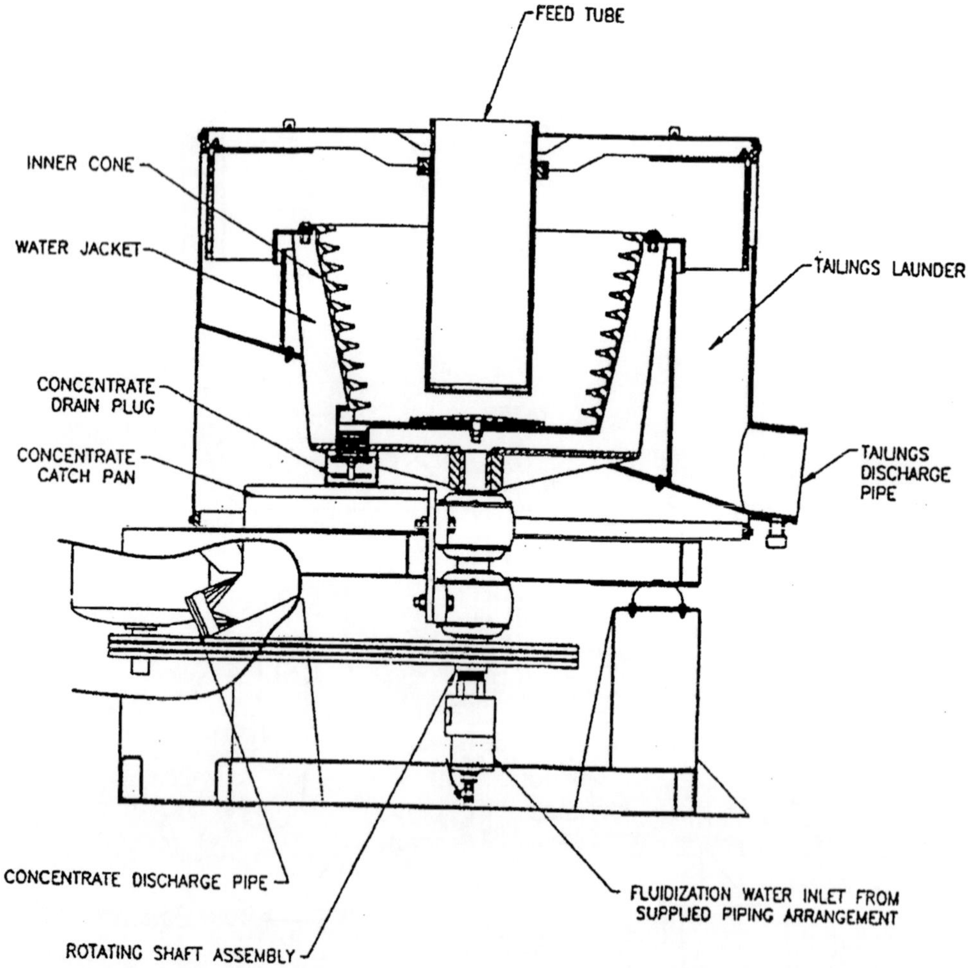

Fig. 20.3. Knelson Concentrator (Courtesy Knelson Concentrators)

The Multi Gravity Separator consists of a slightly tapered open-ended drum as shown in Fig. 20.4. This drum rotates in the clockwise direction and is shaken simultaneously in an axial direction. Inside the drum there is a scraper assembly, which rotates, in the same direction but at a slightly higher speed. Due to the centrifugal force imparted, the heavier particles move against the inclination and wash water flow while the lighter particles are carried away to the other end of the drum and discharged.

Fig. 20.4. Mozley Multi-Gravity Separator (courtesy Richard Mozley Ltd)

Each of the above mentioned Enhanced Gravity Concentration (EGC) technology uses the principles of different traditional gravity technique and has its own advantages and disadvantages. Table 20.1 summarises these aspects and will provide a hint for selecting appropriate EGC technology for a given application.

Froth Flotation

Froth flotation remains as the most efficient and economical beneficiation method for treating complex sulphide ores. Column flotation is the most significant advancement made in this area. Notwithstanding the initial euphoria, the column flotation technology has also struggled to find acceptance by the mineral processing industry. Some operating problems such as plugging of spargers coupled with the need for sophisticated control system have hindered the growth of column flotation. Many improvisations over the past ten years have resulted in different type of column flotation cells, which are finding the place in the industry. Some of the leading and promising column flotation technologies are briefly discussed in the following sections.

Table 20.1 : Summary of existing Enhanced Gravity Concentration technologies

Technology	*Principle*	*Centrifugal Force (max.)*	*Advantages*	*Disadvantages*	*Application (Ore type)*
Falcon Conentrator	Sluice	300 g	High capacity, High "g" forces No wash water	Not efficient in treating slimes	Lead-Zinc Copper, Gold Iron Tin, Coal
Kelsey Centrifugal Jig	Jigging	100 g	Better metallurgical performance due to jigging High capacity	Screen blinding Heavies have to pass through the screen Recovery of ragging material	Chromite Tantalum Alluival-Gold Tin, Iron
Knelson Concentrator	Fluidised Bed	60 g	Use of wash water minimises the misplacement	Higher water requirement	Alluvial-Gold
Multi Gravity Separator	Shaking Table	10g	Ability to treat very fine particles Better metallurgical performance	Low capacity	Lead-Zinc Copper Gold Iron

The Jameson cell is one of the finest refinements made in the column flotation technology (Kennedy, 1990, Clayton, et al 1991). Unlike the traditional column flotation units, the Jameson cell uses shorter column as shown in Fig. 20.5. The feed slurry and air is fed through a down-comer to the bottom of the cell providing co-current flow of air bubbles and particles. In the down-comer the air bubbles tries to rise while the slurry flow pushes them down resulting in better bubble-particle interaction. The feed slurry and air bubbles are discharged at the bottom of the cell and bubbles rise to the lip of the cell. Reasonable froth height is maintained and wash water is used to minimise entrainment of slimes. The Jameson cell provides the following advantages

- Less space requirement.
- Cost effective i.e., low operating costs.
- Treats wide size range feeds efficiently.
- Higher capacity among the column units.

Microcel™ is another variation in the column flotation technology developed by the Virginia Centre for Coal and Mineral Processing, USA (Yoon et al, 1992). The novel feature of this unit is the in-line mixer microbubble generator shown in Fig. 20.6. Apart from using the traditional column features, the Microcel™ uses a portion of the refuse slurry to generate bubbles. At the bottom of the column, a portion of the tailing stream is recirculated using a pump through an in-line mixer, which acts as an aerator. This aeration system produces fine bubbles to enhance the bubble-particle collision and minimises the entrainment of gangue slimes.

The packed column uses stack of corrugated plates as packing to provide small flow passage for intimate bubble-particle interaction. Figure 20.7 shows the schematic diagram

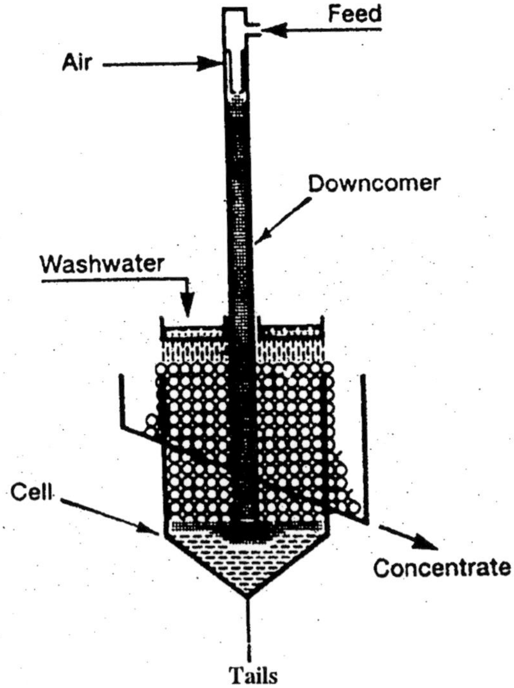

Fig. 20.5 : The Jameson Cell (after Clayton et al, 1991)

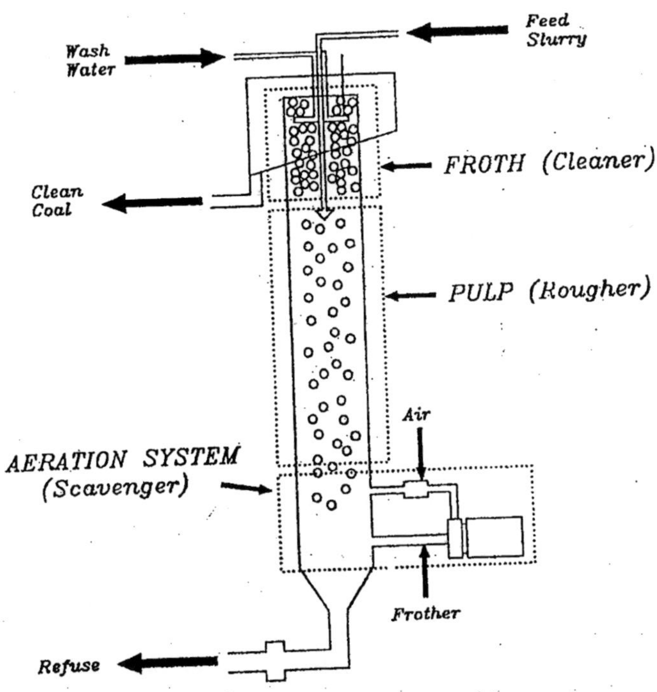

Fig. 20.6. Microcel' along with microbubble generator (after Yoon et al, 1991)

of the packed column unit (Yang, 1990). The corrugated plates are kept at right angle to each other and provide the capillary effects to achieve unlimited froth height. The packing allows the bubbles to break and the particle get attached to another bubble many times before reaching the top of the column resulting in very high separation efficiency. However, need for higher aeration rate and frother dosages along with lower capacities compared to other columns are the drawback of this technology.

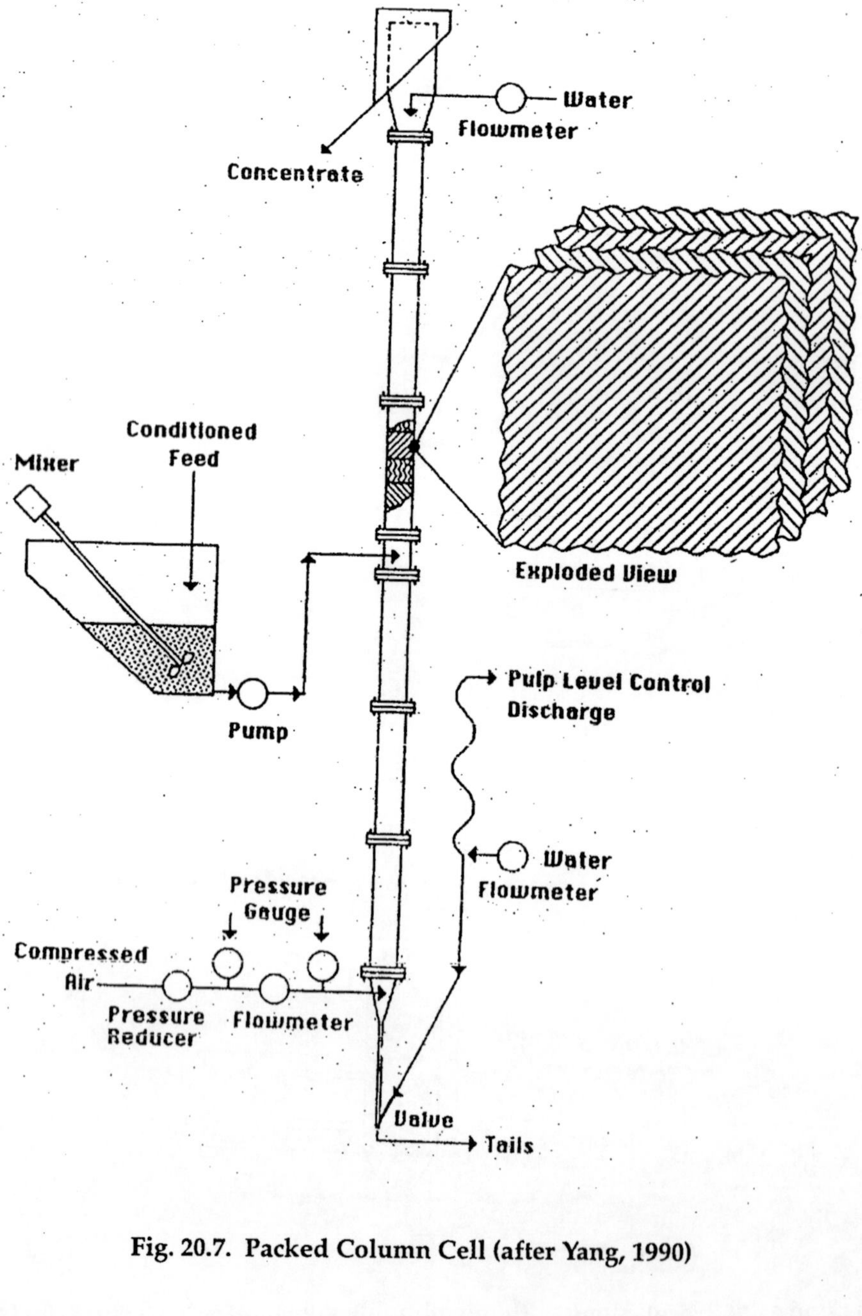

Fig. 20.7. Packed Column Cell (after Yang, 1990)

CASE STUDIES USING THE ADVANCED TECHNOLOGIES

As mentioned earlier, each of the advanced processing technologies has its own merits and demerits. It is advisable to thoroughly investigate these technologies in laboratory for a given application before incorporating in the industry. Regional Research Laboratory, Bhopal has been actively engaged in evaluating the performance of the modern mineral processing technologies and suggesting the suitable techniques to solve long-term problems of the Indian mineral industry. The results obtained from two typical case studies using Enhanced Gravity Concentration at RRL, Bhopal is briefly outlined in the following.

Removal of Graphite from Lead Concentrate

The lead-zinc ore deposits at Rajpura-Dariba and Rampura-Agucha mines occur in two different forms, i.e., Calc-Silicate and Graphite-Mica schist. The ores with Calc- Silicate minerals as impurities are easy to process and the ores with graphite Mica schist are difficult to process. The Graphite-Mica schist in addition to silica contains graphite as a major impurity in the ore. Due to natural hydrophobicity, the graphite reports into the lead concentrate. Graphite above a particular level i.e., 2.5% reduces the lead sinter strength, which is essential during smelting (Patil, et al. 1999).

Several efforts using different flotation methodologies like pregraphite flotation, reverse flotation of graphite etc., though could give good results with high grade ores, they failed with low grade ores containing higher amounts of graphite. Trials using different graphite depressants like nigrocene, depremin-c, quebrecho were found not successful with low grade ores (Gupta et al., 1989).

In an effort to solve this problem, investigation exploiting the wide difference in specific gravities between graphite and lead bearing minerals was taken up along with HZL. Detailed laboratory experiments were carried out on Mozley Multi Gravity Separator using lead rougher concentrate as feed. The results obtained have shown that the graphite content could be reduced to 1.3% from 9.8% with 71% lead recovery. Based on these encouraging results, a double drum MGS unit was installed at Rajpura-Dariba mines, which also produced desired quality of lead concentrate. More recently, another MGS unit was installed at Rampura-Agucha mines to tackle the graphite problem. RRL, Bhopal has also provided support to HZL in optimising the performance of plant scale MGS units.

Removal of Alumina from Iron Ore Slimes

About 18 million tons of iron ore slimes are being discarded as waste by Indian iron ore industry. The Indian iron ore slimes are characterised by high alumina to silica ratio, which causes problem in the blast furnace. It has been reported that one percent increase in alumina content in the feed results in 4% reduction in productivity and 2.25% increase in fuel consumption (Pradip, 1995). Attempts to use the froth flotation and flocculation techniques did not produce desired results as the surface properties of alumina and iron mineral are similar.

RRL, Bhopal has been investigating the possibilities of using the Enhanced Gravity Concentration techniques to treat the iron ore slimes. Samples were treated by different techniques and Table 20.2 presents the summary of the results obtained. It may be noted from these results that all three techniques tried, i.e., Multi Gravity Separator, Knelson Concentrator and Kelsey Centrifugal Jig, were successful in reducing the alumina content. However, the economics of these processes have to be evaluated for selecting a suitable technology for industrial application.

Table 20.2. Performance of Enhanced Gravity Concentrators treating iron ore slimes

Unit	*Feed*		*Concentrate assay*		*%Recovery*
	%Fe	*%Al_2O_3*	*%Fe*	*%Al_2O_3*	
Knelson Concentrator	62.8	3.40	66.0	1.74	49.0
Multi-Gravity Separator	54.9	7.2	63.9	1.86	58.0
Kelsey Centrifugal Jig	63.2	4.2	68.2	1.07	63.7

TECHNOLOGY ASSIMILATION

Absorption of new technologies into the mineral processing plants takes very long time. The technology developers consider that the industry is conservative and reluctant to modify the traditional circuits. On the other hand, the industry blames the technology marketers of promising too much up to the stage of sales and delivering little in the plant floor before eventually washing off the hands with the ultimate excuse that "the feed has changed". Each side must be having many examples to support their statement but the need of the hour is different attitude and approach for technology assimilation. Working together as a team with mutual trust, for mutual growth, through excellence in performance is the only way for survival in the next century.

Mineral processing technologies are much different from other engineering technologies in the sense it need to be fine-tuned for each application and each ore body. Hence selecting mineral processing technology is similar to human resource development, i.e., select a technology with basic capabilities and tune the same to do the required job. Due to these complexities, the technology absorption needs to go through some logical steps to achieve success and jumping off these steps mostly result in disaster.

Figure 20.8 shows a strategy indicating the steps involved in technology development and assimilation envisaging a buffer role for R&D and academic institutes between technology marketers and industry. The methodology shown in Fig. 20.8 is self-explanatory. The technology marketer, industry and R & D institute have to work together to achieve the goal. At different phases of this progress the role of each partner will change and depending on the nature of work one will lead while the others provide support. For example, during technology development, the equipment manufacturer will play the lead role while the industry will control the plant trails phase. Figure 20.8 also shows the anticipated role of each team at different phase of the technology development and assimilation.

CONCLUSIONS

Only some of the technological developments in Enhanced Gravity Concentration and froth flotation are described in this paper and similar developments are happening in other areas of physical separation also. These modern technologies have to be accessed quickly and the suitable ones should be deployed for industrial application without much of gestation period. Concepts of leasing, financing and partnership between equipment manufacturer and mineral processing industry have also to be explored for mutual benefit and growth. Apart from the use of the advanced technologies, the following aspects have also to be given importance in the next century

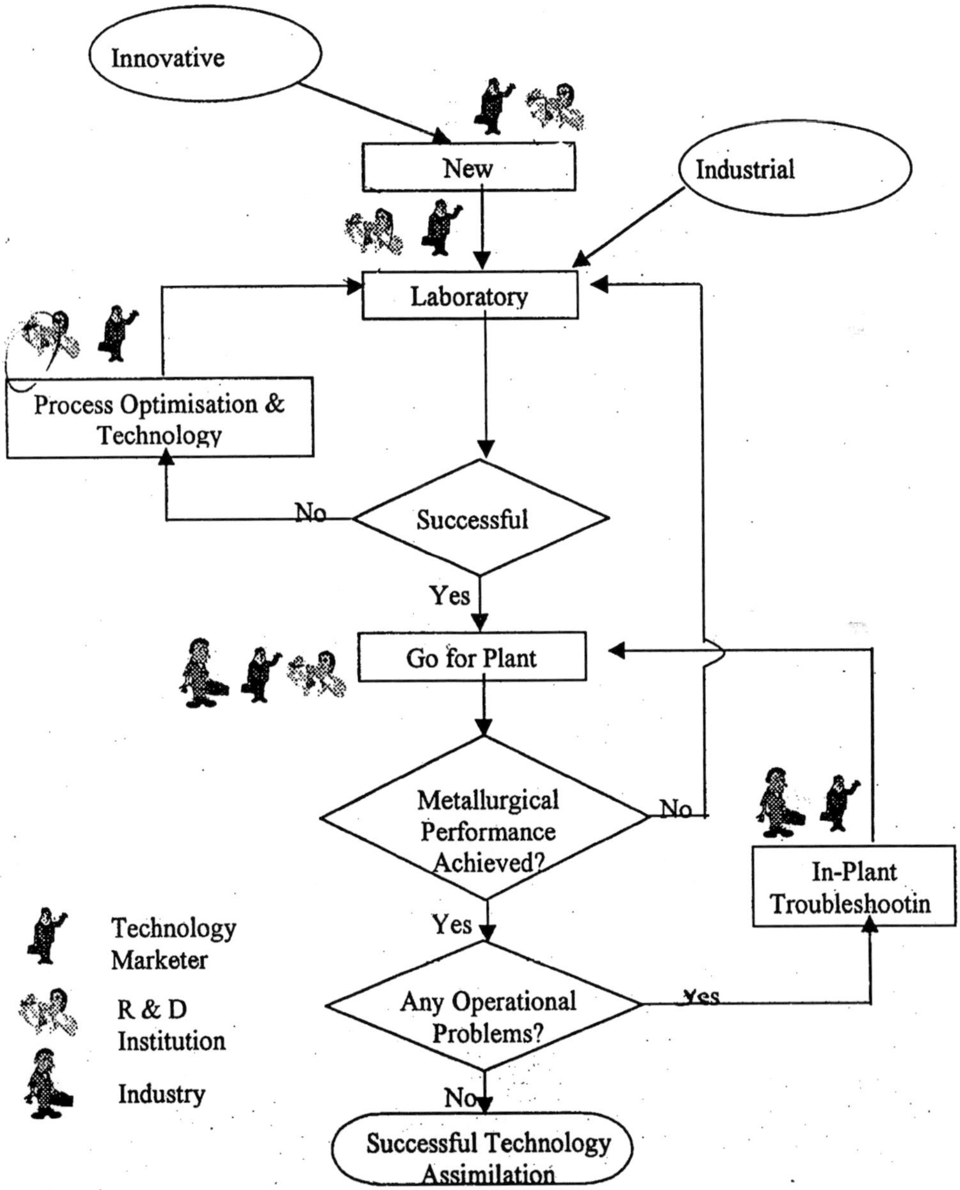

Fig. 20.8 : A strategy for technology assimilation in mineral processing

- Innovative circuit configuration.
- Modular design of processing plants.
- Waste utilization and by-product recovery.
- "Zero Waste" concept.
- Environment friendly technologies.

There are no "ready-made" solutions for the problems of mineral processing industry. With continuously changing feed characteristics and processing complexities, case-specific solutions have to be found in the mineral processing plant floor quickly. Close interaction between industry, equipment designers, R & D and academic institutions will be crucial for the success of the mineral processing industry in the 21st century.

REFERENCES

1. Beniuk, V.G., Vadeikis, C. A. and Enraght - Moony, J.N. 1994. Centrifugal Jigging of Gravity Concentrate and Tailing at Renison Limited. Minerals Engineering, Vol. 4, pp. 577-589.
2. Clayton, R., Jameson, G.J. and Manlapig, E.V. 1991. The Development and Application of the Jameson Cell. Minerals Engineering, Vol. 4, Nos. 7-11, pp. 925-933.
3. Clifford, D. 1999. Flotation Advances. Mining Magazine, November, pp. 235-244.
4. Gupta, B.K., Ravindranath, K. and Chulet, M.R. 1989. Selective Flotation of Galena from the Complex Multi-Metal ore at Rajpura-Dariba. Silver Jubilee Workshop on Flotation, Bhubaneswar, India, Vol. 1, p. II. 4.11.
5. Hall, T.S., 1991. Mineral and Coal Processing, Mining Annual Review, pp. 245-264.
6. Honaker, R.Q., Wang, D. and Ho, K., 1996. Application of the Falcon concentrator for fine Coal Cleaning" Minerals Engineering, Vol.9, No.11. pp. 1143-1156.
7. Holland-Batt, A.B. 1998. Gravity Concentration, Mining Engineering, September, pp. 43-48.
8. Kennedy, A. 1990. The Jameson Flotation Cell, Mining Magazine, October, pp. 281-283.
9. Knelson, B. and Jones, R. 1994. A New Generation of Knelson Concentrators: A Totally Secure System goes on line, Minerals Engineering, 7, (2/3), p. 201.
10. Mozely, R. 1993. Resistance to the introduction and Testing of New Technology in Mineral Processing. Proc. XVIII IMPC, Sydney 15-20, AusIMM.
11. Napier-Munn, T.J. 1997. Invention and Innovation in Mineral Processing. Minerals Engineering, Vol.10, No.8, pp. 757-773.
12. Patil, D.P., Govindarajan, B, Rao, T.C., Kohad, V.P. and Gaur R.K., 1999. Plant trials with Multi-Gravity Separator for the Reduction of Graphite; Communicated to Minerals Engineering.
13. Pradip, 1995. Beneficiation of Alumina–rich Indian Iron Ore Slimes. Metals Materials and Processes, Vol.6, No.3, pp. 179-194
14. Riley, D.M., Firth, B.A. and Lockhart, N.C. 1995. Enhanced Gravity Separation. Proceedings, High Efficiency Coal Preparation; an International Symposium, Kawatra, S.K. (ed.), SME., Littleton, Colorado, pp. 78-87
15. Yoon R.H. Luttrell, G.H. Adel, G.T. and Mankosa, M.J. 1992. The Application of Microcel ™ Column Flotation to Fine Coal Cleaning, Coal Preparation, Vol. 10, pp. 177-188.
16. Yang, D.C. 1990. Packed-Bed Column Flotation of Fine Coal. Part I Laboratory Tests and Flotation Circuit Design. Coal Preparation, Vol. 8 pp. 19-36.

CHAPTER 21

Blasting Technology— For the 21st Century

*S.P. Singh**

ABSTRACT

The mining industry is heading towards a technology driven optimization process. It has been realized that the unit operations such as drilling, blasting, excavation, loading, hauling and crushing are interrelated variables in the total cost equation. The development, advancement and utilization of the innovative technologies are very important for the mining industry to be cost effective.

The drilling and blasting industry is rapidly adopting technology in all forms, in order to improve performance. In the modern mines it is very common to encounter the latest forms of laser measurement technologies, global positioning systems (GPS), communication technology and computer systems. The developments in the areas of planning and design of blasts, drill monitoring, drillhole deviation, drill machine navigation systems and laser profiling systems have been discussed in this paper. The innovative practices in the areas of bulk loading of explosives, controlled blasting, explosive performance measurement and the evaluation of blast outcome and productivity have been outlined. The role to be played by the electronic detonators in the next millennium to improve blasting efficiency and mining economics has been described.

INTRODUCTION

During the past decade, the mining industry has begun to experience a technological revolution. International market pressures and competitiveness ensure that this revolution will continue and accelerate into the 21st century. The development and advancement of innovative technology is essential for the mining industry to be cost effective so that information can flow back into the process for continuous improvement.

*Professor, School of Engineering, Laurentian University, Sudbury, Ontario, Canada, P3E 2C6

The last decade has also seen a dramatic progress in the advancement of blasting technology and the quality and performance of products. Monitoring instruments, measurement technologies and computing tools now have the capabilities to provide a bank of useful information that has previously been the subject of broad assumption. The performance and reliability of explosives and initiation systems are now at a level that allows the distribution and sequencing of explosive energy to be carefully controlled. The objective of this paper is to present state of the art blasting technology, innovative blasting practices and anticipated evolution in blasting trends.

The basic steps in blast engineering are to design, implement and observe the outcome of a blast. Recently developed technologies can enhance and streamline the process for the optimization of blast fragmentation. The major developments in the blasting technology can be grouped according to the blast optimization pyramid (Fig. 21.1). Three main stages of this pyramid are planning, execution and output of a blast.

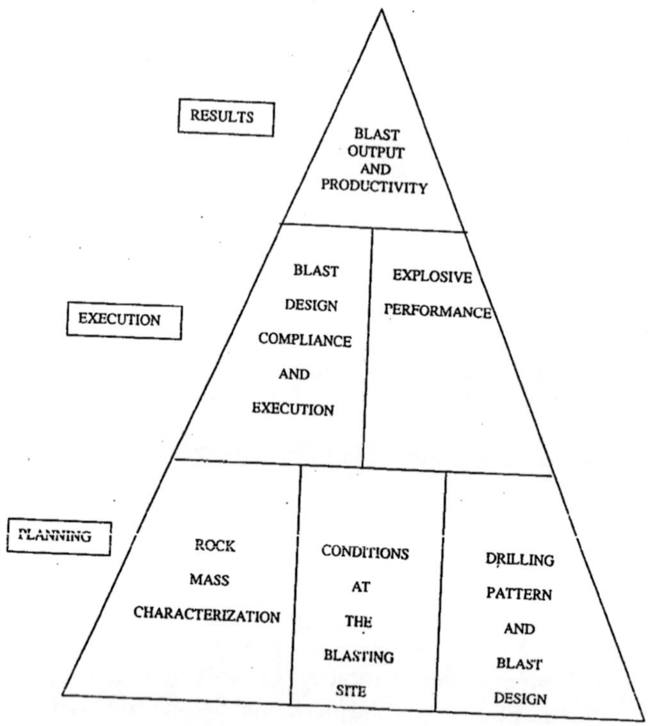

Fig. 21.1. : Blast optimization pyramid.

PLANNING AND DESIGN OF BLASTS

The planning and design of blasts is the foundation of the blast optimization pyramid. Rock mass characteristics and the field controls are the two cornerstones of this foundation.

ROCK MASS CHARACTERISTICS

Rock mass characteristics play a major role in all aspects of blast performance. Explosive energy level and explosive distribution must be matched to geological conditions in order for a blast to be successful. Initiation and timing sequence must be compatible with the rock mass response to explosive loading. The blasting engineer cannot make these adjustments without adequate information and understanding of the rock mass characteristics.

The blasting of the rock involves the interaction between the detonation products of an explosive and the confining rock mass. It is important for us to identify the rock mass properties which influence blasting performance and to modify blast designs to suit different rock mass conditions. Several empirical approaches have been developed to represent the rock mass characteristics in the form of a "Blastibility Index" (Ashby, 1981, Lilly 1986, Scott, 1992, Hino, 1959, Heinen and Dimock, 1976). In order to achieve optimum blasting results, it is necessary to incorporate the most relevant rock mass properties into the blast design process.

DRILLING PATTERN AND BLAST DESIGN

In general, inadequate attention is given to designing blasts to suit the given rock mass conditions. This results in either poor or overblasting. The blasting engineer is, therefore, faced with the challenge of identifying the rock mass properties governing the blasting performance in each situation and to incorporate them in the blast design process. Computer aided blast design techniques, numerical blasting models and expert systems considerably reduce both the cost and time involved in blast design while increasing the quality of prediction. These days computer programs are available with the capability to develop and edit blast designs graphically and optimize those designs using the analysis tools. But the success of these blast designs is strongly dependent upon the accuracy of the input data and the systematic implementation of the blast design.

FIELD MONITORING AND CONTROL

The field monitoring and controls, if properly used, can be valuable tools in the optimization of blasts.

Drillhole Deviation

Drillhole deviation is a common but undesirable occurrence in the mining, tunnelling, construction and exploration industries (Singh, 1996a). The deviation may be defined as the difference between the actual and planned location of the collar, the path and the final destination of the drillhole. The detailed measures to control drillhole deviation have been given in Singh et al. (1998). But in the future, the mining industry needs to utilize modern drilling machines equipped with a hydraulic shockwave reflex dampening system which enables the drill machine to maintain ideal feed force, drilling straight holes, achieving high drilling rates and drill steel economy. Computerized drilling control in modern machines provides the ideal combination of feed force, rotation torque and penetration rates. The drillhole deviation alters the planned burden, spacing and depth of holes thus resulting in sub-optimum results. The typical deviation in a mine has been shown in Fig. 21.2. There is a strong need for developing a reliable technology which can monitor and rectify deviation during drilling.

Drill Machine Navigation Systems

GPS guidance systems have the ability to accurately locate a drilling machine on the planned location and provides benefits in terms of improvement in drill utilization, rock fragmentation and overall mine productivity.

Fig. 21.2. : Drillhole deviation in a mine

Normally, the drill pattern file containing the designed hole numbers, northing and easting coordinates for each hole and the defined horizon to which all holes should be drilled is sent to the drill via a radio network from the planning office. GPS is used to determine the exact collar elevation at each drill location. The drill monitoring system precisely calculates the depth of the drill hole and it significantly reduces over and under-drilling (Hendricks, 1999). The data on actual drill hole positions and length can be uploaded from the drill to mine planning office via the radio network. The knowledge of actual locations and spacings between holes facilitates comparison between actual and planed blast design. GPS-based positioning of the drill provides considerable benefits including improved accuracy and consistency of the blast.

Drill Monitoring

Drill monitors are available to the mining industry since the 1970's. The early monitors were pen strip chart recorders that were developed for the oil drilling industry and adapted to mining (Vynne, 1999). They provide drilling parameters such as depth, pull down pressure, torque, RPM and bailing air pressure. These recorders produced a large volume of paper and minimal amounts of useful data. The data was not summarized and required an engineer to interpret the chart traces. Being a mechanical recorder, the devices were also susceptible to frequent failures.

The type of instrumentation currently available for monitoring of drills has been radically changed by the modern electronics and the development of personal computers and its associated software. The logged data can be analysed, summarized by the personal computer and generated in the form of customized reports to meet the specific requirements of the mine.

In the past, mine management mainly focussed on the Loading and transporting equipment. The situation is changing and there is an increased awareness of the importance of the production drills and their role in the entire mining process. The following benefits of monitoring drill performance are being recognized (Vynne, 1999):

(a) Higher rate of penetration; (b) Reduced overdrilling; (c) Location of weak and strong strata; (d) Optimum blasting; (e) Adequate fragmentation and diggability; (f) Higher bit life; (g) Lower maintenance cost; and (h) Higher accuracy of data and reports.

The drill monitoring systems offer the ability to collect and transmit to the mine office the information such as drill production statistics, ore and waste intersections, exact collar locations, hole positions and depths, maintenance information and consumable tracking. These monitoring systems also determine a Blastibility Index (BI) for each hole. The BI is based upon the intact rock strength as well as structural features such as fracturing, jointing and bedding to provide a composite expression of the overall rock mass competency. The BI values are purely related to the rock mechanical characteristics and are completely immune to drill operator influence. The BI information is very useful for computer based blast design and blast modelling (Hendricks, 1999).

Strata logger systems have also been installed on the new generation drills, providing pull down force, rate of penetration, rotary torque and vibration data (Thompson, 1999).

Laser Profiling

Tight engineering and field control of the front row of holes is of vital importance (Smith and Wiegard, 1999). If the front row burdens are excessive, explosive force will cause motion where there is the least resistance. Improperly designed front row can result in some or all of the following adverse effects:

(a) Inadequate fragmentation; (b) Poor energy distribution; (c) Flyrock (d) Back break; and (e) Coal damage.

Laser profiling of the face is not a new procedure but is quickly catching on in the mining industry. There are several laser systems on the market that allow the operator to measure the front row burden. Without tight control, it is difficult to analyse and optimize the blast design. It has been observed that front row design has a significant effect on the performance of a blast. Fig. 21.2 shows the comparison between old and current technology.

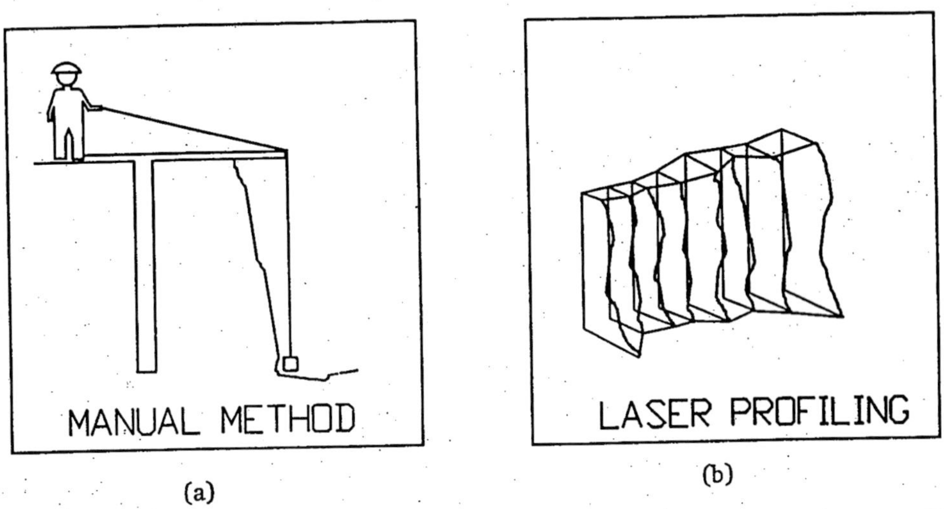

Fig. 21.3 : Assessment of blasthole burden by (a) manual method and (b) laser profiling.

EXECUTION OF BLASTS

The last decade has shown dramatic progress in the advancement of blasting technology and the improvement in blasting performance.

BULK LOADING OF EXPLOSIVES

The equipment utilized to load blastholes has improved in response to more precise blasting products and programs (Watts, 1999). Through new computer-aided design and modular design capabilities comes the ability to manage a bulk loading system which creates records of what was loaded, make data transfers to and from the truck on formulas and densities and to carry out remote control loading.

Previously, there was a serious difficulty in loading bulk explosives in up-holes. But now with the advent of new technology the bulk explosives can be effectively loaded. Fig. 21.4 shows the loading of up holes in a sublevel caving operation with traced ANFO.

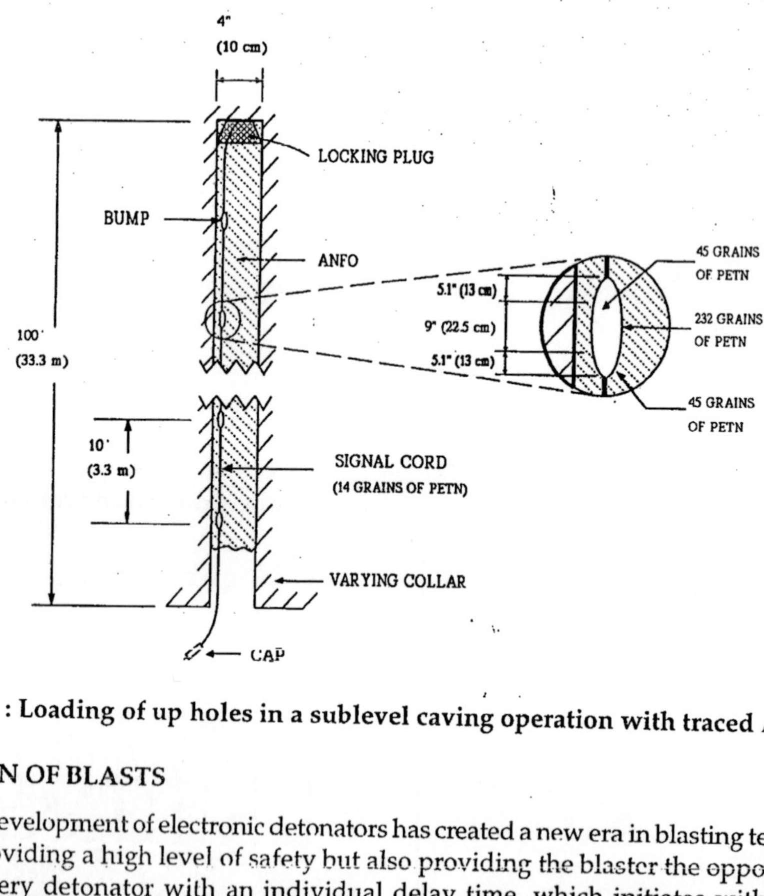

Fig. 21.4. : Loading of up holes in a sublevel caving operation with traced ANFO.

INITIATION OF BLASTS

The development of electronic detonators has created a new era in blasting technology, not only providing a high level of safety but also providing the blaster the opportunity to program every detonator with an individual delay time, which initiates with absolute precision. Currently, several types of electronic systems are being tested and used in the industry, all of which utilize some type of stored energy device to provide energy for their timing and firing circuits. Most of the other detonator technologies use pyrotechnic energy as a means of delay and initiation. The basic components of the shock tube, electric and electronic detonators have been shown in Fig. 21.5.

It is a well known fact that effective control over blast timing will enhance blasting results. Electronic Initiation systems provide the following features which are likely to prove beneficial to the mining industry (Deacon et al, 1997).

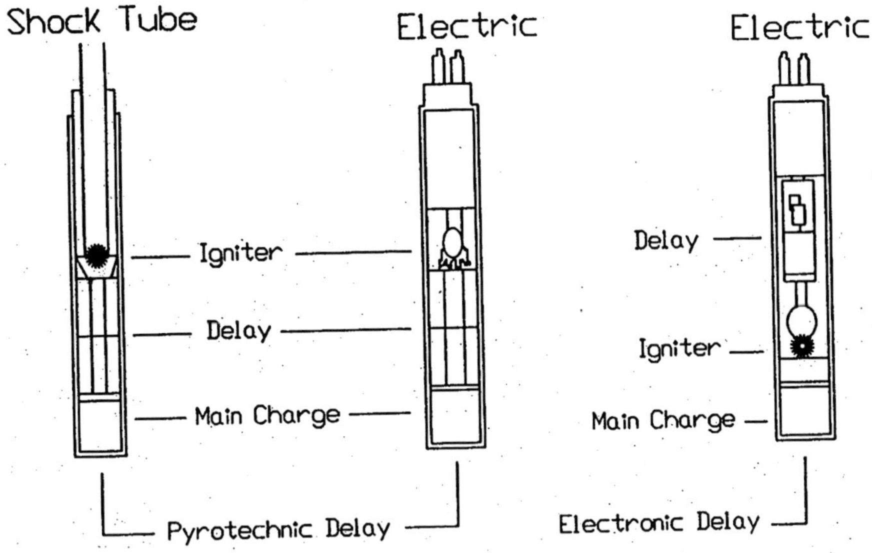

Fig. 21.5. : Comparison between the components of the electronic and conventional detonators.

(a) Increased accuracy in timing; (b) Flexibility in programmed delays; (c) Every detonator is independent with its individual countdown; (d) Testability of the detonators (e) Increased safety

EXPLOSIVE PRODUCT PERFORMANCE

Measurement of explosive performance by velocity of detonation (VOD) monitoring is a time consuming and expensive process. This is a prime example of a diagnostic measurement technology that is not practical for use in every hole of every blast, although it can be a very useful tool for an experienced blasting engineer evaluating product and

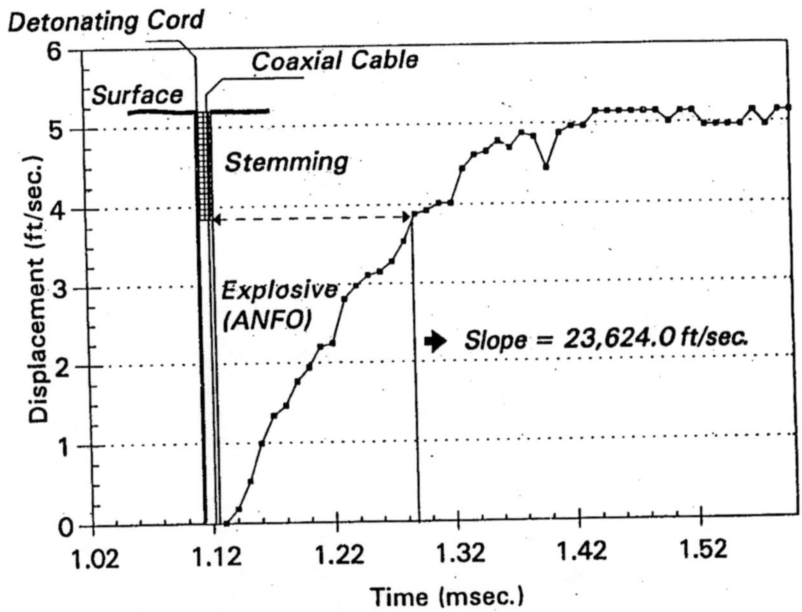

Fig. 21.6. : Continuous VOD monitoring for a blasthole.

blast design changes (Rodgers, 1999). The commonly available instrumentation can be broadly classified into (a) Instant VOD monitoring, which monitors VOD between two points; (b) Continuous VOD monitoring, which monitors VOD for the entire length of the blasthole (Fig. 21.6).

CONTROLLED BLASTING

Blast damage is an inevitable but undesirable product of rock excavation by blasting. The blast damage to the immediate periphery of an excavation may result in dilution and ground control problems (Singh, 1997). Several techniques are used to control blast damage, but the miners feel much at ease with ANFO due to its safety and simplicity of loading. Its lower cost makes it economically attractive to the management. When used alone ANFO causes excessive overbreak. But the tracing of ANFO which is known as tracer blasting, reduces the initial and maintenance cost of supports, provides geometrically precise openings and produces less disturbance to the integrity of the surrounding rock. Tracer blasting has the potential of being very cost effective and safer technique, if properly applied. This technique involves tracing ANFO with a detonating cord prior to loading the main column of explosive (Singh, 1996). The detonator cord is attached to the detonator which is ideally kept in the centre of the blasthole with the help a centralizer (Fig. 21.7).

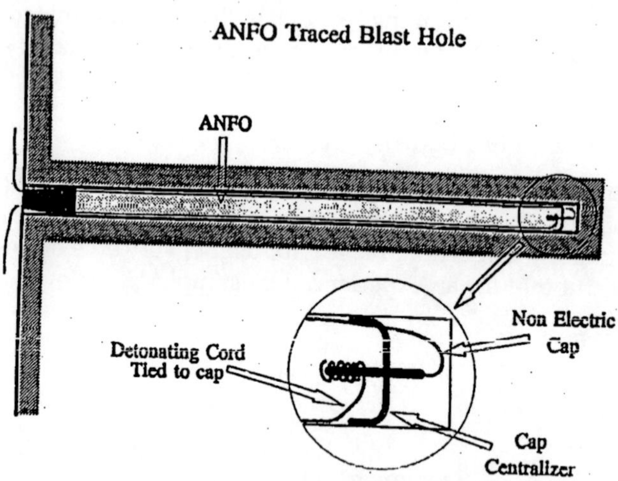

Fig. 21.7. : A horizontal blasthole loaded with traced ANFO.

PHOTOGRAPHY

The systems available for photographing a blast can be classified as follows:

(a) High speed digital computer images;

(b) High speed 16mm camera;

(c) Auto-advanced 35mm camera;

(d) Normal VHS camera; and

(e) High speed VHS camera.

These photography instruments can be utilized to examine face velocity and trajectory, burden displacement, fly rock and stemming ejection. The photography records can be used for assessment of blasts as well as for comparisons between different blasts.

BLAST OUTPUT AND PRODUCTIVITY

GROUND VIBRATIONS AND DIAGNOSTIC MONITORING

It is easier for the blasting engineer to predict and control blast vibrations using advanced digital computer-based and menu driven seismographs. The new generation of seismographs is becoming smaller in size and more portable for use. The new analysis software has the capability to synchronize seismic, acoustic and video signals to better analyse blast vibration data and perform waveform modelling. The seismographs are used not only for blast vibration but also for diagnostic monitoring to determine the cap scatter, misfires, sympathetic detonation. Fig. 21.8 shows the waveform and timing for the first four holes in a cut during drift blasting.

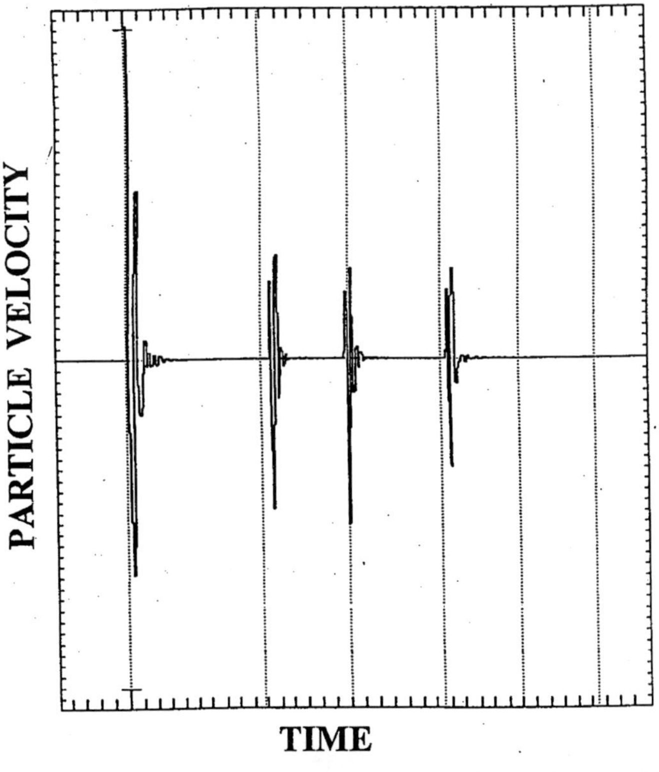

Fig. 21.8. : Waveform and timing for the first four cut holes during drift blasting.

ROCK FRAGMENTATION ASSESSMENT

Historically, blasters have attempted to evaluate the results of their designs by assessing the resulting fragmentation. These measurements have varied from estimates of size difference by observation, time studies of equipment productivity, characterization of muck piles by "objective" observers or manual measurements of fragment sizes in photographs of muck piles. In a very few instances, the muck piles have been sampled and portions of muck piles sieved. Each of these methods (except sieving) gives very subjective and error prone results in even the best controlled conditions. The sieving method is of course very accurate, but is very time consuming, disruptive to ongoing operations and is

The photoanalysis is one of the more recent methods, where photographic images of blasted rock are analysed through computer image processing techniques. Now the blasted rock can be characterized by completely automated and fully integrated, PC based fragmentation analysis system using state-of-the-art photoanalysis techniques. These systems come with high-tech micro-processors, a dedicated image capture board, fast video accelerators and specially designed software to measure the rock fragments and to display and/or print the data in an easy to use form. The results can be displayed as a histogram or cumulative weight percent passing curve (Fig. 21.9). The main advantages of these systems are:

(a) No interference with the mining operations; (b) Reproducible results; (c) Minimizes entry into hazardous locations; and (d) Fragmentation parameters can be included in a mathematical model.

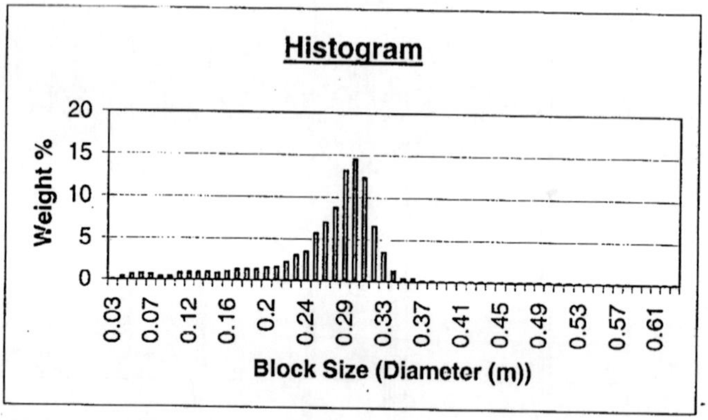

(a)

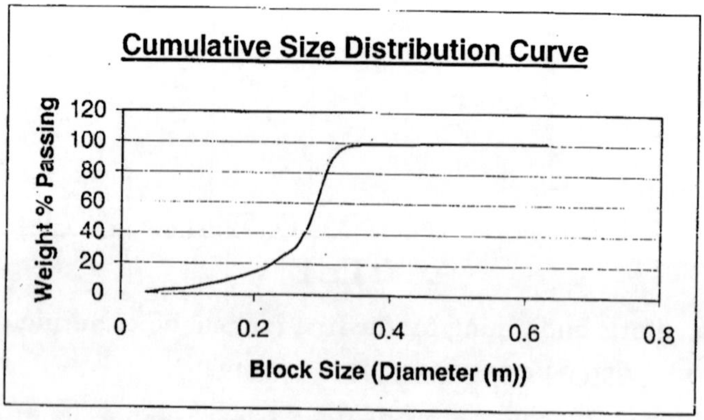

(b)

Fig. 21.9 : Fragmentation assessment results in the form of (a) histogram and (b) cumulative weight passing curve.

ASSESSMENT OF PRODUCTIVITY

This can be done by monitoring the performance of loading, transporting and crushing equipment. A wide spectrum of choices to monitor loaders, shovels and haul trucks are available. Dispatch systems also provide some information regarding excavating and hauling time efficiencies (Rodgers, 1999). Shovel monitoring systems which provide information about the shovelling efficiency in a blasted rock require the operator to indicate the beginning and end of each load.

Shovel performance can be used as an indicator of the muck pile condition or blast efficiency and how the blast effects the post-blasting operations. The performance of a shovel depends upon the rock fragmentation and condition of the muck pile assuming the mine maintains consistent truck/shovel matching and fleet sizing. Normally, resultant fragmentation size distribution is primarily used to rate blast effectiveness. Muckpile looseness would appear to be equally if not more important than fragmentation in assessing blasting efficiency.

The efficiency of the hauling equipment and the degree of crushing needed can also be used as a measure of the effectiveness of the blasting process. The efficiency of a blast should always be assessed in the light of its effect on the downstream operations.

OPTIMUM LEVEL OF TECHNOLOGY

The introduction of new technology, on one hand, reduces the mining cost whereas on the other hand it brings in some new cost depending upon the level of new technology used. The total cost of a mining operation is the cost of technology plus the mining cost. The economically justifiable level of technology varies from mine to mine. The optimum level of technology for a mine is that, which gives the minimum overall cost as shown in Fig. 21.10.

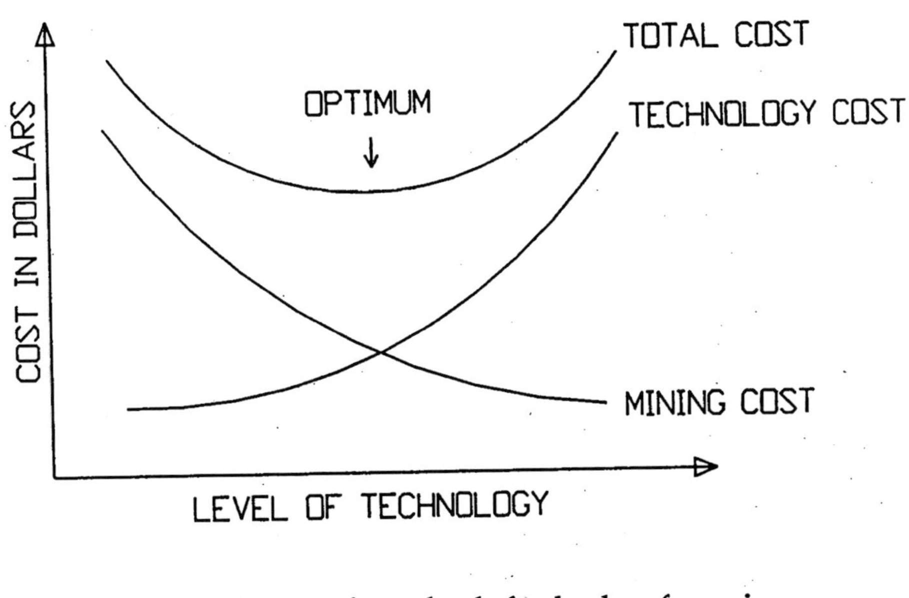

Fig. 12.10. : Optimum level of technology for a mine.

CONCLUSION

The mining and explosive industry is poised on the threshold of some exciting opportunities. The measurement technologies if judiously used, can be valuable tools in the blast optimization process. This is a primary challenge to the vendors of measurement technologies that the measurement and monitoring devices are reliable, user friendly and provide concise usable information.

Drill monitoring systems have the potential to be the most useful technology if this information could be automatically integrated into the mine planning software for use in blast design and explosive selection processes. The advent of electronic detonators has facilitated the controlled sequence of detonation which is one of the critical parameters having a direct effect on overall blast performance in many ways. Explosive energy level and explosive distribution must be matched to geologic conditions in order to achieve optimum blasting results. There is a need for the technology to adequately assess the outcome and productivity of a blast. With the current technology there are no easy answers where comprehensive measurement of productivity is concerned.

REFERENCES

Ashby, John, P. (1981), "Production blasting and the development of open pit slopes", Proc. 3rd International conference on stability in surface mining, Ed. C.O. Brawner, June 1981, Vancouver, pp. 549-572.

Decon, C., Duniam, P. and Jones, M. (1997), "Improved blast control through the use of programmable delay detonators", Proc. 13th annual symposium on explosives and blasting research, Las Vegas, USA, p 55-68.

Heinen, R.H. and Dimock, R.R. (1976), "The use of seismic measurements to determine the blastibility of rocks", Proc. 2nd Conf. On Explosives and Blasting Techniques, Louisville, Kentucky, pp. 234-248.

Hendricks, C. (1999), "GPS positioning and equipment monitoring technology for blasting operations", Proc. MINNBLAST 99, Duluth, Minnesota, USA, pp. 225-238.

Hino, K. (1959), Theory and practice of blasting. Nippon Kayaku Co. Ltd, Japan.

Lilly, P. (1986), "An empirical method of assessing rock mass blastibility", Proc. Large open pit mining conf., Newman (The AusIMM) pp. 89-92.

Rodgers, J.A. (1999), "Measurement technology in mining", Proc. MINNBLAST 99, Duluth, Minnesota, June 1999, pp. 17-23.

Scott, A. (1992), "A technical and operational approach to the optimization of blasting operations", Proc. MASSMIN 92, South African Institute of Mining and Metallurgy, Johannesburg.

Singh, S.P. (1997), "Cost effective blast damage control with tracer blasting", Mineral Resources Engineering Journal, Vol. 6, No. 2, pp. 49-61.

Singh, S.P., Ladoucer, M. and Rouhi, F. (1998), "Sources of drilling errors and their control", Mineral Resources Engineering Journal, Vol. 7, No. 2, pp. 79-92.

Singh, S.P. (1996), "Mechanism of tracer blasting", Geotechnical and Geological Engineering Journal, Vol. 14, No. 1, pp. 41-50.

Singh, S.P. (1996a), "The role of rock-bit interaction in blasthole deviation", Proc. 2nd North American Rock Mechanics Symposium, Montreal, Canada, pp. 583-590.

Smith, J.D. and Wiegard, J.E. (1999), "Laser profiling benefits computer based blast design", Proc. MINNBLAST 99, Duluth, Minnesota, USA, pp. 211-222.

Thompson, D. (1999), "Modern drilling and blasting technology at the Minntac Mine", Proc. MINNBLAST 99, Duluth, Minnesota, USA, pp. 79-96.

Vynne, J. (1999), "Drill monitoring developments and the utilization of the drill data", Proc. MINNBLAST 99, Duluth, Minnesota, USA, pp. 187-196.

Watts, R.T. (1999), "Choosing the correct bulk loading equipment can help optimize your blasting program", Proc. MINNBLAST 99, Duluth, Minnesota, USA, pp. 47- 63.

CHAPTER 22

Technological Implication of Coal Quality Improvement for Power Stations by Beneficiation and Blending

Kalyan Sen, A. Chowdhuri**, S.G. Choudhury**, R. Das Gupta**, S. Biswas*, and N.S. Das**

ABSTRACT

Despite many fold technological benefits that may be accrued from the use consistent, clean fuel and the MOEF's mandate in favour of beneficiation of power coals, not many of the washing plants proposed so far are coming up till date. One of the problems is the reluctance of the consumers to enter into a commercially viable agreement with the coal producers in paying the additional cost of beneficiation. The other, as visualised from the producers' angle, is making the decision on the degree or level of coal cleaning that would impart the necessary qualities to suit the requirements of TPS. It is apprehended that the limit to which a coal needs preparation (i.e. the top size of crushing, specific gravity of cut, moisture level etc.) has to be justified from the specific qualities that are demanded for the boilers.

Beneficiation of high moisture-coking coals by wet processes, particularly treating finer size fraction, means increase in moisture content in the product against mutually agreed limit for the power coals. So far, the washeries are being designed to follow the practice of mixing beneficiated coarse coal with is untreated lower fractions, thereby, maintaining the overall ash of the power fuel below 34%. Limiting only the overall ash content in prepared coal has little commitment on the improvement in combustion properties, because of the possible presence of considerable amount of free mineral matter in the finer fractions.

*Director, Central Fuel Research Institute, P.O. F.R.I., Dhanbad 828 108, India

*Scientists, Central Fuel Research Institute, P.O. F.R.I., Dhanbad 828 108, India

From the beneficiation point of view, it appears that one of the main criteria of preparing non-coking coal is impose a restriction on the quantity of high ash particles in the washed coal. The degree of washing has to be set to such an extent that characteristic ash content at the specific gravity of cut lies below the range of 50-55%. This will also optimise the concentration of total reactive components to a level sufficient to produce good combustion characteristics.

The paper also discussed the implication of blending the imported washed coal with the high ash Indian raw coal from combustion point of view.

INTRODUCTION

Coal continues to be the mainstay of the development of Indian economy through its 2/3rd contribution to the demand of commercial energy. The total consumption of coal during 1996-'97 was 297.6 Mt, out of which Power utilities and captive areas consume about 214.5 Mt of coal. According to the assessment made by the Planning Commission (July 1997), the coal requirement by the end of 9th and 10th five-year plans may be 412 and ad 563 Mt respectively. Only power coal consumption during 1997-'98 is ore than 250 Mt which, as the trend shows, is expected to go up to about 300 and 450 Mt in 2001-'02 and 2006-'07 respectively. Though India has sufficient reserve of non-coking coal to meet this requirement, the inferior quality of such coal with ash content beyond 40-50% poses problem to the power industry.

Because of the unique geological origin, Indian coals show diversified characteristics, some favourable and mostly adverse, towards the power plant utilization. Unlike carboniferous coals of Northern hemisphere, which are enriched in vitrinite, Indian coals (Gondwana drift origin of Permian age) have more inertinite content, even upto 60-70%. The quality is further deteriorating due to the gradual winning of inferior coals from lower horizons and also, due to increased large-scale mechanization of OCP, with net detrimental effects in power plant utilities. One of the detriment is identified as the gradual decline in the GCV of ROM coal (Table 22.1) [1]:

Table : 22.1

	1960-'61	*'70-'71*	*'80-'81*	*'92-'93*	*'95-'96*
Avg. GCV (*Kcal/kg*)	5900	5260	4200	4100	3500

As most of the power plants are designed to consume coal within a range of GCV, such wide variations have adverse effects on the boilers. The problems further extend because of the multiplicity of the supply sources. There are very few power plants (except some pit head plants) getting coal from a single, dedicated source. Most of them are supplied mainly with grade E and F coals (i.e. 85-90%) and partly grade D coals (i.e. 10-15%) [1]. Consequently, the TPS are getting coals of heterogeneous nature, both in quality and size. In absence of proper blending facilities, it is not operationally feasible for the power plants to homogenize the feed coals. Homogenization and quality control being the important ingredients of entire coal development exercise, there is an absolute need for beneficiation of non-coking coals to the extent to meet the qualities required for power generation.

The objective of the preset paper thus, is to provide a basis for identifying the beneficiation strategy to meet the essential requirements for upgrading the raw coal for power generation.

BENEFICIATION-A TUG-OF WAR BETWEEN PRODUCER AND CONSUMER

The technological benefits accrued from the use of consistent, clean fuel is now well established. Coal cleaning becomes one of the important action areas identified for conventional power generation for achieving reduction in CO_2 emission/kWh of electricity produced. The washed coal is more homogeneous in nature, has better handling characteristics and less pollutant but is more cost involving. As there is hardly any incentive for supplying coal of higher heat value covered in the same grade, the producers have no motivation to wash the lower grade coals. From the producer's point of view, he has to meet production target and recovery level and to achieve the optimum in revenue and in resources utilization. All these factors involve additional cost that has to be borned ultimately by the consumer, who is, at present, reluctant to enter into a commercially viable agreement with the producer. The consumer has to appreciate the main benefits of washing, in the light of greater energy content and lower emission level per unit of energy of the clean coal. The main economic advantage of washing the coal lies in the increased utilization of the power plant. A 5% increase in fuel cost is more than offset by a 2% increase in power plant availability. Availability increase of more than 10% can be expected simply by supplying consistent and improved quality coal [2]. However, benefits from the removal of mineral matter will be at least, partially offset by increased moisture content and a compromise needs to be established between the acceptable moisture content and benefits in terms of removal of the non-combustibles.

QUALITY IMPROVEMENT THROUGH BENEFICIATION

The area of non-coking coal beneficiation for use in thermal power plant is quite different from the coking coal upgradation for steel industry. There is no universal flowsheet nor any unique technology. A better approach may be to examine each coal based on its washability and its end-use quality requirement. All currently proposed coal washing facilities for power generation applications have the prime objective of reducing the ash content of the as-mined coal down to a level of 34% in order to reduce the transportation cost and environmental pollution following the MOEF's mandate [3]). As most of the free dirt and banded material that are easily removable, lie in the coarser fractions, the producers aim at beneficiating these fractions and mix the treated coal with untreated smaller fractions, adopting any of the following schemes:

- *Blending of selected Raw coals with clean products*: as most of the washers are used to get feed from multiple sources, the process which was once applied to coking coal beneficiation, seems least promising.
- *Mixing the untreated smaller fractions with beneficiated coarse coal fractions of a raw coal feed*, as is practiced in Piparwar, Bina etc. washeries. Earlier studies on the washability characteristics of some non-coking coals showed that the amount of free dirt/sand having relative density greater than the range 1.80-1.85 may be very high, about 20-45% of the finer fractions (5-10%) of the whole coal) that is included in the prepared coal; this enhances the wear and tear of the boiler tubes and auxiliary units [4,5]. This is further discussed in the light of the present study.

- *Blending the law ash foreign coal (raw/clean) with raw indigenous) coal of high ash*: For example, mixing of 1t of imported coal of 10% ash content with 4t of Indian coal at ash level 40% results in a blend of 34% ash content. Though not practiced in general. This may be attractive in near future, due to the reduction in import duty on thermal coal from 85% in '93-'94 to 10% in '97'-98 [6]. Such step, however, calls for a judicious approach as blending of coals of widely different ranks is likely to cause undesirable differential combustion behaviour in the burners.

- *Mixing the coarser and finer size coals, both beneficiated*: In the light of global interest in Indian power economy and with the advent of modern High Capacity Processor (HCP) like highly efficient H.M. Cyclone circuit, operating in the range of 350-600 tph complete washing circuits for deep beneficiation of both coarse and intermediate fractions are coming up. In the economic analysis for the Bilaspur Coal washery Project, the benefit of using the H.M. Cyclones, though having limitations to go for high density of separation, has been shown to be higher than that in using Jigs [7].

In first two cases, the desired overall ash content and the level of moisture content are maintained but the quality of the washed coal in terms of its end utilization in TPS may not always be assured. It has been discussed earlier that the level of cleaning has to be justified from the specific parameters demanded for the end utilities [4,5,8].

TECHNOLOGY ROUTES FOR BENEFICIATION

The parameters, which need to be considered for selection of beneficiation processes/ routes to achieve cleans of desired qualities, are:

- Top size of crushing.
- Control over fines generation.
- Size fractions of the crushed coal, to be prepared for optimum recovery at desired quality.
- Relative density of cut.
- Near gravity material at cut density.

The ROM coals are crushed to lower sizes mostly for easy and smooth handling and transportation. Pit head washers, particularly of ROM Jig type or H.M. Bath can successfully beneficiate-200mm coal. Else, coals are crushed to lower sizes (e.g., 100, 75mm or even less to suit the top size permissible for the washer like H.M. Cyclone). Studies on the liberation potential of some non-coking coals as a function of comminution have shown that size reduction down to a certain limit, 13 or 10mm does not facilitate desired liberation [4]. The benefits of any liberation of any liberation of mineral matter on further crushing are nulfied by the problems of excessive fine generation, moisture retention etc.

OPTIMUM QUALITY PARAMETER

It is, therefore, felt that the main purpose of beneficiation is to remove the high ash non-combustible particles as far as practicable. In absence of any experimentally determined correlation between the washability characteristics and the actual combustion behaviour of the feed coal, the producers are at a loss to identify the level of cleaning of a coal. There is no clear guidance, except the MOEF's mandate of reducing the ash content down to a level of 34% without commenting of the up-gradation of the washed product. The concept of characteristic ash (λ) which had been established previously as an optimum quality parameter for coking coal washing may be applied to select the specific gravity of cut in power coal beneficiation and to predict the combustion quality improvement thereby. From beneficiation point of view, the characteristic ash content (precisely, the elementary ash of the desired specific gravity fraction in F/S analysis) is one of the inherent properties of coal and washing a coal at a specific gravity determined from (λ) value of 50-55% will minimize the presence of non-combustibles having ash values beyond this limit. Previous study has shown the positive effect of this restriction on ignition properties of some non-coking coals. The properties like relative inflammability, minimum ignition concentrate/ temperature show a gradual change with the increase in specific gravity upto the limit of 1.80-1.85 corresponding to (λ) 50-55%, followed by a drastic change in the pattern above this limit [8]. This also facilitates the improvement in heat value and upgrades the coal to one or two levels. The improved quality and increased price of the upgraded coal may take care of the reduced yield and additional cost of beneficiation.

CASE STUDY

Beneficiation of the finer fractions being one of the pertinent points to be decided in the coming coal preparation strategy, the study has been confined to the coals from different coalfields with varying ash content of their finer fractions: low ash (type A), ash content around 34% (type B, C) and high ash content (type D). Each coal was crushed to nominal top sizes, 100/75/37 mm, deslimed and screened to coarse and fine fractions and subjected to F/S Test. The washability data were simulated on the basis of the following schemes to predict the effect of different levels of washing on up-gradation of the final product/blend in terms of ash and heat values:

(i) Coarser fraction beneficiated (considering Ep = 0.12) and blended with untreated finer fraction, thus simulating the conditions of the existing commercial circuits for a product having ash content 34%.

(ii) Whole coal beneficiated (at Ep = 0.12), simulating the conditions of a jig.

(iii) Whole coal beneficiated (at Ep = 0.04), simulating the conditions of Hm washers.

Simulation has been made for latter two cases (II & III) considering two different levels of beneficiation at specific gravity corresponding to

(a) overall ash content of Cleans = 34%, or

(b) Characteristic ash value (λ = 50-55%) at the sp. gravity of cut

Accordingly, the simulated products are named as I, IIa, IIb ad IIa, IIIb and the results are shown in the Table 22.3. The amount of undesired materials in the washed products is calculated as the % of the sink at sp. gravity corresponding to λ = 55% from the simulated washability data of the washed products. Figure 22.1 represents the typical washability characteristics of the simulated products obtained from coal A. Amounts of undesired particles are marked with asterisks. The amount is found to be less than 5% in the products when the coal has undergone total beneficiation at specific gravity corresponding to λ = 55%.

The results (Table 22.3) suggest that optimum washing by limiting the characteristic ash level at 55% may yield clean coal of ash content varying from 27% to 37%.

Table 22.2: Characteristics of Raw Coals (dry basis)

Coal (top size, mm)	Ash%	Heaart value		Grade	Coarse frac.		Fine fract.	
		GCV	UHV*		Size, mm	Ash%	Wt%	Ash%
A (x100)	36.9	4534	3284	F	+10	40.9	28.3	26.8
B (x 75)	43.0	3744	2496	F	+6	43.8	8.7	34.3
C (x 37)	38.1	4609	3132	F	+13	38.6	24.3	36.8
D (x 37)	46.5	3953	2041	G	+13	47.6	30.5	44.1

* on the basis of equilibrated moisture, 6%.

Table 22.3: Level of Beneficiation and Heat value upgradation*

Coal	Scheme	Condition D-50	Yield %	Ash %	GCV # Kcal/kg	Eq. moist%	UHV Kcal/kg	Grade	Upgr. (level)	Amount % @
	I	2.14	92.6	34.0	4782	6.2	3643	E	1	17.1
	IIa	2.21	92.3	34.0	4802	6.3	3634	E	1	17.5
A	IIb	1.83	73.6	27.8	5350	6.8	4386	D	2	4.3
	IIIa	2.18	93.7	34.0	4800	6.3	3634	E	1	17.8
	IIIb	1.83	77.7	27.3	5400	6.8	4450	D	2	2.6
	I	1.77	64.9	34.0	4568	6.8	3589	E	1	9.1
	IIa	1.83	67.3	34.0	4561	6.8	3589	E	1	7.7
B	IIb	1.85	69.1	34.5	4520	6.8	3524	E	1	8.5
	IIIa	1.86	73.5	34.0	4560	6.9	3580	E	1	4.7
	IIIb	1.85	72.4	33.8	4586	69.9	3605	E	1	4.1
	I	1.70	68.7	34.0	4990	6.4	3625	E	1	7.1
	IIa	1.87	77.5	34.0	5014	6.4	3625	E	1	4.2
C	IIb	1.88	78.4	34.1	5001	6.4	3612	E	1	4.3
	IIIa	1.89	85.8	34.0	5011	6.4	3625	E	1	1.8
	IIIb	1.88	85.0	33.8	5030	6.4	3651	E	1	1.2
	I			34.0						
	IIa	1.72	45.3	34.0	5152	7.3	3543	E	2	5.7
D	IIb	1.87	62.5	37.6	4805	6.9	3117	F	1	9.1
	IIIa	1.79	57.1	34.0	5160	7.3	3543	E	2	1.4
	IIIb	1.87	66.7	36.5	4918	7.0	3250	F	1	3.8

* assuming equilibrated moisture of raw coal = 6%.

\# GCV are given as simulated data on dry basis.

@ percentage of undesired materials in the washed products corresponding to $\lambda = 55\%$.

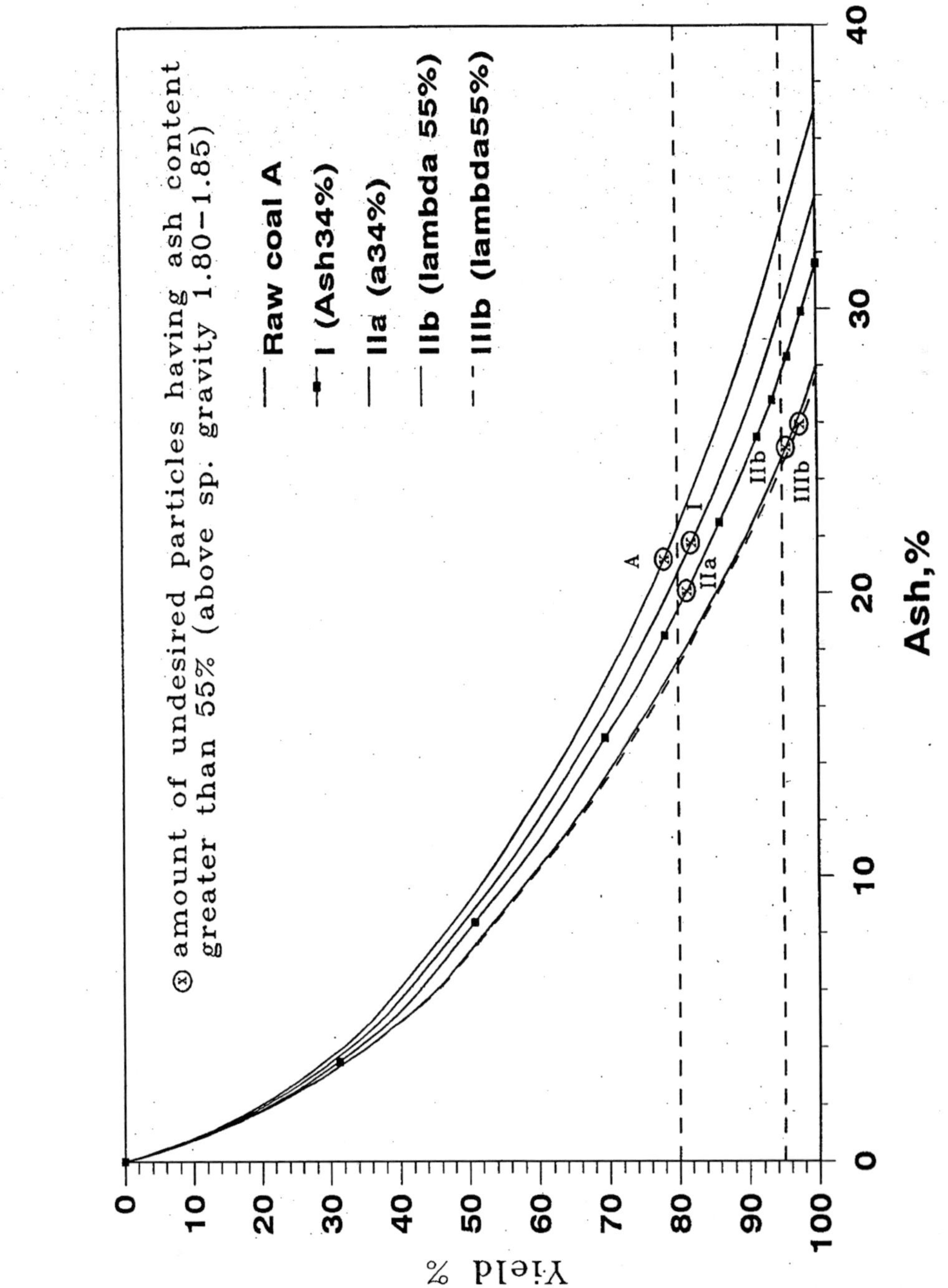

Fig. 22.1: Washability characteristics of simulated products from coal A

Since the finer fraction of coal A is of lower ash value (26.8%), cleaning this coal to an overall ash level of 34% though produces very high yield (92-94%), constitutes very high amount of undesired materials. Attempts to limit the ash content of the particles in the products within 55% through schemes IIb and IIIb reduce the overall yield but improves the grade two levels higher.

In case of coals B & C, beneficiation raises the grade by one level only, though the yields of cleans via schemes II & III are always higher than that obtained from scheme I at the same overall ash level.

For the coals of type D, in which the finer fraction is of high ash content, a product with 34% overall ash may not be achievable by scheme I. The yields from schemes IIa & IIIa can further be increased in case of IIb & IIIb by fixing the optimum parameter, $\lambda = 55\%$. The corresponding overall ash level may be as high as 37%, yet the restriction on λ within 55% limits the presence of undesirable materials to below 4%.

CONCLUSION

The non-coking coal obtained from the large OCP require beneficiation to meet the quality requirements for higher combustion efficiency. In the absence of adequate and authentic quantitative correlation of the quality parameters with the boiler performance it is rather difficult to decide on the level of cleaning. From the producers point of view the beneficiation strategy should focus on restricting the non combustibles to an optimally minimum level irrespective of the total ash content of the product that can ensure upgradation of the raw coal to at least one level higher in the existing grade so that commensurate return on the investments may be achieved for producing value added product. This can be achieved by selecting the sp. gravity of cut corresponding to the characteristic ash range of 50-55 %. It is therefore suggested that in some cases, the beneficiation route should be based on total cleaning of the whole coal in suitable washers with the restrictions on the characteristics ash. Washing the two size fractions in different washers maintaining this restrictions increases the overall yield but may be difficult to implement in heavy medium washers due to constraints on the higher limit of the achievable sp. gravity in such washers.

The suggestions may seem at present to be contrary to the economics, while considering the production target of washed coal (A substantial part of the raw coal is wasted as refuse) and higher cost of total washing (coal to be incurred for deep beneficiation) but may be proved effective in the long run while considering the reduced cost of long distant transportation, increased boiler efficiency, effective boiler design and less cost in flue gas cleaning up & solid disposal. The higher efficiency achieved would in turn reduce CO_2 emissions. It is expected that if half of the country's total thermal capacity were to be provided with washed coal, annual CO_2 emissions would fall by 10% [9].

ACKNOWLEDGMENT

The authors with to thank the colleagues of Coal Preparation for generation of various washability data at CFRI. Special thanks are also due to Mr. S.K. Mitra, Mr. K.K. Sharma, Mr. P.C. Chatterjee and Mr. A. Majumdar for rendering technical assistance and suggestions during preparation of the article.

REFERENCES

1. Economic Intelligence Service : India's Energy Sector, Sept. 1996.

2. Symonds, D.F., 1998. 'Modern Coal Preparation Technologies for Treating High Ash Coals', *Coal Prep. India 98, Int. Con. Coal Beneficiation, New Delhi, 12-13 Jan.*

3. MOEF, Govt. of India Notification, 1997.

4. Sen, K., R. Das Gupta, A. Chowdhury, N.S. Das, S. Biswas and S.G. Chaudhuri, *1998.* 'An approach towards optimum beneficiation of Thermal coals', Coal Prep. India, '98, *Int. Con. on Coal Beneficiation, New Delhi, 12-13 Jan.*

5. Sen, Kalyan, N. S. Das, S. Biswas, S. K. Mitra, A. Sethi, D. Gnana Bharathi, S.G. Chaudhuri, Anjani Kumar and M.M. Bhattacharya, *1999.* "Basic study initiated on imprrovement on coal quality: Beneficiation for cleaner use of non-coking coal", *Int. Symposium on Clean Coal Initiatives (CMRI), New Delhi, 22-24 Jan.*

6. 'Coal Supply To Thermal Power Stations: Qualitative And Quantitative Issues, including Fuel Supply Agreement', 1981. Ed. C. V. J. Verma, *Council of Power Utilities, New Delhi.*

7. Kumar, U., 1998. 'Beneficiation of Power grade Coal-Indian Endeavour', *Coal Prep. India, '98. Con, on Coal Beneficiation, New Delhi, 12-13 Jan.*

8. Sen, K., D.D. Halder, S. Biswas and N.S. Das, *1999.* 'Beneficiation of Non-coking Coals: Basic Concepts and Technology Routes', *Coal for Power Generation–Issues Management vis-a-vis Quality, COALQUAL 99 (DCL), Calcutta, 29-30 Jan.*

9. Reuben, B. S., *1998.* 'Advantages of Coal Beneficiation–AEC's Perspective', *Coal Prep. India,' 98, Int. Con, on Coal Beneficiation, New Delhi, 12-13 Jan.*

CHAPTER 23

Potential of Blast Casting to Improve Economy of Operating Surface Coal Mines in India

A.H. Reddy and M.D. Uttarwar***

ABSTRACT

Surface coal mining in India has been contributing in a massive way to our ever increasing energy demand and that it will continue to rise. The present scale of operations to exploit lower grade deposits with increasing depths has been the key trend in the past two decades. This was mainly possible due to spectacular developments in the field of earth moving equipments like draglines, excavators, drills etc. in a bigger and better way. But during recent times, rising input costs, volatile market due to liberalised regime of economy and growing environmental compliances are posing a major challenge to Indian surface coal mining industry. With the impetus of enhancing the performance and productivity of surface coal mines using the leverage of technology, herein an attempt is made to demonstrate the economic potential of Blast Casting Technique of overburden removal.

1.0 INTRODUCTION

In recent times, the demand for coal has increased enormously due to ever rising energy needs and that the Indian coal mining industry is found to have responded magnificiently to this hike. Such a flamboyant growth in coal production has been made possible through exploitation of large sized surface mines due to their favourable economics, safety, higher recovery, potential for higher production and lower gestation period.

*Assistant Professor, Department of Mining Engineering, Visvesvaraya Regional College of Engineering, Nagpur - 440011, India.

**Assistant Professor, Department of Mining Engineering, Chandrapur Engineering College, Chandrapur - 442403, India.

All India coal production by 1998-99 was 290 Million Tonnes and is slated to rise at 405 Million Tonnes by 2001-02. Owing to the sizeable share of surface mining in coal production, the job of overburden removal will also rise and it will only be possible with the development of most efficient equipment systems.

But during the last decade, the rising input costs due to inflationary effects on costs of equipment, fuel, labour, insurance, growing environmental compliances and fast changing economy have rendered conventional method of overburden removal to be uneconomical leading to reduced productivity and profitability.

In order to offset the dwindling margin of profit and to maximize productivity, there is a need to incorporate innovative methods of overburden removal.

In this scenario, Blast Casting is anticipated to be a cheaper alternative to conventional overburden removal methods. In this technique, chemical energy of explosives is used instead of mechanical energy to move large volumes of overburden directly into decoaled area. The proper application of the technique can either reduce the overall costs of mining or increase the coal production. It can also be used under situations like increasing stripping ratio, fluctuating production demand, for mining of low grade reserves especially located at higher depths etc. Today it is applied as major stripping tool in countries like USA, Canada, South Africa etc.

In the context of today's liberalised regime of Indian Economy, this paper seeks to present a critical appraisal as to potential of blast casting technique based on economic analysis of global case studies with special emphasis on India. An attempt is also made to explore future potential for application in India owing to it's economic benefits. A discussion is also presented on applicability of this technique under different geo-mining conditions and situations.

2.0 FACTORS CONSIDERED FOR APPLICATION OF BLAST CASTING

Despite the significant benefits that may be accrued from blast casting, it should be noted that the method is not amenable to all the stripping operations and has certain limitations. As will be seen below, there are some factors involved that need to be considered in the evaluation process, which may not render it applicable universally.

The practice and experience of various mine operators has revealed that blast casting is suitable in horizontal or flat lying beds or seams and to dipping ore bodies (Dupree, 1987; Rollins and Givens, 1987). It has been successfully applied in mountain top removal, single seam as well as in multiple seam operations. However it is believed that this technique is economically more effective in dipping ore bodies, especially at those areas where the stripping ratios progressively increase to a degree that mining by conventional methods become uneconomical. It allows even lower grade reserves; especially those located at greater depths with high overburden cover, to be worked out economically and at a faster rate. The factors that must be considered before applying the technique at a particular mine site are Pit Geometry which refers to bench height, pit width and the ratio of these two parameters; Overburden Characteristics like depth and hardness; Location of Overburden Dump as to availability of in-pit dumping conditions; Capacity of Stripping and Coal Production Units and Environmental Restrictions etc.

3.0 PURPOSES OF BLAST CASTING TECHNIQUE

Conventional earthmoving methods are not the sole means of handling fragmented overburden, nor they are always the most cost effective. Overburden handling costs in conventional mode do contribute significantly to the overall economics of the mine. Hence, a reduction in these costs can minimize the cost of a ton of coal produced and this happens to be the primary purpose of the method. The other purposes due to which it can be applied under a particular situation in an operating or a new mine are summarized in the following sections.

3.1 For Boosting The Production

When there is sudden increase in the yearly targets of coal, the increased demand of production has to be met with concomitant removal of extra volume of the overburden by existing equipment, which is normally an impossible task. In such situation, blast casting can be a feasible alternative method to remove that extra volume of overburden while utilizing the existing equipment only and a properly executed program whereby stripping rate of overburden excavator is maintained at or near maximum utilization and use the improvement in the % cast over to increase the coal production.

3.2 Replacement Of Equipment

When the principal stripping equipment on job is due for its replacement, it may be advisable to consider equipment that does work well with blast casting. This will lead to procurement of a smaller sized equipment as a result of blast casting doing part of overburden removal, which means considerable savings in capital outlay thereby reducing the cost of production.

3.3 Reducing Capital Outlay for Excavating Equipment

The obvious situation in which casting often proves to be economical is a new surface mine ready to order equipment and in the final planning stages. If the volume of overburden that must be mechanically handled per tonne of coal can be reduced appreciably, there is a consequent decrease in the capital outlay for the excavating equipment. At such times, smaller draglines, fewer support dozers, loaders etc. can be considered which gives scope for reducing the capital outlay.

3.4 Increasing Stripping Ratio

As the pit progresses towards dip, the overburden thickness increases, and that the increased cover over seam is normally handled by truck-shovel combination in case of a dragline pit. Due to poor utilization or inadequate capacity of this system, advance stripping progress may not match the dragline progress thus making it idle or forced to redeploy for advance stripping. This significantly reduces the volume of coal exposure. Under such circumstances, instead of handling that extra volume of overburden by stripping equipment in the form of separate bench, we can use blast casting technique. Thus it is concluded that casting can encourage the mine operators either to plan for higher benches or in other words it allows planning of dragline operations for dipping coal seams. It can also be concluded that casting can open up reserves that have been considered unmineable in the past.

3.5 Improving Productivity of Existing Operation

For a particular operation, sometimes need for raising productivity arises due to either increase in the cost of HEMM maintenance, or stripping ratio or even occurrence of low-grade coal.

Since blast casting warrants use of higher powder factors, fragmentation is greatly improved thereby reducing maintenance costs considerably and in turn longer life of the equipment. In addition to this, an operation reported of using light weight archless bucket for dragline instead of a heavy arched bucket due to better fragmentation (Dupree, 87).

On the other hand in case of occurrence of low grade reserves in some portions of the property with high overburden cover to a degree such that it may prevent economical mining by conventional methods, blast casting can prove to be a economically effective way of overburden removal that too at faster rate.

3.6 Cater to Fluctuating Production Demand

Due to ever changing economic scenario, the demand for coal may fluctuate for various reasons like availability of cheap and good quality coal from abroad due to reduced import duties or increase in the use and finds of oil and natural gas or even switching towards use of unconventional sources of energy like water, winds, sunlight etc.

In situation like depressed coal market especially in countries like India, blast casting can allow an operation to be profitable one because of its flexibility to turn off and on the casting program when to meet the increased demands of the market place without excessive demand on the capital for equipment to handle that increased amount of overburden.

4.0 APPLICABILITY UNDER DIFFERENT GEO-MINING CONDITIONS

In case of strip (open-cast) mining, over-burden must often be blasted so that it can be excavated by a mechanical equipment like dragline i.e. the overburden must be well fragmented and loosened with relatively small lateral displacement. The present status of blast casting can remove over-burden partly while rest of the fragmented rock is removed by draglines or shovels.

The development of this technique started in mines having two benches, one for overburden and another for coal i.e. in a single seam environment. A simple model of the blast casting is shown in Fig. 23.1, which depicts that material can be displaced partly by blasting while the remaining portion requires mechanical removal (Pisaneschi and Frazier,1963; Krapfenbauer, 1990). However, in a South African coal mine, blast casting technique was successfully conducted in double seam operation (Atkinson, 1982). Similarly in multi-seam mines, blast casting of overburden above the top most seam and conventional open-pit operation for removal of interburden and lower seam has also been practiced (Chironis, 1985).

On the other hand, a combination of shovel-dumper can also be used with blast casting. This method may be effective with a moderately shallow overburden and relatively thin coal seam in a single bench operation.

Apart from the methods discussed above, blast casting is equally applicable in rocky mountainous coal mines where coal seam run flat or dip into a mountain/hill side and where a sub-crop or contour pit is usually developed as the pit limit is moved into the

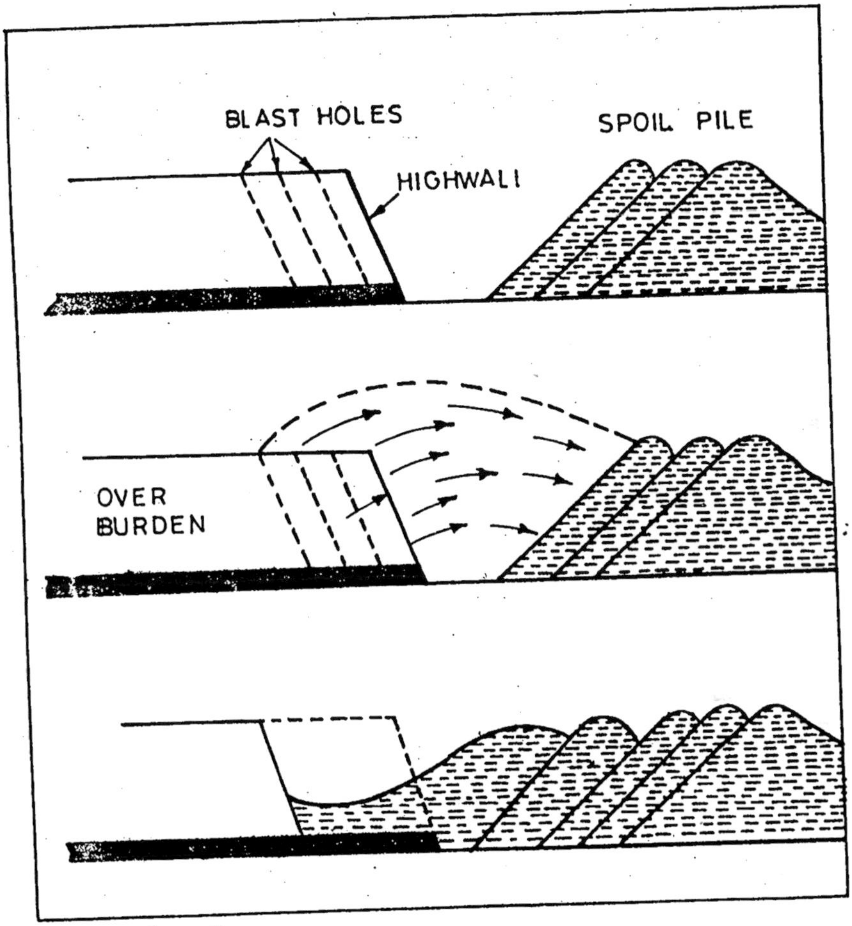

Fig. 23.1 : Overburden Blast Casting in A Single Seam Operation
(After Singh At Al., 1994)

hillside thereby increasing the stripping ratio dramatically. Under such circumstances, a shovel-dumper system in conjunction with blast casting can allow the economic mining of higher strip ratio coal and thus increasing the volume of mineable reserves (Singhal and Naidu, 1989).

The last application includes stripping previously worked deposits/thick coal seams that have been extracted partially by underground methods. Here the security of stripping tool can be in question since it is located over the pillared coal seam. In such a situation, blast casting can eliminate this type of security hazard (Atkinson, 1992). The method can offer special technological advantages for open-pit mining of developed coal seams with special reference to Indian CoalFields where lighter loading equipment can be placed in overburden/interburden benches in conjunction with blast casting.

5.0 OBJECTIVES WHILE PLANNING A CASTING SHOT

The various objectives to be observed while planning a casting shot are magnitude of throw, direction of throw, volume of cast over and desired muckpile profile. The information relating to these objectives is fundamental to the analysis of every blast casting shot and hence it is a must to define them depending upon physical mining situation in which blast casting is to be implemented so that valid economics can be applied. Subsequent sections 5.1 through 5.4 consider each of these objectives with the aim of proving their vitality in the performance of blast casting.

5.1 Magnitude of Throw

It is governed by geo-mining conditions of the site and the parameters of the blast design applied. As hauling operations are practically nonexistent in this method, maximum displacement is the most important criterion whereas good fragmentation is of secondary importance.

5.2 Direction of Throw

Usually the principal direction of throw or rock movement is normal to the free face. Direction of throw affects the blasting sequence and ultimately the quantum of throw. Therefore it has to be considered while planning an optimal blast casting.

5.3 Volume of Cast Over

The desired degree of casting i.e. the volume of cast over in the decoaled are can be achieved by adjusting the H/W ratio, proper location of explosives, (spacing to burden ratio) energy level of explosive, firing pattern etc. commensurate with the level of production and equipment capacity. Here it should be noted that any increase in % cast will reduce the effective stripping rate of equipment and it is therefore careful evaluation is necessary to decide the degree of casting and hence the need to plan for optimal percent cast.

To examine the concept of optimality, it is best to assess two principle-operating scenarios that are possible in conjunction with blast casting. They are : (1) To maintain coal production of mine at present rate and use increase in % cast to reduce cost associated with mechanical overburden removal i.e. optimum blast design based on % cast for the operation. Here it can be primarily concluded that the unit cost of excavation will go on decreasing with increasing percent cast and (2) To maintain stripping rate of overburden excavator at or near maximum utilization and use the improvement in % cast to increase coal production thereby reducing the overall mining cost per unit of overburden shot i.e. optimum casting % based on the economics of operating the existing equipment. Here it can be concluded contrarily to the previous scenario that cost of operations will increase linearly with % cast. The concept is depicted in the Fig. 23.2.

Therefore it can be stated that blast casting can lead to either little economic benefit if the production is held constant at the level before blast casting started with the same equipment in use or no benefit when replacement of present equipment with smaller one is thought about since it would not be justifiable. Under such situation, the obvious choice remains with the operator to use stripping equipment to its full capacity and use blast casting for either increased coal production or in the event of increasing strip ratio of the mine.

Thus summarily it can be concluded that optimum casting design may not necessarily coincide with the optimum economic design.

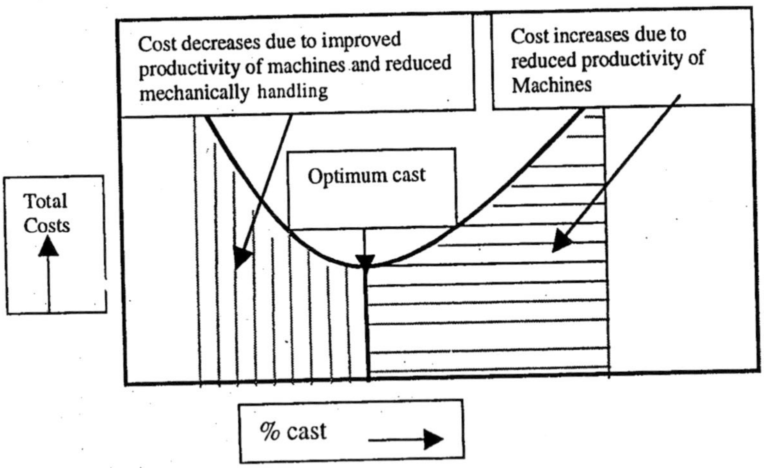

Fig. 23.2 : Concept of optimality for encompassing blast casting in an operating mine

5.4 Muckpile Profile

When laying out blasting operations, the basic requirements are that the rock should be broken to full depth, and that the fragmentation should be within acceptable limits, and safe conditions should prevail. Increasingly, the need to maximize the utilization of capital equipment is calling for an additional criterion; the broken ground should fall in a position and shape that optimizes overall productivity and mining costs. It is because shape and distribution of muckpile affects the time required for stripping due to the fact that 'casting' of overburden into the pit will loosen up rock considerably reducing the time required to fill the excavators bucket. The requirement as to characteristics of muckpile profile is depicted in Fig. 23.3.

6.0 PERFORMANCE ASSESSMENT TECHNIQUES

The performance of blast casting method usually, is quantified by means of Effective Cast Percentage which is defined as the percentage of the broken overburden that does not require re-handling as shown in Fig. 23.4 (a) and 4 (b). In order to perform an economic analysis of blast casting of overburden, the percent cast must be measured. The three most commonly used methods for quantifying the percentage cast are:

(i) Assessment of reduction in the stripping time for the shovel or dragline. (Stach and Hale, 1984)

(ii) Determination of the percentage cast-over by the blast from above the coal (Shaw et al., 1978) and,

(iii) Determination of the percentage cast over in final spoil configuration by blast without being mechanically rehandled (Pisaneschi and Frazier, 1963)

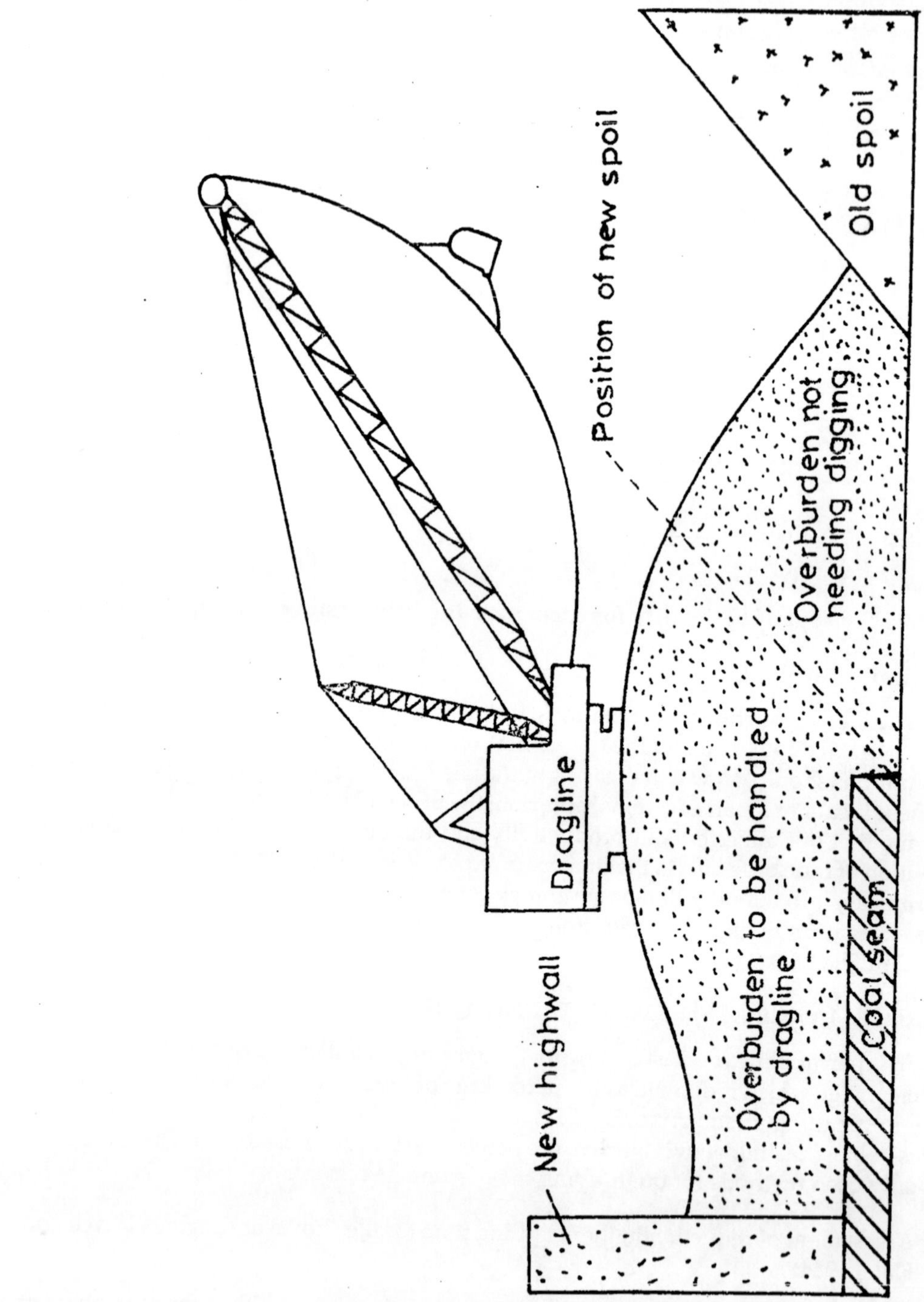

Fig. 23.3 : Effect of shape and distribution of muckpile on cast over (After Cunningham, 1990).

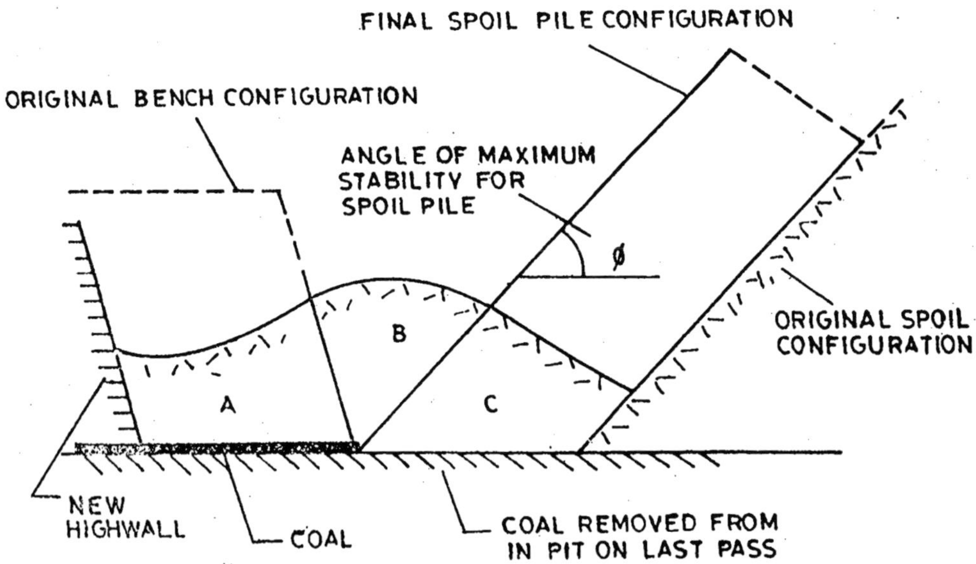

Fig. 23.4 (a) : Cross section of spoil pile configuration immeidately after blast. Illustrating material still left in the space occupied by the original bench (A) material explosively cast in to position in the spoil pile (C) and material in between (C)

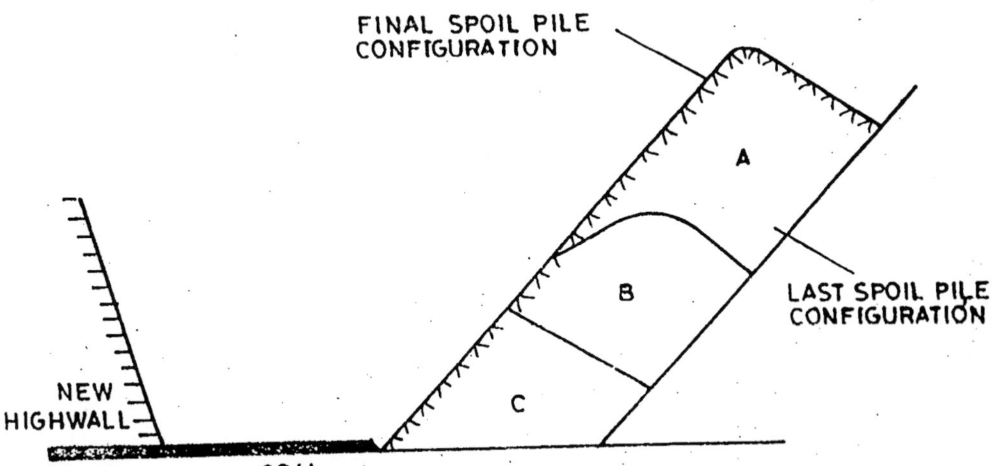

Fig. 23.4 (b) : Final spoil pile configuration (cross section) denoting material explosively cast into spoil pile (C) adjacent material (B) mechanically cast onto spoil pile and material (A) mechanically cast onto spoil pile from above the coal seam. (After Worsey and Giltner, 1987).

7.0 ECONOMICS OF BLAST CASTING

No matter how good the job of design and implementation are done, the bottom line to the success of blast casting is economics. In general, blast casting method adds to the cost of drilling and blasting program. So in order to make it a viable alternative, this added cost must be offset by reduction in overburden removal costs. This section demonstrates the potential of the blast casting technique based on economic analysis of global case studies, economic performance of the technique under different geo-mining conditions with special emphasis on India.

7.1 Economic Performance Of The Method

Actual practice with blast casting has shown that it is an economical method of overburden removal. The primary attraction of this technique is sizeable reduction in stripping costs and ultimately overall costs of mining. In this method, the cost of blasting operations is slightly more than blasting for fragmentation alone, due to use of higher powder factor. But then these costs are more than offset by the reduction in stripping costs. To evaluate the economic potential of the method, a comparative economics of four case studies under different geo-mining conditions are presented in Table 23.1 followed by detailed economic analysis of global case studies in Table 23.2.

Table 23.1 : Comparative Economics of Mines in Conventional Vs. Blast Casting Mode of Operation

Sr. No.	*Name of the Mine*	*Item*	*Conventional*	*Blast Casting*
1.	Thunder Ridge Operations, Canada, (Mountain-top Removal)	Drilling	9.70	13.90
		Blasting	7.90	12.10
		Stripping	82.40	41.00
		Total	100 %	67%
		Net Savings	33 %	
2.	NPR Basin Mine, Missouri, USA (Single –Seam Operation)	Drilling and Blasting	25.78	40.67
		Stripping	74.22	41.88
		Total	100%	82.55%
		Net Savings	17.45%	
3	Bicknell Minerals Inc., Indiana, USA (Double-Seam Operation)	Drilling	15.30	15.30
		Blasting	41.10	45.80
		Stripping	43.60	17.50
		Total	100 %	78.6%
		Net Savings	21.4%	
4	Colowyo Coal Mine, Coloradao, USA. (Multi-Seam Operation)	Drilling and Blasting	39.00	70.00
		Stripping	61.00	20.00
		Total	100 %	90%
		Net Savings	10 %	

Note: All Figures in Percent Costs.

To demonstrate the savings that may be realized as a result of blast casting program, the total production costs of drilling, blasting and stripping for conventional method are considered as 100 % and relative costs are shown under blast casting environment.

As can be seen from Table 23.1, in Thunder Ridge operations, USA, blast casting helped to reduce production cost by as much as 33 % (Rollins and Givens, 1987) over the conventional stripping method. Here the purpose of the technique was to cut across ridge and cast into area excavated by contour and box cuts. A breakup of overburden stripping costs for Bicknell Minerals Inc., USA shows that drilling cost is same for both the methods which implies that blast casting can be applied without increasing the drill capacities while blasting costs are slightly higher for casting (Stach and Hale, 1984). Here the operations were in the close proximity of underground workings and ground vibration levels was the constraint to blast casting program. At the Colowyo coal mine, USA, the net savings were limited to 10 % since drilling and blasting costs were too high for casting due to unfavorable conditions which include water, toe condition and drill scheduling problems. But then this extra cost was still justified because of the added dragline capacity Vs shovel truck cost. Whereas savings were only 17.45% in case of NPR Basin mine, USA, since the dragline was used to its full capacity and improvement in % cast was used to eliminate shovel-truck stripping.

Similarly from economic review of global cases presented in Table 23.2, it can be seen that about 11 mines have reported of cost savings, of which Sasti O.C.P., WCL, India has predicted considerable savings if blast casting is amalgamated in existing operation with few modifications (Vijh and Mathai,1998). In another instance the technique helped to extract a coal seam that could not otherwise be extracted economically (Chironis, 1982). Whereas one such operating mine could improve their cut-off stripping ratio from 1:9:5 to 1:15 because of the technique (Collins, 1987). Another mine reported a 20 % reduction in dragline bucket maintenance costs due to improved fragmentation (Chronis, 1988). In other case, a light weight archless bucket has replaced a heavy arched bucket due to enchanced fragmentation and looser muckpiles (Dupree, 1987).

Contrarily, in spite of decades of experience in surface coal mining, India could not develop or acquire blast casting in their mining operations and hence as such never attained the status of a system of waste removal. But then impetus for the blast casting program in Indian coal mining operations came in order to enhance performance and productivity, which was necessary to maintain unit costs at an acceptable level especially at the time of economic hardship. It is in this context that in the past decade, few dragline pits took pioneering efforts for introducing this technology. Table 23.3 demonstrates the economic performance in these mines.

As can be seen from Table 23.3, the net cost savings in case A and B is limited to 17.10 % and 18 % resp.where the technique was intended to improve the operational efficiency and achieve economic benefits through preparation of large volumes of blasted material for the uninterrupted operation of dragline. Whereas in case C, the net savings was as high as 58.34 % although the blasting costs were significant. This was mainly possible because of higher percentage of cast over which resulted in reduction of stripping costs. Here the objective of the technique was increased coal production and that the option in general to achieve this was to procure additional fleet of equipment for removing that extra overburden cover which means additional capital and that it was not justifiable. Hence introduction of blast casting without need for further investment came as a welcome relief for the project.

Table 2 : Economic Analysis of Global Case Studies

Sr. No.	Name of the Mine	% Cast	Strip Ratio	% Increase in Drilling	% Increase in Blasting	% Reduction in rehandle	% Cost Savings	Remarks	Reference
01	Trapper Mine, Colorado, USA	26	--	--	--	12	--	Heavy Arched Bucket Replaced by light Weight Archless	Chironis'86
02	Trapper Mine, Colorado, USA	30	--	--	--	12	--	As above	Dupree'87
03	Big Run Mine, Penn., USA	60	1:20	--	--	--	--	Economically Viable Due to Blast Casting	Chironis'82
04	Alabama Mine, USA	41	1:23	--	--	--	--	--	Chironis'81
05	Central Pennstate Mine, USA	30	1:17	-	--	--	30	--	--do--
06	Cadiz Mine, Colorado, USA	21	--	--	--	--	--	--	--do--
07	Thunder Ridge Operations, West Virginia, USA	44	1:10	--	--	--	--	--	Rollins and Givens'86
08	--do--	49	--do--	--	--	--	--	--	--do--
09	--do--	65	--do--	--	--	--	--	--	--do--
10	--do--	66	--do--	--	--	--	--	--	--do--
11	--do--	65	--do--	--	--	--	--	--	Rollins and Givens'87
12	--do--	35	--do--	--	--	--	--	--	--do--
13	--do--	55	--do--	--	--	--	--	--	--do--
14	Eagle Mountain Mine, Canada	45	--	15 (Combined)		--	25	--	Singhal and Naidu'89
15	Fording River Operations, # Browny Pit, Canada	26	1:4.5	--	--	--	--	--	Mine report'88
16	--do--	39	--do--	--	--	--	--	--	--do--
17	# Lake Pit, Canada	12	--do--	--	--	--	--	--	--do--
18	--do--	50	--do--	--	--	--	--	--	--do--
19	--do--	33	--do--	--	--	--	--	--	--do--
20	--do--	30	--do--	--	--	--	--	--	--do--
21	--do--	36	--do--	--	--	--	--	--	--do--
22	--do--	48	--do--	--	--	--	--	--	--do--
23	Colowyo Mine, Colorado, USA	48	1:6.5	80 (Combined)		08	10	--	Chironis'85
24	Bicknell Minerals Inc., Indiana, USA	67	1:25	--	10	47	22	--	Stach and Hale'84
25	Samples Mine, West Vir., USA	40	--	--	--	--	--	--	Smith'95
26	Western Pennsylvania Operations, USA	40	1:10	30	33	40	40	--	Anonym'61
27	Sewanee Mine, Tennasse, USA	48	1:34	--	--	--	--	--	Cook and McCutchen'94
28	Lee Ranch Mine, New Mexico, USA	30	1:5.6	--	--	--	--	--	Chung et al., '94
29	NPR Basin Mine, Missouri, USA	25	1:6	58 (Combined)		33	17.45	--	Ewart et al., '93
30	MTR Mine, West Virg., USA	60	1:21	--	--	--	--	--	Chironis'85
31	Ruler St. Open Pit Mine, Sourthen Wyoming, USA	26	1:10	40 (Combined)		25	--		Harder'88
32	Caney Creek Mine, Western Kentucky, USA	50	1:15	--	--	--	--	Strip Ratio 1: 9.5 Before Application	Collins'87
33	Cedrum Mine, Alabama, USA	50	1:11	--		--	--	--	Chironis'84
34	Delta Mine, Ha., Penn., USA	20	--	--	--	30	20	20 % Reduction in Dragline Bucket Maintenance	Chironis'88
35	Chinook Mine, Penn., USA	30	1:7.5	--	--	30	20	--	--do--
36	Highvale Mine, Canada	09	1:5	--	--	--	--	--	Anonym'90
37	Beltona Mine, Alabama, USA	55	1:27	--	--	--	--	--	Hagen'76
38	Western Pennstate Mine, USA	55	1:20	--	--	55	--	--	Chironis'83
39	Rietspruit Mine, South Africa	61	--	--	--	--	--	--	--do--
40	Ramagundam OCP-I Mine, SCCL, India	38	1:3.4	10.2	18.6	23.5	17.1	--	Reddy'90
41	Jayant Project, NCL, India	26	1:2.7	1.7	14.42	26	18	--	Downline issue
42	Sasti OCP, WCL, India	27.5	1:3.7	11	95	18	51.86	--	Vijh'98

Table 3 : Comparative Econimics of Conventional Vs. Blast Casting Method of Overburden Removal for Operating Mines in India

Sr No	Name of the Item	Case A		Case B		Case C	
		Conventional Method	Blast Casting	Conventional Method	Blast Casting	Conventional Method	Blast Casting
1	Number of Holes	108	119	NA	49 (Extra)	49	49
2	Cost of Drilling (Rs.)	588000	649600	15273540	15528060	573300	573300
3	Powder Factor (m^3/Kg)	2.0	1.7	1.8	1.58	2.5	1.8
4	Explosives Consumed (Kg)	90720	106729	424264	485440	51597	75624.20
5	Cost of Explosives (Rs.)	1152144	1366458	19091880	21844800	914136.45	1341627.45
6	Cost of Dlg. & Blstg.(Rs.)	1740144	2016058	34365420	37372860	1487436.40	1914927.40
7	Percentage Cast Over (%)	–	23.50	–	26	–	31.62
8	Dragline Benefit (m^3)	–	59694	–	201995	–	43178
9	Dragline Benefit Cost (Rs.)	–	573062	–	9089775	–	1295340
10	Total Cost of Operation (Rs)	1740144	1442996	34365420	28283085	1487436.40	619587.40
11	Net Cost Benefit (Rs.)	–	297148	–	6082335	–	867849
12	Cost Savings (%)	–	17.10	–	18	–	58.34

(Source: Reddy'90; Downline'90; Sharma & Sharma'98)

Thus it can be said that blast casting technique is an economical way of overburden removal as compared to conventional stripping method. As seen from Table 23.3, in all the case examples, considerable cost savings has been realised due to incorporation of blast casting. As far as purposes while applying the technique of blast casting for particular mine sites in Indian context is concerned, it was found to be used as a tool to enhance the productivity of stripping equipment like draglines whereas in other case it was aimed at increased coal production.

As demonstrated in preceeding paragraphs, thus it can be seen that major savings results from reduction in stripping cost. To conclude with, here is the review of several other operational as well as economic benefits:

(i) Blast casting helps in minimizing rehandling, therefore much smaller sized stripping equipment and fewer auxiliary production machines are required resulting in reduced capital investment, operation and maintenance and subsequently labour cost.

(ii) The throw of the overburden is instantaneous, hence the cycle time is reduced and this may result in an increased production and productivity (Chironis, 1980; Tracy, 1985).

(iii) Better fragmentation because of higher powder factors leads to improved fill factor of the bucket and henceforth faster and easier stripping.

(iv) The technique also helps in improving spoil heap drainage and thereby stability by casting very large rock fragments from the base of the highwall on the floor of the pit to form the base of the spoil pile (Atkinson, 1992).

While blast casting has many benefits, some problems do arise due to its application. It causes undue damage to coal seam and dilution can be prominent. It also results in smaller blasted inventories than buffer blasting. It is more drill and blast dependant; therefore scheduling is more critical. While higher powder factors may create a potential hazard for surrounding environment like ground vibrations, fly rocks etc.

8.0 CONCLUSIONS

Thus summarily it can be concluded that the blast casting is an innovative technique of overburden removal and provides an economical alternative to conventional overburden handling methods. The major economic benefits that can be derived are reduction in stripping costs, saving in capital cost for stripping equipment and subsequently its operating and maintenance cost. Blast casting can allow an operation to be profitable even in the times of a depressed coal market especially in countries like India or an economical extraction of low grade reserves which are located at greater depths with high stripping ratio. The technique when implemented in an operating surface mine can increase profitability and would complete the operation sooner due to increased rate of coal extraction per unit time with a very favourable impact on cash flow.

Thus the summary conclusion is that blast casting should be used to meet the challenge posed by reduced productivity and performance of surface mines. It is also concluded that there exists lot of potential for application of blast casting in Indian surface coal mines in the coming century owing to it's economic benefits.

REFERENCES

Anon, 1961, "Casting overburden with explosives", Coal Age, V66(3), pp 78-82.

Anon, 1990, "Cast Blasting", Downline, Nov., pp 14-15.

Atkinson, T., 1992, "Cast blasting of deep overburden", Section 13.2.3, V(II), SME Mining Engg. Handbook, 2nd edition, pp 1307-10

Chironis, N.P., 1980, "Casting overburden by blasting", Coal Age, V(56), No.5, pp 108-16.

Chironis, N.P., 1981, "Mines Take on Explosive Casting",Coal Age, V(86), No.10 pp 130-35.

Chironis, N.P., 1982, "Blast-Casting pays off at Avery", Coal Age, V(87), No.11, pp 40-43

Chironis, N.P., 1983, "Spoil side stripping succeeds", Coal Age, V(88), No.4, pp 48-53.

Chironis, N.P., 1984, "Angled drilling aids casting", Coal Age, V(89), No.1, pp 64-67.

Chironis, N.P., 1985 (a), "Blast casting succeeds at multi-seam Western mine", Coal Age, V(90), No.12, pp 55-586

Chironis, N.P., 1985(b),"Stripping methods complete at Mountain top removal sites", Coal Age, V(90), No.12, pp 59-62

Chironis, N.P., 1986, "Spoil-side dip line stripping works well with cast blasting", Coal Age, V(90), No.8, pp 40-44

Chironis, N.P., 1988, "Amax dragline dig more overburden with blast casting", Coal Age, V(93), No.6, pp 48-51.

Chung,S.,Mchill, M. and Preece,D.,1994,"Computer cast blast modelling", Proc. Of 20th Conf. On Explosives and Blasting Tech., SEE, Ausin ,Texas,USA, pp 517-21.

Cook, J.D. and McCutches, Dan, 1994, "Timing tips for effective overburden casting", Coal, Aug.,pp26.

Crossby,W.A., Workman,L., Lombardi,J.A. and Simos,J.G.,1982,"Blast designs to improve dragline stripping rates", Clyde-Woodward Consultants, U S Deptt, of Energy 11239-1,Final Report, Nov.

Collins, B., 1987, "How to design high-powered shots for low vibration casting", Coal Mining, March, pp 38-44.

Dupree, P.D., 1987, "Applied drilling and blasting techniques for blast casting at Trapper Mine-potential to save on overburden removal", Mining Engineering, January, V(26), No. 1, pp 13-15.

Ewart, D.L., Richard,M. R.and Pries, S.M.,1993,"Alternatives to shovel/truck overburden pre-stripping;Three case srudies",Innovative mine design for the 21st century, Balkema, Rotterdam, pp. 247-65.

Grippo, A.P.., 1984, "How to get more cast per blast", Coal Age, V(89), No.12, pp 83-89.

Gupta, R.N., Adhikari, G.R., and Singh,B.,1986, "Advances in overburden blast casting", Journal of Mines, metals and Fuels, V(34), pp 487-92.

Hagen, H.D.,1976,"Multiple seam mining in Alabama", Mining Congress Journal, V62(8), pp 22-24.

Harder, Paul,1988,"dozers,Draglines and Cast Blasting", Proc. Of 14th World Mining Congress, Vol.I, pp 493-96.

Krafenbauer, D.K., 1990, "Cast Blasting", Proc of 14th World Mining Congress,V(1), pp 493-96.

Pisaneschi, A and Frazier, H.S., 1963, "Moving overburden with explosives", Mining Congress Journal, V(49), No. 7, July, pp 40-43.

Reddy,B.C.K.,1990,"Economics of technique of blast casting- An approach", National Seminar on Modern trends in Explosives Technology and Applications, Inst. Of Engineers,Nagpur,25-27 Oct., pp 141-49.

Report on Fording Coal Ltd., February,1988.

Rollins, R.R. and Givens, R.W., 1986, "Explosive casting blast round design", Proc. Of 3rd Conf. on Use of Computers in the Coal Industry, A.A. Balkena, Rotterdam, Boston, pp 153-59

Rollins, R.R. and Givens, R.W., 1987, "Blast casting of an Eastern strip mine", SME Annual Meeting, Denver, Colorado, Feb. 24-27, pp 1-6.

Sharma,U.D. and Sharma, P.D., 1998, "Blast casting with SMS explosives-A case study at Sasti Opencast mine", Visfotak, A National Seminar on explosives: safety and technology, Nov., pp 79-82.

Singhal, R.K. and Naidu, H.G., Currie, J.A. and Clausen, G., 1986, "The potential for cast blasting in Canadian surface mines", int. Symp. on Geotechnical Stability in Surface Mining, Calgary, November, pp 145-51

Smith,D.,1995,"U S Mine blasts 6 million yards of overburden every month", Coal International, Nov., pp 229-30.

Stach, G.J. and Hale, N., 1984, "Blast casting in sensitive environment, Proc. Of 10th Conf. On explosives and Blasting Techniques,SEE, Lake Buena, Vista, Fla, pp 218-31

Tracy, J.M., 1985, "Increased production through Cast Blasting in surface coal mines", Journal of Explosives Engg., V(3), No. 3, Sep/Oct, pp 16-23.

Vijh,K.C. and Mathai,John,1998"New mining system with blast casting of overburden at Sasti opencast mine (WCL)", Visfotak, A National Seminar on Explosives: Safety and Technology, Nagpur,Nov., pp. 163-68.

CHAPTER 24

Plants and Equipment for the Beneficiation of Raw Coal

Jürgen Winckler and Gurudas Mustafi***

ABSTRACT

Planning and installation of coal preperation plants require today from the mining companies and their engineering partners along with the consideration of techno/ economical aspects more and more also the responsibility in respect to the environmental compatibility of the projects.

Growing mechanisation of coal mining drastically raises the amount of refuse as well as the amount of fine material in the R.O.M. coal. This leads to an increase and a wide fluctuation of the ash content and to a higher moisture content of the coal. Due to this the amount of raw coal, which is beneficiated, is permanently increasing worldwide. Process steps for the various size fractions of the raw coal have proven their feasibility for the beneficiation of power-station coal and are presented herein.

Aside the beneficiation of the raw coal the improved technologies allow for a recycling of the dumped unwashed coal, which has been wasted in former times due to selective mining methods for example. Even slurries out of ponds, which very often contain material with an ash content of less than 40%, can be beneficiated to make optimum use of the energy, which is available.

1.0 BENEFICIATION OF LUMP SIZED COAL

1.1 General

A great percentage of the power station coal mined throughout the world has been burned so far without preparation or refuse separation, because the preparation costs are too high or combustible substance is lost in the preparation process. Considering, however,

* Dipl. Eng., General Manager Sales - Coal Division, KHD Humboldt Wedag AG, Cologne, Germany.

** B. Sc., Eng. IRIMME Chief of Marketing & Sales Humboldt Wedag India Ltd. Calcutta, India.

the lower wear in the boiler houses and the increased volume of transportation in the case of prior separation of refuse from the power station coal, then a preparation, at least of the lump sized portion should be profitable. Test regarding the abrasive properties of different types of coal have shown that the abrasion and thus the wear are considerably limited due to a separation of refuse from lumpy run-of-mine coal.

The preparation of coarse coal is a process step frequently found in Europe. This has to be attributed to the fact that the proportion of refuse in the raw coal varies with respective grain size. Normally the percentage of refuse in coarse raw coal is higher than in fine granular raw coal. In West Germany about 25 per cent of the raw coal is in the grain size fraction plus 40 mm, and the refuse content of this fraction average 90 per cent, which is nearly one half of all refuse contained in the-raw coal (Table 24.1).

The removal of a certain portion of refuse from the uncrushed raw coal offers great advantages for the preparation process, such as for example

- Decrease of power requirements for size reduction.
- Increased yield of graded coal in the case of domestic coal production,
- decreased yield of slurry,
- decrease of ash content in the raw slurry,
- more favourable homogenizing properties of the raw coal,
- decreased wear in the concentration machines,
- more favourable utilization of the preparation capacity, i.e. increased efficiency of the preparation plant.

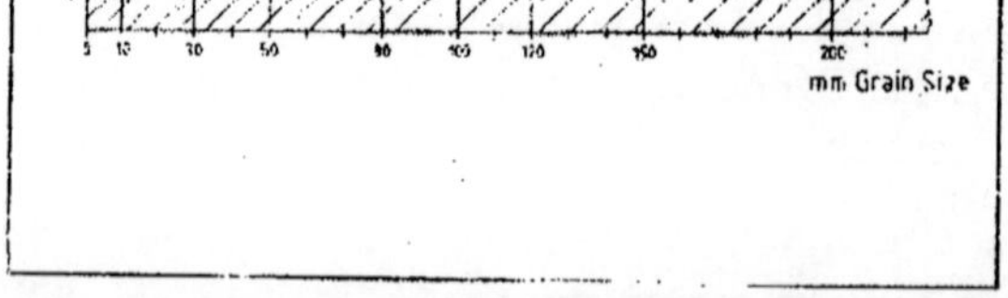

Table 24.1 : Refuse content of a typical German ROM-Coal

The development of the ROMJIG opened the possibility, to use a very simple process for the beneficiation of the lump size coal - the top size is 400 mm and the bottom feed size is 30 mm - by using only one technologically improved machine and some standard auxiliary equipment.

In fact the suggestion to reconsider an economic process for the beneficiation of ROM coal actually came from India. At the first joint meeting of the Indo-German Coal

Group, the German Beneficiation Industry was asked to propose a process, as far as possible dry, to solve the problem of preparing non coking coal.

After many years of development such a plant has been installed in the Bina coal handling plant and is ready for operation. In this plant the raw coal is screened at 30 mm. The coarse coal 250 - 30 mm is fed into the ROMJIG and separated in refuse and clean coal.

1.2 Description of ROM JIG

First of all it had to be ascertained whether lumpy material could also be stratified according to density in water bath like the size fraction of minus 120 mm mentioned before. For this purpose sink tests with a size fraction of 400-40 mm were conducted in water bath which showed that after repeated mechanical lifting and free settlement of the material a stratification according to density is effected, with the refuse settling in the lower area while the lighter coal lumps are overlying.

This knowledge led to the construction of the Movable Screen Jig, designed as single cut jigging machine for the separation of stones from run-of-mine coal (Fig. 20.1).

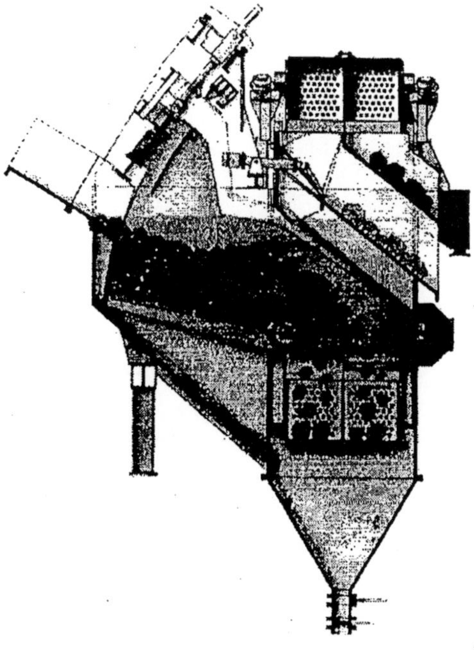

Fig. 24.1 : ROM-Jig

To find a proper name for this kind of jigging machine, we called in "ROMJIG" which is self-explanatory in defining the function of the machine. In Germany we call it also the ..Indian Jig" because the initiation to develop it came from there. In the meantime the word "ROMJIG" has been registered to stand for this type of movable screen jig.

The ROMJIG consists of an unilaterally supported screen panel with a slot width of 15 mm which is hydraulically moved up and down in a water vessel. The movement of the screen panel is effected via an hydraulic cylinder suspended on gimbals, by which the screen panel is lifted. Upon reaching its highest point the oil pressure in the hydraulic cylinder is released and the screen panel freely falls through the water to its lowest dead position.

The kinetic energy is absorbed by a hydraulic shock absorber connected with a central hydraulic plant preventing the building from receiving noticeable shocks.

The temperature of the circulating hydraulic oil is kept within permissible limits by a water-operated head exchanger.

The material fed along the side of the maximum stroke length of the screen panel is stratified according to its density by the lifting movement of the screen panel. Transport is effected due to the inclination of the screen panel and the sliding pressure of the following material. The refuse is discharged by a rotating discharge roll acting as a retaining edge which is set into circular motion via oil motor and chain drive.

The speed of the discharge roll depends on the weight of the material dumped on the screen panel, i.e. on the average content of refuse. This weight acts as a command variable to determine the pressure to be applied to the lifting cylinder by the hydraulic oil; being a sufficiently accurate index for the height of the refuse layer in the stratified material on the screen panel.

The light material resting on top of the refuse layer in front of a weir moves over the weir edge. The discharge of the two products is effected by a twin bucket wheel, sufficiently dimensioned to allow proper de-watering of the products during the lifting process.

Unlike the conventional coarse coal jigs the ROM JIG does not require any backwater to support the jigging process (Fig. 24.2). Only a certain amount of water is withdrawn from the hutch of the ROMJIG through a gate in order to continuously discharge the material accumulating in the hutch. This hutch product is de-watered on a screen with a slot width of 0.5 mm and fed to the raw feed coal for further preparation.

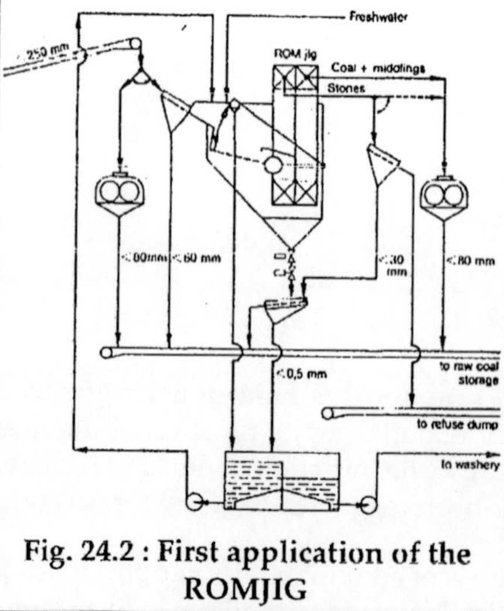

Fig. 24.2 : First application of the ROMJIG

The screen undersize does not need any additional clarification. It runs to a pump sump and is returned directly to the ROMJIG.

Quite generally one can say that the ROMJIG requires so much additional water as it is taken up and carried away by the particle surfaces of the product. As known from experience this comes up to 1-2 per cent or 3 to 6.5 m3/h for the size fraction plus 30 mm.

1.3 APPLICATION OF ROMJIG

Such a plant has been installed in the Bina coal handling plant and is ready for operation. In this plant (Fig. 03) the raw coal is screened at 30 mm. The coarse coal 250 - 30 mm is fed into the ROMJIG and separated in refuse and clean coal.

The refuse discharged from the ROMJIG is separated from the undersize on a stationary bar screen with a slot width of 30 mm and the oversize finally conveyed to the refuse tip.

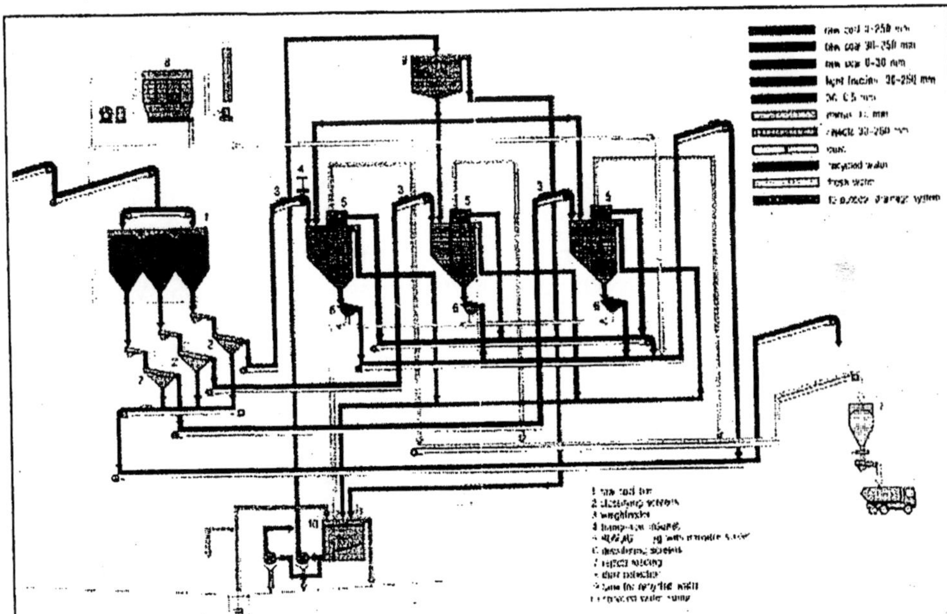

Fig. 24.3 : Flowsheet of the Bina deshaling plant

The light product of the ROMJIG, consisting of coal and middlings, is crushed down to a washable particle size by a double-roll crusher and then conveyed together with the screened out undersize material to the two raw coal storages.

Fine material may accumulate in the water bath having passed through the opening of the jigging surface and will have to be extracted by an occasional introduction of water in order to avoid such accumulations. This hutch product is de-watered on a screen and added to the light material.

The underflow of the de-watering screen is returned to the ROMJIG without any intermediate clarification. From operational data obtained it has been found that, depending on the feed rate, with its low proportion of fines, a balance of solids is formed which in no case exceeds 50 g. The ash content of the circulating solids came up to approximately 60 per cent. This knowledge led to a considerable reduction of the freshwater requirement to 3.5 m3/h, just sufficient for covering the losses of water absorbed by the products.

1.4 Results

At this point I would like to present the predicted results of the non-coking coal beneficiation plant installed in the Bina coal handling plant. This prediction is based on the test results, which have been conducted in the research and development centre of K.HD Humboldt Wedag Ag in Cologne. A five ton sample of the raw coal has been available, to run this test work. The results were as follows:

Screen analysis			*Products of that plant:*		
grain size	Wt. %	Ash %	Product	Wt. %	Ash %
200-30 mm	82.3	44.2	deshaled coal	49.4	28.3
30-0 mm	17.7	37.6	unwashed coal	17.7	37.6
200-0 mm	100.0	43.0	saleable product	67.1	30.8
			Refuse	32.9	68.1

A similar plant has been installed and commissioned in China. The installation was originally a deshaling duty, where the product of the ROMJIG should be sent to the existing washery after having been crushed down to the required size. While commissioning this system the results were so excellent that the plant is selling the product if the ROMJIG as a final product. In fact the imperfection value of the plant has been determined to be 0.09.

1.5 Prospects

The successful first employment of a ROMJIG for large refuse separation of run-of-mine coal opens a wide field of future applications. In view of the favourable secondary conditions, such a low water and power consumption, low expenditure for necessary auxiliary equipment this process stands any economical comparison with other possibilities of large refuse separation and is the ideal process, in case people think of a pit head instalment.

Such a preparation plant for power station coal be exemplified as follows:

The ROM coal is taken to the central crushing station. After having crushed down to minus 400 mm, the coal is classified at 30 mm while the oversize 400 to 30 mm is separated from the refuse in a ROMJIG.

By experience this lump size fraction of the ROM coal has a high content of refuse, so that in this way the greater portion of the refuse is separated and the ash content of the power station coal can be lowered considerably.

Such a preparation will be very simple and economical as it has been exemplified by the use of the ROMJIG for separation of large refuse at the Emil Mayrisch mine.

2.0 BENEFICIATION OF COARSE, MEDIUM AND FINE GRAIN COAL

2.1 General

Whenever the characteristics of the raw coal does not allow, to achieve the required product quality with the ROMJIG by only washing the lump size coal, the grain size of beneficiation has to be reduced. Reasons can be the grain size distribution, when the amount of material < 30 mm is too small and the ash content of the fine material is too high. This leads immediately to the beneficiation of coarse coal by means of Batac® jig.

2.2 Description of the plant

A typical process (Fig. 24.4) for such a plant requires only very few equipments. The raw coal with a top size of 75 (120) mm is at first screened at 13 mm. The coarse product 75 (120) - 13 mm is fed to the Batac jig, where it is separated into refuse and clean coal. The clean coal product is dewatered by means of a fixed screen followed by a vibrating dewatering screen. The dewatered product will be mixed with the unwashed fine coal coming from the underflow of the classifying screen.

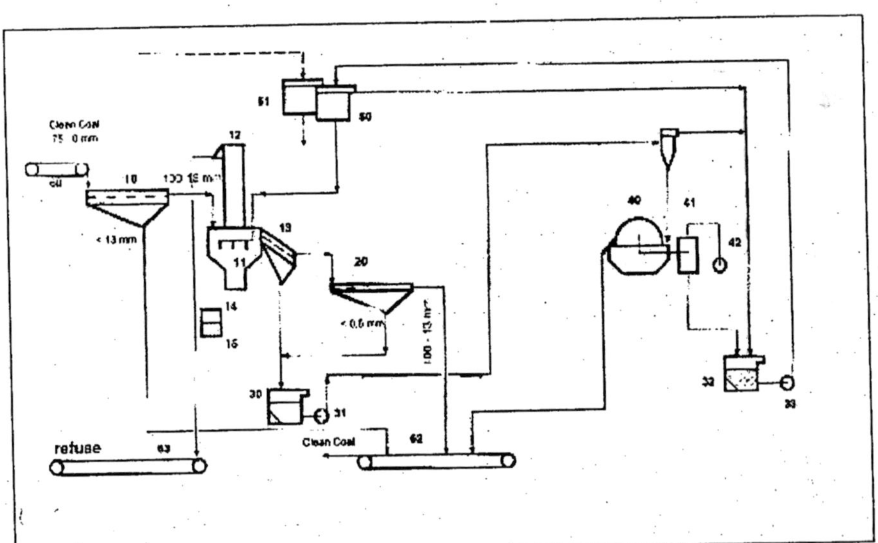

Fig. 24.4 : Flowsheet of a coarse coal beneficiation plant

As it is seen from the picture, the process is closed by means of a combination of cyclones and filters.

2.2 The BATAC® jig for coarse, medium- and finesize coal

The under bed air - pulsated BATAC jig (Fig. 24.5 and 24.6) of reduced weight and constructional volume has been developed for lowering capital expenditure.

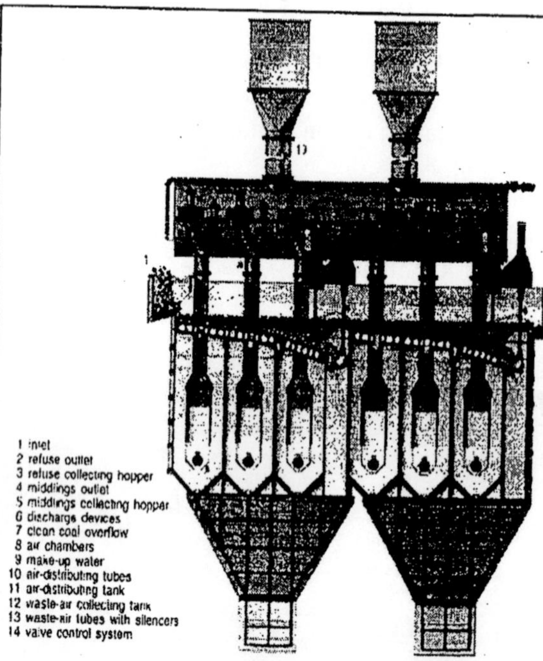

Fig. 24.5 : Batac coarse coal jig

There is one decisive difference between an BATAC jig (Fig. 24.5 and 24.6) and a Baum type jig.

The water current is not generated in an air chamber mounted adjacent to the screen plate but in air chambers underneath the jig bed. These air chamers are intermittently supplied with compressed air by an electronically controlled valve system. That air is intermittently discharged from the system after completion of the upward stroke.

Motion is imparted to the water inside the jig as a function of the pressure generated inside the air chambers.

Moreover, make-up water is added at the lowest location of every jigging chamber to intensify the upward current and to dampen the downward current.

The feed is stratified according to is density by the pulsating motion of the water:

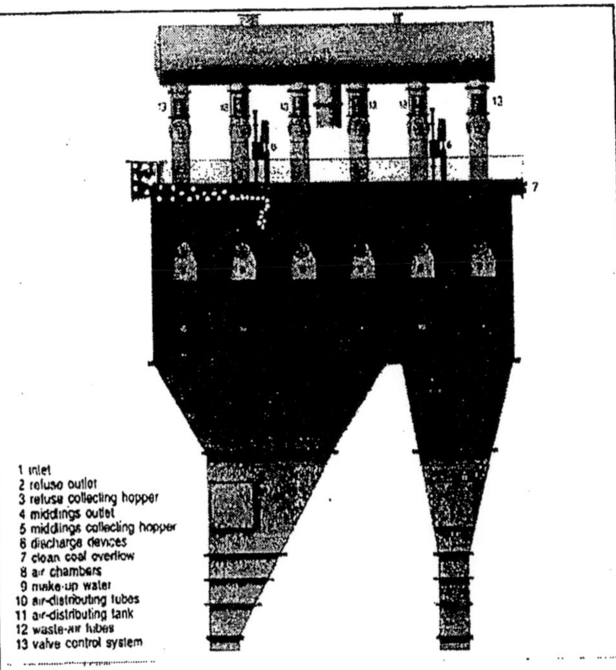

Fig. 24.6 : Batac fine coal jig

- when preparing coal, e.g., in refuse, middlings and clean coal.

The heavy fraction of the stratified raw material is sensed by floats in accordance with the product qualities required. The system provides for controlled withdrawal of the heavy fraction over a discharge device. Jigs of great width are equipped with independently operating discharge devices fitted with separate sensor and hydraulic units. The Magnitude and the kind of kinetic energy admitted for the benefication process are of decisive importance for a successful separation.

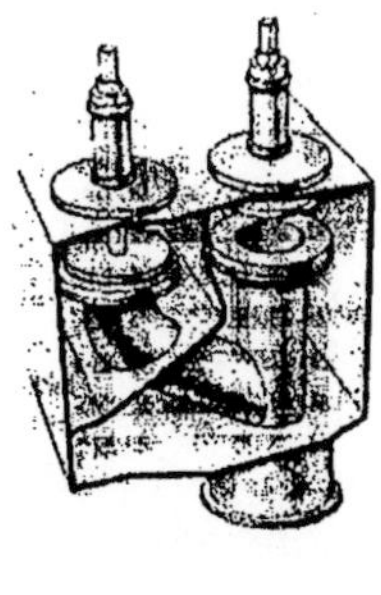

Fig. 24.7 Disk valve

Therefore, special attention has been given to the development of the pulse generators for the controlled admission of the compressed air (Fig. 07,08).

Fig. 24.8 : Rotary flap valve

The quantities of jigging air are controlled electronically. The amount of air can be set individually for each jigging chamber and an identical pulsation frequency (40 - 120 strokes/min) is set for all jigging chambers in the jig PLC (Fig. 24.9).

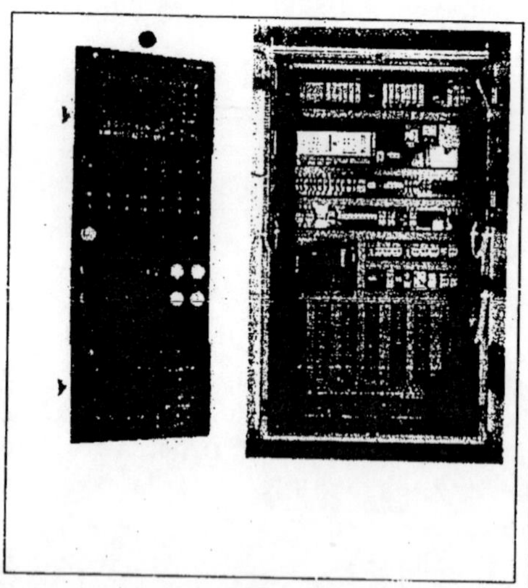

Fig. 24.9 : BATAC jig control system

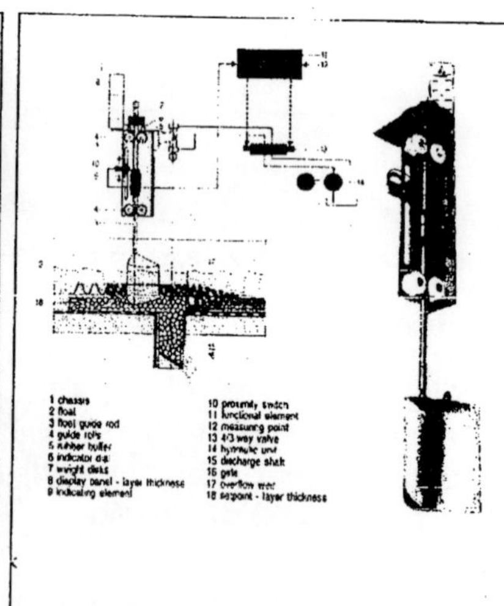

Fig. 24.10 : Discharge control system

The thickness of the coal layers stratified by specific gravity is sensed with the aid of a float equipped with a suitable linkage (Fig 24.10). Formerly employed induced current metering units are today replaced by analog displacement measuring systems operating at ultrasonic velocity. The basic float setting is done by additional weights. The hydraulic systems of the discharge gates (Fig. 24.11 and 24.12) are actuated from the. PID controllers which are integrated in the jig PLC (Fig. 24.9).

Fig. 11 Discharge gate coarse coal jig

Fig. 24.11 : Discharge gate coarse coal

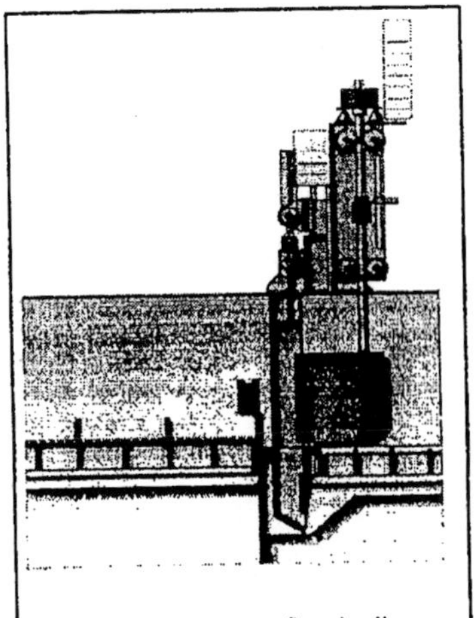

Fig. 12 Discharge gate, fine size jig

Fig. 24.12 : Discharge gate, fine size jig

2.4 Prospects

The Batac jig is so flexible that the bottom washing size can be reduced from 13 mm to 6 mm or even to 3 mm, without a modification of the flowsheet. The only problem is the dry screening at the lower cut points. This can be solved by using the right screen system under these circumstances.

Additionally the depth of beneficiation can be increased by adopting further processes, like the combination of coarse and fine coal jigging and at the end the addition of the flotation system, which is relatively seldom in the beneficiation of a power station coal.

3.0 BENEFICIATION OF DUMPED COAL

3.1 General

In good coal reserves - preferably in open cast mining - a relatively big amount of good quality raw coal is dumped, i.e. wasted, due to the selective mining method. This is the material close to the boarder between coal seam and shale band, which contains upto 50% of saleable coal.

Worldwide plants are being installed to process especially this so called interface material. The process flowsheet of such a plant appears very simple. The heart of this plant is the Batac jig and the depth of beneficiation is again defined by the raw coal characteristics and the product requirements. Due to this the flowsheets of such plants vary widely and I would like to present one system, which can be considered as a standard for this duty (Fig. 24.13).

3.2 Description of the plant

The raw coal is fed in the grain size 50-0 mm to the beneficiation plant. Here it is screened at 0.5 mm first, especially to achieve a good distribution over the width of the following machine and to wetten the material prior to the jig. The overflow of the screen is directly going into the Batac jig, where it is separated in refuse and clean coal.

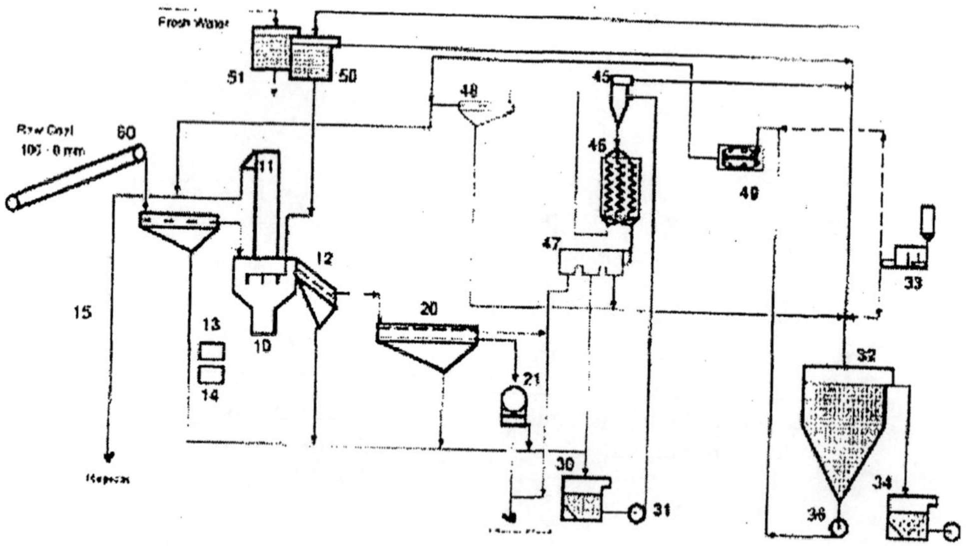

Fig. 24.13 : Flowsheet of a dumped coal beneficiation plant

The refuse are taken out of the jig by a dewatering bucket elevator and fells down onto a belt conveyor, which brings the refuse out of the washery.

The clean coal is overflowing at the end of the Batac jig and is dewatered by means of a fixed screen followed by a double deck vibrating screen. The overflow of the upper deck is dewatered enough and can be sent directly to the clean coal belt conveyor. The overflow of the lower deck is further dewatered in a basket centrifuge, whereas the underflow of the screen is classified together with the underflow of the desliming screen in cyclones, the underflow of the cyclones can be beneficiated in spirals. The spiral concentrate is dewatered together with the jig concentrate in the centrifuge, whereas the refuse are dewatered by small dewatering screen and then sent out of the plant together with the jig's refuse.

The overflow of the cyclone is fed to a thickener, from where the clarified water is recirculated back to the plant. The underflow of the thickener can be either dewatered on filter presses or on centripress centrifuges.

4.0 BENEFICIATION OF POND MATERIAL

4.1 General

In many places in the world slurries are pumped into big slurry ponds. Aside the fact that this is wasting of energy, it also destroys the environment. In future the use of such ponds will no longer be allowed. Technically it is no problem, to close the water circuits of the mines or the preparation plants, which are the main reason, why the ponds exist at all.

This section of the beneficiation plant is unfortunately not only the most expensive - as well in investment as in operation - but at the same time the lass valuable one due to the higher content of moisture and ash. In some countries even the thermal drying is required.

4.2 Description of the plant

A typical flowsheet of such a plant is shown in Fig. 24.14 and consists of the following beneficiation steps.

The slurry is pumped into this beneficiation plant and is at first screened at 3 and at 0.5 mm. The overflow is a check screening to avoid oversize material to enter into this system. The overflow 3.0 - 0.5 mm is separated in spirals into rejects and flotation concentrate. The spiral concentrate is dewatered in screen bowl centrifuges, whereas the rejects are screened and sent out together with the coarse rejects

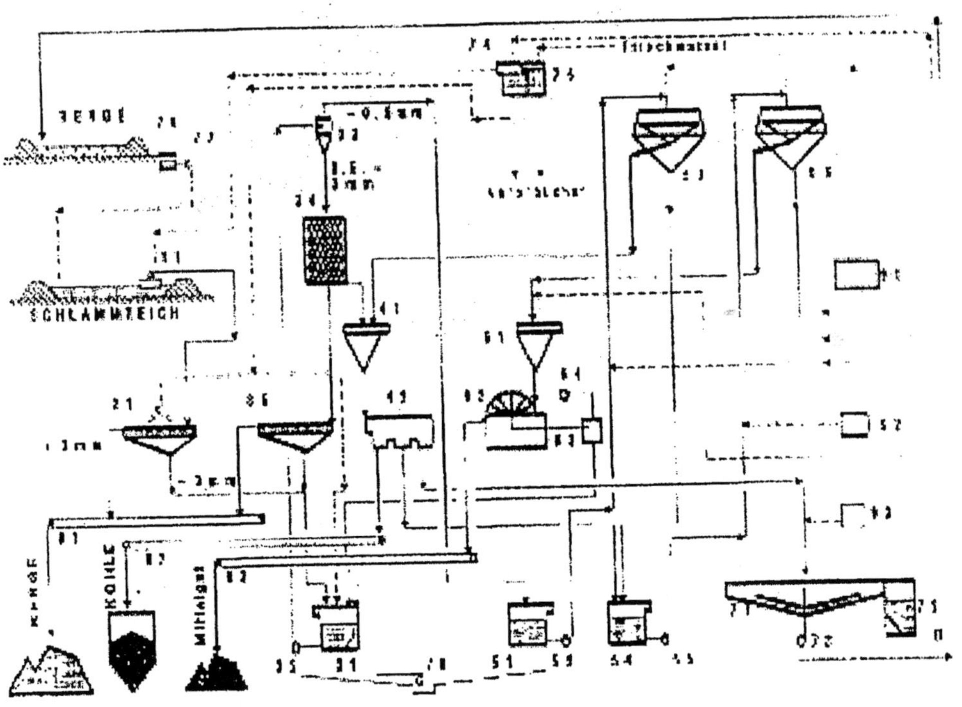

Fig. 24.14 : Flowsheet of a beneficiation plant for slurry pond material

The underflow of the screen is collected in a sump and is being fed to the flotation system. The floatation is the Pneumatic flotation (Fig. 15), a further development of the conventional cell flotation. As the cross section shows, one single central aerator is located on top of the separating vessel. The aerated pulp flows through a pipe into a distributor box near the bottom of the vessel. Orifices in the distributor discharge the pulp vertically, with controlled speed, into the vessel.

A conical froth crowder can be moved up and down on the feed pipe via a hydraulic drive to adjust the size of the froth area to optimise it for various applications. Further control of the concentrate quality is obtained via a wash water system on top of the froth. The economies of the system are further improved by a new type of aerator that does not need compressed air.

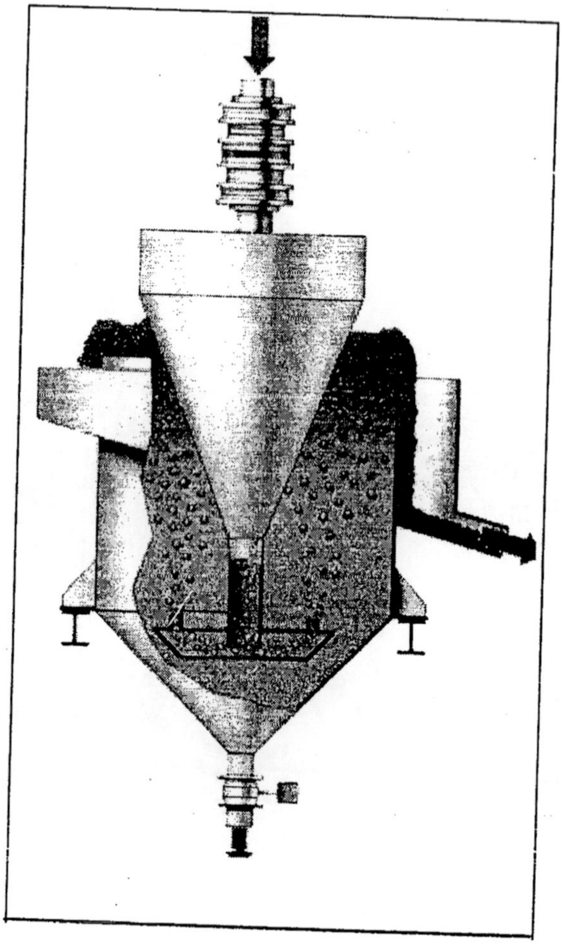

The proven design of the Pneumatic flotation has been further enchanced. The operating features are :

- high recovery, especially for coarse particles
- optimum selectivity due to washing device
- optimum flexibility due to froth corwder
- unbeatable economic efficiency because of
  - low investment costs
  - economical use of spares
  - excellent product quality

In the pneumatic flotation the raw slurry is separated into tailings and flotation concentrate.

The flotation concentrate can be dewatered either together with the spiral concentrate in the screen bowl centrifuge or separately on vacuum filters.

The flotation tailings are sent to a thickener from where the overflow, i.e. clarified water, is pumped back into the circuit. The underflow of the thickener can be dewatered on centripress or filter presses.

5.0 CONCLUSIONS

According to todays standard it is no longer required or necessary to work with open washery circuits and to damage the environment as well as to spoil energy. The described processes for the various applications have proven their flexibility as well as their acceptance in the coal beneficiation industry. In future it will be a must throughout the world, to beneficiate the raw coal before it will be sent to the power station. In the power stations they will appreciate the better and constant coal quality, being fed to the boilers.

CHAPTER 25

Continuous Miner Technology into the 21st Century

*A.D. Johnson**

SYNOPSIS

At the start of the 20th Century coal mining was the most important extractive industry and played a major part in fuelling the Industrial Revolution. Extraction was largely by hand with the help of ponies for haulage.

The introduction of mechanisation in the major western markets of the USA, Australia, South Africa and the UK started in the 1930's and 1940's and resulted in tremendous increases in productivity and safety.

Today mechanised mining equipment is continually being adapted and upgraded in ever shorter cycles leading to an increasing rate of change and associated increasing improvements in productivity, availability, safety and commercial gains.

The Indian underground coal sector, though the fifth largest underground producer in the World with 80 million tonnes per year, still has a low level of mechanisation and massive scope for improvements.

This paper looks at the current continuous miner equipment in use worldwide, examines the tremendous increase in productivity from continuous miner systems in recent years and takes a critical look at the Indian market.

Recently India has introduced semi-mechanised mining with SDL's and LHD's. The scope to introduce new fully mechanised continuous miner systems is examined. Introduction of these systems will increase underground production, productivity and safety, however there is a slowness to introduce these modern proven systems, despite a large shortfall between the energy demand and coal produced. Increasing underground

*Managing Director, Joy Mining Machinery, U.K.

production by continuous miner mechanisation is a positive step to both keep India self sufficient in energy production and reduce mining costs.

Finally, the paper examines several continuous miner projects in active discussion in India and looks at the possible growth of modern continuous miner systems for India during the beginning of the 21st Century.

1.0 INTRODUCTION OF CONTINUOUS MINER SYSTEMS

Coal in the early part of the 20th Century was extracted by hand won techniques only. Powered ventilation and hoisting were developed and improved upon during the early part of the century, but it wasn't until the 1930's and early 1940's, that electrical equipment was introduced initially to load and transport coal and later to undercut, drill and load coal.

The first practical production continuous miners were introduced in the late 1940's to improve productivity in room and pillar mining systems in the USA. As in India, room and pillar mining systems are operated on a place changing sequence whereby one roadway in the system is advanced a certain distance creating a room, after which the mining equipment moves to another workplace while the roof bolting operations are carried out. Under US mining law it is forbidden for an operator to advance beyond the last row of bolts and in conventional equipment sections the unsupported advance, as a function of undercutting and drilling, seldom exceeds 3m to 4m. An advance of this magnitude in the relatively shallow mines of Pennsylvania and West Virginia, and also in many Indian mines, can stand unsupported for a considerable time, enabling the bolting crews to secure the area after the blasted coal is removed.

The introduction of continuous miners enabled the depth of the rooms to be increased to 6m (i.e. the distance from the face to the operator's seat). The ability to mine more coal from a room before moving to an adjacent room brought increased productivity and today, with the introduction of remote control operation, rooms of 12m are taken in many mines with consequent productivity improvements.

The growth of continuous miner systems with rubber tyre shuttle cars and mobile bolting equipment was rapid throughout the 60's, 70's and 80's. The continuous miner system is the first choice for all room and pillar and longwall development operations in the underground coal markets of USA, Australia, South Africa and the UK. Continuous miners have been introduced in and are operating successfully in Russia, China, Japan, Zimbabwe, France, Italy, Mexico and Norway (Spitzbergen). There are now almost 2000 continuous miners operating in over fourteen countries worldwide.

Country	*Estimated No. Operating CM's*	*Estimated No. of Operating Shuttle Cars*
1	2	3
USA	1350	1500
Australia	1903	70
South Africa	275	945

1	2	3
UK	90	40
China	15	10
Russia	7	14
Norway	6	12

2.0 LATEST CONTINUOUS MINER DESIGN

Continuous miners are capable of mining seam thicknesses from 0.9m to 6.0m. Although cutting up to 6.0m is possible, the problems of roof bolting at this height and possible sidewall instability make it preferable at many collieries to mine thick seams in two slices. Cutting up to 6.0m has been carried out at Gloria, Khutala and Matla in South Africa, however the maximum height being cut at present is 5.0m.

Joy Mining Machinery supplies a wide range of continuous miners in its 12CM, 14CM and 17CM series with new designs continuing to replace older models and has dominated the market for continuous miners and shuttle cars for many years.

In most cases room and pillar sections use one continuous miner and two shuttle cars for mineral transport --the electric cable-reeled shuttle cars travel short distances to discharge onto a panel belt conveyor and then make a return journey.

During the 80's and 90's cutter motor power has increased rapidly and continuous miners now have total power ratings from 350 to 900KW. 1000V supply is standard with larger machines now operating at 3300V. Improved bit/pick technology through the use of cad/cam and numerically controlled co-ordinate manufacturing has improved the cutting performance at these high powers.

The basic concept of the continuous miner is of a multi-motor machine with a separate electric motor dedicated for each driven function. A typical machine has one motor to drive each track, two motors to drive the cutter drum, one or two motors to drive the gathering arms and central conveyor, and one motor for the hydraulic pump. Hydraulic functions are kept to a minimum consisting of raise and lower of the cutting head, lift and slew of the swishy tail conveyor (for discharging into shuttle cars), raise and lower of the gathering head (apron) and hydraulic control of the service jacks.

Effective application of the increased cutter forces is optimised through a cutter feedback system which regulates the sump rate of the cutter drums by monitoring the current drawn by the cutter motors.

Electronic systems are now extremely reliable and coupled with the traditional multi-motor design (giving much greater reliability than complicated hydraulic machines), availabilities of 90 to 95% are the norm. A motor and gear case is dedicated for each function (tracks, loading, cutting and pump) and modular construction allow easy trouble shooting and repair. Most models now have on-board electrical diagnostics to allow fast testing and repair of any faults.

During tramming and clean up, the load on the cutter motors is low and traction speed has been increased to improve productivity and reduce tramming times in place changing operations. Gear drive traction systems have replaced the old chain drive, bringing further improvements.

Increased productivity has resulted from the use of dual drives, individually powering two gathering arms of centrifugal spinners. Central conveyors of widths of up to 1.0m are now in widespread use.

Control circuit failures can contribute significantly to down time, but identifying where the fault lies can be very time consuming and can be all the more difficult if investigation with the machine covers off is not allowed. For better control and diagnostics, new computer based electronic control systems are being used. Joy has developed the JNA (Joy Networked Architecture) system to handle radio remote control, machine status, monitoring and machine trouble diagnosis.

3.0 ROOM AND PILLAR MINING

Room and pillar mining is a very efficient form of mining, especially at shallow depths, as the small pillars result in faster place changing and a higher percentage extraction. Its flexibility enables reserves to be mined where geological conditions preclude the use of longwall mining and high outputs are achievable using continuous miners compared with drilling and blasting. Where good roof conditions exist, deep cuts of up to 12m can be mined using remote control before place changing and roof bolting takes place.

Room and pillar sections can have single or double units working simultaneously or separately by one crew, and supersections employing two continuous miners can produce around 4000t per shift. While the operator mines a deep cut, the maintenance crew tram the second continuous miner to the start of the new cut. When one deep cut is completed, the operator walks to the second miner and carries on cutting.

High production outputs in room and pillar sections are achieved with combinations of deepcuts, thick seams, high shuttle car availability, shift changing at the machine, machine swapping and continuous haulage.

Northern Mingo No. 2 Mine in West Virginia, US, which employs shift changing at the machine, produced 19,971t in 24 hours from two continuous miner sections over five machine shifts. Once section uses two remanufactured Joy 12CM12 continuous miners and the other uses two re-manufactured Joy 14CM15 machines.

Pillar extraction using continuous miners is the preferred method to conventional drilling and blasting, as faster extraction of a panel and the extraction of one pillar at a time promotes a safer extraction sequence. Rib pillar mining and its variants, such as the Wongawilli system, have proved very successful in South Africa and Australia. Rib pillar extraction consists of developing long narrow ribs of coal, typically 50-150m long and 7.0m wide, which are extracted on retreat. A Joy Joint Venture at Raspadskaya mine in Siberia is currently mining to an adapted Wongawilli system using a 12CM18 miner. White output is a modest 20,000t per month, this is a considerable improvement on the traditional Russian longwall mine.

4.0 WORLDWIDE MARKETS

It is in the market-led economies where underground coal mining machinery has undergone the most rapid development to enable high productivity and low ownership costs - currently lowest cost per tonne producers of underground coal are the USA, Australia and South Africa.

The cheapest underground coal producers are using heavy-duty longwall installations or a variety of room and pillar/rib pillar extraction methods to compete economically with cheap open cast operations in the world market.

Development for the heavy duty longwall installations is also generally by continuous miner systems operating place changing in 2, 3, 4 or 5 gate roadways.

While introduction of longwall systems has increased dramatically in the 80's and 90's, room and pillar mining still accounts for approximately 50% of underground coal production in the major market economies.

USA

The US coal industry is still the largest market for continuous miners and as such, exerts a strong influence on current designs. In general, US high-seam applications are declining and the market is looking for mid to low-seam continuous miners.

There are approximately 800 underground coalmines in the United Stated. Of these, 53 are longwall mines and the balance are room and pillar operations. However, of the approximate 380 million tonnes of underground coal produced in the US, half of it is from longwall operations and half of it is from room and pillar operations. Underground coal production by mining methods is shown below in graph 1. Longwall output has risen in spite of an opposite trend in the number of longwall faces.

Price pressure has resulted in continuous productivity improvements in both room and pillar and longwall mines, see graph 2 overleaf. These have been realised by the closure of old mines with large infrastructures, closure of low productivity mines, improved management approaches, technological advancements, improved information systems and re-engineering of mining plans (i.e. super sections, increased cut depth). Maximum place change mining cut depth is now 12.2 metres and petitions are in place to increase these depths to 18.3 metres.

The data show that the US industry is mining more coal than ever and is doing so with fewer coal mines, fewer employees and in thinner seams. Continuous miner operations continue to play a major role in the underground industry.

REPUBLIC OF SOUTH AFRICA

In South Africa underground coal production is approximately 110Mt with the very hard thick coal seams close to the surface making bord and pillar mining the most popular method of extraction. Over 90% of production uses either high seam continuous miners or conventional methods in roughly equal proportion.

The thick seams in South Africa have led to exceptional performances. At Khutala Mining Services, which supplies coal directly to the Kendal power station, output is 370,000t/month from three double header sections, two producing approximately 125,000t/month and one at 120,000t/month.

Each double header section consists of two heavy duty continuous miners (1 x 12HM17 and 1 x 12 HM9) and six 20t capacity shuttle cars. Miners have their own conveyor (two per section), located in the 3rd and 9th roadway of a nine entry room and pillar system.

The Joy 12HM21 is capable of cutting up to 6m and is the most popular model in RSA producing up to 100,000tpm.

UK

The introduction of roof bolting techniques in deep mines in the UK has enabled the widespread introduction of continuous miners for longwall entry development. The requirements for narrow roadways and the ability to cut both coal and stone bands, has favoured the 12CM15 machine (see fig 25.1). The UK now produce around 25 million tonne per year from 17 underground mines all of which use longwall systems due to the depth of operation.

AUSTRALIA

In Australia thick regular seams permit longwall mining on a scale not achievable anywhere else but in the US, prompting the use of the drum-type continuous miner, for longwall gate road development. Underground output is now around 75 million tonnes with two thirds from longwall and one third from continuous miners in room and pillar systems.

Oaky North Underground Mine broke the Australian development record for a single unit, mining and roof bolting 3312m during the quarter of July to September 1996, using a Joy 12CM12C5.2. The unit then came back in October of that year with an advance of 1085m. The mining conditions at Oaky North are variable with the immediate roof consisting of heavily laminated silty mudstone with numerous micacaceous partings. The strength of the roof material varies from 1450psi/10Mpa to 7252psi/50Mpa. The seam thickness ranges from 2.5m up to 4.7m.

5.0 UTILISATION OF CONTINUOUS MINER SYSTEMS

Continuous miners today have no problem out-producing the system's ability to support the miner. Removing the mined material from the section and securing the mined area remains the bottleneck, even with deep cut mining made possible by remote control and auxiliary ventilation. Manufacturers are now, therefore, concentrating on improving the machine and reducing delays in the mining cycle, rather than increasing the machine's capacity.

Typically a continuous miner operates at plus 95% availability, but is utilised only 20% - 30% of the time in a 24 hour period. In order to improve this a total system approach is needed in which the extraction, roof support, coal transportation, ventilation, services and materials supply capacities are matched and optimised to the production equipment. New equipment and infrastructure systems are needed to achieve these potential increases in performance.

In addition, haulage with shuttle cars is a cyclic batch system, which involve delays waiting for cars. Over recent years cars have been upgraded by providing larger payloads, introducing fully sealed reliable cable reel systems and the newest innovation is full suspension systems. These allow a more comfortable ride, better handling over poor ground as well as increased payloads and reduced shock loads on the wheel units and maintenance.

The developed countries of USA, RSA, Australia and the UK already have good individual equipment and roofbolting techniques. Markets where development is still taking place, like China, Russia, Poland and India, will all have to adopt fully mechanised roof bolting practices to become truly global competitors.

Major underground coal producing companies worldwide require more rapid gate road development and more productive room and pillar operations. The easiest way to achieve this is through reduced bolting process time. Many manufacturers are now looking at reliable automation to achieve this. Experiments have been carried out with automatic feeds, roofbolt carousels, resin injection, special bolts and a variety of roofbolting rigs from pneumatic to hydraulic. Several manufacturers have recently introduced multi-mast rigs to the market.

6.0 NEW TECHNOLOGIES UNDER INSTALLATION

A new mobile roofbolting rig "the JOY Multi-Bolter" has now been introduced which is one part of the new systems approach

The first was introduced into operation in Capcoal Central colliery in Australia in August 1998. The second was introduced into Sufco mine, Arch Coal, Utah, USA in February 1999.

This machine is a new concept multiple drill mast, platform based, mobile bolter designed, primarily for "place change" continuous miner applications. This design eliminates booms and many other moving parts, reducing wear and repair parts and enhancing the safety and life of the unit.

The wide open platform has hydraulically activated side leafs that lower to create an extremely large working area (typically more than three times larger than other mobile bolters in the same class).

On the front of the machine are four JOY 'HFX' drill rigs with JOY 'HDR' drill motors. The in-board roofbolters provide a side swing of 100 either side of vertical and tilt forward up to 40. The out board roofbolters slide sideways to cover a large area for placement of vertical roof bolts and tilt over 1050 to allow full placement of rib bolts if required.

Extra safety is provided by a two cylinder ATRS (Advanced temporary roof support) which is set to the floor prior to engaging the roof and gives 2 x 10 tonne support in addition to each roofbolter mast's own roof jack yield capacity of 5 tonnes, see fig 28.2.

Due to the open platform and flexibility of positioning of the roofbolter masts at the front of the machine, the multibolter can easily be upgraded with new technologies as they become proven.

Bolting Achievements in Australia

The first JOY Multibolter at Shell's Capcoal Central Colliery, Australia, achieved what is believed to be an Australian bolting record, installing 304 bolts of 1.8m length in just 225 minutes (3 3/4 hours). This was achieved with only two operators. The roadway was 5.2m wide and 2.6m high with four full column resin bolts paced every one metre of advance. The machine can average well over 400 bolts per shift at this mine.

Bolting Achievements in USA

In the Sufco mine, Arch Coal, Utah the JOY Multibolter set what is believed to be an American bolting record. On Friday, February 19 a crew of 2 operators installed 304 roof bolts in 200 minutes. These were 1.5m full column resin bolts on a 1.5m x 1.5m square pattern in a 6 metre wide roadway, 3 metres high. Again averages of well over 400 bolts per shift are well within the range of this machine.

7.0 CONTINUOUS MINER SYSTEMS FOR INDIA

Joy shuttle cars were in general use in India from the 1950's to the 1970's, however this technology was not updated or replaced. Most of India's underground mines now operate room and pillar systems using either manual method (bore, fire, hand load) or semi-mechanised methods (bore, fire, load by LHD or SDL).

From introduction in the mid 1980's SDL technology has grown rapidly with over 744 SDL's and 63 LHD's installed at the end of 1997. The increased use and wide acceptance of SDL technology over recent years has improved productivity, but the 1997 productivity average of 82 tonnes per machine day (96 tonne for SECL) is still only a fraction of that achievable in similar or worse conditions in other markets. Compare this to an average of 2000 to 6000 tonnes per machine day for continuous miners in the USA. Output per man shift (underground) of 0.57 tonnes, leaves India firmly at the bottom of the worldwide productivity league.

Productivity	*Tonnes/Manshift (Approx)*	*Year*
USA	33.24	1996
Australia	21.25	1995
South Africa	14.2	
UK	15.2	1997
Poland	2.4	
India	0.57	1997

Whilst there are reasons for this with a large quantity of low cost labour, there are pressing economic and production pressures which require a change. In the next century several continuous miner systems are already under consideration for introduction. It is interesting to compare the almost constant underground production and productivity in India (see graph 3) with the rapidly increasing productivity and production per mine in the USA (graphs 1& 2)

Joy Mining Machinery have signed a contract with Coal India to supply the first Continuous Miner package in India to SECL's Chirimiri mine. The package consists of :-

- One 12CM15 continuous miner with 2 x 170KW cutter motors, 2 x 37KW DC track motors, 2 x 45KW gathering head motor and a 40KW pump motor.
- Two 10SC32 Shuttle cars 13.7 tonne capacity with 2 x 37KW traction motors, one 18KW conveyor motor and a 15KW pump motor, see picture in fig 25.3.
- One BF14B Feeder Breaker.

- One Joy Multi Bolter, multi drill roof bolter and one Ramtrack single drill bolter.
- Associated Electrical equipment.

The equipment was tendered for in 1955 and, whilst contracts have been exchanged, there are still frustrating delays in getting equipment through the relevant underground and government approvals.

In the CMPDI report long term technology forecast for Coal India Ltd., further large growth of SDL and LHD semi-mechanisation is predicted as well as the introduction of four or five sets of continuous miner equipment per year from 2001. Indeed output for mechanised room and pillar systems is set to increase from nothing to 5mtpa by 2005.

Present tender activity for continuous miners is, 3 sets for SECL (Chirimiri, Pinoura, Vindaya) with a fourth currently on hold (Kapildhara), 3 sets for WCL (2 at Soaner, one at Tandsi), 2 sets for BCCL (Moonidih and Putki) and 2 or more sets for development in ECL at Kottadih and Jhanjra.

The three sets for SECL may start operations in 2001. Joy has formally applied for underground approval for equipment in India. Clearly the range of models available, speed of introduction of technology and acceptance of these fully mechanised systems will depend on how well the first sets operate and on the ability of the Indian approval authorities to accept certification from other overseas testing authorities.

8.0 CURRENT PRODUCTION IN INDIA

Production in India has increased from 73Mta in 1971 (Table 25.1) to a level of 320Mta in 1998, almost by a factor of 4, with a growth rate of 7 % per annum, catapulting India into the position of being the world's third largest producer of coal. The bulk of the increase of production and productivity has come from the rapid expansion of opencast mining which is now contributing almost 75% of the total production. The underground production has remained at a level of 75-80Mta for the last two decades.

Table 1. The growth in production manpower and productivity since privatisation

Year	*Manpower ('000)*	*U/G production*		*O/C Production*		*Total Production*	
		Production Mt	*Productivity T/manshift*	*Production Mt*	*Productivity T/manshift*	*Production Mt*	*Productivity T/manshift*
1971	382	58.5	0.54	17.1	1.00	75.6	0.70
1981	523	76.2	0.55	51.1	2.00	127.3	0.80
1991	554	79.7	0.58	158.1	3.40	237.8	1.32
1998	500	80.0	0.60	240.1	4.8	320.1	1.75

Source : World Coal: April 1999.

This trend towards opencast mining is mainly due to the better economics of opencast, compared to deep mining which, in India, has traditionally been plagued by lack of mechanisation culture.

However, the growth cannot keep pace with the demand and India is already a net importer of over 10mtpa. The figure is set to rise dramatically over the next ten years unless underground mining production and productivity can be greatly increased.

Growth of coal demand (Mt) by terminal year of IXth (2001-2002) and Xth (2006-07 (Xth Plan)

Sector	*1997-98*	*2001-01 (IX Plan)*	*2006-07 (Xth Plan)*
Steel	41.4	51.6	64.0
Electricity	222.0	287.8	447.0
Cement	18.2	21.4	30.0
Others	41.8	51.4	112.0
Total	323.4	412.2	563.0

(Source: Planning Commission, GO1,1997)

Of India's 530 operating mines, 355 are underground accounting for less than 30% of output (80Mta), but employing over 70% of the coal industry workforce. Much of production is still achieved by manual mining, with productivity averaging around 0.60 tonne per manshift.

The introduction of mechanisation has been slow, with semi-mechanised mines and fully mechanised longwall mines contributing less than 25% of underground output.

Through a combination of problems, domestic coal production in India has failed to keep pace with demand. The absence of new capacity may be due in part on over bureaucracy in the coal and utility sectors. Despite moves towards deregulation, the pace is extremely slow.

Most underground operations use bord-and-pillar systems with manual loading. The projected Coal India targets for the period are: 340Mt in 2000, 430Mt in 2003, and520Mt in 2010. However, the precise rate of growth will be dependent on having the capital necessary to develop the reserves and washery capacity necessary and establish an efficient infrastructure.

Like the power sector, the coal sector has been opened up to private investment. Coal India says that 40 mining blocks in different coal fields with estimated reserves of 13000Mt and potential for production of 150Mta have been identified for use by power and steel producers to set up their own mining operations. If underground coal production is going to increase efficiency in the future CIL will have to introduce more fully-mechanised systems with gradual phasing out of the current intermediate technology systems.

In order to provide a framework for investment and upgrading to modern equipment, there is a need to allocate a specific time frame for the evaluation of the proposals for mechanised coal mining equipment and to provide for a fast track for decision making and introduction.

As the result of the nationalisation of the Indian coal industry, there is no coal mining expertise available within the private sector that can be considered financially acceptable to sponsors. The expertise available is only within the public sector, which must have majority control in any partnerships.

To meet the need for the greater financial comfort of foreign sponsors, there is a need for the foreign coal-mining partner to become an equity partner.

It is difficult to attract overseas coal mining companies to India, with all its bureaucratic red tape and delays. Many such companies are reluctant since they perceive that much better investment opportunities lie elsewhere in the world. Therefore, to facilitate rapid private sector involvement in India, there are only two avenues available to make foreign investment more attractive or allow the nationalised public sector to participate in Joint Ventures with private coal companies, with the latter being able to hold a majority control in the JV.

The finances of certain parts of CIL show the size of the problem. In the five years up to 1995-1996, investment in ECL, BCCL and CCL has been Rs42 billion (US$1.184 billion); 52% of CIL's total investment of Rs80 billion (US$2.252 billion). Against this, the aggregate increase in production over the same period in these three companies is only 6Mta, against 42Mta in other CIL subsidiaries. The same money could have been used to develop, promote and guarantee overseas mining houses to run commercially successful operations with modern, world leading technology.

9.0 CONCLUSION

The US mining industry continues to rise to the challenges of thinner, deeper and more difficult reserves by rapid development of equipment to meet the changing market needs. Suppliers of the industry continue to push the technological envelope to improve safety, increase productivity and reduce costs.

Joy Mining Machinery is approaching the future challenges in a systematic fashion. Joy is modifying, refining and re-engineering its individual products so that their output and performance is maximised as part of a system. The 12CM27, the shuttle car with suspension and the Multibolter are examples of technology, increasing safety and productivity. The future for continuous miners will continue to be one of change both technologically and as part of integrated systems.

Diagnostics will increasingly become a standard option for continuous miners in the future. Such systems will be capable of monitoring gearbox vibration, oil contamination, pressure and flow in all hydraulic circuits and give a complete status on electric motors and circuits. Improved diagnostics properly interpreted will reduce machine down time and this will lead to significant production gains.

Continuous miners are part of a whole system involving transportation and roofbolting and, therefore, their overall performance is much lower than their specified cutting rate. Improvements in production from room and pillar sections will be gained through more efficient designs and the operation of outbye equipment such as continuous haulage systems, surge cars, cable-reel shuttle cars and battery powered scoop trams.

There are three critical needs to realise the next level of productivity, reliable micro processor electronics, advanced roof control processors and reliable continuous haulage. These needs are well on the way to being realised. In the USA, the modern continuous miner in a place changing system with an independent bolter providing the roof support system, can mine up to 90M of linear advance per shift.

To reach the next plateau in cost reductions, focus will be on systemic improvements. To maintain these improvements on a ongoing basis, the human element must be reduced through automation. Reliable remote control systems, on board 'intelligence' systems and reliable components, will lead the way to increased automation."

Technology can go so far, the real breakthroughs occur when technology and mining people are in unison. India is lagging far behind the rest of the world in introducing proven fully mechanised continuous miner systems. Underground output has remained static at around 75 to 80 million tonnes per year and the underground productivity is languishing at only 0.57 tonnes per manshift. Whilst there are good reasons for this, the increasing gap between production and usage means India is now a net importer of coal despite having enough coal reserves for well over 250 years. There has never been a better time to liberalise the coal mining markets in India to allow proper investment in modern technology. Increased output and safety from the underground mines could have a major effect on India's balance of payments.

India is at the beginning of the introduction of modern fully mechanised continuous miner systems. There is the potential for rapid introduction of these systems which will dramatically increase production, productivity and safety in the underground mines. As these systems become accepted as the norm, so mining costs will be greatly reduced. However, there is some concern that new projects will not be brought on stream quickly enough. It is up to the leaders in the industry to ensure that India is not left behind. The gap between required demand and supply is widening, yet there are sufficient resources in India to fully meet the supply if the correct infrastructure, equipment and resources are used.

Mr David Johnson has been managing Director of Joy Mining Machinery in the UK since 1990. Prior to this is he was Managing Director of Dosco Overseas Engineering Ltd and worked with Dosco for 30 years. In 1997 he was awarded the Order of Friendship from the Russian Government presented by the Russian Ambassador to the UK and in 1998 Joy UK were awarded the Queens Award for export achievement. David is a former President of ABMEC, the Association of British Mining Equipment Companies.

REFERENCES

Continuous Miners: Machines for Mining, World Coal Aug 98

Continuous Miners: Continuing Story, World Coal Aug 97

Continuous Miner: Continually Adapting: World Coal Aug 96

WJ Graham: Development & Trends in Miners/Bolters: July 95

CMPD1 Long Term Technology Forecast for CIL, 1998

India Focus: World Coal Apr. 99

India Coal: The Future, World Coal Apr. 99

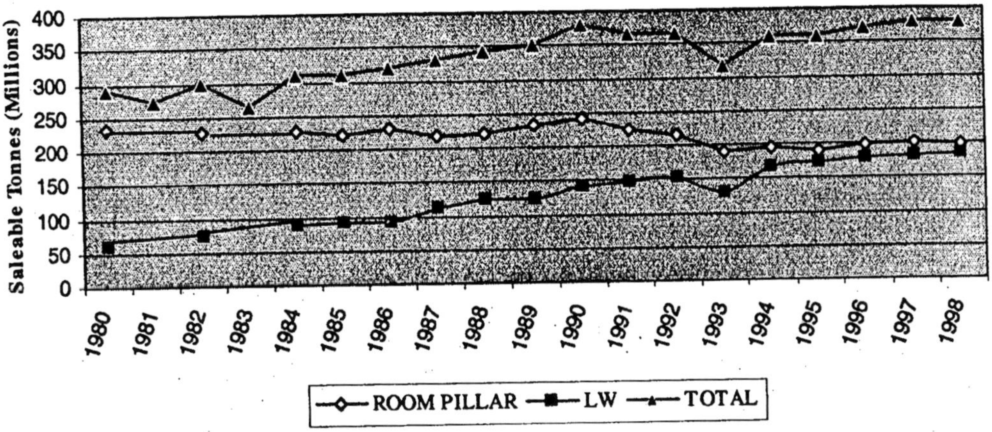

Graph 1 : Underground Coal Production By Mining Method in the US

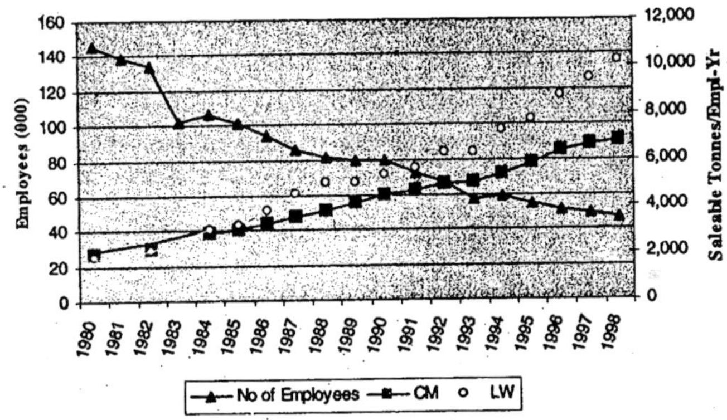

Graph 2 : Average No. of Employees and Saleable Tonnes Per Employee Year Underground Coal Mines in the US

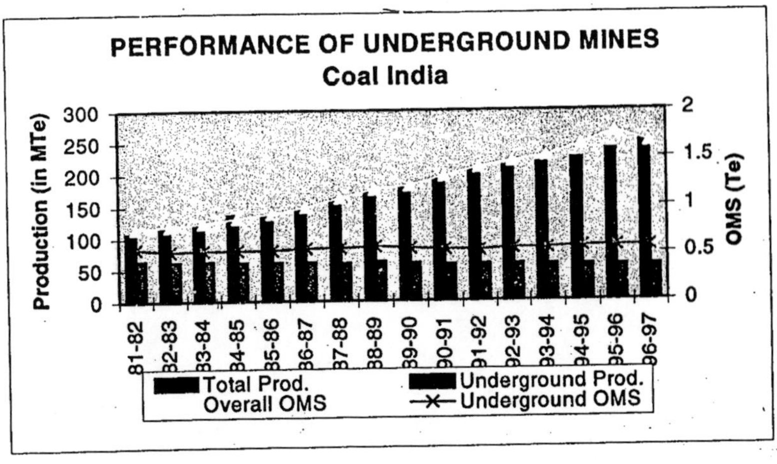

Graph 3 : India Production & OMS

Fig. 1 : Joy 12CM15 Continuous Miner

Fig. 2 : Joy Multi Bolter underground

Fig. 3 : Joy 10SC32 Shuttle Car, UK Coal Mine

Section IV

MINERAL EXPLORATION AND EXPLOITATION

CHAPTER 26

Challenges of Indian Mineral Exploration in 21st Century

*S.D. Prasad**

ABSTRACT

Human race since time immemorial has been in search of mineral wealth, and mineral deposits in turn have dictated the living standards and culture. India has a rich mineral environment and had glorious past in the field of mining. India occupies 2.1% of the total land area of the world and carries over 15% of the world population. The country is endowed with significant mineral resources and the mineral industry constitute an important segment of Indian economy. The modern mining venture in India started with the coal mining in the Raniganj Coalfields in 1774 and gold mining in Kolar Gold Fields (KGF) in 1880.

India produces 84 minerals which includes 4 fuel, 11 metallic, 49 non-metallic and 20 minor minerals. The value of mineral production increased from a level of just US $ 58 million in 1947 upto a level of US $ 10,000 million in 1997. The contribution of mineral sector to the GDP has increased from a level of 0.75% in 1950-51 to 3.5% on date and it accounts for 11% of country's total industrial production.

In 1993, the mineral policy as a part of the liberalisation and globalisation of the Indian economy created conditions favourable for rapid development of both mineral and mining industry. The value of mineral production increased at an average rate of 6% per annum during the VIIIth plan period, i.e., 1992-93 to 1996-97.

The export of mineral and metal are around US $ 7178 millions (1996-97) as against import for over US $ 7808 million which indicate an imbalance. All these factors put together indicate urgent need to accelerate mineral exploration activities.

Mineral exploration is a high risk venture in which chances of making a world class discovery on an average is about 1 in 1000 prospects. Globally the expenditure for mineral

*Chairman-cum-Managing Director, Mineral Exploration Corporation Limited, Nagpur, India.

exploration is of the order of US $ 3.52 billion (1996). While in India it is around US $ 50 – 60 million which is a meagre sum as compared to world standards. Therefore, there is an urgent need for more investments to discover new deposits.

Out of about 2.42 million sq.km. of hard rock area, only 20 – 25% area hold potential for fuel and non fuel mineral resources of which large tract has yet to be fully explored and offer opportunities along with challenges to the geo-scientist of the country.

So far the successful mineral exploration programme for gold, base metal, etc., conducted in the country to date have either centered around old working or near surface deposits by adopting conventional exploration techniques. Such deposits have almost exhausted and search is on for deposits hidden deep below the earth and which are available in complex geological environment particularly in virgin Himalayan region home of rich rock formation of economic importance.

Sustained level of production for various mineral commensurate with the country's demand would required speedy discovery and development of additional deposits. To achieve this goal following challenges are in the foresight.

- Mobilisation of financial resources.
- Discovery of deep seated deposits in geologically complex environment.
- Technological up-gradation in the field of geo-chemical, aero-magnetism, geophysical techniques to reduce expenditure and time in search of new deposits.
- Environmental constraints
- Use of State-of-art drilling equipment for higher productivity.

RICH HERITAGE

The discovery and availability of geological raw material have been crucial in the development of civilisation.

In India the production of minerals dates back to ancient times as witnessed by the remnants of old mining in some parts of the country. Mining and metallurgy have always played an important role in India's history and can be dated back to one of its earliest civilisations, the pre-Harappan period from 4000 to 2500 BC. In "Rig Veda" an ancient Hindu scripture, mention has been made on fire, minerals and gem stones. The Harappan civilisation showed an astounding variety of raw materials and finished products with sharp distinction at different sites such as Mohenjo-Daro, Harppa, Lothal, etc. Harppa having a large number of furnaceses was probably a manufacturing centre whereas Mohenjo-Daro, a capital city showed wider variety of materials. The world famous Delhi-Iron pillar is a marvel of ancient India showing the largest forge-welded object of the era. Recent studies in archaeology in India have confirmed through C-14 dating of samples that there was a flourishing lead-zinc mining activity at Rajpura Dariba near Udaipur (Rajasthan) dating back to 1260-160 B.C. **Chankya**, the politico-economist, who was also known as Kautilaya described in his world famous treatise "ARTHSHASTRA", the significance of minerals in the economy of the state. Detailed description of gossans and ores of gold, copper, silver, lead and tin ore found in "ARTHSHASTRA". The American Society of Metals (ASM) recognised the ancient zinc smelting site at Zawar near Udaipur as one of the international historical land marks for metallurgy. It is obvious, therefore, that

India was on the world map for non-ferrous metals in particular for zinc, many centuries ago. India was also the world's only known home for diamonds for more than sixteen centuries after Christ.

Though the first published reference of coal mining in India dates back to 1774 in Raniganj coalfields, drilling of first oil well at Jaypore in 1866 and gold mining in Kolar gold fields in 1880. Some of the notable discoveries of several large mineral deposits were made, which are being worked even today like lead-zinc in Zawar, copper in Khetri and Singhbhum, gold in Karnataka etc.

POST INDEPENDENCE SCENARIO

The country is endowed with vast mineral resources, it is, therefore, imperative to evaluate potential of these resources for overall industrial development of the country. Despite the glorious past, the mining activities in the country remained dormant over a long period until the beginning of this century when it was revived. The search for mineral gained momentum immediately after independence and adoption of Industrial Policy resolution. Geological Survey of India (GSI) was strengthened and Indian Bureau of Mines (IBM) was established in the year 1948, with the objective of conservation and development of the country's mineral resources and was also entrusted with the task of exploration of minerals. The Mineral Exploration Corporation Limited (MECL) was established in 1972, for undertaking detailed exploration of identified mineral resources, mainly to bridge gap between the initial discovery of deposit and its eventual exploitation. The Directorate of Geology and Mining of various states and other PSU's have also contributed significantly in augmenting the mineral inventory of the country.

As a result, the post independent era registered a rapid growth in the field of mineral exploration activities through various five year plans, resulting in significant increase in National Mineral Inventory as given in Table 26.1.

Table 26.1 : Scenario of Mineral Reserves in India (between 1950-51) (in million tonnes)

S.No.	*Mineral/Ore*	*1950*	*1960*	*1970*	*1980*	*1990*	*1995*
1	2	3	4	5	6	7	8
1.	Barytes	-	0.70	2.54	73.39	70.15	87.00
2.	Bauxite	-	276.00	227.00	2489.00	2525.34	2462.00
3.	Chromite Ore	1.30	4.86	13.86	111.25	88.35	86.00
4.	Copper Ore	-	40.00	244.00	461.56	416.81	416.81
5.	Dolomite	-	302.51	1778.00	4947.47	4967.46	4387.00
6.	Gold Ore	-	4.11	4.90	8.70	17.69	17.79
7.	Iron Ore	-	788.50	8183.00	11478.00	12745.00	13010.00
8.	Lead-Zinc Ore	-	28.45	187.10	368.74	179.13	179.13
9.	Limestone	-	-	49929.00	72229.00	76446.00	76446.00
10.	Magnesite	-	86.00	524.10	-	233.33	245.00
11.	Manganese Ore	-	60.00	128.16	116.64	176.48	167.00
12.	Phosphorite & Apatite	-	-	61.00	132.29	15901	159.01

1	2	3	4	5	6	7	8
13.	Tungsten	-	-	1.68	10.54	39.65	38.11
14.	Tin Ore	-	-	-	-	28.91	28.91
15.	Sillimanite	-	0.25	0.34	12.00	50.69	50.60
16.	Kyanite	-	0.70	3.80	3.03	2.71	2.82
17.	Vermiculite	-	0.01	0.19	0.26	0.31	0.31
18.	Fluorite	-	3.15	12.22	11.89	3.04	2.95
19.	Antimony Ore (t)	-	-	10588.00	10588.00	10588.00	12588.00
20.	Nickel Ore	-	-	14.62	160.26	183.48	183.48

Source: Indian Mineral Year Book : 1997, IBM

The value of mineral production increased from a level of just US $ 58 million in 1947 by leaps and bounds over the last 50 years and has reached the level of US $ 10,000 million. The mineral base has also increased from 22 in 1947 to 84 minerals being produced now-a-days. The contribution of mineral sector to the GDP has increased from a level of 0.75% in 1950-51 to 3.5% on date.

RESOURCE BASE

The development and growth of the present day civilisation depends largely upon mineral resources their consumption and energy. These are infact important parameters for assessing the status of development of a country. India being a large developing country with huge population (1/6th of the world) it is axiomatic that demand for mineral resources will be rising at a very high pace.

Due to industrial revolution all over the world, major changes in the pattern of mineral consumption by industries took place, as a result of which the economy of the most industrialised countries became mineral dependent. It was during World War-II the awareness of mineral resources and need to update the mineral inventory became a reality. In India too, compilations of mineral inventory was initiated in the year 1968 and since then it is being updated on regular basis. The position of *in-situ* reserves and conditional resources in respect of important minerals in India along with its share to world reserves is given in Table 26.2.

Table 26.2 : Reserves of Important Minerals India as on 1-4-95 with percent share in world reserves (*Reserves in million tonnes except where otherwise specified*)

Sl. No.	*Mineral*	*Total Indian Reserves*	*Total World Reserves*	*India's Share %*
1	2	3	4	5
METALLIC MINERALS				
1.	Iron Ore	13460.00	230000	5.5
2.	Bauxite	2462.00	28000	9.0
3.	Copper metal	4.37	610	0.7

1	2	3	4	5
4.	Chromite Ore	86.23*	7400	1.16
5.	Manganese Ore	176.48	5000	3.7
6.	Lead metal	2.32	120	1.8
7.	Zinc metal	10.08	330	3.0
8.	Gold metal (t)	67.90	51000	0.1
9.	Tin metal (t)	3272.00	10000000	-
10.	Tungsten metal (t)	86532.00	3300000	2.3
NON-METALLLIC MINERALS				
1.	Rock phosphate & Apatite	159.00	34000	0.5
2.	Ilmenite	90.15	440	20.48
3.	Barytes	87.00	500	14.0
4.	Kyanite	2.82	108	2.5
5.	Graphite	4.58	380	8.2
6.	Fluorite	2.95	311	6.9
7.	Asbestos	9.30	Large	--
8.	Limestone	76446.00*	NA	--
9.	Dolomite	4387.00	NA	--
10.	Quartz & Silica sand	983.52*	NA	--
11.	Kaolin	1068.00	NA	--
12.	Fire clay	696.72*	NA	--
13.	Gypsum	238.00	NA	--
14.	Sillimanite	50.69*	NA	--
15.	Felspar	31.29	NA	--
16.	Calcite	10.57	NA	--
17.	Diamond (000 carats)	982.00	NA	-
18.	Mica (t)	60000.00	NA	--
19.	Wollastonite	6.29	NA	--
20.	Magnesite	24.51	340.00	7.20
21.	Granite (Cu.m)	102.74	NA	--
22.	Marble	82.46	NA	--
23.	Bentonite	36.55	NA	--

(i) World Reserves are as per USBM Mineral Commodity Summaries 1996

(ii) NA – Not available.

(iii) * Reserves are as on 1.4.90.

Based on the reserves availability, the relative abundance and scarcity of various minerals in India is given in Table 26.3.

Table 26.3 : Adequcy of Mineral Resources in India

Sl.No.	*Grouping*	*Abundant*	*Adequqte*	*Deficit*	*Scarce*
I.	Fuel	Coal	Lignite	Coal (Coking)	
II.	Metallic Minerals				
	(a) Ferrous	Iron Ore	Chromite (Metallurgical grade) Manganese Ore	Chromite Refractory grade)	Nickel, Tungsten Cobalt, Molybde-num, Vanadium
	(b) Non-ferrous	Bauxite (Metallurgical grade)	Zinc	Bauxite (Chem. & Refractory grade) Copper, Lead	Anitmony, Gold Platinum group of metals, Silver, Tin
III.	Industrial Minerals	Barytes, Ballclay, Bentonite Calcite, Dolomite (excluding low silica grade), felspar, fire clay, Fuller's earth/ Garnet, Gypsum, Ilmenite & Rutile, Kaolin, Lime-stone, Magnesite, Mica, Pyrophyllite, Pyrite, Ochre, Quartz & Silica Sand, Quartzite, Beach sand, Sillimanite, Zircon, Steatite	Corundum, Graphite, Rocksalt, Vermiculite Wollastonite	Apatite & Rockphosph-ate, Asbestos, Fluorite Kyanite	Andalusite, Borax, Diaspore Diatomite, Native Sulphur Potash
IV.	Precious Stones				Diamond, Emerald, Sapphire, Ruby
V.	Minor Minerals	Granite (Dimension Stone) Marble.			

Though a total 84 minerals are being produced, the country is self sufficient in 35 minerals only which constitute primary raw materials to basic industries like, Iron and Steel, Ferro Alloy, Aluminum, Cement, Refractories, Glass and Ceramic, Foundries, Abrasive and broad spectrum of inorganic chemicals.

India is by and large self sufficient in coal and lignite, bauxite, chromite, iron ore, manganese ore and almost all industrial minerals. There is however very significant gap in respect of 17 minerals and metals for which country is dependent largely on imports. A few other important mineral commodities are not yet known in India in economic proportion such as PGE, REE, tin, antimony, cobalt, nickel, mercury, potash, sulphur and some rare minerals used in high technology. However, a significant resource for a number of such deficient/poor minerals locked up in low tonnage/low tenor deposits which could only be exploited by adopting advanced processing technology and or/modified managerial and economic frame.

In the global scenario the production is usually below 10% for most of the important minerals. The resources of the minerals are also negligible in comparison to world resources.

The existing gap between the demand and supply position for the most minerals is expected to widen adversely by the end of the century.

The value of mineral production excluding atomic and minor minerals increased from an insignificant level of about US $ 59 million at the time of Independence to US $ 10000 million during the "Golden Jubilee year of Independence". It is evident that country's own expertise and resources in mineral exploration, acquired over the last 50 years played a vital role in increase in the value of mineral production and diversification of different minerals.

NEED OF EXPLORATION

The systematic and planned efforts in mineral exploration by GSI, MECL, State Governments and other exploiting agencies in the central and state sectors during the last plan periods have resulted in building up of mineral resource base, resulting a sea change in industrialisation and modernisation of the country. Since independence the value of mineral production as well as number of minerals in production increased considerably and in respect of 35 minerals, the country is wholly or largely self-sufficient. The fuel minerals account for 85% of the value where as 15% was accounted for by metallic and non-metallic minerals almost in equal proportion.

With the envisaged annual growth rate in GNP at 6% during VIII plan and beyond, the requirement of minerals is expected to be more than double in next 10 years, while chronic shortage of some important base metals, strategic minerals, fertiliser minerals, noble metals, and elements required in high technology is also anticipated. Therefore, the search and evaluation of the new deposits through systematic and intensive exploration are necessary to boost up the production of minerals and reserves.

Exploration is extremely risky and expensive. The discovery of new deposits which are economically viable and can be of international importance is rare, usually with success ratio of 1:1000. In India where most of the surface deposits have been delineated, major discovery may involve only concealed prospects, the exploration of which needs advance technical inputs as well as higher investment.

In spite of the risk in exploration, the expenditure incurred for discovery of a specific deposit is insignificant in comparison to the tonnage and value of the minerals proven and also the socio-economic upliftment of the area.

India is largely a country of small mines where 95% operating mines produced about 50% of the value of the minerals. In order to reduce dependence on import and trade deficit in mineral sectors private investment need to be encouraged and there is an urgent need to accelerate exploration activities for the minerals specially base metals, gold, diamond, rock phosphate, tin, coking coal etc. in order to reduce import bill of the country.

The progressive economy as well as the population growth demands increased quantities of minerals to be used in next decade and beyond, leading to greater depletion of the minerals inventory in shorter time. One of the most interesting factors is that most of the successful mineral exploration programmes conducted so far have either centered on ancient mine working or near surface deposits by adopting conventional exploration techniques. We have now reached a point where we have exhausted such near surface occurrences. Now we are forced to search deposits which are hidden deep below as well as those which occur in complex geological environment. Hence utilisation of advanced integrated exploration techniques which could lead to a discovery of world class sub-surface deposits, would be an immediate steps in this direction.

CHALLENGES

It is an established fact that no country in the world is endowed with all the requisite mineral resources and it can never attain self sufficiency in all the minerals due to limited conducive and tectonic milieu. India also lacks adequate resource base in respect of a number of minerals, for which the country has to depend largely on imports. The total value of minerals and metals imported has been far in excess of that exported and the share of minerals and metals in the total merchandise imported has throughout remained much higher than exported. Hence the trade deficit stood at staggering total of Rs. 25578 crores in 1996-97. India having 2 % of the global area sustains 16 % of the world population and our per capita consumption of mineral resources is one of the lowest in the world. The present trend indicates that the demand for minerals would continue to increase rapidly. Being the second largest populated country, India would definitely play an important role in influencing the balance of foreign trade of minerals and metals in the years to come. It is, therefore, imperative that while we should try to make the best use of available mineral resources we have also to intensify our efforts in exploration so that we could add to the currently known resources of the country. To maintain sustained level of production for various minerals commensurate with country's demand would required speedy discovery of and development of additional deposits. To achieve this goal various **challenges** in foresight are discussed below :

1. ***Mobilisation of Financial Resources*** :Exploration is an extremely high risk venture. A saying in the oil industry is that **"It is a gamble with nature where millions are spent before billions are won"**. While successes of scientific interest are not rare, the chances of making commercially viable discoveries are extremely low. A study of US exploration activity between 1942-67 showed that the chances of economic discovery resulting from any exploration programme was 1:300. In another study made by a giant multinational organisation like Cominco over forty years period showed that out of 1000 prospects initially taken up for investigation resulted in identification of 80 prospects worth exploration in detail of which 18 were taken upto exploitation stage, of which finally only 8 were found commercially viable over a long duration.

However, success ratios in India was considerably higher in the past but it is bound to go down in the future as we have more or less exhausted the easily identifiable prospects. Such a retardation in success at the present juncture of the Indian economy will have far reaching consequences. The only viable solution to tackle this problem would be to intensify the exploration efforts and to usher in the state of the art technologies in the exploration industry to discover new deposits located at greater depths and in geologically complex situations.

Hence, a substantial investment in exploration is a must and is sure to benefit the country's industrial growth in the medium and long term perspective. It is because if one takes a look on a global scale, India stands no where in terms of investment in exploration programme annually as compared to the other countries. Whatever meagre investment is made by several exploration agencies in India, major share goes on salaries and wages. This is the one area where serious efforts are to be made by the Government and other private sector companies in India to face the future challenges in the mineral industry.

The new National Mineral policy, 1993 allows induction of foreign participation in exploration and mining. As already mentioned the investment in mineral sector, may it be mining or exploration involves risk capital. Given the fact, the progress in this area will always be slow and takes some time before the results are visible. Thus, to attract private investment for mobilisation of financial resources and to create self of deposits, Govt. would have to continue risk investment at exploration stage. It is in this context, that the need for governmental funding and encouragement for exploration efforts in the country needs not only to be maintained but in fact substantially intensified. Simultaneously to this Indian agencies should be encouraged to form "Joint Ventures" with foreign investors for generation of additional funds for exploration, technological up-gradation and procurement of high tech equipments.

2. *Discovery of Deep Seated Deposits in Geologically Complex Environment* : In the country till date, majority of the exploration programmes were centered to old working or near surface deposits by adopting conventional exploration techniques. It is a known fact that the base metal deposits specially, Cu, Pb, and Zn, which were out cropping or having surface indication in from of old working gossans etc. have all been explored. Now the time has come when a systematic approach is required to locate new deposits in concealed as well as those occur in geologically complex situation. The concealed or blind deposits are very difficult to find without concepts and techniques that have been developed in recent times. Space and air-borne appliances permit large areas to be surveyed very quickly and the most promising zones could be selected for further detailed ground follow up.

3. *Technological Up-gradation* : It is of foremost importance to adopt integrated technique in search of hidden mineral deposits. Remote sensing forming a part of the space air borne surveys, deals with the study of satellite imageries of different nature and aerial photographs or stereo imageries. With the launch of IRS-IC and ID, India has become world leader in acquisition of remote sensing data. It is essential that the full potential of the remote sensing data be harnessed so that more and more hidden deposits are found.

   Apart from remote sensing techniques, advanced techniques of surface and airborne geophysical surveys, geo-chemical surveys, geo-botanical surveys, tectonic concepts etc. needs attention and its application as modern concepts in the field of mineral exploration. This modern approach could reduce expenditure and time in search of new deposits. The integrated approach involves handling of voluminous data, which are heterogeneous in scale, time of acquisition and precision levels. Management of such huge data manually is not possible and needs handling through computers. Thanks to the development of GIS and GPS system, management of such data is possible now. GIS is very useful in integrating, modelling, analysing different types of data.

The use of GIS and GPS system in the country has started recently and its extensive application in the mineral search is the need of the hour. Though the data processing of exploration has been made faster using various computer software, but it can be further extended to make pre-feasibility and feasibility reports making the exploration more meaning full and attractive for investor in mining and further down stream activities.

4. ***Environmental Constraints*** : In the country like India, which sustains as much as 16% of the world population on nearly 2.5% of the landmark environmental preservation and improvement attains a special significance and needs to be assigned priority. In mineral sector its relevance is of greater appeal as this sector is quite sizeable and that a good part of it is located in environmentally fragile areas comprising tribal habitat, wild life sanctuaries, archaeological monuments etc. and has a very little chance to over come environmental constraints. The only viable alternative to overcome this constraint, either wholly or partially, would be to invest in exploration on a larger scale so that a number of mineral deposits could be discovered in different geographical locations, which in turn could offer the investors a choice of deposits to choose from not only with reference to environmental impact, but also, in the bargain, take into account the availability of infrastructure, scale of investment required, quality and size of the mineral deposit etc. Further, mineral industry articulate regarding the implementation of national laws relating to environment and forest, simplification of procedures to address the requirement and development at one end and conservation of environment on the other. This aspects needs immediate attention as the country still continues to export a number of minerals in raw form which in turn fetches very low values not commensurate with the efforts and money spent and the environmental degradation it causes. It is high time that the country goes in for more and more value addition for export so that we could realise higher value for our inputs. In this connection, the exploration industry will have to reorient its approach by compulsorily conducting a wider range of tests both in the laboratory scale as well as on the industrial scale so that the prospects of value addition in individual mineral deposits could be ascertained at the initial stage itself.

5. ***State of Art Drilling Equipment for Higher Productivity*** : Though there are many advanced techniques of mineral exploration, the art of drilling continues to be the ultimate back bone of any mineral exploration. Drilling helps in three dimensional subsurface delineation of ore body. Like other exploration techniques, drilling techniques have also undergone rapid technological advancement. Highly sophisticated state of the art of prospecting rigs are in use today in foreign countries like Australia, Canada, Africa etc. In India drilling techniques need many fold improvement to come to world standard, i.e., Hydrostatic drills with wire line system of core retrival, fast and deep drilling, Reverse circulation (RC) drilling, down the

hole hammer drilling (DTH) vaccum suction drilling etc. With these advanced drills, controlled and directional drilling, underground exploratory drilling and large diameter drilling can be achieved by using additional drilling tools. The high tech drilling rigs are manufactured mostly in the western countries. In order to compete in global market, highly sophisticated drilling rigs have to be adopted in the country without giving second thought. The exploration of mineral deposit by advanced drilling rigs with higher productivity leading to generation of quality data with less expenditure in a very short time.

CONCLUSION

To sum up, India is known to have high geological prospectivity for a variety of minerals. The country is very attractive to exploration companies for exploration/ investment in mining for a number of reasons such as high geological potential for new mineral discoveries, available of a reliable geoscientific data base, well developed infrastructure, large technical manpower, rule of law, political stability, well organised banking system and dynamic financial institutions. The national agencies like GSI & MECL, fully equipped with data base and technology, are prepared to share the efforts of interested parties in the field of mineral exploration. All these positive aspects offer exciting and attractive opportunities in the mineral sector for the potential investors.

The total leased mineralised area for all the major and minor minerals is about 31000 sq.km. i.e. around 1% of the total land area of India. However, the actual land area occupied by actual mining operation including overburden dumps etc. may not exceed 0.2% of the land area.

The geological prospectivity comparable with mineral rich countries such as Canada USA, South Africa, Australia etc. exist in the Indian sub-continent and there is ample scope for mineral development. These potential areas need intensive and extensive multi disciplinary exploration inputs.

Thus, unlimited opportunities available in the mineral sector needs attention of the private entrepreneurs which could be en-cashed by over coming the various **challenges in foresight during the 21st century.**

REFERENCES

1. Report on the working group on Mineral Exploration (other than coal & Lignite) 1996, Ministry of Mines, Govt. of India.
2. B.B. Tandcn, 1998, An Overview of Indian Mining Sector and Investment Opportunities. Paper in "The conference on Mining Sector 1998." Johannasburg, South Africa.
3. Sachdeva, O.P. 1995, An Overview of Mineral Exploration in India Paper in "The 1995 Mining Symposium".
4. Indian Mineral Year Book 1997 : Indian Bureau of Mines, Nagpur.
5. Statistical Profile of Minerals 1997-98, Indian Bureau of Mines

CHAPTER 27

Diamond Potential in India and Exploration Strategies to be Adopted

Abhijeet Mukherjee and K.S. Rao**

ABSTRACT

A number of new kimberlite bodies have been reported recently mostly by the Geological Survey of India, Directorate of Geology and Mines, M.P., Atomic Minerals Division and other organizations from Dharwar and Bastar Cratons of India confirming the fact that many more pipes are yet to be discovered. The potential Indian cratons with kimberlite clusters are Bundelkhand - Aravalli (Panna cluster), Bastar (Raipur and Tokapal cluster), Dharwar (Wajrakarur, Chigicheral, Kalyandurg and Maddur - Narayanpet cluster). Overall there are over 60 kimberlite/lamproite occurrences in India. This indicates that there is lot of scope for finding new diamond bearing rocks in these potential cratons. Application of recent geological, geophysical and geochemical techniques are required to further enhance the search and evaluation of new kimberlite pipe/bodies. These advanced techniques are used in area selecton, finding the target and prioritization and evaluation.

INTRODUCTION

India offers an exciting challenge to the diamond explorationists and the possibility for the successful discovery of kimberlite/lamproite pipes which are the host rocks for diamond. In order to develop India's diamond prospectivity the initiatives that should be taken into consideration are assessment of known diamond localities, area selection for primary kimberlite exploration, definition of appropriate laboratory and field testing equipments (both geophysical and geochemical) initiation of primary diamond testing programmes.

AREA SELECTION

There is substantial gap between production (~31000ct) and requirement (over 30 million carat) of diamonds in India. This quantum is minimized through imports of roughs.

*National Mineral Development CorporationLimited,Khanij Bhavan, Masab Tank Hyderabad - 500 028.

However, the possibility of discovering new sources of diamonds is bright as vast tracts/belts of potential areas remain to be explored in detail. A number of new kimberlite/lamproite occurrences have been reported by Geological Survey of India recently from Dharwar and Bastar Cratons confirming the fact that many more pipes are yet to be discovered. The primary sources of diamonds in India are kimberlites/lamproites (K/L), emplaced during the Upper Proterozoic (~1100Ma). The different K/L fields in India is given in Table 27.1.

Table 27.1 : The different K/L fields in India.

Sl.No.	*K/L Field*	*Craton (Intruded into)*	*District*	*No. of pipes/ bodies reported*	*Age*
1.	Panna Diamond Belt	Bundelkhand - Aravalli	Panna	02	Upper Proterozoic
2.	Raipur Kimberlite Field	Baster	Raipur	05	-do-
3.	Tokapal Kimberlite Field	Bastar	Bastar	04	-do-
4.	Wajrakarur Kimberlite Field	Dharwar	Anantapur	13	-do-
5.	Chigicherla K/L Field	Dharwar	Anantapur	04	-do-
6.	Kalyandurg K/L Field	Dharwar	Anantapur	03	-do-
7.	Chelima - Ramannapeta Lamproitic Field	Eastern Margin of of Cuddapah Basin (Mobile belt)	Guntur	03	-do-
8.	Maddur-Narayanpet K/L Field	Dharwar Craton	Mahboobnagar and Gulbarga	28	-do-

Area selection looks at large scale tectonic and geophysical concepts and their application to exploration i.e., geotectonic controls on the occurrence of diamonds and recognization of areas that might preserve deep lithospheric roots in the Indian Cratons mostly Archons (mentioned in Table 27.1) favourable for diamonds. Finding the target requires the role of airborne magnetic, electromagnetic and radiometric surveys and remote sensing techniques. Remote sensing helps in arriving at the structural controls on the intrusion of K/L and stressing the role of linear structures which may be visible in satellite imageries (Fig. 27.1). K/L are essentially controlled by major faults/lineaments in ancient cratonic regions (Fig. 27.2). While these form the general criteria for search of K/L, a more precise prognostication requires understanding of the various structural features such as structural domes, intersection of faults, reactivation events of major faults/lineaments as their emplacement if facilitated by the splaying of faults around closures of structural domes and resultant fractures of major faults. These are interpreted through geological, satellite imageries, aerial photographic and regional Bouguer gravity data. The conceptualised exploration strategy has resulted in the discovery of 18 K/L pipes in Maddur - Narayanpet area and 3 K/L pipes in Wajrakarur area.

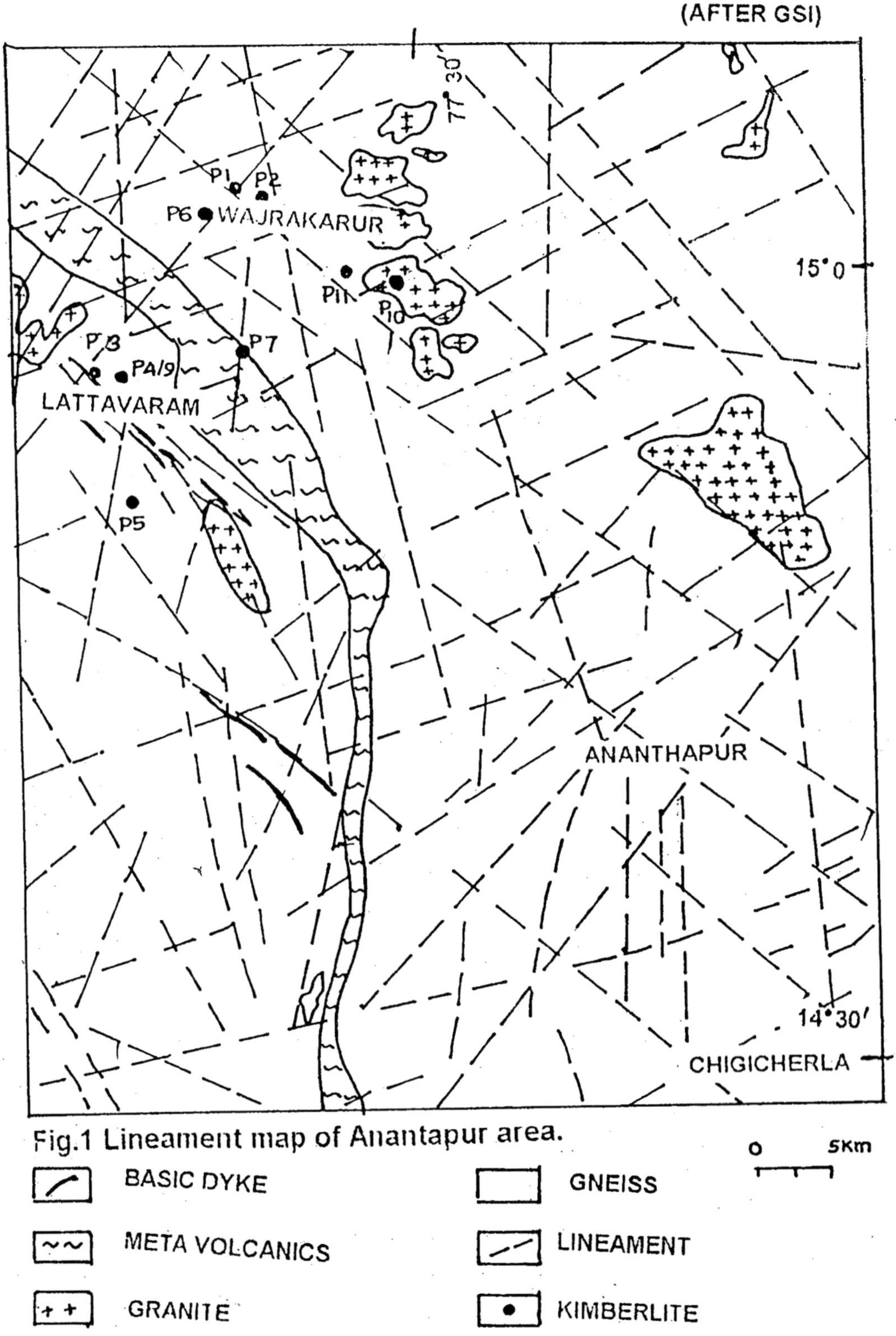

Fig. 27.1 : Lineament map of Anantapur area

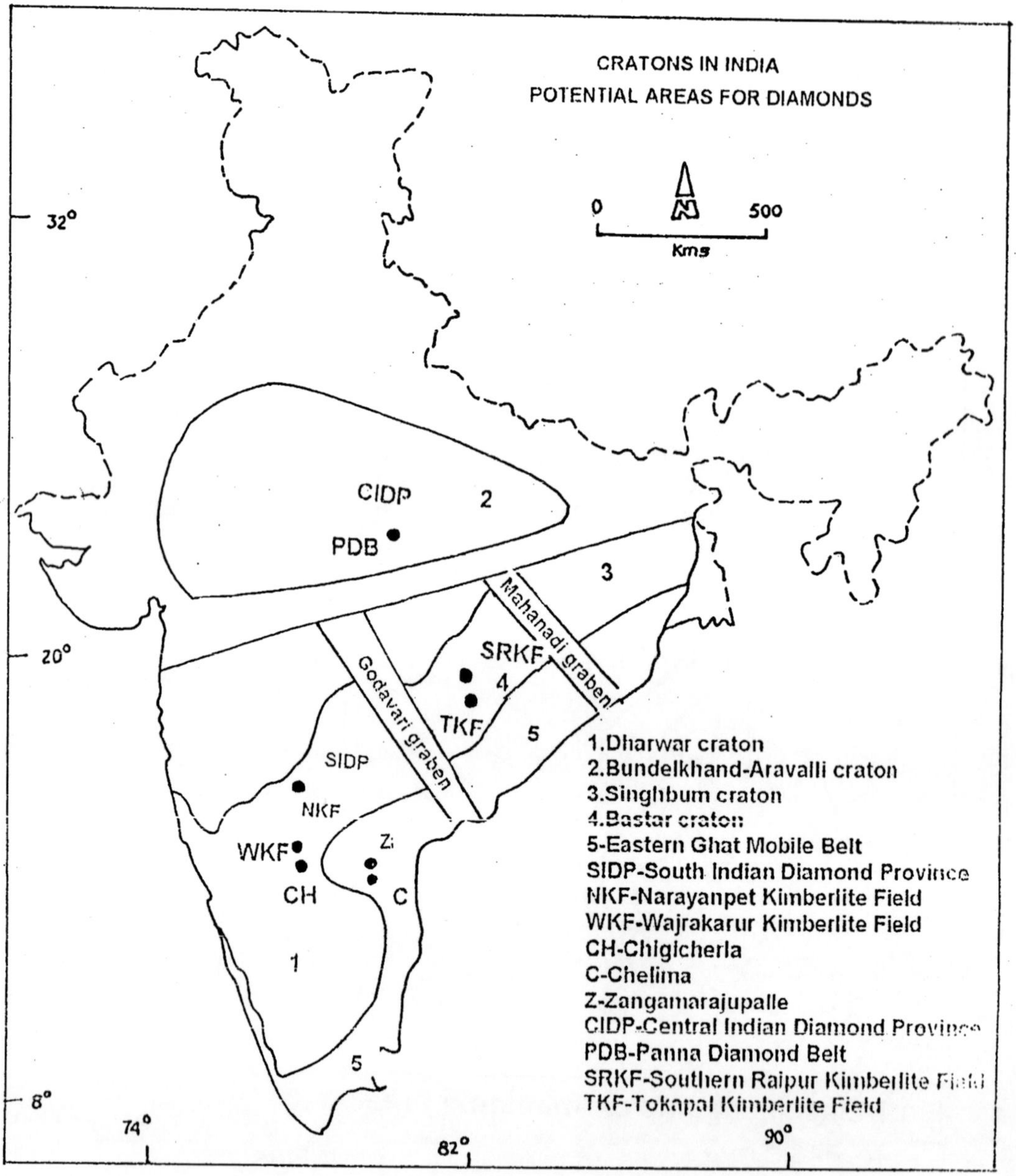

Fig. 27.2 : Potential areas of diamond in India showing different cratons

Table 27.2 : The lineaments that control the emplacement of kimberlites and lamproites in India.

Sl.No.	*Area*	*Important lineaments*
1.	Panna Area	NNW - SSE NE SW
2.	Raipur Area	WNW - ESE
3.	Tokapal Area	NW SE
4.	Wajrakarur - Lattararam Area	ENE - WS W and NW and SE
5.	Chigicherla Area	NE - SE and NW - SE
6.	Maddur - Narayanpet Area	WNW - ESE, NE - SW and NW - SE
7.	Chelima - Zangamrajupalle Area	NE - SW
8.	Kalyandurg Area	ENE - WSW and NNW - SSE

The areas with low heatflow preferably 40mw/m^2 within the cratons are preferred with deep seated lineaments.

GEOPHYSICS

The magnetic susceptibilities for pipes reportedly range from 100 to 6000 × 10^{-5} SI units emu (Fig. 27.3). The Air Borne Magnetic Survey requires flying height of 50-100m with a spacing of 250m. These methods have certain disadvantages like

(i) The anamoly caused by trap rocks over pipes.

(ii) Slightly magnetic pipes are barely distinguishable from general background.

(iii) Small pipes are missed.

(iv) ASGN magnetometer recorded non-existent anomalies.

Present Air Borne Magnetometers are capable of providing magnetic readings at intervals of <0.2 seconds (i.e., 12m intervals along line). These methods are sometimes hampered by certain reasons like complexity of the area due to weathering (oxidation of hematite, magnetite) and multiple intrusion. Non-magnetic sediments obscure/deminish the magnetic signatures.

The gravity method is generally used as a follow-up method. Negative responses are obtained over pipes and the amplitudes are < 1 milligal (Burley and Greenwood,1972 and Carlsen et al., 1984).

Electrical methods are limited to follow-up study delineating pipe outlines. Resistivity values of 20 Ω were obtained over kimberlites in Zaire(Magnee, 1950).

Clay minerals cause low resistivity values of 200 Ω over kimberlites. Electromagnetic methods werre applied in several South African pipes and Ellendale area of Australia. EM anamolies show high amplitudes and slow decay rates.

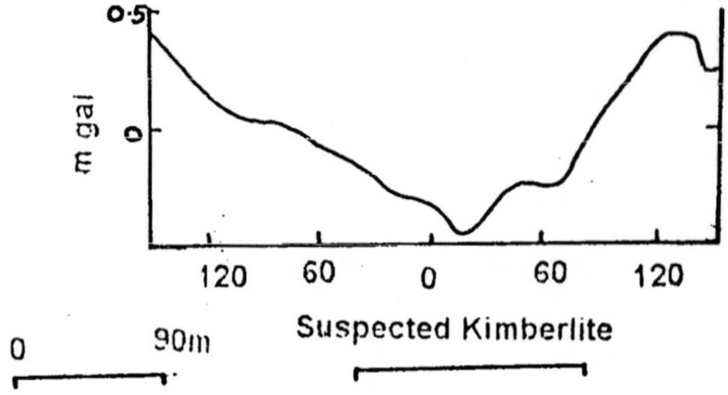

Gravity survey of Maxell No.1

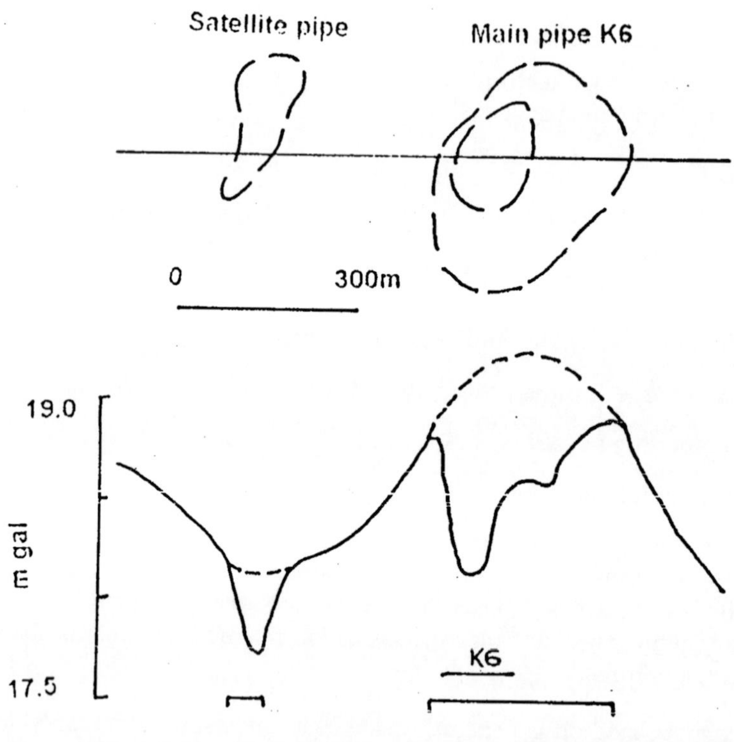

Fig. 27.3 : Gravity survey of kimberrlite pipe-Sekameng, Lesotho

Radiometrics are used to distinguish dolerites/basalts and kimberlites where magnetics are ambiguous, K (potassium) upto 10% were observed in leucite bearing lamproites. Twenty six lamproites were recognized in Ellendale (Australia) by this method.

HEAVY INDICATOR MINERAL STUDIES

Indicator mineral studies of minerals like garnets, ilmenite, spinel, etc., from stream sediments is another method which helps in locating primary diamond deposits. Prioritization and evaluation of the kimberlite pipes by interpreting the major and trace element data of these indicator minerals is essential because less than 1% of the kimberlite pipes found are likely to be economic and viable exploration programme requires the early rejection of uneconomic finds. Trace element micro analysis of indicator minerals and better understanding of the genetic relationship between micro and macro diamond populations will improve the precision of target evaluation.

NMDC has carried out preliminary studies with respect to heavy indicator minerals, thermometry, barometry and oxygen fugacity on some of the occurrences from different Indian cratons namely Bundelkhand - Aravalli (Majhgawan pipe, Panna), Bastar (Payalikhand, Behradih and Tokapal) and Dharwar (Venkatampalle i.e., pipe - 7, Anumpalle i.e., pipe - 10 in Wajrakarur and Kotakonda in Mahboobnagar district of Andhra Pradesh (Abhijeet Mukherjee at al. 1995 and 1997).

Heavy indicator minerals (HIM's) are the best tools used the world over for early recognization and rejection of targets. HIM studies have been applied to different diatremes/ pipes. The presence of "GIO" garnets indicate the presence of diamonds in a pipe (Fig. 27.4). These "GIO" (harzburgite) garnets are found in Majhgawan, Payalikhand and Venkatampalle pipes (Abhijeet Mukerjee and Rao, 1998a and Abhijeet Mukerjee and Kimothi 1998b). In case of eclogitic garnets the Na_2O% is greater than 0.07 wt%. The garnets with similar composition indicates that these eclogitic xenoliths are from diamond stability field and they carry diamonds. Four out of fifty garnets analysed in Bahradih show $Na_2O > 0.07$ wt% indicating its sampling from the diamond stability field, also confirmed by the presence of diamonds. S1 and S2 spinels also indicate that the emplacement of the diatreme is from diamond stability field and confirms the presence of diamonds. Anumpalle (pipe - 10) in Wajrakarur has S1 and S2 spinels which confirms the presence of diamonds in that pipe (Abhijeet Mukherjee et al., 1998).

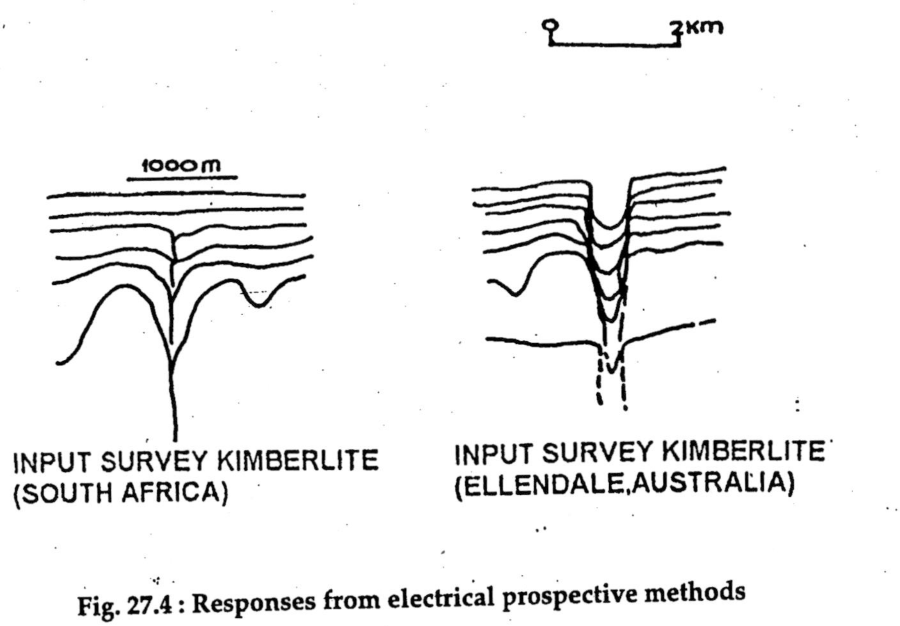

Fig. 27.4 : Responses from electrical prospective methods

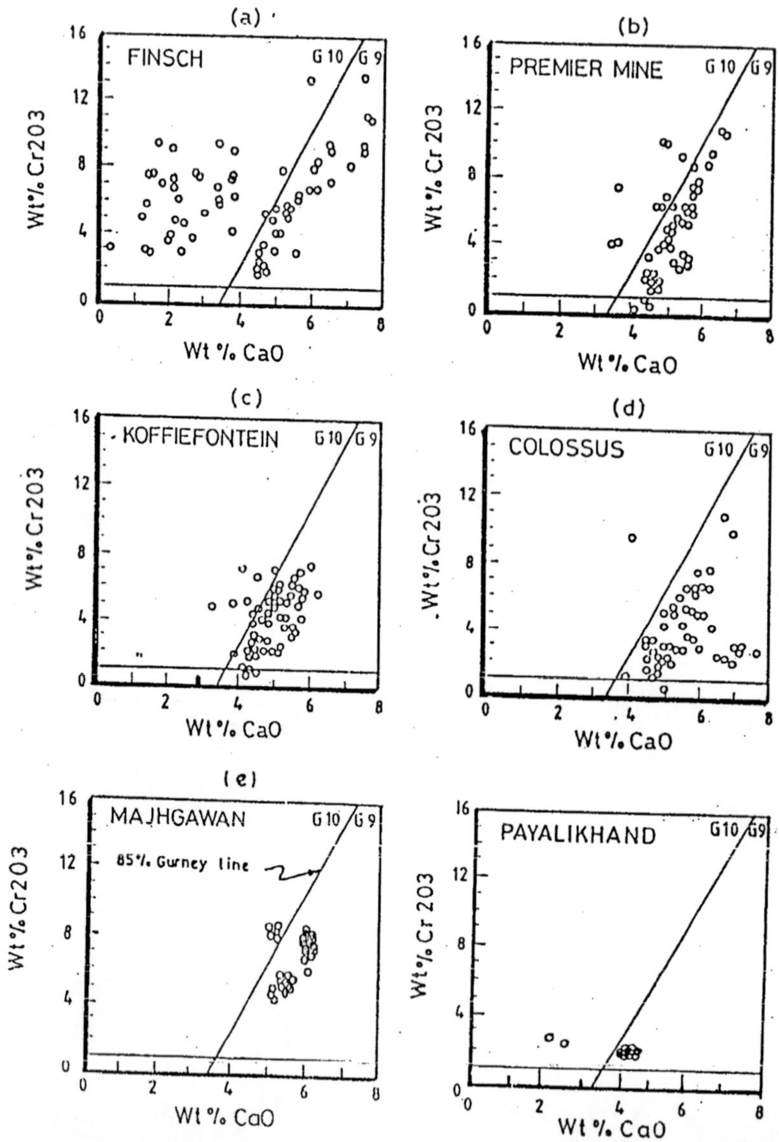

Fig. 27.5 : CaO vs Cr2O3 plots of garnets (a) high grade (b) medium to high grade (c) medium to low grade (d) barren (e&f) Majhgawan and Payalikhand plots

The impetus to develop these methods was first generated by the discovery in Botswana of kimberlite pipes under deep sand burden which made them too expensive to evaluate by bulk sampling.

The methods developed by Gurney (1984, "GIO" - garnet), Griffin et al., (1992, nickel thermometry), Sobolev (1974) and Zhou et al.,(1992, S1 and S2 spinels) have been very useful for prioritizing the targets.

DISCUSSION

National agencies/institutions which possess the capability to undertake air borne geophysical survey are NRSA, AMD, and AMSE (GSI). Laboratory facilities available with these organizations have to be further augmented to cater the needs of diamond exploration. NMDC is operating the only mechanised diamond mine in the contry at Majhgawan in Panna district, Madhya Pradesh. The present level of diamond production is 34,000 carats/anum. The diamond treatment plant includes Heavy media separation and X-ray sorter units. The Majhgawan diamond mine and the plant are being expanded to increase the diamond production to 84,000 carats/anum by next year.

Studies on the indicator minerals help in taking timely decision of selection of promising areas and this study could be taken up by NMDC, if prospecting licences are given to it in Andhra Pradesh and Madhya Pradesh. Exploratory drilling may be required to be further supplemented by evaluation drilling and explatory mining and NMDC is possessing adequate drilling facilities and expertise. Bulk assaying facility is also available with NMDC at Panna and recently NMDC has acquired ITPH HMS mobile plant for the treatment of bulk samples for correct assaying of exploration samples.

The liberalized Mineral Policy of Government of India has not yielded desired results in the field of diamond exploration and exploitation due to various licencing procedural problems. The multinational companies are not willing to auction the diamonds mined in India within the country. At this stage creation of National Diamond Authority is required for pooling the expertise and scientific know-how available with GSI, NRSA, NMDC and NGRI and DGM Orissa and Madhya Pradesh to give a thrust to exploration and exploitation of diamonds in India. The chairman of CGPB 99 (precious and semi-precious stones) emphasised the need for better interaction between State Governments and GSI for locating potential areas for exploration. Diamondiferous rocks do not come by easily and concerted and sustained efforts are essential. The exploration technology needs to be constantly reviewed, refined and updated in tune with the developments elsewhere, for achieving best results. In the recent years many kimberlites were discovered in Canada and valuable information on these kimberlites is accessible through internet. This information might help us improve or refine our current exploration strategy. For this purpose Government should identify a nodal agency preferably NMDC based on its rich experience in the field of exploration and exploitation of diamonds in order to achieve quick results in a given time frame.

India has large diamond cutting industry especially for small diamonds and India imports about 70% of the world production of the rough diamonds. The country also dominates the exports of cut and polished diamonds. 7 out of every 10 diamonds are cut and polished in India. Since this trade almost entirely is dependent on imported rough diamonds, there is an urgent need to boost indigenous diamond production.

ACKNOWLEDGEMENTS

AM and KSR thank the CMD and Director (Technical) , NMDC for their constant encouragement and permission to publish this paper.The views expressed in this paper are those of the authors and do not represent that of the organization.

REFERENCES

Abhijeet Mukherjee, Kimothi, G.P. and Babu, E.V.S.S.K. 1998b. Geochemical correlation of garnets from Majhgawan and Payalikhand diatremes, Madhya Pradesh. In Intr. Conf. On Kimberlites and Diamond Exploration. Cond. by AEG. pp. 9-10.

Abhijeet Mukherjee and Rao, K.S. 1998a. Mantle minerals in kimberlites/lamproite bodies and their significance in diamond exploration with special reference to Majhgawan pipe, Panna district, Madhya Pradesh. In Intr. Conf. on Kimberlites and Diamond Exploration. Cond. by AEG, pp. 9.

Abhijeet Mukherjee, Rao, K.S. and Babu, E.V.S.S.K. 1998. Study of mantle minerals from few kimberlite/lamproite bodies from different Indian cratons. Extr. Abstr. 7th Intr. Kimb. Conf., Cape Town, South Africa (unpaginated).

Abhijeet Mukherjee, Rao, K.S. and Bandyopadhyay, D. 1995. Geological and geochemical characteristics of Majhgawan diamondiferous pipe, Panna district, Madhya Pradesh. Intr. Symp. On Diamond Exploration and Gemstones. Cond. by SGAT, Bhubaneshwar, pp. 27-38.

Abhijeet Mukherjee, Rao, K.S., Bandyopadhyay, D., Roy, G. and Chatterjee, A.K. 1997. Geothermometry and oxygen barometry of co-existing iron-titanium oxides of Majhgawan diamondiferous pipe, M. P. Jour. Geol. Soc. of India, vol. 49, pp. 55-60.

Atkinson, W.J., Hughes, F.E. and Smith, C.B. 1984. Areview of the kimberlites of Western Australia. In Kornprobst J., ed., Kimberlites I : Kimberlites and related rocks, Elsevier, Amsterdam, pp. 195-224.

Burley, A.J. and Greenwood, P.G. 1972. Geophysical surveys over kimberlite pipes in Lesotho. Inst. Geol. Sci. London, Geophys. Dir., Ref. 540, 1069/72.

Carlson, J.A., Johnson, R.B., Mccallum, M.E., Campbell, D.L. and Padgett, J.P. 1984. Evaluation of geophysical techniques for diatreme delineation in the Colorado Wyoming Kimberlite Province. In Kornprobst J., ed., kimberlites : Kimberlites and Related Rocks, Elsevier, Amsterdam, pp. 21-32.

Griffin, W.L., Gurney, J.J. and Ryan, C.G. 1992. Variations in trapping temperatures and trace elements in peridotite suite inclusions from African diamonds : evidence for two inclusion suites and implications for lithosphere stratigraphy. Contrib. Mineral. Petrol. vol. 110, pp 1-15.

Gurney, J.J. 1984. A correlation between garnets and diamonds in kimberlites. In Glover, J. E and Harris, P. G., ed., Kimberlites occurrence and origin : a basis for conceptual models in exploration. The Univ. West. Aust. Publ., vol. 8, pp. 143-166.

Magnee, I.D. 1950. Delimitation geo-electrique du premiere pipe de kimberlite de couvert dans les champs diamantiferes du Kasai (Congo Belge) Proc. 18th Intr. Geol. Congr. 5, pp. 52-58.

Sobolev, N.V. 1974. Deep seated inclusions in kimberlites and the problem of the composition of the upper mantle (in Russian : English Transl. By D. A. Brown, AGU, Washington, D. C.

Zhou, K., Zang, A., Wang, W.M., Xie, X. and Guo, L. 1992. Spinels as indicator for diamonds. In : Ext. Abstr., 5th Intr. Kimb. Conf., CPRM. Spec. Publ. Brasila, vol. 2, pp. 208-211.

Section V

ISSUES—ENVIRONMENT AND MANAGEMENT

CHAPTER 28

Coal Slurry Pond Reclamation —A Contribution to Pollution Control

*Michael Schütze**

INTRODUCTION AND BACKGROUND

Most of the Indian Coal contain higher percentages of inorganic impurities compared to the coals of the western world and Australia. These inorganic impurities present in Indian coals are intimately mixed with the coal matrix. Regarding washabilty characteristics, this means that Indian coal is difficult to very difficult to wash. According to Central Fuel Research Institute (CFRI) nomenclature, the washabilty number (WN) of Indian coal is in the range between < 25 to 75 compared to WNs of 90 to 150 for Western and Australian coal (the lower the number the more difficult to wash). Another characteristics to express the washability of coal is the Near Gravity Material (NGM); in most cases, NGM for Indian coal is between 25 to 30%, sometimes as high as 50%, whereas for Western coals NGM is in the range of 15 to 20%.

Especially for the washing of coking coal with an ash content in the raw coal of up to 35%, these washability characteristics require a special treatment. Therefore, coking coal washing here in India may be characterised by the following features:

Total washing is required, i.e. all the material between 0 and 150 mm has to be washed,

- In most cases, the ash content in the "Clean Coal" must not exceed 17% (compared to figures around 10% for Western coal).
- Washing to 17% ash requires finer crushing for better liberation causing.
- Large quantities of fines which are difficult to treat.

These features require three product washing, i.e. clean coal, middlings and rejects and complex washing circuits have to be used. The requirement of complex washing circles for the washing of coking coal is not an unique Indian feature, but this is common practice

*Montan-Consulting GmbH, Germany

in almost all coking coal producing countries. All around the world, in many cases the bottleneck in the coking coal washing plants is the treatment of the ultra-fine material minus 1 mm, either in the process of sorting and/or the in the process step of dewatering the fine products.

What can the operator do in a case where the grain size distribution of the raw material is getting worse and worse, i.e. the amount of fines is increasing, and, on the other hand, the capacity of the equipment is limited. In many cases in the past, the only solution was and is sometimes still today the disposal of the ultra-fine material in ponds. It is estimated that the amount of fines which are disposed of is in the range of 1 to 2 % of the raw coal plant feed. During the overall lifetime of a washing plant, the amount of slurries that has been collected and stocked for decades may be in the range of some million tonnes. There are reports from Russia and South Africa about overall quantities of several hundred million tonnes of slurries stocked in ponds which cover enormous amounts of land.

The ponds represent a source of relatively cheap energy if the slurries are reclaimed and the calorific value is high enough. Out of experience with coal washing, and slurry disposal and reclamation it is known that the ash content of the slurries may be in the range of 12 to 15% up to more than 60%. In cases where the ash is relatively low, it is common practice to reclaim the slurries, to expose the slurries to the atmosphere in order to reduce the moisture content and then to use the slurries for heating or power generation. If the caloric value of the slurries is too low, there is no possibility to use this potential of cheap energy unless the ash content of the pond material is decreased and the calorific value is increased.

On the other hand, the slurry ponds are of a certain danger to the environment. Not only that large quantities of land are consumed, but coal dust may also pollute the air during longer dry periods; sometimes there is also the possibility of self ignition. There is a further danger to the environment because the groundwater may get contaminated.

The figure gives an impression how these ponds look like, they often look like dead lunar landscapes (Fig. 28.1).

SLURRY POND BENEFICIATION PLANT

To use the pond material as a fuel the calorific value has to be increased or the ash content has to decreased. For this purpose, a process was developed which is a combination of proven technology and equipment. Before the process flow sheet was designed, a large-scale programme for sampling and testing was worked out, core samples of the pond were taken and the washability was tested. It should be noted that the thorough preparation and implementation of this sampling and testing programme is one of the most important steps for the success of the beneficiation plant.

The beneficiation process can be divided into three stages:

After the raw material has been exploited in the ponds and transported to the plant by trucks, in the first stage, lumps of slurry and agglomerates are dissolved in a type of conditioner drum thus changing the consistency of the material into a suspension. At the same time, foreign bodies like stones, plastic, wood and metal with a size of more than 35 mm are removed from the suspension by a rotating screen (Figs. 28.2, 28.3, 28.4, 28.5). The next figure shows the simplified flow sheet of the process (Fig. 28.6).

After having passed the conditioner drum, the slurries are classified into different grain sizes and then fed to spirals and flotation for further treatment The coarse material (0.3 - 2.5 mm) is treated on a two-stage spiral system, where clean coal and rejects are produced. The ultra-fine material minus 0.3 mm is separated on a two-stage pneumatic flotation system into clean coal and rejects.

The last stage of the process is the dewatering of the products. Clean coal from the spirals plus the flotation concentrate from the first stage are dewatered on a vacuum belt filter, the flotation concentrate from the second stage is treated on a vacuum drum filter. The flotation tailings are dewatered in a chamber filter press.

The entire washery water system in the plant is designed as a closed system. Under normal operating conditions, neither make-up water for the process is needed nor is surplus water emitted from the water circuit. Solids-free water from the overflow of the thickener is recirculated to the plant. The closed system can nearly be assured, as the amount of water which gets into the plant with the feed material is approximately the same to the amount of water which leaves the plant with the washed products. Therefore, this technology can also be interesting for regions with water scarcity or adverse climatic conditions.

The entire process is completely automatically and is controlled from a control room. The plant is equipped with measurement and control devices necessary for the process as well as for the recognition and location of malfunctions. For the operation of the plant, only three people are required.

The next two figures show a view into the plant and give an impression of the consistency of the washed products (Figs. 28.7 and 28.8).

So far two plants, one in Poland and one plant in Germany have been planned and commissioned. The capacity of the plants is in the range between 55 and 75 TPH and is limited by the size of the conditioner drum. In both cases, flotation tailings with an ash content between 40 and 50 % are treated. In case of the German plant, the clean coal was used as PCI-coal, in Poland the clean coal product serves as fuel in a near-by power plant.

The Table 28.1 (Fig. 28.9) shows the process results of the German plant which had been operating for nearly 7 years.

Table 28.1 : Process Results of the Slurry Preparation Plant

	Feed	*Concentrate Spirals+Flot.1*	*Concentrate Flot.2*	*Rejects*	*Foreign Matter*
(m-%)	100	40 – 45	5 –10	45	5
Moisture (m%)	22 –25	22 –24	24 – 28	20 – 22	
Ash (m%)	40 – 50	9 – 11	18 - 22	>70	
Grain Size Range	--	2.5 – 0 mm	0.35 – 0 mm	2.5 – 0 mm	

The process results for this raw material show that a coarse concentrate with an ash content in the range of 9 - 11 % at a yield of 40 to 45% and a finer concentrate with an ash content between 18 to 22 % could be achieved. Tailings quality with more than 70% ash can be regarded as excellent. After blending both the concentrates a final ash content of around 11 to 12 % at an overall yield of 50% is achieved. The moisture content is in the range of 22 to 25 %; depending on the use of the product a further reduction of the moisture content by thermal drying may be necessary.

ALTERNATIVES

The process described so far was designed for the treatment of flotation tailings with high ash contents. The washability characteristics required two stage spiral and flotation systems to get a sufficient yield. In the case of raw material with better washability characteristics one-stage systems may be sufficient and in certain cases, flotation may not be required at all. Especially if the raw material consists of slurries which have not been treated in a flotation, the process may be simplified.

A further possibility to expand the use of this type of preparation plant is designing the plant as a mobile unit. Normally the slurry ponds are scattered over the mining regions and the amount of slurry in the respective ponds often does not justify the expenditures of a stationary plant. In such cases, the plant can be designed in a modular way where the equipment is built into 30´-ISO- norm-containers and where the containers are connected with simple coupling systems. This lay-out leads to savings in time and cost.

CONCLUSIONS

The last two figures (Figs. 28.10 and 28.11) show a pond site before and after rehabilitation, these figures explain themselves.

The beneficiation process described is using proven equipment with well-known technology minimising the risks to a potential operator. The beneficiation of pond slurries is a good example for an ecological and economical process to produce good quality low cost fuel from material that was formerly regarded as waste. The reclamation of pond material is an active contribution to pollution control as hazards to the environment may be mitigated and former pond sites may be converted to agricultural or forestry land-use.

Fig. 28.1 : Slurry Pond, Capacity 2,5 Mt

Reclaiming of Slurry

Montan-Consulting GmbH, Germany

Coal Slurry Pond Reclamation
- A Contribution to Pollution Control -

Picture 2

002-a4.ppt

Fig. 28.2 : Reclaiming of Slurry

Loading of Slurry on trucks

MC Montan-Consulting GmbH, Germany	Coal Slurry Pond Reclamation - A Contribution to Pollution Control -	Picture 3

003-a4.ppt

Fig. 28.3 : Loading of Slurry on trucks

Fig. 28.4 : Feeding Station of the Slurry Beneficiation Plant

Fig. 28.5 : Dissolving Drum for Slurry and Separation of Impurities (wood, iron, cables, stones)

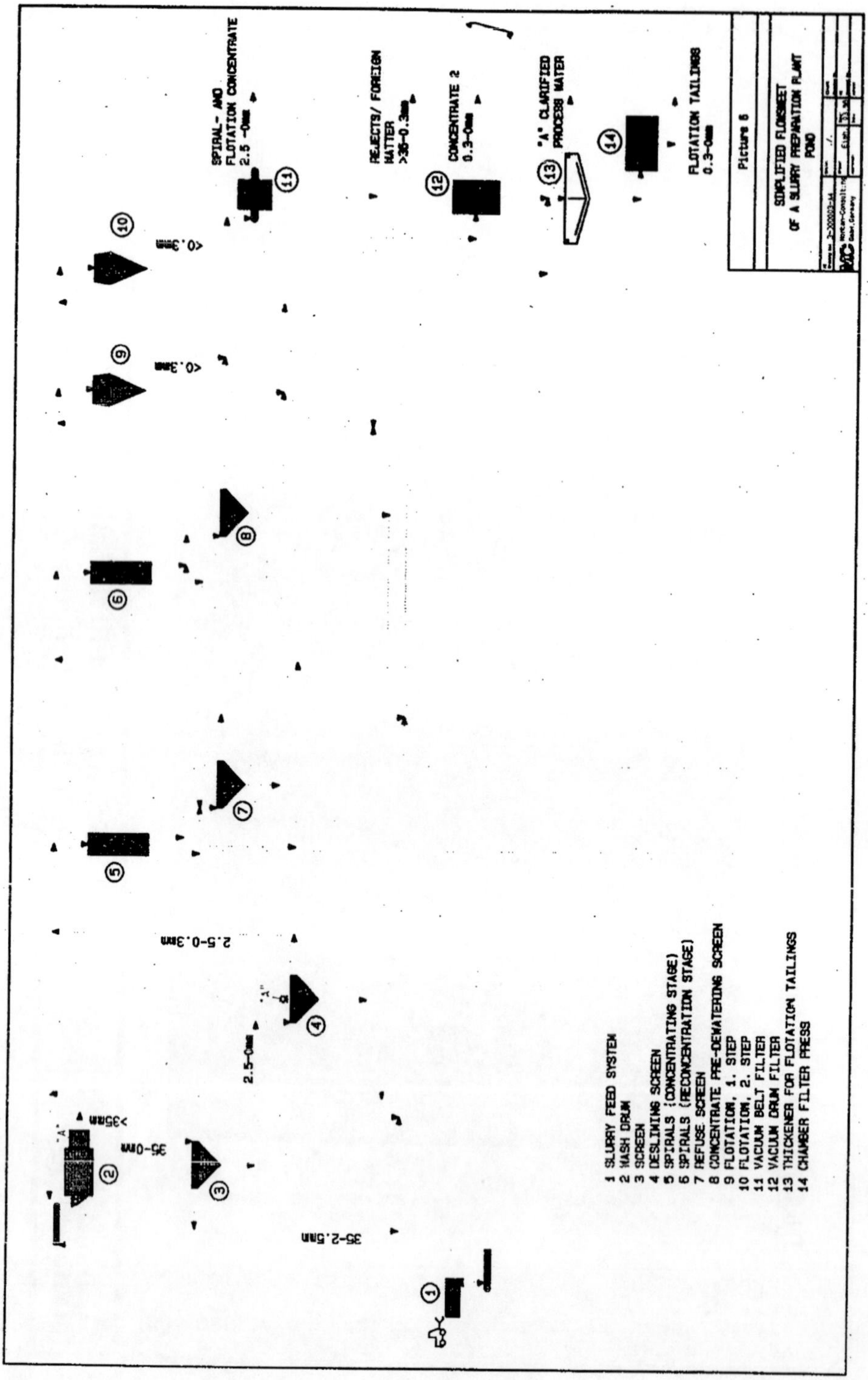

Fig. 28.6 : Simplified Flowsheet of A Slurry Preparation Plant Pond

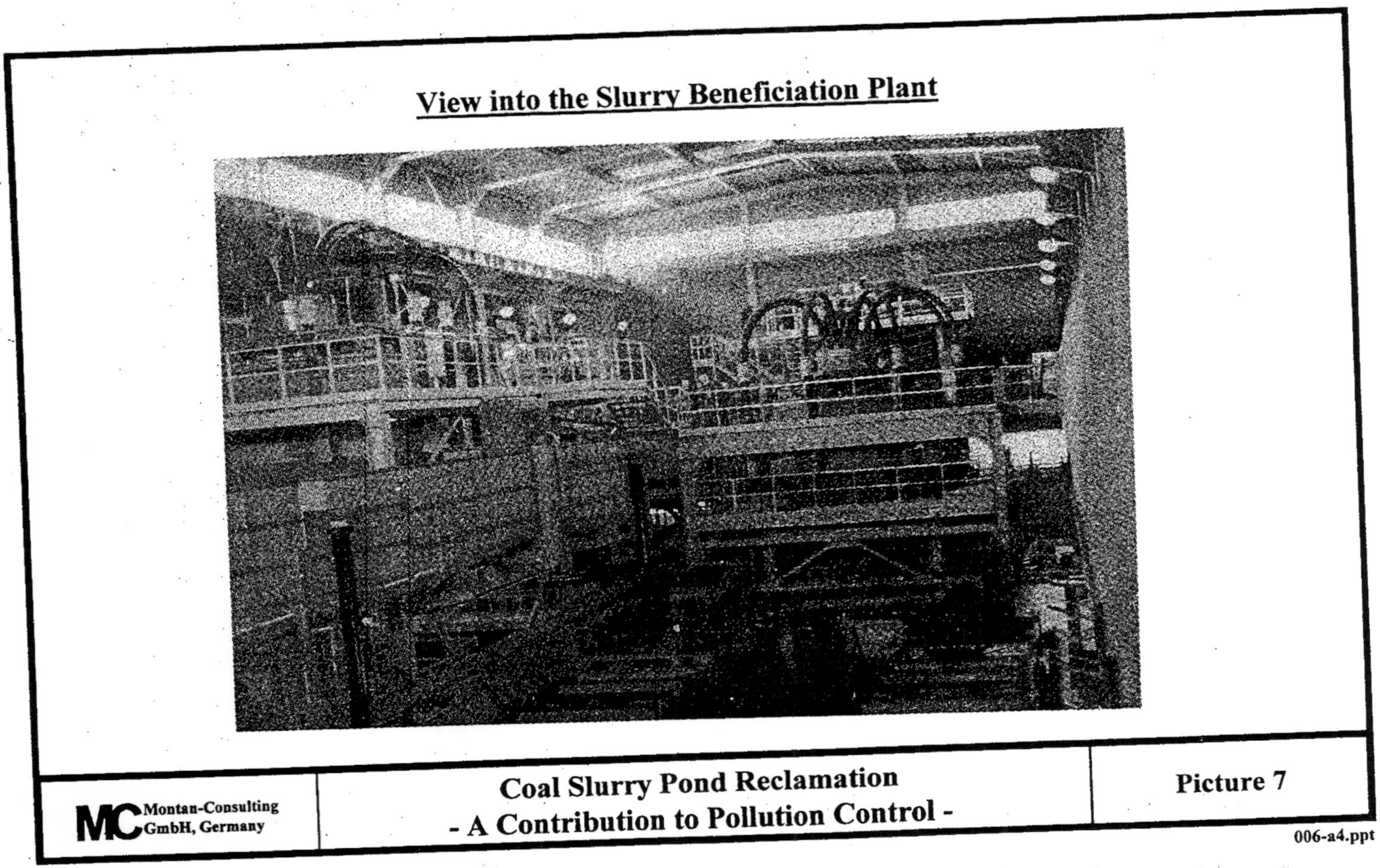

Fig. 28.7 : View into the Slurry Beneficiation Plant

Final saleable Products, low-ash concentrate (right bulk) and middlings (left bulk)

MC Montan-Consulting GmbH, Germany	**Coal Slurry Pond Reclamation - A Contribution to Pollution Control -**	**Picture 8**

007-a4.ppt

Fig. 28.8 : Final saleable Products, low-ash concentrate (right bulk) and middlings (left bulk)

Slurry Preparation Plant
(Fechner, Germany)

	Feed	Concentrate 1 Spirals + Flot. 1	Concentrate 2	Rejects Spirals + Flot. 2	Foreign Matter Rejects, Wood, Iron, Plastic etc.
Mass distribution (m-%)	100	40 - 45	5 - 10	45	5
Moisture (m-%)	22 - 25	22 - 24	24 - 28	20 - 22	--
Ash (m-%(d))	40 - 50	9 -11	18 - 22	75 - 78	--
Grain size range	--	2.5 - 0 mm	0.35 - 0 mm	2.5 - 0 mm	> 2.5 mm

MONTAN-CONSULTING GmbH, Germany

Fig. 28.9 : Slurry Preparatioin Plant (Fechner, Germany)

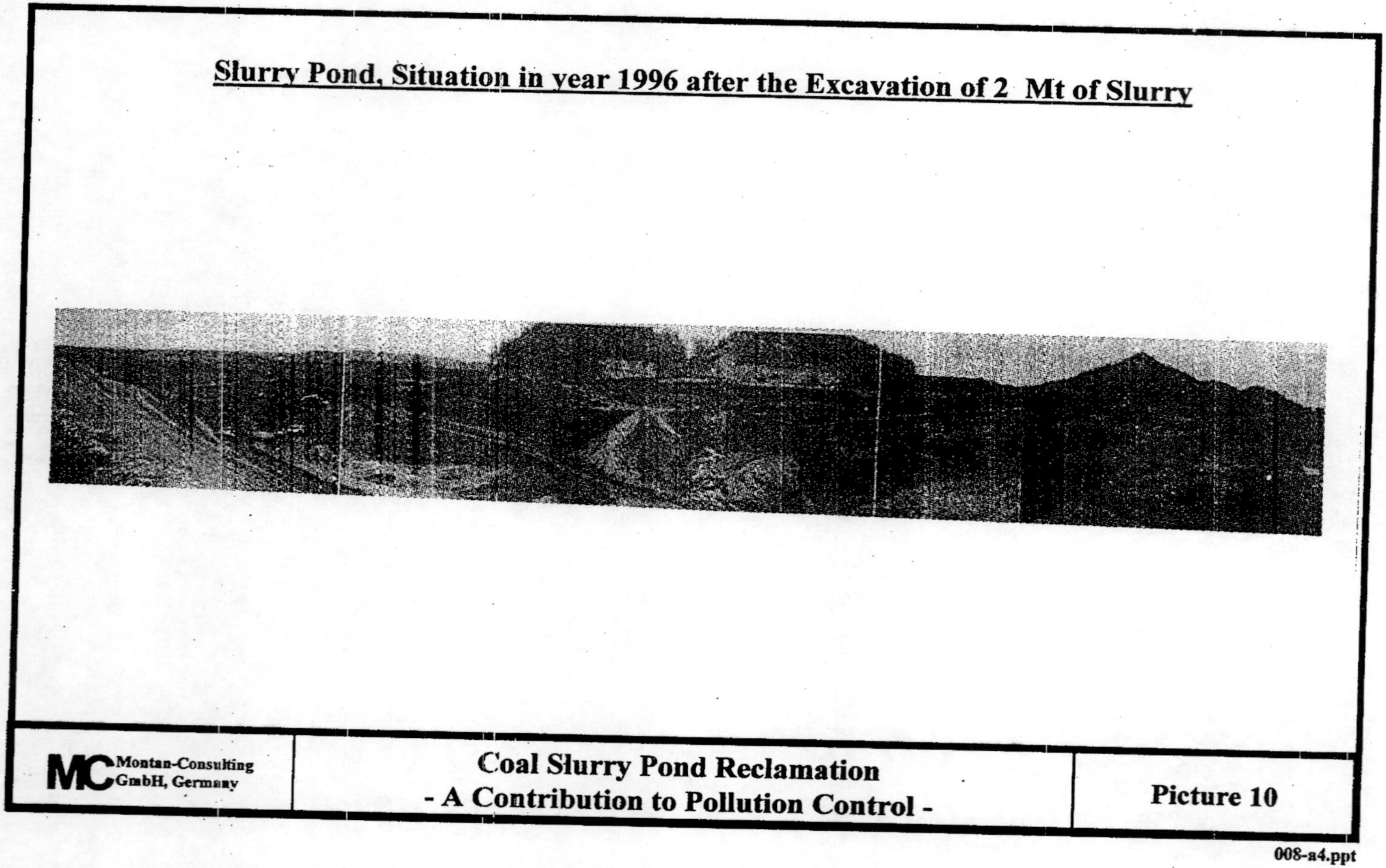

Fig. 28.10 : Slurry Pond, Situation in year 1996 after the Excavation of 2 Mt of Slurry

Fig. 28.11 : Rehabilitated Pond Site

CHAPTER 29

Ground-water Management in the Lausitzer Braunkohle Aktiengesellschaft (LAUBAG)

*Ulrich Höhna**

GROUND-WATER LOWERING – RE-ESTABLISHMENT OF GROUND-WATER LEVEL

Beginning with the exploration and exploitation of the 1st Lusatian seam 200 years ago, the conflict between mining and industry and water reserves had been also effecting the Lower Lusatian region. First, the withdrawal of water which was necessary to extract coal from this seam had only minimal effects on the environment, but the beginning of the mining operation from the 2nd Lusatian seam resulted in a new aspect: The lignite seam had to be extracted deeply within the range of the ground-water area of the ice-marginal valleys. Increasing water quantities were pumped and the static, i. e. the non self-renewable, ground-water reserves were claimed. Subsequently, the ground-water level was lowered extensively. This had serious consequences for the water resources of the Lower Lusatian area. The water supply from surface wells was endangered and the flow-off behaviour of the waters inside the cone of depression was strongly influenced. Furthermore, this caused adverse effects on the population of the varied indigenous flora and fauna and a change in the structure of agricultural sites.

The effects remained restricted to the catchment area of the Schwarze Elster river until the mid of the 50ies. But beginning with the development of new opencast mines, the depression cone increasingly influenced the catchment area of the Spree river and to a less extend also the Lusatian Neisse river. In 1989, the ground-water cone of depression reached its maximum extent in the Lower Lusatian region covering an area of approximately 2,300 square kilometres. This was also reflected by the pumped water quantities of 1,200 million cubic metres per year, coming from the 16 opencast mines at that time. The total ground-water deficit amounted to 13 billion cubic metres.

*LAUBAG-Consulting, Ingolf Arnold, Lausitzer Braunkohle Aktiengesellschaft.

At present, LAUBAG operates 5 opencast mines in the Eastern part of Lower Lusatia, which is mostly within the catchment area of the Spree river.

The company's activities with regard to an environmentally acceptable use of the water resources focus on the following:

1. Ground-water pumping is limited to a required minimum extent. Only that quantity of water is to be pumped that is needed for a safe operation.
2. The pumped ground-water is preferably used for preserving the natural reserves and to guarantee the minimum run-off of rivers.
3. We support efforts made by the Government and the states for re-establishing the ground- and surface water systems.

The requirements of the Federal Mining Law, Nature Conservation Law and Water Resources Law – aiming at minimising the consequences of lowering the ground-water – are stipulated in the licenses and water supply permits granted by the mining authorities. LAUBAG carries out different measures to saving the ground-water resources. Model calculations with efficient computers make it possible to find optimal solutions for the placing of pumps and wells to be used and, therefore, the quantities of pumped water. Ground-water balances for the catchment areas of the rivers are worked out, the large-scale ground-water development is determined and protective measures are prepared.

To simulate the underground flow processes, the Lusatian lignite mining industry uses program systems which are both based on the Finite-Difference-Method (which is the most frequently used method applying the program GEOFIM) and the Finite-Element-Method.

Within a legal framework, the mining company is not only obliged to minimise the consequences of the unavoidable withdrawal of ground-water. Moreover, sensible areas are to be protected and compensation or substitution measures have to be taken, respectively, as far as possible. Reports on ecological requirements and expert opinions as well as investigations are essentially supporting the decisions about water compensation measures. The mining licenses and water supply permits necessary for the operation of an opencast mine contain numerous detailed measures.

Each of the measures requires careful consideration of important aspects, such as:

- Quantity of water
- Quality of water
- Efficiency of the measure and
- Sustainability.

MONITORING SYSTEM

All the protective measures require a reliable and continuously working monitoring system of the surface and ground-water systems. This includes the collection of basic hydrological and data of the abiotic and biotic stock of the uninfluenced pre-mining condition of the natural areas within the range of the current and forecasted depression cone. Areas with ground-water levels close to the surface, which are endangered due to the lowering of the ground water level, can be identified by means of terrain photo-surveying and mapping. Based on this, hydrological and biological monitoring systems are installed to collect data of the initially uninfluenced situation of those ground-water dependent wetlands in the surrounding of the opencast mines. Experts from different fields and several Federal States are co-operating interdisciplinary in this field.

The results of the monitoring system enable us to decide upon measures to protect and maintain those areas.

PROTECTIVE AND COMPENSATING MEASURES WITH REGARD TO WATER RESOURCES

Sealing walls

Depending on the hydrological, geological and technical conditions, an additional measure is applied to support the conventional drainage by filter wells – the so-called sealing wall.

Two methods are used:

— the continuous slot cutting process with direct flushing, and

— the slot grab excavation method with self-hardening flushing.

The principle of the sealing wall bases on the idea that on the "water side" the natural ground-water table is maintained whereas on the opencast mine side the ground-water level is lowered to the necessary extent. Parallel to keep the deposit free of water, this measure is used to protect waters, wetlands and marshy areas in the direct vicinity of an opencast mine.

The longest sealing wall produced continuously so far has been established along the Lausitzer Neisse river at the eastern boundary of the Jänschwalde opencast mine. It has an average depth of 75 metres and a current length of 7.8 kilometres. The total length of the sealing wall will come to approximately 11 kilometres. At present such a sealing wall is unique. It avoids the lowering of ground-water onto the territory of the Polish Republic and protects the river landscape of the Lausitzer Neisse river.

Ground-water infiltration plant

Once, the "Laszinswiesen" (a moist meadow area) were the gateway to the "Spreewald" area. Just as in this protected area, a natural network of watercourses runs through these moist meadows. Centuries ago, the resident farmers unravelled the maze of watercourses and arranged them according to their needs; in the 70ies of this century, the watercourses were step by step extended to a system totalling more than 60 kilometres of straight ditches. The ground-water level in this wet area is only 0.5 to 1 metres below the surface level. The central area of the "Laszinswiesen" is protected, the area is cultivated in accordance with the responsible nature conservation authorities.

During the next years, the Jänschwalde opencast mine will come close to this area. Without taking adequate compensation measures, the ground-water level would be lowered by about 7 m due to mining drainage. The meadows would fall dry and rare birds and plants would loose their habitats.

Various measures to protect the meadows have been initiated

A so-called dam concept has been jointly developed with the farmers and conservationists. Pit water from the Jänschwalde opencast mine will be purified, fed to the watercourses and from there distributed into the ditches via a pumping station. From 2002, approximately 2.5 million cubic metres of water will additionally flow to the "Laszinswiesen" in this way.

In contrary to the open infiltration carried out so far, the main part of the dam concept is the closed infiltration. Between meadow area and opencast mine, there will be installed a filter well system. Another 5 million cubic metres of water will be injected annually into the soil step by step via wells and percolation pipelines until 2010. This measures aim at establishing a so-called closed water wall to prevent the meadows from falling dry.

The feasibility of the seepage and infiltration concept has been verified by thorough studies. This method has also been successfully applied in the Rhenish area "Schwalm-Nette".

At the end of 1998, a pilot plant was put into operation to test different infiltration methods at site and to investigate the raw water treatment for applicability as infiltration water in closed infiltration plants. The planned test period is two years. Based on the results of this pilot project, a plant will supply water to this moist area during the opencast mine operation close to this area for about 10 years.

INTERESTING FACTS ABOUT THE PILOT PLANT

The plant consists of a pumping station, a water treatment plant, an infiltration plant including measuring equipment and ground-water monitoring pipes, and a 1.7 kilometre long pipeline between the two plants. Approximately 40 cubic metres of water flows through the pipeline. Via a 40 metres deep infiltration well, the water is fed to a horizontally laid 40 metres long seepage pipeline. The water quantity can be variably distributed. The plant operates automatically. Important data are constantly measured, stored and periodically analysed.

Users of water

Lowering the ground-water has consequences for the waterworks and operators of private dug wells. LAUBAG provides adequate compensation by financing the extension of existing wells or installing new wells or the connection to the public drinking water network. Additional expenses are compensated for by granting favoured costs for irrigation water.

Branitzer Park

Feeding point Petershainer Fließ

Steinitzer Quelle

Rothwassergraben

Welzow-Süd opencast mine, distribution of water

PIT WATER PURIFICATION

Due to the special geological condition in Lusatia, the ground-water has a high natural iron content. When directly feeding pit water into receiving rivers the reaction of solved iron with oxygen is caused and thus the formation of precipitable iron-III-hydroxide. This results in a clearly visible ochre coloration of the water courses. It is therefore the task of pit water purification plants to remove the iron and therefore counteract the formation of iron hydroxide in the natural water courses.

The iron content in the pit water varies. It is between only a few to 70 mg/l, in extreme cases even higher. This water mainly comes from dumps. Moreover, it has a low pH-value, that means it is relatively acid.

By adding limestone the pit water is neutralised in the purification plants. In a subsequent ventilation stage the water absorbs oxygen, the iron oxidises and afterwards the iron hydroxide flocculates in the settling tank. Clear water containing less than 3 milligram of iron per litre leaves the purification plant and is fed to the natural water courses.

DISTRIBUTION OF PIT WATER

LAUBAG is not responsible for distributing the pit water quantities, this is the task of the water authorities. They grant water law permits according to the water resources law and licenses, respectively.

More than half of the pumped ground-water quantity flows into rivers and rivulets after purification. Therefore, the mining company compensates for the lack of natural inflow arising from the use of the pumped ground-water.

The supply of pit water to local wetlands is limited during the next four to five decades due to balancing the Spree-run-off with public welfare in mind. The minimum run-off of the Spree must be guaranteed because this river supplies water to the Spreewald biosphere reserve and to the Federal capital of Berlin.

At present and also in the future, LAUBAG uses more than 15 per cent of the pit water to supply wetlands. Annually, more than 44 million cubic metres of water are returned to the water circulation by feeding the rivulets, ponds and swamps. Within the framework of a careful consideration process involving also the responsible nature conservation authorities, the areas that are considered worth protecting are selected on the basis of water resources balances. It goes without saying that all additional expenses are taken over by LAUBAG.

The remaining part of the pit water is used in the opencast mines and for the process of power generation. Treated to industrial water, the pumped ground-water from the opencast mines satisfies the demands of the Lusatian power plants of Jaenschwalde, Boxberg and Schwarze Pumpe.

Pit water from the Welzow-Sued opencast mine with a good quality is treated in a LAUBAG-own plant to adhere to drinking water qualities. Via long-distance pipelines the water is supplied to villages and towns within a radius of 30 km.

Modern drinking water treatment plant at Schwarze Pumpe

Diagram of water distribution

CHAPTER 30

Coalbed Methane : Challenges Ahead

P. Karunesan and U. P. Singh***

ABSTRACT

Coalbed Methane is a by-product of underground coal mining operations. It is a mining hazard. Its presence makes the mine gassy and so the mining equipments to be used in these gassy mines need to be flame proof or intrinsically safe. Thus, the presence of CBM in the mine not only makes the mining operations difficult and risky but also makes it costly. Therefore, it must be controlled at the working faces and also at other points in mine layout. This can be achieved by well-designed ventilation system but its ventilation in the atmosphere adds to greenhouse gas and therefore contributes to global warming. However, CBM is a remarkably clean fuel when burnt. Therefore, if this methane can be trapped from coalbeds economically and burnt properly as fuel, it is as an additional source of energy. Moreover, reducing both, the escape of methane gas to the atmosphere and mining hazard. This concept led to development of Coalbed Methane (CBM) resources and attracted attention of the scientists. In the recent years it has been developed as an alternate clean energy resource in many parts of the world including India.

Indian energy scenario reveals that India will continue to be largely dependent on coal, for its energy requirements, which will cause negative environmental impacts. At the same time India's Natural Gas reserves are going to last hardly for 20 years and even during this duration, its supply is not going to meet the projected demand for the gas. Thus, first and foremost challenge ahead of CBM industry is to (i) trap the CBM economically, and (ii) successfully supplement natural gas resources with CBM, partially reducing the dependence on coal.

Though, CBM has been identified as an alternate clean source of energy, however, the CBM activities in the country is still in its infancy stage, therefore, typical of an immature and evolving technology and concept, series of issues are unresolved, forcing the implementers to face many challenges and contradictions, which are of technological, regulatory and financial nature.

* General Manager (Production), Coal India Limited, Calcutta.

** Dy Chief Engineer (New Technology), Coal India Limited, Calcutta.

The unconventionality of CBM leads to the question of its availability, identification of potential basins, collection of various geological data and generation of data bank, evaluation of gas content and Gas In Place (GIP), reservoir engineering, drilling techniques, ascertaining the damage to coal seam due to CBM operation, etc. Even the engineering technologies and machineries required for the development of CBM technology are not adequately available in the country and the domestic oil field services are yet to adopt to suit the need of the CBM industry.

CBM is not covered under any of the statutes of the country. Who is owner of the gas; what are the constituents of CBM (its definition); act under which this resource has to be prospected and extracted; regulatory issues related to the safety of the concerned mines and proper discharge of produced water; etc. are some of the issues yet to be resolved.

Indian coal companies have large expansion plans for their coal mining activities and, hence, are not in a position to even carry out the initial work of identification and assessment of CBM potential in different coalfields of India. CBM development project is capital intensive with very high risk.

Apart from these, transportation of CBM is one of the very important issues need to be addressed properly. In absence of national pipeline grid, new ways to utilise CBM near the point of production itself, have to be found out.

However, one should not get disheartened. These questions are the new-age cousins of questions people asked whenever a new resource came into existence. First the high power new engines were invented and then only came the seat belt.

Government of India (GOI), through an MoU between Ministry of Petroleum and Natural Gas (MOPandNG) and Ministry of Coal (MOC), has initiated a step towards resolving some of the regulatory issues pertaining to ownership of the gas and act under which CBM has to be prospected. However, the law of the land does not support it. Involvement of private national and international agencies in development of CBM is expected to bring with it not only adequate technologies but sufficient finance also.

Participation of CIL in a jointly funded project with UNDP/GEF for identification, design and implementation of programs to recover CBM in safe, cost effective and environmentally acceptable manner from the working mines may, at the end of the project resolve some of the issues envisaged above.

Since development of CBM technology and production of CBM gas is a time consuming venture of very high risk, it is imperative to address all the issues so that a sincere integrated effort for development of CBM technology in the country may be put.

INTRODUCTION

Coalbed Methane is a by-product of underground coal mining operations. It is a mining hazard. Its presence makes the mine gassy and so the mining equipments to be used in these gassy mines need to be flame proof or intrinsically safe. Thus, the presence of CBM in the mine not only makes the mining operations difficult and risky but also makes it costly. Therefore, it must be controlled at the working faces and also at other points in mine layout. This can be achieved by well-designed ventilation system but its ventilation in the atmosphere adds to greenhouse gas and therefore contributes to global warming.

However, CBM is a remarkably clean fuel when burnt. Therefore, if this methane can be trapped from coalbeds economically and burnt properly as fuel, it is as an additional source of energy. Moreover, reducing both, the escape of methane gas to the atmosphere and mining hazard. This concept led to development of Coalbed Methane (CBM) resources and it attracted attention of the scientists. In the recent years it has been developed as an alternate clean energy resource in many parts of the world including India.

Indian energy scenario reveals that India will continue to be largely dependent on coal, for its energy requirements, which will cause negative environmental impacts. At the same time India's Natural Gas reserves is going to last hardly for 20 years and even during this duration its supply is not going to meet the projected demand for the gas. Thus, first and foremost challenge ahead of CBM industry is to (i) trap the CBM economically, and (ii) successfully supplement natural gas resources with CBM, partially reducing the dependence on coal.

CBM constitutes a clean fuel having heating value of approx. 8500 K Cal/Kg compared to 9000 K Cal/Kg for the natural gas and is generally of pipeline quality and therefore, it can be directly fed to the national grid of pipeline.

ISSUES

The CBM activities in the country is still in its infancy stage, therefore, typical of an immature and evolving technology and concept, series of issues are unresolved, forcing the implementers to face many challenges and contradictions, which may be of technological, regulatory and financial nature. Some of the major issues to be resolved are:

Technological issues

India is estimated to have around 2 Trillion Cubic Meters of CBM reserves. The estimation, which has been made about the availability of CBM gas in the country, is based on mainly the international experience in this regard and hence it can not be accepted with a very high confidence level. Therefore, some of the foremost tasks are:

- To identify the probable coal and lignite basins enriched with CBM gas, and
- To ascertain the quantity of gas in these basins.

Assumptions on ideal gas content may be possible from coal isotherms or from coal rank, reservoir depth and maceral content, however, to understand the recoverable reserves it is necessary to study permeability, gas saturation, structural and stratigraphic variables, etc. Various agencies, concerned with development of CBM technology in the country, need to put their efforts in this direction. Government and private agencies are expected to come forward with coordinated activities in this regard.

Coal Companies

Coal companies are expected to carryout the following activities:

- *Basins identification*: By a study conducted by ARI, U S A, it is estimated that India is having around 2 Trillion Cubic Meters of CBM reserves in the various Coal Basins of India. However, these estimates need firming up through exploratory drilling. Further to this, one of the most important tasks is to identify these potential basins itself for CBM extraction.

- *Development of Data bank*: Exploration Data needed for CBM Development are: Total Area, Number of seams and their thickness, Proximate and Ultimate analyses, Rank of Coal, Geological Structure, Hydrogeology, Geothermal gradient, Reflectance, Maceral Composition, Mineral Matter, Dispersed Organic matter, Gas composition and percentage, Permeability and porosity, Cleats and fracture pattern, Adsorption Isotherms, Quality of adsorbed water, etc. Apart from these, Reservoir Engineering and modelling, Open hole drill stem testing, Side wall pressure core testing, etc. is also required to be carried out. Coal rank and burial depth determine the gas content where as net coal thickness gives an idea of gas in place. Over and above, Coal diffusion properties, absolute and relative permeability, etc. need to be evaluated for knowing the coal reservoir properties of the probable CBM potential coal blocks.

  — Extensive data is available in respect of Total area, Number of seams and their thickness, Proximate and Ultimate analyses, Rank of Coal, Geological Structures; and

  — Data is available in sporadic manner for Hydrogeology, Geothermal gradient, Reflectance, Maceral Composition, Mineral Matter, Dispersed Organic matter, Gas composition and percentage covering all coalfields of India.

  — However, no data is available for rest of the parameters i.e. Permeability and porosity, Cleats and fracture pattern, Adsorption Isotherms, Quality of adsorbed water, Reservoir Engineering and modelling, Open hole drill stem testing, Side wall pressure core testing, etc. This is so, because for coal mine planning and designing, all the tests that need to be conducted, for generation of gas related data were not considered necessary.

- Therefore, extensive exploration have to be carried out once again, even where detailed exploration has already been carried out for mining projects, in our coalfields to know about these parameters.

- This will facilitate in identifying the potential CBM basins, its gas reserves and flow characteristics, which are the major determinants of commercial viability of the CBM Project.

- In Indian coalfields all locations, where coal is occurring at depth (200 to 1500 m) and having high rank coal or evidence of igneous body intrusions, should be made target for CBM exploration, such as Damodar, Sone, Kanhan Valley coalfields, etc. But these conclusions are based on very limited data available and for firm assessment; extensive drilling and tests have to be carried out in these coalfields.

National/International Players

Once the coal, its properties, its storage/release mechanisms are well understood for different coalfields, outside agencies, which are technically and financially sound, may be allowed to participate in Coalbed Methane Exploration and Exploitation in the various blocks. They are expected to look into the matter of reservoir engineering, recommendation and adoption of best drilling techniques, completion, stimulation, and spacing strategies for the wells and ultimately to the economics and scale of Coalbed Methane production.

Legal or Regulatory issues

The concept of commercial utilisation of Coalbed Methane (CBM) is new in this country and therefore, CBM is not covered under any of the statutes. The unconventionality of CBM gas leads to the question of its definition, ownership, act under which this resource has to be prospected and extracted, regulatory issues related to safety of the concerned mines and discharge of produced water, etc.

Ownership

The concept of Coalbed Methane gas as an energy source did not exist when Acts and Regulations; such as The Oil Fields (Regulation and Development) Act, 1948; The Petroleum and Natural Gas Rules, 1959; The Coal Bearing Areas (Acquisition and Development) Act, 1957; The Coal Mines (Conservation and Development), Act, 1974; and Mines and Minerals (Regulation and Development) Act, 1957 were framed. Thus, CBM is not covered under any of the statutes including acts and rules governing oilfields and coal mines.

Therefore, the question arises who is the owner of Coalbed Methane? Is it Coal Producing Companies? Yes, feel the Coal Producing Companies, mainly due to these reasons:

- The gas is inherent in the coal, and
- A mining company's right to remove Coalbed gas through ventilation or drainage, is obviously a necessary safety precaution without which coal can not be mined, has been well established over long years of mining practice. This infers that CBM is their property.
- Moreover, as per the Petroleum and Natural Gas Rules, 1959, "Petroleum" means naturally occurring hydrocarbons in free state whether in the form of natural gas or in a liquid. But we know that CBM is absorbed on the internal surface of the coal and hence it is not in free state and therefore not covered under petroleum definition and thus, it should not be termed as natural gas, which is a petroleum product.

But non coal companies, especially Petroleum and Natural Gas companies, who are dealing with the extraction of the natural gas in the country, argues that:

- As CBM is a separate mineral resource, rights granted for mining of coal do not cover other materials, such as gas or oil. This means that the mining right for coal not necessarily includes the rights to extract CBM also. Thus, the existing laws do not clearly establish the rights of coal companies to extract CBM.
- CBM is a Natural Gas and hence claim for exploration and exploitation falls under their jurisdiction.

All the above arguments were raised by either Coal companies or by Petroleum and Natural Gas companies to get administrative control over exploration and exploitation of CBM in the country.

Therefore, which ministry should administer the CBM Projects? Whether it should be with MOC or should it go to MOPandNG. Since the inception of the CBM project in the country, in the year around '90, the matter remained unanswered.

Who is owner of the CBM gas in the country was a burning issue till recently. This confusion caused a lot of problem not only to GOI but also to interested national and international investors and operators. This was one of the main reasons that all out efforts were not put for development of CBM industry in the country.

Initially, under a letter from PMO, Ministry of Coal was entrusted the job to develop CBM technology in the country. But of late a Memorandum of Understanding (MoU) has been signed on 9th September 1997 between Ministry of Petroleum and Natural Gas (MOPandNG) and Ministry of Coal (MOC) for resolving some of the issues facing the CBM industry including ownership issue. Arrangements and procedures on exploration/ exploitation of Coalbed Methane (CBM) in the country have been covered briefly in this MoU. It has been agreed that:

- The Coalbed Methane (CBM) is a natural gas and, therefore, the Ministry of Petroleum and Natural Gas (MOPand NG) is required to administer CBM projects under the existing Oil fields Act and Rules (Regulation and Development Act, 1948 and Petroleum and Natural Gas Rules, 1959).
- Since CBM is an associate of Coal and therefore it is required to be extracted in a manner so that the interests of coal and lignite industry are not getting affected adversely as regard to their present or prospective mining.
- Preparation of fiscal and other terms for a "Model Contract for Production of CBM" by the MOPand NG.
- The procedures, to deal with CBM related activities, need to be so organised that the prospective investor is required to interact with only one organisation and not with several organisations under different ministries, which would lead to confusion and serious difficulties in coordination, etc.

Though, the signing of the above MoU is an effort in the direction of resolving ownership issue of CBM between MOC and MOPand NG, however, yet it is not covered under any act and thus, it is not supported by the law of the land.

Definition of CBM

Once the ownership issue is resolved, the second most important issue to decide is: whether the CBM gas includes all the gases and condensate being discovered below surface, which might have migrated on the associated rocks and sand earlier.

However, this issue has lost its relevance as now, as it has been decided that Ministry of Petroleum and Natural Gas (MOPand NG) will administer the activities of CBM in this country and therefore, MOPandNG, which is empowered to decide about migrated gases and condensate, may decide that associated gases will also be extracted by the same agencies who have been entrusted the job of CBM extraction.

Mine Safety and Water Disposal Regulation

- Present mine safety regulations are inappropriate to deal with gas recovery safety and, therefore, it is necessary to incorporate gas recovery safety with mine safety regulations.
- During coalification process huge quantity of water is produced. For production of CBM, wells have to be dewatered first. It may happen that initially (may be from few months to few years) only water is being produced in very large quantity. The water produced has to be discharged properly without adversely affecting the crops, water bodies, water table, etc., of the area. Production of water is an essential activity for CBM extraction and therefore, it is essential that regulations for proper discharge

of produced water is developed to avoid any future conflict with those who are getting adversely affected.

Financial Issues

Indian coal companies have large expansion plans for their coal mining activities and hence are not in a position to even carry out the initial work of identification and assessment of CBM reserves in different coalfields. Therefore, even the exploratory work need to be carried out by outside agency.

Government of India (GOI) through the Directorate General of Hydrocarbons (DGH), Ministry of Petroleum and Natural Gas, is finalising the policy framework for development of CBM in the country. The DGH has drafted a model contract with terms and conditions for undertaking the CBM development through competitive bidding process. The qualifying agency needs to arrange its own finance for the whole operation.

Though the model contract includes the exploratory work also, however, if the exploratory work is being carried out initially itself, it will help in assessing:

- Total CBM reserve in the country, and
- Have better energy planning for the country.

The coal companies, however, do possess a large volume of mining data and also have substantial technical competent manpower. Hence joint ventures with coal companies may be an alternative to take the development of CBM Industry in India.

Commercial issues

In USA, the development of CBM was encouraged by special incentives, which was not available to the natural gas industry. In India, the demand for such incentives have been raised anticipating the expected higher cost of production, mainly, because much larger number of wells are required in case of CBM extraction and even after that there may not be considerable production of CBM during initial years and hence delayed Return On Investment (ROI). However, in USA, CBM has to compete with cheap gas but in India, the competition would probably be with imported gas, which would be costly. Thus, the question of special incentives can be settled only when we have more knowledge about the production costs of CBM.

Transportation of gas

As gas pipeline infrastructure, except for HBJ pipeline from Maharastra to U.P., is not well developed in the country, especially in the eastern part (where bulk of the coal resources of the country occur and hence potential CBM basins), simultaneous action needs to be taken for building up this important infrastructure. GOI has taken certain steps to improve this infrastructure in the country. The status of major existing gas pipelines in the country is:

- The 2300 km long HBJ gas pipeline capacity has recently been increased to carry from 18.2 to 33.4 MMSCMD gas from the Western coast of India to the Central and northern parts of country with the addition of 560 km loop line.
- Gas Authority of India Limited (GAIL) is further looking into the aspect of transportation of additional quantity of natural gas through the HBJ pipeline by

doubling its capacity from 33.4 MMSCMD to 60 MMSCMD by putting an additional pipeline from Hazira to Vijaipur.

- Apart from this GAIL also owns 725 km of existing pipelines that had been constructed and operated by ONGC earlier. Gas customers have also laid 320 km of pipeline themselves, whereas GAIL has laid an additional 600-km of pipeline in different areas of India. The regional pipelines vary in length from 0.5 to 75 km and were generally installed to meet the needs of a single or a group of customers.
- The smaller pipeline networks of GAIL supply gas to industries from offshore and onshore gas fields in the western, northeastern and southern parts of the country. The regional pipelines are in the areas of Mumbai, Gujrat, Rajasthan, Andhra Pradesh, Tamil Nadu, Pondicherry, Assam and Tripura.
- GAIL is considering having trans-national gas pipelines from eastern side of the country and this would provide a significant growth potential for the future.
- GAIL has also launched LPG pipeline project involving laying of a 1250-km pipeline from Kandla to Loni to transport 1.7-2.5 MMTPA of LPG.

Since, the development of pipelines initially may be a very costly preposition and, therefore, new methane utilisation technologies may have to be studied for consumption or utilisation of produced gas in the producing area itself, such as putting up captive power generation plants, chemical plants, fertiliser plants, industrial gas, town gas and use of gas as transportation fuel (CNG or LNG), etc. New Technologies have been recently developed for converting gas into various liquids, which can be transported with ease, compared to gas transportation.

Engineering

Engineering technologies and machinery required for the development of CBM technology are not adequately available in the country. The domestic oil field services yet to adopt to suit the needs of the CBM industry.

UTILISATION OF CBM

Methane gas having concentration of more than 95% (Pipeline quality) can be directly fed to the national grid of pipeline whereas mine gas having methane concentration of about 50% (taking into consideration the gob gas, etc.) may be utilised as a fuel for coal drying, etc. in the mine itself. There are various ways in which methane can be utilised. **(Refer Fig. 30.1 : Some Methods of Coalbed Methane Utilisation)**

Combustion

Undoubtedly, the most common method of utilising Methane liberated, as a by-product is the combustion. Methane is environment friendly non-toxic combustible gas having heating value of approx. 8500-8900 K.Cal/M^3 compared to 9000 K.Cal/M^3 for the natural gas. It constitutes a rather clean fuel taking into account the products of its combustion owing to modern technologies applied to natural gas combustion (where methane is main component), the emission of 30-60% lower carbon dioxide can be obtained when producing the same amount of energy compared with oil and hard coal. The contribution to the greenhouse effect by methane combustion is much smaller than the effect of releasing the same amount of methane in the atmosphere.

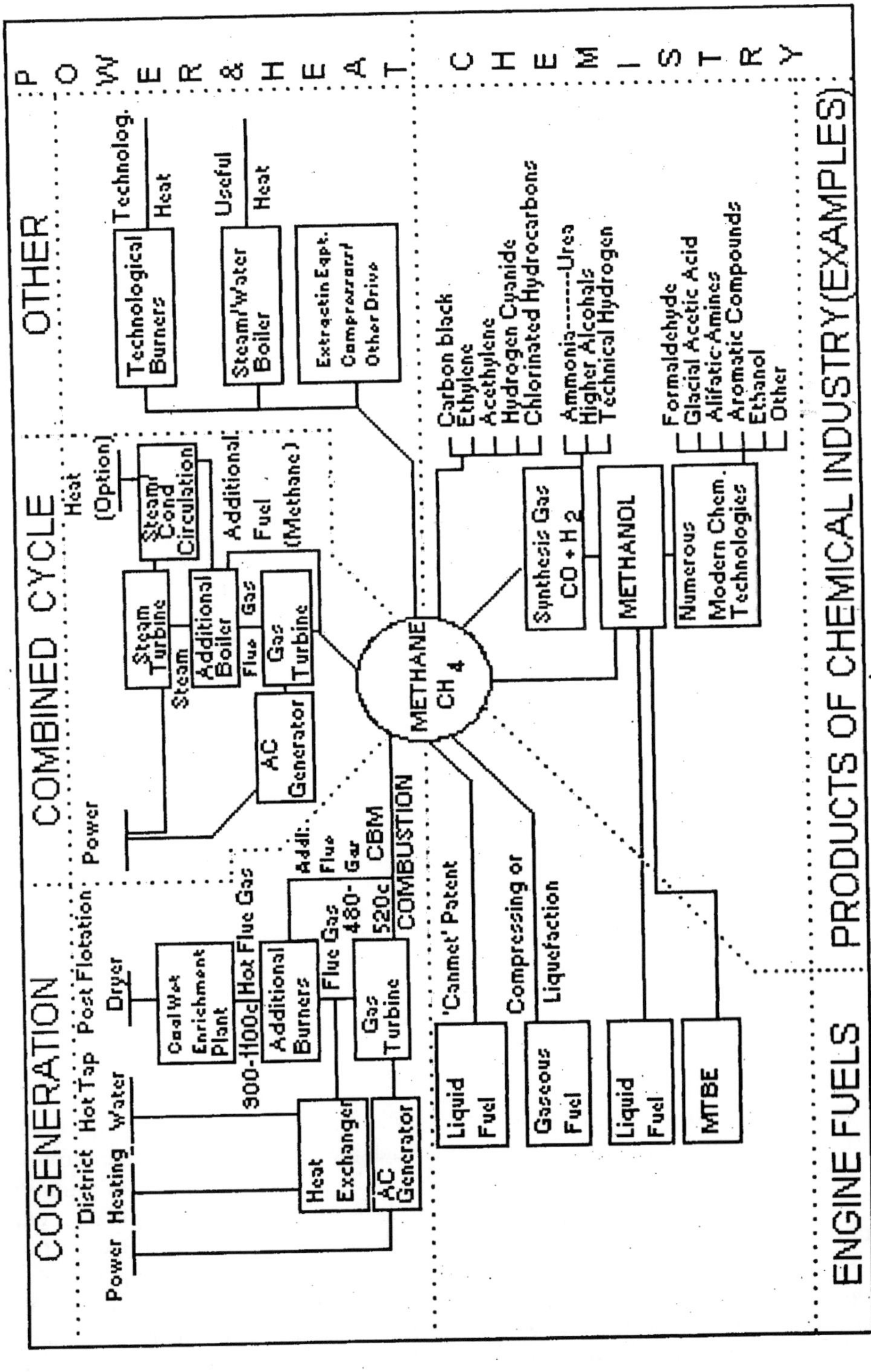

Fig. 30.1 : Some methods of Coalbed Methane utilisation

- It can be utilised as a local energy resource to dry coal out put from the coal preparation facility, generate captive power, or for other local uses. CIL is having a fleet of HEMMs and other vehicles, which can be powered by compressing the gas.

Chemical Energy

More modern and elegant methods of energy production from CBM are applied worldwide enabling more complete utilisation of methane chemical energy.

- Pipeline quality methane is used to produce acetylene, Carbon disulphide and Carbon Black as a feedstock for tyre production. Even some of the chemical processes developed utilise 40% mixture of methane and air.
- Methane can be converted into synthesis gas, which is a mixture of Carbon Monoxide and Hydrogen, to be used as feedstock for Methanol, aldehydes and alcohol synthesis. Some of the other chemicals can be obtained from methane are: Methyl chloride, chloroform, Tetra chloride carbon, etc. to be used commercially further.
- Pipeline quality methane can serve as an automotive fuel for both Cars and City Buses after being compressed (CNG) and liquefied (LNG) and pumped into fuel tanks.
- Some of the other uses of methane are in Plastic production, conversion of methane to gasoline range hydrocarbons via isobutene. Methanol is highly regarded as a potential source of clean energy and methanol derivative (MTBE) as an Octane-booster substitute for lead in Petrol.

Thus, we see that methane is having large utility, the challenge before the CBM industry in India is to adopt such technology, which can be appropriately applied in our situation.

RISKS INVOLVED IN CBM PROJECT

Investment in CBM projects is highly risky, especially in India, because it has to deal with various unknowns and uncertainties.

Untested Geological Settings

Unknown geological data, potential basins, gas content and its extraction, flow rate, etc. is incorporating a high degree of risk in estimating the viability of the project. Whatever estimate about the gas reserve, potential basins, etc. have been made are based upon the, experience in the foreign country which may not be directly applicable in Indian conditions.

Project cost

Project cost estimation may be a wild guess. The ability of any coal to produce methane depends on a variety of geological parameters such as: general geological environment, specific geohydrologic conditions, characteristics of the target coal, etc. These parameters vary in their applicability to particular coalbeds, for example geohydrologic data will be essential for water-saturated seams, but will be far less significant with respect to gas-saturated coalbeds. Therefore, even for a well-conceived and designed reservoir engineering for a particular type of coal, it is very critical. The conventional 'rules of thumb' about well spacing and stimulation design rarely hold and therefore situation remains uncertain and hence the cost. It is the adoption of different techniques and technologies as per the requirement of the situation, make a project successful and hence initial estimation of project cost is not possible, and the project cost goes on changing.

- High volume of water production may push back the gas production in terms of time and hence the return on investment may jeopardise the cost estimation.
- Earlier experience of methane drainage in India is discouraging (Muslia Ghusick and Amlabad). However, since then a lot of technological advancement has taken place as well as higher scale of operation is expected to take place and hence the past experience may not hold true and therefore one should not get discouraged for the development of CBM industry.
- Since the quantity of the gas, which will be available from a particular block or region at a particular time, cannot be predicted well in advance, therefore, no advance planning can take place to utilise the gas, which will be available at a later date, and hence market can not be developed in advance and need to be developed only after result of exploration phase as well as pilot production, is available which may cause delay in the cash inflow in the project.

CONCLUSION

Since development of CBM technology and production of CBM gas is a time consuming venture, it is necessary to address all the issues to attract outside agencies put a sincere coordinated effort for development of CBM technology in the country so that gap between demand and supply for gas can be met. Strategic alternatives for development of CBM market should include linkage of gas to existing gas-fired power plants of the region and also to captive power plants for mining and other industries from its inception itself.

Note : Views expressed in the paper are that of the authors and are not necessarily the views of CIL or GOI.

CHAPTER 31

Indian Coalbed Methane

R. Bhaskaran & U.P. Singh***

ABSTRACT

Coalbed Methane (CBM) is a newly developed energy source and its exploitation started over a decade back. It has attracted attention in the recent years as an alternate clean energy source in many parts of the world including India. CBM is a colourless, odourless gas trapped in layers of coal seams by water pressure and overlying rock formations. Thus, It is inherent in coal beds and emits from coal seams. The quantity of CBM stored in coal is closely related to the rank and depth of the coal. With fracture of coal, pressure on the coal decreases, the methane desorbs from the coal, the released gas diffuses through a coal matrix - to the coal cleat - to the well bore and finally it comes out from the drilled well. The diffusion of gas depends upon the concentration of methane, the desorption isotherm, the distance methane gas has to travel to reach a cleat or a fracture, the relative permeability of the cleat and fracture system, etc.

Mainly there are two approach in practice to extract CBM from coal seams; (i) to extract CBM from the virgin blocks prior to mining operation, where coal mining may take place after 20-30 years of CBM operation; and (ii) integrated approach: to extract much of the methane from the coal seams before, during and after mining operation from the virgin area, working mines as well as from the gob area simultaneously.

So far extraction of CBM from the virgin blocks is concerned, the Government of India, as a policy, has decided to offer blocks for development of CBM in the country to prospective operators. Ministry of Petroleum and Natural Gas has been entrusted with finalising of the policy framework. The Directorate General of Hydrocarbons (DGH) has drafted a model contract with terms and conditions for undertaking the CBM development through competitive global bidding process. However, a few blocks have already been allotted to the interested agencies and techno-economic feasibility of the project has already been established in one of the coalfields.

* Director (Technical), Operation, Western Coalfields Limited, Nagpur.

** Dy Chief Engineer (New Technology), Coal India Limited, Calcutta.

Of late, Coal India Limited has also participated in a jointly funded project with UNDP/GEF for identification, design and implementation of programmes to recover and use Coalbed Methane in safe, cost effective and environmentally acceptable manner during and after mining operations. Once commercial viability of this project is being established, it will give a major boost to adoption of integrated approach of methane extraction technology in the country.

India has many CBM potential coalfields. Gas content in these coalfields ranges from 3 M^3/Tonne to 40 M^3/Tonne. Based on thickness, burial depth and rank of coal, the Damodar basin is expected to have the greatest CBM potential in India. Permeability, gas saturation, structural and stratigraphic variables are quite favourable in this basin. CBM Gas In Place (GIP) potential of the un-mined portions of Damodar basin coalfields is estimated to range from 0.5 to 1 Trillion M^3 whereas the total CBM potential of India is estimated around 2 Trillion M^3.

By the year 2009-10, the estimated demand for the Natural Gas in India is expected to be 110 Billion M^3 whereas supply is only expected in the level of 35 Billion M^3. Further India is having a growing demand of approx. 10,000 MW of electric power every year and thus, every year increase in gas supply is required at the rate of 50 Million M^3/day. Therefore, it is obvious that CBM is having assured market in India.

WHAT IS COALBED METHANE?

Coalbed Methane (CBM) is the gas inherent in coalbeds and emits from coal seams. It is associated with coal as a by-product of the coal formation process. It is trapped in coalbeds and gets released during and after mining.

It exists as a monomolecular layer, absorbed on the internal surfaces of the coal matrix, except for the small amount of free gas that is sometimes found in the cleat system of a coal seam.

As coal has a very large internal surface area, over 100 Million Cubic Meter/Tonne of coal, and as Methane molecules can be very tightly packed in this molecular layer, a large quantity of methane can be held in coal seams, may be up to 80-100 M^3/Tonne of coal.

The Coalbed Methane Industry aims to economically exploit and utilise the methane associated with coal. This industry has adopted the techniques practiced in the traditional Oil & Gas industry and has been successfully developed in USA since the 1980s.

Association of Methane in Coal

Coal originates from vegetal matters. The plant debris gets converted into coal in two stages; first by biochemical process or humification process, which converts the debris into peat and then by, geochemical and coalification process by which peat is transformed into higher ranks of coal.

During the vegetal decomposition process at the time of coalification, a series of by-products are also generated including water and methane. The pressure exerted by naturally formed water keeps the methane adsorbed on internal surface of coal. As the coal progress from peat to anthracite, about 160-170 M^3 of methane is expected to be generated per Tonne of coal.

The amount of methane produced during coalification process greatly exceeds the capacity of the coal to hold the gas. Additional methane may naturally get released from the coal as its burial depth decreases, due to surface erosion or uplifting. The excess gas migrates into the surrounding rock strata and onto any traditional sand reservoirs that may overlay the more deeply buried coals. **(Refer Fig. 31.1).**

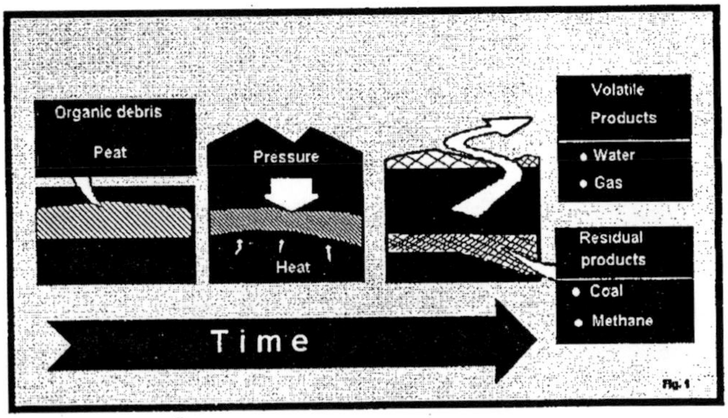

Fig. 31.1 : Key steps in Coalification process

Migration

The gas, which escapes from the coal and migrates into the surrounding rocks, strata and sand reservoir, is termed as **"Natural gas"**, whereas the gas still remaining in the coal seams is termed as **"Coalbed Methane"**.

Liberation

The amount of methane stored in coal is closely related to the rank and depth of the coal. The higher the coal rank and the deeper the coal seam is presently buried, the greater is its capacity to hold gas. The gas is liberated through the cleat system in coal when the pressure on coal is reduced. **(Refer Fig.31.2).** The reduction in pressure is generally achieved

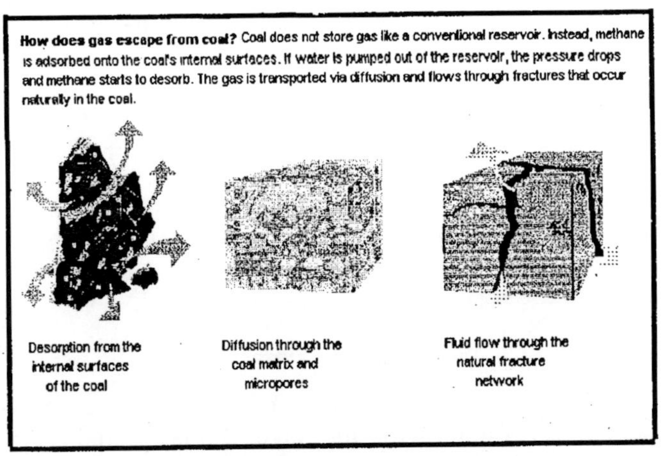

Fig. 31.2 : Liberation of methane from Coal

by lowering the water pressure i.e., dewatering of the coal seams. The amount of gas released is controlled by an adsorption isotherm, which is highly nonlinear.

Extraction of CBM from Virgin Blocks

For production of CBM from coal seams, vertical boreholes are drilled and coal seams are being fractured hydraulically by injecting fluids, water, foam and may be fine sand into the borewell to create cracks and fractures in the Coalbed, which liberates water and methane. Dewatering process further accelerates the desorption of methane from the coal seams. After the methane desorbs from the coal, the released gas must diffuse through a coal matrix until it reaches a coal cleat, the natural fracture network in coal. The gas then flows through these cleats or fractures and reaches to the Wellbore. **(Refer Fig. 31.3).**

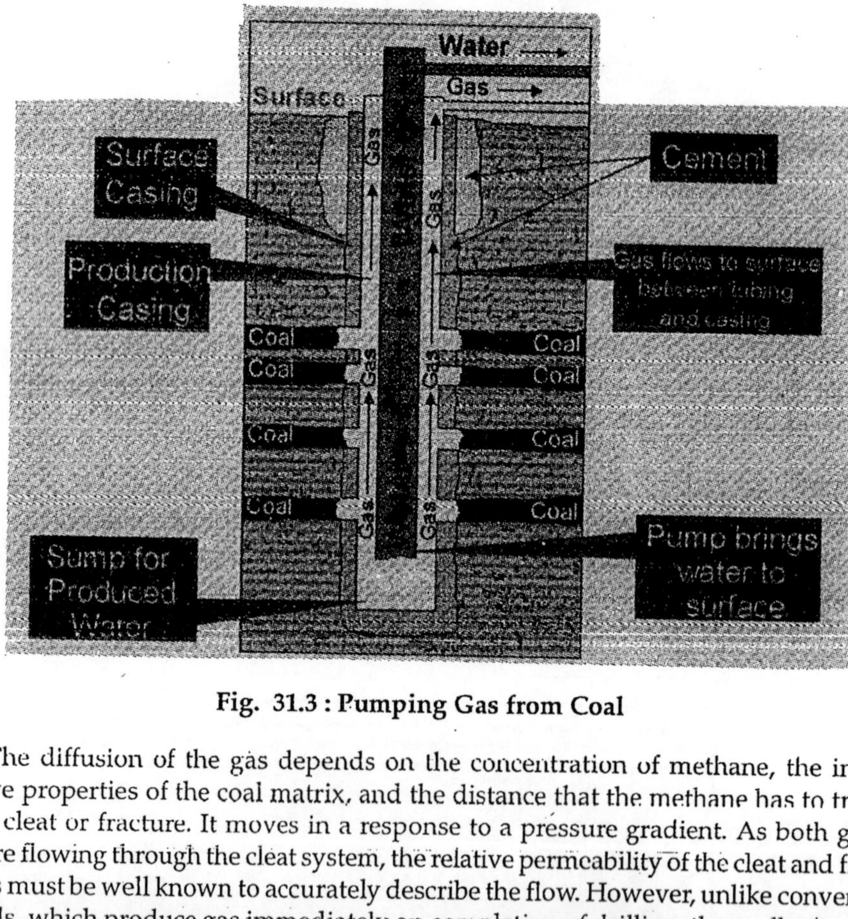

Fig. 31.3 : Pumping Gas from Coal

The diffusion of the gas depends on the concentration of methane, the inherent diffusive properties of the coal matrix, and the distance that the methane has to travel to reach a cleat or fracture. It moves in a response to a pressure gradient. As both gas and water are flowing through the cleat system, the relative permeability of the cleat and fracture systems must be well known to accurately describe the flow. However, unlike conventional gas wells, which produce gas immediately on completion of drilling, the coalbed methane wells pump out only water for considerable period of time (may be a few months/years) before methane is produced.

Integrated Approach of Coalbed Methane Extraction

Under the mine regulation and safety rules, methane must be controlled at the working faces and at other points in the mine layout. This has traditionally been performed using a well design ventilation system. However this task is becoming more difficult to achieve economically as mining reaches deeper levels, bigger and more powerful ventilation equipment is required. Sometimes even this process is ineffective, necessitating the use of supplementary methods, such as mine degasification to extract much of the coalbed methane from the coal seams before, during, and after mining. In addition methane released to the atmosphere is a major green house gas, second only to carbon dioxide in its contribution to potential global warming. Thus, all out effort need to be put to capture released methane effectively and convert this nuisance into a resource by selling it for profit. Integrated approach for methane extraction is an effort towards this direction. **(Refer Fig. 31.4)**. It is fully associated with coal mining operation. It incorporates various separate and distinct methods, such as:

(i) Degasification in advance of mining by drilling vertical holes and practicing standard techniques followed in producing coalbed methane in a virgin area.

(ii) Degasification by horizontal holes drilled into a coalbed from active workings within a coalmine.

(iii) Degasification by gob holes drilled from surface into the caved out area behind longwall faces.

(iv) Degasification by cross-measure or inclined boreholes drilled from existing mine entries.

(v) Degasification by horizontal holes from special purpose shaft, etc.

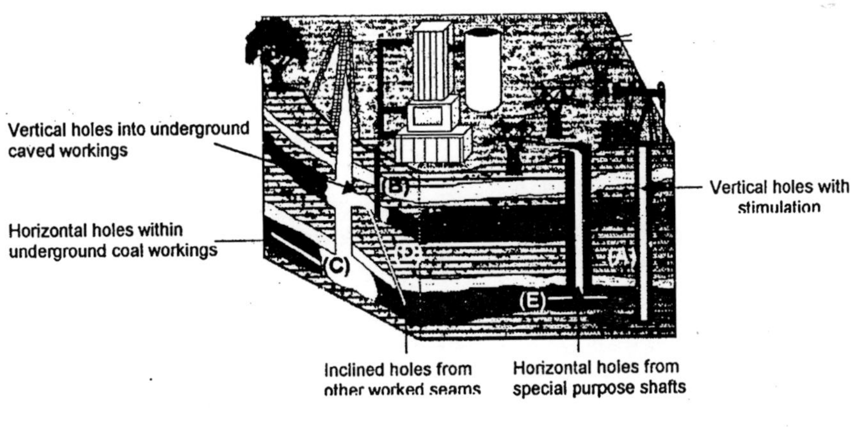

Fig. 31.4 : Integrated CBM Recovery Strategy

Methane Degasification Methods

Some of the following distinct coalbed methane extraction methods, incorporated with integrated approach, are in use in various part of the world:

(i) *Vertical Wells*: The term "vertical well" is generally applied to a well drilled through a coal seam or seams and cased to pre-drain the methane prior to mining. The wells are normally placed in operation 15 to 20 years ahead of mining and the coal seam is hydraulically fractured to remove much of the methane from the seam. For this purpose both vertical and horizontal fractures are produced in the coal seam.

To enhance the flow of gas from a vertical well, either hydraulic fracturing or open-hole cavity completions are generally used. These fractures generally do not seriously affect the mineability of the coal seams if the roof is competent.

Vertical wells can recover up to 70% of the gas content of the coal. Vertical wells offer an advantage over other methods because they can be applied to multiple coal seams simultaneously.

(ii) *Gob Wells*: The "gob well" refers to the type of coalbed methane recovery well that extracts methane from the gob areas of a mine after the mining has caved the overlying strata. Methane liberated from coal seams above and below the mine fill the voids and bed pressure pushes the gas out in to mine gob through the fractures. Gob wells allow the methane to be removed from the gob to the surface before spilling over into ventilation system. Gob wells are drilled to a point 10 to 50 feet above the target seam prior to mining, but are operated only after mining fractures the strata around the wellbore.

It can recover up to 50% of methane emissions depending on geological conditions and the number of gob wells within the panel. Gob wells are the most effective method of reducing methane content in rapidly moving longwall faces.

(iii) *Horizontal Boreholes*: Horizontal holes are drilled into the coal seam from development entries in the mine. They drain methane from the un-mined areas of the coal seam shortly before mining, reducing the flow of methane into the mining section. The recovery efficiency of this technique is low. Normally, about 10% to 20% of the methane is recovered from the drilled area. The drainage efficiency using horizontal holes less than 1000 feet is generally less than 20%. As a result, these horizontal holes are used in longwalls only where methane concentrations are high and permeabilities are relatively low.

(iv) *Cross-Measure or Inclined Boreholes*: Cross-measure boreholes are drilled at an angle to the strata, normally from existing mine entries. The boreholes are strategically placed above areas to be mined with the goal of pre-draining the overlying strata and exhausting gas from the gob area.

(v) *Other Degasification methods*: A variety of additional methods for drainage of methane from coal measures has been attempted or used on a production basis throughout the world. These include the horizontal holes from special purpose shafts, the packed-cavity methods, the superjacent methods and the method of using directional drilling, etc.

The pattern of methane release from the coal seam or seams will ordinarily be controlled primarily by the mining methods, the location of the gas in the seams and the surrounding strata, and the permeability of the relevant geological materials. If the integrated approach is being applied, the gas extraction can be both adequate (about 80-90% recovery) and cost efficient. **(Refer Table 31.1).**

Table 31.1: Methane Recovery and Utilisation Strategy

Considerations	*Enhanced Gob Well Recovery*	*Pre-Mining Degasification*	*Ventilation Air Utilisation*	*Integrated Recovery Combined Strategies*
Recovery Techniques	**• In Mine Boreholes** **• Vertical Gob wells**	**• Vertical Wells** **• In Mine Boreholes**	**• Fans**	**• All Techniques**
Support Technology	• In Mine Drills and/or Basic Surfage Rigs • Compressors, Pumps, and other support facilities	• In Mine Drills and/or Advanced Surface Rigs • Compressors, Pumps, and other support facilities	• Surface fans and Ducting	• All Technologies • Ability to Optomise Degasification Using Combined Stragegies
Gas Quality	• Medium Quality (11-29,000 kJ/m^3) (300-800 Btu/cf) (approx. 30-80% CH_4)	• High Quality (32-37,000 kJ/m^3) (900-1000 Btu/cf) (above 90% CH_4)	• Low Quality (1% CH_4/ or usually below 1%)	• All Qualities
Use Options	• On Site Power Generation • Gas Distribution Systems • Industrial Use	• Chemical Feed stocks in addition to those uses listed for medium quality gas	• Combustion Air for On Site/Nearby Turbines and Boilers	• All uses
Capital Requirements	• Low	• Medium/High	• Low/Medium	• Medium/High
Technical Complexity	• Low	• Medium/High	• Low/Medium	• High
Applicability	• Widely Applicable • Site Dependent	• Technology, Finance and Site Dependent	• Nearby Utilisation • Site dependent	• Technology • Finance, and Site Dependent
Methane Reductions	• Upto 50%	• Upto 70%	• 10-90% recovery	• 80-90% recovery

Benefits

The recovered coal mine methane provides various types of benefits to a mining company. These benefits include, but are not limited to, conversion of a nuisance into a resource, reduced ventilation costs, downtime costs, production costs, development cost, and shaft development costs; increased extractable reserve, mine safety; reduced dust problems, water problems; worker comfort; etc. All of these benefits are achieved internal to the mining operation and can be easily analysed by the mining company.

Apart from this companies can also be benefited by its ability to use the recovered gas as an energy source, either at or near the mine site or by injecting it into a commercial gas pipeline system. It can be utilised as a local energy resource to heat buildings, dry coal output from the coal preparation facility, generate electrical power, power vehicles by compressing the gas, or other local uses. The value of the methane as energy resource may be very large and it can make a significant contribution to profits.

CBM STUDY

The major technical challenges Indian CBM industry facing is lack of knowledge about the gas content and permeability of the coal available. Assumptions on ideal gas content may be possible from coal isotherms or from coal rank, reservoir depth and maceral content, however, to understand the recoverable reserves it is necessary to study permeability, gas saturation, structural and stratigraphic variables, etc. Though no co-ordinated effort has been put till date to identify the potential basins of India for the CBM, however, some knowledge has been gathered by various agencies. They have put their, though limited, efforts in this direction. Some of the findings are as follows:

Experiences at Parbatpur Block

In 1990-91, Central Mine Planning & Design Institute, Ranchi initiated methane emission and desorption studies on core samples obtained in the process of coal exploration from Parbatpur block in Jharia coalfield, Bihar, where deep mines are to be opened and which are expected to be gassy. The tests were carried out by CMRI, Dhanbad. The test results indicated that methane content was generally higher than 8 M^3/Tonne, which increases with depth at the rate of 1.3 M^3/Tonne per 100-metre depth. The highest recorded value of methane content was 14.93 M^3/Tonne at a depth of 800 metres.

Amoco's study

AMOCO analysed several samples of coal from Damodar Valley coalfields for maturity, coal quality, maceral content and sorption. The coal samples, analysed and used in the study, were obtained from the year 1990 to 1994. Their findings in this basin are that:

- Coal rank varies from high volatile bituminous C to medium volatile bituminous, the rank is highest at Jharia and East Bokaro and decreases both towards the east and west.
- There is some influence around the sills and dykes but the affect is limited to the immediate area around the intrusive. Although the coal quality varies greatly, the coals can be generally classified as high ash coals ranging from 10 to 40 percent ash.
- Based on the rank, thickness, quality and sorption isotherm data the CBM gas in place (GIP) potential of the unmined portions of these coalfields could range from 0.5 to 1 TCM and assuming a 50% recovery rate, recoverable reserves could range from 0.25 to 0.5 TCM. These estimates provide a rough guess potential for gas in place and direct measurements are necessary for accurate GIP used for economic evaluation.
- Maceral content varies dramatically from vitrinite rich to inertinite rich and the variation occurs both laterally and vertically. Amoco had detailed study of the maceral and was intended to know the absolute permeability of coals of Damodar basins. For this basin, only maceral data is available and therefore to infer absolute permeability, use of maceral and absolute permeability data from coals in San Juan, Black Warrier and Sydney basins were made. **(Refer Table** 31.2 **&** 31.3**).**

Table 31.2: Maceral Composition by Basin

Basin	*Coal Age*	*Percent Vitrinite*	*Percent Exinite*	*Percent Inertinite*
San Juan (USA)	Upper Cretaceous	67 - 95	0 - 4	4 - 29
Black Warrior (USA)	Carboniferous	75 - 94	1- 10	5 - 27
Sydney (Australia)	Permian	23 - 85	0 - 80	14 - 77
Ordos (China)	Permo- Carboniferous	35 - 54	1 - 18	35 - 60
Damodar Valley (India)	Permian	37 - 83	0- 24	7 - 53

Table 31.3: Maceral and Permeability Data Used in the Study

Basin	*Maceral Analysis*	*Core Face Cleat Permeability*	*Well Test Permeability*
San Juan (USA)	41 by core Lab, 250 Counts, 1990 + 1992	10 Plugs, Dry Helium in CMS-300 @ 600 PSI, by Core Lab, 1990	Well tests and History Matches for hundreds of Wells
Black Warrior (USA)	32 by Core Lab, 250 Counts, 1992	N/A	Well tests and History Matches for hundreds of Wells
Sydney (Australia)	49 by Core Lab, 250 Counts, 1992 + 1993	7 Whole Core, Humidified Helium @ Reservoir Net Confining Pressure by Core Lab, 1992	Pressure falloff Tests in 3 Wireline Slimhole Coreholes
Ordos (China)	11 by Core Lab, 250 Counts, 1993 + 1994	N/A	N/A
Damodar Valley (India)	40 by CMPDI (India), 1993. 4 by Core Lab, 250 Counts, 1994	N/A	N/A

- The Permian coal samples from the Damodar Valley exhibit a range of vitrinite compositions from the lower end of the San Juan and Black Warrior compositions through the upper end of the Ordos Basin composition. Coals of the Damodar Valley reach a thickness of over 300 feet. There is a wide range of maceral composition in the various seams. The relative lower vitrinite (higher inertinite) content of the coals suggests that the coals will be poorly cleated, having a low "basal" permeability. However, Some of the coals are cleated. Interpretation of the maceral and permeability data lead to the following conclusions:
  - ♦ There is no clear correlation between maceral composition and permeability.
  - ♦ Coals have a low "basal" permeability. Secondary geological features must be present to provide permeability of economic significance.
  - ♦ Geological factors other than maceral composition have a significant effect to either enhance or reduce permeability.
  - ♦ Geological factors which enhance permeability are:
    - — An increase in the number of regional or tectonic fractures,
    - — Shallower depth of burial,
    - — Intermediate coal rank (0.7-1.9% R_o),
    - — Over-pressured coal or aquifer.
  - ♦ Geological factors which reduce permeability are:
    - — Mineralization occludes cleats and fractures,
    - — Greater depth of burial,
    - — Coal rank either too low or too high (<0.7 or>1.9 % R_o),
    - — Under pressured coal or aquifer.

- The maceral composition of coals from the Damodar Valley does not preclude the possibility of having good or excellent permeability, provided geological factors, which enhance permeability, are present. Permeability may be similar to the best coals in the San Juan or Black Warrior basins, given the presence of the proper geological factors.

- Therefore, Amoco felt that CBM exploration in India should be continued with an emphasis on locating areas, which are enhanced for permeability and to collect maceral and permeability data to build on this initial preliminary work.

- More maceral and permeability data is required to better understand the maceral-to-permeability relationship. Exploration for CBM should be concentrated in areas where geological factors, which enhance permeability, can be found.

GEECL Experience

M/s. Great Eastern Energy Corporation Limited (GEECL) has been offered 225-sq. km. area in Ranigunj coalfield for exploration and exploitation of CBM. They conducted techno-economic feasibility study and have found that gas contents in the primary target seams are well within the productive range.

They drilled two exploration core holes at Surajnagar and Poradiha. This initial drilling program in conjunction with desorption testing, the determination of a characteristic desorption isotherm, and chemical analysis of several coal samples provided basic data on gas contents and its variation with depth and coal chemistry. Their observations are as follows:

- The Surajnagar desorption results are mixed but generally favourable. From the base of the Panchet to a depth of approx. 500 meters the coals are undersaturated and generally contain less than 100 scf/t of CBM. Below depths of 500 meters, however, the contained gas increases to 150-300 scf/t of gas.

- Results of the test indicate that at depths and pressures likely to be encountered in the Ranigunj area, the maximum anticipated gas content on a dry-ash-free basis is approx. 287 scf/t. The Langmuir volume or the maximum sorptive capacity of a dry-ash-free coal at high pressures was calculated to be 390 scf/t.

- A large amount of in-seam variability was evident in the samples from the Surajnagar well. Gas contents near the top and bottom of the seam are in the range of (+)250 scf/t whereas the central part contains less than 300 scf/t. A clear correlation between coal grade and gas contain was not immediately evident.

- At least some portions of the deeper seams are very near to saturated.

- Extensive vertical natural fracture has been noted in the several wells of the area. Unstimulated flow to the aforementioned core holes indicates that permeability is good, at least in some areas or some stratigraphic intervals.

DEMAND FOR CBM

Indian Energy scenario

In 1995, coal accounted for 63.3% of India's primary energy production, whereas petroleum accounted for 18.6 percent, hydroelectricity 8.9 percent, natural gas 8.2 percent, and nuclear power 1 percent. The current fuel mix is expected to change slightly through the forecast period ending in 2010. Coal (65%) is projected to remain roughly the same as in 1995, while hydro (14%) and natural gas (10%) will have higher shares of total production. Oil production will decline sharply, however, to only a 9 % share. The increasing trend in natural gas share is mainly because gas has superior characteristics as fuel and is also an environment-friendly fuel. Moreover, there has been an accelerated development of gas-based technologies in the power generation sector and thus, natural gas is likely to be India's most important potential alternative to coal. India is planning to make widespread use of natural gas in power generation and in the industrial and residential sectors.

Coal is the major exhaustible resource of the country and has a life expectancy of over 240 years. The R/P ratio for natural gas, though it has declined over the years, is still at a higher level than that for crude oil and is expected to be about 28.7 years **(1994/95, TERI yearbook 1998/99)**. Natural Gas has the proven reserves of approx. 0.7 Trillion M^3 and gross production of approx. 29 Billion M^3. The projected demand for the natural gas in the country in the year 1999-2000 is estimated to be 50 Billion M^3 and is expected to rise to 70, 110, 200 Billion M^3 in the year 2004-05, 2009-10 and 2019-20 respectively. However, supply is expected to be not more than 35 Billion M^3 by the year 2009-10. A huge deficit, therefore, haunts the Indian gas scenario.

To bridge this huge shortfall and meet the rising gas demand, the alternative with the country is to explore the possibility of other sources of gas. The options are either to bring the LNG and cross border pipeline gas through imports or to exploit its unconventional gas resources such as CBM, UCG, gas hydrates, and shale gas.

The option of large-scale imports of natural gas through pipelines from neighbouring countries, require political and diplomatic initiatives, which need to be more assertive.

CBM constitutes a clean fuel having heating value of approx. 8500 K Cal/Kg compared to 9000 K Cal/Kg for the natural gas and is generally of pipeline quality and therefore, it can be directly fed to the national grid of pipeline. By a study conducted by ARI, U S A, it is estimated that India is having around 2 Trillion Cubic Meters of CBM reserves in the various Coal Basins of India.

Projects under consideration (Import)

To bridge the gap of demand and supply Government of India is increasing imports of natural gas, mainly from Oman, Iran and Afghanistan. GAIL is planning to expedite Liquefied Natural Gas (LNG) shipments to India from suppliers such as Qatar, Malaysia, Oman and Thailand. GOI has taken up some Mega Projects to import the gas such as:

- Construction of Oman Oil Company-India Pipeline at the cost of 5 billion US $ which will pass through the deep sea covering a distance of approx. 1150 Km. It is proposed to have two 24" pipeline to supply around 30 Million M^3 of gas each day from each of the pipeline.
- Construction of Iran - India subsea pipeline.
- Import of LNG from Oman and Qatar: The Qatar will supply 7.5 MMTPA of LNG to India. The Heads of Agreement (HOA) to this effect has already been signed. The HOA provides for supply of 5 MMTPA of LNG for Dahej and 2.5 MMTPA of LNG for Cochin for a period of 25 years providing an initial build up period. The LNG to be supplied contains higher hydrocarbons and the HOA provides for exclusivity to

Petronet LNG Limited for the state of Gujrat and Kerala. The receipt of LNG is likely to commence by the end of 2002.

Even if all the projects under consideration are getting commissioned as per schedule and supply commences in time, then also there would be vast gap in demand and supply of the gas.

Indian Power Position

The present Installed Power Generation Capacity of India is approx. 90000 MW and with this generation country experiences the energy shortage of about 11.5% and Peak Power shortage of 18%. The load is increasing almost 10-13% per annum and the country is having the growing demand of approx. 10,000 MW every year and thus, every year increase in gas supply is required at the rate of 50 Million M^3 /day.

Thus considering the demand and supply position of gas as well as present power position in India and its growing demand, it is certain that CBM is going to play the important role in bridging the gap.

Gas Price

Natural gas price has been deregulated in the country w.e.f October 1, 1997 and the consumer price of the gas, at landfall point, was linked to the price of a basket of LS/HS fuel oils. The price varies between the floor price of Rs. 2150 per 1000 SCM and ceiling of Rs. 2850 per 1000 SCM and is adjusted on a quarterly basis. Supply of gas in north eastern states is at a concessional price varying between a floor price of Rs. 1200 and a ceiling price of Rs. 1700 per 1000 SCM respectively. This price corresponds to a calorific value of 10,000 K calories per SCM. All taxes etc. are extra. In addition to the above price the consumers along the HBJ pipeline are required to pay transportational charges @ Rs. 1150 per 1000 SCM along HBJ line linked to a calorific value of 8500 K Calories per SCM. The cost of transportation in other areas is site specific and varies as per the infrastructure required for transporting gas to customer's premises.

The gas price applicable to domestic consumers is indexed to the administered retail-selling price of LPG, which is increasing regularly. The present price is Rs. 6.75 per SCM of natural gas (Domestic). Under the present circumstances CBM will be able to fetch a market determined remunerative price for itself.

GOI'S INITIATIVE

The Directorate general of Hydrocarbons (DGH), Ministry of Petroleum and Natural Gas, Government of India (GOI) has been entrusted with finalising the policy framework for development of CBM in the country. The DGH has drafted a model contract with terms and conditions for undertaking the CBM development through competitive bidding process.

- It has been announced that New Exploration Licencing Policy (NELP) will guide exploration for CBM, recognising the fact that CBM cycle differs from Oil & conventional gas where considerably longer periods are necessary to determine the commercial viability before proceeding to full development investment.
- Better terms for CBM investment have been envisaged using concessional agreement rather than Production Sharing Contracts. The new concept of Production Level Payments (PLP) on sliding scale would govern the CBM contracts. These PLPs are made bidable. The salient features of the draft model contact are:

- ♦ Work Programme: Consisting of 4 phases
  - — Phase I: Exploration, 3 years;
  - — Phase IIA: Pilot Assessment, 3 years;
  - — Phase IIB: Market confirmation, 2 years; with 2 years extension for frontier areas
  - — Phase III: Development, 5 years
  - — Phase IV: Production, 25 years
- ♦ Relinquishment: No relinquishments at the end of phase I, it begins at the end of phase IIA. Second at the end of first year of phase III, then again after 3 years of phase III when all non-producing areas are to be relinquished.
- ♦ Exit Options: At the end of phase I, IIA & IIB.

The bidding round will be open for both domestic and foreign companies with up to 100% equity and no government participation. Some of the provisions for successful bidders are:

- A nominal commercial bonus of Rs. 1.0 crore will be payable on declaring the discovery commercial.
- The CBM producers will be able to sell gas at market-determined prices in the domestic markets.
- Subject to approvals, infrastructure status has been proposed for the CBM projects, which allows a 5-year tax holiday from the date of production of gas. The government is considering extending the tax holiday period by additional 2 years.
- All production saved will go to the CBM producer after payment of 10% royalty to the respective state governments. Additional production linked payment would be payable to the central government on a sliding scale based on production levels. Likewise, in Oil & Gas PSC's no custom duty would apply.

The DGH in its brochure has indicated 10 blocks for CBM exploration in six coalfields for offer in the first round. However, the no. of blocks on offer may be sized down by the time bid round is announced as a result of reviews and updating of the information.

It is expected that with final approvals to the CBM policy envisioning better fiscal terms such as gas sales at market price and infrastructure status to CBM industry would encourage domestic and foreign investors and set the stage for initiating the CBM development process in India.

CIL's Initiative

Of late, Coal India Limited has also participated in a jointly funded project with UNDP/GEF for identification, design and implementation of programmes to recover and use Coalbed Methane in safe, cost effective and environmentally acceptable manner from the working mines. Apart from this, it is likely to execute demonstration project at the Moonidih and Sudamdih coal mines located in Jharia coalfield for the recovery and use of mine associated Coalbed Methane.

This project is expected to address major issues that have presented barriers to adopting measures that effectively drain, recover and use methane during mining:

- The methane that is released during mining is often viewed as a nuisance and not a valuable resource since most of it is released in the ventilation air and although the volume of gas that is released is large the concentration is low. Prior attempts to drain gas for the purpose of increasing safety were attempted in mines where the mining face advance rate was low and attendant relaxation of the adjoining strata was not significant. Furthermore, the attempts were not long-lived and the lack of immediate success thwarted additional attempts. Due to the low initial production rates there was no attempt made to recover and use the gas. The drainage technology proposed for this project are advanced, and provisions have been made to study the behaviour of the strata that releases the methane during and after mining; so that fact based approaches to location and design of the underground drainage bore holes and surface wells into virgin coal and gob areas are optimised.

Expected situation at the end of the project:

- A power generation facility installed at the Moonidih mine will be operational. The facility will be fuelled by methane recovered from coal and adjoining strata ahead of mining and from the gob areas left as mining passes through a mine.
- Capacity will be expanded and strengthened in CMPDIL so that this institute can provide services ranging from conceptualisation of CBM recovery projects through planning, staging, performing, drilling, testing, evaluation and development of CBM deposits.
- Capacity will be expanded and strengthened in BCCL for planning and executing undergrround drilling projects using state-of-the-art underground drilling equipment to drain gas from coal and surrounding well proven strata. These strategies for recovery will be incorporated into mine planning for extraction of gassy coal deposits.
- Expanded numerical reservoir and strata relaxation modelling capabilities and understanding of the economics of coalbed methane development.
- Expanded laboratory facilities in which advanced technology for measuring the sorption characteristics of coal can be tested. Permeability tests will also be performed under in situ stress conditions allowing reservoir characteristics to be used in project evaluation.
- There will be a coalbed methane clearinghouse located on the CMPDIL campus that will serve as an interface for GOI concerned agencies and a source of information and contacts for national and international parties interested in development of coalbed methane in India. The clearinghouse will also act as a library and Internet gateway connected via Internet with linkage to Ministry of Coal, CIL, and Regional CMPDIL offices.

CONCLUSION

Methane, a newly found source of energy trapped within the coal, till recently was being considered nuisance for the mining industry particularly and for the society in general. Its emission in the mine adds to the mine safety & ventilation costs while released gas to atmosphere is a major green house gas, second only to carbon dioxide in its contribution to potential global warming. Therefore, it is imperative that it is being extracted timely, not only to generate revenue from its sale but also to lower down the ventilation costs, downtime costs, production costs, development costs, shaft development costs, etc. in a mine.

CBM constitutes a clean fuel and is generally of pipeline quality and thus is ideal substitute for Natural Gas. India is having potential for CBM. There is a huge gap between demand and supply of natural gas. CBM can play the important role in bridging this gap. The competition it expects to face is with the imported gas & LNG, which are costly. Therefore, CBM is not only having assured market but is expected to fetch a remunerative price also for itself. Moreover, GOI's initiative in CBM field makes investment in CBM development a more viable proposition.

However, in the absence of national grid of pipe lines, especially in the methane potential eastern region of the country, new methane utilisation technologies need to be found out for consumption or utilisation of produced gas in the producing area itself.

Strategic alternatives for development of CBM market should include linkage of gas to existing gas fired power plants of the region. Most of the coalfields, especially in eastern region, face problem of power shortage and thus the produced CBM can be used in the Captive Mine Power Plants at the pit head itself.

Note : Views expressed in the paper are that of the authors and are not necessarily the views of CIL or GOI.

ACKNOWLEDGEMENTS

We especially want to thank Coal India management for allowing us to publish this paper as well as for permitting us to use the data available with CIL.

REFERENCES

Environmentally Sound Energy Development - India Pilot Project, prepared by Tata Energy Research Institute, New Delhi, Dec - 1994, p VII.10-VII.11.

TERI Yearbook 1998/99

Pandey S. K., Prospects of Coalbed Methane and its introduction in India, Minetech Vol 17 No. 1, Jan - Feb., 1996, pp. 14-23.

Foster Robert A, Maceral Permeability Analysis, Amoco Production Company, U S A, p 4-5.

Coalbed Methane of the Damodar Valley Coalfields - Bihar and west Bengal, India by J. L. Cooper, B. B. P. Srivastava & P. Kumar, Intergas '95, May 15-19, 1995, The University of Alabama, Tuscaloasa, Alabama, USA.

Project Report on the Geological Evaluation of CBM Gas in Raniganj Coalfields - Prepared by Great Eastern Energy Corporation Limited, Dec-1997.

CHAPTER 32

Opportunities for Coalbed Methane Exploitation in India

*A.K.Singh**

ABSTRACT

Coal Mining is an anthropogenic source of atmospheric methane, which plays an important role in greenhouse effect and hence the global warming. Projected increase in coal production to meet growing energy demand in the 21st century will lead to considerable increase in emission of methane to the atmosphere. During the next millennium, the wave of interest would be to stabilise greenhouse gas concentration in the atmosphere.

Recovery and utilisation of coalbed methane (CBM) are expanding quickly which would increase mine safety and reduce the atmospheric load of methane. The present state of art in the field of CBM with special reference to Indian coalfields is discussed in this article.

India has approximately 60 coalfields having coal reserves of about 196.9 billion tonnes. Although coalbed gas resources in most of the coalfields of the country are yet to be established, some of them appear to have good potential for commercial exploitation of the gas. Several reservoir properties such as coal seam thickness, in-situ gas content, sorption characteristics, maceral composition and adsorption isotherm have been determined in the three primary Indian coalfields that will be key factors for coalbed methane resource development. Methane emissions to the atmosphere from coal mining activities and potential methane recovery level have been presented. Proposed CBM projects in the country would not only restrict ever-increasing emission levels, but also, put a value on an energy resource being wasted until recently.

INTRODUCTION

Energy is an essential component to standard of living, economic prospects and performance of any country. The challenge of the 21st century is the sustainable development of energy resources on cost-effective basis and protecting the global environment too.

---

*Central Mining Research Institute, Dhanbad 826 001, India.

In India, coal provides about two-third of total commercial energy needs and will continue to be a major energy source over the foreseeable future. Despite ambitious intentions of the Government of India to maximise hydro electricity generation, presently almost 72 percent power generated is through the thermal component (Palit, 1999). The all India coal production in 1997-98 was 297.169 million tonnes (CCO, 1998). Of this, the share of open cast production was 228.141 million tonnes (76.8% of the total production) and that of underground production was only 69.028 million tonnes (23.2% of the total production).

Based on an exercise carried out for estimation of coal demand for the year 2009-10, coal production in the country will have to be programmed at 690 million tonnes. With fast depletion of shallow coal deposits, possible decline in coal production from the open cast mines is anticipated. The country will have to depend mostly on underground mining of deeper and gassy seams for meeting its increasing coal demand.

One of the problems relating to exploitation of deeper coal seams with improved mechanised methods will be increased rate of methane emission in the underground workings. Methane, an explosive gas in concentrations of 5.4 - 14.8% by volume in air is known as a major safety hazard ever since the beginning of underground mining. In the history of Indian underground coal mining, forty percent of disasters have been caused due to explosion of firedamp alone killing 839 persons altogether (Singh et al., 1999).

On the other hand, methane is a radiatively important greenhouse gas that accounts for about 15 percent of the current increase in commitment to global warming (USEPA, 1990). Atmospheric methane levels are increasing presently at the rate of 0.9 percent per year with concentration increasing from 0.79 ppmv in the pre-industrial period to 1.72 ppmv at present (Crutzen, 1998). According to a recent estimated annual global emission strengths, nearly 530 million tonnes of methane are produced every year (Tyler, 1999) which are much beyond the cleaning levels of the sinks.

In June 1992 at Rio de Janeiro, 152 countries signed the United Nation Framework Convention on Climate Change (UNFCCC) which aims to 'stabilise atmospheric greenhouse gas concentrations at the level that would prevent dangerous anthropogenic interference with the climate system'. The UNFCCC Third Conference of the Parties in December 1997 at Kyoto adopted the Kyoto Protocol aimed at an overall 5.2 percent reduction in greenhouse gas emission below 1990 levels in the five year period 2008 - 1012.

It is essential therefore, that greenhouse gas emission in the next millennium must be restricted to minimum in the process of exploitation and utilisation of coal. Ventilation of underground coal mines drains out huge amount of methane present in the coal seams to the atmosphere. Methane is emitted to the atmosphere from surface mining of coal and post mining operations as well. About 27 million tonnes of methane is emitted every year due to coal mining (Tyler, 1999). Technological options for reducing methane emissions from coal mining are available and pre-mining degasification from the surface vertical wells is one of them. The gas so produced can be and is being successfully used as a non-conventional source of energy. In the present paper, it has been evaluated if opportunities for commercial recovery of coalbed methane in various Indian coalfields exist or not.

ORIGIN AND CONTROL PARAMETERS OF METHANE IN COAL

Methane is native in coal seams and has its origin in the process of coal formation. Plant materials carried away by streams were deposited in river basins. Over time, these ancient plants were covered by increasing overburden and were compressed by weight of the sedimentary layers. Eventually, the biochemical and geochemical transformation of plant materials begins with peat and ends with anthracite through lignite, sub-bituminous and bituminous coal due to the intense pressure and heat present across the sediments.

In the early biochemical stage of coalification process, the easily degradable compounds such as chlorophyll, cellulose and protoplasm are converted to water, ammonia, carbon dioxide, methane and other gases by bacterial decomposition. The gaseous products migrate through the pores and fissures in the sedimentary rocks to the atmosphere as the peat is deeply buried. Further burial of peat gives rise to the later geochemical stage with the increase in temperature and pressure. In this stage, hydrogen and oxygen content of the woody matter decreases forming methane, water and carbon dioxide. Large volume of gaseous by-products is generated during coalification process. Most of the gas so formed either moves into overlying strata or is emitted to the atmosphere. However, a significant amount of gas (mostly methane) is stored on the internal surfaces of micropore network of coal. Because of its very large internal surface area and high pressure exerted by the overburden, coal normally stores substantial quantities of methane in adsorbed state over internal surfaces of the micropores.

Of the several parameters that determine resource potential for exploitation of CBM, Gas content, sorption capacity, coal rank and depth, the maceral composition, permeability of the coalbeds and seam gas constituents are very important.

Gas content of the coal is the most important parameter for coalbed methane production commercially. Higher gas contents are preferred for economic exploitation of CBM. Though the gas content relationship is basin specific, it generally increases with the coal rank and depth of the coal seams. Pressure controls the quantity of methane adsorbed in coal.

Transport of methane through the coal matrix does not occur as the fractures or cleats that permeate coalbeds are often filled with water. Methane is released and its movement through the coal matrix begins when pressure on the coal seam is partially lowered by natural erosion or mining activities or by removing water from the coalbed. Gas flow rate is governed by permeability of the coalbed that may vary from 0.01 millidarcy to 100 millidarcy. For good permeability, the coal needs to have a well-developed and open cleat and joint systems. Three distinct phenomena occur simultaneously as and when coalseam pressure is reduced by dewatering to degass the coal seam: (1) desorption of gas from the coal surface with pressure lowering; (2) an increase in effective stress; and (3) two phase flow through the coal seam fracture system with continuous reduction in water saturation. When methane desorbs from the coal surface, the coal matrix shrinks because of an increase in free surface energy resulting in widening of cleats, fissures, and increase in permeability. It has been reported that increase in permeability is pronounced at high stress (Harpalani, 1991).

Cumulative gas recovery and the efficiency of recovery are highly influenced by the gas saturation level of the coal. The most effective method of investigating sorption of methane by coal and its gas saturation level is the determination of adsorption isotherm which show the manner in which the amount of sorbed gas varies with pressure at a constant temperature. At low pressure, the sorption capacity increases almost linearly with pressure. As pressure continues to increase, the coal's ability to adsorb gas decreases, eventually reaching a pressure level at which very little added gas can be stored. Higher rank coals with larger surface areas have higher sorption capacity.

No systematic study exists in India that rigorously provides these important parameters regarding the potentiality of CBM production. Selected data assembled by the author is presented in this paper for some coalfields of the country.

GAS POTENTIAL OF COAL BEARING BASINS OF INDIA

Knowledge of the amount of methane liberated per tonne of coal produced in working mines is useful for preliminary examination of gassiness of coal seams. Based on mine specific measurements of rate of emission, all the underground coal mines in India have been categorised into Degree I, Degree II and Degree III of gassiness by the Directorate General of Mine Safety (DGMS, 1967). More than 66 percent of the underground coal mines in the country are Degree I gassy mines where the rate of emission is less than 1 cubic metre per tonne of coal produced. Nearly 27 percent of the underground mines have been categorised into Degree II of gassiness as the rate of emission in these mines lies between 1 and 10 cubic metre per tonne of coal produced. The remaining 25 underground mines are Degree III mines as the rate of emission exceeds 10 cubic metre per tonne of coal produced. Rate of emission of methane in mines gives an insight into the gas content of working seam as it has been observed that in general the emission rate is 3 to 5 times the gas content of the seam. It is however emphasised that the coal seam, which records low emission rate in the mine, does not necessarily have low gas content as the rate of emission depends on several geo-mining conditions. In case of some mines, there is significant emission of gas from the roof and floor strata of the coal seam being worked. This must be taken into consideration for estimating gas content of coal seam from rate of emission. In absence of detailed information of mining method vis-à-vis prevailing geo-mining conditions, gas content estimation from rate of emission may result in misleading conclusions.

Of the 25 Degree III mines, only four mines belongs to tertiary deposits in Assam. Remaining 21 Degree III gassy mines are located in the Jharia, Raniganj and East Bokaro coalfields of Damodar River Valley. Details of the gas desorption parameters determined by the author in some gassy blocks in different coalfields of the country are presented in the following sections. An attempt has been made to quantify the gassiness of the coal seams of these coalfields. Preliminary investigations are based on the available data on rate of emission of methane in different underground mines. However, gas desorption data determined by CMRI in few specific target areas have also been presented.

Jharia Coalfield

Jharia coalfield which is well known for the development of high rank coal seams, covers an area of 450 sq. km lying between latitudes 23°37' and 23°52' and longitudes 86°06' and 86°30'. Roughly sickle shaped in plan (Figure 32.1), Jharia coalfield is located mainly in Dhanbad district of Bihar province with a small western portion in Giridih district of Bihar. The most important coalfield in the Damodar Valley, Jharia is also the premier coalfield of India as it is the only store house of prime coking coal in the country.

There are about 380 collieries out of total 900 for the whole country. In spite of the similarity of chemical and petrological attributes and almost equal depth of occurrence, the particular seam exhibits wide variations rate of emission in different mines possibly due to varying geo-mining conditions and hydrogeological regimes in operating mines.

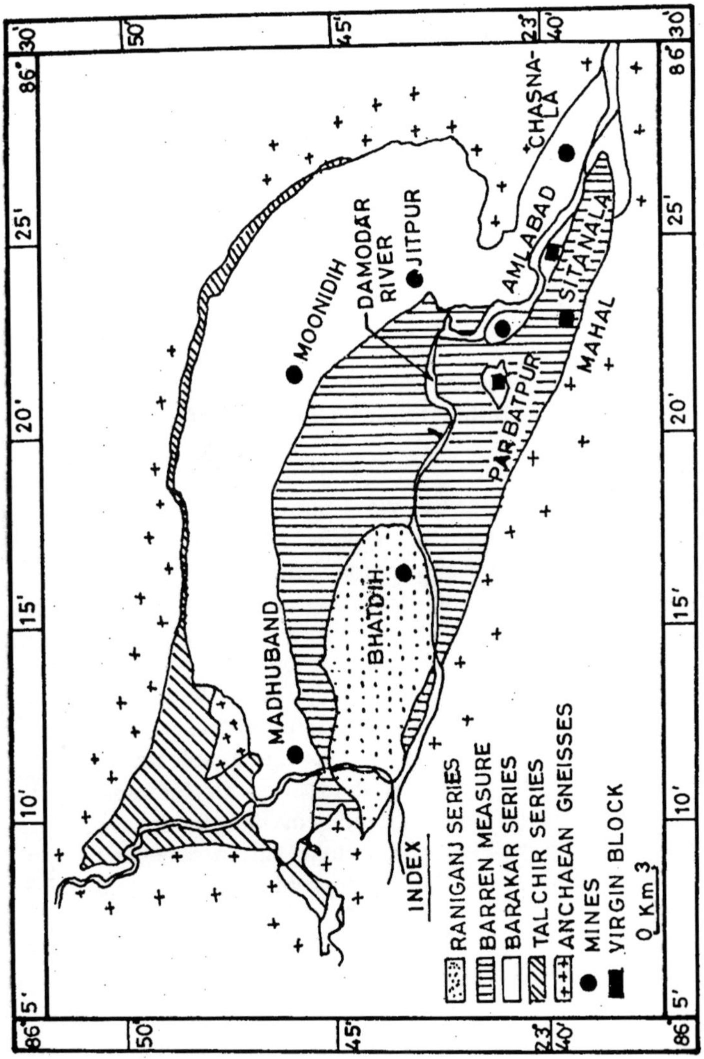

Fig. 32.1 : Geological Sketch Map of Jharia Coalfield

Some of the underground mines as Moonidih, Amlabad, Jamadoba, Bhowra, Sitanala, Chasnala, Jitpur, Sudamdih and Bhatdee etc. have been categorised as Degree III mines where the rate of emission is more than 10 cubic metre per tonne of coal produced. It is emphasised that the Degree III mines in the Jharia coalfield are very close to thė Damodar River which traverses the coalfield from west to east.

Archaean with unconformity in Jharia coalfield form the basement of the Lower Gondwana formations, which consists of the rocks of the Talchir and Damuda series. The Damuda consists of Barakar Measures, the Barren Measures and the Raniganj Measures. The Barakar and Raniganj formations contain the coal seams of workable thickness. Sills and dykes of dolerite and mica-peridotite of lower Cretaceous age are met through out the coalfield but mainly confined to the disturbed areas. These dykes and sills have not only pyrolitised some of the coal seams but have sometimes contributed to local accumulation of gas in the pockets.

The Barakar Measure contains nearly 25 coal seams of 1.22 m and above in thickness out of which eighteen coal seams numbered I to XVIII have regional extent. The Raniganj Measure which is developed only to a limited extent in south western part contains about 9 coal seams of 1.22 m and above in thickness. Total in-situ reserves of coal in Jharia coalfield have been estimated to be 19.4 billion tonnes (Coal Atlas of India, 1993).

The geothermal gradient of 34°C to 53°C per km in Jharia coalfield is very favourable for hydrocarbon generation. From the view point of proximate analysis and vitrinite reflectance, the Barakar coal of Jharia is classified as low to medium volatile bituminous coals that is very favourable for gas generation (Murray, 1996). R_0 values for Barakar coals of Jharia are mostly around 1.0% and range from 0.9 to 1.4 but comparatively lesser R_0 values between 0.86 to 0.9 have been obtained for coals of Raniganj Measure.

Central Mining Research Institute (CMRI), Dhanbad has carried out extensive investigation on the determination of in-situ gas content of several virgin blocks by Direct Method (Diamond and Levine, 1981). A synthesis of available data broadly indicates high concentration of methane at Moonidih, Amlabad, Chasnala, Sitanala and Parbatpur blocks mainly in the seams IX to XVIII. There is a paucity of data on the gas content of older seams I to VIII. A general trend of increase in gas content with increase in the depth of the seams has been observed. It has been found that gas content increases by 1.30 m^3/t per 100 metres of increase in depth. The highest value of 14.3 m^3/t of gas content has been measured at XIVA seam at a depth of 795 m in Parbatpur block. The results for seams XVIII to XIV encountered at different depth levels have been found more than 9 m^3/t for all the seams. Relatively low gas content of 6.3 m^3/t to 8.90 m^3/t has been observed in Amlabad, Sitanala and Chasnala areas. This may be due to lower hydrostatic pressure in these areas. The sorption capacity determined by adsorption isotherms has been determined for Barakar coals. It has been found to be in the range of 10.6 m^3/t to 18.0 m^3/t at a pressure of 40 kg/sq. cm for XIV to XVIII seams. The isotherms predict good saturation level of coal. The Barakar Measure in Jharia Coalfield especially south-eastern part of it covered by Barren Measures appears to be ideal targets for commercial production of CBM where gas in place has been estimated to be 1.5 trillion cubic feet.

Raniganj Coalfield

The Raniganj coalfield fails within Burdwan district of West Bengal province and a small western portion in the Dhanbad district of Bihar province. It is situated between latitudes 23°34' and 23°52' and longitudes 86°36' and 87°17'. It extends over an area of 1550 sq. km. The field is the birth place of coal mining in the country. The easternmost part of the Damodar Valley Coalfields, Raniganj remains one of the premier coalfields of India even after continuous coal exploitation for nearly two centuries.

Raniganj coalfield is well developed and is second in the country after Jharia in respect of coal quality and production. There are more than 300 collieries producing about one quarter of the total coal production of the country. Some Degree III underground coal mines at Seetalpur, Sodepur, Chinakuri, Parbelia, Ranipur, Ghusick, Damra, Tirat, Kalidaspur and Madhujore etc. are located near the Barakar and Damodar rivers. Although the deepest coal mine of India at Chinakuri is located in this coalfield, mostly the top seams are being exploited in this coalfield and large reserves in the lower seams are locked up like Jharia coalfield. It is worth mentioning that the deeper seams may be more gassy as the gassiness generally increases with depth.

As in the case of Jharia coalfield, Lower Gondwana formations are resting on the Archaean with unconformity. Archaean is exposed on all the directions of the coalfield except the east. A thin patch of Talchir rocks as shown in Figure 32.2 is exposed in the north western part. The rocks of Barakar, Barren Measure and Raniganj stage of the Damuda series occupies a larger part of the coalfield. The Barakar having irregular exposure along the northern boundaries of the coalfield consists of twelve coal seams. The rocks of Raniganj Measure are well exposed in the southern half, which contain eleven coal seams of 1.2 m and above thickness. Total in-situ coal reserves of this coalfield have been estimated as 22 billion tonnes (Coal Atlas of India, 1993) of which 20 billion tonnes are non caking coals. The next younger sediments viz. Panchet and Supra Panchet are exposed in smaller areas of south-western part. Numerous sills and dykes of mica-peridotite and dolerite have been injected in the ages from Jurassic to Cretaceous.

The geothermal gradient in the Raniganj coalfield is 25°C to 52°C for 1 km, which is slightly low in comparison to the Jharia coalfield. Generally, the Barakar coals of Raniganj coalfield is of low moisture, low to medium volatile and of caking type. The Raniganj coals of this field are normally high moisture, high volatile and non-caking type. The R_0 values of coals from Raniganj Measures range from 0.6 to 0.9 and slightly higher for Barakar coals. These values are reasonable for CBM potential. The highest R_0 value obtained by the author is 0.94% for a coal sample retrieved from the depth of 1478 m in an exploratory borehole in Durgapur area.

Gas content measurements have been made on coal core samples from mine faces as well as from exploratory boreholes in the virgin blocks. The gas content of Dishergarh seam at Chinakuri colliery, which is at a depth of about 630 m from the surface, has been obtained as 5.93 m^3/t. In Ghusick area, a lump of coal sample from R-VIII seam has been found to have methane content of 6.1 m^3/t. The gas content of Barakar coal core retrieved from a depth of 1478 m in Durgapur area has been found to be 7.62 m^3/t but more than 90 percent of the gas has been obtained as residual gas in this case. Maximum sorption capacity of the coals at a pressure of 40 kg/sq. cm are observed to lie in the range of 9.5 to 11.8 m^3/t. These values when compared with estimated in-situ methane content of the coals revealed substantial undersaturation of the coals upto the depth of 400-500 m. The distribution of gassy seams is highly uneven and local concentration of methane is observed in some pockets. There is an evidence of gas flare in an abandoned coal exploration borehole in the Surajnagar bock in the south western part of Raniganj coalfield. However, gas desorption parameters in this virgin block has to be carried out carefully.

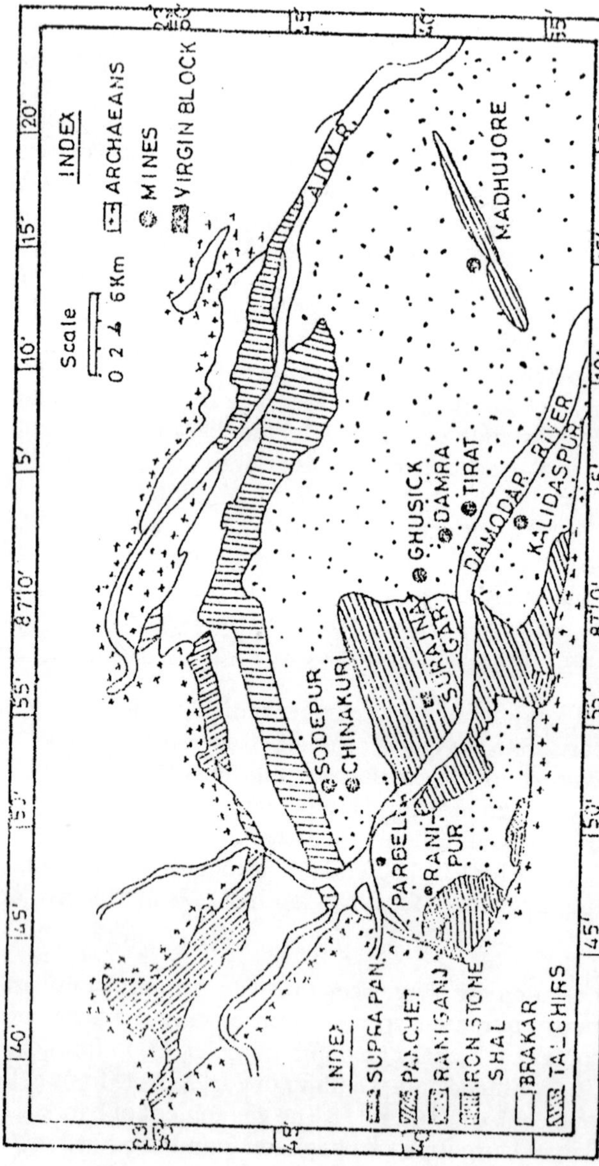

Fig. 32.2 : Geological Sketch Map of Raniganj Coalfield

East Bokaro Coalfield

The East Bokaro coalfield, which extends over an area of 237 sq. km, is a major source of medium coking coal. Enclosed between the latitudes 23°44' and 23°48' and longitudes 85°40' and 86°04', this coalfield is situated in the Bokaro and Hazaribagh district of Bihar province.

There are only a limited number of coal mines (about 25) in this coalfield, producing about 15 million tonnes of coal every year. The underground coal production is only 1.2 million tonnes per year and the rest is being produced from open cast mines. Of the 10 underground mines, Dhori, Jarangdih and Sawang etc. are Degree III gassy mines.

As shown in Figure 32.3, rocks of Lowermost Talchirs to Panchet and Supra Panchets of Lower Gondwana system are present here. These formations overlie the Archaean metamorphics exposed in the northern, eastern and southern boundaries of this coalfield. The Panchets and Supra Panchets are observed in the western part of the field near Lugu hill. The Talchir is very scantly exposed on the eastern boundary. The rocks of the Barakar stage, Barren Measures and Raniganj of Damuda series are developed extensively. There are altogether 21 coal seams in Barakar and 8 coal seams in Raniganj formation with a total reserve of about 5.6 billion tonnes (Coal Atlas of India, 1993). About 2.8 billion tonnes of coal occur at a depth of less than 300 m. Sills and dykes mainly Lamprophyre and mica-perditite are present as intrusions throughout the coalfield especially in the western part. The coal seams have been affected by these intrusions at places.

The thermal gradient in east Bokaro is similar to that of Jharia Coalfield. The Barakar coals of all the seams in the field are characterised by low moisture and low to medium volatile bituminous coal with higher caking index. The Kargali seam is the most important seam in respect of its wide extent, thickness upto 31m and strongly caking quality. The East Bokaro coals have a R_0 value in the range of 0.9 to 1.18, which is quite favourable for CBM.

Methane content of coal seams in Asnapani block does not exceed 5.0 m^3/t, which is not very encouraging. This block is near to working mines at Jarangdih. However, gas content of the coal core samples retrieved during exploratory drilling from depths of 600 m in the virgin Koiyotanr block, is more than 8.0 m^3/t which, is quite good for CBM exploitation. The Koiyotanr block is far away from any working mines in the field. The sorption capacity has been determined for only two samples of Kargali seam that is about 14 m^3/t at a pressure of about 40 kg/sq. cm. Based on the limited gas desorption data, it can be said that East Bokaro Coals are also ideal for coalbed methane extraction.

CONCLUSION

Mine specific measurements of rate of methane emission and methane content of determination of coal seams in some virgin blocks of Jharia, Raniganj and East Bokaro Coalfileds suggest suitable sites for exploitation of coalbed methane. Production and subsequent utilisation of CBM in these virgin blocks will not only fulfil the growing needs for energy but also help to make safe the prospective mines from gas hazard and to control the atmospheric load of this potent greenhouse gas.

ACKNOWLEDGEMENTS

The author is grateful to the Director, Central Mining Research Institute, Dhanbad for his kind permission to present this paper. Thanks are also due to various organisations viz. CMPDI, ONGC, MECL and GSI for sponsoring the study in different areas.

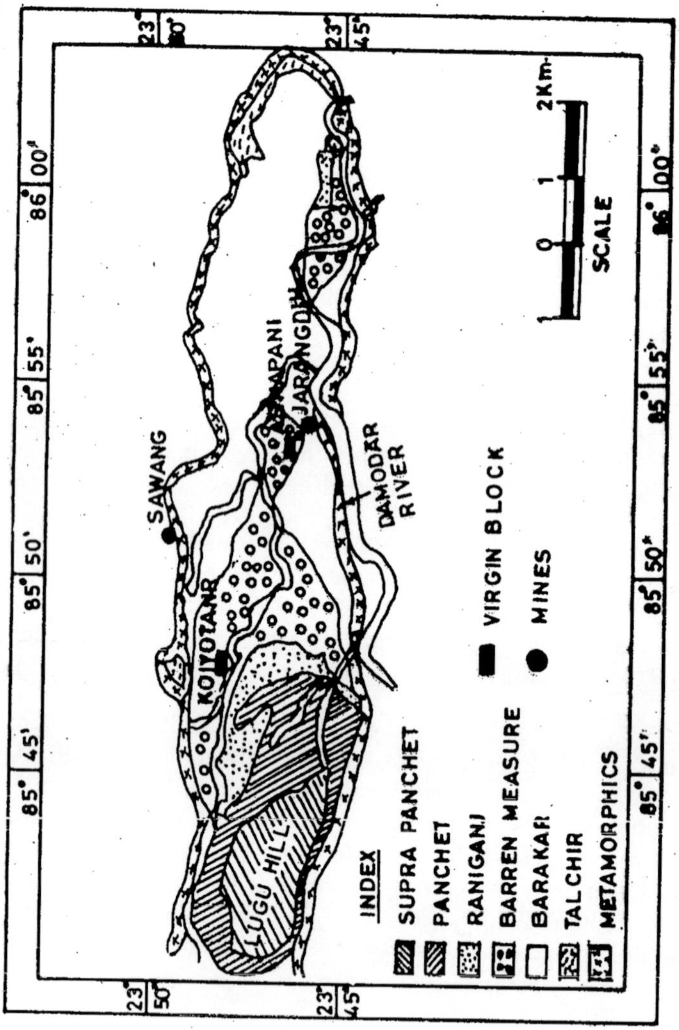

Fig. 32.3 : Geological Sketch Map of East Bokaro Coalfield

REFERENCES

CCO, 1998, Annual Review of Coal Statistics 1997-98, Government of India, Ministry of Coal, Coal Controllers Organisation, Calcutta, p. 7.

Coal Atlas of India, 1993, Published by Central Mine Planning and Design Institute Limited, Ranchi on behalf of Coal India Limited, Calcutta, India.

Crutzen, P. J., 1998, Global Environment Chemistry, Narosa Publishing House, New Delhi, pp. 1-11.

Diamond, W. P. and Levine, J. R., 1981, Direct method Determination of Gas Content of Coal: Procedure and Results, US Bureau of Mines Report of Investigation RI 8515.

DGMS, 1967, The Coal Mine Regulations, Directorate General of Mine Safety, Dhanbad, Bihar, India.

Harpalani, S., 1991, Production of Methane from Coalbeds, Quarterly Review of Methane from Coal Seams Technology, **89(4)**, 43.

Murray, D. K., 1996, Coalbed Methane in the USA: Analogues for Worldwide Development, Coalbed Methane and Coal Geology, Gayer and Harris (eds), Geological Society Special Publication No. 109, London, UK, pp. 1-12.

Palit, A., 1999, Environmental Aspects of Coal Power Plants in India and Clean Coal Technology, Proceedings of International Symposium on Clean Coal Initiatives, 22-24 January, New Delhi, India, pp. 431-445.

Singh, A. K., Kispotta, J., Singh, H. and Mendhe, V. A., 1999, Indian Scenario of Coalbed Methane, Proceedings of International Symposium on Clean Coal Initiatives, 22-24 January, New Delhi, India, pp. 729-737.

Tyler, S. C., 1999, Measurements and Modelling of Atmospheric Methane Using Stable Carbon Isotopes, IG Activities Newsletter, Issue No. 16, March, pp. 3-7.

USEPA, 1990, Methane Emissions and Opportunities for Control, Workshop Results of Intergovernmental Panel on Climate Change, United States Environmental Protection Agency, Co-ordinated by Japan Environment Agency, September, p. 7.

CHAPTER 33

Mining in Fragile Areas - The Challenges : A Case Study of Captive Mine of Ambuja Cement in Himachal Pradesh, India

Y Raghavendra Rao, Vijay Kumar** and Sanjay Kumar****

ABSTRACT

Mining is one of the core industries which plays a significant role in economic upliftment of the country and is vital for sustenance, well being and comforts of human beings. India produces about 84 minerals and many of them are located in sensitive and fragile areas of our country. Mining operations induce both positive and negative changes on pre-mining environmental and socio-economic status of the surrounding regions. Opencast mining in the fragile areas in general, has the ill effects viz. visual intrusion, landscape degradation, land slides, deforestation, migration of fauna, noise and vibration due to blasting, dust and air pollution and hydrological imbalances.

The paper highlights the endeavours of Ambuja Cement towards sustainable mine development at its mine located in Solan district of Himachal Pradesh. At the mine, base line environmental data collection studies are undertaken and impact assessment as well as environmental action plans are prepared. Mining operations adhere to guidelines based on an environment management programmes approved by Statutory Bodies. All out efforts are made to adopt environmentally sound technologies for winning and conveying the limestone. The company promotes partnerships with NGOs and local communities to ensure better environmental awareness and management.Stricter and more vigorous

*Joint President

** Manager (Geology)

***Dy. Manager (Mines)

Gujarat Ambuja Cements Limited , Ambujanagar.

environmental standards are adhered to and regular monitoring and auditing is resorted to. The principles of sustainable mine development based on scientific mine planning are followed for the protection of environment giving due consideration to the socio-economical aspects.

INTRODUCTION

Himalayas, the youngest chain of mountain on the earth at the northern crest of Indian sub-continent form a significantly important region from the environment point of view owing to its fragile nature, sensitiveness to seismicity and other minor disturbances. The **ecologically fragile and sensitive region host immense quantum of mineral wealth like limestone, magnesite, phosphorite, gypsum, soapstone, baryte, dolomite, rocksalt** etc. Many of the above minerals are being actively mined in the region mostly by opencast method of mining using mechanised as well as manual means. There are numerous working leases of various minerals owned by Central and State Govt. undertakings as well as private companies/individuals.

Gujarat Ambuja Cements Ltd popularly known as **Ambuja Cement,** a public limited company engaged in manufacturing cement since 1986, holds a mining lease for limestone and shale in Himachal Pradesh. During a short span, Ambuja Cement has set up several bench marks in various fields, viz. **productivity, environment improvement, energy efficiency, quality and profitability** etc. and won prestigious awards including **Pollution Control Award from the Prime Minister of India. Ambuja Cement** believes that industrialisation and environment protection should go hand in hand. The company is committed to protect environment around its mines and practices eco-friendly mining techniques for mineral raising apart from suitable reclamation and rehabilitation measures. **Kashlog Limestone Mine** of **Ambuja Cement** in Himachal Pradesh was awarded **Sita Ram Rungta Memorial Award for Social Awareness for the year 1998-99 by Federation of Indian Mineral Industries (FIMI).**

Kashlog Limestone Mine

Limestone is the chief raw material for cement manufacture. The total requirement of Limestone for our Himachal cement plant is met from the captive mine located near **village Kashlog in Arki tehsil of Solan District.** The mine is operated with best available technological means to ensure the highest standards of Safety, Productivity and Environment Protection. Crushed limestone from mine to plant is transported by a 3.2 kms long closed overland belt conveyor system in an eco-friendly manner.

MINING OPERATIONS

At **Ambuja Cement,** mining operations are undertaken keeping corporate environmental objectives in mind. Emphasis is given to eco-friendly scientific mine planning. Adequate pollution control techniques are adopted while winning the raw material. Compliance to statutes is our commitment. Social welfare measures are duly undertaken for workers and local communities.

The exploitation of limestone is carried out by highly mechanised open-cast mining system. The deposit has been opened up at two different locations and this facilitates simultaneous excavation of various grades of limestone. This approach helps in conservation of minerals by proper blending and solves the problem of slope stabilituy by avoiding formation of a series of benches at a particular place.

Mine Planning

Mining operations are undertaken in a safe and scientific way in accordance with an approved mining plan from Indian Bureau of Mines. Mining plans comprising inherent environmental management components are prepared using computer aided facilities at the mine planning division.The 10m height working benches are planned to have a slope of 70^0 with an ultimate pit slope of $45^{0.}$

Drilling

Drilling is carried out by deploying hydraulic drills for primary blasting purpose. It is resorted to at locations which are difficult to work by Ripping and Dozing technique. Around 60-70% of the production requirement is met through Drilling and Blasting. The drill has inbuilt system for flushing water to suppress the dust at the source itself. The inbuilt compressor and remote controlled travel switches provide additional flexibility and safety to the machine. Wet drilling operations make the operations practically dust free. Drills have also in- built dust extraction system for arrest of particulates.

Blasting

Blasting operations are carried out in a controlled manner to minimize fly rock generation for safety of civil structures, machines and nearby habitation and agricultural fields. Due care is taken to keep the ground vibrations and air blast levels to the lowest possible limits to avoid any adverse impacts on the surrounding environment.

- *Proper Blast Design* : The blasting parameters like spacing, burden, depth, sub-grade, charge per hole, delay pattern, firing pattern etc. have been established after careful studies at the mine by concerned experts from Universities/Research Institutes. The operations are carried out under the supervision of Mining Engineer, well versed with technique to ensure quality and safety in the work.
- *Bottom Hole Initiation System for Blasting* : In this system of blasting, the detonating cord in the hole is replaced with a non electric detonator (Exel). During firing of the blast, the explosive charge is initiated at the bottom of the hole and this ensures that most of the explosive energy is used in actually breaking the rock, thus reducing the fly rock considerably. Even on the surface of the hole, the delay between different holes is provided by non electric connectors ensuring correct firing as per the blast design. The process reduces the maximum instantaneous charge per delay which finally reduces the ground vibration levels. Furhter, noise less trunk delay systems are used to reduce air blasts.
- *Blasting Mat for Fly Rock Control* : For arresting fly rocks, where ever needed, mats are used for covering the blastholes. The mats were designed in consultation with Central Mining Research Institute (CMRI).This restricts fly rock and shows our commitment towards environment protection.
- *No Secondary Blasting* : The common practice of secondary blasting for breaking of large size boulders has been totally eliminated. The hydraulic rock breaker is being utilised for breaking the same. The process helps in avoiding the common problems of fly rock and noise associated with the secondary blasting. The endeavours are also being made to use hydraulic breakers fot primary breaking.

- *Ripping and Dozing* : As the limestone deposit is rippable at certain locations, a giant Ripper - Dozer has been deployed to produce limestone from such places. The adoption of the ripping and dozing technique at mine has considerably reduced dependence on blasting and its associated problems.

Raising and Transportation

The ripped/blasted and fragmented limestone is raised using 4.5 m^3 hydraulic excavators and 5.74 m^3 wheel loader and loaded to the 35 tonnes capacity dumpers. The crushed material from crusher to plant is conveyed through 3.2 kms long closed concyor belt system having adequate carrying capacity. The conveyor passes through three tunnels with varying lengths and gradients along three major ridges falling between plant and mine area. **Load cells** have been installed in the tunnel to monitor creep displacement. The conveyor is simple to operate and consumes less power and there is no possibility of dust generation and thereby the system is termed as **eco-friendly transport system**. The conveyor is designed after detailed geo-technical investigation of the area by experts from Indian Bureau of Mines.

ENVIRONMENTAL MANAGEMENT

Top Soil Management

The negligible soil over limestone is carefully scrapped and collected manually and stacked separately. The soil is used in re-vegetation/plantation schemes. Massive afforestation work has been undertaken in the non-mineral bearing areas within as well as out of the lease using available soil from the area.

Air Environment

Continuous monitoring of air quality using Respirable Dust Samplers(RDS) is undertaken at different places in mining area and its surrounding on regular basis. To measure Respirable Particulate Matter(RPM) below 4 microns, the state-of-the-art equipment **Dust Trak** is also used to measure particulate concentrations. Apart from this the parameters like SO2, NOx, CO, dust fall and free silica are also measured during auditing studies. Crusher stack monitoring study is also undertaken on regular basis.

The preventing measures comprise use of **dust collectors/water infusion systems in drills, atomised water spray system in the stone crusher,** water sprinkling over roads and on blasting faces/sites, effective maintenance of machines etc. Crusher and surge hopper are equipped with bag filters. The mine is equipped with two water tankers of 7 and 12 KL capacities. Apart from this a 10 tonnes capacity road roller is deployed for road maintenance and rows of trees have been planted along haul road sides for arresting propagation of dust.

Noise and Vibration

Ambient noise levels in the mining area and surroundings are regularly measured apart from measurement of levels within and outside of HEMMs.Good maintenance and operational practices are adopted to contain the levels within permissible limits. Further,the operators are provided with ear muffs and plugs and they are trained to use them regularly.

All the blasts at the mines are **monitored in three different directions using blast vibration equipment Minimate DS-077.** The regular monitoring helps in improving the blast design for reducing ground vibrations. The observations at the mine reveal that both ground vibrations and air blast levels are well below the safe limits.

Water Regime

The mining operations do not disturb the ground water regime. However, surface and groundwater studies are underatken at six locations in the mine and adjoining areas.The studies reveal that water is potable.

Check Dams

To control wash off and exclude silts from the run-off storm/rain water in the gullies, more than 30 nos. of check dams have been built at proper sites at suitable intervals on the surface water courses i.e. valleys of the active mining areas. These check dams are preceded by interception ditches. Some of the the check dams have been constructed using scrap steel and these are strong enough to withstand the impact of rolling boulders. The silt arrested at check dams is utilised for horticultural works.

Check Filter

Three nos. of Check filters have been built at the culminating points of water courses. A number of steel pipes of 150 mm dia have been fixed in cement masonry wall built across the flow channel. The pipes are filled with sand, gravel and pebbles to act as inverted filters. This arrangement allows water to flow through but arrests suspended particles. After each monsoon the steel pipes are cleaned and refilled with filtration media.

Oil/grease trap

For used oil recovery from the mine workshop, oil-water separation tanks have been constructed. The used oil is collected after filters and used as furnace oil in the D.G sets situated in the plant.

Apart from above, to treat the mine run-off, one on stream sedimentation pond has also been designed to finally discharge the clear water to natural water courses as well as to aid in the recharge of ground water regime.

Vegetation/Afforestation

The mining area is covered with little natural vegetation and is rocky in nature. However, the afforestation programme in and around the mining area was started prior to commencement of mining activities. Till date more than 75000 saplings of broad leaved and coniferous species have been planted in the mining and surrounding areas. The saplings have been selected on the basis of their suitability to the local conditions.While selecting the species, adequate consideration is given to ecological aspects and economical importance. A nursery has also been developed near mining area.The areas selected for plantation include non-mineralised areas in the leasehold, areas out of the lease, sides and slopes of the roads, colony area and plant side etc. The future proposed locations comprise peripheral portions of mine, along the check dams/check filters (along water courses) etc. A study of the growth rate of selected species revealed that Himalayan cyprus has got highest growth followed by moderate growth of Leucaena and lowest of poplar species.

COMMUNITY DEVELOPMENT

Community development, socio-economic welfare and environmental conservation measures comprise construction and financial aids to schools, medical care facilities for locals through hospital/mobile dispensary, imparting training to village women in cottage industries, dairy and animal husbandry, assistance in development of orchards, organisation of cattle camps, soil conservation, water harvesting, construction of biogas plants, waste land development etc. Apart from this, various infrastructural facilities like pucca approach roads to villages, play grounds, storage tanks for drinking water, water ways etc. have been established for the welfare of locals. The company also liberally sponsor sport events and arrange cultural fairs and participate in various festivals.

OTHERS

The Domestic Garbage Management System has been introduced in the township. Colony garbage is segregated into bio-degradable and non bio-degradable fractions.The former is used for manure preparation after decomposting, while the latter is recycled. To improve general aesthetic beauty of the plant and colony areas, gardens and lawns have been developed at various ocations.Green belt has been developed over an extensive area surrounding plant and colonies. Well built asphalt roads are maintained in plant and residential areas. Well designed water drainage and sewage facilities have been provided in the campus. Full fledged recreational facilities and well maintained play grounds are provided.

MONITORING

For effective Environment Management in the mine, cement plant and surroundings, an environment monitoring cell comprising of experts from various expertise has been created. A micro-meteorological station has been set up for regular data collection and monitoring. Regular monitoring is undertaken with respect to air, water, noise etc. to ensure that they conform to the standards laid down by the Indian Bureau of Mines, Central and State Pollution Control Boards and MOEF.

Remotely sensed satellite data is utilised for continuous monitoring and assessing the environmental impact with involvement of experts for change detection landuse/ landcover studies pertaining to 10 kms buffer zone of mine. The study conducted at an interval of six years revealed that built up land increased from 0.26 to 1.98%, cultivated area 18.20 to 33.53 %, forest green cover 47.73 to 50.02% and water bodies 0.68 to 0.72% while on the other hand waste land decreased from 33.13 to 13.75%. No adverse impact on the landuse is revealed while significant barren land has been put to productive use resulting in overall green cover in the buffer zone. This reveals overall positive impact after commencement of developmental project.

ENVIRONMENTAL AUDITING

It has been planned to conduct Mine Environmental Audit to have a systematic, documented, periodic and objective evaluation of environmental management system including equipment performance with the aim of safeguarding and protecting environment to meet regulatory requirements and to promote good relations with local communities. The first such study undertaken during 1998-99 for evaluation of environmental performance after the commencement of mining operations in the area involved the technical analyses of various environmental parameters and check sampling by an independent and competent agency (Indian Bureau of Mines) to assess the efficacy of the existing mitigation measures and suggest improvements wherever necessary.

CONCLUSION

GACL is conscious of its responsibility towards environment and has proved the environmental performance at its captive mines in Gujarat and Himachal Pradesh. In Gujarat, the entire limestone production is achieved using eco- friendly equipment Surface Miner. Parts of mined-out areas in Gujarat have been reclaimed and rehabilitated by constructing cement units, a sewage water treatment plant, afforestation, agrifarming and orchard development apart from creating water reservoirs.Mines have received several awards and laurels for better mining practices, reclamation and rehabilitation of mined-out areas and environment improvement of surroundings. Our Gujarat unit has been accredited with most coveted Environmental Management System : ISO - 14001 certification for plant and mining area.

Substantial amount of environmental pollution and hazards have been witnessed by fragile Himalaya in the past by unscientific mining in the area. Sustainable mine development based on appropriate and scientific mine planning is the need of the hour and requires balancing the protection of natural environment with the need for social and economical development. The case of Kashlog Limestone Mine is one such example.

CHAPTER 34

Procedure for Environmental Clearance of Mining Projects

R.K. Suri

INTRODUCTION

Environmental clearance from the Ministry of Environment and Forests (MoEF), Government of India, was earlier an administrative requirement intended for very large mines or new mega-projects undertaken by the government or public sector undertakings. After the enactment of the Forest (Conservation) Act, 1980 and subsequently the Environment (Protection) Act, 1986, the requirement of environmental assessment and appraisal was extended to mining projects where diversion of forest land is involved. However, with a view to control and prevent environmental degradation arising out of developmental projects, MoEF, Government of India, issued a Notification referred to as Environmental Impact Assessment (EIA) Notification on 27th January, 1994 (as amended on 4th May, 1994), which makes EIA a statutory requirement for mining projects (major minerals) with leases for more than five ha or for the expansion or modernisation of any existing mine. MOEF have further amended the above notification making it mandatory to conduct public hearing for all projects covered in Schedule-I of the EIA notification dated 27th January, 1994 with effect from 10th April, 1997. Environmental clearance will not be required at the time of renewal if there is no increase in the originally sanctioned lease area.

PUBLIC HEARING

Public hearing has to be conducted by the respective State Pollution Control Boards (SPCBs) as per the procedure contained in the Notification dated 10th April, 1997 (Annexure-I). After Public hearing is conducted by the SPCB, the proposals are sent to MoEF alongwith public hearing reports for obtaining environmental clearance. In this context, the following clarifications are made:

*Ministry of Environment & Forests CGO Complex, Lodhi Road, New Delhi - 110 003, India.

- For site specific projects, such as mining (major minerals involving more than five ha) and prospecting and exploration of major minerals in areas above 500 ha public hearing should be conducted before the site clearance applications are forwarded to MoEF. In those projects, where the MoEF has already accorded site clearance, the proposal for environmental clearance will be considered only after public hearing is conducted.

- As per procedure laid down, public hearing has to be conducted in all such cases where SPCB has either already issued No Objection Certificate and the concerned proposals have not been received in MoEF before 10.4.1997.

SITE CLEARANCE FROM ENVIRONMENTAL ANGLE

In the case of mining projects (major minerals) involving more than 5 ha of land or expansion/modernisation in which an existing mining lease in excess of five ha is proposed to be further expanded or prospecting and exploration projects of major minerals in areas above 500 ha are required to obtain environmental clearance from the MoEF. Project proponents will be required to furnish an application alongwith the prescribed environmental questionnaire, details of public hearing and consent to establish from the respective SPCB, to MOEF through the State level agency dealing with mines. In the case of strategic minerals like fuel minerals, atomic minerals, etc., the concerned Central Government agency will forward such applications with its recommendations. MoEF will communicate its decisions to the forwarding State/Central Government agency which will be requested to ensure that the conditions governing the site clearance are complied with while granting the lease. The said site clearance shall be granted for a sanctioned capacity and shall be valid for a period of two years.

PROJECT ENVIRONMENTAL CLEARANCE

Only new mining projects (major minerals) with leases for more than 5 ha or expansion or modernisation of any existing mine are required to obtain environmental clearance from MoEF. Further for every mine, big or small irrespective of size of lease area, operating or new, environmental clearance from the MOEF is to be obtained if it is located in ecologically sensitive/fragile areas notified under the Environment (Protection) Act, 1986, which are given below:

- Doon Valley in Uttar Pradesh, Dhanu Taluka in Maharashtra and Aravalli Ranges in Gurgaon District of Haryana and Alwar District of Rajasthan.

- Identified coastal areas as per the Coastal Regulation Zone Notification, forests, wildlife, mangroves, biosphere reserves, hill and mountain area, etc.

Under the EIA Notification, 1994, it is mandatory for any person who desires to undertake any new mining project (major mineral) with a lease area of more than 5 ha to submit.

- An application in the prescribed proforma alongwith filled in questionnaire.

- A Project Report which should include *inter-alia* EIA/EMP (Environmental Management Plan).

The proposals, where forest land is a part of the project site, approval under Forest (Conservation) Act, 1980 is essential.

The reports submitted with the application shall be scrutinised by M0EF and discussed in a meeting of the Expert Committee set up under the provision of EIA Notification, 1994. MOEF will take a decision on according clearance within three months from the receipt of the required document. Often, it is found that proposals lack necessary details and essential information. In such cases, MOEF asks the project authorities for details and missing information. Under the present procedure, the project authorities are required to submit the missing information/action plans within two months. If furnished, the case is reconsidered for a decision. If full information/action plans are not furnished within two months, the case is rejected for non-furnishing of information. After preliminary scrutiny of the essential information/action plans and other details, the project proposal is placed before the Expert Committee. Project authorities are invited for a brief presentation and providing necessary clarifications to the members of the Expert Committee and wherever necessary; site visits are made for on-the-spot assessment of environmental aspects.

On the basis of such an exercise, the Expert Committee makes its recommendations for approval or rejection of the projects on merit. While recommending approval of a project, the committee also suggests certain environmental safeguards in specific cases. After recommendations of the Expert Committee the proposals are then processed by the MoEF for approval/rejection. The Ministry has also set up six regional offices to carry out monitoring of compliance of environmental stipulations.

ANNEXURE-I: PROCEDURE FOR PUBLIC HEARING

(1) Process of public hearing:

Whoever apply for environmental clearance of projects, shall submit to the concerned SPCB twenty sets of the following documents namely:

(i) An executive summary containing the salient features of the project both in English as well as local language.

(ii) Form XIII prescribed under Water (Prevention and Control of Pollution) Rules, 1975 where discharge of sewage, trade effluents, treatment of water in any form is required.

(iii) Form I prescribed under Air (Prevention and Control of Pollution) Union Territory Rules, 1983 where discharge of emissions are involved in any process, operation or industry.

(iv) Any other information or document which is necessary in the opinion of the Board for their final disposal of the application.

(2) Notice of Public hearing:

(i) The SPCB shall cause a notice for environmental public hearing which shall be published in at least two newspapers widely circulated in the region around the project, one of which shall be in the vernacular language of the locality concerned. SPCB shall mention the date, time and place of public hearing. Suggestions, views, comments, and objections of the public shall be invited within thirty days from the date of publication of the Notification.

(ii) All persons including bonafide residents, environmental groups and others located at the project site/sites of displacement/sites likely to be affected can participate in the public hearing. They can also make oral/written suggestions to the SPCB.

Explanation

For the purpose of the paragraph person means,

(a) any person who is likely to be affected by the grant of environmental clearance;

(b) any person who owns or has control over the project with respect to which an application has been submitted for environmental clearance;

(c) any association of persons whether incorporated or not likely to be affected by the project and/or functioning in the field of environment;

(d) any local authority within any part of whose local limits is within the neighbourhood, wherein the project is proposed to be located.

(3) Composition of Public Hearing panel:

The composition of Public Hearing panel may consist of the following, namely:

(i) Representative of SPCB;

(ii) District Collector or his nominee;

(iii) Representative of State Government dealing with the subject;

(iv) Representative of Department of the State Government dealing with the Environment;

(v) Not more than three representatives of the local bodies such as Municipalities or Panchayats;

(vi) Not more than three senior citizens of the area nominated by the District Collector.

(4) Access to the Executive Summary

The concerned persons shall be provided access to the Executive Summary of the Project at the following places, namely:

(i) District Collector Office;

(ii) District Industry Centre;

(iii) In the Office of the Chief Executive Officers of Zila Parishad or Commissioner of the Municipal Corporation/Local Body as the case may be;

(iv) In the head office of the concerned SPCB and its concerned Regional Office;

(v) In the concerned Department of the State Government dealing with the subject of environment.

Section VI

ISSUES — SAFETY, PRODUCTIVITY & SOCIO-ECONOMIC DIMENSIONS

CHAPTER 35

Mine Safety- Challenges of the 21st Century

B. Bhattacharjee and S.N. Padhi ***

INTRODUCTION

The current millennium, which is now almost at its end has witnessed some of the greatest changes in the recorded history. The millennium began in the middle ages, then progressed through the industrial revolution, the age of science and the age of electricity and is now rapidly moving towards the information age. The eve of the new century will be time of introspection. For the first time in the history, a new era, a decade, a century and a millennium will draw simultaneous, a time everybody will like to celebrate. However not long after the celebrations stop, the realities of 21st century will set-in. The mining industry like many other industries of the world will have to face a new type of challenge which will be unique to the social, economical and environmental order of the new century. The issue is whether we are prepared to face the new challenges. While drawing analogy from other industries, we find the computer industry is fighting hard to meet the challenges of Y2K problem. The chemical industry is battling to meet the stringent demands of protection of the environment.

The engineering industry is preparing to face its challenge of quantum technology up-gradation and quality improvement. But when we talk of mining industry, where challenges are nothing less, the question is whether we are preparing ourselves enough.

Efforts for preparing the mining industry to meet the challenges of the next century started early, perhaps a decade and a half back. Seminars, Workshops and discussions were held to identify the safety needs also. The thrust area was however on technology upgradation and development to meet the projected high mineral output indigenously, where man and the environment were not in the focus. The drastic change in the global political and economic scenario thereafter, resulted in complete shifting of priorities.

*Director of Mines Safety.

**Director-General of Mines Safety.

Demand pattern of minerals also changed abruptly calling for matching changes in the means of production and hence the safety needs. Man and environment came into sharper focus and so also occupational health and safety.

The steep demand for minerals has to be met through quality assurance, protection of environment, stringent social demand of safety at work place in a highly competitive environment where only the best and the very best can survive.

PROFILE OF MINING

India has a rich heritage in mining and metallurgy dating back to 4000 BC. It is also the home of the world's earliest zinc technology. Since independence, Indian mining industry has shown rapid growth from production of 22 minerals valued at Rs.70 crores in 1950 increasing to 64 minerals valued at about Rs.33,500 crores in 1996-97. Currently the country produce 4 fuels, 11 metallic and 49 non-metallic minerals, besides about 20 minor minerals. The total number of persons employed in the mineral industry is over 1 million on an average daily basis.

Currently there are 592 coal mines, 32 oil projects and a large number of non-coal mines. Projections of out put for most minerals in the 9th and 10th five year plan also shown steep increase. Mining is likely to extend to more complex areas, including the sea bed and hostile regions bringing it with a new challenges in exploitation. The progress of the Indian mining industry had been phenomenal in terms of production, productivity which has been achieved through intensive mechanisation and adoption of new technology. Indeed never before in history mineral industry has experience such revolutionary changes at accelerated pace. The demand for the next century will be still more great which make the safety challenges equally stiff.

SAFETY AND HEALTH

In this century we have been able to bring down the fatal accidents and the fatality rates within certain limits but unfortunately it has almost plateaued thereafter without showing any signs of break-through. Whereas some of the developed country are fighting in terms of reducing lost time injuries, our priorities has been limited to reduction of fatal accidents and even some times to multi fatal accidents. With growing awareness and flow of information, it is unlikely that such status of mine safety will be acceptable in the next century.

In the field of occupational health, the Indian mining industry continue to remain in a stage which is much below the acceptable standard set up by the developed world. The recent legislative changes in respect of occupational health and the recommendations of the Conference on Safety in Mines are only tips of the Iceberg. Total elimination of silicosis and other occupational diseases are still a distant dream.

CHANGING SCENARIO

The last decade of this century has seen different changes in technical system, economical and political scenario of the world. With the policy of liberlisation, globalization and removal of trade barriers the challenges of competition is also fast entering the Indian mining scenario. The industry can no longer operate in an insulated market devoid of external competition. There is increasing demand of products of high quality at competitive price. Availability of quality product is no longer a distant dream for people living even in the remote corner of the globe. This has been possible because of awareness of the customers and competition amongst the producers. International standards have been set and the products need to reach these standards to be acceptable not only internationally but also nationally.

A revolution has taken place in the field of information technology. Mines which are even worked in remote and backward places can no longer have the dubious advantage of being insulated from the general world. Information worth making news about any incidence in a mine, however remotely it may be located is immediately transmitted to the world. Mine accidents, therefore, is likely to get more publicity than what it was in the past.

There has been increasing awareness towards the need of protection of the environment. Mining has been branded as an economic activity which results in degradation of the environment. The challenge of winning minerals without causing appreciable damage to the environment is another new challenge to the mining community.

Individual health and safety is likely to find greater priority in view of changing social priorities. The responsibility of health and well being of mine workers will however continue to remain with the mine operator but monitoring and surveillance may not be limited to only to the operator and monitoring and government agencies but extend to other external agencies. Some non-government organisations are already taking up the issue of protection of miners health and safety as also the cause of protection of environment of the mining community. Changes in the social and economic priorities had taken place very fast, the magnitude of which probably could not be foreseen clearly by the planners. There is a need to modify the plans by foreseeing the changes that are likely to come. However, economic and profitability will play the key role in the survival of the fittest.

New productive technology will hold the key role to survival of the mining companies. The future will see emergence of new technology.

MINING OF NEXT CENTURY

It is necessary to visualize the changes that are likely to take place in the mining in the mining industry of the next century before the safety needs are identified. In fact the last decade of this century is itself witnessing changes which was perhaps not within the stretch of common imagination shortly before. The industry will be highly competitive, trimmer, efficient, flexible and thoroughly business like where only the fittest will survive will be less. Obsolesce will be very fast and each organisation has to be ever prepared with contingency measures against evolution of new systems and products which will make them redundant. This constant fear of redundancy will results in increase of work related and social stress which will add a new dimension to occupational health and safety. Specialist services will be increasingly used by the parent oraganisation. Thus there will be greater use of contractors and contract related work. The work force of the contractors may have to move from place to place and may be deployed in remote places for carrying out work. There will be greater use of longer periods of work extending to twelve hours a day and twenty eight day at a stretch followed by the same period of rest. This will result into a new type of stress and may give new dimension to occupational health and safety.

Geo-mining condition is likely to become more difficult as exploitation extends to deeper horizons. Problems of outburst, strata control in highly stressfield, problems of heat and humidity, hauling from long distances and winding from greater depths will need to be addressed. Our in-house expertise in solving these problems has to be forged to get the desired result. There will be ever increasing need of sharing and gaining experiences in these areas on a global basis.

This decade is witnessing inroad of new technology and new methods of exploitations. Coal bed methane (CBM) drainage and other new methods of exploitation which are in budding stage may gain prominence. There will be therefore need to evaluate the new technology and methods to determine the safety needs. New technology is often brought in without assessing the social and organisational needs. The results is generation of stress in the work-persons and their social environment. Many of the developed countries had undergone such experiences in the process of change over and the Indian mining industry may gainfully utilize these experiences while switching over to extensive mechanisation. There is need for advance socio-technical evaluation of mechanisation giving weightage to technical benefits vis-à-vis social costs. In many cases such evaluation have been made at a later stage when the adverse consequences had surfaced and are therefore post-mortem in nature when the damage has already been set in. It is necessary to foresee these challenges of occupational health and safety and device corrective measures.

Electronic and computer technology is making very fast inroads in our daily life and even in traditionally conservative mining industry. This technology is changing so fast that almost every day better system are being marketed. In this race, standardization of the product is becoming a poor second and is trailing far behind. This fast track development of information technology have pitfalls that the user may not be able to determine the safety needs of the system and the consequence of the failure of the system or misinterpretation of the results therefrom, against which sufficient checks and balances have not been provided, may lead to disastrous consequences.

Emergency response would be another critical area in the future. Efforts need to be directed to prevent disaster in mines and also to mitigate its effect, should one take place, besides preventing its effects spreading to the surrounding. As mines become safer, expertise and experiences of dealing with emergency will become scarce. This gap can cause serious hurdles in dealing with mine emergencies in the future, unless skills are constantly updated and sharpened and tools are provided to effectiveness and responsiveness of the emergency plans.

Mining engineering today is no longer an attractive profession. The profession has perhaps not been able to fulfill the aspiration of the younger generation and there is a fast erosion of its popularity. This is likely to result in a bankruptcy of talents in the profession. The younger labour force has also an alienation towards the rigours of physical work in mines. Skilled and dedicated manpower will therefore be a scarce commodity.

Resources will become scarce and our field of ;operation will appear to be a play ground of abundance and waste. There will be constant need for innovation and evolution of new technology to win minerals probably. Thus occupational safety and health initiatives will require constant development. Globalisation will lead to entry of multi players offering goods and services to the mining industry. Standardisation and testing therefore will have increasing role to play.

CHALLENGES OF MINE SAFETY

The traditional means for improving safety is perhaps too slow and ineffective to meet the demand of growing standards of occupational safety. There is need to go for quantum of jump to meet the demand of society for better work place safety as well as to achieve the international standards. The issue is how to achieve these objectives. Some of the ideas to meet the objectives are given below:

Concept of quality in safety

Starting from the beginning of this century in the climate of complete laissez faire, the mine operators did not believe that protection of persons at work in mines against occupational hazard was their responsibility. At a later date, acceptance of their responsibility was extended to only within the ambit of the legislative provisions. In the later part of the century tripartite mechanism gained some footing. All these efforts resulted in decline of fatal accidents but did not really make the work place safe. In an attempt to give greater emphasis to safety, creation of institutional mechanism like Internal Safety Organisation and Safety Offices took place. Though the mechanism had initial bouts of success, in the long run it perhaps drove a wedge between safety and production, with the accountability of safety slowly shifting towards it, defeating the very purpose of its creation. It is a reality that safety can not be and should not be separated from the process of production. It has to be interwoven in the matrix of production. Further, like any other process of production where quality is the mainstay, so also quality should be ingrained to safety for its success. Thus be it a safety audit, a risk assessment or even a safety awareness programme there is need for achieving quality of highest standards. In the past even when proven technology had been introduced without paying due attention to quality at each step, the result had been far from satisfactory. Roof bolting technology, workers health surveillance programmes, powered support long wall faces etc. are a few examples in the long list where lack of quality had been the main cause of failure to achieve the desired results.

Proactive initiatives

In our country safety initiatives had been mostly reactive in nature. It is an accident or dangerous occurrence which initiates preventive measures, after learning from the experience. Sometimes repetition of such incidences can only pave the way for safety initiatives. This process, which arises out of findings of inquiry ,has its own limitations as whole hearted participation of many concerned is never possible in an environment of conviction and self defence. With availability of standards and pool of rich experience, proactive approach for identifying the safety needs and designing corrective measures should form the basis for any safety programme.

The thought process in mines safety has been perhaps restricted by the artificial barrier of prescriptive safety legislation. Traditional emphasis on adhering to the legislative framework was relevant in the development stage but with maturing of safety initiatives the goal need to be far ahead .With development of standards and availability of more and more safe tools and methods, with which legislative amendments cannot kept pace with, there is scope for improving safety. Many more hazards can be identified and corrective measures brought in which are not covered by the safety legislation. In developed countries as well as in some other industries within the country the concept of risk assessment and safety management plans has been already developed and practiced. This process gives unlimited scope to improve safety in mines. Each mine and in turn each activity of the mine can be analysed for its risks followed by its assessment and finally providing control measures against the risks through the development of safety management plans. This is a concept which has tremendous potential and should come in a big way in the mining industry. We have to prepare ourselves to use those tools for achieving the best results. The biggest strength of the tool is the scope of sharing experience on safety matters at grassroots level through involvement at different levels. The safety management plan derived from the assessment of risks again fixes and quantities responsibilities at different level to make the mine safe. The principle however is built on the concept that it is the primary responsibility of the operator to ensure safety where different levels have responsibilities based on capability and influence within the organisaitonal framework.

Selection of technology

In the world of globalisation and access to information, availability of off-the-self technology is no longer the limitation. The problem is the identification of the requirement of the technology and thereafter its selection from the shelf. This requires in-depth understanding of the problems inherent to the mines. Scientific studies by the external agency is only an aid and by no means the answer to selection of technology. Selection of proper site specific technology will perhaps be a big challenge to the mine operators but once it is successfully done , technology will not only be productive but also safe, where operators will not have sleepless night later.

Testing and standardization

With globalization and liberalisation there will be large number of vendors providing goods and services to the mines. Each vendor will project the best of his product, claiming it to be superior to the others. There is need for the standardisation of the product and determine whether they adhere to the standards. Many of the products used in the mine still do not have national or international standards. Testing of the inputs as well as the output as to whether they conform to the required standard is another item critical to mines safety and production which ensure quality of the end product. It will be necessary to standardize the system of testing and quality control as well as lay down the criteria of acceptance and rejection.

Training

Training has been always thought to be the key area in improving safety. The conventional training system has probably reached its level of competency. The present system of giving fixed training in the class room or even on the job has its limitation and is not proving to be effective enough. Training has to be made inter-active. Computer based training program with facilities for continuous upgradation and up-datation and capabilities of being moulded to specific requirements hold promise. On line computer based unit learning system which are interactive has the advantage where the trainer and the trainee has the scope to vary the contents of training based on site requirement and see the results of training. There is flexibility of modifying the training programme based on needs.

Development of Virtual Reality (VR) offers significant potential to improve safety training for mining personal in activities which are to risky to be simulated physically. This technique is used in developed countries in training of pilots in aviation industry, nuclear maintenance and battlefield training in military. Research group around the world are developing prototype training systems to evaluate the potential for safety applications using VR for mining personal.

Information technology in safety development

Information technology is making deep inroads into every sphere of our lives and mine safety is no exception. In the field of assessing, transferring, modifying and disseminating information and developing data base, information technology will have ever increasing role to play. It is for us to use this technique for development of protocol for safety management.

Research and development

Traditional R&D initiatives had been a closely guarded secret. This may result in duplication of research efforts with different agencies working in allied subjects. With easier accessibility of information in future research initiative of different countries and different institutes can be pooled together for the best possible results.

While the research initiative in a competitive atmosphere will be more applied, economic driven and result oriented, there will be fear of fundamental research suffering in the process. There will be however need to continue with fundamental research to form the foundation of mine safety. Laying down the research priorities will be another key area of stress.

CONCLUDING REMARKS

The next century will continue to witness dominance of workforce and the environment in the process of production. The workforce will not remain a mere tool to serve the more privileged section of the society but will become a separate entity which will also share the fruits of toil. Technology will be a means to and not the end to achieve the objectives of higher output. Success will depend on a perfect synthesis of efforts of workforce and technology, with man and environment remaining in the centre. But a word of caution need to be added that planning should be flexible to absorb paradigm shift. Experience has taught us that any paradigm shift can have a revolutionary influence in the whole industry. We have to remain vigilant to survive even in the ever changing world.

ACKNOWLEDGEMENT

The views expressed in this paper are those of the authors and not necessarily of the organisation they represent.

The authors are faithful to the Director-General of Mines Safety for permitting to write and present this paper.

CHAPTER 36

Coal Production with High Performance Shearer Loader

*Dipl.-Ing. Karl-Heinz Rieser**

INTRODUCTION OF EICKHOFF MASCHINENFABRIK GmbH

Eickhoff Maschinenfabrik GmbH, is a privately owned company and was founded in the year 1864 (Fig. 36.1, Eickhoff Products). Eickhoff started out as a foundry, specialized in products for the coal mining industry. The first Eickhoff coal cutting machines for longwall mining were put into operation in Germany in the year 1914. Since that time Eickhoff was involved in the introduction of mechanised longwall mining in almost all coal producing countries. Many innovations in the shearer technology were introduced by Eickhoff. Today we are world wide known as manufacturer of high quality and high performance shearer loaders.

Besides the shearer loaders Eickhoff manufactures the following products (Fig. 36.1, Eickhoff Products):

— Roadheaders for mining and tunnelling

— Gearboxes for mining application and all other industrial aplications

Many longwall face conveyors, even those of our competitors are equipped with Eickhoff gearboxes. Another large market outside the mining industrie are gearboxes for wind generators. Our gearboxes are well accepted by the manufacturers of wind-energy-generators because of their compliance with the high requirements in regard to reliability and noise level.

— Belt conveyors

— Electrically driven lokomotives for underground and surface operations (Manufactured by our 100 % Eickhoff subsidiary named "Schalke")

*Eickhoff Maschinenfabrik GnbH Bochum.

— Coking plant equipment (Manufactured by our 100 % Eickhoff subsidiary named "Schalke")

FACTORS INFLUENCENG LONGWALL PRODUCTIVITY

The following factors determine the performance of a longwall (Fig. 36.2, Longwall):

Geological conditions

The geological conditions are generally a given fact. The effect of the geological conditions on the longwall operation can only be influenced within certain limits by the mine- and panel layout, the type of equipment and the mode of operation. Parameters such as depth of the seam, seam thickness, gas content, coal hardness, inclination, rock intrusions, roof and floor conditions greatly affect the productivity of a mine.

Mine layout and panel planning

The general mine layout depends in many respects on the geological conditions Depth, inclination of the seam and geological irregularities determine the entire structure of a mine such as panel and roadway layout. Panel length and width influence the productivity of a face and the frequence of face moves. The optimum panel layout can differ from mine to mine.

Training of the mine staff

Wrong assessment of the geological conditions or inattention underground can cause severe production delays and cost or even life threatening danger.

Poor maintenance, carelessness, inefficient fault diagnostics or incorrect operation of the equipment can lead to major production delays and increased cost for repairs and rebuilds of the equipment.

For a highly mechanized, high production face it is essential to have a well trained staff of machine operators, technicians and electricians.

Capability of the equipment

For reaching a high production rate it is essential to operate equipment that is capable of working continuously reliable on a high performance level.

The performance of the shearer loader as the first and perhaps most important link in the production chain of a longwall system depends on the following criteria:

- cutting power
- haulage power
- portal clearance
- availability
- ease of maintanace and repair

TECHNICAL FEATURES OF A HIGH PERFORMANCE SHEARER LOADER

The modern high performance shearer loader is an all-electric shearer with cutter motors mounted to the ranging arms and with electric haulages. Fig. 36.3 shows the drive concept of a modern all-electric shearer versus a hydraulically driven shearer.

The all-electric shearer provides more usable power for cutting due to separate drives, higher efficiency of the electric motors compared to hydraulic drives and less friction losses because of shorter geartrains. Fig. 36.4 shows the drive concept of an Eickhoff SL 500 shearer.

The production of the shearer loader is not only limited by the installed power. A very importend criteria is as well the clearance under the shearer body, to allowe the produced coal to pass under the shearer. Therefore a modern shearer loader has to offer a strong but compact design. Fig. 36.5 shows the cross-section of an SL 300 shearer loader.

High installed cutting and haulage power combined in a compact machine, requires a very high mechanical stability of all shearer components. The improvement of the strength of the individual shearer components and the complette shearer is reached by computer aided design (Fig. 36.6, Load Simulation by FEM), intensive testing (Fig. 36.7, Load Testing Shearer Body), high quality standards in material and production and many years of experience in shearer design. Fig. 36.8 shows the tie rod design of the SL 500.

A modern shearer loader can considerably increase the hourly production rates of a face provided that the conveying system can keep step. Tables 36.1 and 36.2 show a comparison between different types of shearer loaders.

Table 36.1 : Technical Comparison

	EDW 170/340 L	*EDW 230 2L 2W*	*SL 300*	*EDW 300/380 LH*	*EDW 450 L*	*SL 500*
Operating Range (m)	1,5—3,0	1,8—4,0	1,4—4,0	2,0—5,0	2,0—5,0	2,0—6,0
Shearer Weight (t)	32	35	40	50	55	65
Shearer Length (mm)	9206	8868	10620	9056	8950	11970
Continuous Cutting Power (kW)	2 × 170	230	2 × 300 or 2 × 360	1 × 380	1 × 450	2 × 500
Haulage Power Installed (kW)	approx. 32	23	1 × 35 or 2 × 80	Approx. 50	2 × 25	2 × 54 or 2 × 80
Voltage (V)	1000	1000	1000/3300	1000	1000	3300

Table 2 : Comparison of Producitivity

	EDW 170/340 L	*EDW 230 2L 2W*	*SL 300*	*EDW 300/380 LH*	*EDW 450 L*	*SL 500*
1	2	3	4	5	6	7
Cutting Speex Max.	4,2 m/min.	6,4 m/min.	10,0 m/min.	5,9 m/min.	6,6 m/min.	13,0 m/min.

1	2	3	4	5	6	7
Production Max. Down Hill	700 t/h	1200 t/h	1900 t/h	1100 t/h	1300 t/h	2500 t/h
Cycle Time	162 min	100 min	73 min	101 min	94 min	71 min
Shift Production Max. (ROM)	2100 t	3400 t	4600 t	3400 t	3600 t	4800 t
Difference	0	+ 62%	+ 35%	0	+ 6%	+41%

The production comparison is based on following operational parameters:

Face/Operational Parameters

Face Length:	220 m
Seam Thickness:	2,70 m
Coal Hardness:	medium
Inclination:	5 degrees
Drum Diameter:	1,80 m
Web Depth:	0,8 m
Cutting Time Per Shift (8 h):	4 h (50 %)

The replacement of worn-out and unreliable equipment for modern high performance machinery seems to be a very expensive proposal at first sight, but it can be an investment which is very quickly recovered in the form of a considerable increase in production.

Another very efficient approach to increase productivity is the increase of the overall longwall availability, that means the time the longwall produces coal.

The longwall availability depends on the downtimes caused by geological problems and the downtime caused by brakedowns of the equipment.

In the production chain, the shearer loader in particular needs more attention. The speed of the shearer on the conveyor, the large amount of moving parts, the loads imposed on the shearer due to dynamic loads, vibration and pushing loads from the shield supports are all factors which substantially contribute to a greater risk of failure.

New, modern shearer loaders reach availabilities between 90 and 100 %, proper maintenance and rebuilds provided. The availability of the older hydraulic shearer generation can be considerably lower.

EXAMPLES OF HIGH PERFORMANCE LONGWALLS

In the years 1994 and 1995 Eickhoff commissioned two newly developed high performance shearer types, the SL 500 and the SL 300, at two German mines. The SL 500 was delivered to Ensdorf Mine of Saarbergwerke AG (today DSK) while the SL 300 was delivered to Walsum Mine of Ruhrkohle AG (today DSK).

In both cases the mines intend to increase longwall production considerabely to reduce the total number of faces. Ensdorf Mine installed a completely new face system consisting of roof supports, A.F.C. and shearer. Walsum Mine replaced the shearer only. Tables 36.3 and 36.4 show the operational parameters and the main specifications of the new and the old shearers.

Table 36.3 : Operational parameters Ensdorf Mine

Face Length:	335 m
Seam Thickness:	3 m
Coal Hardness:	medium - hard
Rock Intrusions:	0,3 m rock layer
Inclination:	10° - 15°
AFC:	1000 mm, 2 x 400 kW

	Eickhoff SL 500	**Eickhoff EDW-450-L**
Cutting Power	2 × 480 kW	1 × 450 kW
Voltage	5000 V	1000 V
Haulage Power	2 × 60 kW	2 × 25 kW
Portal Clearance	650	500
Maximum Speed	13m/min	6.4 m/min.
Daily Production Average (ROM)	9500 - 14000 t	6000-9000 t

Table 36.4 : Operational parameters Walsum Mine

Face Length:	300 m
Seam Thickness:	1,9 m
Coal Hardness:	medium - hard
Rock Intrusions:	yes
Inclination:	0 - 5°
AFC:	880 mm, 2 x 315 kW, 1 x 400 kW

	Eickhoff SL 300	**EDW-230 2L 2W**
Cutting Power	2 x 300 kW	2 x 230 kW
Voltage	3300 V	1000 V
Haulage Power	2 x 35 kW	2 x 23 kW
Portal Clearance	600 mm	420 mm
Maximum Speed	14 m/min	9 m/min
Daily Production (ROM)	6500 - 8000 t	5500 - 7000 t

In January the Ensdorf Mine reached a record production of 23.540 t on one day (24 H) during a performance test. Fig. 36.9, shows the Eickhoff SL 500 of Ensdorf Mine

Fig. 36.10, shows the Eickhoff SL 300 operaring at Walsum Mine.

Both mines could increase production with their new equipment. Bearing in mind that both shearer types were prototypes at that time this performance is considered a great success.

Saarbergwerke AG ordered a further four SL500 shearers to cut down the existing 5 production faces to only 3 high performance units.

Ruhrkohle AG ordered a total of 17 additional SL 300 shearers for Walsum Mine and other mines operated by them.

CONCLUSION

This paper summarizes the important factors whiche have an effect on the performance of a longwall. For reaching a high production rate it is essential to operate state of the art equipment. The shearer loader is the first link in the production line of a longwall and because of the loads imposed on it during operation and it's complexity the shearer loader is one of the most critical machines in the longwall face.

In closing I would like you to note that we cannot overemhasize the importance of state of the art machinery for the save and profitable longwall mining.

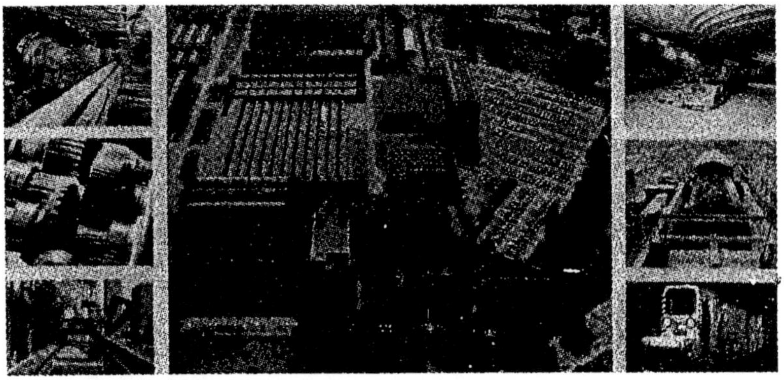

Fig. 36.1 : Eickhoff Products

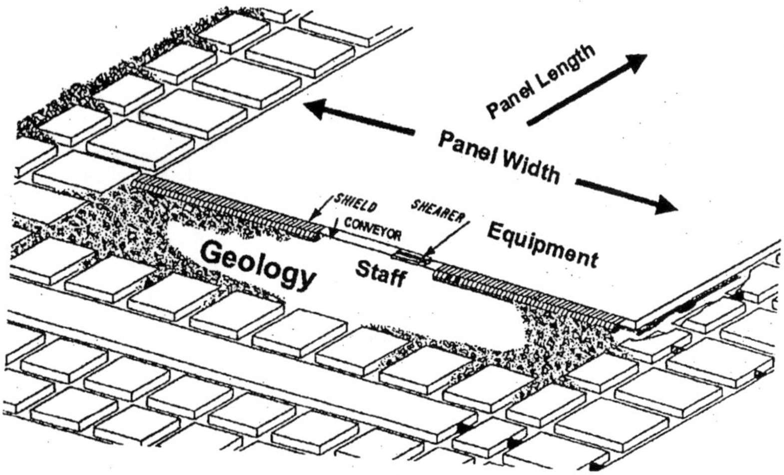

Fig. 36.2 : Longwall

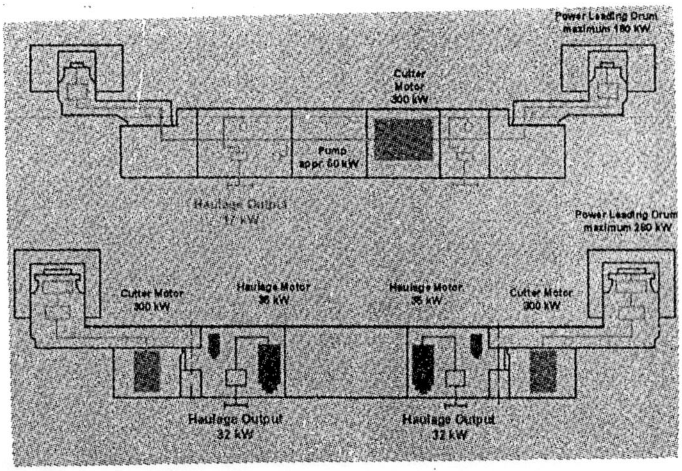

Fig. 36.3 : Shearer Loader Comparison

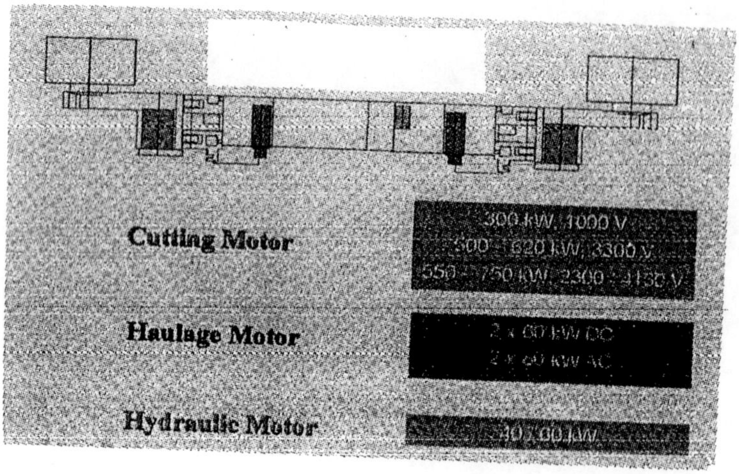

Fig. 36.4 : SL 500 Drive Concept

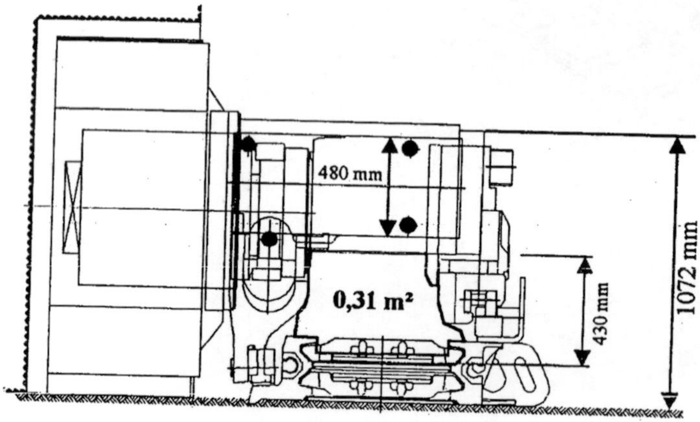

Fig. 36.5 : SL 300 Cross-Section

Fig. 36.6 : Load Simulation by Finete Element Method

Fig. 36.7 : Load Testing Shearer Body

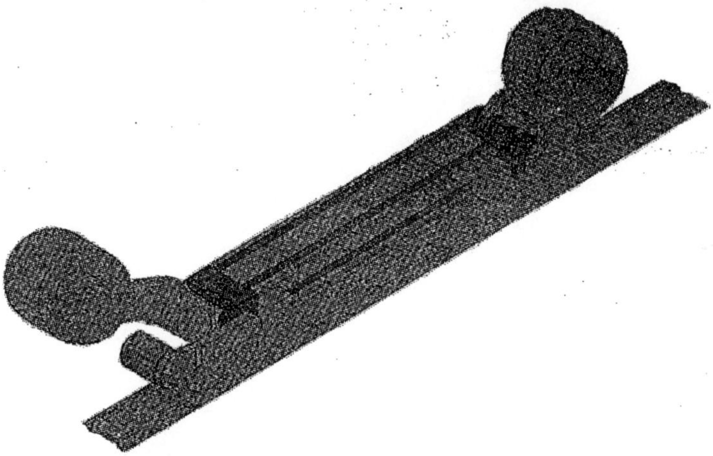

Fig. 36.8 : SL 500 Tie Rod Design

Fig. 36.9 : SL 500 Ensdorf Mine

Fig. 36.10 : SL 300 Walsum Mine

CHAPTER 37

Mining and Legislation Present Scenario and Future Prospects

*V.S. Rao**

1. INTRODUCTION

1.1 It is often said that India is a mineral-rich country. This is true to the extent that the geological record indicates that the Indian Precambrian Shield hosts a wide array of formations with exciting mineralogical possibilities. India has large reserves of coal, iron ore, bauxite, barytes, dolomite, manganese ore, mica, zircon, granite, marble etc.

1.2 However we are not a mineral-rich country in the sense that, despite a mining tradition spanning over a century, India is still a relatively insignificant producer of minerals by global standards, barring a few minerals such as iron ore, bauxite, and limestone. We produce about 84 minerals whose total contribution to GDP is less than 4%. Our total mineral production in 1997-98 was of the order of US$ 9.3 billion, though the contribution of non-fuel minerals was only about US$ 1.5 billion. By contrast, Australia's mineral exports alone totalled approximately US$ 26 billion in 1996-97. (1)

1.3 The level of exploration activity in a country at any given time is a good indicator of the long term mineral development potential of a nation. According to the data collected by the Metals Economic Group (MEG), the total planned expenditure on exploration in 1997 was US$ 5.1 billion. While nobody is aware how much India spends on exploration, a figure of US$ 10 million is mentioned in certain circles. This figure, when compared with the amounts being spent in other countries, as shown in Table 37.1, is obviously woefully inadequate for any serious development.

1.4 It is important that we not labour under the misconception that we have achieved any real success in attracting investment. In various fora it is proudly stated that the total foreign investment commitments till date in Indian mining total Rs.3,158 crores

*President, Federation of Indian Mineral Industries

(approximately USD 0.75 billion). However the reality is that the total inflow of foreign capital as on 30.11.98 was only Rs. 74 crores (approximately USD 17.6 million). In comparison the direct foreign investment in Chile's mining sector, in 1997 alone, was of the order of USD 1.63 billion. (2)

Table 37.1 : Leading Exploration Targets - 1998

	Country	*Exploration expenditure (US$ million)*
1.	Australia	494.6
2.	Canada	308.0
3.	U.S.A.	242.7
4.	Chile	177.0
5.	Indonesia	136.4
6.	Peru	136.1
7.	Mexico	127.2
8.	Brazil	121.9
9.	Argentina	76.0
10.	Papua New Guinea	59.9

2. LIBERALISATION - TOO LITTLE, TOO LATE

2.1 The main reason for the comparative low level of mineral development in India is that our laws and procedures have done very little to encourage investment in the development of mineral resources. Historically the mining of most metallic, fuel and precious minerals was reserved for the public sector. As a result, even today the Indian mining industry is dominated by the public sector. Public sector mines, while comprising only about 10 per cent of the total number of mines in the country, account for over 90 per cent of mineral production by value. It should be noted however that this public sector bias in the minerals sector is not a phenomenon unique to India. In most countries there was a feeling that minerals are a strategic national asset the control of which should be vested in the State.

2.2 However the mid 1980s saw the start of a regulatory reformation in mining sectors around the world. Recognising that they lacked the technical and financial wherewithal to fully develop their mineral sectors, nation after nation began modifying their legal and regulatory frameworks in a bid to attract private sector investment. Since 1985 more than half the world's nations have either adopted a new mining legislation, made a major revision to the existing laws, or are currently working on a new draft legislation. The limited capital available promoted a fierce spirit of competition which saw many countries making swift and sweeping changes to their laws in order to attract move investment.

2.3 In this global race to attract mineral development funds India was a late starter, and for all practical purposes, has remained a non-starter till date. It was only in 1993 that we amended our National Mineral Policy followed by legislative amendments in January 1994. As these amendments did not result in any significant inflow of investment at the ground level, a Committee was formed in February 1997 to review our mining legislation and suggest how to streamline and simplify rules and procedures. This Committee was headed by the then Secretary of the Ministry of Mines and included representatives of the Central and State governments as well as from the Federation of Indian Mineral Industries (FIMI). Till date the Committee's Recommendations, which were focused on making India's mineral sector a much more attractive destination for private investment, have yet to be implemented.

2.4 Today the global mineral exploration scenario is going though major changes. Depressed commodity prices, surplus supplies and an abundance of known deposits waiting to be developed have resulted in a slow-down in world-wide exploration expenditure. Between 1997 and 1998 global exploration spending fell by 31% and is expected to fall further in 1999. Global mining majors have made huge cutbacks in staffing and expenditure. For example, Billiton reduced exploration and development staff by 80%. In 1993 when the liberalisation of our mineral sector was initiated there was a boom in the global exploration industry. Had we acted quickly and decisively at that time, we would have succeeded in attracting large numbers of investors who, once committed, would have remained in India even through the present lean phase. As it is now we face an uphill battle to attract investment.

2.5 Today's market for mineral development funds is a buyer's market and investors have multiple highly competitive options. If we fail to recognise this fact we will not only fail to attract foreign capital, but we will not even be able to attract Indian companies to invest in mining. We are already seeing reputed Indian companies, such as Binani Industries, Hindustan Zinc, NMDC Ltd. and Sterlite Industries, venturing into exploration and/or mine development overseas. We are also seeing FDI bypassing India and going to less prospective but more investor friendly countries.

2.6 While we are still struggling to fully implement the first phase of reforms, some of the world's most successful investment destinations are already embarking on new waves of reforms. A good example of this is Peru, which saw investments to the tune of US$ 1.3 billion in its mining sector in 1996 (including US$ 300 million in exploration and prospecting). Even now the Peruvian Ministry of Energy and Mines continues to work closely with their Ministry of Finance to try and improve the existing tax regime for new investments in mining. [3]

3. THE NATIONAL MINERAL POLICY

3.1 Perhaps one of the reasons for the slow rate of change in our mineral legislation is that there is an inherent conservatism in the National Mineral Policy, 1993. This policy, which was designed to guide the country's mineral industry into the competitive era of the globalised economy, is based on the premise that "minerals are exhaustible and non-renewable resources" and that "their exploitation has to be done keeping in view not only the present but the long-term needs." These statements pre-suppose that we know exactly what is beneath the earth and there is nothing more to be explored, which is not the case. Exploration by modern methods and state-of-the-art technologies will certainly further enhance the resource position.

3.2 Further the perception that mineral resources are finite and strategic commodities that must be carefully conserved and rationed for future use runs contrary to the objective of a vibrant, growing mineral industry. It is important to recognise that from an economic point of view, the value of resources exploited today is greater than that of those exploited tomorrow. In addition minerals that are left in the ground so as to conserve them for tomorrow's needs may ultimately become sterile and never be mined, on account of technological advances making today's uses obsolete. A good example of such a situation is the sheet mica industry. Thirty years ago the vacuum tube industry was one of the major consumers of sheet mica. However with the development of semiconductor technology, the importance of the vacuum tube industry diminished considerably and with it the fortunes of the sheet mica industry.

3.3 The NMP, 1993, further states that :

Conservation of minerals shall be construed not in the restrictive sense of abstinence from consumption or preservation for future use, but as a positive concept leading to augmentation of resource base through improvement in mining methods, beneficiation and utilisation of low grade ore and rejects, recovery of associated minerals, reduction in their requirements of minerals per unit of material output etc.

3.4 This attitude expressed in the National Mineral Policy towards mineral conservation is primarily a defensive one where it is assumed that there are only a fixed quantity of reserves available and that conservation efforts should focus on improving productivity of mining, processing and utilisation of minerals, so as to make these reserves last longer.

3.5 The proactive role of exploration as a means of significantly augmenting the country's reserve base is not given the emphasis it deserves. The NMP, 1993 seems to envisage exploration as being predominantly a public sector domain, with "the initiative and cooperation of the private sector being drawn on as required". Since modern mineral exploration is a high-risk, capital-intensive activity requiring state-of-the-art technology it would seem logical that public sector participation would be minimised and that the financial and technological muscle of the private sector would have to be roped in.

3.6 The notion that it is best if control of mineral resources vests with the state, either on the grounds of strategic importance or as a result of their revenue generating potential, will only serve to slow down the development of this sector. This theme of the NMP is reflected in sections of the existing mineral legislation such as Section 17A of the MMRD Act which deals with reservation of areas by Government for the purpose of conservation of minerals.

4. ATTITUDE OF THE STATES

4.1 Creating and maintaining an investment-attracting climate in today's highly competitive global scenario entails more than a one-off revision of legislation. Apart from an on-going process of legislative analysis and fine tuning in order to maintain and enhance competitiveness, it also requires attitudinal changes on the part of the States where the ground level activity takes place. The continuous changes in attitudes and mineral legislations in nations around the world, even those that are already considered as highly favourable mineral investment destinations is indicative of the keen competition prevailing in the minerals sector.

4.2 Ultimately, whether or not investment flows into India's mining sector will depend largely on the attitudes of the individual State Governments at the ground level. The

States that frame transparent laws and time-bound procedures will succeed in attracting investment. Rajasthan is an excellent example of this fact. As a result of its progressive attitude, Rajasthan's exploration industry is years ahead of any other State and it is only a matter of time before this momentum is translated into new mines coming on stream. On the other hand, many mineral rich States such as Andhra Pradesh, Bihar, Karnataka, Madhya Pradesh and Orissa, have still not been able to work out a clear-cut policy for mineral development. Consequently despite their mineral potential they have attracted negligible investment. There seems to be a mistaken impression among certain States that their geology is unique and that investors will come flocking to them. Nothing could be further from the truth. The reality is that investors are flocking to many destinations but India, perhaps with the exception of Rajasthan, is not one of them. In Brazil for example, in 1998 the Government issued 12,000 exploration permits for an area of 13.3 million hectares.[(4)]

5. FISCAL FRAMEWORK

5.1 The operating fiscal environment is one of the main criteria considered by private sector mining companies while assessing investment attractiveness. Exploration is the business of finding mineral deposits that can be commercially mined. The risks are high and the odds are heavily stacked against success. Mineral prospects (high concentrations of minerals) are extremely rare, and less than one prospect in a hundred becomes a mine. To test these low odds requires a lot of time and money, commonly in the range of three to ten years, and between $US 10 million and $75 million, depending upon the deposit type and location. It is therefore very important that the deposits that are found provide a return to the investor that reflects the risk in investing funds during the exploration programme.

5.2 Thus before an exploration programme can commence there must be a reasonable certainty that if an economic resource is found the investor will be allowed to make an adequate return. Mine development involves large, high-risk investments, which are made against a backdrop of uncertain market and geological conditions. This high level of uncertainty demands a higher return on investment than a risk free situation. As a high incidence of fiscal levies reduce the return on investment, if the tax burden is perceived to be too high, investment will not take place, regardless of geological potential. An investor-friendly tax regime is therefore crucial for attracting sustained investment in the Indian mining sector.

5.3 The Indian mineral sector is currently governed by a multiplicity of taxes at the municipal, State and Central levels. Unfortunately due to a tendency, in some quarters, to view the minerals sector as a cash cow, all kinds of new financial conditions and levies are being imposed under some excuse or the other. It is crucial to understand that the long-term interests of the industry are also in the best interests of the State. Given room and incentive to grow, the mining industry has the potential to deliver far more, by way of employment generation, infrastructural development, and increased revenue, than if the industry's growth is stunted by a heavy fiscal and regulatory burden.

6. FOREST/ENVIRONMENT CLEARANCES

6.1 India has comprehensive environmental laws which can result in significant delays to mining projects particularly those located in forest areas and/or near wildlife sanctuary areas. There are cases on record where clearances have taken as long as

2-3 years. Potential cost escalations as a result of these "green tape" delays serve as a major disincentive to mining investors.

6.2 There is a need for the unique role and characteristics of the mining industry to be taken into consideration. Contrary to popular belief, mining is not the main cause of forest destruction. In fact it is responsible for less than 5% of the total diversion as on date. Mining is an indispensable part of the nation's industrial framework and provides the raw materials for economic growth and development. Besides providing considerable revenue to the Government it also generates wealth, provides employment and develops physical and social infrastructure in many remote and backward areas. In light of its essential and site-specific nature, mining should be readily permitted in any part of the country that is not declared as ecologically fragile. Furthermore the diversion of land for mining should not be viewed as a short term activity but rather should be considered for periods ranging from 50-100 years till the exhaustion of ore in the deposit.

6.3 Normally a mineral deposit extends over a certain area. Thus once MOEF agrees in principle for its diversion the entire zone, consisting of clusters of adjoining mining leases, should be considered for diversion instead of granting approvals on a lease by lease basis. A standard set of conditions may be applied for the diversion of such an area/zone. In this regard each State Government can prepare a district-wise list of existing mining leases, and wherever such areas coincide with forest land, they may be considered for diversion en bloc. Within this zone in which diversion has been agreed to in principle, the State Government should be allowed to grant and/ or renew leases as and when they come up for renewal subject to the lessee complying with all specified conditions, including compensatory afforestation requirements.

6.4 Developmental projects in India invariably face a tug-of-war between the economic benefits of the project and issues of concern to local populations. In a bid address this issue legislation was introduced, under the Environment Protection Act, making public hearings mandatory. However for smaller mining leases, say less than 100 ha., the requirement of public hearing is neither necessary nor feasible from a logistical and economic point of view. Similarly in the case of renewal of a mining lease, no matter what the size, a public hearing should not be necessary as the displacement and various impacts caused by the project have already been taken into consideration when the lease was granted. This is also kept under constant scrutiny by the Indian Bureau of Mines during each of the five-yearly reviews of the mining plan, which includes an EMP.

7. CONCLUSIONS

7.1 Till date India's mining industry has been predominantly in the public sector. However it is very clear that with constant demands for government funds from various social sectors, the future growth of the minerals sector will hinge on the level of involvement from the private sector. In order to truly encourage private sector investment in minerals and mineral-based industries there is a need for urgent reform of investment-inhibiting laws and regulations, such as Section 17A of the MMRD Act. Similarly serious commitment is required on the part of Government, particularly at the State level, to promoting large-scale private investment in mineral exploration.

7.2 As we enter the 21st century we must make a firm commitment to developing and maintaining an internationally competitive mineral investment climate in India. The primary objective of our mining legislation must always be to promote and develop mining activity, with regulation and control as incidental functions. It is only through a legislative/regulatory framework that is continually evolving in response to market forces, that we will be able to attract the flows of private sector capital and state-of-the-art technology required to develop a globally competitive world-class mining industry in the new millenium.

CHAPTER 38

Increased Performance in Underground Drifting due to Mechanization with DH-Equipment

*Andreas Schroth**

ABSTRACT

The Deilmann-Haniel (DH) group of companies provides specialized technical services to the international mining industry, and is one of the leading specialized underground contractors in the Federal Republic of Germany 100 years of experience and utilization of the latest technology have enabled DH to maintain the highest standard of service and level of success. With a staff of more than 6000 people, its engineering/ development projects include shaft sinking underground development construction of mining machinery civil engineering tunneling and road construction.

The application of ground freezing design and construction of absolutely watertight shaft lining systems utilization of road headers and full-face tunnel bring machines as well as shaft dolling are techniques with which DH has much experience. Tunneling and shaft sinking activities are supported by DEILMANN-HENIEL Maschinen-u Stahlbau GmbH which is specialized in design and manufacture of machinery and systems such as a hydraulic sidetipping loaders for a wide range of applications Crawlertype drill jumbos are other typical examples of the large variety of products manufactured by DEILMANN-HANIEL Maschinen-u Stahlbau GmbH Other products find their application in shaft sinking i.e., winches hoists, shaft drill jumbos, grabs working platforms, special structures. For completed shafts DH plans, designs, constructs and supplies complete shaft installations, i.e. hoisting equipment. DH's unwelded steel guide system, shaft stations, road junctions, bottom frames shaft stations special structures.

* Dipl. Eng., Deilmann-Haniel Maschinen-und Stahlbau GmbH. Dortmund.
Tel:-49-231-2891-470. FAX--49-231-2891-314. e-mail: DHSchroth@t-online.de

In the realm of international cooperative experience. DH has successfully executed or has under way projects in Canada, Chile, Korea, Slowenia, Indonesia, India, Morocco, France, the United States, China Spain, Burkina Faso, Austria and the C.I.S.

INTRODUCTION

The Deilmann-Haniel (DH) group of companies provides specialized technical services to the international mining industry, and is one of the leading underground contractors in Germany and worldwide. 100 years of experience and utilization of the latest technologies have enabled DH to maintain the highest standard of service and level of success. With a staff of more than 6,000 people, its domestical and international engineering development projects include design and manufacture of mining machinery, shaft sinking, underground development, drilling boring technology above and below ground.

A Deilmann-Haniel uses all of the latest drifting methods, we have always played a prominent role in the development of such procedures. The full facing machine technology used in the rock drivage of the hard coal mining industry was introduced by us. We were also involved in a major way in helping to achieve the present daily performance and safety standards for road heading machine in the German hard coal mining industry.

We invented and developed the shaft freezing method for unstable waterbearing strata with a highly load resistent and absolute watertight shaftlining system and exported that knowhow and equipment even to China.

SUCCESS THROUGH EXPERIENCE GAINED FROM OWN SITES

The machines for drilling and blasting technique, for shaft sinking works and any necessary special constructions are designed and manufactured in our offices and workshops.

Deilmann-Haniel Maschinen-und Stahlbau has the good fortune of being able to manufacture mining machinery, which benefits from the experience and know-how gained in our contracting sites.

The result of this is that our machines have achieved a market share of well over 70% in Germany, and our machines are exported to many mining countries throughout the world.

These machines have the highest degree of efficiency, reliability and quality. As a result the customer saves costs for maintenance, spareparts and downtime.

All Deilmann-Haniel equipment is robust and sturdy and built to withstand the tough underground treatment during operations.

In addition to the most modern technology, which ensures the highest level of performance of DH machines, it is the durability and the reliability of our machines that guarantee the greatest drifting performance at the lowest operating costs.

DEFINITION OF HIGH-PERFORMANCE HEADING

The "High performance heading" demands the efficiency of high quality machinery, which is a necessary prerequisite for positive operational results in roadway drivage.

Without a suitable operations technique and an individually-tailored operational organization system, however, the performance potential cannot be realized.

An optimal result can only be achieved with an individually-styled optimum operational arrangement with sufficient space availability.

Only in a heading system with a well-coordinated procedural technology can the machines used fulfill any performance expectations.

The requirements for optimized roadway drivage are high performance capacity at the lowest possible costs.

The high-performance heading is complicated and complex system. Numerous influencial factors, such as geometric factors, operational sequences consisting of partial operations which must be coordinated with one another, process technological contexts, external interference, such as logistics or power supply, and also geological conditions, rock precautionary measures and kind of roof support, all have an effect on the results of the work.

Not to forget the high risks in uncertain or unknown geological situations and the enormous logistic demands of high speed drivages.

TRANSFERABILITY TO INDIA

The before mentioned requirements of a high-performance-heading may be valid as well for the Coal Mining Industry in India, however, with different priorities. In India, increase of advance rates and higher technical and economical efficiency might be first priorities. This aims can be achieved by mechanizing the drilling-and loading operations with our adequate machinery.

Obvious interdependancy of these prerequisites leads to the conclusion, that any specific situation for a drill and blast heading has to be analyzed individually and carefully in order to find the optimum layout. Detailed knowledge on potential and restrictions of machines and installations is essential for optimization of equipment and procedures.

Again it must be emphasized, that equipment for mechanization is expensive and has to lead directly to higher profitability and has to contribute together with optimum organization to higher performance.

EQUIPMENT FOR MECHANIZATION OF DRILL-AND BLAST HEADINGS

Drilling

The simplest and in certain cases the optimum solution is the mechanization of drilling work with pneumatic high-performance hand-field drifters and pneumatic legs. Modern handheld drifters can have an advance speed of 0.3 to 1 m/min producing only modest noise emission of about 100 to 105 dBA. But electro-hydraulic drilling jumbos are being used more and more due their better capabilities. The output of the hydraulic drifter is approx. 17 KW.

For the hard coal mining industry, DH builds single-boom and two-boom drilling jumbos on a crawler type undercarriage.

The advantages are:

- With fully hydraulic drifter hammers, drill holes with diameters of 32 to 63 mm and more can be made at high speeds.
- In medium-hard rock, a net drilling speed of at least 2-3 m min should be standard.
- The completion of one pull with e.g. 2.-m length and 12 m 2 section area should be less than 1 hour by a single boom drilling jumbo.
- Less physical and noise stress to the miners.

The minimum width of a single boom jumbo is only 1.200 mm and the height is only 1.300mm. These dimensional data are very important, because in heading works machines have to be at the face and proper maneuverability is essential, even in galleries of 4.5 m width or less.

By the modular design of the Deilmann-Haniel drilling jumbos with fixed and telescopic drilling booms with varying types of feeds, including telescopic feeds, customer shaped solutions for all types of drilling problems in roadway drivage can be found.

Loading

Crawler-mounted side-dump loaders made by Deilmann-Haniel are known as essential component of highly performing drilling-and blasting systems in mining and tunneling.

By now, more than 800 units have been delivered to the mining industry all over the world. Electro-hydraulic side dump loaders that load the blasted rock or coal directly into the mine car or onto a chain conveyor are standard in high performance headings.

By choosing optimum machinery, it should be regular to clean the face in any cross section area within 1 to 2 hours, provided that the linked haulage system has an adequate capacity.

For matching in the best possible way all components of a conventional tunnelling system and for allowing optimized performance and cost-effectiveness, a whole range of types was developed by DH. This range comprises eight sizes of dump loaders with shovel capacities of 500 1 to 2000 1, telecopic arms with a stroke of 600 mm to 1.5 m and slewing ranges from 15 degrees to 30 degrees. Their loading capacities range from 30 to 200 m3 per hour depending on the travelling length during loading works.

Suitable for the section areas in the coal industry of Columbia could be electro-hydraulically driven side-tipping loader on crawler undercarriage types L 513 or DH250T with 30 KW to 55 KW drive rating and a shovel capacity ranging between 500 and 600 1.

This loaders are predominantly used for cross sections of 6 to 12 m2.

These machines with its small dimensions were designed for use in smallest roadway cross sections offering a maximum mechanization of loading

However, the lowest possible time for the loading cycle can only be achieved, if the subsequent haulage of blasted material can be performed simultaneously and with adjusted capacity.

With the sidedump loader, various other attachments can be used such as drilling unit, ripper unit or even a roadheader unit. That makes the sidedump loader a extremely versalite machine.

Support Setting

This generally most time consuming and hardest labor can be eased by using the side dump loader for following operations:

- the arch-segments can be transported to the face in the bucket.
- with an attachment at the bucket the roof segment can be lifted to the roof and laid upon the forepoling girders one after the other, with the sidedump loader travelling slowly forward and the forepoling girders being extendes to the face.
- from the up-lifted bucket, the arch connecting locks can be fixed, tightened and the lagging material be installed.

RELIABILITY

Deilmann-Haniel machines for drilling, support setting or loading have an availability of 90 to 95%. That of course depends on regular maintenance according to regulations. As standard, DH-machines are in permanent use in underground conditions for many years without being recondition in a surface workshop, which is the result of highest possible design and manufacturing quality.

CONCLUSION

Specific for drill and blast headings are their flexibility, comparatively low capital cost and wide field of application with regard to geological conditions and rock characteristics. Due to thorough mechanization, which basically is possible for each individual working sequence, they are now-a-days of comparable performance at mostly lower specific cost.

Reliability and a high degree of availability, are besides of a high power potential, the reasons for the utmost efficiency at lowest costs of Deilmann-Haniel equipment in drill- and blast headings.

Thus the main objectives of mechanization, namely

- minimization of cost
- increase of efficiency
- reduction of manpower at the face
- increase of advance rate
- improvement of working conditions
- dramatic improvement of safety standard and drop of accidents will be achieved.

CHAPTER 39

Improving productivity in mining sector–Future visions with special reference to coal sector

*R.B. Mathur**

1. Productivity and Quality are key factors for survival and growths in the global competitive mining environment of today embracing both coal and metal mining industry in India and abroad.

2. The economic success of the country is achieved through contribution of people at work and labour productivity is an important parameter in the achievement of economic success in the technological and information age of today and 21st century.

3. Technology has affected to a great extent labour productivity and that is the reason for an upsurge in opencast coal mines of our country. New economy is emerging from scientific and technological achievements.

4. The impact of technological innovations and improvement in the mining sector can be enhanced by integrating the application and adoption of technology appropriately with existing mining systems, both in O/C and U/G for optimising the output and minimising costs for achieving best productivity. Introduction of higher and higher capacity dumpers, shovels and draglines and surface miners in a few cases in O/C mines and introduction of mass production technology in U/G are the best examples of new approach which managers of today are wanting to adopt.

5. In recent years there has been a major upsurge of interest and activity in technology improvement all over the country in Industrial Sector. The Government of India has expressed time and again deep anguish over poor in different poor productivity sector including Core sector. The fundamental cause of slow process of development in our country is poor capacity utilisation. Capacity utilisation in power sector is less than 70% (PLF). Similar is the situation in the steel sector. But, there is no cause for complacency as there is still wide scope for improvement.

---

*Ex CMD, Wel, Coal India.

6. During the pre-nationalized period, there was no concept of system capacity utilisation in Coal Sector. Presently, CMPDIL assesses on annual basis the capacity of each mine and monitors the performance on that basis. But, I have no hesitation in expressing that still many persons are not aware of the intricacies involved in calculation of productivity in mining sector future vision has been chosen by me. Further this subject matter has also assumed importance in view of surplus coal scenario in case of coal mines lack of funds with mining sector PSUs, liberalisation/ globalisation and import of minerals which is posing a threat to indigenous mining industry.

It is there, imperative that cost of production of minerals is brought down by proper system capacity utilisation. The coastal based industries have already stated relying on imported coal because of economic considerations are quality. Coal is no longer a sheltered section of economy and the very survival of coal industry depends on efficient utilisation of capacities created in 7th, 8th and 9th plans and improving productivity.

Year	*ECL*	*BCCL*	*CCL*	*WCL*	*SECL*	*NCL*	*MCL*	*CIL**
1980-81	63	49	101	70	-	-	-	71
1985-86	61	71	84	85	82	74	-	78
1990-91	46	64	87	62	79	102	-	79
1995-96	58	65	55	80	91	80	78	73
1996-97	66	79	56	89	94	85	84	79
1997-98	70	83	61	96	98	93	92	85

* NEC included in CIL

Leadership role in improving productivity vision

Leadership:–Leadership is an influence process. The leader is different from other people in the ability to get them to perform activities that lead to desired outcome or goal. In the Mining sector the leader has to command over masses.

Vision:–Leaders see what other's can not. Somehow the leaders of industry or mining sector develop a vision or a belief as to what is possible and then act to achieve the same. However, there is a between a PSU leader and a private sector leader. What Bhargava could do for MARUTI, his successor could not and whether public sector could reach commanding heights of economy as envisaged earlier, is any body's guess.

In coal sector, what is the vision for U/G productivity faced with very large manpower which can not be dispensed with? This is a challenge for mine managers.

7. The effective utilisation of money, material energy and other inputs interact directly or indirectly on labour productivity and vice versa. All this leads us to realize that the productivity of each resource is interdependent on the productivity of other resources in the wealth generating process. It may some times happen that the productivity of one resource may show improvement or otherwise at the cost of, or due to change in productivity of other kindred resources. This may be examplified while considering labour productivity in a man, machine system–which is the way

most systems operate today. A change in technology could be the major reason for a change in labour force productivity. Likewise, if machines and equipment are functioning at an optimum level to utilization, extra effort put in by the labour force for productivity improvement may not result in enhanced labour productivity. It may be possible to change the labour productivity frontier only with improvised or new technology. The human resource is the most important resource in wealth creation. It is widely acknowledged, and is also true that, even if all other resources are taken away, and the human resource is left to perform, people will be creative, dig out material from mother earth and convert ores into useable metals and coal to energy.

The start of such a process may be slow. However with the right technology and other appropriate resources in conjunction with the labour force, and with productivity and quality, true wealth can be created.

8. Methodology–Global Concept

A. *Labour Force Productivity* of an economic sector at the national level is quantified as the ratio of output expressed in terms of wealth generated by a particular economic sector, to the quantum of labour input, in respect of the same economic sector under reference, as follows:

Labour force Productivity Measure of a country's economic sector for a period of time (say one year)

$$\left[\frac{\text{GDP1 in monetary units of economic sector for a particular year}}{\text{Labour Force employed in the economic sector under reference during the same year}}\right]$$

This gives the wealth produced by a country's economic sector, per capita of labour force employed for the year under reference.

B. Labour Wage Productivity of an economic sector at the national level is quantified as the ratio of GDP generated by a particular economic sector, to the quantum of computed wages for the labour force employed. This may be exemplified as follows:

Labour Wage Productivity Measure of a country's economic sector for a period of time (say one year

$$\left[\frac{\text{GDP in monetary units of an economic sector for a particular year 1.}}{\text{Computed Labour Wages for the particular sector reference during the same year 3.}}\right]$$

The Labour Wage Productivity measure gives the wealth produced expressed as GDP in monetary units per unit money value of computed wages in respect of a country's economic sector under reference.

The above 2 methodologies give an idea of productivity measured in terms of GDP, which in fact embraces a wider concept of production achievement rather than a mere tonnage.

9. To understand capacity Utilization and concept of productivity as is popularly known in coal sector we have to be first familiar with a few definitions, related to working of HEMM in o/c mines.

a. PERCENTAGE AVAILABILITY

$$\frac{\text{HRS WHEN THE MACHINE IS FIT TO BE PUT TO ITS DUTIES}}{\text{TOTAL SHIFT HOURS}} \text{ X } 100$$

$$\frac{\text{SHIFT HRS - (MAINTENANCE HRS+ BREAKDOWN HRS)}}{\text{SHIFT HOURS}} \text{ X } 100$$

$$\frac{\text{WORKED HOURS + IDLE HRS}}{\text{TOTAL SHIFT HOURS}} \text{ X } 100$$

b. PERCENTAGE OF UTILISATION SHIFT

$$\frac{\text{HRS. WHEN THE MACHINE IS USED}}{\text{TOTAL SHIFT HOURS}} \text{ X } 100$$

$$\frac{\text{WORKED HOURS}}{\text{TOTAL SHIFT HOURS}} \text{ X } 100$$

10. International / CMPDI Norms for % Avail. and utilisation of HEMM

Main Hemm	% Avail of shift Hrs		% Utilisation of shift Hrs.	
	International	CMPDI.	International	CMPDI
Walking Draglines	85	85	72.25	73
Ele. Rope shovels	80	80	68	58
Hydraulic Shovels				
Rear Dumpers	-	80	-	61
- Electrical	80	-	68	54
- Mechanical	80	67	68	50
Drills Dozers	-	70	-	45

11. Breakdown Hrs. per worked Hr. of HEMM in CIL-Figures relate to the year 1995-96

EQUIPMENT	*BREAKDOWN HRS. PER WORKED HR*
* DRAGLINES	0.27
* ROPE SHOVELS	0.69
* HYD. SHOVELS	1.38
* REAP DUMPERS	2.42
* DRILLS	1.76
* DOZERS	2.38

12. Maintenance Hrs. per worked Hr. for HEMM in CIL-Figure relate to the 1995-96

EQUIPMENT	*MAINTENANCE HRS. PER WORKED HR.*
DRAGLINES	0.11
ROPE SHOVELS	0.26
HYD. SHOVELS	0.32
REAP DUMPERS	0.36
DRILLS	0.35
DOZERS	0.38

13. Global Opencast Technology Trends

13.1 Modernization of opencast mining operations is taking place continuously in the world coal industry. The major areas in which developments have taken place are:

(i) Upgradation in equipment sizes

(ii) Use of advanced electronics and computer control systems for equipment reliability.

(iii) Improvement in the operational efficiency of the equipment and system.

(iv) Alternatives to wheeled transport for coal and overburden i.e. crushing and conveying of coal and overburden.

(v) Steep angle conveying of coal.

(vi) Selective mining of thin bands using special equipment.

(vii) Continuous mining methods.

13.2 Almost all the major dumper manufacturers have launched high capacity models in the 170-220 T range. Recently a few companies have introduced 240-T and 300-T dumpers commercially. Probably, the 300-T dumper may become standard equipment in world coal industry.

13.3 Higher capacity excavators are also becoming available to work with these large capacity dumpers. Rope shovels of 46 Cum bucket and hydraulic shovels of 50 cum bucket are in regular production. Wheeled loader of 28 cum size with an 1800 hp diesel engine with a loading rate of 2800 tph is now being used with a 240-T dumper.

13.4 Performance monitoring systems have already been applied to excavators. These enable detailed analysis of digging cycles and plant and operators performance. They can lead to reduced downtime, optimization of digging cycles and the minimization of power consumption. The Dragline Performance Monitor also tracks the movement of walking dragline, allowing automatic transfer of the path onto the mine plan. This enables rapid identification of inefficient digging patterns. Weighing facilities are allowing shovel operators to load a dumper to a per-set payload. This result in a better dumper life and efficiency.

13.5 Drilling equipment has already advanced to a semi-automatic operation. It is possible that it will be the first to yield to full automation. At present, emphasis is placed on the derivation of blast design related data from the rig during drilling. Optimization of blast design to yield the desired fragmentation with minimum fly rock and explosive consumption is a proven route improved efficiency.

13.6 Machine health monitoring units using on-board computers, a variety of sensors, and electronically activated devices are available for shovels, loaders, dragline, dozers and off-highway trucks. They monitor engine, drive train, hydraulic and tube systems to provide immediate warning of impending failure. The can also performance. Periodic system checks can also be made, alerting the operator to potentially catastrophic conditions.

13.7 Automatic haul truck dispatching has been adopted in a number of large and complex open pits. Some of these systems are capable of monitoring the position of trucks within the pit automatically and checking the status of loading equipment and dump points by digital radio links.

13.8 The most significant developments in the automation of opencast mining have occurred in the fields of mine design and operations planning. Improvements in data gathering and processing power are leading towards rapid and realistic designs. Automation of plant and operations is at an early stage. There is a continuing trend towards the use of operational monitoring and control systems as a means of enhancing productivity and efficiency. The problem delaying the wider use of automatic system are not all technical.

13.9 A pit utilizing truck haulage will often be designed and scheduled differently from a pit employing conveyors. Conversion from one haulage mode to the other is likely to be costly. It may be either very time consuming or very disruptive or both. Nevertheless, some operations have chosen to minimize truck haulage by switching party to conveyor transport. This is done by installing an intermediary, in-pit crusher. The crushing unit can be fully mobile, semi-mobile to fixed, depending on site-specific requirements. The economics of belt conveying are particularly attractive in large volume operations, those where steep climbs are involved are where haul distances are longer than a few kilometers. Hindering their introduction to surface mines has been the apparently cumbersome nature and rigid configuration of large conveyor installations. It is mainly these problems that recent technical developments have addressed with a good deal of success. A second problem is that, for conveyor transportation, the mined material has to be within certain size limits. In hard rock mines that means crushing prior to conveying and tends to limit conveyor haulage to the coal friction. Here, too, advances in mobile equipment designs using rubber tyres and crawlers have introduced a flexibility hitherto unavailable.

14. Opencast Technology in India

14.1 The opencast mining systems adopted in India so far have been based on shovel-dumper system for coal extraction and overburden removal. Draglines have been deployed for overburden removal wherever conditions were suitable. No major change is likely in these technologies upto the end the Ninth Plan. A beginning has been made with mobile in-pit crushing and conveying of coal in Pipewar project. Crushing and conveying of overburden has been introduced in Ramagundam OC-II project of Singareni Collieries Co. Ltd. Recently some surface miners have been introduced in opencast mines of Mahanadi coalfields.

14.2 So far as the equipment size is concerned, Draglines with bucket size of 30 cum. Rope shovels with bucket size of 20 cum, 170 T Rear Dumpers, 770 HP Dozers and 311 mm bore-dia Drills are already in use. These sizes are unlikely to be exceeded during the Ninth Plan.

15. Man Productivity (O.M.S)

15.1 The productivity in India coal industry has been very low as compared to the world standards. This is more pronounced in the case of underground mines, especially of CIL, where the man productivity has been stagnation at around 0.55 tonnes per manshift over the last ten years. The man-productivity in CIL opencast mines, however, has registered a significant increase from 2.24 t/manshift in 1986-86 to 4.36 t/manshift in 1994-95. This has enabled the overall man-productivity in CIL to increase from 0.92 t/manshift in 1986-86 to 1.75 t/manshift in 1995-96.

15.2 The actual man-productivity achieved by CIL and SCCL in the first four years of the Eighth Plan, and the projections for 1996-97 (terminal year of the Eighth Plan) and 2001-02 (terminal year of the Ninth Plan) are tabulated in Table 39.1.

Table 39.1

Sl. No.	*Company*	*UG/OC*	*92-93*	*93-94*	*94-95*	*95-96*	*96-97*	*01-02*
1.	CIL	UG	0.6	0.6	0.6	0.6	0.6	0.7
		OC	3.8	4.0	4.4	4.7	4.9	6.0
		OVERALL	1.5	1.5	1.6	1.8	1.8	2.2
2.	SCCL	UG	0.7	0.7	0.7	0.7	0.7	0.8
		OC	4.5	4.4	4.6	3.7	4.3	5.5
		OVERALL	1.0	1.1	1.1	1.2	1.2	1.4

16. Machine Productivity

16.1 The equipment utilisation in opencast mines of has also been quite low. There is substantial scope for improvement of machine productivity especially in opencast mines. The actual performance of HEMM in CIL opencast mines in 1995-96 and the achievable performance, which can be deemed as the target for 2001-02, are tabulated in Table 39.2.

Table : 39.2

Sl. No.	Equipment	CIL Actual (95-96)	CIL Achievable (2001-02)	SCCL Actual (95-96)	SCCL Achievable (2001-02)
1.	Roper Shovels	35	47	-	-
2.	Hydraulic Shovels	31	39	55	57
3.	Rear Dumpers 35-85 T	22	26	-	-
	Rear Dumpers-120-170 T	40	42	46	55
4.	Draglines	65	67	79	98
5.	Dozers	23	26	64	65
6.	Drills	21	30	38	40

17. Productivity Improvement

17.1 Higher man-productivity is a resultant of a combination of measures based on in-depth studies of the conditions prevailing in different areas and individual mines. Insofar as the productivity in opencast mines is concerned, this is basically dependent on increasing the utilisation of Heavy Earth Moving Equipment. Measures should be taken to ensure that procurement of equipment is duly synchronized with land acquisition and availability of basic infrastructure like workshop, trained manpower etc. More attention should be paid to the phasing of equipment in the Project Report. If necessary, this part of the Project Report should be reviewed from time to time and phasing of equipment re-synchronized.

17.2 HEMM equipment are progressively becoming technologically more sophisticated and more expensive. It is, therefore, of utmost importance that the manpower deployed for maintenance of these equipment should be highly skilled, property trained and experienced.

18. Suggestions for the Improvement of systems capacity utilization - (Equipment Maintenannce)

- WORKSHOP FACILITIES
- MOBILE MAINTENANCE VAN FOR FIELD EQUIPMENT
- PREVENTIVE MAINTENANCE
- CONDITION MONITORING
- QUALITY SPARES
- SPARES MANAGEMENT - COMPUTERIZED INVENTORY CONTROL
- STAGE INSPECTION AT MANUFACTURER'S WORKS
- TRAINING BY MANUFACTURERS IN MAINTENANCE SKILLS
- REHABILITATION

19. Suggestions for the improvement of capacity utilisation system (operational)

(I) TIME MANAGEMENT

- ENFORCE SHIFT OVERLAP OF HALF AN HOUR
- MOBILE CANTEENS-ELIMINATION OF EXTENDED TIFFIN BREAK
- HOT SEAT CHANGE

(II) OPERATIONAL ASPECTS

- OPERATIONAL PLANNING AND IMPLEMENTATION
- GOOD FRAGMENTATION
- LARGE BLASTS
- LARGER INVENTORY OF BLASTED MATERIAL (2 WEEKS)
- MOBILE DIESEL FILLING
- MINE COMMUNICATION SYSTEM (WALKIE-TALKIE)
- OPERATIONS TRAINING

India today stands at the threshold of the 21st century with the economic scenario undergoing a change in the context of globalisation of economy and ongoing reforms. In order to keep pace wit the changes which are taking place in the rest of the economy, the mining sector needs to be overhauled for maximizing system capacity utilisation. I am happy to state that the process has already been initiated, by this seminar.

As you know coal is a very important fuel and plays an important role in meeting the energy demand in the country. The growth of the coal mining industry was very slow till nationalisation in the early seventies. From about 78 mt in 1973, the coal production rose to about 236 mt in 1997-98. India now ranks 3rd amongst the coal producing countries in the world.

Productivity in India Coal Mines is far lower than what is found in some other parts of the world including countries like China, be it machine productivity or labour productivity. The major challenge before the coal industry is to improve its labour productivity by reducing excess manpower and improving the skill levels of the work force.

RBI has recently said that the industrial performance as measured in the index of industrial production recorded a growth rate of 3.6% during April-Oct 98 substantially lower than 6.2% for the same period last year. This trend needs to be reversed as quickly as possible. The need of the hour is to identify our strengths and weaknesses so that we can prepare ourselves for the opportunities, which the next millennium will present before us not with standing the threats.

CONCLUSION

It is to be noted with satisfaction that CIL and its subsidiaries are fully conscious of the urgent need for improving the System Capacity utilization and have on their own initiated several meaningful steps to bring about improvement which include strengthening of weak units, their regular monitoring, rehabilitation programme of HEMM, continuing dialogue with the workman associated with various activities. In this connection, I am reminded of an excellent practices at Jayant Projects (10 million capacity) where the General Manager had fixed 1/2 day for a dialogue with key workmen for redressal of their grievances and for listening to their suggestions. This practice was existing in the year 89-90, 90-91-92 and Jayant Project was judged as one of the best projects in Coal India. It is in this context that I feel that the suggestions made in earlier paragraphs will be of assistance in supplementing the efforts being made.

CONCLUSION

1. What Gets Measured get done?
2. If mining companies are measuring productivity in terms of OMS, they will keep a tag on manpower numbers, may be at the expense of inefficiencies else where. Hence equipment productivity and investment productivity are also important.
3. Coal India can easily increase profit mergin by raising selling price and not improving productivity. It may survive in this situation on a short term basis. In the long term there will be competition from Pvt. Sector and imported coal.
4. Productivity = Good management.

Total productivity measurement systems are required to be adopted. There must top management support. The measurement should be inderstable.

The vision of "Productivity" in the next millennium in the mineral sector has to be improved with dedicated efforts so as to compete with world market.

CHAPTER 40

Productivity Gains through GPS Technology : An Opportunity for Indian Mining Industry

P.S. Upadhyaya and H.C. Dahiya***

1.0 SYNOPSIS

Satellites offer chance to a wide range of industries to improve their operating efficiency through better communications. After it's proven success in communications, I.T. Industry and Maritime Sector, the satellite technology offers the Indian Mining Industry a key to potential productivity gains.

GPS Technology is one such application, which is extensively being used in surface-mines world over. Indian Mining Industry is new to this technology, but does not lag behind the international mining community. In India, M/s. National Aluminium Co. Ltd. (A Public Sector Mining Co.) is about to commission one such system at its fully mechanised surface Bauxite Mine, targeting the potential gains through better Fleet Management and Equipment Health Monitoring.

Surface mining contributes a major share of the total mineral production in our country. A substantial production of coal, copper, lead and zinc and almost total limestone and lignite comes from surface mines.

Productivity gains through application of GPS technology in a typical surface mining environment is discussed in this paper.

2.0 GPS AND ITS FUNCTIONING

2.1 What is GPS?

GPS i.e. Global Positioning System is a satellite based navigational, positioning and surveying system, which measures the position and velocity of a target on the earth's surface

*DGM, Engineers India Ltd.

**Dy. Manager, Engineers India Ltd.

(say in a mine) by using radio- signals emitted by multiple satellites. The system basically calculates the distance of the target from various satellites in the system based on Triangulation Principles. A minimum of four satellites are required to provide the data to solve a four-variable equation involving x, y, z and Time.

Two such systems available globally today are USA's Navigation Satellite Timing and Ranging (NAVSTAR) and RUSSIA's Global Navigation Satellite System (GLONASS) giving coverage all over the globe regardless of time or weather conditions. USA's NAVSTAR system became fully operational in 1993. It is maintained by US Dept. Of Defence, and consists of 24 satellites. Each satellite completes an orbit every 12 hrs, and are so spaced that at least five are visible to users at any point of time. Similarly, GLONASS is maintained by Russian space forces. This system also consists of 24 satellites, and operates in a slightly lower orbit thus completing orbits approx, every 11 hrs. The spacing of GLONASS satellites also ensures five satellites in view at a time.

The use of above GPS systems ensures positional accuracy right down to the centimetre. The combined NAVSTAR and GLONASS application (48 satellite constellation) ensures that 11 satellites are usually available anywhere in the world. More availability of satellites means better reliability and accuracy than using a single system.

2.2 How GPS Works? (Refer Fig. 40.1)

There are different ways in which GPS is employed: single point v/s relative positioning, static v/s kinematic positioning, and real-time v/s post positioning. A mining environment typically requires relative positioning or differential technique for greater accuracy, kinematic positioning for moving equipment, and real-time positioning for instantaneous feed-back. In a typical Mining GPS application, a stationary ground base station/reference is set-up, which is a precisely surveyed point i.e. its latitude and longitude are known accurately. This base-station relates each satellite's position and time to a common reference. Thus, the GPS receiver at base station can calculate the errors in the signals it receives from satellites and transmits corrected co-ordinates to field receivers.

For a mining application, the equipment tracking is made possible by creating 'virtual' beacons for the key mine-locations like various benches, road - intersections, crusher, maintenance- shop, etc. Each virtual beacon is a database entry that contains X, Y and Z co-ordinates of a location and a defined circle of influence surrounding the location to let the system know when equipment enters or leaves the location area. The information of virtual beacons' network is stored in local memory on-board the mining equipment.

During the operation, GPS receivers on the equipment continuously 'position' themselves using the GPS satellite signals. The on-board communication processor transmits the positional data at regular interval (say every 30 sec) to the central computer at control room or base-station, which uses the data in its assignment decisions.

The positional data is stored in a data base, which can be further used for:

- showing real-time (current) equipment location.
- taking historical scatter-plots.
- full-shift replay of equipment movement.

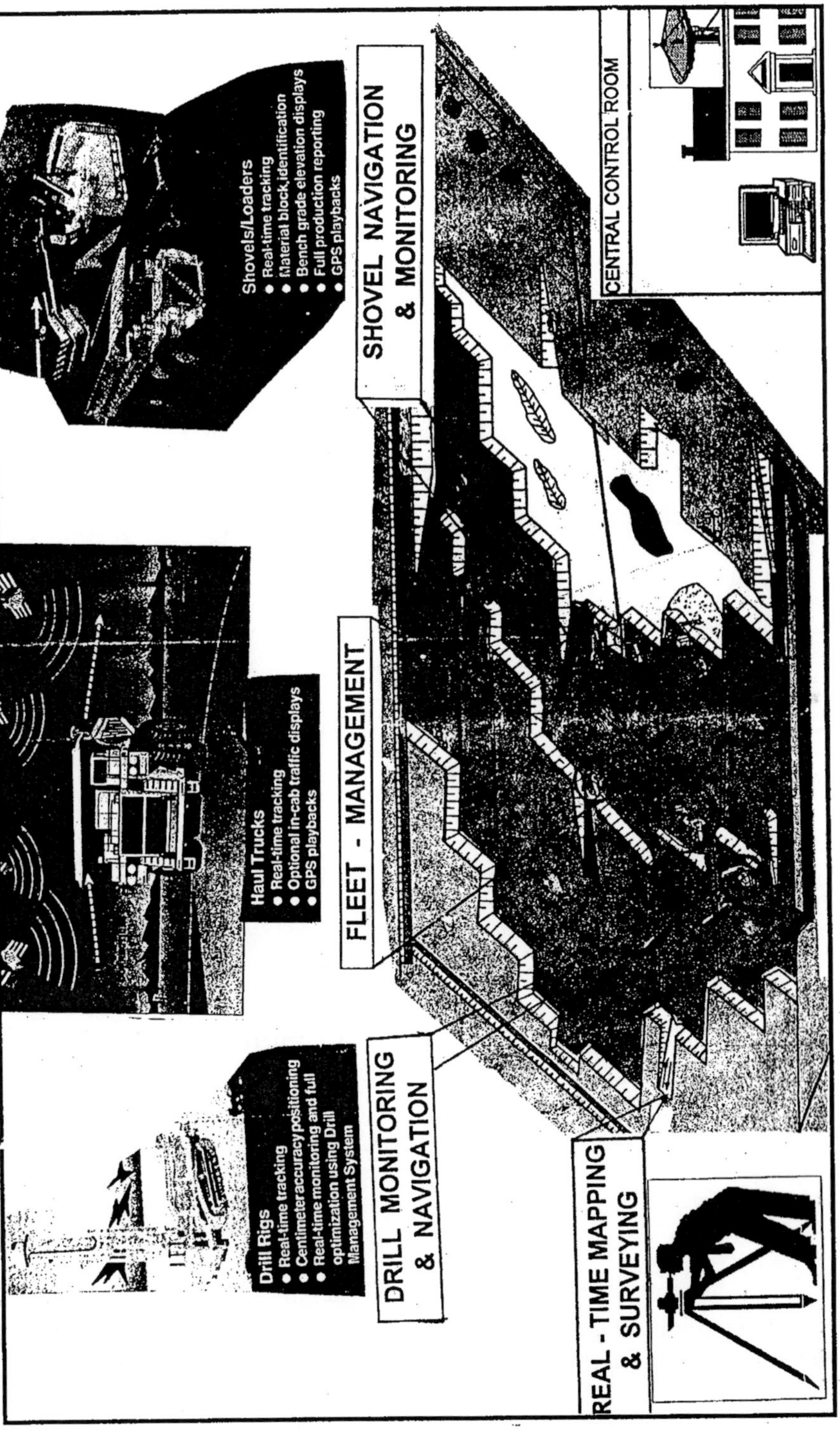

Fig. 40.1 : GPS Application in a Surface Mine

The above information is used for planning of optimal auto-dumper assignments, shovel-advances, dumping patterns, grade control, road-use planning, accident investigations and analysis, equipment-maintenance and comprehensive production reporting.

3.0 AIMS AND AREAS OF GPS APPLICATION IN SURFACE-MINING

Mining is a tricky and capital intensive business, where investors cannot afford inefficiencies. The Management would like it's 4M-resources (i.e. Man, Machine, Material and Money) to be utilised as efficiently as possible.

Now-a-days, technical advances are changing the business strategies of mining companics. Global Positioning Systems (GPS) is one such application, which has become an invaluable tool to mining operators to achieve the objectives of better efficiency and productivity in their operations.

As discussed earlier, the use of proven GPS technology gives a reliable, accurate and real time 3-D position of mobile mining equipments like dragline, dumper, dozer, motor-grader and drills in a typical surface-mine.

The **AIMS of GPS APPLICATION** in Surface-Mines are to achieve :

- Improved resource/equipment utilisation and productivity.
- Accurate blending and quality control to meet requirements of the process selected.
- Improved safety standards.

GPS TECHNOLOGY has been applied in the following mining areas :

- Performance and Health Monitoring of Equipments
- Navigation
- Mapping and Surveying
- Simulation Studies.
- Customised Reporting

In summary, the user can expect one of the two benefits :

— a substantial productivity gain with a fixed equipment fleet

or

— a significant reduction in the equipment fleet to meet required production, apart from better quality control and safety.

Performance and Health Monitoring of Equipments

Production monitoring systems are used to measure the performance of operating equipments and personnel in real time. The on-board electronic-sensors and software record and transmit the equipment performance data to base station/control room of the mine. For example, Shovel sensors give load data for each bucket pass, per- shift or per-day performance. Drill performance monitor gives data for bore-hole location, depth, penetration rate, specific energy, meterage drilled per shift and cumulative along with logging details of rock-types encountered. Penetration rate of the drill and shovel-digging can be controlled using the monitored data as explained later at para 4.0.

The equipment health is monitored by placing sensors at strategic locations, which detects the functioning parameters. A typical engine-health monitoring system on HEMM can give the following information/signals:

— Engine- coolant Temperature

— Engine- oil Pressure

— Engine rpm

— Transmission- oil Temperature

— Hydraulic- oil Temperature

— Fuel Level

Vital signs data and alarms are automatically collected and transmitted to base station for real-time analysis. This data gives an advance warning for planning of preventive maintenance of the equipments and hence, reduces time and production losses due to untimely breakdowns of production/support equipments. Thus, GPS makes 'Predictive-Maintenance' a reality.

3.2 Navigation

This application is typically used in conjunction with computerised truck despatching to track movements of mobile equipments during the operation-shift. Knowing the precise equipment-position is important for generating optimal dumper-allocations. This application can also be used to check travel routes and work progress of all auxiliary mining equipments like dozers, graders, water-sprinklers or maintenance trucks. From safety point of view, portable temporary GPS units can be installed on external vehicles entering the mine, so that the exact position of the same is known during its stay inside the mine.

The navigation system enables the machine operators to locate the machine via a real time display on screen. Thus, it allows accurate navigation from blasthole to blasthole for drills, or maintaining a predetermined bench gradient or face position for shovel/loader as per the mine-plan/schedule, with an accuracy to within few centimeters of design co-ordinates. The tracking or positional data is stored in the database, which can be used for mine-planning purposes like dumping-patterns, road-use planning, water-sprinkling requirements, etc.

3.3 Mapping and Surveying

Mapping and Surveying is an integral(and the most important) activity in any mining operation, especially its' speed and accuracy. In general practice, periodic surveying for mine plan-updation and reconciliation takes many days for field and table work involving the entire survey-crew. The decision maker/planner has to wait for the survey to be done, brought back to office, process the data and feed to the mine-planning package. This process is time consuming, disruptive to the survey schedule and often too late for the actual field requirement. Using GPS technology, a single person can operate the survey package with centimeter level accuracies in real-time. As line of sight is not necessary with GPS, surveying is possible in fog, snow and even darkness. Such systems are fully compatible with various mine-planning softwares i.e. field data collected can be directly fed into the mine-planning softwares for plan-updation, etc. This results into increased speed of data acquisition, updation and co-relation of mine-progress database with minimum personnel for field and table work.

3.4 Customed Reporting

Accessing and using key production information is critical to achieving efficiency and success of any mining operation. GPS can provide a comprehensive array of standard/ customised production reports at any point of time or periodically for management-reviews and decisions, which help in taking timely decisions.

4.0 USES OF AND GAINS THROUGH GPS

As mentioned earlier at para 3.0, GPS has the potential to revolutionize mining productivity through optimal utilisation of the 4 M-resources (i.e. Man, Machine, Material and Money). Uses of and Gains through GPS in some of the mining areas is elaborated in the following paragraphs.

4.1 Drilling with GPS

Use of GPS practically eliminates the requirement of surveying for drilling operation. It enables the drill operator to precisely position drill over blastholes without initial surveys. Blasthole layouts are downloaded directly into the On-Board- system. The operator is then guided by the on-board computer display for hole to hole movements. It provides accurate 3-D positions with respect to the ground surface, and compensates for irregular bench floors. This leads to reduction in over/under drilling resulting into better fragmentation and even looser muck-profiles.

Drill on-line monitoring system enables automatic transfer of drill-hole data like hole-position, hole-depth, specific energy requirements, rock-profiles etc. directly to the explosive charging trucks. This results into efficient blast designs and reduction in time and manpower requirements, thus increasing the overall productivity of drilling and blasting operations.

4.2 Loading and Hauling with GPS

The use of GPS along with computerized truck - despatching leads to optimisation of the dumper-shovel-loader utilization for given production targets and hence productivity of the fleet is improved. The shovel-application leads to better control over digging patterns in terms of selectivity based on ore grades or limits. The shovel-positioning system gives precise information for each block it is digging. This allows the operator to verify the correctness of the material at all times, at night or in inclement weather. This means less dilution of ore at boundaries or contact zones. Thus, quality control and blending requirements can be managed by using GPS. Also, due to real-time colour graphic display of elevation of the equipment on-board in relation to target bench elevation, the operator can maintain a predetermined bench-floor level or face-position. GPS navigation/tracking application is used to physically locate the machines and optimise the dumper assignments to waiting shovel/loader, thus lowering the queuing time and increasing the overall fleet productivity.

Using the on-board load sensor with GPS, the accurate measurement of volume excavated per pass, per shift and accumulative can be directly recorded and transmitted to the data base for automatic updating of mine face-position and mine-plans. The direct transfer of mine plans from mine-planning software to the machine's on-board computer gives access to current mine plans within minutes of their creation/updation.

All these information are used to optimise and control the machine's activities, eventually leading to robotics and autonomous operation of mining-equipments.

4.3 Auxiliary Activities

The mining application of GPS is not limited by the type of surface mining equipments. GPS can be utilized for most of the auxiliary mining activities like dozing, grading, water-sprinkling on haul roads, explosive-charging and maintenance. The dozer-positioning system allows operator to orient in the work-area, displays all key operating parameters e.g. planned v/s actual slope of the road under preparation. The system can be used in accident investigations by being able to show times and locations of involved equipment.

5.0 AVAILABLE GPS SYSTEMS AND THEIR APPLICATIONS

GPS technology is fast becoming an important mine-management tool in surface mines of developed/developing countries e.g. USA, Canada, Australia, South Africa, where competition and cost of manpower is high. A few systems along with their applications are as hereunder :

(i) M/s MODULAR MINING SYSTEMS INC. offers "DISPATCH" , which includes equipment tracking; shovel, drill and dozer positioning; equipment health monitoring; and customised reporting systems or modules. It manufactures it's own hardware. Its application-sites are spread world over e.g. CODELCO (Chile), Palabora Mining Co. (S.A.), CVRD (Brazil), and NALCO (India).

(ii) M/s CATERPILLAR offers Computer-Aided Earthmoving System (CAES) for its equipments, which uses machine-mounted computers in conjunction with GPS receivers, data-radios and displays. CAES uses systems from M/s Trimble, Aquila, Micom and its own development.

(iii) M/s Integrated Mining System(IMS), which is an alliance of M/s Condor Earth Tech., M/s Gemini Positioning Systems, M/s Thunderbird Mining Systems, M/s Tritronics, offers an integrated package including navigation systems for different equipments (e.g. ORENAV, DRILLNAV, DOZERNAV, GRADENAV); positioning systems, and real-time mapping and surveying systems.

Australian mines (Howlick Blackwater, Tarong, Blair, Athol) are using Tritronic's Fleet Management System(FMS), which is now part of Integrated Mining System(IMS). Peabody's Rochelle Mine in Wyoming will be the first integrated mobile equipment monitoring/navigation site.

Indian mining industry is not lagging behind the international mining community. The first Indian mining site to use GPS is NALCO's Panchpatmali bauxite mine, where Modular Mining System is being commissioned for the new equipment fleet (i.e. for Dumpers and Loaders) under it's expansion programme.

6.0 CONCLUSION

It can be seen clearly that GPS, which was rarely mentioned in the mining industry a few years ago, is fast becoming an indispensable part of today's open-pit mines environment. It is an ideal solution for open-pit mines seeking to gain better control over their operations. Hence, it can be concluded that GPS has the potential to revolutionize mining productivity through optimal utilization of 4 M's- i.e. Resources of Man, Machinery, Material and Money.

REFERENCES

1. Mining Magazine - May, 1998.
2. Mining Magazine - June, 1998.
3. World Mining Equipments- Dec, 1997.
4. In house data.

CHAPTER 41

Taking Stock of Near Misses–Real Concern of the Hour

*A.K. Acharya**

The Oil Industries in India are always susceptible to serious hazards in the form of accidents, fire explosions, property and man hour loses due to the very nature of handling, storing and transportation of petroleum/petroleum products. Day by day the technologies and the process systems have become more complex and that is getting further complicated due to change in human behaviour at the same time. The man-machine relationship has become more critical, making the whole system into a complex dynamic shop floor. Of-course, disasters or accidents may not occur on a regular basis, but definitely there are certain indications or early warnings in the systems which clearly establish the causes for future major occurrences We therefore must not overlook those events considering them as minor issues. These areas should not be neglected or dealt with purely on casual sense. These "near misses" are pointers to the inefficiency of the system or inadequacy of the safety measures. Moreover, they serve as indicators/signs of major accidents/disaster to take place the site. The smaller the ear misses, the greater is the likelihood of something serious occurring at a later stage.

"Accident" is a happening which takes place for not having the foresight and usually results in various types of personal injury and/ or damage to the equipment and property. Although there are many reasons attributable to accidents, the three main causes are the improper attitude unsafe acts and unsafe conditions. Similarly, the "near misses" are also caused due to similar factors. On account of any one of the parameters or a combination may lead to the "near misses" in the system. Various case studies show that "near misses" account for approx. 40% of total incident occurred in India during the past 8 years period. This paper deals with various aspect of lapses in the "MAN-MACHINE INTERPHASE', which results in "near misses".

It is worth mentioning here that near misses always indicate the failure of the system i.e. man-machine interphase. The most important and common aspect in maintaining the

*Suptdg. Engineer (S & E), Oil India Limited

best relationship between man and machine is QUALITY. *Only quality persons and quality machines can ensure a safety system either in running or standby condition.* The message on quality should be passed down the line, making every person accountable for the quality. Quality control can not be the matter of concern for a particular group only. It should be everybody's personal responsibility. Building quality in every aspect of the job is an integral part of the whole job. This should be treated as a crucial factor for deciding the performance of the concerned man. Individual potential should be assessed based on the success or failure in managing not only safety but also taking consideration of near misses happening under his/her jurisdiction. The near misses almost indicate that quality is not being considered to be in the forefront for the people under whose control the near misses have appeared. This approach will be definitely a major contributor to improve the quality at work, thereby averting the near misses. All the concerned people regardless of their position or job description must understand that they have all responsibility for maintaining quality of the system for safe operation avoiding any near misses.

Experience of the people plays a vital role to decide the suitability of the people for the required quality of work. If at all near misses have occurred, experience of lapses on the part of the concerned people will help others not to repeat similar lapses. The more people learn from each other's mistakes, the better. New thinking and new initiatives must be taken up to arrest the future near misses.

Just accepting human error as the cause of the near misses is not enough. This human error must be treated as failure in the man-machine relationship. It must be ensured that it dones not repeat in future. Motivation and training are the real answers to this problem. The Management group must be committed enough to take necessary measures to create or maintain a culture in which constant safety awareness is fostered among the people as a continuous process. The training imparted to the people should be need-based instead of being simply incentive based.

(d) Unsafe Acts

Besides the above, negligence, carelessness and complacency on the part of the people also give rise to many a near-misses, which are termed as "unsafe acts'.

Always there are certain well established operating procedures and practices in the system. However, these are not followed strictly and religiously–may be due to lack of awareness of the consequences of doing the work in a wrong or short cut methods. Most of the temporary nature of works for achieving desired goal set by higher officials lead to unwarranted near misses. People who hurry for meeting targets often land up in a safety state at a later stage. This type of make do acts becomes a major contributor of the future near misses in the system.

(e) Non-Compliance of Proper System

Non-compliance of Hot Work Permit, Cold Work Permit and HT/LT Clearance systems may result in the near misses of the system. While using such permit system, continulus supervision is a must during deployment of contractor's labour at site. Simply issuing such permits for the sake of safety/statutory compliance is not sufficient. The ethics of issuing such permit should be taken in the true spirit. People should never forget the past better experience of the Piphr Alpha Disaster in 1984.

(f) Assessment of Health of the Equipment

Healthiness can only be ensured by proper quality of jobs under taken for the equipment. Specific and periodical checks to confirm the health standard must be followed. Condition monitoring of the equipment should be done to find out the various parameters during running/operating condition of the machines. These findings, test results and checks tell us regarding the recommended repair, overhauling or maintenance of the equipment. Hence quality consideration of the equipment only at the time of installation does not guarantee the safe operation of the equipment for the entire life span. This over-confidence on the machines may lead to the near misses situations. Hence, much more though should be given before the arrival of the near misses. Proactive measures only can lessen the major blows. Prevention is not only better, but cheaper than cure.

(g) Analysis of Failure Rate–an early signal for nearmisses

Equipment failure rate at every point of time during its useful life indicate the reliability of the system. This life is divided into three separate periods. Usually these periods are distinct but sometimes they do overlap on each other.

The are–

(i) Wear-in period

(ii) Normal operation period

(iii) Wear-out period.

Let us analyse one such typical "*BATH TUB shaped failure rate curve*" that an equipment experiences during its lifetime. (See Fig. 41.1)

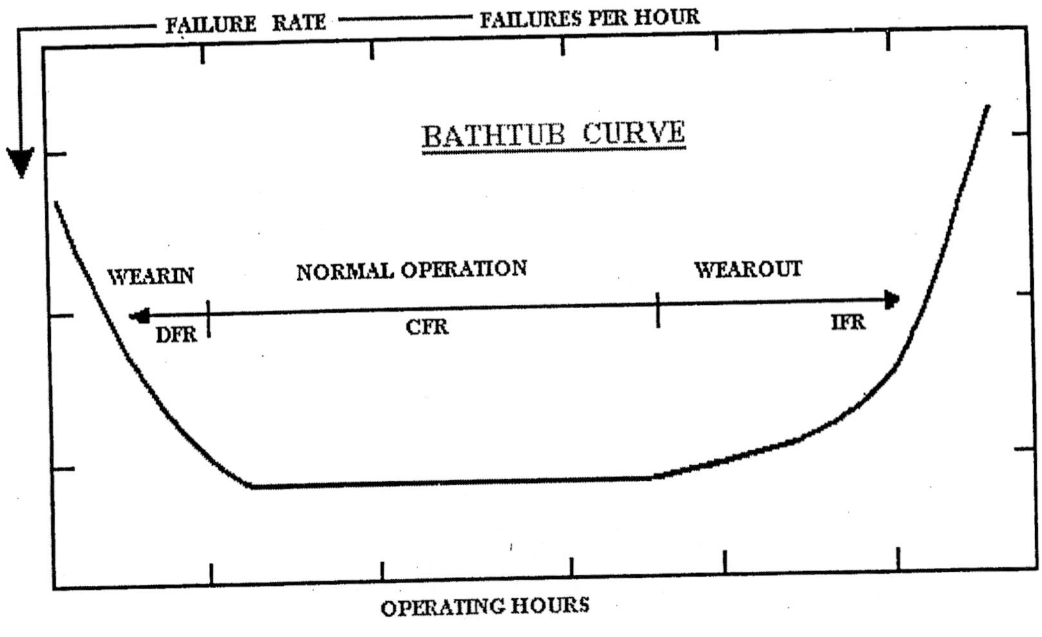

Fig. : 41.1

The wear-in period : (Also known as infant mortality period).

As time passes, the wear-in period is characterised by a decreasing failure rate (DFR) i.e. r(t) decreases. Tomorrow's probability of failure will be less than that of today. This period is usually meat for start-up period of the equipment. Normally the failures here signifies material defects and human errors during assembly or the equipment after being subjected to overhauling has been put into service beginning with a new life cycle with another wear-in period.

The Normal operation period : This period covers the major part of an equipment's life and is characterised by a constant failure rate (CFR) i.e. r(t) dones not change as the equipment ages. The probability of failure w.r.t. time remains constant (See Fig. 41.1).

The Wear-out period : As the time passes, the equipment wears out and material strength degrades. After a long CFR normal operation period, r(t) beings to increase which is represented as Increasing Failure Rate (IFR). Normally the equipment should be overhauled when the same enters into the wear-out period.

Wear-out failures occur when the failure rate is almost inn up ward treat. They are IFR. There is probability of failure tomorrow if it has not happened today. *There are more chances of "near misses" to take place in this wear-out period.* The main causes of wear-out failures are:

(a) Fatique

(b) Erosion

(c) Corrosion

(d) Corrosion–Erosion

(e) Mechanical wear

The failure process can be described by a mathematical equation called weibull distribution :

i.e. r(t) = (k/v) (t/v) k-I

where r(t) : Weibull failure rate

V -> *location parameter* i.e. characteristic age at failure.

For different equipment/items/components operating in a given environment, larger the valve of 'V', the greater the reliability.

Conversely, for a given equipment/item/components operating at different environments, smaller the valve of 'V', more severs is the environment or operating conditions.

An increase in 'V', implies improved reliability *K -> Shape parameter*. Used to Calcuate the variance

i.e. K <I, DFR Wear-in

K =I, CFR Normal operation.

K >I, IFR Wear-out.

For K >I, r(t) increases as 't' increases Fig. 41.2 shows some curves for V = 100 and different values of K.

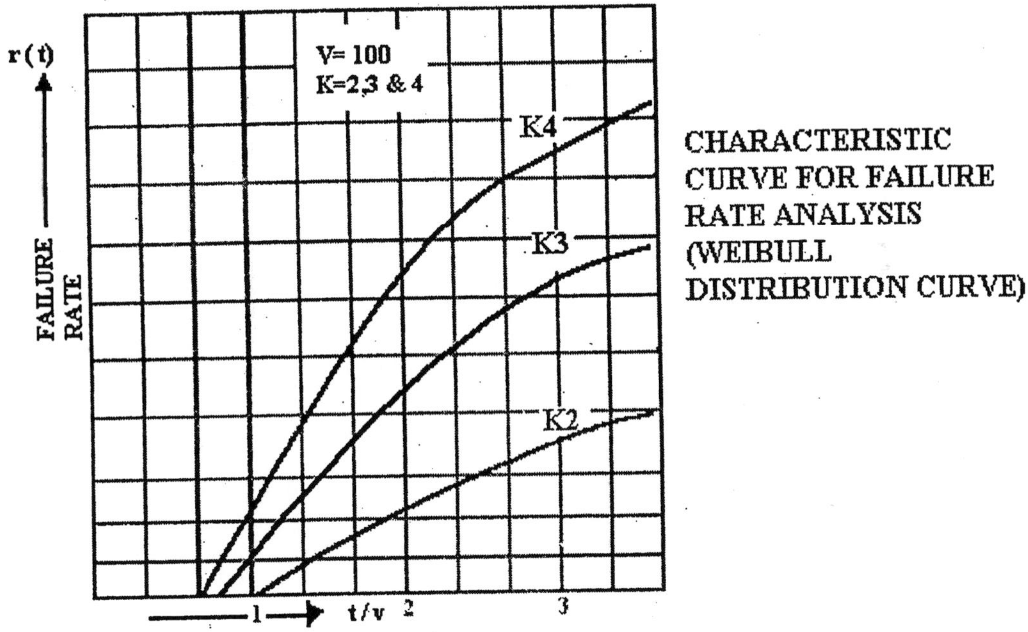

Fig. 41.2

(K = 2, K = 3 & K - 4)

Wibull, a Sweedish engineer originally obtained the above formula from empirical fatigue failure data.

Pipeline leaks also can be represented by the Weibull distribution.

Let us consider that V = 29,000

K = 3.5

then r(t) = (k/v) (t/v) K-I

= (3.5/29,000) (16,094/29,000) 2.5 = 0.0000278 failures per hour

(t) = equipment has not failed constantly till 16,094 hrs, after which replacement was done.

The paper deals with a few examples and the curves which give a idea that how important a role maintenance plays on deciding the factors of near misses. The calculation of failure rate for a specific equipment signifies the risk or chances of probability of failures per hour which can be safely tolerated. This tolerance limit will be equipment/component specific only as per the application/utility and purpose for which the equipment serves.

Always there is a chance of creation of hazardous atmosphere and dangerous situations, especially in the high risk and complex oil industries. Short cuts to meet the targets should not be a practice in following safe operating procedures and adopting Safety Management technique on Oilfields areas. Further, more and more inexperienced Contractors for Man Management services who are quoting for lowest tender price ultimately make the quality of service poor with many lapses. Safety Officers who occupy the back seat can not always fight a losing battle with the Management to allow near misses to go unabated, Maintenance of Equipment should never be treated as a costly affair. Hence, proper failure analysis with a Scientific approach and study of the equipment failure rate must be done as a pre-requisite to installation and operation. Ultimately goal being to control the near misses and to avoid accidents during operations.

Section VII

DEMAND AND SUPPLY PERSPECTIVES AND FINANCING MINERAL SECTOR

CHAPTER 42

An Overview of India's Iron Ore Industry - Strategies for Future

P.R. Tripathi*

1.0 India with its Territory of 3.29 Million Sq.Kms is the Seventh largest country in the World and has vast mineral resources. It ranks Sixth in the Iron Ore production after Australia, Brazil, combined USSR, China and USA with its production of around 70 Million Tonnes in the year 1998 making 7.0% of the total World Production of 1,060 Million Tonnes.

There has been a continuous increase in the production scenario of iron ore in the 90s and it is expected that from the production level of around 54.87 Million Tonnes in the year 1990, India may end with a production level of around 80 Million Tonnes in the year 1999.

2. RESOURCE POSITION

India has abundant reserves of Iron Ore of good quality. The recoverable reserve of Direct Shipping Grade Haematite ore are nearly 9600 Million Tonnes and that of Magnetite around 3100 Million tonnes making a total recoverable reserves of 12.7 Billion Tonnes. There is an ample scope for further increase of these reserves since in many of the ore bodies, there are geological conditions favouring the possible extension of ore at further depth. The Haematite ore is distributed mainly in Bihar, Orissa, Bailadila, Dalli-Rajhara-Rowghat in Madhya Pradesh and Bellary-Hospet Sector in Kamataka and major occurrence in Goa, Reddi area, while the Magnetite is mostly available in Karnataka and to some extent in Andhra Pradesh and Goa area.

3. STATUS OF PRESENT PRODUCTION AND CONSUMPTION

Production of Iron Ore in the country is being achieved through the combination of large mechanised mines (mostly in Public Sector) and smaller mines operated on manual or Semi-mechanised mines (mostly in Private Sector). These operations can be broadly grouped under the following three categories.

*Chairman-cum-Managing Director, National Mineral Development Corporation Ltd. (A Govt. of India Enterprise) Khanij Bhavan, Masab Tank, Hyderabad - 500 028 A.P., India.

a. Captive Production Mines owned and operated by Steel Plants mainly for their own consumption.

b. Mechanised mines owned and operated by Central and State Government Undertaking in Public Sector mainly for export and internal consumption of merchant Steel Plants.

c. Smaller mines owned and operated by Private parties mostly by manual and semi mechanised methods of mining for export as well as internal consumption.

3.1 The production of iron ore which was around 42.2 million tonnes in 1980-81 increased to 54.87 million tonnes in 1990-91, out of which around 40% was consumed for internal domestic use and the remaining was exported. After 1990-91, with the liberalisation of the economy and changes being affected in the Steel Industry of India, when it has been decided that the steel plants could be put up by Private Sector, the ratio of possible domestic consumption and export is likely to have a change. Details of production, internal consumption and exports of iron ore during the last 19 years are given at Table 42.1.

Table 42.1 : Production/Consumption of Iron Ore (1980-81 to 1998-99)

(Figures in Million Tonnes)

Year	*Production*	*Export*	*Domestic Consumption*
1980-81	42.167	22.402	13.847
1981-82	41.085	23.676	15.786
1982-83	42.357	21.737	15.183
1983-84	39.038	22.073	15.249
1984-85	43.131	25.527	15.618
1985-86	44.955	30.150	16.325
1986-87	52.254	28.226	17.069
1987-88	50.924	29.410	17.600
1988-89	49.911	33.040	19.331
1989-90	55.437	35.558	19.740
1990-91	54.870	31.580	23.290
1991-92	57.460	31.540	25.920
1992-93	58.120	28.520	29.600
1993-94	58.670	31.500	27.170
1994-95	64.500	28.000	36.500
1995-96	66.580	31.340	35.240
1996-97	66.600	31.700	34.900
1997-98	75.720	35.610	40.110
1998-99	70.680	31.270	39.410

4. REQUIREMENT OF DOMESTIC STEEL INDUSTRY

The existing production capability of domestic steel industry through primary steel making as well as through alternatives routes of Electric Arc Furnace, Sponge Iron etc combined is around 30.0 million tonnes of crude steel. Out of this, iron ore is basically being consumed in the primary steel making industry to produce hot metal through integrated blast furnace route and the remaining route is consuming iron ore in the form of Sponge Iron/DRI/ HBI.

The combined installed capacity of the integrated steel plants of Steel Authority of India Limited, TISCO and VSP under RINL is around 21.5 million tonnes of hot metal for which the estimated iron ore requirement would be around 34 million tonnes. The remaining domestic requirement is for the capacities of various sponge iron units which at the moment have installed capacity of 5.76 Million tonnes including coal based as well as gas based plants needing iron ore of around 9 million tonnes of specific grades in the case of gas based and normal good grade lump in the case of coal based plants. Number of mini-blast furnaces have also been set up during this decade which have a combined capacity of around 3.5 million tonnes of Hot Metal needing around 5.5 million tonnes of Iron Ore.

To meet the projected increasing demand of steel, number of steel plants have been planned utilising many normal blast furnaces, DRI, Electric arc as well as through mini blast furnaces. The Table 42.2 gives a list of projected plants in the various parts of the Country.

Table 42.2 : Projected Steel Plants in India

S.No.	*Capacity (Million tonnes)*	*Company*	*Status*
1	2	3	4
1.	1.80	Ispat Industries Limited	Under implementation
2.	2.00	ISMT	Under planning
3.	1.25	Neelachal Ispat Limited	Under implementation
4.	4.50	Larson & Toubro Ltd	Under planning
5.	10.00	Tisco, Gopalpur Plant	Under planning
6.	1.15	MESCO Steel	Under implementation
7.	1.25	Malvika Steel, UP	Under implementation
8.	0.38	Bellary Steel&Alloys Ltd	Under implementation
9.	0.25	Elkay Steels Co.Ltd	Under Planning
10.	1.25	Jindal Vijayanagar Steel	Under implementation
11.	0.26	Dalton Steels Ltd	Under planning
12.	0,25	Bhuwalka Ispat Ltd	Under planning
13.	1.50	Mukund Vijayanagar Steel	Under Planning

1	2	3	4
14.	0.70	Mittal Steels Ltd	Under planning
15.	0.18	Dalmia Cement(Bharat)Ltd	Under planning
16.	1.00	Raunaq Industries Ltd	Under planning
17.	1.50	RD Steel Ltd	Under planning
18.	1.00	Kirloskar Steel	Under planning
19.	0.30	Canara Steels Ltd	Under planning
20.	0.20	K.I.O.C.L.	Under implementation
21.	2.00	Usha Iron & Ferro Metals	Under planning
22.	0.60	Sathavahana Ispat Ltd.	Under planning
23.	0.12	Kirloskar Ferrous Ltd.	Under implementation
24.	0.22	Kalyani Steels Ltd	Under planning
25.	0.26	KAP Steel Ltd.	Under Planning
26.	0.30	Kanishka Steel Industries	Under planning
27.	0.14	National Engg.Co.Ltd	Under planning
28.	1.25	NECO Engineering,Raipur	Under implementation
29.	0.60	SJK Industries, Tadipatri	Under implementation

5 REQUIREMENT OF SPONGE IRON AND DRI PLANTS

The details of the existing sponge iron plants and their installed capacity, location and iron ore requirement is given in the Table 42.3.

Table 42.3 : Existing Sponge Iron Plants by 1996-97

S.No.	*Plant*	*Location*	*Capacity*	*Quantity*	*Source*
1	2	3	4	5	6
A GAS BASED DR PLANT					
1.	Essar Gujarat Ltd.	Hazira	1,320,000	L 1,800,000 P 800,000	Bailadila KIOCL Import Maharastra
2.	Grasim Industries	Raigad		L 750,000 P 800,000	Bailadila Import
3.	Ispat Industries	Raigad	1,000,000	L 1,200,000 P 800,000	Bailadila
	Sub-total - Gas Based Plant		**3,070,000**	**L 3,750,000** **P 2,400,000**	

1	2	3	4		5	6
B. COAL BASED DR PLANTS						
1.	Sponge Iron India	Kothagu-dam Andhra Pradesh	60,000	L	120,000 Planned	Bailadila
2.	Hindustan Electro Graphite	Borai, M.P.	60,000	L	120,000	Bailadila-U/I
3.	Tamilnadu Sponge	Salem Tamilnadu	30,000	L	60,000	Bellary-U/I Hospet
4.	Goldstar	Viziana-garam Andhra Pradesh	440,000	L	800,000	Bailadila-U/I
5.	Bellary Ispat	Bellary Karnataka	60,000	L	120,000	Bellary-U/I Hospet
6.	Bihar Sponge Iron Ltd	Chandil, Bihar	150,000	L	300,000	Barajamda-U/I Banspani
7.	Prakash Industries	Champa, MP	375,000	L	840,000	-do-
8.	Sunflag	Bhandara, Maharastra	150,000	L	300,000	Bailadila Planned
9.	Orissa Sponge Iron Limited	Keonjhar, Orissa	150,000	L	300,000	Barajamda Banspani -Planned
10.	Jindal Strips	Raigad, MP	600,000		1,050,000	-do- U/I
11.	Nova Iron Steel	Champa, MP	150,000	L	300,000	-do- U/I
12.	Monnet Ispat	Raipur, MP	300,000	L	600,000	-do- U/I
13.	Kumar Metallurgical	Nalgonda, AP	60,000	L	120,000	Bailadil-U/I
14.	Raipur Steels	Raipur	30,000	L	60,000	-do- U/I
15.	Vandana Ispat Ltd	Raigarh, MP	75,000	L	150,000	-do- U/I
	Sub-total for Coal based plants		**2,690,000**		**5,240,000**	
	Total		**5,760,000**	**L** **P**	**11,440.000** **2,000,000**	

U/I indicates Under implementation

P indicates Planned.

6. IRON ORE EXPORTS

India has been exporting a substantial quantity of around 30 million tonnes which makes up around 8-10% of the total Sea Borne Trade of around 400 million tonnes in the International Market. Because of its geographical advantages being closer to major consuming countries like China, Korea, and Japan, India's position in these countries has been very well established. Taking into consideration the huge reserves of iron ore available and distributed in various parts of the country, it is felt prudent to continue to maintain its share of export since it provides additional export earning in the present competitive age of global economy. It is also important to mention that since the domestic Steel Plants proposed are at various locations, the logistics of the plants may not permit consuming only domestic ore as some of them may find it more economical to import the iron ore from other countries like Australia, Brazil etc. With the present globalisation scenario and liberalisation of the economy, the situation is further going to lend credence to this competitive action and therefore it is essential for the Indian iron ore producers to not only maintain their presence in the established market but also explore avenues to increase the export to new markets which may prefer the indian iron ore brands because of better quality and subject to competitiveness being established.

7. The iron ore industry is directly affected by what happens in the global steel industry. Steel has always been a tough and hard industry because it is at the sharp edge for any major economic change. Since the steel has been highly fragmented and has been under the majority level of government ownership and protection, it has led to widespread capacity and therefore it will be going through a lot of restructuring in the future years. With the above scenario, the Iron Ore Producers in India have to gear themselves for improving the productivity and reducing the cost of the ore to not only their export customers but also domestic customers.

8. FUTURE OF IRON ORE MINING - DEPENDENCE ON INFRASTRUCTURE

There is a good future for the iron ore mining because of increasing steel production at domestic level as has been outlined earlier. If the demand projected by the Steel Ministry is able to materialise, the expected domestic consumption requirement would be around 119 million tonnes in the year 2001-02 and around 125 million tonnes by the year 2005-2006. The existing production capability of the mines could be summarised as in Table 42.4.

Table : 42.4

Mines	Capability
Public Sector	
1. <u>NMDC</u>	
i. Bailadila, M.P.	13.00 Million Tonnes
ii. Donimalai, Karnataka	4.00 Million Tonnes
2. Steel Authority of India Ltd	
Bihar, Orissa, & M.P.	23.70 Million Tonnes

	Mines	Capability
3.	Kudremukh Iron Ore Ltd	
	Magnetite Concentrate & Pellets	7.00 Million tonnes
Private Sector		
1.	Goa	18.00 Million tonnes
2.	Tisco & Others	
	Bihar, Orissa, & Karnataka	19.30 Million tonnes
	Total	**85.00 Million tonnes**

Taking into consideration the above demand of 125 Million tonnes and assuming 90% utilisation factor, a capability of built up of around 140 million tonnes would be required for the domestic consumption itself. If the level of export is maintained even at the current figure of 30-35 million tonnes, the total capacity of around 180 million tonnes would be required to be established in the country.

Sensing the expected increase , various mines are being targetted for opening mainly by the Public Sector and some of the deposits are identified and expected to be exploited are as in Table 42.5.

Table : 42.5

S.No.	*Area*	*Mine*	*Expected capacity*
1.	Madhya Pradesh	Bailadilal 10 & 11 A	5.50 MTPA
2.	Madhya Pradesh	Bailadila 11B	5.00"
3.	Madhya Pradesh	Rowghat	6.00"
4.	Bihar	Chiria	10.00"
5.	Orissa	Malangtoli	10.00"
6.	Orissa	Daitri	4.50"
7.	Orissa	Sundergarh	10.00"
8.	Bellary Hospet	Kumaraswami	7.00"
9.	Bellary Hospet	Ramandurg	10.00"
10.	AP	Ongole Magnetite	3.00"
	Total		70.00 mtpa

The production capabilility if developed above would require a tremendous improvement in the infrastructure covering rail transport and port handling facilities in various parts of the country.

10. INFRASTRUCTURE IMPERATIVES

The economics of iron ore trade largely depends upon the efficient infrastructure support for the movement of iron ore from mine head to domestic consumers as well as for overseas consumers. Nearly 70% of the landed cost at the consumer end is comprising of Railway freight and Port handling cost while the mining cost is only around 30%. Excepting some of the steel plants located close to the iron ore reserves, this analysis will be more or less applicable to most of the newly upcoming plants as well as export trade, barring the deposits of Goa where costly railway transport is substituted by River/Sea transport.

10.1 Railway facilities

Railway facilities which is primarily being operated by the Indian Railways are extremely costly as compared to the similar railway facilities being operated by Iron Ore Producers in Australia, Brazil and South Africa and therefore there has to be definite change in not only providing efficient service but also cutting the cost if the steel industry has to remain competitive. The rail facilities in the Eastern sector of Bihar and Orissa is extremely lacking since the rail line is still not upto Paradeep Port and therefore the export activities of the Eastern Sector iron ore producers is very poor making operation of these mines uneconomic. Situation in the Bailadila Sector as well as Bellary-Hospet Sector is also very unfavourable because in Bailadila sector the railway is charging 30% extra inflated mileage because of the legacy of installation of this line in special Ghat Section with the development of Bailadila mine and the initial traffic being very much limited. The freight charges are increasing every year while the average price in the international market are repeatedly getting reduced so much so that the actual price of the iron ore in US dollars is not firm even what it was in the year 1982. The development of the mine would therefore depend on suitable condition for infrastructure improvement and this matter would be most crucial affecting the development of iron ore mine since otherwise it may become cheaper to import the iron ore from other countries by various steel plants/sponge iron plants.

10.2 Port facilities

Role of the Port having adequate draft to accommodate largest possible vessels is important for reducing the cost in transportation to long distance markets. Currently four ports namely Visakhapatnam, Chennai, Goa and Paradeep are handling iron ore export from India.

While Paradeep can only handle panamax size cargo vessels(60 thousand capacity), Chennai Port and VPT are able to handle cape size vessels(13 0,000 to 140,000 tonnes capacity). Goa port is able to handle bigger vessels beyond cape size since it has arrangement of middle stream loading where the larger vessels after being partly loaded could be topped up. However, with these constraints of limitation of the size of the vessel and further constraints of rate of loading, the competitiveness of the Indian ore for export is severely lost.

It is understood that Chennai Port is likely to be closed for iron ore export and therefore it becomes essential to develop another port in Western Coast where Belikeri could be the most preferable location since it could take care of existing draw backs and provide for bigger vessels upto 250,000 capacity with modem loading arrangements so that the ore from Bellary-Hospet sector could become most competitive.

Improvement has also to be made through Paradeep Port which is under consideration with the private sector participation of Rio Tinto in development of the major mining in Malangtoli and independent operation and opening of separate loading jetti in Dimri.

11.0 TECHNOLOGICAL IMPROVEMENT IN MINING OPERATIONS

Though the infrastructure cost is considerable, still there is a scope for reducing the cost of mining and improving the quality of the product by adopting suitable technology involving proper excavation and suitable beneficiation methods so that the resource position could be further enhanced by reducing the existing 55% cut off grade of Iron values further to lower values. The technological changes could be implemented only when the basic size of working mine is increased. Presently the 5 million tonne per year mine being operated in the India is considered satisfactory size while internationally the mines of 20 to 30 million tonnes like Shishen in South Africa, Carjas in Brazil, Mt.Newmann and Hamersley in Australia are the normal size operations. It is essential to increase the size of the mining operations by sharing infrastructure at the mine head like transportation by conveyors, beneficiation plant etc. which is possible only by combining number of ore bodies adjacent to each other in one mining complex. The main constrain in this direction is because of existing small leases being granted to various small mine operators by different states. This is one of the major impediment in not being able to modernise the mining operations since the lease holdings could not be amalgamated because of ownership reasons. It will therefore be essential to rectify the situation to be able to reap the benefit of higher technological options available to industry.

12.0 CONCLUSION

Iron ore mining in India is a major wealth generating industry for the country. It has a large scope for further improvement both as regards the value as well as the quality of downstream industry covered by the steel. Though some action is being planned for achieving the expected increase, special impetus need be given to the following areas:

(i) The mines have to be upgraded by improving the technology content in mining, mineral dressing and transportation in the existing mines. Though this could be achieved easily in the existing large mines being operated by various public sector units, it will not be possible to do the same in a large number of small mines which are being operated by semi mechanised methods. The authorities responsible for guiding the mineral development like Ministry of Mines (IBM) as well as the State Governments responsible for granting the leases would have to act upon and insist that the iron ore leases have to be amalgamated to ensure an adequate size of the mine block for introducing the latest technology methods for gettng the desired results in respect of quality and cost of the iron ore.

(ii) The railway transportation from the mine to the port especially for the export purpose even as well as for domestic needs, will have to be reviewed in respect to capability and the cost of delivery. It may require major changes in management systems since the transportation by Indian Railways is the costliest as compared to all the existing business operations in the world.

(iii) The new ports may have to be located (like Belikeri in West coast) both in East as well as in the West side with proper ship handling and other capabilities to ensure the least cost and efficient operations.

It is hoped that all those concerned with the growth of the Iron and Steel industry and the economic well being of the country, would be able to give due consideration to the iron ore mining and related issues, some of which have been touched upon in the above paper and strategies will be made and implemented to achieve the real plan position for this sector.

CHAPTER 43

Rock Phosphate Resources in India—An Overview

*T.N. Jaggi**

ABSTRACT

Present food production in India is about 203 million tonnes/year which has to increase to 240 million tonnes/year by the year 2010 to feed the increasing population. Thus fertilizer applications to the crops has to be increased to produce more feed from the same cultivated area which is stagnant since several years.

Phosphate Rock is the only source of P_2O_5 for the production of Phosphate Fertilizer. At present production of Phosphate Fertilizer is only 8.2% of P_2O_5 from the indigenous Rock Phosphate. Thus to meet the demand of P_2O_5, indigenous exploitation and exploitation of Rock Phosphate resource has to increase or else the import of Rock Phosphate will increase to meet. The Rock Phosphate requirement will reach to 6.0 million tonnes by the year 1999-2000 (presently 2.5 million tonnes/year).

India has reserves of 14.7 million tonnes of high grade rock phosphate (+30% P_2O_5) and about 190 million tonnes of low grade rock phosphate with an average of 12% P_2O_5. Brief description of the various rock phosphate deposit, in India has been given with their reserves, locations, beneficiation and utilization patterns and studies. These low grade resources can be beneficiated to produce rock phosphate for chemical processing to produce Phosphate Fertilizer. There is also ample scope of use of new fertilizer products made out of Indian Rock Phosphate to meet the P_2O_5 need of the country.

Apart from these reserves there are number of geological occurrences of Rock Phosphate deposits which could become a big resource. There is ample scope for conducting detailed geological exploration to firm up the various occurrences of Rock Phosphate and exploitation to meet the assured demand of Rock Phosphate of the Indian sub-continent.

*Consultant/Chief Advisor, Oswal Chemicals and Fertilizers Limited. Ex-Chairman and Managing Director, Pyrites, Phosphates and Chemicals Ltd. (A Govt. of India Enterprises) and Past President of Mining Geological Metallurgical Institute of India.

The Techno-Economic use of these rock phosphates for production of Phosphate fertilizer has also been discussed. A lot of business opportunities are feasible.

1.0 PREAMBLE

The present food production level is at **203 million tonnes/year**. India has to substantially increase its food production to **240 million tonnes/year by the year 2010 A.D.**, to feed the expected **1.2 billion population**. This has thrown up challenge of increasing the food production as increase of cultivable area is not feasible.

Present consumption of Fertilizers ($N+P_2O_5+K$) is 16.75 million tonnes of all the nutrients in the (year 1998-99).

One of the major inputs for increasing the food production is to increase the fertilizer input. We are consuming only 84.0 Kg. of NPK/ha (1997-98) whilst the average of the world is 87.1 Kg./ha.

Phosphate fertilizer application is a must for balanced nutrient application for maintaining the soil productivity and fertility. The ideal ratio recommended for India Council of Agricultural Research is 4 : 2 : 1. The present ratio is around 8.5 : 3.5 : 1 (year 1998-99).

Thus in future years, demand for phosphate fertilizers will increase very largely not only for higher demands of fertilizer application but also for correcting the imbalance. Various agencies have estimated phosphate fertilizer of 6.8 million tonnes P_2O_5 in the year 2010, when our food production requirement will reach 240 million tonnes/year.

2.0 RAW MATERIALS FOR PHOSPHATE FERTILIZER

The essential raw material for phosphate fertilizer is Phosphate Rock, which provides the 'P' or P_2O_5. Sulphuric acid, Nitric acid, Hydrochloric acid or Electric energy (electrothermal process) are the other essential materials/inputs for extracting P/ P_2O_5 values from the Phosphate Rock.

Phosphate rock

Phosphate rock is nothing but the name given to mineral phosphate rocks, which in addition to tri-calcium phosphate mineral, has in combination, other substitutes like fluorine chlorine, magnesium, sodium etc. Its usefulness for phosphate fertilizer production depends on its chemical and mineralogical composition. Generally phosphate rock with +31% P_2O_5, Fe_2O_3 and Al_2O_3 below 2%, SiO_2 below 5%, fluorine 3.5%, MgO 0.5% and chlorine below 100 ppm is considered good raw material for production of phosphate fertilizers namely, Single Super Phosphate (SSP) (0-16-0), Di-Ammonium Phosphate (DAP) (18-46-0), NPK complexes and Nitro-Phosphate (NP).

3.0 PRESENT STATUS OF ROCK PHOSPHATE IN INDIA

3.1 Fertilizer Industry in India is utilizing the Rock Phosphate both from indigenous sources and also from imports. The scenario over the last 3 years is given in the Table 43.1.

Table 43.1 : Production and Imports of Rock Phosphate

No.		Year 1994-95 Quantity	Year 95-96 Quantity	Year 96-97 Quantity	Year 97-98 Quantity
1.	Indigenous rock phosphate	883	1065	1022	1159
2.	Imported rock phosphate	2501	2601	2396	2693

Source : Fertilizer Association of India - Fertilizer Statistics

3.2 Demand of Phosphate Fertilizer in India

Demand of Phosphate Fertilizer has been estimated in Table 43.2. This is terms of M.T. P_2O_5 per year comes as below :

Table 43.2 : Production and Imports of Rock Phosphate

Year	Consumption/Demand of P_2O_5
ACUTAL	
1996-97	2977
1997-98	3917
1998-99	4100
PROJECTED	
1999-2000	4305
2009-2010	6800

Source : Actual Figures are published in Fertilizer Association of India - Fertilizer Statistics Project Figures are based on growth rate of 5%.

3.3 Contribution of P_2O_5 from Rock Phosphate

It will be seen from the data of Table 43.1 and 43.2 that contribution of P_2O_5 from indigenous Rock Phosphate is 0.32 million tonnes P_2O_5 whilst total consumption/demand was 3.92 million tonnes P_2O_5 in the year 1997-98. *Thus the contribution of P_2O_5 from indigenous Rock Phosphate was only 8.2%.*

This in the year 2009-2010 projected additional requirement of P_2O_5 comes to 2.88 million tonnes of P_2O_5 i.e. 9.0 million tonnes of 32% P_2O_5 of Rock Phosphate. Thus there is plenty of scope in exploitation of indigenous rock phosphate for the production of Phosphate Fertilizers.

4.0 RESERVES OF INDIAN ROCK PHOSPHATE

Rock phosphate deposits in India are located at 18 places. The total reserves come to about 217.2 million tonnes. The reserves at various locations are given in the Table 43.3.

Table 43.3 : Reserves of Indian Rock Phosphate

('000 tonnes)

Sl.No.	Name of the Deposit and Location		Reserves*	Grade % P_2O_5
1.	Jhamarkotra (H.G.) (A,B,C,D,E and F Blocks)	(Rajasthan)	14.7	30
2.	Jhamarkotra (L.G.) (A,B,C,D,E F, G and A-Ex Blocks)	(Rajasthan)	51.8	15-20
3.	Jhabua - Cherty	(Madhya Pradesh)	14.1	20
4.	Jhabua - Carbonate	(Madhya Pradesh)	12.5	12
5.	Hirapur (Resource)	(Madhya Pradesh)	22.0	7-10
6.	Hirapur	(Madhya Pradesh)	18.7	23
7.	Mussoorie (Medeota and Durmala)	(Uttar Pradesh)	16.0	18-22
8.	Mussoorie (other blocks)	(Uttar Pradesh)	25.0	16-18
9.	Maton	(Rajasthan)	6.2	22-24
10.	Lalitpur	(Uttar Pradesh)	6.3	15-18
11.	Purulia	(West Bengal)	10.5	20-25
12.	Singbhum	(Bihar)	2.5	7-25
13.	Kasipatnum	(Andra Pradesh)	0.5	25-42
14.	Cumbum (Cuddapah)	(Andra Pradesh)	3.5	20-30
15.	Kanpur Group	(Rajasthan)	8.0	15-25
16.	Jhamarkotra (H. Block)	(Rajasthan)	1.8	22-25
17.	Biramania	(Rajasthan)	2.9	10-18
18.	Pondichery	(Tamil Nadu)	0.20	10-15
	Total		**217.2**	**Million tonnes**

(*Source* - IBM-Statistics, GSI Reports, PPCL Estimates and Personal communication)
* Reserves of all categories - Proved, Probable, Possible and Resources.

Classification of Reserves has been done by Indian Bureau of Mines, which gives the position in Table 43.4.

Table 43.4

Fig. In million tonnes

	Apatite	*Phosphorite*	*Total Phosphate*
Proved category	0.6	88.2	88.8
Probable category	11.6	25.8	37.4
Possible	1.0	33.0	34.0
Sub-total	**13.2**	**147 .0**	**160.2**
Resources			57.0
Grand total			**217.2**

4.1 Jhamarkotra Rock Phosphate Deposits

This deposit is located at Jhamarkotra, about 25 kms from Udaipur City (Rajasthan). It has reserves of over 66.5 million tonnes of rock phosphate, out of which about 14.7 million tonnes only is high grade i.e. +30% P_2O_5 and balance of 51.8 million tonnes are low grade - 15-20% P_2O_5.

High grade (+30% P_2O_5) reserves are being mined at the rate of 600,000 tonnes/year and the low grade, associated with the high grade are only being extracted. High grade ore is being crushed to minimum 12 mm size and is soled to the fertilizer industry for producing phosphate fertilizer.

Low grade phosphate rock is beneficiated to produce concentrate with 32-34% P_2O_5, CaO 50.4%, SiO-2 4.5%, Al_2O_3 0.5%, Fe_2O_3 0.3%, MgO 1.4% and F 3.2 + Chlorine 100 ppm.

Mining is being done by Rajasthan Mines and Minerals Limited, by open cast method. Beneficiated ore production of +34% P_2O_5 has started. Plant capacity is 240,000 tonnes per year. This plant capacity is being expanded to 320,000 tonnes per year.

4.2 Jhabua Rock Phosphate Deposits

This is situated in Jhabua district of Madhya Pradesh, about 20 kms from Maghnagar railway station, (about 50 kms west of Ratlam railway station on Delhi-Mumbai highway railway line).

There are two type of deposits at this place. One is Carbonate phosphate rock and second is Cherty rock phosphate. Reserves of carbonate phosphate is 12.5 million tonnes (12.5% P_2O_5) and Cherty phosphate is 14.1 million tonnes (20.0% P_2O_5).

Mining is being done by M.P. State Mining Corporation Limited. Present production is around 100,000 million tonnes/year which contain +24% P_2O_5.

Beneficiation tests have been done both on bench scale and pilot plant scale by IBM (Indian Bureau of Mines). The rock concentrate contain 32% P_2O_5, 45.58% CaO, 12.04% SiO-2, 0.84% Fe_2O_3, 0.46% Al_2O_3 and 1.9% MgO. Recovery of P_2O_5 is 85% in the beneficiation process.

Cherty phosphate, bench scale and pilot plant scale tests have given rock concentration with 35.36% P_2O_5, 46.46% CaO, 10.28% SiO-2, 1.77% Fe_2O_3, 1.24% Al_2O_3, 0.21% MgO. Recovery of P2O-5 is 90%. Flow sheet of a beneficiation plant has been drawn. No beneficiation plant has yet been set up.

4.3 Hirapur Rock Phosphate Deposits

This deposit is situated about 80 kms from Sagar railway station (Centraly Railway) in the district of Chatarpur of Madhya Pradesh State.

The reserves has been estimated by Geological Survey of India. Deposit has a mineable reserve of 18.7 million tonnes, with about 23% P_2O_5 and there are also low grade ore reserves of 22.0 million tonnes with about 7-10% P_2O_5. Both the varieties of ore have got high Fe_2O_3 4 to 10% and Al_2O_3 3 to 6% and SiO_2 - 10 to 30%.

Bench scale beneficiation studies have been done, which shows the low grade ore with 18% P_2O_5 can be beneficiated to produce concentrates with +32% P_2O_5, 40-45% CaO, 10% SiO_2, 0.8% Fe_2O_3, 0.5% Al_2O_3.

Mining of this deposit is being done to produce selectively ore of +30% P_2O_5 with 4% Fe_2O_3, 3% Al_2O_3. This is done by hand sorting. Annual production is about 30,000 tonnes. Mine capacity is 45,000 tonnes/year.

4.4 Mussoorie Rock Phosphate Deposits

Mussoorie rock phosphate deposit situated around Mussoorie town in the districts of Dehra Dun and Tehri of the State of Uttar Pradesh contain a total of 41 million tonnes of rock phosphate with an average content of 15 to 18% of P_2O_5. The deposit occurs at various location like Madeota, Durmala, Bhusti-Jhalikhal, Chipildih etc. The reserves in Maldeota and Durmala total to 16 million tonnes (18-22% P_2O_5) other blocks total to 25 million tonnes (16-18% P_2O_5).

The deposit at Maldeota and Durmala are being worked by underground mining method by Pyrites, Phosphates and Chemicals and Limited (Govt. of India Undertaking). Annual production rate is 130,000 tonnes (+18% P_2O_5). Rock phosphate mineral is entirely being used as straight phosphate fertilizer, after grinding it to 100 mesh power (Application of the type of fertilizer is mostly for crops grown in acid soils).

4.5 Maton Rock Phosphate Deposit

This deposit is situated at Maton, 15 kms from Udaipur city in the state of Rajasthan. Total reserves are about 6.2 million tonnes with 20-24% P_2O_5. Mining by open cast method is done by M/s Hindustan Zinc Ltd. (Govt. of India Undertaking). Present mining rate is 50,000 tonnes/year +24% P_2O_5, however the mine has capacity of production of 100,000 tonnes/year of Phosphate ore.

Beneficiation plant of 600 tonnes/year capacity is also setup. This is not being worked at present, as it is un-economical to work the plant, although process is well established to upgrade the run-of-mine grade (24% P_2O_5) to +32% P_2O_5.

4.6 Lalitpur Rock Phosphate Deposit

This is situated at Lalitpur about 80 kms from Lalitpur railway station (Central Railway) in the state Uttar Pradesh. Reserves are about 6.3 million tonnes (15-18% P_2O_5). This deposit is being mine by U.P. State Mineral Development Corporation Limited. Production is about 15000 tonnes/year. Beneficiation studies (both bench scale and pilot plant scale) show that this ore can be upgraded to +32% P_2O_5 with SiO_2 below 3%, Fe_2O_3 0.5%, Al_2O_3 0.4%. This process has been developed with the aid of Canadian Government and pilot plant tests by IBM has confirmed the technology.

4.7 Purulia Rock Phosphate Deposit

This deposit is situated at Beldih in the district of Purulia, in the State of West Bengal, The reserves total to 10.5 million tonnes (20 to 25% P_2O_5). This is being mined by West Bengal Mineral Development and Trading Corporation Limited. Present production rate is around 6,000 tonnes per year. This rock is entirely being used for direct application as fertilizer in acid soils.

4.8 Singbhum Rock Phosphate Deposit

This deposit occurs in the district of Singbhum of the State of Bihar. This is the form of pebble and is of igneous origin. The occurrence is patchy and is not economical for exploitation. It is not being mined at present. Reserves are about 2.5 million tonnes.

4.9 Kasipatnam Rock Phosphate Deposit

This deposit occurs near Kasipatnam village near in Vizagaptanum in the state of Andhra Pradesh. This is of igneous origin and occurs in the form of veins 0.2 to 0.4 meters thick, inclined at 60° to 70°. The reserves are estimated around 0.5 million tonnes. This is being worked by Andhra Phosphate Limited. Production is about 3,000 tonnes/year.

4.10 Cumbum Rock Phosphate Deposit

Deposit of rock phosphate at Cumbum (Cuddapah) in the state of Andhra Pradesh are indicated at 8.0 million tonnes (15-25% P_2O_5). Mining is being done by Andhra Pradesh Mining Corporation Limited, but as it is low grade, it is not being worked.

4.11 Kanpur, Jhamarkotra (H Block) and Birmania Rock Phosphate Deposit

These are located in the state of Rajasthan and are being exploited by M/s. Rajasthan State Mines Development Corporation Limited. The grade of the deposit is about 10 to 25% P_2O_5. At places where the ore body is high grade i.e. +25% P_2O_5. Total production does not exceed 40,000 tonnes/year.

4.12 Rock Phosphate Resources

Geological Survey of India has reported 16 more occurrences of Rock Phosphate at various places (given in the Table 43.5). These are low grade and further exploratory work is required to assess their economic exploration.

Table 43.5 : Rock Phosphate Resources

1 2	3	4
1. Assam	Bandarigiri, Garo Hill Distt. Batwa-Borghat Khaddum-Lama, Jaintia and Khasi Distt.	Phosphate Nodules - 10% P_2O_5 Phosphate Nodules - 15% P_2O_5
2. Bihar	Bhavanathpur, Palamau Distt.	Phosphate shales - 2.5% P_2O_5
3. Gujarat	Godpur, Jhura Kutch Distt.	Phosphate Nodules - 2-7% P_2O_5
4. Himachal Pradesh	Nigali-Dhar-Korgal, Sirmur Distt Dati Deab, Mahasu Distt.	Phosphate Nodules-10-30% P_2O_5 Phosphate rocks - 2-3% P_2O_5
5. Jammu and Kashmir	Poonch, Rajori, Udhampur Distt.	Phosphate rocks - 2.5% P_2O_5
6. Lakshadwip	Aminidive Islands	Guano Phosphate - 10% P_2O_5
7. Meghalaya	Gari Gill Distt.	Phosphate Nodules-5-10% P_2O_5
8. Orissa	Mayrbhanj and Sundergarb Distt.	Phosphate rocks - 5% P_2O_5
9. Rajasthan	Newania Apatite, Udaipur Distt. Kerpura, Distt. Sikar · Fathehgarh	Apatite 34% Apatite 40% Phosphate - 5-15% P_2O_5

1	2	3	4
10.	Tamil Nadu	Tirchirapalli Distt.	Phosphate Nodules - 21-26.5% P_2O_5
11.	West Bengal	Chirugora deposit Panrkidih deposit	Apatite 6.7% Apatite 20%

5.0 PROSPECTS OF UTILIZATION OF INDIAN ROCK PHOSPHATE

Rock phosphate from various deposits have been tested for various uses. The potential of their use for Phosphate Fertilizer production is as below :

5.1 Single Super Phosphate (SSP)

Production of SSP requires rock phosphate with at least 31% P_2O_5, which on acidulation with Sulphuric Acid, produces SSP containing 16% P_2O_5 (Grade-I). Rock will 28% P_2O_5 will give 14% P_2O_5 SSP (Grade-II). (These specifications for Grad-I and II are fixed for SSP in the Fertilizer Control Order by the Government of India).

5.2 Phosphoric Acid

Röck Phosphate from Mussoorie, Jhamarkotra, Maton, Jhabua, Lalitpur have been tested for production of Phosphoric Acid. These are suitable for production of Phosphoric Acid but the purity has to be increased to at least 28% P_2O_5 with (Fe_2O_3 and Al_2O_3) to be less than 3% and MgO below 1.5%.

Beneficiation Studies done upto pilot plant scale show that rock concentrate of the required purity can be produced.

Techno-Economic Feasibility Studies done by M/s Trebhawan Consultants and Associates, New Delhi for production of Phosphoric Acid from Jhamarkotra, Jhabua and Lalitpur Rock Phosphate show that it is feasible to produce the Phosphoric Acid at the same cost as imported rock around US $ 60 per M.T.(CIF). Since the reserve are low, plants of smaller capacity have to be put up. (Studies were sponsored by M/s Rashtriya Chemicals and Fertilizers Ltd., Modi Enterprises, Lurgi India, M.P. State Mining Corporation and U.P. State Mineral Development Corporation Ltd.)

5.3 Nirtro Phosphate

Low grade rock phosphate from Hirapur and Jhabua have been tested by Norsk Hydro in Norway. Although these rock contain more Fe_2O_3 and SiO_2 but still are suitbale for this route.

Techno-Economic Feasibility Studies done by M/s. Trebhawan Consultants and Associates, New Delhi (Sponsored by M/s National Fertilizer Ltd.) shows that it is feasible to produce Nitro Phosphate (NP) from Hirapur and Jhabua Rock Phosphate at the same cost as than from imported Rock Phosphate at US $ 60 per M.T.(CIF).

5.4 Elemental Phosphorus

Rock Phosphate from Lalitpur and Hirapur (in chips form) are suitable for production of Elemental Phosphorus by thermal process. Rock Phosphate from other deposits are low grade and beneficiated powder is to be pelletised before use. This is not economical.

5.5 Direct Application Fertilizer

Rock Phosphate production, ground to 100 mesh is suitable for use in Acid Soils as phosphate fertilizer. Rock Phosphate from Mussoorie, Purulia, Jhamarkotra, Maton have been tested and approved by Indican Council of Agricultural Research and Government of India as suitable source of 'P' for agriculture and plantation crops in acid soils of PH 6.00 and below.

5.6 Partial Acidulated Phosphate Rock (PAPR)

This is a phosphate fertilizer product, which is produced by partial acidulation of rock phosphate with sulphuric acid. The acidulation is done to the extent of 25%-50% of normal stiometric requirement of acid for complete acidulation. This contains 25-50% the P_2O_5 in the water and citric acid soluble form.

PAPR products prepared from Mussoorie and Maton rock phosphates have been agronomically Tested by agricultural research institute and found that these product can be used in phosphorus deficient soil at any PH as a phosphate fertilizer in traditional agricultural crops etc. High yielding varieties of crops will require completely water soluble phosphate fertilizer, such as SSP, DAP, TSP and Nitro Phosphate. PAPR provide a cheap source of P_2O_5 to the crops. Its use is increasing in many countries of the world.

It will be seen from the foregoing paragraphs that Indian Sources of Rock Phosphate are :

1. Generally poor/low grade in P_2O_5.
2. Suitable for beneficiation, to produce concentrates of 28-30% P_2O_5.
3. Deposits are spread out at various places and thus cannot support a large fertilizer complex.
4. Rock deposits are located inside, away from the sea coast and thus if fertilizer products are produced at or near the deposit for utilization within 300 to 400 kms, of the deposits, cost of P_2O_5 fertilizer from indigenous rock phosphate become cheaper or comparable to use of imported rock phosphate.

6.0 INVESTMENT OPPORTUNITIES FOR ROCK PHOSPHATE MINING

Indigenous rock phosphate reserves in Jhamarkotra, Jhabua and other places have a total reserves of 210 million tonnes (about 30.9 million tonnes P_2O_5). These sources can easily sustain a production of 1.5-2.0 million tonnes P_2O_5/year.

To utilize economically the low grade rock phosphate, following suggestions are made.

1. Beneficiation the low grade rock phosphate to produce concentrates with 28-32% P_2O_5 with P_2O_5 recovery not below 84 to 90%.
2. Utilize the concentrates for production of
   (a) Single Super Phosphate
   (b) Partially Acidulated Phosphate Rock

(c) Rock Phosphate Powder for Direct Application

(d) Nitro Phosphate

(e) DAP/NPK Fertilizers

Techno-Economic Evaluation shows that the above routes are suitable for rock phosphate of Jhabua, Hirapur, Lalitpur, Mussoorie, Jhamarkotra and Maton.

Investment opportunities are possible for the following :

1. Direct Application of Rock phosphate
2. Production of Single supper Phosphate
3. Production of Partially Acidulated Phosphate Rock
4. Production of Nitro Phosphate/DAP Fertilizers

CONCLUSION

It will be seen from the foregoing paragraphs that there is huge requirement of rock phosphate for production of Phosphate fertilizer. The demand will increase by 9.0 million tonnes in the year 2010 AD.

Indian rock phosphate deposits are generally low grade but most of the deposits can be beneficiated to produce concentrates of +31% P_2O_5. These concentrates if used locally for production of Phosphate Fertilizer can be competitive in cost to the imported rock phosphate. Recently Techno-Economic Study done in this regard confirms the financial viability. This approach gives a good scope for business/investment opportunity for exploitation of low grade Indian Rock Phosphates.

CHAPTER 44

New Challenges ahead Beyond 2000 AD in Mining State-of-the-Art for Limestone Mines in Indian Cement Industry

P. Paul, F. I. E., D. K. Panda*, G. C. Mishra* & S. J. Raina**

ABSTRACT

Limestone mining operations in India have been in vogue for quite a long time. The demand of limestone has increased manifold since the last 50 years, especially when the cement and steel industries came into limelight. To feel the importance of limestone mining quantum in our country, this industrial mineral occupies the 2nd highest position next to iron ore in terms of production capacity and 2nd in order next to coal in terms of resources (reserves). The demand of raw material is so immense that unless the technology upgradation for their exploitation is brought forth at a faster pace, the said core sector industries may not be able to achieve their ambitious targets. The technology upgradation so conceived has to be gradually inducted to achieve higher scales of productivity, economics, quality improvement and environmental safety. Technology upgradation state-of-the-art in mines and quarries comprises mainly the continuous surface miner technology, input crushing (mobile/semi-mobile crushers), electric overhead trolley dump-truck haulage, high angle belt conveyors transport, plasma blasting etc. meant for high productivity, increased economics and reduced environmental hazards.

1. 0 INTRODUCTION

The mining sector in our country has been systematically in progress for around 60 -70 years. Coal, iron ore, bauxite ore, manganese, base metals, building materials, clays, limestone and dolomite and other allied materials were being produced on an increasing trend, though some of the major minerals were being mined on a large scale for mass consumption and for export as well.

---

*National Council for Cement and Building Materials New Delhi

Limestone mining in our country has been in steady progress on a regular basis since the core sectors, the main users of limestone, started quite sometime back, even though some small scale mining operations in limestone were in vogue much earlier to cater to the calcined lime kiln industry - a small scale sector operation. The limestone demand is mainly concentrated in the cement industry which consumes around 65-70% of the total limestone production in India whereas the steel sector requires around 15-20%. the balance being consumed in the chemical and other small scale industries. Limestone production in our country constitutes the second largest raw material sector after the coal mining sector. Maintenance of grade of raw material has become the major issue as the high grade reserves of limestone are getting depleted gradually and a strict mineral conservation policy is necessary in the near future either through optimal inpit blending schemes of low and high grades by computerised mine planning techniques or through rational beneficiation methods which though expensive for a low-value mineral like limestone may have to be resorted to in the near future quite scrupulously and under compulsion.

2. 0 BASIC CONCEPT

With the above background in mind, the upgradation of mining technology for limestone mining is more so necessary since the quantum of production requirement of limestone in our country is increasing, at the same time the demand of qualitative requirement is also on the rise with times. In order to meet the ever increasing quantitative and qualitative requirements of limestone for the steel, cement and other sectors, the technology upgradation in the limestone mining sector of our country has already started, though at a belated stage, but it is expected to pick up the momentum by the beginning of the next millennium

3. 0 AVAILABLE MODERN TECHNOLOGY FOR LIMESTONE MINING

In the international scenario, technology upgradation has taken place in several fields, namely :

(a) Continuous Surface Miner (CSM) operation.

(b) Inpit mobile/semi-mobile crushing - conveying system.

(c) Electric overhead- trolley dump truck haulage.

(d) High angle conveyor (HAC) transport. (e) Plasma blasting technique.

Whereas the surface miners, inpit crushers and the hydraulic rock breakers are being gradually introduced in the limestone mines of our country for continuous rock breaking, the other state-of-the-art technologies such as large sized electric overhead trolley-assist dump trucks for high speed, long haul and energy-efficient operations and the high angle belt conveyors for belt length reduction are yet to gain momentum in the mining industry of India. For beneficiation state-of-the-art, dry beneficiation like differential grinding and screening, magnetic separation using high intensity magnetic separators, electrostatic separation, optical sorting and bio-leaching are being tried in some of the limestone mines of our country. Infact a particular cement plant in our country has introduced high intensity magnetic separators as a dry beneficiation technique for separating off-grade material (granite) from limestone, thereby upgrading its quality subsequently.

The subsequent paragraphs in this article will try to elucidate briefly the salient features of some of the above technologies which may be made applicable in the large mines of our country.

3.1 Continuous Surface Miner (CSM)

In the early 70's, road millers in construction work were being used in USA for cutting concrete roads. In coal mines, the operators were using coal ploughs or coal-cutting machines known as cutter-loaders or power miners for cutting coal. Some US companies also developed a few prototype cutting machines by merging the above two equipment, having a frontal cutting or milling drum with cutting tools (picks) all around the circumference of the drum. Thus the lengthy experience in construction equipment manufacture of road millers for cutting concrete/asphalt and the technical know-how obtained from the power miner and undershearer/overshearer technology for coal cutting led to the development of a suitable continuous mining equipment for the 21st century—the versatile Continuous Surface Miner (CSM) for opencast mining operations of which we are much knowledgeable and will be hearing very often in the coming years ahead.

3.1.1 *Application in Limestone Rock Formations*

The continuous surface miners being presently used on commercial basis all over the world are able to cut soft to medium hard consolidated rocks having compressive strengths of upto 50 Mpa, such as coal, bauxite, gypsum, lignite and limestone. For fractured jointed unconsolidated limestones having compressive strengths upto 80-100 Mpa, the surface miners are also able to work quite efficiently. However, research is under progress for cutting slightly harder consolidated rocks having compressive strengths of 60-80 Mpa by suitable modifications in cutting tools (picks) design of the milling drum and also by changing the metallurgical properties of the cutting pick metal.

3.1.2 *Classification of Continuous Surface Miner (CSM)*

Continuous surface miners which are meant for uninterrupted rock-breaking operations, namely breaking/loosening, loading and discharging/conveying the broken material as well as ground surface leveling/dozing all by one single machine are classified under three categories at present

(a) Cutting drum CSM (Wirtzen, Germany and Easi-Miner, USA)

(b) Bucket Wheel type CSM (Krupp, Germany)

(c) Boom type CSM (Voest Alpine, Austria)

(a) Continuous Surface Miner (Cutting Drum Type)

The cutting drum continuous surface miner which was developed in Germany and USA based on road milling technology for concrete and asphalt cutting has a frontal helical-shaped cutting (milling) drum with tungsten carbide cutting tools (picks) fitted on the outer edges of the drum lips. The type, number and spacing of these cutting picks are very important for the cutting operation and speed which again depends on type and hardness of rock. The cutting width of the drum of one particular model which we are using in our country, at present varies from 1. 9 in to 4. 2 in and the cutting height/depth from 0. 15m to 0. 6m, depending on the model, size and capacity of the equipment. The cutting speed varies from 2-4 m/min for soft and medium hard fractured limestones to 7-8 m/min for coal, lignite and other similar minerals in softer strata.

The rock after being cut by the milling/cutting drums picks is referred by the helical (spiral) shape of the drum towards the center of the machine where the cut material falls

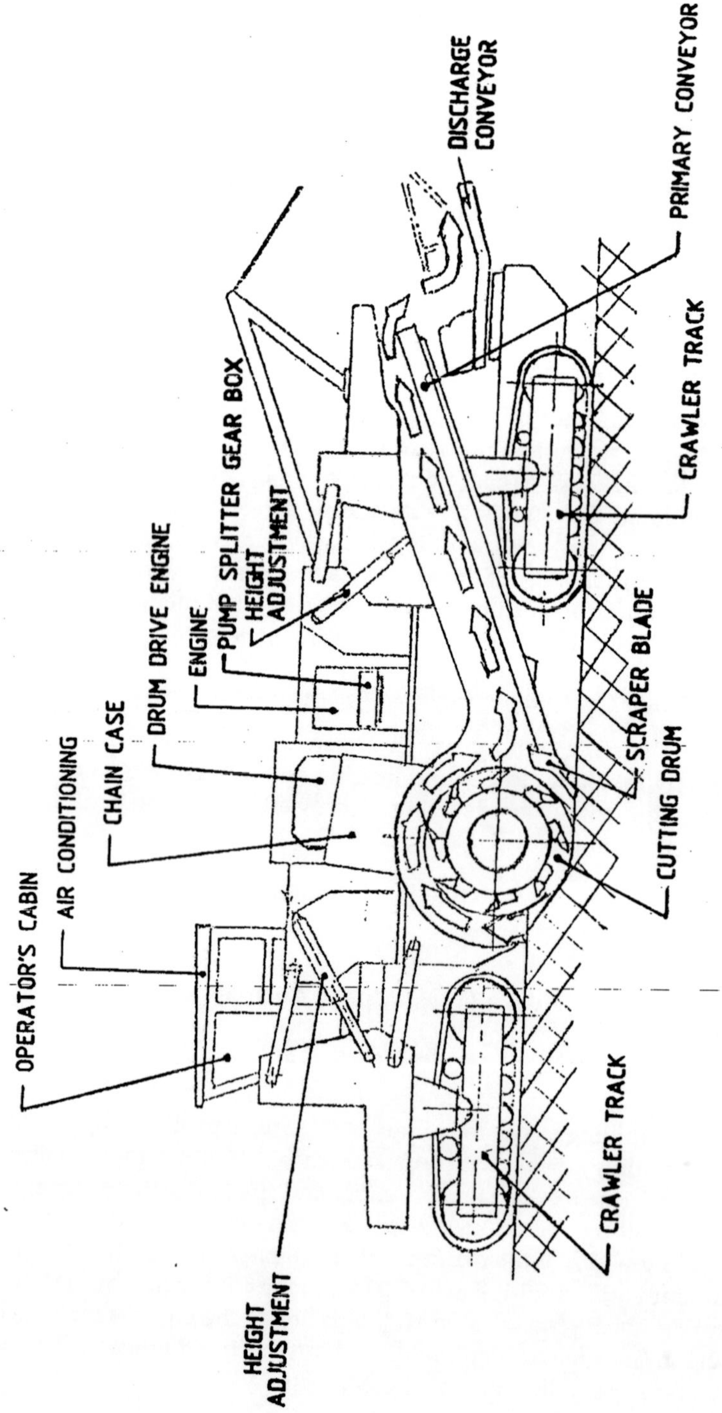

Fig. 44.1 : Surface Miner in Operation

through a chute onto a loading conveyor underneath the equipment main frame from where the material is reported to the end discharge belt conveyor for being off-loaded into the waiting trucks or belt conveyor transport system which comprises a set of mobile transfer conveyor, shiftable belt conveyor and a stationary belt conveyor to convey the material to the user plant. The equipment has 3 to 4 travelling crawlers, each mounted on hydraulically-operated cantilevered arms and a laser-operated radar for night operations to distinguish between mineral and non-mineralised rocks. Fig. 41.1 illustrates the cross-sectional view of a cutting drum surface miner in operation.

(i) Optimum Capacity of Cutting Drum Surface Miner

The optimum capacity of the surface miner depends mainly upon the cutting width, cutting height/depth, cutting speed, rock hardness and type, pick changing time and degree of maintenance and repair and can be determined in the following manner for various types of transporting system :

$$Q_c = \frac{60 \times Q_k \times K_c}{\dfrac{Q_k}{S \times h \times v} + t_i + t_o} \ \text{m}^3\text{/hr (bank)}$$

Truck Transport

where,

Q_k = truck payload (m^3)

S = cutting width of surface miner (m)

h = cutting height/depth of SM (m)

v = cutting speed of SM (m/min)

t_i = waiting for/positioning of trucks = 0.50 min

t_o = manoeuvering of SM = 1.2/n min (minm.)

n = no. of trucks loaded per machine pass

K_e = availability of system

$$Q_c = 60 \times S \times h \times v \times K_c \ \text{m}^3\text{/hr (bank)}$$

Belt Conveyor Transport

where,

S = cutting width (m)

h = cutting height/depth

v = cutting speed (m/min)

K_e = availability of system

The effective capacity of a cutting drum surface miner in various types of rocks is given below :-

- For soft consolidated limestone (less than 500 kg/cm² strength) : Range
  40 m³/hr (smallest model)
  350 m³/hr (largest model)

- For medium hard fractured limestone (less than 800-1000 kg/cm² strength) : Range
  30 m³/hr (smallest model)
  250 m³/hr (largest model)

- For coal, lignite, gypsum, schistose clay, bauxite etc. (less than 300 kg/cm^2 strength) : Range 100 m^3/hr (smallest model) 900 m^3/hr (largest model)

(ii) Method of Working

For techno-economic advantages (high productivity and low operating costs), continuous surface miners are best suited for mineral deposits of longer strike length, preferably more than 500 metres at a stretch due to its typical continuous operation of working. Two box-cuts, slightly larger than the cutting width and length of the surface miner and with a face height equal to the cutting height of the equipment are developed initially one at either extreme end of the deposit. In the initial box-cut No. 1, the surface miner first makes horizontal cuts 1,2 and 3 in transverse direction to the strike length of the deposit (normal direction of face advanced) so as to make adequate space for the equipment to manoeuvre and turn at ease (Ref. Fig. 44.2).

The surface miner then cuts a horizontal strip of rock No. 4, working parallel to the strike length of the deposit in a continuous operation with a cutting width and height equal to the corresponding operating dimensions of the equipment, before it reaches the box-cut No. 2 at the other end where it makes horizontal cut Nos. 5, 6 and 7 in transverse direction to the strike length so as to make adequate space for easy manoeuvring and turning. The surface miner then makes horizontal cut Nos. 8, 9 and so on, strip by strip in up and down movement by fleeting along the strike length of the deposit. The effective cutting speed of the surface miner during operation varies from 2-4 m/min in soft rocks (less than 50 Mpa strength) and medium hard fractured rocks (less than 80-100 Mpa strength) to 7-8 m/min for softer rocks like coal, lignite, gypsum, bauxite etc (compressive strength less than 10 Mpa).

To maintain continuity of operation, it is preferable to use for material haulage a set of belt conveyor modules comprising a number of mobile refer conveyors working in tandem with a shiftable belt conveyor which is laid parallel to the movement of the surface miner along the strike length of the deposit and a stationary belt conveyor laid transversely to the shiftable conveyor near box-cut No. 1 for final material transportation out of the pit (Ref Fig 44.2.)

The mobile transfer conveyors which are of shorter lengths can be effectively used to bridge the gap between the shiftable belt conveyor and the surface miner or between the stationary belt conveyor and the miner. Once the bridge distance between the shiftable conveyor and the surface miner becomes too long for the mobile transfer conveyors, the shiftable belt conveyor is then shifted to its new position (Ref Fig. 44.2). In case of truck transport the discharge belt of the surface miner loads the broken materials onto the dump trucks standing near the miner.

In India, typical block size for a surface miner operation whose capacity is around 200 tph in soft limestones would be 400 m X 200 m. Whilst cutting a block, a rib of 15 m on either side of the block is to be left uncut for easy machine turning which can be later on removed by drilling, blasting and dozing. The SM takes around 3-4 minutes for turning round or steering for the reverse cut. Fig. 44.4 shows a typical working for a quarry by SM at a cement plant in the south. Salient features of the performance data of continuous surface miner for 3 different quarries in India are given in Annexures 1, 2 and 3.

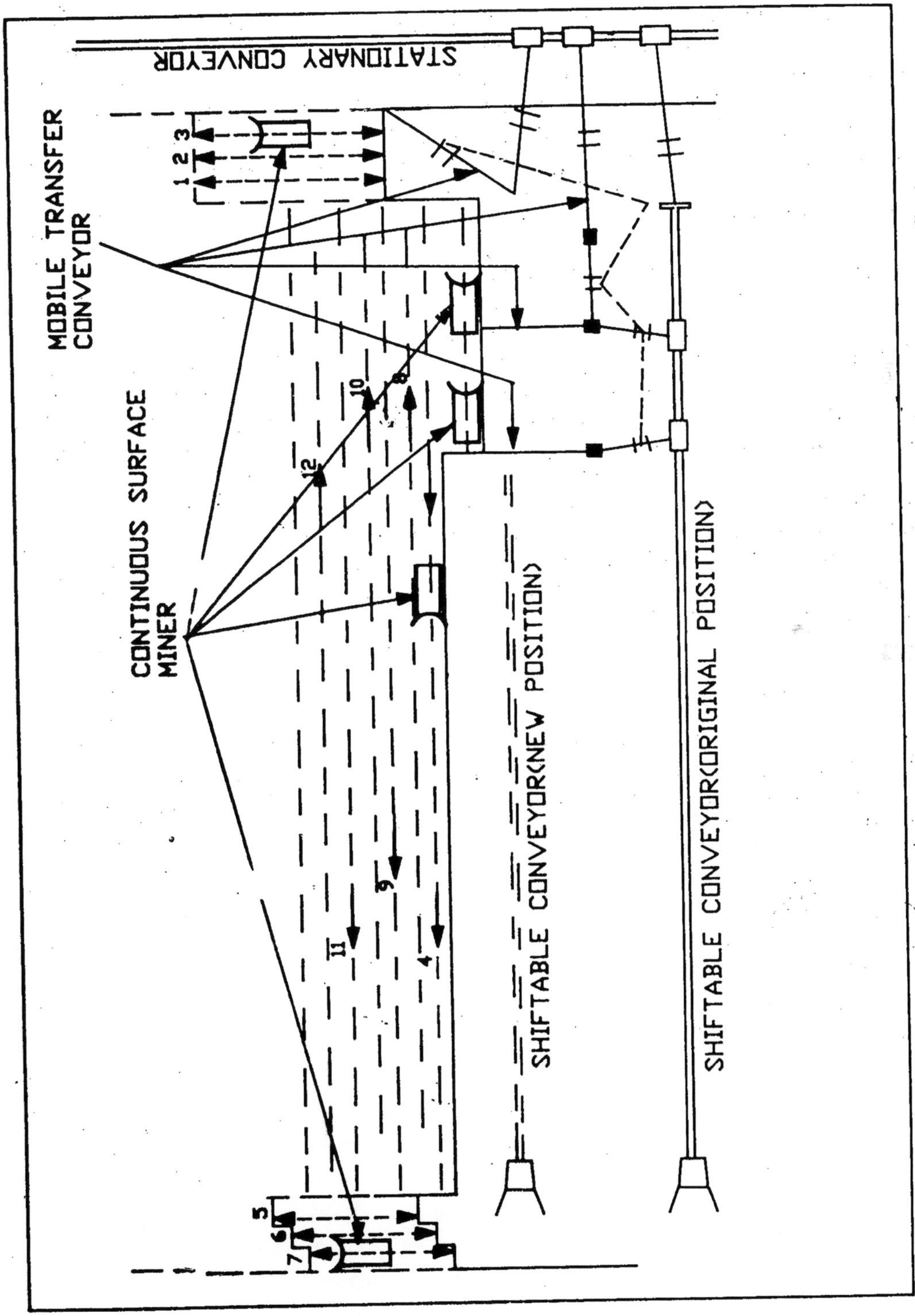

Fig. 44.2 : Selective Mining Using Cont. Surface Miner and Conveyor (Box-cut Method)

(iii) Merits of Continuous Surface Miner Technology

The surface miner is also a good selective mining equipment for quality control purpose. It can cut thin strips of graded limestone upto 25-50 mm thick strata without any contamination/dilution with the contact zone (non-mineralised patch) which can be cut separately and then thrown as rejects. In case of limestone deposits having clay intercalations or high magnesia or silica patches occurring as vertical or inclined intrusions, these can be easily circumvented by the skilful manoeuvering of the surface miner (Ref. Fig. 44.3). The limestone can be cut at first wherever possible and afterwards the waste rock intrusions can be removed at leisure. In case of conventional mining by shovel-dumper combination, such finer selective mining is not possible.

The surface miner primarily cuts material upto 300 mm size pieces, sometimes less than 100 mm size, thereby eliminating primary crusher and sometimes even secondary crusher requirement. The ground surface after being cut by the surface miner is level and trafficable, thereby eliminating use of dozer for road making. The drilling and blasting operations are completely absent in surface miner technology because of direct rock cutting—a big environmental hazard protection. But the major advantage is the large saving in owning and operating costs of surface miner with belt conveyer transport over the conventional mining system to the tune of about 30-35%. Since the surface miner has a long economic life (I 2-15 years) and the belt conveyer system has also an elongated life (20 years), the payback period for the equipment is quick and the return on investment is faster. Another interesting fact of cost economics of continuous surface mining equipment when used with belt conveyor transport is that the total capital investment of the system is marginally lower by around 8-10% than the conventional mining and crushing equipment required for the same production.

(iv) Cost-Benefit Analysis

An analytical study has been made for cost-benefit analysis to compare the capital investment as well as the owning and operating costs of continuous surface miner system with the conventional mining equipment system. The following major equipment were included for our study in both the systems:-

Continuous Surface Miner	*Conventional Mining System*
1. Continuous Surface Miner	1. Hydraulic Excavator
2. Set of Belt Conveyors	2. Rear dump truck (2-2.5 kin length of haul)
3. Bulldozer	3. Front-end loader
4. Workshop service centre etc.	4. DTH drill m/c
5. Secondary Crushing Plant (if necessary)	5. Bulldozer
	6. Workshop equipment
	7. Primary Crushing Plant
	8. Secondary Crushing Plant
	9. Hydraulic Rock Breaker
	10. Portable Air Compressor
	11. Infrastructural facilities like haul roads, service centre, auto garage etc.

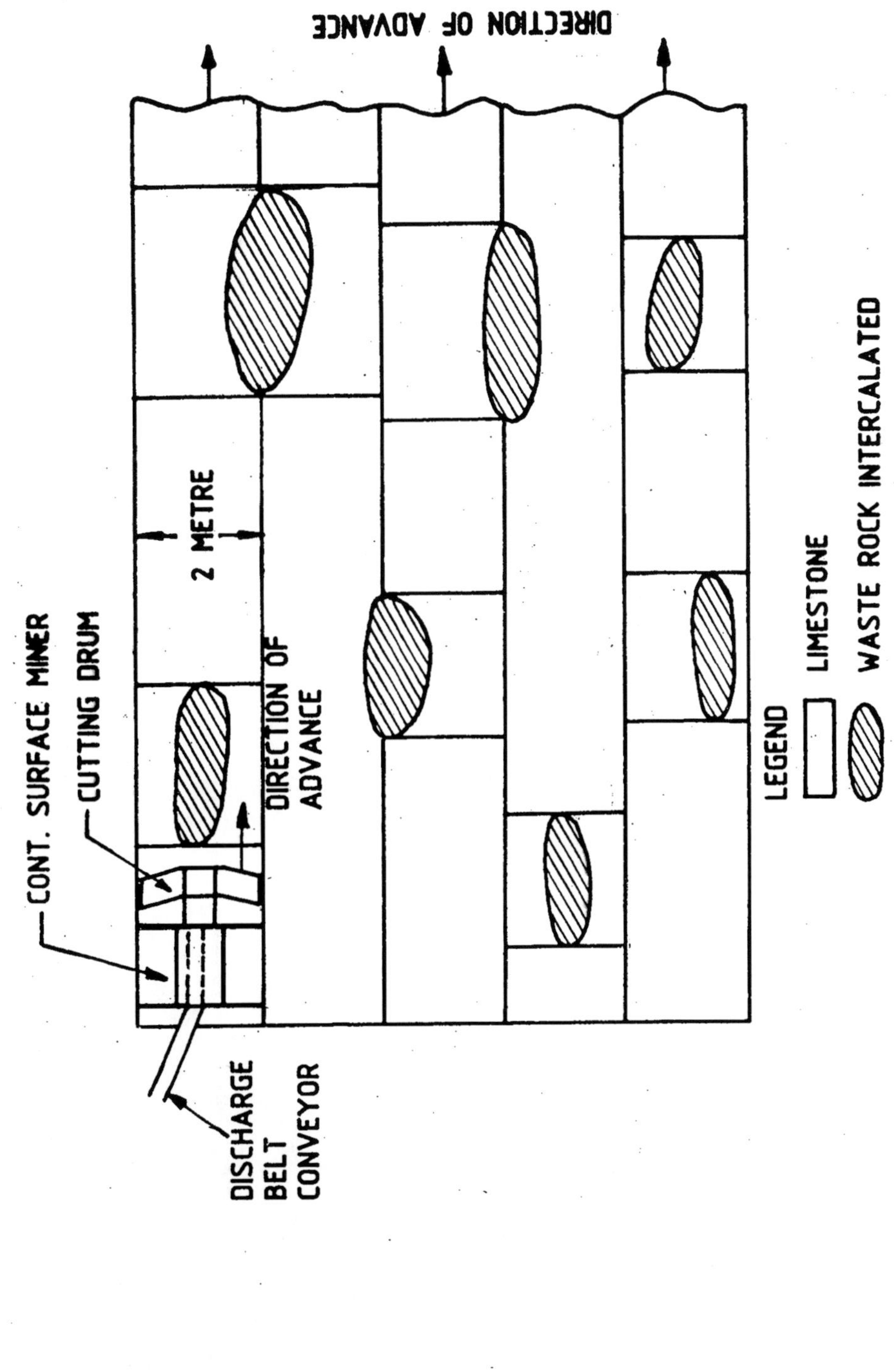

Fig. 44.3 : Plan view of limestone blocks and waste rock intercalated sub-blocks with width of block/sub-block matching with cutting width of surface miner

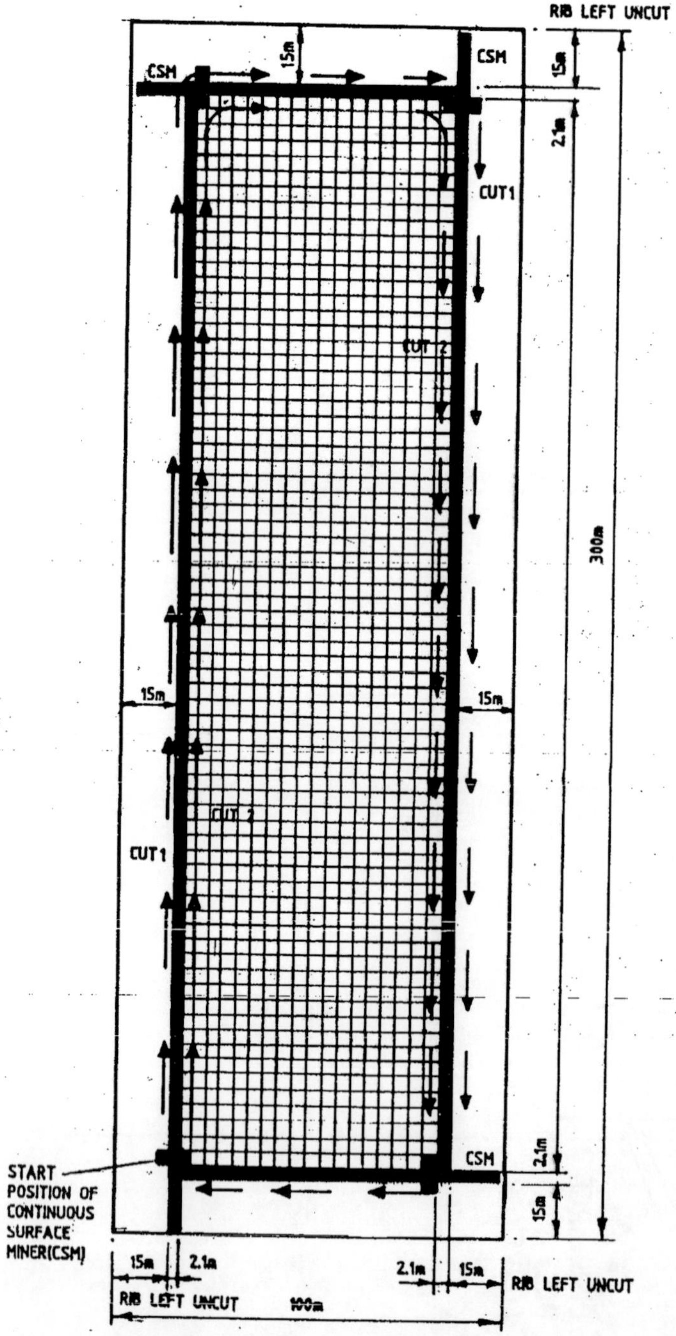

NB: RIB OF 15m IS LEFT UNCUT FOR EASY AND SAFE TURNING OF M/C SO AS TO HAVE ENOUGH SPACE FOR MANOEUVERING

Fig. 44.4 : Method of Cutting by Surface Miner

The savings in capital investments is marginal around 9-10% over the conventional mining equipment system, whilst the savings in owning and operating costs of CSM system is high around 30-35% over the conventional mining system. Thus the payback period and return on investment for the surface miner system is quicker over its longer economic life period of around 12-15 years. Details of an analytical cost-benefit study for large mine producing around 2.0 million tonnes/year of ROM limestone are given in Annexure-4.

(b) Continuous Surface Miner (Bucket Wheel Type)

This type of surface miner is similar to bucket wheel excavator except that it has four (4) large bucket wheels (3.5-3.7 m dia) in parallel in the front with several buckets around each wheel and 5-10 bucket teeth on each bucket. The cutting width is also quite large varying from 5.5 to 7.0 m depending on the model, type and size (Ref. Fig. 44.5). These surface miners of Krupp, Germany are larger than the cutting drum type of Wirtzen, Germany and Easi-Miner, USA and are much more expensive (investment). The nominal cutting height of a mining face is around 2.3 to 2.8 meters depending on model and size of equipment, even though the diameter of bucket wheels are larger.

Bucket wheel surface miners are mounted on 2 to 3 single-floating track crawlers, each fixed on hydraulically-operated cantilevered arms. While the travelling speed is 2-3 km/hr, the cutting speed is around 1-5 m/min (aver--3). The largest model of bucket wheel surface miner can produce around 3 000 m^3/hr whereas the smaller version can produce around 1500m^3/hr. The miner can work in rocks having compressive strength of 50 Mpa and in fractured rocks having compressive strength of 100 Mpa. The loading of material is carried out by two lead-in gathering belt conveyors at the two outer bucket wheels, the main loading conveyor underneath for carrying the broken material of the two inner bucket wheels and the gathering conveyors of the outer bucket wheels and the main discharge conveyor at the end. The drive can be either electric, diesel-electric or diesel-hydraulic.

Method of Working

The method of working is by ramp-cut method, i.e. by cutting a ramp at one end of a mineral deposit, then by cutting horizontal strips in parallel at a height equal to the nominal cutting height of the surface miner (70-75% of bucket-wheel diameter) and then going up a ramp at the other end of the deposit. Hauling units are placed at the higher ground surface level for loading purpose whereas in case belt conveyor transport, the conveyor set modules composing mobile transfer conveyors, shiftable belt conveyor and the stationary belt conveyor are relocated according to the disposition of the surface miner. For mining at next lower level, the surface miner cuts the horizontal strips in a similar manner like the initial cut except that the hauling or conveying units are located on the diate upper level.

The merits and advantages are similar to those of the cutting drum surface miner.

(c) Continuous Surface Miner (Boom Type)

This type of surface miner comprises a long boom, 900 mm long in the front with a spiral-shaped and oscillo-rotating (clockwise) cutting drum having tungsten carbide cutting teeth on its outer circumference lip edges. The cutting width is around 5 meters in one slice with a cutting height equal to 5-6 meters. This surface miner operates its cutting operation from top downwards rotating clockwise and oscillates laterally on one side marginally. The rock after being cut from the top falls onto a set of spinners which carry the material

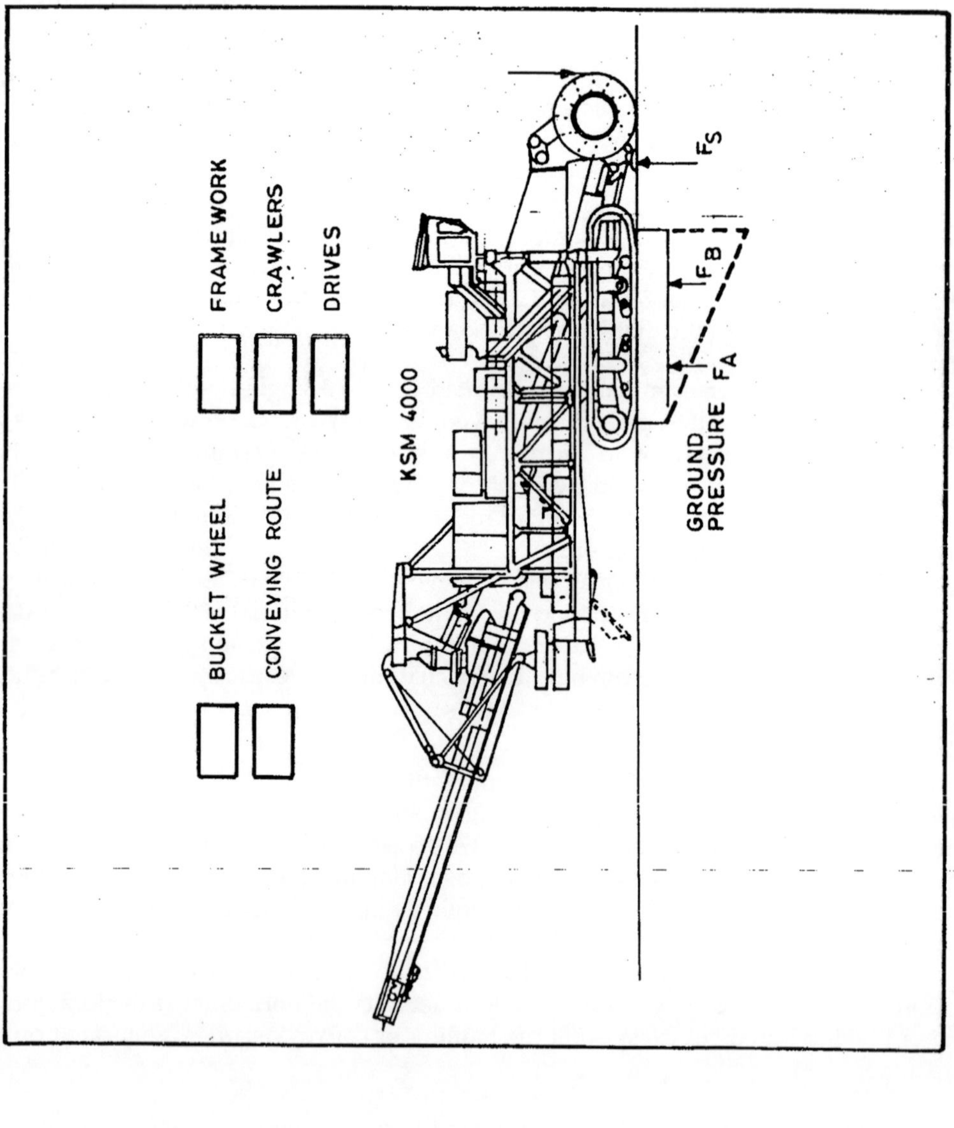

Fig. 44.5 : KSM Main Assembly Groups and Load Distribution

onto a loading belt conveyor which in turn loads the broken rock onto a discharge conveyor at the end. This machine is mounted on crawlers, has a programmable logic controller and is provided with remote-controlled mechanism.

The surface miner can work in rocks having compressive strength of 50 Mpa but is able to cut weak, fractured rocks having compressive strength of 80- 100 Mpa. Its cutting speed in soft consolidated rocks is around 1-3 m/min. It is easily adaptable to an existing (drill and blast operation) quarry where the mining faces are already available for the machine to start- cutting from the existing bench top. The merits are similar to those of the cutting drum type miner.

3.2 Inpit Crushing – Conveying Technology

Inpit crushing-conveying system in open-pit mines for hard rock crushing has become a part of the total quarry planning strategy since it has a close relationship with the excavating and hauling equipment movement and is also economical to operate due to the larger fuel saving on account of less trucks being used due to improved truck cycle operations.

3.2.1 Applicability of Mobile/Semi-mobile Crushing System in Quarry

Inpit crushing can be either mobile, semi-mobile or permanently-installed crushing plant inside a pit. Where the strike length of a limestone deposit is quite long and the grade is quite uniform, mobile crushers (Ref. Fig. 44.6) can be used in conjunction with mobile transfer conveyer, inclined shiftable belt conveyor and stationary overland conveyor for continuous mining operation vis-a-vis high productivity. Where the mine working faces inside a pit are quite a many, it is preferable to use semi-mobile crushing plant (Ref. Fig. 44.7) which can be temporarily set up by powerful hydraulic jacks at one end of a deposit quite safe from the blasting operations, on a lower level so that dump trucks working inside the pit move with material payload down gradient and the mineral reserves on the upper working levels can last for at least 4-5 years. From the semi-mobile crusher location, the inclined shiftable conveyor and the main overland conveyor are set up.

3.2.2 Structure of Module and Merits of the System

The mobile/semi-mobile crusher module comprises the normal mineral processing/ sizing equipment (hopper, feeder, crusher, belt-conveyer, screen etc.) and an expensive transporter carrier on which are mounted the processing equipment. However, this high cost of transporter carrier is offset by minimal infrastructure requirement like civil/ structural, absence of long haul roads for dump trucks, smaller auto-garage facilities due to less dump trucks usages etc. Thus the overall capital investment for mobile/semi-mobile crushers are much less as compared to that for conventional stationary crushing, plant-dump truck system. Also the owning and operating costs of mobile/semi-mobile crushing system are lower by about 35-40% when compared with similar costs of conventional stationary crushing plant.

At third category of which there are several numbers working in our country, especially the cement industry is the permanently-installed crushing plant inside an open pit at the farthest end of a limestone deposit on a non-mineralised zone. From the inpit crushing plant, a belt conveyer carries the material to the user plant, thereby reducing dump truck haulage movement considerably and hence less fuel consumption, less haul road requirement, less manpower etc. In one of the largest limestone mines of a very large

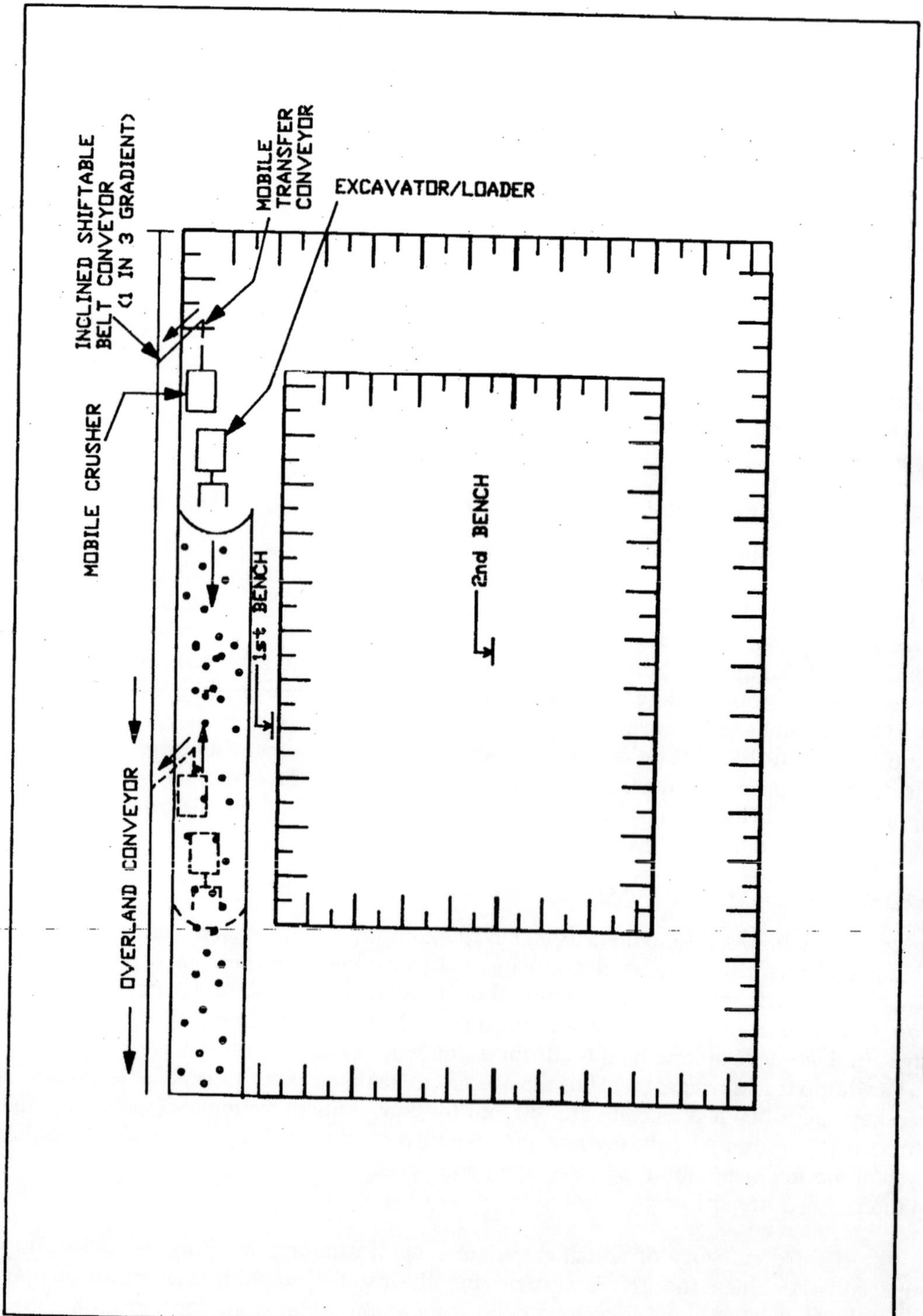

Fig. 44.6 : Mobile Crushing unit (direct loading by excavator/loader)

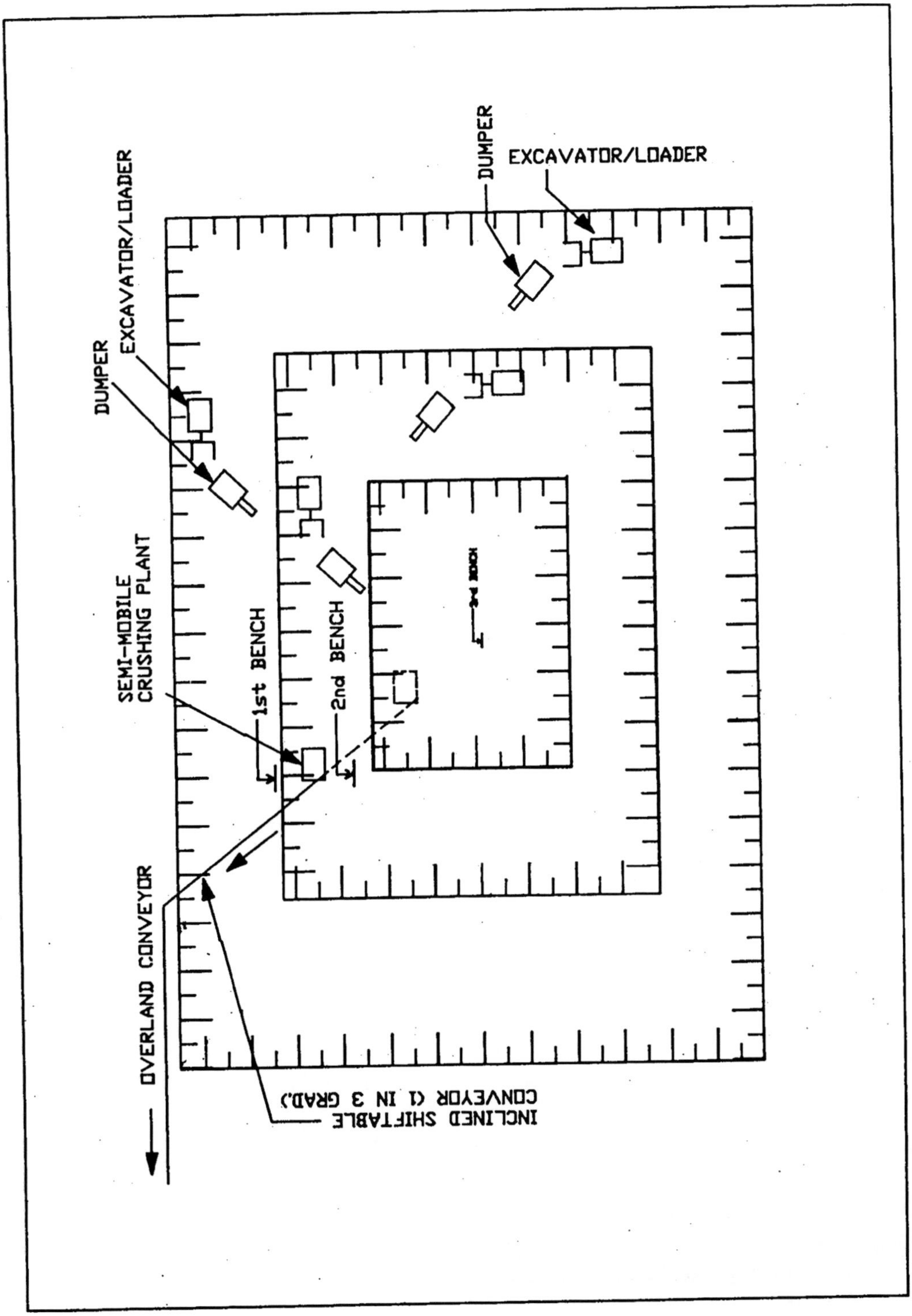

Fig. 44.7 : Semi-Mobile crushing unit (loading through dumper)

cement plant in our country, the production of limestone is around 15,000 tonnes/day at present, working round the clock in 24 hours in a day. This was made possible by installing the inpit crusher which has not only increased the production rate due to the highly productive continuous operation of belt conveyor transport to the plant from the crusher but has also reduced the hauling cost due to less requirement of dumpers. The haul distance of dumpers inside the pit has been completely reduced from 2.5 k.m. to 1.3 k.m., thereby reducing the dumper requirement, fuel consumption and ultimately the hauling cost. Even though the investment on belt conveyors is high, its operating cost is very low (being electrically operated) and the overall operating cost on hauling and transporting by belt conveyor is much lower by around 35-40% than the- cost of hauling the limestone from the pit to the crusher being carried out earlier by dump trucks.

3.2.3 Categories of Transporter Carrier Mountings

The transporter carrier mountings of the mobile/semi-mobile crusher module can be of 4 types, namely skid-mounted (up to 50 tonnes weight of crusher); wheel-mounted (upto 200 tonnes weight); crawler-mounted (upto 1000 tonnes weight); and walking shoe pad-mounted (more than 1000 tonnes weight). Wheel and crawler-mounted carriers are more popular and easily mobile.

These transporter carriers are very capital-intensive and are a critical item of the total modular structure. A comprehensive maintenance schedule is necessitated for this capital intensive unit of the mobile/semi-mobile crusher.

3.3 Electric Overhead Trolley Dump Truck

The electric overhead trolley-assist dump-trucks (Ref. Fig. 44.8) basically comprise a combined diesel-cum-electric mode of operation so that at the moving mine faces the diesel mode is in operation whilst on the main haul road section the DC-controlled electric trolley mode functions freely and at high speeds of over 30 km/hr (average), thereby reducing the truck cycle time considerably, consequently the number of trucks and hence reduction in fuel oil consumption. These trucks are economically suitable for long haul distances of 12-14 lans on steep gradients and ramps and for large tonnages hauled and are thus economically advantageous over the conventional diesel dump trucks of today whose economic, haul distance is not beyond 2.5 - 3 kms and which find difficulty in their movement on steep ramps and slopes.

The electric overhead trolley dump trucks which are available today in capacities of 150 tonnes and above operates a unique static control system of using diesel/alternator and electric overhead trolley control simultaneously without any switchgear both of which are coupled, whilst the DC traction rear wheel motors of the trucks are powered (2900 KW) via a thyristor control panel from the alternator output. This control regulates the flow of power from the trolley line to the alternator so that the engine is also driven, essentially shutting off the need for diesel fuel. Thus the wheelmotors and auxiliary functions like steering, blowers etc. are operated through trolley line source via alternator, hence diesel usage is virtually zero. Another advantage of thyristor control is that the diesel engine is also driven during dynamic braking on downhill haulage, as a result of which the alternator being supplied with AC power acts as a synchronous motor driving the engine. This reduces the load of the diesel engine considerably.

Fig. 44.8 : Electric O/H Trolley Dump Truck

The merits of the electric overhead trolley-assist dump trucks are manifold and are stated below:-

(a) high speeds of electric mode over steep gradients and ramps,

(b) reduced cycle time and reduced number of trucks,

(c) high saving in fuel consumption (over 90%) on steep gradients and ramps due to electric power usage mostly,

(d) excess rear wheel motor capacity,

(e) increased engine and wheelmotor life.

(f) deeper pit operation,

(g) more economical over longer haul distances (12-14 Kms) as compared to the economical haulage distance of 2.5-3.0 Kms of the conventional diesel dump trucks.

Salient features of the operating parameters of electric overhead trolley-assist dump trucks are given in Annexure-5.

3.4 High Angle Conveyor (HAC)

Normally, belt conveyors can carry material loads up a gradient of 1 in 3 to 1 in 4. But today even inclinations of 60° slope for short conveying distances and lifts have been made possible, thanks to the development of high angle conveyor (HAC), also called "sandwich belt conveyer". These belt conveyors are very useful for mobile/semi-mobile crushing systems inside a pit where the crushed material has to be transported out of the pit up the pitwall sides for shorter lifts and lengths.

The system (Ref. Fig. 44.9) comprises two belt conveyors, the bottom belt called carrying conveyor conveying the material and the top belt called floating conveyor as a part of an endless, short-belt conveyor system which hugs the bottom belt carrying the material. The carrying belt is supported on closely-spaced troughing idlers and the floating conveyor belt of the endless conveyor system is softly pressed onto the conveyed material of the carrying conveyor belt so as to prevent back spillage/rolling by closely-spaced fully-equalised standard impact rolls on spring-loaded pressing modules that press down on the return belt of top conveyor.

There are several advantages such as reduced length of belt conveyor thereby reducing investment cost, maximum utilisation of energy for material conveying, use of cheaper electric power as compared to the expensive diesel fuel of dump trucks, longer economic life as compared to truck life (6-8 years), smaller service roads requirement and less maintenance as compared to the larger haul roads requirement and maintenance for dump trucks etc.

3.5 Plasma Blasting

Recent work on an electrical method for breaking rock shows that such technology will be commercially viable at an earlier date in the near future. The method, called plasma blasting, is based on the fast delivery of stored energy to fracture the rock. In the plasma blasting process, the energy is released in the form of an electric discharge into a small amount of liquid in a hole drilled in the rock. The electrical discharge rapidly vaporizes

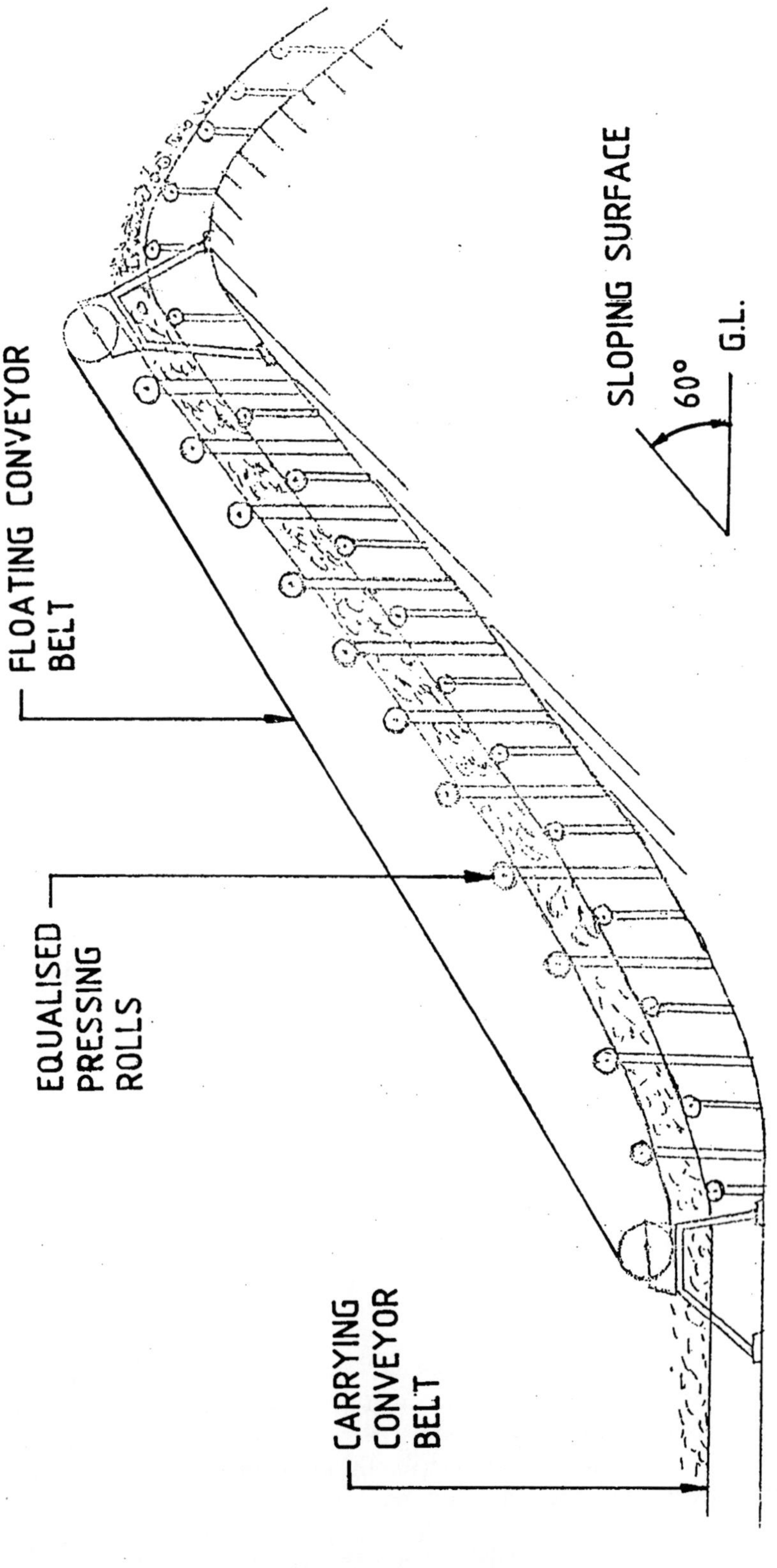

Fig. 44.9 : Side view of high angle conveyor belt

and disintegrates the liquid, thus creating a high pressure plasma confirmed in the rock. The expanding hot gas similar to chemical explosives detonation generates a shock wave that in conjunction with cracks propagating the blast zone fractures the rock.

In this method, electrical energy is stored in a capacitor bank and is suddenly released by a spark gap switch. The power is conducted to the blast region by a heavy dumper low resistance coaxial cable which basically comprises an assembly forming the two electrodes between which the electrical breakdown occurs, thereby terminating the system. This coaxial assembly called blasting probe is inserted in a liquid-filled hole previously drilled in the rock. Ordinary tap water can be used as the blasting fluid in vertical or sufficiently inclined holes, but suitable additives must be added to increase the viscosity when blasting in horizontal holes.

In the first few micro-seconds of electrical breakdown in the liquid, a conductive channel is formed between the inner and outer conductors of the blasting probe. Heat generated in an electric arc vaporizes the surrounding water and induces the system to high temperature. Since the discharge is very short, a steep pressure rise takes place on the walls of the hole, generating a shock wave in the rock. From, studies conducted, the generated pressure pulse is similar to the one as in detonation of chemical explosives. Thus an electrical blast of a given energy will produce equivalent mechanical energy as a mass equivalent to the explosives representing the same energy.

4.0 CONCLUSION

Mine owners and operators in the world are opting now-a-days for higher degree of mechanisation and automation in the mines and quarries, however keeping in mind the human resources factor. The latest mining state-of-the-art and technology that are being introduced for adoption in the long run are being encouraged more and more because of their inimical high productivity and economical parameters, not to forget, ofcourse, their environmental and safety considerations. One such hi-tech of which we shall hear more often by the end of this century is the introduction of the Continuous Surface Miner state-of-the-art and its increased usages in the limestone mines of India, especially for soft and fractured medium hard rocks, so as to increase productivity manifold, reduce operating costs substantially, eliminate some of the environmental hazards like dust, gaseous fuels, noise, flyrocks and ground vibrations generated during drilling and blasting operations. Perhaps this equipment may find its utility also in other minerals, in our country such as friable iron ore, bauxite, coal, etc. Though such equipment are being used in coal, lignite, bauxite, gypsum, limestone. etc. in other countries of the world, in our country the equipment usage has been restricted to limestone at present.

Another upgradation in technology could be the usages of more and more inpit crushing and conveying systems (mobile and semi-mobile crushers) in the mines/quarries, thereby leading to less usages of diesel-driven dump trucks and consequently less fuel consumption, reduced haulage costs etc. Introduction of belt conveyors in mines/quarries reducing thereby the dump truck usages to a minimum especially when used with continuous surface miners or inpit crushers will reduce the haulage/transportation costs considerably because electrical power is cheaper than diesel. A hitech method which may catch up in the earlier part of the next century in our country could be the introduction of electric overhead trolley dump trucks which has the unique system of utilizing both electric and diesel modes wherever necessary. It remains to be seen, however, how our mineral and mining industry would adopt these hi-tech state-of-the-art in the various mines and quarries for greater socioeconomic benefits n the long run.

ACKNOWLEDGEMENT

The authors extend their thanks to the Director General, NCB for giving permission to publish this paper.

REFERENCES

1. D. Bullivant, (1987). Current surface mining techniques, Bulk Solids Handling 7(6), pp. 827-830.
2. R. Sagner and H. Willnauer, (1991). Krupp surface miner - the solutions for continuous surface mining, Bulk Solids Handling 11(1), pp. 185-191.
3. Annon, (1991). Technical Information Literature on Wurtzen Continuous Surface Miners. Wurtzen International-first edition and Wurtzen Brochures, Germany, 25-10.01.0486, pp. 1-7.
4. Annon, (1986). Surface Miners move into the Big League, World Mining Equipment, pp. 44-46.
5. Michael P. Sassos, (1984). In-pit crushing and conveying systems, E&MJ 4, pp. 46-49.
6. Idem. Ibid. J.A. Dos Santos, (1988). High angle conveying - HAC for surface mining, pp. 117-125.
7. Carel Kruger, (1994). ISCOR leads the way with in-pit electric haulage system, Bulk Solids Handling 14 (1), pp. 71-74.
8. Rozgonyl Tibor and Golosinski Tad. Continuous Surface Mining-Equipment Operation and Design.
9. Pasemechmetoglu A.G. and Karpuz C. Mine Planning and Equipment Selection 1994.
10. Wirtzen and Krupp. Technical Literature Continuous Surface Miner.
11. Paul P., Mishra G.C., Panda D.K. : Computer-aided Mine Planning for Highly Mechanised Quarry Operations - Paper presented at AFCM International Technical Conference on Cement Manufacture at Kaulaoampur, Malyasia, 1995.

ANNEXURE-1

Salient Features of Performance Data of Continuous Surface Miner
QUARRY-I (SM 2100 Capacity 180 tph)

1.	Total size of Limestone Deposit	= 800 × 200 m
2.	Workable size of Block for SM	= 300 × 100 m
3.	Rib size left for safety on either side of block	= 15 m
4.	Compressive Strength	= < 400 kg/cm^2
5.	Turning Time for SM for each turn	= 3-4 minutes
6.	Time taken by SM (capacity = 180 tph) (for 300 × 100 m Block)	= 65-70 days in hard rock
7.	Cutting Speed of Drum	= 3-4 m/min
8.	Pick Consumption (High pick consumption due to harder rock condition)	= 3-5 per 1000 tonnes
9.	Cost of Pick	= Rs. 450/- to Rs. 500/-
10.	Working Slope of existing flat terrain (Can work upto maximum 1 in 6 gradient)	= 1 in 100
11.	Working Shifts	= 3 shifts
12.	Maintenance period per shift	= 1-1.5 hrs.
13.	Water spraying on top of cutting-drum for dust suppression	
14.	Capacity of water tank	= 4180 litres
15.	Time taken to load one dump truck (20-t)	= 4 minutes
16.	Size of Product	= 70% less than 75 mm 20% 75-150 mm 10% > 150 mm (Max-300 mm)
17.	Manpower required	= 2 Operators-cum-Mechanic

ANNEXURE-2

Salient Features of Performance Data of Continuous Surface Miner QUARRY-II (SM 2100 Capacity-200 tph)

1.	Total size of Kankar Deposit	= 600 × 200 m
2.	Workable size of Block for SM	= 300 × 100 m
3.	Rib size left for safety on either side deposit	= 15 m
4.	Compressive Strength	= < 300 kg/cm^2
5.	Ratio of Co,pressive Strength to Tensile Strength	= < 10 (Workable upto 13)
6.	Turning Time for SM for each turn	= 3-4 minutes
7.	Time taken by SM (200 tph) for 300 × 100 m Block	= 60-65 days
8.	Cutting Depth of Drum	= 0.2 mm
9.	Cutting Speed of Drum	= 3-4 m/min
9.	Cutting Width of Drum	= 2.1 m
11.	Pick Consumption	= 1-2 (average) per 1000 tonnes
12.	Cost of Pick	= Rs. 450/- to Rs. 500/-
13.	Time taken to load one dump truck (20-t)	= 4 minutes
14.	Working Shifts (Effective working hrs. = 6 hrs./shift)	= 2 shifts
15.	Maintenance period per shift	= 1-1½ hrs.
16.	Manpower required	= 1 Operator + 1 Helper

ANNEXURE-3

Salient Features of Performance Data of Continuous Surface Miner QUARRY-III (SM 2100 Capacity = 200 tph)

1.	Area of Operation (Block Size)	= 400 × 200 m
2.	Time taken by SM (200 tph) for 400 × 200 m Block	= 140-145 days in soft rock
3.	Compressive strength	= < 300 kg/cm^2
4.	Seismic Wave Velocity	= 800-2800 m/sec
5.	Cutting Speed of Drum	= 4-4.5 m/min
6.	Cutting Depth of Drum	= 0.15-0.2 m
7.	Cutting Width of Drum	= 2.1 m
8.	Cutting Drum Speed	= 92 RPM
9.	Drum Dia	= 900-1050 mm
10.	Pick Consumption	= 1-2 (average) per 1000t
11.	Diameter of cutting bit on pick	= Either 12.5 mm or 15.5 mm or 17.5 mm
12.	Spacking of Picks	= 76 on the drum
13.	Pick holders life	= 3000 hours
14.	Pickholder is fitted to a plate which is welded to the drum lip	
15.	Cost of Pick	= Rs. 450/- to Rs. 500/-
16.	Loading Conveyor	= 4 ply or 3 ply rating
17.	Fuel Consumption (600 HP Engine)	= 40-60 litre/hr.
18.	Working Shifts	= 3 shifts
19.	Maintenance Period per Shift (Sunday - Preventive Maintenance)	= 2 hrs. per shif
20.	Time taken to load one dump truck (15t) Time taken to load one dump truck (25t)	= 3 minutes = 4-4½ minutes
21.	Size of Product	= 85% Less than 75 mm
22.	Life of Equipment	= 12-15 years
23.	Manpower Required	= 2 Operator

ANNEXURE-4

Cost-benefit analysis for comparison between continuous surface miner system and conventional mining system for a large mine producing 2.0 mt/year rom limestome

		Continuous Surface Miner System Using Belt-conveyor Transport (2 Shifts Working) (Major Equipment Considered)	Conventional Mining System Using Dump-Truck Transport and Crusher (2 Shift Working) (Major Equipment Considered)
1.	Capital Investment (Rs. in Million)	145	160
2.	Savings in Capital Investment by using CSM	15.00	9-10%
3.	Anual Owning & Operating Costs (Rs. in Million) including Royalty and Cess	100.00	140.00
4.	Savings in Annual Owning and Operating Costs by using CSM	40.00 ((30-35% maximum)	

ANNEXURE-5

Salient Features of Performance Data of Electric Overhead Trolley-Assist Dump Trucks (Source : A Coal Mine in South Africa)

1.	Payload (Present day)	:	150 t and above
2.	Gross Weight	:	260 t and above
3.	Drive Mode	:	Diesel/Electric (Alternator Thyristor Control)
4.	Power Source	:	Diesel/AC and DC Electric Overload
5.	Trolley line Voltage	:	1200 - 2000 volts
6.	Output of diesel engine	:	1600 - 2000 HP
7.	Output of real wheel motors	:	2900 KW
8.	Haul distance capability (economical)	:	beyond 2.5 kms and upto 13-14 kms.
9.	Average speed of electric mode (Twice that of diesel mode)	:	21-28 km/hr.
10.	Truck cycle improvement	:	Around 26%
11.	Fuel saving (on ramps and gradients)	:	90-95%
12.	Percentage increase in deeper pit operation	:	40-100%

CHAPTER 45

Future Trends in WCL—Transition from Opencast to Underground Coal Mining

*N.P. Bhati**

INTRODUCTION

Western, Southern and part of Central India basically depends for its coal requirement on Western Coalfields Limited (WCL). Due to its strategic location, there has always been a pressure of coal demand on WCL. At the time of Nationalization (1973-74) the existing command area of WCL had produced about 6.5 Mt (0.755 Mt from opencast and 5.775 Mt from underground). During 1998-99 production of WCL from the above referred command area has reached to a level of about 32 Mt indicating nearly 5 times increase in about 25 years. Average compound growth has been 5.5%.

Total coal reserves in WCL command area (as per the inventory of GSI) as on 1.1.99 is 9.6 Bt against the grand total of Indian reserves of 208.7 Bt. It is observed that the share of reserves of WCL is only 4.6%, whereas its contribution in production is about 11%. This clearly indicates that the utilization of reserves in its command area is quite high and relatively will exhaust earlier.

FUTURE SCENARIO OF OPENCAST MINES IN WCL

Wardha Valley coalfield in Maharashtra State is the main contributor of opencast production in WCL. Opencast Mines in WCL has been mostly planned for a stripping ratio of 3 to 4 Cum/t indicating a depth range of 80 to 100 m in Wardha Valley Coalfield. The existing operating mines in Wardha Valley Coalfield have reserves of about 175 Mt and are contributing about 17 Mt per annum. It can be easily seen that even at the current rate of production the reserves are likely to get exhausted in next 10 years.

Intensive exploration has been done in Wardha Valley Coalfield to locate new areas for exploitation and also to find the possibilities of extending the existing mines to greater depths. The detailed exploration inferred that by deepening some of the mines and opening

*Regional Director, CMPDI, RI-IV, Nagpur

new areas, it would be possible to add about 450 Mt to the inventory of opencast reserves. These additional reserves would have to be extracted with a higher stripping ratio ranging between 6 to 8 Cum/t. Since, these mines are not likely to be viable on the notified prices, therefore, commercial agreements may have to be done with the customers. Considering an optimistic scenario, where all these mines will be opened, the production from all the opencast mines of WCL in 2016-17 would be little less than the production of 1998-99 i.e. around 20 Mt.

Technically it is feasible to further deepen some of the mines to a depth of 200 m, which may add additional 200-250 Mt of reserves to the opencast inventory indicating an increase of about 10 years life.

The above statement is summarized in Table 45.1.

Table 45.1 : Opencast Production Scenario of WCL
(excluding the OC depth range between 200-250 m)

(All figures in Mt)

OC PRODUCTION	*98 – 99* ACTUAL	16 – 17	36 – 37*
Existing Projects	22	2	0
New Projects	0	18	10
TOTAL	22	20	10

*Forecast beyond 16-17 is likely to change with the proving of new reserves.

It is worth taking note that other coal companies (likely to supply coal to the same consumers), like Mahanadi Coalfields Limited (MCL), and South Eastern Coalfields Limited (SECL) are working at a lower stripping ratio as compared to WCL and would continue to work at much lower stripping ratio because huge shallow reserves with them. Rail transportation cost from Korba (SECL) and Ib Valley (MCL) to say Chandrapur Thermal Power Station is about Rs.520/t and Rs.610/t respectively. With the further increase in the stripping ratios in WCL mines this advantage will get neutralized soon.

In a long term scenario it will be difficult for WCL to compete with other coal producers national or international by working opencast mines at a stripping ratio of around 10 Cum/t and producing "E" grade coal.

FUTURE SCENARIO OF UNDERGROUND MINES IN WCL

Production from the underground mines of WCL command area at the time of Nationalization was 5.8 Mt, which has now reached to about 9.7 Mt indicating a compound growth rate of nearly 2% per annum.

Total extractable reserves in the currently operating underground mines of WCL area, which are amenable for underground mining, as on 1.4.99 are about 430 Mt. Underground mines of WCL are currently producing at the rate of about 10 Mty and with this rate the existing mines would last for nearly 43 years. Since, the production of opencast mines is likely to tapper down, as indicated in the above para, it would be essential for WCL to step up its production from underground mines to make up the loss created by the exhaustion of opencast reserves. Therefore, production from underground mines will have to be increased. On a long term planning, geological exploration was spread to locate new areas and also the possibilities were examined for extending the life of the existing mines. The detailed analysis indicates that it would be possible to add about 750 Mt of additional extractable reserves to the inventory of underground mines.

Barring a few good blocks, the underground mining in WCL command area can be considered as a difficult mining. Therefore, the cost of coal produced from underground mines is likely to be higher. As such, opening of underground mines would depend on the cost benefit analysis of the consumers. In an optimistic scenario where the proposed new mines would come into operation, the production from underground mines would reach to 17 to 18 Mt by 2016-17. This indicates a compound growth rate of 3.7% per annum.

The above scenario is given in Table 45.2.

Table 45.2: Underground Production Scenario of WCL

(All figures in Mt)

UG PRODUCTION	*98 – 99* ACTUAL	16 – 17	36 – 37*
Existing Projects	9.7	7	5
New Projects	0.0	11	10
TOTAL	9.7	18	15

*Forecast beyond 16-17 is likely to change with the proving of new reserves.

OVERALL SCENARIO OF WCL

Based on the above analysis an optimistic scenario, which can be achieved with a long term strategy and planning, is given in Table 45.3 & 45.4.

Table 45.3 : W C L Production Scenario

(All figures in Mt)

	98 – 99 ACTUAL	16 – 17	36 – 37*
Existing Projects	31.7	9	5
New Projects	0.0	29	20
TOTAL	31.7	38	25

*Forecast beyond 16-17 is likely to change with the proving of new reserves.

Table 45.4 : Share of Underground & Opencast Production

	98 – 99 ACTUAL	16 – 17	36 – 37*
U G	30%	50%	60%
O C	70%	50%	40%

DEMAND vs. PRODUCTION POTENTIALITY

Provisionally coal demand on WCL has been estimated to be about 34 Mt in the year 2001-02 and about 43 Mt in 2006-07.

Assuming a subsequent growth of even 4% per annum the demand would rise to about 63 Mt by 2016-17.

Since the potential of production in WCL is only around 38 Mt in 2016-17, there will be a shortage of about 15 Mt by that time.

Since potentiality of increasing production from opencast mines would be decreasing, therefore, more and more stress would have to be given for the improvement in the share of underground technology.

This analysis indicates that WCL would have to start a transition from opencast to underground mines.

STRATEGIC PLANNING

It is clear that there will be always a demanding market for WCL coal. Long term strategy for both opencast and underground mines needs to be clearly spelt and implemented sincerely.

Strategy for Opencast Technology:

The above analysis clearly indicates that –

1. The cost of excavation in opencast mines would have to be brought down so that deeper reserves can be mined at the rates acceptable to the consumers.
2. Care has to be taken to conserve the coal reserves, which may become opencastable in future.

   Excellent environment management has to be done for the opencast mines so that society accepts their existence in deepening.

   Quality control in the opencast mines has to be strengthened to satisfy the needs of the consumers.

Strategy for Underground Technology:

1. Cost of production from underground mines would have to be reduced and the selling price would be maintained at a level, which is acceptable to the consumers.
2. Mass production technology like Continuous Miner and Longwall would have to be introduced at selective places.
3. Special status and incentive has to be given to the underground management team.
4. R & D work has to be taken up to manage the problems being faced in the underground mines.
5. Strong R & D support would be essential for the development and absorption of the technology.

6. Since, lot of coal in Wardha Valley Coalfield would have to be extracted by using sand stowing and there is a scarcity of sand reserves, therefore, technology to use the substitute of sand has to be established.

7. Innovative managerial and operating techniques to reduce cost of production and increase the productivity would have to be tried and established.

CONCLUSION

The survival and existence of WCL depends on meeting the above challenges by improving performance in all the disciplines. There is a clear time available during which the situation has to be managed and author is confident that WCL will be able to maintain its commanding position in this area.

To achieve the above goal following actions have already been initiated:

1. Coal resources on the deeper side of the existing opencast mines have been kept reserved for future deepening.

2. Deepening of the opencast mines have been started.

3. Long term agreements with the consumers are being initiated.

4. Efforts are being taken to reduce the cost of excavation.

5. Technology for substitute of sand has been developed on pilot plant basis.

6. Global bids have been called for introduction of mass production technology in selective underground mines.

7. Environment Team in WCL has been strengthened. Permanent arrangement of environment monitoring has been established at CMPDI, RI-IV to have incompany scientific analysis and necessary corrective action.

ACKNOWLEDGEMENT

The author is thankful to the Chairman-cum-Managing Director of CMPDI for the encouragement and kind permission for the presentation. The views expressed in the paper are of author only and not necessarily of the organization to which he belongs.

CHAPTER 46

Financing the Coal Sector—New Modalities

*Shrikant Kulkarni**

This paper examines the changing face of the Indian coal industry, and analyses its impact on the nature of investors and investments in the sector.

A number of sources of funds exist for investments in the infrastructure sector in India. These include:

1. Central and State Government budgets (i.e. tax receipts of the governments)
2. Development Financial Institutions (State Industrial Development Corporations, State Finance Corporations, sector specific funding agencies like REC, etc.)
3. Multilateral Lending Agencies
4. Equity from the Indian capital market (Indian corporates, Shareholders)
5. Debt from the Indian capital market (Financial Institutions, Banks, Bondholders etc.)
6. Equity and Debt from Foreign Investors (Foreign Direct Investment by corporates, foreign banks, GDR holders, etc)

While some of the sources are well developed, others are expected to emerge in the near future as the Indian capital market becomes deeper. The sources have different risk appetite, and therefore different return expectations. A host of factors impact the risk return relationship for investments in a sector, and therefore determine the nature of investors who would fund the requirements of the sector. In the context of infrastructure sectors, the government policies, the sector structure, nature of the players, nature and growth of the market, etc. are some of the key factors driving the nature of investments and investors.

1. CHANGING FACE OF INDIAN COAL INDUSTRY

Control over the production, distribution and prices of coal was first imposed in 1944. The controls were as much to protect the interests of the consumers against soaring

*Head — Energy, Crisil Advisory Services

prices under scarcity conditions as to safeguard the labor from being exploited. Controls were also to safeguard against mines being subjected to unsafe mining practices and to prevent scarce reserves of high-grade coals being wasted away through their wrongful utilisation.

In July 1967, prices of coal were decontrolled and were to be determined by market forces. Railways exercised the domineering role in this scenario regulating the dispatches. Thus, the sellers could not sell to whomever they wanted to and hence at the best price.

With the nationalisation of coal industry in 1973, formal regulations on coal were once again introduced. The Government controlled the production, pricing and distribution of coal. The Planning Commission estimated the requirement of coal for the entire country. Based on these estimates, the Ministry of Coal decided the Long Term Linkages for each of the consumers. There were Linkage Committees for each industry - Power, Cement, Fertiliser, Steel, etc. - with appropriate representation of the concerned industry, GoI, and Coal Industry. The Linkage Committees had quarterly meetings to decide on the short-term linkages. The prices, for the various grades of coal, were the same across the country, and were decided by the Ministry of Coal. A Coal Price Regulation Account was maintained, where the low cost producers placed their profits to be utilised for making up the losses of the high cost producers. As decisions were centralised and competition was totally absent, the companies neither paid for their inefficiency nor reaped the benefits of their efficiency.

The focus on economic liberalization in the country and the increasing demand for coal, especially from the power sector, have raised a number of issues related to the coal industry. The most critical issue is the ability of the existing players to meet the increasing demand at competitive prices. The other critical issue is arranging for the large quantum of investment required which cannot be met by budgetary support alone. In order to address these issues, the GoI has initiated steps for the deregulation of the coal industry. These include:

1. Amendment of the provisions of the Colliery Control Order 1945, according to which the price and distribution of coking coal and grades A, B, C & D of non coking coal, hard coke and soft coke have been deregulated. Further, CIL and SCCL have been allowed to fix prices of E, F & G grades of non-coking coal every six months till January 2000 on the basis of the BICP formula. It is envisaged that pricing would be market determined subsequently.

2. Mining of coal from a captive coal mining block has been permitted, subject to the following provisions:

   (a) A company engaged in thermal power generation or iron and steel production or in any other end uses notified by the GoI from time to time, can mine coal from a captive coal block provided such a company is registered under section 3 of the Companies Act, 1956.

   (b) A company engaged in thermal power generation or iron and steel production or engaged in any other end uses notified by the GoI, from time to time, can mine coal from a captive coal block through an associated coal company, provided both the companies are formed and registered under section 3 of the Companies Act, 1956. The associated coal company is required to be registered with the objective of supplying coal on an exclusive basis to the end-user company. Further, the share capital clause of the Memorandum and Articles of Association of the end-user company should provide that it will hold, at all times, and in any case for the entire duration of the mining lease over the captive coal mining block, not less than 26% of the equity shares having voting rights in the associated coal company. The share capital clause of the Memorandum and Articles of Association of the associated company should provide that not less than 26% of the equity shares having voting rights would be held by the end-user company at all times.

(c) There can be a holding company with two subsidiaries i.e. (i) a company engaged in thermal power generation, iron and steel production or in any of the end uses notified by the GoI and (ii) an associated coal company formed with the objective of mining coal and supplying the coal on an exclusive basis from the captive coal block to the end-user company mentioned in category (i). All the three companies should be formed and registered under section 3 of the Companies Act, 1956.

In line with the liberalisation philosophy, CIL has become commercially oriented. This is manifested in its insistence not to open any new mines which would not yield an 18% IRR at the current prices. Further, a cash-and-carry payment mechanism has been instituted for sale to SEBs to check the increasing receivables position of CIL.

The Government of India has significantly reduced the customs duty on import of coal. Coupled with the significantly low prices in the Asian market, import of coal has become an attractive option to a number of consumers, especially those located in the Western part of the country.

Illustration: Policy Changes in Coal Sector during last twenty-five years

1972	Nationalisation of Coking Coal Industry
1973	Nationalization of Non-Coking Coal Industry
1976	• Captive mining by private players in iron and steel industry allowed • Sub-lease of mines to private parties allowed in certain cases
1993	• Quantity restriction of Coal import removed and coal included under the Open General License Category • Captive mining allowed for power industry • Private investment for washeries allowed • Process of reduction in import duty of coal initiated
1996	• Partial deregulation of coal industry : • Pricing & Distribution reforms of coking coal & high grade (A, B, C & D) steam coal • Captive mining allowed in the cement sector • Further reduction in import duty of coking & non-coking coal • Certain CIL subsidiaries referred to the Disinvestment Commission, Government of India Issue of notification by Ministry of Environment & Forest, Government of India, making it mandatory for thermal plants located over 1000 km from the pithead to use washed coal with ash content not more than 34 %.

2. WIDER FINANCING SOURCES AVAILABLE TO THE SECTOR

Coal India Ltd., the monopoly coal mining company in the country, has traditionally been the vehicle for most of the investments in the sector. The source of funds has been a mix of Government of India and internal accruals of CIL. CIL has raised debt from Financial Institutions and Banks and also by issue of Bonds. The World Bank has also been a provider of funds to the sector.

The part liberalisation of the sector by permitting captive mining of coal envisaged investments in the sector by the end users, and the capital market sources that the end users may be able to tap. CIL had identified 47 coal blocks for non-coking coal and 9 blocks of coking coal for captive mining purposes. Captive coal mining blocks have been allotted to 15 companies involved in power generation and 5 companies in iron and steel. These include State Electricity Boards (e.g. TNEB, PSEB) and Independent Power Projects (e.g. Bhadravati Power Project promoted by Ispat group. North Chennai Thermal Power Project promoted by Videocon group).

Captive mining by end users could be a good opportunity of attracting commercial funds to the sector which may not be coming in through CIL. This could be on account of limited borrowing capacity of CIL and also investors who may be more comfortable lending to the private sector end user companies based on the balance sheet strength of these companies and/or the fundamental strengths of the coal mining venture. However, the progress on these blocks has not been encouraging. CIL has been accused of offering relatively inferior quality of the coal blocks to the end users. The restriction of selling to only the user company could also be an issue as it could require the mine to operate at a sub optimal production level.

Though the captive mine for Bhadravati Power Project appears to be dormant at the moment, the fate of the captive mines linked to a number of other Independent Power Projects (Videocon, Jayamkondam, Srimushnam) is linked to the fate of the power project itself. The issues of relatively weaker credit quality of the purchaser, being faced by the Independent Power Projects, also apply to the power sector linked captive coal mining ventures.

The public sector end users, namely the State Electricity Boards, have also attempted to act as a conduit for commercial funding of the sector. The SEBs, given their relatively poor financial position and lack of mining expertise, are not well suited to undertake captive mining. Certain SEBs have therefore envisaged formation of Joint Venture Companies with a 26% SEB ownership (to satisfy the captive mining criteria) for undertaking coal mining. The private sector (74%) partner would be responsible for development, financing and operations of the mining venture. The Joint Venture route has the potential to attract direct investments by mining majors - i.e. from companies interested in the development and operation of mines, and not just in funding them. Of course, the issue of credit quality of the coal purchaser (as in case ofIPPs) continues be a hurdle.

At least two SEBs have experimented with the Joint Venture route. WBSEB has entered into a joint venture with a Private Company, M/s EMTA, and formed a joint venture company called Bengal EMTA Coal Company. The private company holds 74% equity and WBSEB and West Bengal Power Development Corporation Ltd. (WBPDCL) share the balance 26% stake. It is understood that the company has commenced mining, and is selling coal at a discount to the CIL price. Punjab State Electricity Board (PSEB) floated a Request for Proposal inviting private sector entitiés for formation of a Joint Venture Company. It is understood that PSEB has received bids from three Indian firms, which are presently being evaluated.

3. IMPERATIVES TO ATTRACT COMMERCIAL FINANCING

As mentioned abcve, financing of a sector is affected by, and in turn affects, various other attributes of the sector including government policy, sector structure, operating efficiency, nature of the players, nature and growth of the market, etc.

Post 1973, the Government policy dictated public sector financing of coal mining. However, as the country has changed tracks on its economic principles, two essential imperatives have emerged for the coal sector to play its role in the larger national economy.

1. Large investment requirements to increase production so as to meet the anticipated increase in demand
2. Efficiency improvement closer to world standards so as ensure reasonable price of coal while ensuring financial viability of the sector

Government policy should be focussed on addressing these imperatives and the sector structure, financing sources, nature of players, etc. need to be aligned with both objectives. At present, the policy appears to be dictating dominance of a deregulated monopoly supplier. While this may *prima facie* appear attractive to commercial investments, the situation is not sustainable and could pose a significant risk to the investors. The partial reform of the sector does not appear to be adequate to attract investments or efficiency to the sector.

The traditional financing sources - Government of India, World Bank funds, internal accruals of CIL and commercial borrowings by CIL - are clearly not going to be sufficient for the investment requirements of the sector. Nor does the present structure impose competitive or shareholder pressures on the operators to optimise the efficiency level. However, as a short-term measure, CIL could perhaps raise funds through an public equity issue (resulting in dilution of Government holding in CIL), which could be further leveraged to raise debt funds. The relatively healthy balance sheet of CIL, as reflected in its investment grade rating, indicates that this could be possible. Of course, there would be a limit on the quantum of funds that CIL could raise on its balance sheet, either as equity or as debt.

The Indian capital market - specifically the Corporate sector made up of large business houses - appear to have the appetite for investing in large infrastructure projects. A number of international mining companies have also been interested in India, and have been waiting in the wings for the appropriate opportunities. Both the Indian and foreign private sector would be able to attract appropriate debt funding from Indian and overseas markets.

However, the coal sector, in its present form, would not be attractive to commercial investments. Some of the international mining companies have shown interest in the development of coal mines jointly with CIL. This could be a good starting point, since it could ease some of the procedural hurdles, and could result in quick investment and technology inflow. Joint venture efforts with coal consumers may only partly address the issue. To fully utilise the private sector investment potential and efficiency improvements, commercial mining of coal appears to be imperative. The Chari Committee has envisaged total deregulation of mining, pricing and distribution. The following key steps would be essential to successfully attract large-scale investments into the sector:

1. Mining of coal to be permitted to private sector investors (and not only end-users). Mining companies to be able to select their customers. This is essential for two reasons: (i) the economic size of a mine is typically larger than the requirement of a single user, and (ii) freedom to choose purchaser may be seen a means of mitigating purchaser credit quality risk.

2. Allocation of blocks to be done by an independent agency, and not by one of the sector players.

3. Modification of sector structure to induct competition; independent regulation until appropriate competition has developed. (Improvement of Railway and Port infrastructure would be essential to promote effective competition).

It is appreciated that a number of issues need to be addressed while implementing the above measures. However, it is felt that the coal sector has to set itself the two objectives discussed above for it to play its critical role in the Indian economy. The policy makers and policy implementers need to gear up to address whatever issues arise in achieving the said objectives.

CHAPTER 47

Business Excellence in Mining

Anindya K Pal, Sriram Srinivasan** and TN Gunaseelan***

INTRODUCTION

Business of mining is not a simple matter of 'digging minerals' but like any other business, it is a complex interaction of several functions, which work in harmony for a common goal or purpose. These functions include domain specific activities like Mine planning, Mine operations, Processing plants, Logistics, Environment management, etc., besides other areas like Strategic planning, Finance, Human resource management, Materials management, Marketing, etc. In today's highly competitive scenario, the demands of the market are exacting and complacency would very surely wipe out any company. To survive and flourish in the long term, every organisation has to pull up their socks and strive for overall effectiveness and seek 'Excellence'. Certain basic questions that come to mind at this juncture like – Are we willing to seek excellence? Are we prepared or ready to start the journey towards excellence? The former question establishes the organisation's intent while the latter questions the readiness and capability. It's a bit like climbing a mountain. One has to have a serious desire or intent for climbing the mountain and secondly one has to be physically fit, learn the skills involved to undertake the ardour. Without both these ingredients, the quest for summit is meaningless.

GETTING READY

Consider the following. A mining company thrown in the midst of competition decided to take action to revitalise itself. The top management started a cost cutting drive. The performance of a division was to be measured by how much cost it could cut compared to the previous years' targets. The Table 47.1 was the immediate result

Table : 47.1

(Cost in Rs per tonne)

Department	*Avg. Previous Year*	*Apr*	*May*	*June*	*July*	*Aug.*	*Sep.*
1	2	3	4	5	6	7	8
Mining	68.80	68	65	58	52	50	47

*Tata Steel (FAMD), Presently studying in IIM, Calcutta.

**SRG Services & Consultancy (P) Ltd., Calcutta,

1	2	3	4	5	6	7	8
Maintenance	72.50	72	70	72	75	80	92
Crushing	65	64	64.8	66	70	75	79
Total	**206.3**	**204**	**199.8**	**196**	**197**	**205**	**218**

The mining people got a lot of praise and the others got stinkers in September. What really happened? The mining in charge increased the burden and spacing marginally, to reduce drilling and blasting cost, deferred OB removal to bring down non-productive mining activity, overloaded dumpers to transport more per trip, neglected road maintenance, etc. With this, they could achieve a short term cut in cost. On the other hand, even after the efforts taken by the maintenance and crushing personnel, finally the poor work practise lead to higher costs of maintenance and crushing. This is a clear case of misalignment of goals at the micro level because of lack of shared vision and commitment to common goals. It was an internally destructive competition caused by a short sightedness. The costs in future months are likely to be much more due to handling of backlog OB removal, major repairs, replacement of wear parts in crushers, etc.

So although the company wanted to improve their performance by cutting costs, the way the goal was perceived, interpreted and implemented by various constituents of the organisation, left much to be desired. Instead of functioning as a team, the organisation functioned as a group of individuals, each moving in a direction they perceived as the goal. The resultant was contrary to the company goals. This is what a lack of cohesiveness can do to an organisation. It's like having a cap lamp on your head to see the path but no map to lead you to the working face in an underground mine. Unless you have the correct map, know where you are and where you want to go, the light on your head is really not removing the darkness around you.

The 'warm up' routine involves a clear understanding of some fundamental questions:

1. Who are we? (Characteristics of the organisation and people thereof)
2. Where are we now? (Current status and business position, current level of competition, etc.)
3. What does the future look like? (General perception of opportunities and threats in the business environment in future)
4. Where do we want to go? (Objectives of the organisation in a given future scenario)

Who are we?

An organisation, like individuals, also has a set of guiding principles and tenets. These *timeless guiding principles* remain implicit behind every strategy and action the organisation pursues. These are *Core Values* of the organisation that must be conveyed to everyone in the organisation through communications and through the actions of the leaders. For example, if a core value of a mining company, is 'safety first' then it is not enough to print pamphlets, celebrate safety week and/or display slogans, but in addition, supervisors and managers should follow safety rules by wearing helmets and approved boots, take care to mitigate unsafe conditions, discourage unsafe practices, punish errant employees and so on. If 'customer satisfaction' is a core value, the mine manager should strive to maintain timely supplies of required quality and quantity, even if the cost shoots up temporarily. *The key is not what core values an organisation has, but whether it has core values at all.* These values are the conscience of the organisation that guides the people in the organisation to take the right actions.

Like any human being, every organisation has its purpose for 'living', in other words, its 'aim in life'. These are very important to the people inside the organisation. These are not goals or any business strategy. Such fundamental reasons for existence of the organisation in the industry is its *Core Purposes*. For example, in case of Coal India Limited, its core purpose may be 'to produce such quantity and quality of coal so as to maintain India's self-reliance in coal for power and metallurgical needs'. Thus primary role of core purpose is to guide and inspire the members of the organisation.

Where are we now?

Business excellence is about achieving quality and improved performance across the organisation with a firm focus on all the stake holders of the organisation like customers, suppliers, share holders, employees, the community, etc. The emphasis is on integration and balance of key criteria (like leadership, planning, resource and people management, processes) leading to long term business results and competitiveness. Thus the *where are we?* introspection must take a holistic view of the current position, rather than a narrow view of profitability, market capitalisation or such business related parameters alone. It is not enough that a mine is the highest producer of iron ore, what also needs to be seen is, what does it do for environmental protection, what does it do uplift the lives of the people in the area, etc.

It is necessary to look at our own performance vis-a-vis other organisations within India and globally. The purpose of this is to generate a scale of measurement and place us at some position with respect to somebody else's performance. This is essentially the process of bench marking.

What does the future look like?

This essentially is the process of visualisation of possible future opportunities and threats in the overall mining business environment, with special attention to the specific mineral the company is involved in. The visualisation should not only be with respect to the market for the mineral but also from the point of view of possible changes in legislation, problems due to possible adverse environmental and social impacts, changes in technology, etc.

Where do we want to go?

We now know our current status and we perceive how the future looks like. It is now necessary to know the destination in the next 20 – 30 years. The organisation should define clearly worded and measurable objectives. Consistent with the concept of business excellence these should cover:

- Business goals – Like Production, Costs, Profitability, Shareholder value, etc.
- Customer satisfaction goals – Like quality, timely delivery, reduced customer complaints, etc.

- People goals – Like development human resource skills, improving working environment, etc.
- Social and Environmental impact goals – Like contribution in development of the area, environment mitigation measures, etc.

Figure 47.1 summarises this process of 'getting ready'. This is the process by which the organisation will be bonded into a single entity, with a single-minded purpose.

FIGURE 47.1

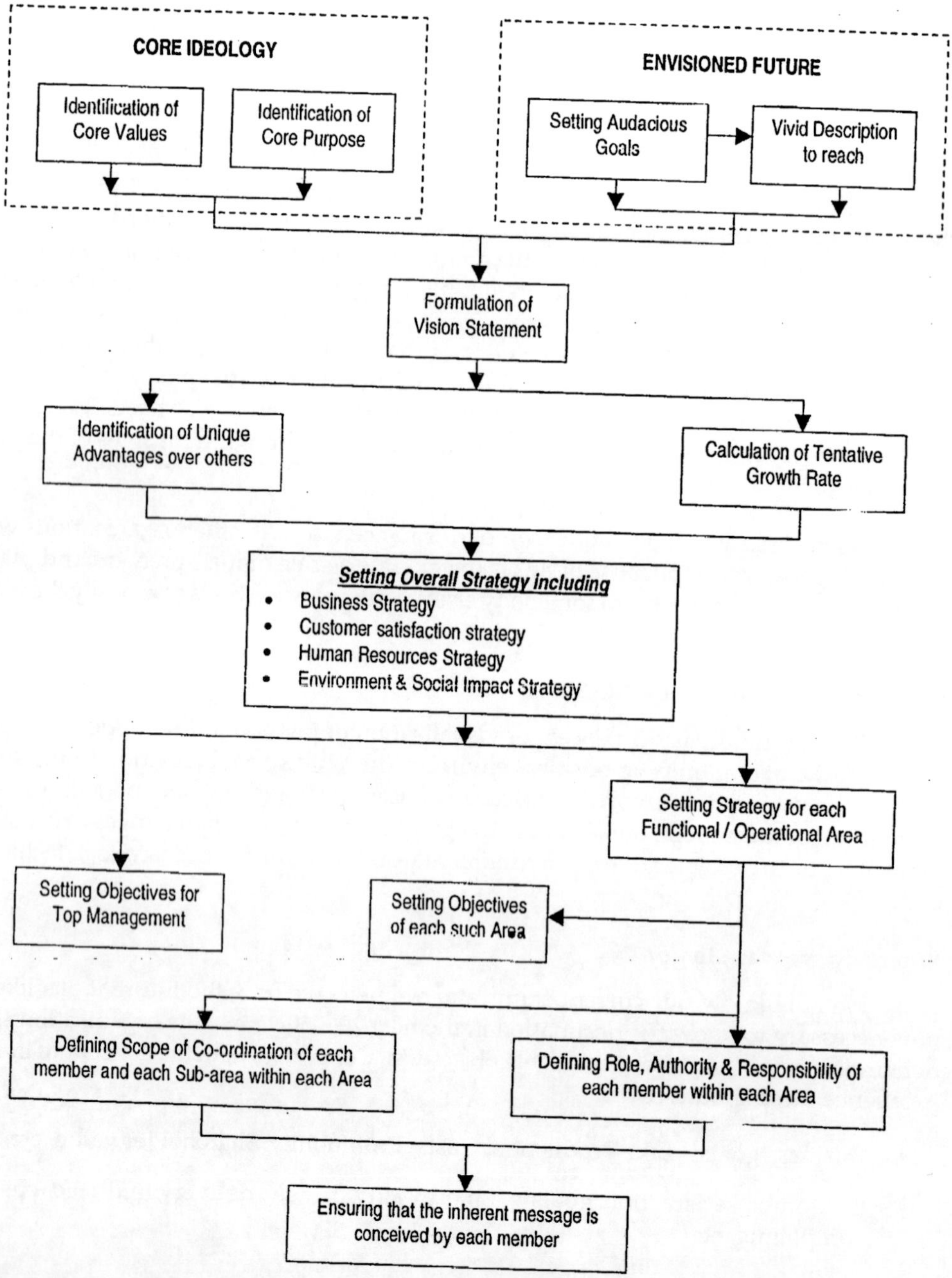

THE JOURNEY

The journey towards excellence is the process of implementation of the goals and aspirations of the organisation with faithful adherence to the core values and core principles of the organisation. Excellence is something that is superior, that has good qualities in a high degree. This definition is inherently suggestive of measurement. While measurement is understandable, but

- What should be measured and how to go about it?
- Achievement of what standards should constitute excellence?

What and how to measure?

An improvement is basically a distance between two points. One extreme is unanimously the 'starting point', i.e., the present status/performance level of the organisation. The other point is the destination set out in our organisational goal. The organisational goal is normally quite broad and would need to be translated into specific terms for each involved department so that the target is related to the specific function being performed by each department. For example, one business goal of an organisation is to reduce cost of mining by 10% after adjustment for inflation. The specific targets for individual departments may read something like :

Drilling : (a) Improve bit life by 15%
(b) Eliminate hydraulic oil spillage
(c) Improve drilling rate by 15%
(d) Increase utilisation of drills to 90% etc.

Blasting (a) Maximum boulder size below 1000mm
(b) Reduce time for blasting to 40 mins, etc.

However, to arrive at such targets it may not be a straightforward exercise of setting departmental goals. It is important to understand how the various activities are related and how various departmental goals will collectively lead to achievement of the desired objective. It may be easy to reach a conclusion to reduce OB removal for 2 years to reduce mining cost, but its implications after the 2 year period may be disastrous. A reduction in drilling and blasting cost may cause the cost of shovelling, crushing to go up. Hence a systematic approach is required to convert the business attributes (BA), customer satisfaction attributes (CA), people attributes (PA), social impact attributes (SA) into measurable engineering and management actions to achieve maximum impact.

The 'House of Quality' matrix

Consider an example of a limestone mine supplying a cement plant. Based on a broad target to optimise raw material cost to the plant, the following action groups could be categorised:

- Adherence to ROM specification
- Adherence to Quantity specification
- Reduced cost of mining

Generally it may be difficult to improve all the BAs in a group / bundle. Though sometimes a creative solution can be found that satisfies all, usually the organisation have to trade off one benefit against another. Hence, the best way is to measure relative importance to the organisation of all BAs. Such weightings are based on team members' direct experience with customers or on organisational goals and objectives. In Fig. 2, weightings are displayed next to each BA – usually in terms of percentages, a complete list totalling 100%.

Group / Bundle	*Business Attributes (BAs)*	*Relative Importance (RI)*
Adherence to Product Specification	More control over quality supplied	15%
	More control over size range supplied	7%
	....	
Adherence to Quantity Requirement	More control over total quantity supplied	10%
	More control over effective quantity supplied	12%
	....	
Reduced cost of mining	Decrease in Production Cost	12%
	Increase in Equipment Productivity	18%
	....	
	A Complete list totals	100%

Fig. 47.2 : Relative Importance Weights of Business Attributes

Organisation that wants to match or exceed other similar operations' performance, must first know where it stands relative to the BAs. So, we list evaluations of our own operations vis-à-vis that of other similar operations (say, Mine A and Mine B) or in comparison to domestic or international industry standards.

Group / Bundle	Business Attributes (BAs)	(RI) (%)
Adherence to Product Specification	More control over quality	15
	More control over size range	7
	....	
Adherence to Quantity Requirement	More control over total quantity	10
	More control over effective quantity	12
	....	
Reduced cost of mining	Decrease in Production Cost	12
	Increase in Equipment Productivity	18
	....	

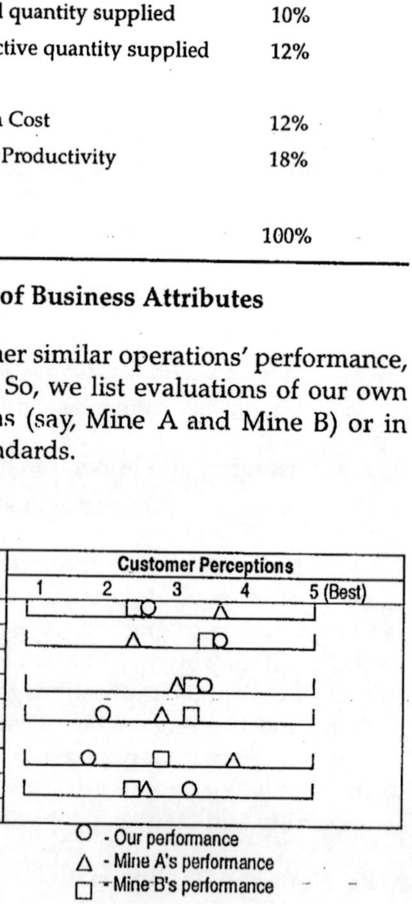

Fig. 47.3 : Evaluation of present performance

Now we need to describe the important attributes in the language of the Mining Engineers. Figure 47.4 displays a matrix where the BAs are interpreted in terms of the contributing *Mining/Engineering Characteristics (MCs)* that are likely to affect one or more of the BAs. The positive sign means the mining engineers need to increase the MC from the present level in order to achieve the concerned BA. The MC should be described in measurable terms.

		Mining Characteristics	Product Spectn.					Quantity Targets				Cost Parameters				
			(+) Sampling on face	(+) Blending	(-) Dilution	(-) Product size	....	(+) Equipment Availability	(+) Equipment Utilization	(-) Handling loss	....	(-) Variable cost	(-) Overhead cost	(+) Equipment Productivity	....	....
	Business Attributes (BAs)	RI (%)														
Adherence to Product Spectn.	More Control over quality	15														
	....															
	....															
Adherence to Quantity Reqmt.	More Control over effective quantity	10														
	....															
	....															
Reduced mining cost	Decrease in production cost	12														
	....															
	....															

Fig. 47.4 : Mining Characteristics

The matrix now needs to be developed further by evaluating the relationships that exist between BA and MCs regardless of whether the MC was selected for that BA or not. That is to say, how each MC affects each BA. The relationships can be categorised into positive, strongly positive, negative and strongly negative. This evaluation should be based on the team's own experience and also on wide discussions with relevant people. For example, in Figure 47.5, the mining characteristic of increasing sampling frequency has a strong positive relationship with the business attribute of 'more control over quality'.

Once the relationships between the mining characteristics and business attributes have been linked, we now have a set of mining characteristics for which we need to evaluate our current performance levels in these identified mining characteristics and also the levels of performance that is being achieved by Mine A and Mine B for the same areas. This is the stepping stone to setting actual performance goals for the operations. These are basically objective measures or bench marks at the present. Figure 47.6 demonstrates this process.

Relationships

Symbol	Relationship
☑	Strong Positive
✓	Medium Positive
☒	Strong Negative
×	Medium Negative

		Mining Characteristics	Product Specification					Quantity Targets				Cost Parameters				
			(+) Sampling on face	(+) Blending	(-) Dilution	(-) Product size	....	(+) Equipment Availability	(+) Equipment Utilization	(-) Handling loss	....	(-) Variable cost	(-) Overhead cost	(+) Equipment Productivity	....	....
	Business Attributes (BAs)	RI (%)														
Adherence to Product Spectn.	More Control over quality	15	☑	☑	✓	✓										
	....															
	....															
Adherence to Quantity Reqmt.	More Control over effective quantity	10						☑	☑	✓				☑		
	....															
	....															
Reduced Mining Cost	Decrease in production cost	12	×	✓	✓	✓		✓	☑	✓		☑	☑	☑		
	....															
	....															

Fig. 47.5 : Relation between Mining characteristics & Business Attributes

The final stage in developing this 'house of quality' is the 'roof matrix', and setting of targets to achieve. Development of a 'roof matrix' is the process of understanding the inter-relationships amongst the various MCs if such relationship exists. For example, as in Figure 47.7, equipment utilisation has a strong positive relation with equipment productivity, but has no relation to sampling frequency.

This matrix now provides an understanding of the need for improvement in various MCs based on the BAs. What characteristics need to be improved? How does it impact other characteristics? What should be undertaken first? and finally what should be the specific target set at this point in time? The entire matrix, *the House of Quality*, helps the team to visualise the linkage between the mining activities performed and the broader business results. This picture enables proper understanding leading to setting meaningful targets for each individual activity taking into consideration the effort and resource required to achieve the target.

Relationships

Symbol	Relationship
☑	Strong Positive
✓	Medium Positive
☒	Strong Negative
×	Medium Negative

	Mining Characteristics		Product Specification					Quantity Targets				Cost Parameters				
			(+) Sampling on face (no.of samples / day)	(+) Blending (% Deviations from target)	(-) Dilution (% of gangue/ low grade material)	(-) Product size (% of oversize & undersize)	....	(+) Eqmt Avlblty (% of standard availability)	(+) Eqmt Utilization (% of eqmt availability)	(-) Handling loss (% of production)	....	(-) Variable cost (Rs. per unit production)	(-) Overhead cost (Rs. per unit production)	(+) Equipt. Productivity (% of rated capacity)	....	....
	Business Attributes (BAs)	**RI (%)**														
Adherence to Product Spectn.	More Control over quality	15	☑	☑	✓	✓										
	....															
	....															
Adherence to Quantity Reqmt.	More Control over effective quantity	10						☑	☑	✓				☑		
	....															
	....															
Reduced mining cost	Decrease in production cost	12	×	✓	✓	✓		✓	☑	✓		☑	☑	☑		
	....															
	....															
Objective Measures	Our performance		16	5	5	10		75	80	2		60	38	75		
	Mine A's performance		25	1	2	4		75	80	5		50	30	95		
	Mine B's performance		13	3	4	8		70	75	3		55	35	80		

Fig. 47.6 : Objective Measures

THE DESTINATION (?)

Now that we understand what and how to measure, achievement of what standards should constitute excellence. Do you have to be the best in all areas in the world to be excellent? Can organisation be called excellent on mining only? We must reflect on the definition of business excellence again. Business excellence is about holistic excellence, it is about balance. It is about a process, a journey rather than a destination. Aristotle wisely

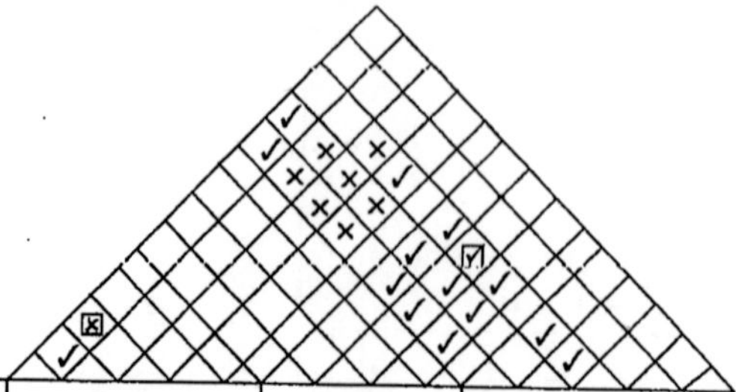

Relationships

- ☑ - Strong Positive
- ✓ - Medium Positive
- ☒ - Strong Negative
- × - Medium Negative

Mining Characteristics			Product Specification					Quantity Targets				Cost Parameters				
			(+) Sampling on face (no.of samples /day)	(+) Blending (% Deviation from target)	(-) Dilution (% of gangue/ low grade material)	(-) Product size (% of oversize & undersize)	....	(+) Eqmt Avlblty (% of standard availability)	(+) Eqmt Utilization (% of eqmt availability)	(-) Handling loss (% of production)	....	(-) Variable cost (Rs. per unit production)	(-) Overhead cost (Rs. per unit production)	(+) Equipt Productivity (% of rated capacity)	....	....
	Business Attributes (BAs)	RI (%)														
Adherence to Product Spectn.	More Control over quality	15	☑	☑	✓	✓										
	....															
	....															
Adherence to Quantity Reqmt.	More Control over effective quantity	10						☑	☑	✓				☑		
	....															
	....															
Reduced Mining Cost	Decrease in production cost	12	×	✓	✓	✓		✓	☑	✓		☑	☑	☑		
	....															
	....															
Objective Measures	Our performance		10	5	5	10		75	80	2		60	38	75		
	Mine A's performance		25	1	2	4		75	80	5		50	30	95		
	Mine B's performance		13	3	4	8		70	75	3		55	35	80		
Technical / Managerial Difficulty			1	3	3	4		4	2	2		5	5	4		
Estimated Resources (% of total required)			2	10	1	15		15	1	1		1	1	10		
Targets			25	3	2	6		80	90	1		55	35	85		

Fig. 47.7 : House of Quality

said " We are what we repeatedly do, excellence then is a habit not an act". So as in the analogy of a mountain climber, as you prepare, undertake the ardour and reach a summit, you will discover new summits to conquer.

REFERENCES

Hauser, John. R and Clausing, D, 'The house of quality', *Getting the product right*, pg 99-109

Ethos Management Limited, *Business Excellence Model – Self assessment using assess rapidscore workbook*, pg 2 – 32

Prof. Banerjee, Ashish and Dr. Das, Ranjan, IIM Calcutta, *Lecture notes on product management and strategic marketing*, unpublished

Covey, Steven, 1994, *Seven habits of highly effective people*, McPhersons Printing Group, Australia, pg 46.

Section VIII

R&D INITIATIVES

CHAPTER 48

R&D Challenges for Indian Coal Sector

R.K. Chechani & S.N. Sahay***

1.0 INTRODUCTION

In India oil and gas reserves are limited. Although there is considerable potential for hydel and nuclear power, their use on a large scale is also limited due to environmental and safety considerations. Therefore, with estimated reserves of over 200 billion tonnes, coal occupies a dominant place in the energy scenario of the country. Currently about 62% of the country's commercial energy is produced from coal.

With 16% of the global population India consumes hardly 4.5% of world energy. Still the country has to rely on non-commercial sources, like firewood and animal waste, to a large extent. About 40% of the total energy consumed comes from such fuels. The share of commercial energy has been rising and needs to go up substantially in coming years so as to reduce reliance on fuels like firewood and minimize deforestation.

Against this backdrop India will continue to rely on coal as the primary source of energy well into the next century. The challenges facing the coal sector are not likely to lessen, but can only multiply. R&D activities will not only have to address issues relating to productive and safe mining in increasingly difficult geomining conditions but also focus on greater utilisation of coal as an energy resource including its use for replacement of depleting oil wherever possible.

2.0 R&D SCENARIO : PAST AND PRESENT

In the coal industry the R&D scenario is marked by two distinct time domains : the early period which as isolated efforts to improve operational standards, and the post-nationalisation period characterised by coordinated research activity.

2.1 Pre-Nationalisation Period

Early R&D activities before independence of the country were mostly limited to transplantation of technology from the UK with some modifications to suit Indian

*Director (RD&I), Central Mine Planning & Design Institute Limited, Ranchi.

**General Manager (R&D), Central Mine Planning & Design Institute Limited, Ranchi.

conditions. Such transfer of technologies was carried out by the British companies having coal mining interests in India. There was very little concerted effort to improve the performance of the industry through organised R&D and indigenous development.

Mining, Geological and Metallurgical Institute of India (MGMI) was among the first to take up research in mine safety, fires and explosions. The investigative studies carried out by the First and Second Subsidence Committees (1913 and 1927) also rank among the major efforts of the time.

Following the recommendations of the Coal Mining Committee (1937) and Indian Coalfields Committee (1946), Central Fuel Research Institute (CFRI) was set up near Dhanbad in 1946. This institute carried out pioneering work in coal beneficiation, combustion, recovery of by-products and formed coke technology among others.

After independence of the country, organised R&D began in several key sectors with the establishment of the Council for Scientific and Industrial Research (CSIR). A chain of laboratories under CSIR was set up across the country to meet the R&D needs of various industries. For the mining sector Central Mining Research Station (now CMRI) was established at Dhanbad in 1955. This institute along with Regional Research Laboratories at Bhubaneswar, Hyderabad, Jamshedpur and Jorhat carried out useful work on coal in the following years.

National Coal Development Corporation (NCDC), founded by the government in 1956, gave a major thrust to mechanisation of mining operations and trial of hitherto untried mining methods for thick and steep coal seams in India. A few large mines in the private sector too carried out innovative trials in the 1950s and 60s involving stowing, supports and machinery. The NCDC also laid stress on centralised mine planning and exploration which later played a key role in stepping up production in the nationalised coal sector. Despite these moves in technological upgradation, however, coordinated R&D activities exclusively for the coal sector had to wait until nationalisation of the industry.

2.2 Post-nationalisation Period

Following the nationalisation of coal mines in two stages in 1971 and 1973, Coal Science and Technology Plan was initiated by the government in 1975. The same year Central Mine Planning and Design Institute (CMPDI) was established as a subsidiary of Coal India Limited (CIL). CMPDI was entrusted with the task of coordinating R&D activities under the S&T Plan as the nodal agency for coal and lignite sectors.

The Coal S&T Plan is administered by a high-level Standing Scientific Research Committee (SSRC) of the Ministry of Coal. The SSRC is assisted by four technical sub-committees on :

- Production, productivity and safety.
- Coal beneficiation.
- Coal utilisation.
- Environment and ecolog.

The primary objective of the industry-funded research under this S&T Plan is to develop technology and systems suitable for Indian mining conditions and their successful integration with mining operations and coal utilisation. Since these operations involve multi-dimensional complexities, a wide range of problems from coal exploration to extraction, from combustion processes to health and safety of workers can be investigated for overall benefit of the industry and its surrounding areas.

Over the past two decades a total of 174 sponsored research projects under the Coal S&T Plan have been completed by various academic and scientific institutions and the coal producing companies. Many of these projects have contributed greatly to production, productivity and safety in mines along with improvements in coal utilisation and environment.

In addition to the Coal S&T Plan a Research and Development Board has recently been constituted by CIL, the largest coal producer in the country. Since 1997 this Board has started funding R&D projects. Currently there are 11 projects being executed. This funding by the R&D Board of CIL is likely to facilitate full-scale experimental trials, which otherwise may involve financial risks for an operating mine. A total of Rs. 125 crore has been ear-marked by CIL to finance R&D activity over a five-year period.

3.0 S&T PROJECTS

Since the inception of Coal S&T Plan in 1975 the number of research projects funded by it in various fields of study of shown in Table 48.1.

Table : 48.1

Subject Area	*Number of projects*	*Project costs (Rs. in Lakh)*
Exploration	17	633.8
Underground mining methods	11	1298.4
Underground coal gasification	1	322.0
Underground machinery and coal workability	9	232.9
Blasting in mines (underground and opencast)	4	62.1
Strata control and roof supports	22	221.5
Mine ventilation, methane drainage and fire control	25	279.5
Transport	6	117.1
Stowing	3	261.1
Subsidence	5	256.2
Opencast mine planning and slope stability	3	35.8
Corrosion studies and condition monitoring of equipment	3	26.18
Accident analysis and analysis of projects	2	13.5
Coal beneficiation	20	1368.3
Coal utilisation	31	802.8
Environment and ecology	12	229.8

The total funding involved in the completed projects was nearly Rs. 62 crores. As can be seen in Table 48.1, underground mining methods and beneficiation account for a substantial part of the total R&D outlay. However, among the 11 projects on underground methods, a project on hydraulic mining alone involved Rs. 8.19 crores. Similarly, among 20 projects in coal beneficiation two projects involving washeries account for Rs. 8.16 crores. In exploration also just two projects out of a total of 17 account for over 60% of the R&D expenditure in that area.

The bulk of the projects, except a few involving underground coal gasification and stowing were essentially small studies with low financial inputs. Many of the projects were laboratory studies only, and some involved just theoretical work.

In a few critical areas like strata control, roof supports, ventilation, methane and fire control, though as many as 47 R&D projects were executed, the total expenditure made amounts to only Rs. 5.01 crores. In major areas of concern like subsidence and stowing, only 8 R&D projects were undertaken involving Rs. 5.2 crores.

Since the inception of Coal S&T Plan financial outlay has never been a constraint. Allocations have often remained unutilized due to R&D proposals not addressing the requirements of the industry. In a six-year period from 1992 only 59 proposals out 190 submitted were considered appropriate for funding.

Efforts are underway to indentify and address R&D effort such that benefits to the industry can be maximized. Currently, a total of 42 research projects involving a financial outlay of Rs. 58 crores are in various stages of execution.

4.0 FUTURE R&D CHALLENGES

From a level of about 70 MTY at the time of nationalisation of the industry in 1971-73 coal production has risen to 299 MTY in 1998-99. R&D activities during this period have made significant contributions in the rise in production, safety standards and environmental matters related to the industry. There have been notable improvements in geophysical exploration techniques, hydrogeological investigations, mechanisation of workings, replacement of timber support by roof bolting, coal beneficiation and utilisation processes. Mining methods like depillaring with cable bolting and blasting gallery for thick seams have been successfully introduced. R&D on environmental issues like reclamation of quarried land for revegetation has proved useful.

In the light of experiences gained and the ground which has been covered, R&D efforts need to be intensified in certain areas. Production from underground mines has stagnated around 55 MTY since nationalisation. Modification of conventional methods like board and pillar, improvements in roof support practices, introduction of machinery or systems to improve productivity call for extensive field trials and mine oriented studies.

Studies specially targeted towards improvement in longwall productivity, indigenous development of spares for longwall equipment and for removal of underground transport bottlenecks are like to prove essential for capital intensive longwall projects. Trials and introduction of condition monitoring equipment and systems for underground machinery, powered supports and also HEMM for opencast mines may be increasingly necessary for optimum utilisation of equipment.

Management of hard roof strata calls for intensive studies and trials of remedial measures. Warning systems for water danger, collapse of ground in populated localities over old workings, failure of dolerite/basalt sills and spread of old mine fires may prove particularly useful in the Indian mining context.

Environmental issues like noise, air and water pollution are likely to need increasingly greater R&D effort to improve operational standards and reduction in costs. In the area of coal utilisation also extensive R&D trials are called for in mobilising coal for greater use in urban and rural areas as a replacement of other fuels.

As mentioned at the outset coal is likely to play an increasingly greater role in the energy scenario of the country in the coming decades due to depletion of indigenous reserves of other fuels. The coal industry is not only required to geat itself for increase in coal demand but also for capacity utilisation of investments in the changing economic scenario. Increasingly difficult geomining conditions and greater social and environmental awareness are likely to add further dimensions to the complexities that future R&D efforts will be required to handle and resolve.

CHAPTER 49

Coal Mine Planning with Due Consideration of Strata Control and Environmental Protection

W. Keune and J. T.E. Kook**

ABSTRACT

The presented planning system

- is based on many years of experience in hard coal mining with winning coal on different seams in one of the most complicated deposits in the world,
- applies computer technology that handles the vast amounts of required data to describe a coal mine,
- reduces the planning times by approx. 50%.

The system is based on standard hard and software such as Oracle™ or MSOffice™. It runs on ordinary PCs and laptops to ensure world-wide availability. Special software such as the finite difference method is also used, if required, as is the physical modelling technique.

1. INTRODUCTION

Considering that a mechanised mine needs roughly US $ 100 to 150 million of investment, planning is very important for minimising the risks. A distinction can generally be made between seven types of risks:

- Political risks.
- Social risks.
- Geographical risks.

*Deutsche Montan Technologie GmbH. Essen, Germany

- Legal risks.
- Risk of Deposits.
- Technical risks.
- Financial risks.

Taxation, customs and fees are not fixed for a long-term project, instead they may be decided upon within political processes.

Foreign investors usually provide their own management. Religious or ethnic conflicts may cause problems.

Mining activities often take place in geologically active formations. High floods or lack of water are definitely geographical risks.

Uncertainties in legislation on mineral ownership and changing of legislation on mining and the environment are further risk factors.

Although the first four items are of great importance, this paper will focus on the three other risks mentioned above.

2. COMPUTER AIDED MINE PLANNING AND DESIGN

To minimise the further risks, a large amount of data has to be evaluated. The aim of planning is to combine the deposit, mine infrastructure and capital input in a cost-optimised way. To reach this target, different scenarios need to be developed. For that purpose, these computer aided mine planning systems are very important.

For developing scenarios, different soft- and hardware tools are required. Fortunately most tasks can be solved using standard software like spreadsheet or CAD-programs. Special tasks such as strata control or subsidence prevention require special software.

All engineering disciplines involved in mine planning need to retrieve data from a unique source. Working within one framework makes great demands on communication [1]*.

2.1 Deposit

The assessment of the deposit requires information about, e.g., the quantity and quality of coal, the amount of waste, thickness, depth, dipping of the seams and tectonical structures.

The result of the exploration illustrates the amount of data. Figure 49.1 describes a small deposit. It covers an area of roughly 4 km^2. The deposit is explored by about 100 drillings ranging up to 250 m in depth. This means 15 km core description and additional data depicting the surface and infrastructure has to be handled.

Unfortunately, the coal seams were not named during the sedimentation period. Therefore geologists have to identify the coal seams by using their experience with key strata. The data needs to be interpreted to derive a 3-dimensional model of the deposit. Therefore, data sections and cross sections are conventionally generated using graphical programmes. After correlation of coal deposits ISO-patches and maps depicting floor-contours are to be created.

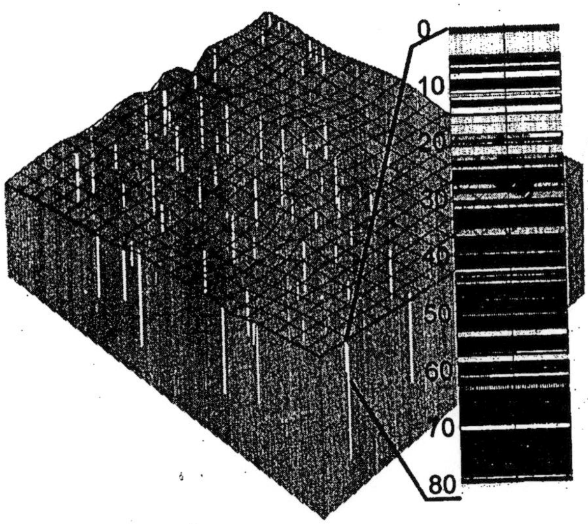

Fig. 49.1 : Surface and drillings

Computer-based planning is able to perform an analytical approach. Identification of coal seams from drillings works are as follows:

- Based on the data an inquiry is generated straight away. For instance, to find the top seam, all data describing coal at the upper position within each drilling is separated.
- From this data all drillings outside the outcrop are removed.
- To take the technical performance into account a minimum thickness is defined.
- A length-dipping ratio is calculated to find tectonical structures that distinguish between extractable parts of the deposit.

The results of analytical and conventional identification of the coal seams are compared and show that more than 95% of identified resources are identical. Remaining uncertainties have to be clarified geologically. Using the analytical approach, the time needed to derive a model of the deposit from the exploration results can be reduced by more than 50%.

Due to the fact that this approach is based on a full 3D-model of the underground and surface, cross-sections, volumes, quality distribution, basically all types of ISO-lines may be generated. This means that the depiction of the deposit e.g. in maps, sections and 3D-views and all dependent data is fully included.

2.2 Technique

After fixing the deposit, engineering will proceed in mine layout. Conventional planning takes border conditions into account such as dipping, thickness, characteristics of seam and host rock etc. This is done by interpreting maps, sections and other bases containing the necessary information. The results are shafts, roadways and panels as shown in Figure 49.2. Analytically one divides that task into two subtasks.

The 1st step reduces the deposit geometrically to the mineable area. That area has been derived from the planning system, regarding the border conditions. Due to the fact

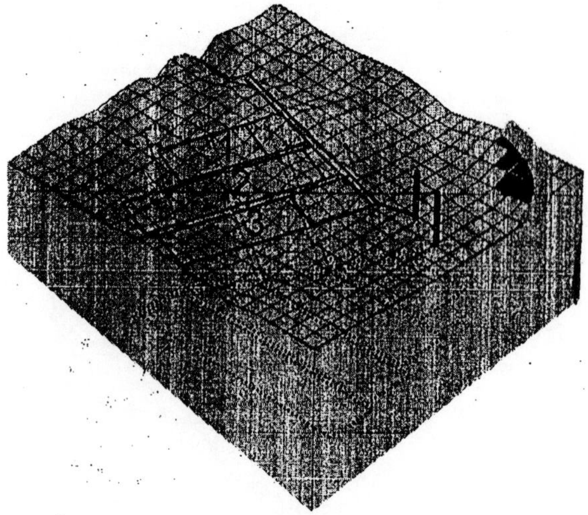

Fig. 49.2 : Mine Layout

that one has to mix two ISO-planes (e.g. seam -thickness and -level) the computer will decrease the time to fix that area dramatically.

The 2nd step covers the technical approach to fit the goal to extract the deposit into the frame, using the suitable extraction method, the machinery availablilty, transportation and so on.

A key criterion in underground mining is the control of the strata. The aim is to maximise the exploitation of the deposit regarding the stability of all excavations. Regarding roadways, the support and maintenance expenditure are to be minimised.

The German mining industry has more than thirty years of experience in strata control. Due to the average extraction depth nowadays of roughly 1000 m German mining investigation began in the early sixties with R&D programmes to develop a system of strata control.

Strata control is indispensable in the planning of underground mines. In Indian deposits with shallow depths and thick sandstones in the roof, strata control has the same importance as in deep mining.

The system of strata control includes the following elements:

- Measurements and observations in galleries and faces, estimation of the behaviour of support and strata, verification of the planning and developing the prediction methods.
- Rock pressure calculations and geologic mapping for optimising the mine layout.
- Physical and numerical modelling for the behaviour of support and strata in faces and galleries research.
- Testing rigs for investigation and improvement of gallery and face support.
- Test benches for analysing the properties of rocks and building materials [2].

The prediction tool of the system uses geometrical data and the representation tool of the mine planning system.

After preparation of a mine layout, including panel positions and time schedule one has to take into account that mining will influence the environment. For instance a river as shown in Figure 49.2 can be influenced by subsidence. Normally a waste dump will cover a certain area on the surface. It is needed for dumping extracted rock material. Generally the mine operator has to reclaim the affected area before leaving it [4].

Therefore, the engineering workforce has to predict all influences of mining on the surface and to prepare an environmental impact plan (EIP). On the one hand measures have to be taken to prevent damage. On the other hand subsidence might help keeping e.g. a harbour open. The deposit of mud may make the floor rise and thus turn the harbour inoperable. Subsidence can decrease the floor. The EIP is based on an exact prediction of all those impacts. For that reason a numerical prediction system for ground movement is one tool of the workbench the above described mine planning system represents. This system uses a stochastic approach to calculate ground movement. After calibration it is applicable under all mining conditions [3].

2.3 Finance

The success of a mine largely depends on low costs and high revenues. This requests the control of costs right from the start of the planning process.

Conventional finance and technical planning run more or less parallel. There is no online data exchange. Instead it is only applied at fixed points, e.g., for the selection of new equipment. Planning is -in this case- restricted to fewer planning scenarios, because it is too time consuming to consider a cash flow analysis for each modification introduced, e.g., in case of a new pressure or subsidence prediction.

The mine planning system provides methods and actual data to each member of the planning team. Therefore an engineer calculating subsidence is able to estimate the influence of modifications e.g. in panel length or time sequence on cost .

Figure 49.3 shows the mentioned multi-use of the same data. This figure combines the mining sequence with the mine layout and the rate of advance in relation to expenses and revenues. 90 % of all information shown has been generated with standard software. The data on which the figure is based, comes from the data base management system.

3. CONCLUSION

Under consideration that the removal of an error while planning charges a split part of the same while mining, planning becomes of higher importance in modern mining. In fact planning means exchange and processing of data between different mining disciplines. The ventilation system and requirements depend on the mine layout. But an optimal ventilation system may increase the costs beyond certain limits or make extraction impossible. The same data as mentioned before is used to estimate the length of transportation routes. Every change in the layout is automatically provided to all mine engineering disciplines and can be thus considered.

Five years of computer based planning taught us that the planning system may provide tools to accelerate the work, to avoid mistakes and to develop different scenarios

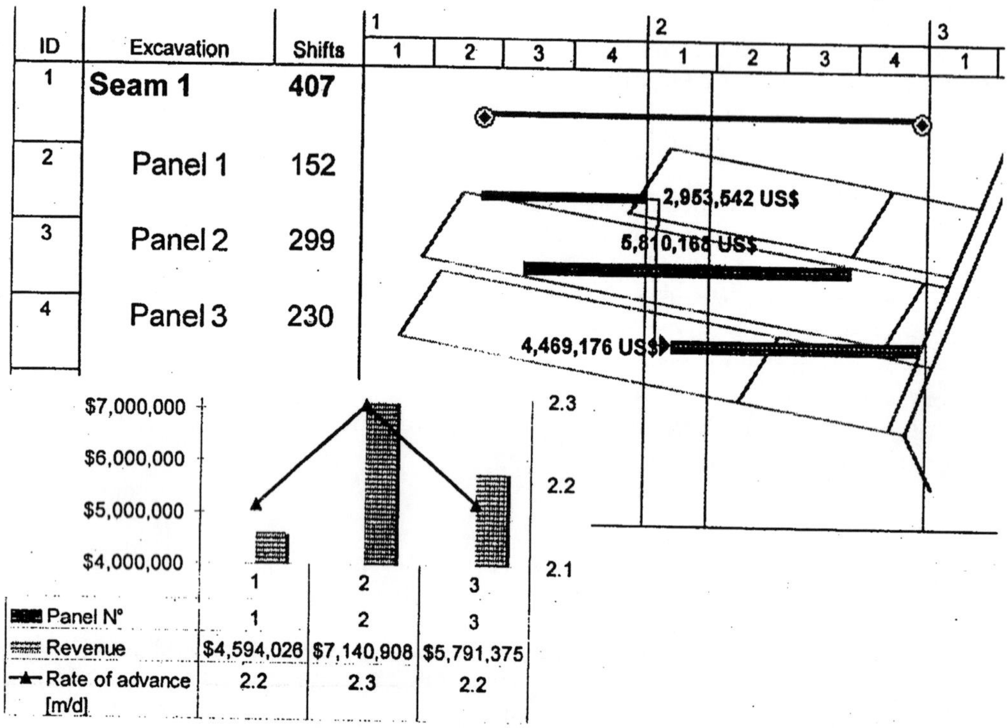

Fig. 49.3 : Basic parameters for cash flow analysis

easily. But successful planning mainly depends on the use which is made of those tools. On the one hand knowledge of mining operations and mine planning issues is of importance, on the other hand the knowledge of handling a very complex system guarantees success.

REFERENCES

1. Keune W.: A mine information system based on an OMT model of the coal mine and deposits in: R.V. Ramani, (Hrsg.): 26th International Symposium Application of Computers and Oparations Research in the Mineral Industry. Society for Mining, Metallurgy and Exploration Inc. Colorado, USA 1996, S. 347-353.

2. te Kook J. and W. Keune: The German system of Strata Control and a Mine Information System to calculate subsidence and earth movement. 2nd seminar on ground control in mining. S.K. Sarkar (ed). Oxford and IBH Publishing Co. Pvt. Ltd. New Delhi, Kalkutta 1996. S. 341-356.

3. Keune W. and V.P. Talwar. Subsidence prediction techniques including modelling. Journal of Mines, Metals and Fuels. Vol XLV, Calcutta 1997. pp. 49 – 56.

4. Keune W. et al: Enlargement of a deep mine information system (IS) with a model of the surface. An application for Bobovdol in Bulgaria. In: Minning in the 3rd Millenium. Xth International Congress of the International Society for Mine Surveying. Congress Proceedings. Promaco Conventions Ltd. Freemantle (WA) 1997. pp. 649 – 658.

CHAPTER 50

New R&D Initiatives in Coal and Non-Coal Mineral Sector for the Next Millennium

*Gupta. R.N**

ABSTRACT

Increasing demand of power will call for higher production of fuel minerals. The coal industry itself has set high target of production by the turn of the century, out of which a large chunk is projected to be achieved through intensive mechanisation whereas production from manual bord and pillar workings is projected to remain almost constant. Thrust has been given to mechanised longwall method of extraction and other mechanised bord and pillar methods. In the opencast sector, high capacity mines using state-of-the art technology are to be developed. The demand projections for most of the minerals show substantial increase in production by several folds. This will be achieved through high capacity opencast mines using large equipment. Mechanisation has been envisaged in some of the metalliferrous mines including copper, zinc and lead. These mines use state-of-the art technology. Blasting is likely to remain the primary method of breaking rocks and therefore increased production would call for larger blasts. Workings are likely to be made in difficult geo-mining areas, which were unthinkable in the past. Many workings may approach near built up areas, which may bring new problems in safety.

NIRM is a Research and Development organisation. The mission of NIRM is to work with Civil and Mining Industry to identify opportunities and deliver solutions through outstanding science and technology. This paper gives a brief information about new R&D initiatives in coal and non-coal mineral sector for the next millennium.

1. NEW TRENDS IN MINING EQUIPMENT DESIGN

It is possible to predict what the future mine will look like by extrapolating from the present mining equipment development trends. While major paradigm shifts through the

---

* Director, National Institute of Rock Mechanics, Champion Reefs (P.O), Kolar Gold Fields – 563 117, India E-mail: gupta_rn57@yahoo.com

introduction of new extraction systems cannot be ruled out, gradual but steady advances in automation and equipment reliabilities may be as effective when they are sustained over a sufficiently long period of time.

Unit cycles in mining have not changed for many years. Substantial productivity increases over the past three decades have mainly been achieved by using larger and heavier machines. However, the investment in size has probably reached the point of diminishing returns and increased productivity in the future will have to be generated by other means.

- Introduction of new mining systems which will radically change the shape of mining or at least transform certain parts of the mining cycle, e.g., the continuous rock excavation for hard rock mining.
- Improving the performance of existing equipment by smart sensors and automation
- Increasing machine availabilities by better maintenance.

It may be illuminating to smart by trying to visualize what a typical mine site would look like in twenty years. This is far enough to expect some significant technology leaps but not too far for them to be unrecognisable. In twenty years, most of us would still be alive and, some us may even be working in that typical site and it is a fair bet that they will not find anything we now cannot understand – no nano-liberators nor little humanoid machines.

2. PLANNING AND DESIGN

Planning forms the backbone of any mining project all problems that are likely to be met need to be taken care of during the planning itself. Failure to identify the trouble spots and lack of co-ordination between different implementing agencies may lead to failure of project. There is need to achieve the protective environment, health and safety of the miners. The social and economical benefits accruing from the project may not be kept limited to the company and the consumers but extended to surrounding inhabitants also. There is a need to make the local inhabitants develop sense of belonging and pride for the project itself.

2.1 Standardisation, Testing, Monitoring and Environment

Globalisation emphasis in standardisation of mining equipment and even the operations, testing and monitoring will play greater role in future. There is a need to develop and standardise such facilities not only in the test houses but also at the manufacturing units and users end to develop national accreditation to ensure quality and assured after sales and service of the products used in the mines.

The need for improving the environment both at work and outside needs no elaboration. With increasing mechanisation the problems of respirable dust and high noise levels are likely to surface soon. Realistic strategies for sampling of airborne respirable dust under some conditions need to be evolved through adaptation and suitable modification methods available elsewhere.

3. GEOLOGICAL INFORMATION SYSTEM

A Geological Information System (GIS) is being used within this research effort to standardise the data collected and provide a common graphical and tabular format and analysis system for researchers to access information on underground coal mines. In addition, GIS's can perform relational analysis on multiple map and tabular data sets, which are not possible through map-creation or CAD package or through databases alone.

In addition to standardising ground control analysis, there is an increasing need for rapid analysis of ground of ground control data in connection with increasing production rates at longwall mines. The examination of a large number of interrelated parameters and their impact on time constraints imposed by rapid production requires computer-aided techniques. A GIS provides the means to quickly, quantitatively, and reproductively evaluate mine-wide conditions.

For example, a typical data set that would be acquired through a mine-wide monitoring system would be pillar and panel loading data taken during mining. These data can be read automatically into a GIS-compatible format and related to other data sets, such as surface subsidence measurements, shield leg pressures, roof characteristics, or time and location of bumps and roof falls in and around a panel. In addition, textual data, such as roof control plans and accident reports, and graphical information, such as shield designs can be attached to appropriate locations or areas of the mine and subsequently retrieved in relation to particular parameter or search for those locations (e.g., in response to a critical pressure value being reached in a nearby pressure cell or shield).

4. INTELLIGENT ROCK MECHANICS

Non-linear relationship of rock mechanical parameters, development of interaction mechanism of geological features and construction, self-learning of rockmass constitutive model, symbolic representation of experience knowledge, development of expert system, integrated intelligent analysis method, estimation of rock burst risks in opening and deep gold mines, optimal design of underground openings, slope stability analysis, and real-time prediction of roof pressure in coal mines.

Artificial Intelligence is a science of simulating human intelligence by using computer technology. Expert system is one of branches of artificial intelligence. It is currently applied in many fields. Intelligent Rock Mechanics (IRM) is a new branch of rock mechanics. Based on the latest developments of artificial intelligence, expert system, neural network, fuzzy mathematics and the ideas of system science, the intelligent rock mechanics is an interdisciplinary science aimed to solve the complex problems in rock mechanics and engineering.

4.1 Expert System for Coal Mine Roof Bolt Design

In geo-mechanics problems the empirical approach is required as the rock is not homogeneous and the exact relationships among various design factors are not clearly understood. The empirical methods usually have some fudge factors whose values are to be determined depending on site conditions from some table or graph or from the experience of the user. It is not even possible for a user to assimilate all this information generated by different domain specific experts. The expert system will help the user to gain access to this huge knowledge base where all the information is stored in terms of simple rules. This system will not be biased and will come up the some answer all the time. This may not be the case for an expert who may get influenced by knowledge and perceptions that do not pertain to the specific problem being analysed. The expert system will not jump to a conclusion and can be used conveniently any time whereas the experts are normally scarce as well as costly and perishable.

4.2 Expert system for blast design

An expert system to design and evaluate blasts has been expanded to include a fragmentation prediction function and an algorithm to select blasting detonator delays. The delay selections are based on realistic mean values and variability estimates.

Because the number of factors, which can cause poor blasting results is very large and each factor can have many levels, a simple procedure to identify the cause of a blast problems is not possible. In addition, symptoms of a poor blast are usually described qualitatively because of the difficulty of obtaining quantitative measures. Therefore, it is impractical to develop a totally mathematically based blast evaluation procedure. Since the diagnosis of blasting problem relies heavily on personal knowledge and experience of an expert to reach a conclusion, expert opinion is probably the only practical approach to solve such a multivariable problem. This expertise is embedded in an expert program, which prompts the user to describe the blast results and the computer reaches conclusions based on the stored information. This process is consultative expert system, and the rules from a knowledge base.

4.3 Expert system for Seismic Hazard Analysis

The purpose of seismic hazard analysis is to define the maximum expected magnitude of event and to estimate probabilities of the occurrence of an event of this magnitude over a designated future time period. The system links both numerical and symbolic processes, providing a tool more powerful than either can achieve in isolation. The inherent structure of the program enables it to handle complicated geophysical and mining information and reach reasonable conclusions speedily.

4.4 Robotics in Mining

There are the objective preconditions for use of automated multifunctional machines in mining. Some methods for evaluation of mining robotisability are offered. Functional modeling and petri nets were used for synthesis and dynamic simulation of mining robotics systems. It enables to construct some robotics-based technologies for underground mining. The trial of rail robocar and tele-loader in the Siberian ore-mines showed it is necessary to have a post-learning of control system.

There are some objective preconditions for robotics use in mining: increase of expenses on a miners labour, limited possibility of traditional mechanisation and automation for a changing mining environment, deterioration of miners labour conditions as the depth of mines, experience accumulation of robotization in machinery. However, a transfer of machinery robotization in mining is impossible owing to working place mobility a changeability of both technological conditions and operations, an unreadiness of mining infrastructure to robotization.

Mining robot is an autonomous mining machine with flexible control that provides the multifunctional use of working head during mining. There are three forms of robotics in mining: creation of technological robots on abase of mining machines with adaptive control, use of remote controlled manipulators for mechanization of manual actions, introduce of information robots in case of accidents. Now there are some prototypes of mining robots, such as automatic drilling machines, roof –bolters, shortcreting machine, tele-loaders

5. HIGH-SPEED VIDEO CAMERA APPLICATIONS

A typical blast lasts on an average 1 to 2 seconds. The event can be broken down into parts with duration ranging from a few hundred microseconds (the passage of explosively induced stress waves through the rock), through several tens of milliseconds (the distribution of the resulting fragments into the muckpile). To assess the results of a production blast, it is necessary to obtain quantitative information on the size of fragment in the muckpile and the rate and velocity at which the fragmented rock material is through a way from the bench face. This call for the use of highly sophisticated monitoring instrument like-high speed video camera.

6. IMAGE PROCESSING TECHNIQUES

In this, the rock muckpile are photographed with a 35-mm camera, medium format camera, or video camera. The photograph images are the digitised and stored in the computer memory. Next, the digitized photographs are processed on the computer to enhance the rock fragments and detect the edges. Once the outlines of the individual rock fragments are defined, the size of fragment can be determined. The size of fragment can be related to the minimum screen size through which the fragment would pass. The final step involves calculating the fragment distribution, converting fragment areas in specified size intervals to pass into the same intervals.

The system can be developed to be fully automatic. It is much faster and care must be taken to check and maintain the accuracy of overall process. Black and white images of high contrast are preferred for this work in addition to avoid shadows whenever possible.

7. LASER PROFILING TECHNIQUES

Laser profiling technique was developed in England. It is remote surveying method used for precisely measuring a face. The results are computed and printed out giving a series of cross sectional profiles of the face and the degree of burden. The portable laser profile system includes a theodolite with electronic distance measurer. It is used to energize and send a pulsed infrared laser beam to the face where it bounces back to the theodolite. The flight time of the send/return beam is recorded, computed and translated into the precise distance between the theodolite and the face. The procedure is replicated in a grid fashion for every 5 to 6 ft until the complete face area has been surveyed.

All recorded data is downloaded to computer/print out system which is set up in the office for obtaining comprehensive read out information and isometric schematics from this information blaster can design a shot with preciseness that was not possible before. We can even combine computer data with geological conditions.

Another important use of the equipment is muck profiling. It is difficult to obtain an accurate survey of muckpile using conventional technique because some one has to climb the blasted rock to provide target points. With laser equipment this is not needed and many points can be measured safely

8. BLAST MONITORING BY VELOCITY OF DETECTION MEASUREMENT

Reliable VOD system allows one to examine how the explosive performs as used in the blasthole. By measuring the VOD of explosive charges in the blasthole, mine operators are able to assess if the explosive is detonating at the rate stated by the manufacturer and is compatible with the properties of the material being blasted. An appreciable difference between the claimed and measured VOD would be a cause for concern and could be due to (1) poor mixing of the explosives, (2) inadequate priming, (3) desensitisation of the charge like penetration of stemming into the charges or by the dynamic pressure generated by adjacent earlier firing charges, and (4) extravagant claim by the explosive manufacturer.

Explosive quality problems, water attack and inadequate priming are examples of problems that can be identified using this equipment. The delay timing accurately can also be determined. The amount of deck stemming needed to avoid cross propagation of independently delay decks can be determined as well. In hole, VOD measurements can also help to identify the effect of sleep time on explosive performance. These are three types of VOD systems.

- Time Domain Reflectometry (TDR)
- Continuous Resistance Wire Technique
- Fiber Optics System

VOD signal is measured with a resistance wire probe and displacement (m) versus (ms). The slope of the signal is a measure of the velocity at which the probe is connected and is calculated without applying any filters or mathematical algorithms.

9. ROOF LOGGING SYSTEM

For effective support it is important to gain an understanding of the geological and geo-mechanical conditions around the roadways-especially to obtain information about the boundary layers, changes in rock types and separation of the strata. Such data make it possible to select the optimal rockbolt type and bolting density. To obtain in-situ data before rock bolting a real-time logging system is being developed for implementation during drilling of the rockbolt holes and field trials with the prototype system were conducted in an underground coal mine. At present, the system for mechanical data acquisition and a data-logger with a memory card.

Information on the roof rock, such as rock type, rock strength, strata constitution and distribution of discontinuities, is indispensable for effective rock bolting. Currently in coal mines core drilling is carried out to determine the RQD (Rock Quality Deformation) and to make a log sheet. Mechanical tests of the physical properties of the rock are also conducted when necessary. These data are used to set a standard of the rock bolting, which regulates the bolt type, bolt length, bolt setting density, etc. Much time is needed. However, for the core drilling and mechanical tests.

A new logging system is, therefore, being developed to obtain information from the performance of a drilling machine, such as torque, thrust, revolution and stroke. All the data were recorded during drilling on a memory card and analysed off-line by the software system.

10. CONTINUOUS MINING TECHNIQUES FOR UNDERGROUND METAL MINES

There are basically two systems of continuous mining in underground mines. One of them, which are used to mine soft minerals such as coal and potash, depends much on a continuous miner. It combines the individual operations of cutting, drilling, blasting, and leading into one operation by this single machine. Another system is chiefly for mining hard metal ores. Based on intermittent and continuous ore breakage, it forms the continuous mining operation line by using the continuously operated ore drawing and haulage equipment.

11. PATTERN RECOGNITION TECHNIQUES

The detection technique of minor faults with seismic parameters and pattern recognition methods has been studied. It provides a new idea in which the computer technique, fractal technique and pattern recognition technique are used in seismic data interpretation, for the structural interpretation of coal seismic data and has two main properties: (1) using more seismic information (particularly dynamic information); (2) using computers with a higher resolution to quantify and analyse seismic data and to locate fault points without influence of interpreters. With his new method, the distribution of minor faults of about 5 m throw (other minor structures) in mining areas can be detected in the area of good seismic geological conditions, and the precision and reliability of interpretation are clearly improved. This creates great economical benefits for the coal mining industry.

REFERENCES

V. Konyukh, Technological applications of robotics in mining, Proceedings International Conference on Mining Science and Technology, 1996, PP 465-470

Hongto et. al, Underground Optical Fibre Industrial TV Monitoring Systems in the Coal Mine, Proceedings of International Conference on Mining Science and Technology, 1996, PP 499-502.

CHAPTER 51

R&D Initiatives for Coal Sector : A Perspective for the 21st Century

*O.P. Rao**

ABSTRACT

Coal and Lignite resources available in the country can meet the energy requirements through out the 21st century at the present rate of production. Mining and utilisation of these resources require R&D support. Indian coal sector has been plagued with chronic problems like low productivity, subsidence, mine fires, and low end use efficiency, most of them are after effects of unscientific mining carried out prior to independence of the country and nationalisation of the coal industry. The chronic problems are adversely affecting the efficiency of mining, environment and efficiency of coal utilisation. Therefore, finding effective solutions to these problems in order to minimise the adverse effects will be a major R&D challenge in the 21st century. Though India is the fourth largest coal producer in the world, technologically lagging behind the advanced countries. India need to adopt the latest technologies in mining as well as energy sectors in order to meet the quantitative and qualitative demand and share of energy from coal and lignite in the 21st century. The concepts of energy security and sustainable development may have to be followed to fulfil the energy demands. The R&D perspective for coal sector in the 21st century, as suggested in the paper, is based on these concepts.

INTRODUCTION

Coal and Lignite are the major constituents of the primary energy resource of the country. India accounts for about 0.8% of the geological reserves and 5.7% of the proven reserves of coal and lignite in the world. The geological coal and lignite reserves in the country are estimated at 204.65 and 27.45 billion tonnes respectively as on first January 1997[1]. More than 70% of the nation's power generation is based on these 2 resources. Coal is an important raw material in the production of steel accounting for about 13% of total coal produced in the country. The estimated coal production in the terminal years of

*Scientist, Council of Scientific and Industrial Research, New Delhi

IXth, (2001-02) Xth (2006-07) and XIth (2011-12) Five Year Plans is 364,418 and 572 million tonnes respectively [2]. However these projections are being revised now.

The projected production figures indicate that coal continues to dominate the energy scenario in the country well into the 21st century. Therefore proper development of these resources forms the main thrust of the country's energy strategy for the 21st century. This requires adequate, appropriate and timely R&D support in finding new or upgrading existing reserves, increasing the production of coal as well as utilising it efficiently, economically, safely and in an environmentally benign manner. Based on this Philosophy a perspective for R&D initiatives for coal sector in the country for 21st century is formulated in this paper.

R&D PERSPECTIVE FOR THE 21ST CENTURY

The review of eighth plan achievements and ninth plan proposals are given in the Annexures I and II. The R&D perspective for 21st century is likely to depend on several issues: Slippage in the achievements during the eighth five year plan (1992-97), targets for the ninth five year plan (1997-2002) and chronic problems in mining of coal and its use, developments in energy sources specially the renewable and their application - including inter substitution of energy sources, adequate availability of oil and natural gas, foreign exchange position, economical, technical and political factors etc. Accordingly the R&D perspective for the 21st century has been discussed in the following paragraphs.

Exploration and Coal Resources

As per the report of the Chari Committee, the present coal resource position is adequate to meet the projected demand in the 9th plan. However, only 3 billion tonnes of recoverable reserves are presently available for meeting partly the 10th plan demand and have been projectised. In order to meet the full projected coal demand of the 10th plan, there is a need to add about 6 bt reserves annually (3 bt on the account of depletion of reserves by mining and 3 bt required for new mines) to the proved category during the 9th plan period [3]. These figures may be revised.

The foregoing highlights the fact that the established coal resources that can be considered for project planning are not adequate and exploration efforts need to be intensified urgently. In order to achieve this the thrust needs to be laid on upgrading the reserves under the categories of "inferred" and "Indicated" to the "Proved" category. Consequently, concerted efforts are required to increase both the regional as well as detailed drilling to meet the future demand projections and increase the coal reserves. Preference should continue to be given for locating opencastable coal and lignite in order to sustain the present production mix of underground and opencast and to meet the demand of coal in the power sector.

In order to increase the quality and the confidence level of exploration data for avoiding geological surprises during implementation like faults, dyke or unforeseen water seepage etc. which could stall mining and cause-huge losses, the adoption of latest technologies is essential. Efforts were made through sustained and continuing R&D activity at Central Mine Planning and Design Institute to increase the pace of detailed exploration and make it cost effective with multi parametric, multi disciplinary data for a better geological modelling of the coal sources. Many such R&D programmes have been used gainfully, and are being employed on routine basis for geological exploration. These efforts need to be accelerated through induction of latest technologies for faster and reliable data.

Exploration activities are necessary not only for augmenting the coal reserves, but also for acquiring comprehensive geo-technical information which is useful for taking investment decisions and to minimise exploitation risks. In terms of financial inputs the cost of exploration is insignificant; varies from 3 to 5% of the total investment [3]. Additional investments in exploration can provide the much needed assurance of availability of reserves, improvement in the quality of exploration inputs and eliminate avoidable delays in mine planning and implementation of projects.

Considering the long gestation period involved in implementing coal and lignite projects, it is necessary to take up a comprehensive exploration programme on a long term basis. A shelf of projects can then be prepared and specific projects can be taken up depending upon the requirement and cost.

It is necessary to give equal emphasis on upgradation of reserves as well as regional exploration work, whereby new deposits could be established.

Extraction of Coal from Existing/old/Abandoned Mines

The unscientific mining or slaughter mining practised prior to independence of the country and nationalisation of coal mines had left millions of tonnes of good quality coal unmined which is a resource to be tapped now. Examples are coal locked in pillars of multiple and thick seams, underneath aquifers, fragile ecology, and surface structures etc. The quantum of coal still available for extraction can be estimated by the following examples.

The prevalent method of Bord and Pillar mining accounting for major production from underground mines is characterised by low production and productivity in addition to blocking of large reserves in thick and multiple seams. Thick coal seams account for nearly 60% of the total coal reserves. The extraction has been deplorably low to the level of only 30% resulting in nearly 2500 million tonnes of good quality coal getting locked up in developed Bord and Pillar workings [4]. In Jharia coalfields alone 900 million tonnes are locked in pillars [5]. Mechanised depillaring need to be given emphasis in 21st century for extraction of thick seam standing on pillars.

A large number of unknown, unstable, abandoned and water logged workings exist in Raniganj and Jharia Coalfields. Approximately 363 million tonnes of coal are estimated to have been locked up in the virgin seams below these old workings. Sprawling townships have developed and are still continuing to grow over these unsafe workings [6].

Identification of such old workings is a burning problem for the coal mining industry because accurate mine plans are not available for locating them to do the repair work in order to make them safe. In absence of proper information, during the development of present workings there is always apprehension of intersecting with old workings leading to major disasters. Hence, at most of the places an excess barrier is left as a safety measure to avoid possible disasters. Determining barrier pillar thickness is a challenge. Thus identification, delineation, proper stowing and extraction of coal below abandoned workings and extraction of coal locked up in pillars are challenges in the 21st century.

Technologies have been developed and applied for optimal liquidation of pillars underneath aquifers and fragile ecology, for example the wide stall Method through which 40% recovery of coal is possible[7]. However further R&D work is essential to increase the recovery of coal from pillars safely under different geomining conditions.

Mining Technology

The total projected production in the ninth plan and subsequent periods is envisaged to come from opencast and underground mines around the ratio of 75:25 [1]. This trend is fast depleting the reserves amenable for opencast mining and neglecting the development of underground mines where large potential exists for meeting the future demand. For example only 30% of coal reserves in Coal India Ltd. are amenable to opencast mining, while 70% by underground mining. But at the moment 75% production is coming from opencast mines [2]. This trend needs to be reversed and more emphasis is required in improving underground mining production and productivity in the 21st century.

Bord and Pillar method of mining remains to be the most popular method of underground mining: 88.7% in 1996-97 and 84.4% in 2001-02, while Longwall technology formed 5.5% during 1996-97 and 12.9% in 2001-02. In underground coal mining, the Longwall technology is used upto 100% in China, UK, and Germany and Australia produces about 65% of underground production from Longwall faces [2]. In India, the production from longwall is not even 10%. The main constraint in increasing the production from Longwall, appears to be that the geomining conditions in India are not favourable for large scale introduction of mechanised Longwall mining [2]. It will be necessary to overcome this constraint to practice Longwall on a larger scale in the 21st century to improve the production from underground mines.

Continuous developments are taking place worldwide in the mining technology and India has to catchup with them to meet the demands of the 21st century. Several developments have already taken place in the modernisation of opencast mining. Higher capacity equipment are being used with performance and health monitoring systems. Operational efficiency of the equipment has been improved. There is a continuing trend towards the use of operational monitoring and control system as a means of enhancing productivity and efficiency. Automation of operations is being contemplated. The most significant developments in the automation of surface mining have occurred in the mine design and operational planning etc. Surface miner is available for selective mining [2]. It is possible that robots may be used for mining of coal in the hazardous mines.

Millions of timber props and wooden sleepers are used annually in Indian coal mines, which is disturbing the ecological balance due to deforestation. Adequate availability of timber will be a constraint. Moreover with the use of timber, roof fall and to some extent side dislodging are still the major contributing factor for fatal accidents. At present huge amount of coal is locked up in developed pillars of Bord and Pillar workings. Depillaring provides a huge stock for good quality coal. Wooden supports create problems for manoeuvrability and mechanisation in the depillaring work. In view of the above alternate roof support systems are required. Roof bolts appear to be the suitable and easily available alternative for the purpose. The roof bolting technology is being implemented in Indian coal mines. But further development is required to apply roof bolting under different geomining conditions. The aim of 21st century could be "Timberless Mining".

Mining economics depends upon efficiency of blasting. Hence, the emphasis in the 21st century in the area of blasting has to be on high degree of production, productivity and safety with the utilisation of minimum resources of men and machines, materials and money and also to lower down the cost of blasting.

Sand is the principal stowing material. Availability of sand is a constraint. Flyash as a stowing material is being tried. Many other alternate material to sand have to be tried and established for stowing in the 21st century. This topic assumes importance since the share of coal production from the underground mines need to be substantially increased, which results in large requirement of stowing materials and old and abandoned workings have to stowed for safety and for extraction of coal locked up below these workings.

Lignite Development and Utilisation

The important known occurrence of lignite in India is confined to the states of Tamilnadu, Rajasthan, Gujarat, Kerala, Jammu and Kashmir, and Pondicherry. The geological reserves assessed as on 1.1.97 are 27.45 billion tonnes of which 94% are in Tamilnadu and Pondicherry [1].

Power sector remains to be the major consumer of lignite accounting for 78.5% of production during the terminal year of IX plan (2001-02) [2]. The demand of lignite is likely to increase in view of increasing gap between production and demand of coal. Besides power generation, lignite is being used as fuel in chemical, paper, cement, textiles and sugar industries.

The demand of lignite based power generation is increasing substantially and acceptability of lignite in other industries such as cement, textiles, chemical and brick industries is gaining momentum. Hence it is estimated that demand for lignite could continue to be there during IX and Xth Five Year Plans and whatever quantity of lignite is produced could be utilised. Thus there is both the need and rational for stepping up the production and utilisation of proved reserves of lignite by increasing the production and productivity from existing mines and by increasing the area of exploitation.

The following are emerging technologies for utilisation of lignite.

- Underground gasification wherever geomining conditions and economics permit.
- Integrated Gasification Combined Cycle for power generation.
- Lignite oil tar slurry to substitute petroleum fuel in boilers.
- Use of lignite in blast furnace for production of liquid iron.
- Manufacture of humic acid from lignite in an economical way and use the humic acid as organic manure in agriculture.

Coal Bed Methane

India is a major producer of coal and is expected to produce about 346 million tonnes by 2001-02. Due to fast depletion of shallow coal deposits, extraction from the deeper mines will be necessary to meet the future requirements. Winning of coal from the deeper seams will involve greater emission of methane in the underground. Methane is a pollutant and a potent greenhouse gas contributing to global warming. India has substantial coal reserves about 206 billion tonnes. Coal bed methane (CBM) associated with coal reserve could be a major source of energy. Commercial extraction of coal bed methane is being done in USA.

The CBM industry in India lags behind primarily because of the knowledge gap of its resource potential, non availability of suitable technology and inadequate funds. Some work has been initiated in India by oil producing companies for exploration of CBM. CMPDIL jointly with UNDP proposed a project for coal bed methane recovery from Indian coal mines and its commercial utilisation. The cost of the project is estimated to be Rs.69.21 crore. The objective of the project is to degassify coal seams for safer extraction and to demonstrate viability of harnessing CBM in India [8].

Favourable conditions for occurrence of coalbed gases are generally limited to areas where bituminous coals occur. In India, such areas are mostly located in coalfields of Damodar Valley namely Raniganj, Jharia, East Bokaro, West Bokaro, Ramgarh, South Karanpura, North Karanpura Coalfields [8].

The Development of coalbed methane has particular relevance to India on account of the following: [8]

- A study of the location of coal and lignite fields vis-a-vis petroliferious basins in India would indicate the production centres of coal are mostly located in the Damodar-Son-Mahanadi valleys and are far away from oil and gas producing centres. CBM from such areas acts as natural gas.
- There is overall shortage of electricity in the country. The shortage is particularly acute in the Bihar - Bengal region where coal resources are available in abundance and possibility exists for availability of coalbed methane. Thus, the existing shortage of electricity may at least be partly bridged using this unconventional source of energy.
- Coalbed methane is a major contributor towards the green-house effect. As a matter of fact, many countries, including India, which are poor in oil/gas resources and have relied heavily on coal burning as primary energy source resulting in serious air and water pollution, can draw some benefit from coalbed methane.
- Thick seams have been encountered in many major coalfields. These seams are being presently quarried in the shallower regions. Exploitation of thin, average, medium and thick seams beyond certain depth have their own limitations. At depth beyond certain limits mining problems of strata control, methane emission, fire hazard etc. also multiply leading to limitations in exploitation and economic mining. Extraction of coal bed methane under such conditions helps in taking out a part of the stored energy and in some cases may also lead to removing the hazard of emission due to degasification.
- Deeper coal deposits which otherwise are uneconomical to mine at present day economics and/or are not likely to be mined in the fore-seeable future can be accessed for CBM recovery and thus effectively release a part of the stored energy in coalbeds.

Man and Machine Productivity

The productivity in Indian coal mines has been very low as compared to the world standards which is more pronounced in the case of underground mines. The productivity from underground coal mines continues to be low and stagnated and whatever increase achieved in OMS, is due to higher productivity from the opencast mines. The figures for productivity during eighth and ninth plans are given in Table 51.1 in Annexure-I. This trend is somewhat disturbing as 75% of the workforce is deployed in the underground mines and their contribution is only 27.5% of the production resulting in very high costs which may not be sustainable. With the rapidly depleting reserves of coal amenable to opencast mining, the country's coal production may therefore plateau at 500-600 million tonnes per year, if not at lower levels, unless new deposits are located [3]. This production may be low in the 21st century.

The low productivity of mines of Coal India Ltd. have contributed largely towards the adverse working results, however the large manpower which has been inherited by CIL has affected the productivity adversely. OMS for CIL is targeted at 2.24 tonnes (UG-0.C9 t, CC-6.53 t) in 2001-02 against the actual achievement of 1.86 t (UG-0.57 t, OC-5.12 t) in 1996-97 [1].

Though during the eighth plan period, the targets of machine utilisation have been achieved when compared with the norms fixed by CMPDIL for various Heavy Earth Moving Machinery (HEMM), there is a scope to enhance machine utilisation. Productivity for HEMM in terms of machine utilisation in CIL is targeted at 72% for shovels, 60% for dumpers, 63% for dozers, 65% for drills and 93% for draglines against the CMPDIL norms of 58%, 50%, 45%, 40% and 73% for these machines [1]. There is an urgent need to review the capacity utilisation as well as the norms to increase the productivity from HEMM in the 21st century, since the equipment are progressively becoming technologically more sophisticated efficient and expensive.

Subsidence and Fires

The Indian coal mines have a chronic problem of subsidence and fire since 19th century mainly because of the unscientific mining. A large number of studies and investigations have been carried out to understand the subsidence behaviour of coal measures and in preparing norms establishing relationships between subsidence parameters and subsidence predictive models. The emphasis in 21st century could be on development and implementation of subsidence control measures and management methodologies.

The number of fires in 1994 in all Indian coal mines was 196, including 65 active fires in Jharia coalfield alone. The coal blocked due to fires in India has been roughly estimated as 1900 million tones and coal burnt due to fire is about 45 million tonnes [9]. Till March 1996 an amount of Rs.75 crore was spent on dealing with the fires [10]. Though many fires have been controlled with the technologies developed indigenously, much work is required to be done in the 21st century to fully tackle mine fires.

Reclamation of Land

Mining activity displaces people and affects the environment and ecology of area. Rehabilitation of people affected due to mining is becoming a social, economical, legal and political problem and will be an important issue in the 21st century. Reclamation of mined out areas is an effective solution to the problems and a challenge to achieve sustainable development. Development of a reclamation methodology for the abandoned and worked out coal mine site is very necessary for restoration of ecology of the area and to restore the degraded land for some useful purpose.

In the period prior to Nationalisation due to selective mining and uncontrolled caving, large tracts of subsided areas are found in older coalfields like Raniganj, Jharia and South

Karanpura. In Raniganj area the land affected by mining was estimated to be 74.23 sq.km consisting of 143 surface subsidence, 223 abandoned coal quarries and 6.4 sq.km of fire [11].

Ecosystem destruction by opencast coal mining is inevitable. During mining, not only the vegetation is destroyed but there also occurs loss of soil, The flora and fauna, hydrological relations and soil biological systems are drastically disturbed. Huge amounts of infertile overburden materials are dumped on adjacent lands in addition to pit scarred landscape.

Slopes of external dumps are an important area to be considered for proper and effective reclamation. The slopes of external dumps are to be suitably, graded/terraced for plantation. Scientific studies have to be undertaken for this purpose, whenever required.

For effective bio reclamation of disturbed land, scientific studies may have to be carried out to select suitable species for each coalfields and also sustainable sequence of reclamation from grass to shrubs.

Thus, the research activities for dump reclamation need to address the following objectives: maintain longterm stability of slopes, stabilise soil against wind and water erosion, control air and water pollution, maintain and enhance aesthetics of mining areas, preserve the biodiversity and finally to put the land to productive use at minimum cost.

Coking Coal Washeries

The performance of coking coal washeries has not been satisfactory, over the years, the capacity utilisation and yield have been low, achieving 65.2% and 45% respectively in 1997-98 [12].

The washeries are also under utilised due to shortage of raw coking coal of suitable quality. CIL washeries commissioned during sixties, were designed to wash raw coal of 22-25% ash content, but ash content gone up to as high as 30% which is a factor responsible for poor performance of washeries. The ash content in washed coal being supplied to steel plants is around 21% [1].

Other reasons are:

- Feed coal being from a number of sources has inconsistent quality, size and washability characteristics.
- Feeding of oversized materials cause jamming of washery circuit resulting in poor yields.
- Presence of shale, stone cause problems.
- Delay in modernisation/modifications of washeries as recommend by Altekar Committee.

Poor performance of washeries increases the cost of washing and leads to higher imports of coal.

Maximisation of use of indigenous coals for steel sector

Coking coals constitute only 15% of the total coal deposits in India and only 18% of the coking coal reserves are of prime coking variety. India's imports of coking coal are steadily going up on the ground that the domestic coking coals are deficient in quality and quantity. If energy security is policy of the country, then import of coal need to be discouraged to allow maximum use of indigenous coals. There are technologies for the use of non coking coals and sub standard coking coals in steel making, namely partial Briquetting of Coal Charge, Stamp Charging, Selective Crushing of Coal, and Coal Dust Injection into blast furnace. These technologies have been introduced in the Indian steel plants and if they are pursued vigorously with R&D support for adaptation can lead to greater use of domestic coals in place of imported coals. Another indigenous source to be tapped is Low Volatile Medium Coking Coal (LVMC).

The proven reserves of LVMC coal are 5.7 billion tonnes and constitute about 34% of the total coking coal reserves in India. These coals are characterised by high ash content (32 to 36%) and very difficult washability characteristics. However, studies indicate that when beneficiated to 17% ash they exhibit good coking properties and can be used as an excellent blend. The present production of LVMC coal, which is around 20 MT per annum, is being fed to power houses and other non-metallurgical purposes. Washing of LVMC coal either in existing washeries by suitably modifying the circuits or by establishing a new washery would help in augmenting domestic coking coal availability. Existing washeries which are designed for upper seam coals can not produce techno-economically acceptable product from LVMC.

Beneficiation of Power Grade Coal

The power grade coals presently supplied to power plants have average ash content of 45% (range 30-55%) and average calorific value of 3500 kcal/kg (range 3000 to 5000 kcal/kg) and are associated with unwanted non-coal materials like shales, stones, and iron pieces etc [13]. The quality of coals in future are likely to deteriorate further with ash content going above 50%. Use of high ash coals is affecting the performance of thermal power plants in terms of lower efficiency, low PLF (average 60%), higher pollution and higher maintenance cost. Beneficiation of coals to a limited extent can reduce the problems. The need for beneficiation of coals for power generation, its technoeconomic advantages have been the subject of debate for more than 2 decades. Though the techno economic advantages of use of beneficiated coal in the power plant have been established through studies and practical trials, the commercial application is yet to begin.

The Ministry of Environment and Forests has issued a notification making it mandatory to use washed coal with 34% ash content in the power plants located at 1000 kms from coalfield or in sensitive areas from 1st June, 2001. The R&D studies have shown that the ash content of 34% can be achieved through limited beneficiation. This needs to be established through demonstration plants before large scale commercial applications of power grade coals which is unavoidable in the 21st century. The beneficiation techniques can be dry or wet which depend upon technoeconomics of washing.

Clean Coal Technology

This term entered the energy vocabulary in 1980s. It represents a new generation of advanced coal utilisation technologies, that are environmentally cleaner and more efficient and often economical than conventional burning processes. The clean coal technology (CCT) is meant for 21st century to meet the efficiency and environmental needs for utilising coal on a large scale. CCT are appropriate to utilise inferior grade coal efficiently and cleanly, hence suitable to India. CCT can be applied all along the fuel chain from coal mine to its end use. CCTs have been developed/under development in USA, Europe, China and Japan etc. in the four market sectors.

- Advanced electric power generation.
- Environmental control devices.
- Coal processing for clean fuels.
- Industrial applications.

The integrated gasification combined cycle (IGCC) and pressurised fluidised bed combustion (PFBC) are the 2 advanced electric power generation systems under development which are suitable for Indian coals. IGCC has been demonstrated in the capacities ranging from 117 MW to 350 MW and the technology is poised for commercial application. The feasibility studies carried out with Indian coals have shown that IGCC is technically and environmentally much superior and economically comparable with the conventional pulverised coal fired (PC) power plant technology presently adopted in India. Therefore IGCC has the potential to replace the PC technology in India in the 21st century. Supporting R&D and demonstration are required for adaptation of IGCC technology to Indian coals prior to commercial application.

IGCC enables to use coal as a raw material and, therefore, offers multiple choices to improve the economics of coal utilisation. The fuel source for IGCC is the synthesis gas produced from coal gasification. Besides power generation, the synthesis gas can be used for production of chemicals, fertilisers, and liquid fuels offering the options for cogeneration and coproduction. Super critical steam condition in coalfired boiler can improve the efficiency but does not provide flexibility of coal use and higher efficiency.

There is a technology gap in PFBC which is hot gas cleanup. Indian coals contain high ash and also the ash is highly abrasive which can lead to faster erosion in combustor and fuel path. In view of these constraints, PFBC may find application in India much latter to IGCC. However the fluidised bed technology is suitable for high ash coals.

The efficiency of IGCC at present is about 45% which is projected to go up to 60% with the developments taking place in the gas turbine technology.

Technologies have been developed as well as under development for control of emission of particulate, and oxides of sulfur and nitrogen from the pulverised coalfired power plants in the advanced countries.

At present in the Indian power plants, only the particulate matter is controlled through the incorporation of electrostatic precipitator. Emission of sulfur and nitrogen will need to be controlled in the 21st century. The technologies are available in the advanced countries and can be imported and applied to the Indian plants. R&D work is not required. In addition to these, emission of particulate matter below 2.5 micron and trace elements like mercury also need to be controlled in the 21st century.

In India, the present reserves of oil and natural gas are likely to last for 25 to 30 years. It is experienced that imports of oil can get affected by political reasons. Energy security is going to be a policy for many countries in the 21st century. If this concept is followed in India there is no option but to go for conversion of coal to oil which requires extensive demonstration with Indian coals. In any case it will be a good strategy to develop and demonstrate the coal to oil technology on semi commercial scale in the country and be prepared to face any eventuality. There are many technology gaps which need to be addressed to reach the semi commercial scale.

In the area of industrial application, the high concentration coal water slurry containing 70% solids and 30% water can replace pulverised coal, fuel oil and furnace oil etc. in the boilers and furnaces. In Indian coals, the carbonaceous matter is finely mixed with the mineral matter and needs finer grinding for beneficiation. Dewatering of fines in the beneficiation process is inefficient and expensive. Use of coal as slurry eliminates the need for dewatering. An alternate transport system to rail transport is required for coal movement which is possible through pipeline transport of slurry. Hence coal water slurry has a special relevance to India. Extensive R&D is required in the beneficiation of coal, slurry preparation and additives, pipeline transport and slurry combustion for application of coal water slurry technology in the country. Work on pilot plant scale of 2 tph capacity was carried out in India.

The concept of coal refinery similar to oil refinery is proposed in the advanced countries which will be suitable to India in the 21st century.

CONCLUSION/STRATEGY

Coal and Lignite continue to be the main primary energy resources and can meet the energy requirement of the country for the whole of 21st century. The quantity and quality of coal and lignite have to meet the energy needs. For this, R&D support is essential for exploration, mining, environment control during and after mining, safety, reclamation of land and rehabilitation of people, beneficiation and utilisation of coal and lignite etc. The concept of Sustainable Development will play a major role in extraction and utilisation of coal and lignite in the 21st century.

Indian coal sector is beset with chronic problems such as low productivity, and mechanisation in mining, mine fires and subsidence, abandoned mine workings, low priority on reclamation of mined out areas and environment management, low yield from coal washeries, and low priority on introduction of clean coal technologies. All these problems need to be solved with the support of appropriate R&D for full exploitation of coal and lignite in the 21st century.

The coal sector requires speedy and reliable exploration techniques to make available enough reserves to meet the demand. Present trend of mix of coal production from under ground and opencast mines need to be reversed and higher proportion of coal production has to come from underground mines. Man and machine productivity has been very low as compared to the world standards. This needs to be tackled on high priory by improving the performance of both man and machine. Environmental issues will assume greater importance in coal mining as well as coal use. Adequate attention and resources will be required on reclamation of land, and rehabilitation of people and restoring ecology and biodiversity.

At present only the coking coals for steel sector are beneficiated. The benefits of use of washed coal for power sector have been established. The coal producer, power producer and environmentalist are in favour of use of beneficiated coal in boilers. The first priority should be on setting up of non coking coal washeries followed by R&D to further improve techno economics of washing.

Coal and lignite are now utilised mostly through combustion route which cannot meet the efficiency and environmental norms of the 21st century. The role played by coal and lignite has to change from a mere energy resource to utilising it as a raw material. The clean coal technologies can fulfil all these requirements. They are capable of attaining efficiencies upto 60%, control the emission of particulate and oxides of nitrogen and sulfur to very low levels economically and use coal and lignite as raw material.

Summing up, coal and lignite will constitute major raw material for energy and industrial sectors and will be utilised more efficiently, cleanly and economically in the 21st century.

REFERENCES

1. Ninth Five Year Plan (1997-2002), Volume II, Planning Commission, New Delhi.
2. Report of the Working Group on Coal and Lignite for the Ninth Plan, (1997-2002), Planning Commission.
3. Report on Integrated Coal Policy 1996, Chari Committee, Planning Commission.
4. Optimal Extraction of Coal Locked in Pillars of Multiple and Thick Seams, National Institute of Rock Mechanics, KGF, 1998.
5. Development of a Model vis-a-vis study of parameters Influencing Abutment Loading of Pillars etc, Project proposal Central Mining Research Institute, Dhanbad, 1998.
6. Seguin K.M, N.Sen and O.P. Gupta, 1995, Final Report, Project on Identification of Abandoned Unknown and Unsurveyed Underground Collery Workings in Raniganj Coal field, Geophysical study.
7. Optimal Extraction of Coal Locked in Pillars Underneath Aquifers and Fragile Ecology, Project Proposal, Central Mining Research Institute, Dhanbad, 1998.
8. Coalbed Methane Recovery and Commercial Utilisation, Central Mine Planning Design Institute, Ranchi, 1998.
9. Background Material on Environmental Measures and Subsidence Control in Mines, Department of Scientific and Industrial Research, New Delhi, 1997.
10. Proposal for Dealing with Fire and Subsidence at Rajapur/South Jharia Colliery, Central Mine Planning and Design Institute, Dhanbad, 1999
11. Restoration of Abandoned Coal Mines in Eastern Coalfields, Central Mine Planning and Design Institute Ltd, 1995.
12. Varma S.K., 1999, Demand and Availability of coking coal for the Indian Iron and Steel Industry. Proceedings of the International Conference on coking coal.
13. Report on Scheme for Dealing with Fire and Subsidence at Rajapur/South Jharia Colliery, CMPDIL, Dhanbad, 1997.

Annexure-I

REVIEW OF EIGHTH PLAN ACHIEVEMENTS

Major objectives for coal sector during eighth plan were: scientific evaluation of coal and lignite resources; lignite development; a coordinated approach towards production, transportation and pithead stock utilisation to meet the demand; improvement in productivity; maximisation of the use of indigenous coking coal for steel making; beneficiation of non coking coal; introduction of effective environmental management plan and making the coal industry self supporting.

A statement comparing the targets and achievements in the Eighth plan and giving Ninth plan targets is available in Table No. 51.1 and the salient features are indicated below:

The coal exploration target was aimed at meeting the demand for coal over a 15 years perspective through intensification of regional exploration in new areas and promoting recent discoveries. As a result of exploratory efforts the geological coal reserves have increased from 196 to 204.65 billion tonnes, while the lignite reserves have increased from 6.47 to 27.45 billion tonnes during the plan period. But the proved coal reserves have increased from 61.14 to 72.73 billion tonnes which corresponds to only about 35% of the available geological reserves. Main reason is that only about 75% of exploration drilling target was achieved [1].

The average annual rate of growth in the production of coal fell from 6.08% to 4.5%, which can be attributed to the decision of the Ministry of coal not to takeup projects yielding less than 16% of IRR resulting in delay in implementation of new projects. This had also resulted in the decline of underground: Opencast production ratio from 28:72 to 25:75 [1].

Productivity from the underground mines continues to be low and the improvement in output was mainly confined to opencast mines.

Though targets for utilisation from heavy earth moving machinery were achieved, there is a need to review the capacities and norms in order to achieve optimum utilisation of machinery.

The problem of higher ash content in the coking coals continued and the ash content in washed coals supplied to steel plants was 21%. In order to maintain the desired ash level of 17% plus or minus 0.5%, the steel industry is increasingly depending on imports which have increased to 9 million tonnes during 1996-97 against 3.5 mt originally envisaged [1].

The proposed utilisation of Low Volatile Medium Coking Coals (LVMC) to increase the availability of domestic coals for steel sector did not comeup to the required extent. No major R&D project was undertaken during the eighth plan period and the plan expenditure was Rs.19.78 crore (23%) against an approved outlay of Rs.87 crore earmarked for coal S&T projects by the Ministry of Coal [1].

Table 51.1: Targets and Achievements in Coal Sector [1]

Item	*Eighth Plan Period*		*Ninth Plan Target*
	Target	*Achieved*	
Regional Exploration Drilling, meters	403400	299658	10,00,000
Coal production, average annual growth rate, %	6.08	4.5	5.3
Ash content in the washed coking coal, %	17%	21%	17%
New Coal projects sanctioned and started, million tonnes	72	43	62
Productivity (CIL)			
-Output per Manshift (OMS)	1.65 t	1.82 t	2.24 t
-(Underground^{UG}, opencast^{OC})	(UG 0.66 t OC4.32t)	(UG 0.59t OC4.88t)	(UG0.69 t OC6.53 t)

Annexure-II

NINTH PLAN PROPOSALS (1997-2002)

The R&D priority areas identified were as follows: [1,2]

- Measures for augmenting domestic coal production to reduce the huge gap between demand and supply since large scale imports are not feasible due to lack of facilities at the ports and constraints of foreign exchange.
- Improvement in productivity of men and machines.
- Improvement in mining technology.
- Expediting coal and lignite exploration to match the reserves with the projected demand/production.
- Development of equipment/machine suitable to Indian conditions.
- Productive techniques for prevention and control of mine fires.
- Stabilisation of worked out areas.
- Alternatives to sand for stowing and replacement of timber for supports.
- Environmental management in coal mining areas.
- Utilisation of coal Bed Methane as a source of commercial energy.
- Recovery of coking coal from washery middlings, slurry and fines.
- Beneficiation of non coking coal for power generation.
- Development of clean coal technologies.
- Demand side management.

CHAPTER 52

Behaviour of Fractured Roof Strata over a Longwall Panel and Determination of Capacity of Powered Roof Supports

*Samir Kumar Das**

SYNOPSIS

Selection of powered roof supports in the longwall face must be done very carefully under any roof conditions particularly under the envelop of fractured roof strata since the latter behaves very erratically and offer many abnormal situations. The behaviour of this fractured roof strata was minutely observed in the F_2 longwall panel of XVI combined seam of the Moonidih colliery of B.C.C.L. situated in the Jharia coalfield of India. In many occasion as soon as the shearer used to cut coal the fractured roof strata, in between the face and tip of the canopy use to cave down instantaneously causing tremendous amount of strata control problems. This strata control problems tuned the face powered roof supports to behave very erratically with lowering down efficiency. By intense field study it was also observed that the healthy medium strength massive roof strata present over the thicker fractured roof strata, only offer the load due to it's own deflection. The details regarding the investigation, examination and analysis of behaviour of the total roof strata and the powered roof supports of the above mentioned F_2 longwall face are highlighted in this paper.

An intensive field study and experimental works were carried out in number of longwall panels in various mechanized coal mines situated in various coalfields in India and actual behaviour of coal measure roof strata are evaluated and new strata mechanics concepts around the longwall panel developed which prompted author to develop a very simple but realistic mathematical model to determine the optimum capacity of the powered supports. This mathematical model considered the total loads offered by the detachment of strata layers/composite layer blocks right from the face and also loads due to the deflection of strata layers/composite layers situated above the detached roof rockmass. This formula along with the other associated developed formula by the author calculated

* Associate Professor, Department of Mining Engineering, Indian Institute of Technology, Kharagpur–721 302, West Bengal, India

the optimum capacity requirement of the powered supports (chock-shield supports), to be used in XVI combined seam of the said mine which is 567 tonne (for support spacing 1.5m) i.e. only 84 tonne higher compared to the existing design one. It is also seen that under the fragmented roof condition, the horizontal strata pressure always tend from the face towards the goaf side. Therefore, all the legs of the chock-shield supports shall be inclined towards the face so that they can offer higher horizontal reaction forces from the goaf towards the face. It is also found that the split base designed powered roof supports are unsuitable in Indian mining conditions.

1. INTRODUCTION

Mechanised coal wining in medium thick coal seam by the sophisticated split base powered roof supports had been first introduced in the F_1 panel in the XVI combined coal seam of Moonidih colliery of BCCL which is situated in the Jharia coal-field of India. A detailed strata mechanics investigation and examination and compatibility of the powered supports had been carried out in this panel as well as in the next F_2 panel. Similar investigations were carried out in numerous longwall panels in various mines situated in various coal-fields in India where mechanised single pass longwall system are being practised but all such faces incorporated rigid based powered supports. All these investigation yielded new thoughts on actual strata behaviour over the longwall panels and a very realistic mathematical model for the determination of the capacity of the powered supports has been formulated. This formula is based on the combined loads offered by the detachment of strata layer/composite layer blocks right from the face and also loads offered due to the deflection of strata layer/composite layers situated above the detached roof rockmass. This formula along with the other associated developed formulate by the author prompted to determine the optimum capacity requirement of the powered supports to be used in the XVI combined coal seam of Moonidih colliery. In India coal seam of less than 1.5m thick is considered as a thin seam, from 1.5m to 4.8m is the medium thick whereas above 4.8m is considered to be a thick seam.

2. DETAILS OF THE PANEL

The pertinent details of the F2 longwall panel (Fig. 52.1) are as follows:

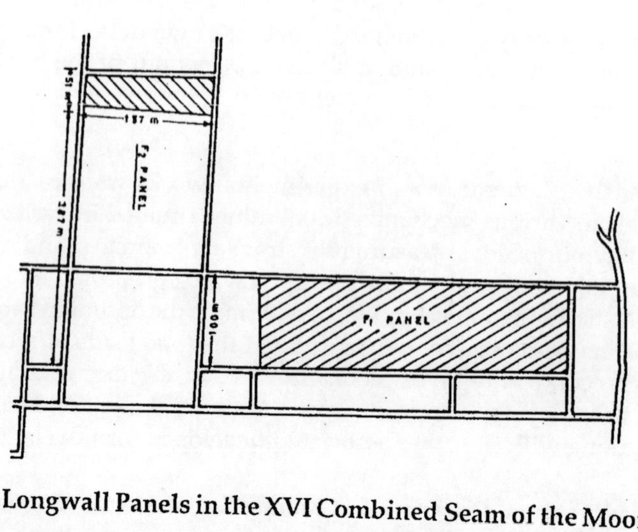

Fig. 52.1 : F_1 and F_2 Longwall Panels in the XVI Combined Seam of the Moonidih Colliery

Name of the Panel	:	F_2
Location	:	In between 400m and 500m horizon at XVI combined seam
Depth of working	:	Minimum : 470m and Maximum 500m
Face length	:	157 m (centre to centre)
Panel length	:	338m
Seam thickness	:	3.0 to 4.2m
Average thickness	:	3.5m
Height of extraction	:	3.0 to 3.5m

There are numerous stone bands in the seam. Apart from these stone bands, the panel is affected by two faults with throws around 0.4 m and 0.7m respectively. There are also numerous igneous intrusions like sills, stone boulders and also jhama along the floor level.

Gradient of Seam	:	6^0 to 9^0

2.1 Roof and Floor

Immediate Roof : As per borehole litholog No. NCCJ-16, it is clearly indicated that roof above the powered supports consists of 7.6m intercalated sandstone and shale. Above this intercalated sandstone and shale layer there is 12.8 m medium to coarse grained sandstone layer and afterward there is 7.6m and 8.6m intercalated sandstone and shale layer and medium grained sandstone layers respectively. There is no uniform consistency of immediate roof strata over the XVI combined coal seam. The roof rocks are fractured and intercepted by numerous slips. Fig. 52.2 is the NCCJ-16 borehole litholog of Moonidih colliery.

Overlying coal seam from XVI Combined Coal Seam

XVII Bottom coal seam	-	Parting : 80m
XVII Top coal seam	-	Parting : 120m

Above this F_2 panel XVII Top coal seam was extracted whereas XVII Bottom coal seam was partially worked out.

Immediate Floor : Immediate floor consists of 1.6m carbonaceous sandy shale and thereafter there is a layer of 1.5m medium grained sandstone. Underlying coal seams in the panel are virigin.

Size of the gate roads are 4.8m in width and 2.75m is height (approx.).

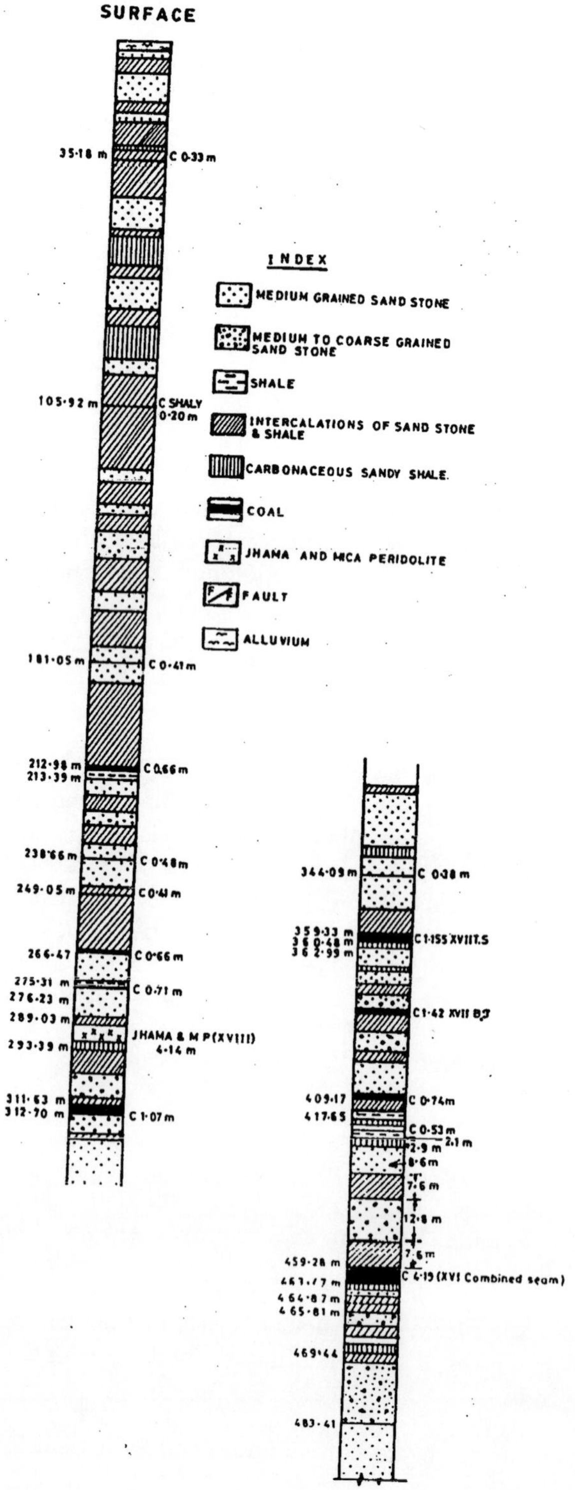

Fig. 52.2 : N.C.C.J-16 Borehole Litholog of Moonidhi Colliery

2.2 Brief about the Face Machineries

Shearer

Type : KGS-320W/2 BPH Narrow Web Double Ended Ranging Drum shearer

Drum diameter	:	1.8 m
Web Depth	:	0.63 m
Maximum cutting height	:	3.75 m
Power	:	2 x 132 KW

Armoured Flexible Chain Conveyor

Type	:	225/BP/160/IND
Power	:	2 x 160 KW

Stage Loader

GROT	:	225/750/IND
Capacity (maximum)	:	700 tonne per hour
Power	:	160 KW

Powered Supports (Fig. 52.3)

Type	:	FAZOS-17/36-PO_2/SW-1.5 Chock Shield Powered Roof Supports
Closed height	:	1.7 m, Open height: 3.6 m
Support centres	:	1.5
Number of legs in each support	:	4
Face inclination	:	Upto 35^0 (longitudinal) ± 15^0 (transverse)
Setting load	:	4 x 104 tonne
Yield load	:	4 x 121 tonne
Setting resistance	:	61.6 to 67.5 tonne/m^2
Yield resistance	:	84.00 to 90.22 tonne/m^2
Mean floor bearing pressure	:	214 tonne/m^2
Mean roof bearing pressure	:	86 tonne/m^2

Support advance	:	Upto 0.7m
Base	:	Split base type

Effort of Advancing Ram

For shifting face conveyor	:	19 tonne
For advancing support unit	:	34 tonne
Weight of each unit	:	14.3 tonne
Type of control	:	Hydraulic adjacent control
Supply pressure	:	240 to 290 bar

Power Pack

Type	:	T-100/32, Capacity 100m³/min
Working pressure	:	160 to 300 bar
Maximum pressure (for short duration)	:	340 bar
Motor power	:	55 KW
Tank capacity	:	1000 litre

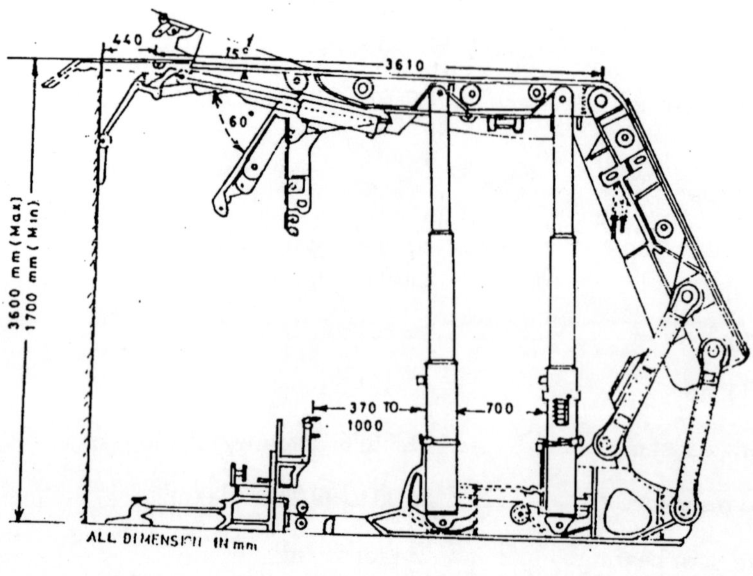

Fig. 52.3 : 4 × 484 Tonne Chock-Shield powered roof supports being used in the Moonidih Colliery.

Maximum pressure at the hydro-accumulator	:	360 bar
Application for face inclination	:	Longitudinal : Upto 35^0 Transverse : ± 15^0

3. OBSERVATIONS OF ROOF STRATA BEHAVIOUR AND ITS IMPACT ON THE CHOCK-SHIELD SUPPORTS

3.1 Overall observations

- The panel was worked out with retreating longwall system. Half face method of shearing was normally practised in this panel. Because of maximum extended height of the powered support is 3.6m excess coal around 0.7 to 1.2m used to be left on the floor although its maintenance with respect to consistency was very difficult. Coal floor is quite competent to hold the powered supports. Caving of roof in the goaf was excellent. No roof weighting was felt in the face nor any periodic weighting was noticed during the extraction process.
- Overhang in the main gate road barrier pillar edge was found 3m and the overhang in the tail gate road barrier pillar edge was 4m. Behind 1 to 67 No. chock shield supports, there was practically no overhang during the observation period because of fractured intercalated sandstone and shale roof layer just above the powered supports.
- Caving of contact roof was nice. Again it was observed that the sizes of the caved boulders were medium at the floor level due to the fractured intercalated sandstone and shale but were very large at the top most point of the muck-pile. Caving of roof rocks were in orderly manner. More than 85% of the rock layers were piled up one after another uniformly and only less than 15% of the total goaf stones were scattered. Void was seen in between the caved roof layer and roof over the goaved out area.
- Average face to canopy tip distance prior to cut was around 45 cm.
- Heavy spalling of coal in the face was observed. Around 75% of the face was affected by face spalling.
- The 6m to 7.5 roof above the powered support was quite bad right from the 3m away from the edge of main gate road wall to the tail end of thee face. Roof was generally fractured and used to rest over the 55% length of the canopy towards the goaf side.
- No chock shield supports became solid in the healthy and normal condition throughout the run of the panels during the observation period.
- Stone intrusions around 0.8m to 1m thick were found in patches all along the face for which drilling and blasting method was practised for its fragementation.
- No water procolation was noticed in the longwall face.
- Average zonal caving angle found was around 75^0 to 80^0. Height of caving near the drive-head was found around 8m and near tail end it was found 6m. Cracks were also found in the intact rock-mass above the caved portion of rock.
- The powered roof supports were suffering from the following problems:

- Tendency of toppling while lowering and advancing the supports since the centre of gravity of the powered support unit is not as close to the base. This is because of the split base design and heavy structural weight. The tilting (tilting angle goes upto 20^0) of the powered supports used to reduce the contact area of the canopy with the roof resulting in poor roof supporting and as a consequences, it always used to cause a small roof falls. However, in the disturbed roof area problems were highly aggravated because of severe tilting of the chock shields and a roof height around 6m to 9m used to cave down forming a huge cavity over the chock shield supports. Negotiations of each phenomenon were very difficulty, time consuming and involved safety of oth men and machines. A heavy cycle of operations for repositioning and straightening of powered supports were done throughout the extraction process in F_2 panel. Tilting of the chock shield also posed severe problems during advancing the support.
- Damaging of pressure gauges, legs and legs seals, side rams, advancing ram pins, front canopies, etc.
- Bypassing of fluid in the yield valve and legs, leakage in legs and side rams, etc.

- A normal fault of 0.7m down-throw and several patches of geologically disturbed zones were egotiated in both the panels.

In the faulted area roof cavity used to be formed in most of the occasions due to the falling of roof stone. In these geologically disturbed zones, the stable superincumbent zone was disturbed. To stabilize these zones roof bolting was done in the longwall face. The brief about the roof bolting system was as follows:

Average number of bolts in a row	:	35 to 40
Length of each bolt	:	1.5 to 2.0m
Inclination of bolt on the roof of the face	:	30^0 to 40^0 (from horizontal)
Average number of bolts/m^2	:	2.2 to 2.3

The bolted stable roof was then used to be caught by the canopy of the chock shield powered roof supports. In most of the occasions wooden chocks were also erected in these geologically disturbed zone.

- The roof of the layer-II was very stable. The shape of the roof in the goaf was looked like a hyperbola (actually caving edge line was a fourth degree parabolic shape).

3.2 Pressure Survey

Initially all the powered supports were provided with pressure gauges but latter on most of those gauges were damaged. From these pressure gauges the pressure variation were measured directly. Pressure readings for all the supports and in their legs were taken for a couple of months directly from the pressure gauges.

For analysing the loading pattern on the face, leg pressure recorded for all the supports were then averaged based on the different zones. The number of supports fallen under 7 zones are as follow :

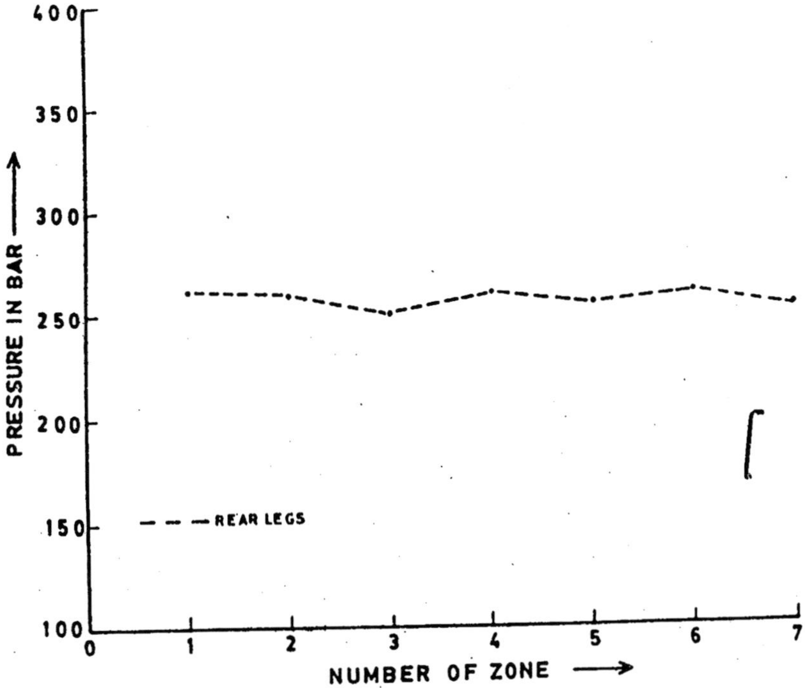

Fig. 52.4 : Average pressure on the rear legs of the powered supports in the different zones

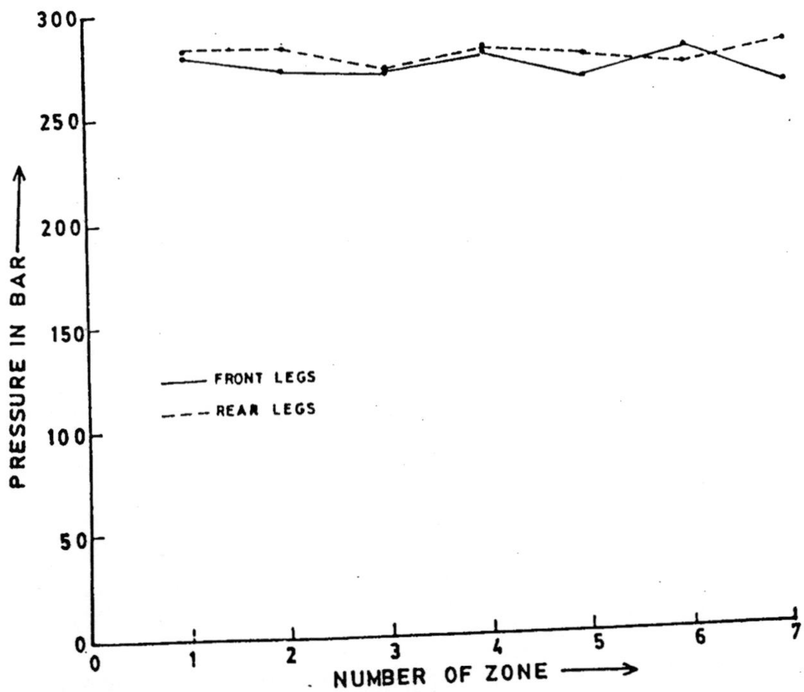

Fig. 52.5 : Average pressure on the powered supports in the different zones

	Zone-1	Zone-2	Zone-3	Zone-4	Zone-5	Zone-6	Zone-7
Chock Shield No.	1-10	11-20	21-30	31-40	41-50	51-60	61-67

Chock shield No. 1 was located near to the bottom gate road whereas the chock shield No. 67 was located near to the top gate road. Table 52.1 depicted the average maximum and also the zone average pressure of the chock shield supports fallen under the various zones. Fig. 4 and 5 show the average pressure in different zones in various days in the F1 panel of XVI combined seam of the Moonidih colliery.

Table 52.1 : Average Pressure in Bar

Zone No.	*Maximum*		*Average in the zone*		*Standard deviation*	
	Front	*Rear*	*Front*	*Rear*	*Front*	*Rear*
1.	250.00	260.00	239.65	247.93	56.07	42.39
2.	305.00	300.00	255.85	271.52	23.68	11.46
3.	290.00	280.00	255.80	240.53	19.23	36.72
4.	320.00	273.33	276.00	247.53	3.81	42.29
5.	300.00	300.00	274.95	261.32	11.80	12.13
6.	305.00	290.00	255.53	255.36	32.34	22.51
7.	277.50	273.33	261.66	253.13	25.92	6.70

3.3 General Inference

It has already been stated that no chock shield powered roof supports became solid while wining coal in both the F_1 and F_2 panels as well as in the pre-salvaging period. From the Table 52.1 it is very clear that except in zone-1 (where roof was stable), the average maximum front leg pressure was higher compared to the rear leg pressure and in case of the zone average pressure the front legs pressure were more compared to the rear leg pressure. This is because of the roof layer-1 above the powered supports (i.e. the highly geologically disturbed layer of intercalated sandstone and shale) was giving the horizontal thrust from the front to the goaf side. The stable superincumbent roof was disturbed and there was practically no overhang throughout the longwall face in both the panels. The second sandstone layer i.e. the layer No. 2 may partially offer the deflection load if the face remain standstill for a longer time. Because of the fractured roof, the setting pressure was adjusted at various level from 240 bar to 290 bar. From the pressure reading one can find out that the powered support leg pressure were more or less near to the level of setting pressure. The maximum front leg and rear leg pressures in the zone-4 was 320 bar and 273.33 bar respectively whereas both the maximum front leg and rear leg pressure in the zone-5 was 300 bar. The maximum load over the powered supports offered by the superincumbent roof was around 210 tonne to 250 tonne.

3.4 Convergence in the Main Gate Road

Convergence recording had been taken regularly in the main gate road ahead of the longwall face. Around 2.9cm convergence was noted around 76m ahead of the longwall face and the maximum convergence recorded around 11.3cm at 4.0m ahead of the longwall face (Fig. 52.6). In the Fig. 52.6 convergence curve of a particular point in the main gate road (ahead of the face) has been shown. During that time the longwall face was standstill for around 40 days. The convergence value increased in an asymptotic manner showing a peak around 5.70cm. The convergence value at the starting point was 3.5cm. Hence, average rate of increase convergence value

$$= \frac{57 - 35}{40} = 0.55 \text{ mm / day}$$

3.5 Surface Subsidence

Since XVII Top seam was worked out long back and XVII Bottom seam was being extracted (during this experimentation period) a maximum surface subsidence around 1.4m was observed over the F_1 panel. More subsidence was also gradually occurring during that time. In the underground, the actual spot of this maximum surface subsidence observation point was at the gate road of F_1 panel superimposed over the middle of working C_2 panel in the XVII Bottom seam. It means the superincumbent strata above the F_1 panel was disturbed. The angle of draw of the subsidence profile was 20^0 and the rate of surface subsidence was 0.66 mm/day.

4. BRIEF ABOUT THE DETERMINATION OF THE CAPACITY OF THE POWERED SUPPORTS

In brief the proposed methodology of support rating determination shall be done based on logical borehole litholog analysis and also based on weighting zone under the fourth degree pabolic curve concept, which is as follows:

- Identify the rock layer/layers above the coal seam.
- Identify the plane of weakness
- Group the various layers/composite layers
- Estimate and classify the stability of those layers/composite layers (Table 52.2 and 52.2B).
- Estimate magnitude of overhang of various layers/composite layers (Annexure-1)
- Estimate the amount of caving angle of various layers/composite layers (Table 52.3).
- Identify layers/composite layers of the immediate roof rock which may detach from the face and also identify the layers/composite layers which will deflect and provide load over the powered supports. To determine the capacity of the powered support the lower most strata layers or composite layers must be taken into consideration that it will detach right from the longwall face freely.
- Draw strata layers and then weighting zone based upon the drawn caving angle. The fourth degree parabolic curve can be drawn by touching the various points cut at the two extreme points of every layers/composite layers as previously defined. The loading zone will be in between these two curves.
- Determine the load to be offered by the various layers/composite layers over the powered supports considering that the total lower layers/composite layers will be

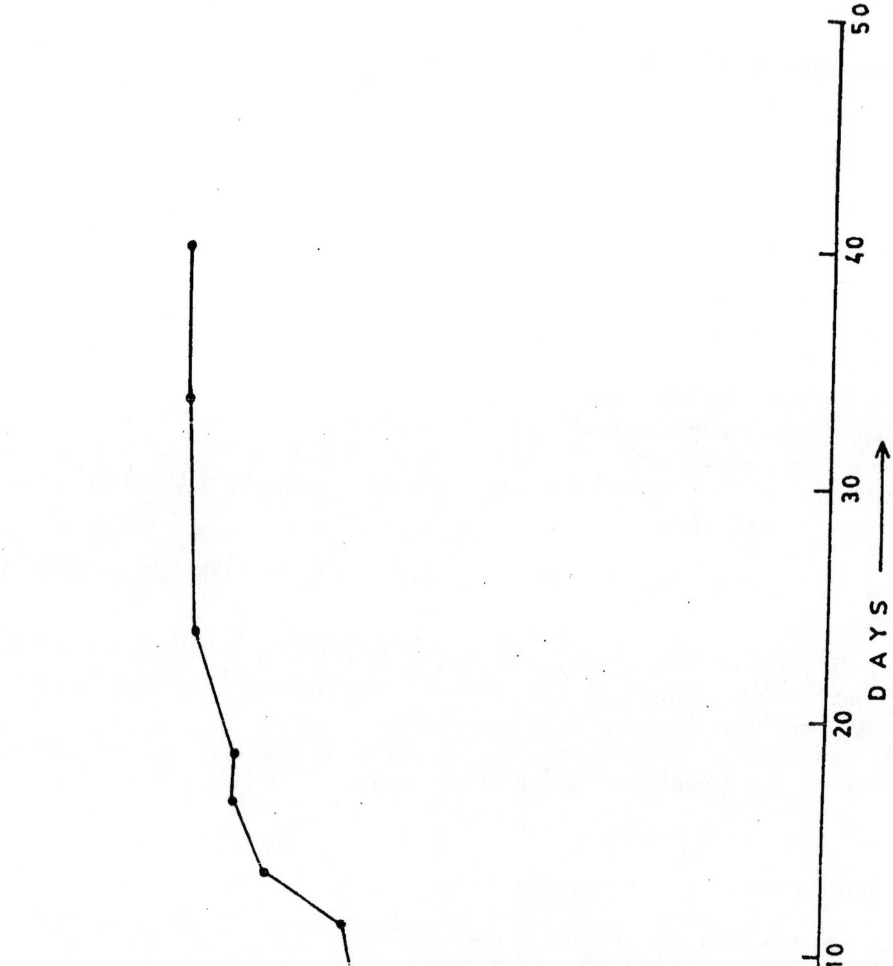

Fig. 52.6 : Shows the Convergence of a particular point when face is standstill (in main gage road)

detached right from the face and will offer free loading based on moment equation.

- Find out the layers/composite layers above the detached layres/composite layers which will deflect and offer loading due to deflection. This loading will also be solved by the moment equation. (Table 52.4 may be consulted).
- Find out the total loading by adding all the above mentioned loading (loading by detached layers/composite layers and also the loading due to deflection) and based on the moment equation find out the resistance to be offered by the powered supports to resist the same total loading.
- Find out the capacity of the powered supports based on the following formula:

$$T_y = \frac{R x S_t}{\gamma_i x \gamma_l}$$

where,

T_y = Capacity of the powered support in tonne

R = Resistance of be offered by the powered supports in tonne

S_t = Factor of safety

γ_i = Efficiency factor of the powered support due to the inclination of legs etc. γ_i is considered to be 0.85 for chock shield support and 0.8 in case of shield supports but $\gamma_1 = 0.90$ in case all legs of the chock shield are vertical.

γ_1 = Efficiency factor of the powered support due to the leakage in the pipelines and valve system and also other mechanical defects. γ_1 is considered to be 0.9.

$$R = \frac{W_1\left(\frac{L_1}{2} + \frac{H_1 Tan\alpha_1}{2}\right) + W_2\left(\frac{L_1 + H_1 Tan\alpha_1}{2} + \frac{H_2 Tan\alpha_2}{2}\right) + \frac{3}{8} W_3\left[\frac{1}{3}(2L_1 + H_3 Tan\alpha_3) + H_1 Tan\alpha_1 + H_2 Tan\alpha_2\right] + \cdots}{P}$$

Where W_1 W_2 W_3 etc. are the weight of the 1st, 2nd, 3rd etc. immediate roof layer/ composite layer blocks respectively starting from the top of the powered support.

H_1 H_2 H_3 etc. are the thickness of the 1st, 2nd, 3rd etc. immediate roof layer/composite layer blocks respectively starting from the top of the powered support.

L_1 = Distance between the longwall face and the caving edge.

$90° - \alpha_1$, $90° - \alpha_2$ $90° - \alpha_3$ etc. are the caving angle of the 1st, 2nd, 3rd etc. immediate roof layer/composite layer blocks respectively starting from the top of the powered supports.

P = the distance from the face to the centroid of resistance offered by the powered support.

5. DETERMINATION OF SUPPORT CAPACITY FOR THE XVI COMBINED COAL SEAM

Based on the compressive strength, RQD percentage, plane of weakness, etc. the litholog as shown in Fig. 52.2 can be divided into various layers which are shown in the Table 52.5. It has been calculated and estimated that the layer-1 and sandstone layer (layer-II and III) of the roof rock above the chock shield powered support comes under the Class-II and Class-IV category of rock respectively (Table 52.6). Hence these layer-I and layer-II are highly cavable and good cavable respectively. Based on the experiences of the roof strata in the XVI Top seam of the same Moonidih colliery, the sandstone layer has been divided into 2 parts viz. bottom layer-II, 5m thick and the top layer-III, 7.8m thick. The layer-II is expected to offer the deflection load over the powered roof supports. The chock shield powered roof supports are rightly selected but instead of split base, the rigid base would have been the best choice.

5.1 Determination of Capacity of the Powered Roof Supports if the Existing Type Chock Shield Supports are Used (Fig. 52.3)

From the field study, it is very clear that the layer-I (intercalated sandstone and shale layer) comes down very easily as soon as the powered supports are advanced and it offers the load of it's detached rock-mass. But as the longwall face advances at a faster/moderate rate, parts of the layer-II (sandstone layer) shall only offer the deflection load. It has been seen that in the normal case and under such fractured roof condition, the average length of overhang of layer-I behind the powered support was 1.0 and the average caving angle was around 75^0.

Average density of roof rock	=	2.5 t/m³
Average inclination of seam	=	1 in 7 i.e. 8.13°
Hence, Cos 8.130	=	0.9895
Support spacing	=	1.5 m

The distance of the goaf edge from the longwall face (as per Fig. 52.3) is $L_1 = 4.65 + 1.0 = 5.65$ m (considering the web depth of cutting is 600 mm). Fig. 52.8 shows the details regarding this.

The distance of the centroid of resistance by the chock shield support from the face is 3.686 m. Therefore, weight of the layer-I:

$$W_1 \quad = \quad 5.65 \times 7.6 \times 1.5 \times 2.5 \times 0.9895 - 153.33 \text{ t}$$

The distance of the centre of gravity of this layer-I (load = W_1) from the longwall face:

$$C_1 = \frac{5.65}{2} + \frac{7.6 \text{ Tan } 15^0}{2}$$

$$= \quad 3.943\text{m}$$

Therefore, the moment of the load W_i taken from the face is:

$M_1 = 159.33 \times 3.943 = 628.27$ t-m

The weight of the sandstone layer-II offering deflection load over the chock shield support. The length of overhang = 5.57m

Therefore, $W_2 = 10.22 \times 5 \times 1.5 \times 2.5 \times 0.9895 = 189.61$ t

The moment of this load W_2 taken from the face as per the given formula:

$$M_2 = \frac{3}{8} \times 189.61 \times \left[\frac{1}{3}\left(2 \times 10.22 + 5 \times \text{Tan}50^{\circ}\right)\right]$$
$$= 625.68 \ \text{t-m}$$

Therefore, the resistance to be offered by the clock shield support against all these loads :

$$R = \frac{628.27 + 625.68}{3.686}$$

$$= 340.\ 19 \text{ tonne}$$

So, the unit capacity of the chock shield should be:

$$T_y = \frac{340.19 \times 1.15}{0.9 \times 0.9} = 482.98\text{t} \approx 483 \ \text{tonne}$$

(1) Factor of safety'S_f' under such a roof condition = 1.15

(2) Efficiency factor of the chock shield support due to the inclination of leg

$\eta_i = 0.90$

(3) Efficiency factor due to the leakage, etc. in the system

$\eta_1 = 0.9$

[Note : Normally the supports were withstanding a load less than 265.05 tonne because practically there was no overhang in the layer-I. The caving angle of this layer was around 75^0

Therefore, the weight of the layer-I

$W_0 = 4.65 \times 7.6 \times 1.5 \times 2.5 \times 0.9895$

$= 131$ tonne

The distance of centre of gravity of this layer-I (load = W_O) from the longwall face

$$C_0 = \frac{4.65}{2} + \frac{7.6 \text{ Tan } 15^0}{2}$$

$$= 3.34 \text{ m}$$

Therefore, the moment of the load W_0 taken from the face is :

$$M_0 = 131.0 \times 3.34 = 327.96 \text{ 5-m}$$

Considering more or less the same deflection load to be offered by the layer-2, the resistance to be offered by the chock shield support :

$$R_0 = \frac{437.96 + 625.68}{3.686}$$

$$= 288.56 \text{ tonne} \approx 289 \text{ tonne}$$

which is around 51.19 tonne less than R.

5.2 Determination of Capacity of the Powered Roof Supports if the Recommended Type Chock Shield Powered Supports are Used (Fig. 7)

The distance from face to the goaf edge of the canopy of the chock shield (with a cut around 0.6m web depth, average tip of the canopy to the face distance around 0.4m and considering the same overhang length i.e. 1.0m behind the powered supports). Fig 52.8 shows the details regarding this.

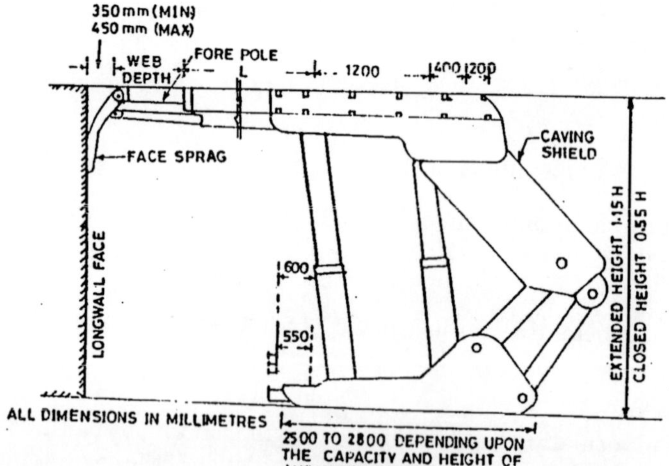

WHERE H = EXTRACTION THICKNESS OF SEAM
L = DISTANCE FROM FACE TO THE CENTRE OF THE FRONT LEG- (350 mm + Web depth)
= 1450 mm (APPROX) FOR WEB THICKNESS 600 mm WITH WIDTH OF AFC PAN CROSS SECTION 762 mm
= 1650 mm (APPROX) FOR STANDARD WEB ATHICKNESS 800mm WITH WIDTH OF AFC PAN ACROSS SECTION 762 mm
= 2000 mm (APPROX) FOR WEB THICKNESS 1000 mm WITH WIDTH OF AFC PAN CROSS SECTION 1000 mm

OVERALL CANOPY WIDTH	:	1450 mm
OVERALL BASE WIDTH	:	1400 mm
SUPPORT CENTRE	:	1500 mm

Fig. 52.7 : Recommended chock-shield powered roof supports to be used in the XVI combined seam of Moonidih Colliery

$L'_1 \quad = \quad 5.45m$

Support spacing = 1.5m

The mass of the layer-I offering the total detached load over the face supports :

$$W'_1 = 5.45 \times 7.6 \times 1.5 \times 2.5 \times 0.9895$$

$$= 153.69 \text{ tonne}$$

The distance of the centre of gravity of this layer-I (load W'_1) from the longwall face:

$$C'_1 = \frac{5.45}{2} + \frac{7.6 \text{ Tan } 15^0}{2}$$

$$= 3.743m$$

Therefore, the moment of the load W'1 taken from the face is :

$$M'_1 = 153.69 \times 3.743 = 575.29 \text{ t-m}$$

The mass of the sandstone layer-II offering the deflection load over the chock shield support. Length of overhang = 5.57m

Therefore, $W'_2 = 10.02 \times 5 \times 1.5 \times 2.5 \times 0.9895 = 185.90$ tonne

The moment of this load W'_2 taken from the face as per the given formula :

$$M`_2 = \frac{3}{8} x185.90x\left[\frac{1}{3}\left(2x10.5 + 5\text{Tan } 50^0\right)\right]$$

$$= 604.14 \text{ t-m}$$

The distance of centroid of resistance of the chock shield support from the longwall face is 3.13m (considering inclination of the legs at a height of 3 to 3.5m is around 5^0). Therefore, the resistance to be offered by the unit chock shield supports as designed and shown in the Fig. 52.7:–

$$R` = \frac{575.29 + 604.14}{3.13}$$

$$= 376.81 \text{ tonne}$$

Therefore, the capacity of the chock shield support per unit

$$T'_y = \frac{376.81x1.15}{0.85x0.9}$$

$$= 566.45 \text{ tonne} \approx 567 \text{ tonne}$$

Considering that the :

(1) Factor of safety 'S_f' under such a roof condition = 1.15

(2) Efficiency factor of the chock shield support due to the inclination of leg,

$n_i = 0.85$

(3) Efficiency factor due to the leakage, etc, in the system

$n_1 = 0.9$

The capacity of each unit of this chock shield support shall be only 84 tonne higher compared to the existing design one and also it will be the best suitable for the XVI combined seam of the Moonidih colliery. All the 4 legs of this support will offer resistance against the load, the horizontal component of which tends from the face towards the goaf.

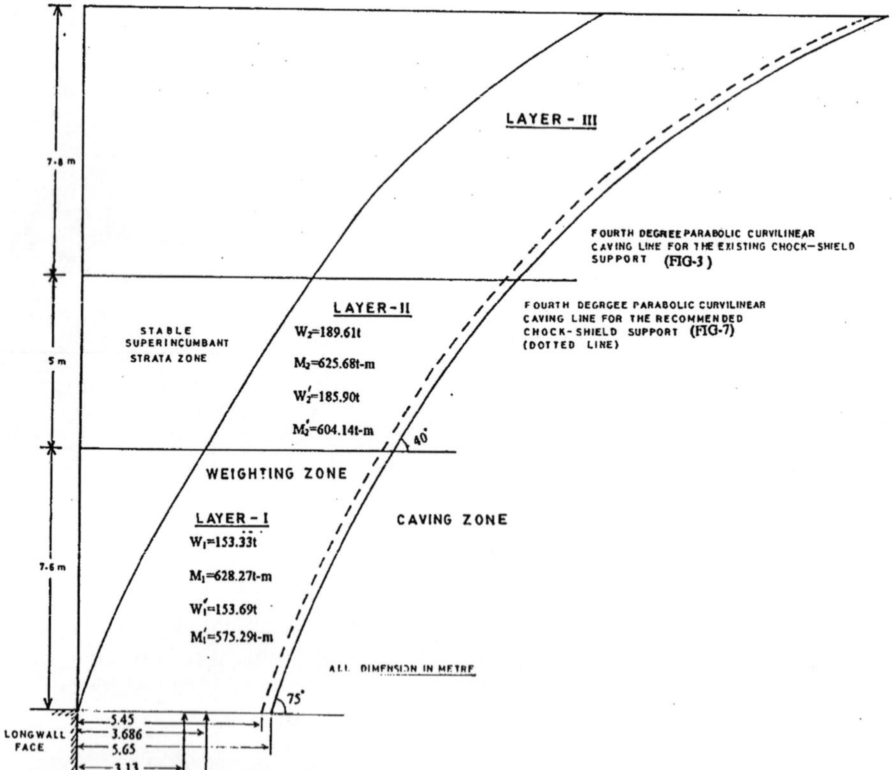

$W_1W'_1$ are the weight of first composite layer block in the weighting zone w.r.t. various types of powered supports and $M_1M'_1$ are their respective moments taken from the longwall face.

$W_2W'_2$ are the weight of the second layer block in the weighting zone w.r.t. various types of powered supports and $M_2M'_2$ are their respective moments taken from the longwall face.

Fig. 52.8 : Various Roof Layers Offering Loads over the powered supports over the longwall face of XVI combined coal seam of Moonidih Colliery.

From this experiment and study, it is very clear that although the capacity of the existing but instead of the split base, the rigid base design would have been the right choice. In general, under such a fragmented roof condition, the horizontal strata pressure shall always tend from the face to the goaf side. Therefore, the legs of the chock shield supports shall be inclined towards the face so that they can provide higher horizontal reaction forces from the goaf towards the face. In Indian mining condition the split base of the powered roof support is found unsuitable.

6. CONCLUSION

The immediate thick roof strata (composite layer number-1) above the powered roof supports and also over the entire F_2 longwall panel of the XVI combined seam of Moonidih colliery of Jharia coalfield, India, was highly fractured but the medium strength layer-2 roof rock (massive sandstone) was healthy which never offered any surge loading weighing in the face. This indicates that the layer-2 roof rock only used to offer deflection load and thereafter it used to cave down beyond the "weighting zone" offering no surge load over the powered supports. In other words presence of thick weaker strata as an immediate roof rock over the powered support is a boon of safety against a disaster of support system due to sudden caving of massive roof present above the farmer provided that the capacity of the support is the optimum. However the fractured roof rock on many occasions caused very difficult situations due to instantaneous caving of the same in between the freshly exposed longwall face and the tip of the canopy. The outcome of this phenomenon is the poor production, lower longwall face advancing rate, lower support efficiency, higher frequency of face equipment stoppage and breakdown, wastage of time, money and manpower to support the roof over the powered supports by the wooden cogs, bars and bolts, higher consumption of materials and spares, higher chances of accidents, etc. The horizontal force under such a fractured roof condition tends from face to the goaf side. Therefore, to negotiate such a phenomenon all the rear and front legs of the chock shield supports should be inclined at an angle of 5^0 to 17^0 towards the face so that they shall offer higher horizontal force from goaf to the face side. The determination of the capacity of the powered support must be done scientifically by the logical borehole litholog analysis, geo-technical parameters of roof rock and based on 'weighting zone' under the fourth degree parabolic curve concept which has been done in this case. The optimum capacity of the chock shield support per unit to be used in XVI combined seam of Moonidih colliery of Jharia coalfield, India, shall be 567 tonne.

REFERENCES

1. Das Samir Kumar, Ph.D. Thesis, on "Optimized Selection of Powered Supports in Mechanized Coal Mines in India", Indian School of Mines, Dhanbad, 1994.

Table 52.2A(t): Roof rock Classification and Ratings of Indian Coal Measure Rocks in Longwall Panels Based on Concept of Stability of Various Roof Layers or Composite Layers

Sl. No.	(A) Rock parameters	Class					
		I	II	III	IV	V	VI
1	2	3	4	5	6	7	8
1.	Mean weighted uniaxial compressive strength. kg/cm²	0-100	>100-200	>200-300	>300-500	>500-800	More than 800
	Ratings : (Max rating 25)	0-3	>3-6	>6-9	>9-15	>15-24.9	>24.9-25
2.	Mean weighted RQD %	0-30	>30-40	>40-50	>50-70	>70-85	>85-100
	Ratings: (Max. rating 25)	0-6	>6-8	>8-10.5	>10.5-16	>16-20.5	>20.5-25
3.(a)	Type of Rocks	Weak & soft carbonaceous shale, lamination of shale in argillaceous sandstone, very soft fractured intercalation, splinter shale, etc.	Sandy shale, fissile rocks, weak and soft carbonaceous shale, very soft & fractured intercalation coal, etc.	Most of the types of shale, weak sandstone. Alternate layers of various kinds of rocks at a close succession. Intercalated sandstone & shale.	Hard shale, Medium strength sandstone, etc.	Quite strong sandstone, hard shale etc.	Very strong sandy shale, competent fine grained sandstone with shale laminations, thick massive sandstone, dolerite etc.
	Ratings : (Max. rating (10)	0-3	>3-5.5	>5.5-7.5	>7.5-8.5	>8.5-9.5	>9.5-10
(b)	Absence of severe cracks, fissile beds, splitting, slip planes etc. (presence of a minor fault of more than 0.5m average throw in a longwall panel in case of seam thickness of more than 2.0m should be avoided. There shall not be any fault in a seam of less than 2m thick.)	Presence of very high density cracks, coarse grained sandstone with carbonaceous intercalation around 0.05 to 0.08m or less in thickness and fissile sandy layer of less than 0.15 m thick, zone near to the minor fault	Presence of moderate cracks, fissile beds like carbonaceous material bed thicker than 0.1m, clayey rocks, etc. more than 0.2m thick, zone near to the minor fault.	Presence of moderate cracks, fissile beds like carbonaceous material, bed thicker than 0.1m, clayey rocks, etc.	Cracks may be present but can not do any adverse affect. If fissile beds of very fine grained sandstone are present moderately then their thickness may vary in the range 0.1-0.3m.	Slight cracks may be present if fissile beds of very fine grained sandstone are present (rarely) their thickness may vary in the range 0.3 to 0.6 m	Not present

1	2	3	4	5	6	7	8
	Ratings (Max. rating 10)	0-3	>3-5.5	>5.5-7.5	>7.5-8.5	>8.5-9.5	>9.5-10
(c)	Absence of groundwater	Severe water problem	Water under moderate pressure	Water under less pressure	Moist only	About completely dry	Completely dry
	Ratings (Max. ratings 10)	0-4	>4-6	>6-7	>7-8	>8-9	>9-10
4.	Thickness of prominent bed layer (m)	0.0-0.35	>0.35-1.0	>1.0-2.0	>2.0-3.0	>3.0-4.0	Above 4.0
	Ratings (Max. ratings 20)	0-1.5	>1.5-5.0	>5.0-10.5	>10.5-16.0	>16.0-19.9	>19.9-20
5.	Total ratings 100	0-20.5	>20.5-36	>36-52	>52-72	>72-93.3	>93.3-100

Table 52.3[1] : Behaviour of Layers/Composite Layers of Immediate Roof Rockmass in Longwall Panels (Face Length : 100 m to 150 m)

Sl. No	Parameters	Class					
		I	II	III	IV	V	VI
1.	Stability nature	Extremely unstable	Highly unstable	Stable	Good Stable	Very good stable	Highly stable
2.	Caving nature	Extremely cavable	Highly cavable	Very good cavable	Good cavable	Poor cavable	H.ghly uncavable
3.	Caving angle	>81° - 85°	>75° - 81°	> 65° -75°	>50° - 65°	> 30° - 50°	30° and less
4.	Overhang length (m)	0	> 0 - 0.5	> 0.5 - 1.5	> 1.5 - 3.0	> 3.0 - 6.0	>6 and more
5.	Longwall system favourable	Caving system	Caving system	Caving system	Caving system (induced blasting may sometimes be necessary for initial caving)	Induced caving or stowing	Stowing system
6.	Support requirement	Composite layer/layer of rockmass may be dislocated right from the face edge which will offer higher pressure at the front rows of the legs of the powered supports. I.F.S. working mode support system (in case of thin seam) and combination of conventional and I.F.S working mode along with powered for poling should be selected in all other cases. All legs of the powered supports shall be inclined towards the face	Composite layers/layer of rockmass may be dislocated right from (near to) the face edge which will offer higher pressure at the front rows of the legs of the powered supports. I.F.S working mode support (in case of thin seam) and combination of conventional and I.F.S working mode along with powered for poling should be selected in all other cases. All legs of the powered supports shall be inclined towards the face.	Composite layers/layers of rockmass will offer more or less equilibrium loading over the powered supports. I.F.S. working mode support system (in case of thin seam) and combination of conventional and I.F.S working mode along with powered fore poling should be selected in all other cases. All legs of the chock shield powered supports shall be either in V' orientation fashion or shall be kept slightly inclined towards the face	Composite layer/layers of rockmass will offer more loading condition over the rear side of the powered supports particularly those which are installed in the middle of the face. For other cases the higher pressure will be felt in the rear rows of the legs due to higher overhang. Conventional working mode (in case of thin seam) and combination of conventional and I.F.S working mode along with powered forepoling should be selected in all other cases shall be used. All legs of the chock shield supports shall be either oriented in 'V' fashion or shall be kept slightly inclined towards the goaf	For regular caving of immediate roof induced method of roof fracturing like blasting, etc. may be practised which is not only time consuming, cumbersome, costly and less productive process but also it will not give any guarantee of caving of upper layers/composite layers which may cause severe loads on the powered supports. High pressure will be generated in the rear row of the chock shield supports. Very compact conventional design 4 leg chock shield supports with all the legs inclined towards the goaf shall be used. Capacity of the powered supports should be very high as far as practicable. Single telescopic legs should be used. However, longwall with stowing method will be the best in this case	Tremendous amount of load will be generated by the roof strata at the time of failure in the case of caving system of longwall mining which may be the far beyond the highest capacity of the powered supports so far developed in the world market. Longwall method can be practised along with the combination of stowing.

ANNXURE-1[1]
LENGTH OF OVERHANG

Based on vast field statistics and trial and error method a formula has been derived to find out the length of the overhang of the immediate roof rock behind the powered supports which is as follow

$$C_{oh} = \sqrt[3]{\frac{h\sigma_c K}{\partial}}$$

Where, C_{oh} = Length of overhang in metre

h = Thickness of overhang team in metre

σ_c = Compressive strength of rock beam in kgf/cm^2

∂ = Density of rock in t/m^3

k = A class constant

Value of k is as follows:

For sandstone = 0.217

For other composite layer with various rock combination and also for coal = 0.0131

However, it is observed that value of C_{oh} = 0 if the strata is highly geologically disturbed.

The calculated value of C_{oh} will will be reduced by 30% to 40% if water problem in the strata in present and also it depends upon the degree of watery condition.

The calculated value of C_{oh} will again be reduced in case of lower most bed, if a very high support setting pressure is selected and if severe rupture is given in the immediate roof layer by the powered supports.

Table 52.4[I] : Relationship Among Strata Formation, Caving Angle and Mean Average Compressive Strength

Strata Formation	*Caving Angle*	*Mean average compressive strength (kgf/cm²)*	*CTRAT*
Extremely weak strata including very weak composite layer, very weak sandstone layer, etc. i.e., extremely cavable strata	>81°-85°	Less than 100	0-20.5
Very weak sandstone, weak and soft carbonaceous shale, very soft and fractured intercalation, etc, i.e., highly cavable strata	>75°-81°	>100-200	>20.5-36
Most of the type of shale, shale sandstone, weak sandstone, intercalated shale and sandstone, etc. i.e., moderately stronger very good cavable strata	>65°-75°	>200-300	>36-52
Hard, grey shale, medium strength sandstone, etc. i.e., for good cavable strata.	>50°-65°	>300-450	>52-67
Quite strong sandstone, medium strength massive sandstone, etc. i.e., for poorly cavable strata	>35°-50°	>450-600	>67-79.1
Very thick massive sandstone, very strong sandy shale, competent fine grained sandstone with shale lamination i.e., for highly uncavable strata.	>30°-35°	>600-800	>79.1-93.3
Very thick strong massive sandstone i.e., for extremely uncavable strata.	>=30°	>800	>93.3-100

Note: [However, it is observed that the massive sandstone or thick moderately massive sandstone behaves peculiarly. In a very weak sandstone roof, the caving angle was found around 75° to 90° whereas in moderately stronger massive sandstone roof caving angle was only in between 27° to 35°. Therefore, caving angle in case of massive sandstone roof varies in a indistinct manner. But the composite layer beam consisting of various strata has some consistency regarding the formation of caving angle and the angle is generally in higher order compared to a single layer of strata of the same thickness. The major role play to determine the caving angle is the mean weighted average compressive strength, tensile strength, thickness of strata, presence of fractured plane, density or rock etc.

CTRAT: Converted total rarting into 100 under a given condition. For moe simplification when MWREC (Mean weighted rating of total composite layer or layer) is considered then the equivalent value of CTRAT shall be considered against it.

Table 52.5[1] : Roof Height Above the Powered Supports Affected Near the Face

Colliery	*Average height of Extraction (m)*	*Total height of Roof Affected above the Powered Supports (m)*	*Ratio Between Total Affected Roof Height To Extracted Thickness*	*Caving height (Caves within short time after supports are advanced (m)*	*Ratio Between immediate caving height to Extracted Thickness*
Moonidih					
1. XVII Bottom Seam	1.4	9.77	$\frac{9.77}{1.4} = 6.97$	4.75	$\frac{4.75}{1.14} = 3.39$
2. XVI Top Seam	2.5	11.25	$\frac{11.25}{2.5} = 4.5$	8.22	$\frac{8.22}{2.5} = 3.288$
3. XVI Bottom Seam	1.95	12.94	$\frac{12.95}{1.95} = 6.635$	5	$\frac{5}{1.95} = 2.56$
4. XVII Top Seam	1.8	10.40	$\frac{10.40}{1.8} = 5.77$	4.01	$\frac{4.01}{1.8} = 2.22$
East Katras					
X Seam	2.4	14.40	$\frac{14.40}{2.4} = 6$	8.8	$\frac{8.8}{2.4} = 3.66$
GDK-11 A Incline					
1 Seam	2.8	23.27	$\frac{23.27}{2.8} = 8.31$	16.62	$\frac{16.62}{2.8} = 5.93$
Churcha West					
V Seam	3.2	12.4	$\frac{12.4}{3.2} = 3.875$	5.5	$\frac{5.5}{3.2} = 1.718$
Jhanjra Project					
R-VII Seam	3.5	17.62	$\frac{17.62}{3.5} = 5.034$	12.62	$\frac{12.62}{3.5} = 3.60$
Scale of searching for the total affected roof height and caving height			4 to 8.5 times the Extracted Thickness		2 to 6 times the the Extracted Thickness

Table 52.6 : Roof layers/Composite Layers Above the Chock-Shield Powered Supports with Their Approximte Physico-Mechanical Properties Over the XVI Combined Seam near F_2 Panels (as per NCCJ-16 Borehole Litholog) of Moonidih Colliery

Rock Formation	*Layer/Composite Layer Approx. thickness (m)*	*Estimated Mean Weighted Compressive Strength (kgf/cm²)*	*Estimated Mean Weighted Tensile Strength (kgf/cm²)*	*Estimated Mean Weighted RQD Percent*	*Average Caving Angle*	*Layer/Composite Layer No.*
Intercalated sandstone and shale	7.6	376.39	40.70	62.45	75^0*	I
Medium to coarse grained sandstone layer	12.8 — ** 5m	400.00	43.50	67.41	58^0	II
	12.8 — 7.8m	400.00	43.50	67.41	58^0	III

* The averave cavinng angle was 75^0 because of the geological disturbances

** The 12.8m thick medium to coarse grained sandstone layer has been divided into 2 parts viz. Layer-II : 5m thick and Layer-III : 7.8 m thick

Table 52.7 : Ratings, classification, overall classification and other parameters of the Roof Rock above XVI Combined coal seam of the Moonidih Colliery

Parameters	*Composite Layer/ Layer-I*	*Composite Layer/ Layer-II*
1. Mean weighted uniaxial compressive strength (Rating)	11.29	12.00
2. Mean weighted RQD% (Rating)	14.27	15.58
3. (a) Type of rocks (Rating)	7.0	7.5
(b) Absence of severe cracks, fissile beds, splitting, etc. (Rating)	3.0	6.0
(c) Absence of groundwater (Rating)	7.0	7.0
4. Thickness of prominent bed layer ((Rating)	20.0	20.0
Rating out of 100	62.56	68.08
Class of roof rock	IV II*	IV
Amount of α_1	43^0	50^0
To be followed	15^{0}*	50^0
Amount of overhang	2.46m	5.578m
To be followed	1m*	5.578m
Overall rating of roof rock-mass	66.02 less than 52*	
Overall class of roof rock	IV III*	
Overall α_1	48^0 Less than 30^{0}*	
Overall amount of overhang	1m* and 5.578m	

* Because of roof above the powered support is highly fractured and it was physically probed.

Section IX

INFORMATION TECHNOLOGY IN MINERAL INDUSTRY

CHAPTER 53

Truck Despatching Systems in Surface Mines in the 21st Century—Journey Towards Complete Automation

*C. Amitabh** , *Ajoy K Ghose*** *& Paul L Richard****

ABSTRACT

In the foreseeable future upto 2050, surface mining will continue to dominate the mineral production scene with mega sized mines for economies of scale. Truck dispatching has now become an integral part of any large surface mine for improving machinery availability and utilization, management control, real time control and has proven its economic viability in over 100 installations in the world. The paper presents a synoptic overview of the current state-of-the-art of truck dispatching systems.

A typical case study of the potential benefits of the application of truck dispatch system in the Indian surface mining milieu is presented. The need for further upgradation of the system for further cost-cutting and additional features which could help realize the ideal of a completely automated mining system are discussed. The integration of the truck dispatch system with mine planning and materials control and the concepts of telemining are presented, which hopefully would emerge as the dominant system in the surface mining industry.

1.0 GLOBAL SURFACE MINING INDUSTRY – THE FORESEEABLE FUTURE

With the mining industry poised to enter the new millennium – a new era is being ushered in with heightened competitiveness and the imperative to achieve higher levels of profitability by the mining corporations. It needs to be realized, however, that technological change alone will not be the criterion for overcoming the current and future challenges of

*Manager – Enterprise Solutions, Wipro Limited – Software & Services, India

**Formerly Director, Indian School of Mines,

***President & CEO, Wenco International Mining Systems Limited, Canada

competitiveness. Technological change does not imply merely the implementation of Heavy Earth Moving Machinery, hardware and software packages in the mining industry. The key issue for remaining competitive in the evolving dynamic scenario is to implement a total system, either as an integrated whole or in delineated parts, which would enable the industry to plan its resources efficiently, design the different operating systems and operate the same based on plans, which would be able to accommodate deviations in mine operational dynamics. This paper aim at identifying the needs of the 21st Century surface mining with special emphasis on management of shovel dumper fleet resources as well as to provide an enabling information technology based solution.

The surface mining sector, being extremely resource intensive in terms of capacity and population of HEMM in deployment, needs extreme care in managing the processes governing efficient utilization of the same. The design aspect of an expert system has to be planned in such a way so as to successfully make it work in a demanding environment as well as to ensure higher ROI at an optimized payback period. This has to consider strategic enhancements in fleet management techniques as a key performance indicator in futuristic surface mining horizons. The paper will also examine the concepts underlying the need to incorporate new technology in the arena of fleet management – as a tool for guiding and controlling the mobile resources deployed in mines. Profitability and competitiveness in surface mining operations are directly impacted upon by the performance statistics of the deployed resources as reported by management information systems in real time.

2.0 TRUCK DESPATCH SYSTEMS—A PHASED GRADUATION

The history of fleet management goes back to as far as the late 70's when it was realized that the capital investments made towards procurement and operation of shovels and trucks did not validate the expected returns over time. It was observed that despite theoretical verity in a particular shovel/truck combination for achieving a target production, in practice, the same did not necessarily hold good. The performance differential between plan and actual was reportedly 80 – 85% . The reasons for such shortfall could never be ascertained owing to frequent changes in mine dynamics and non -availability of monitoring tools. The differential was principally attributed to machine breakdowns, idling due to non-availability of blasted material, idling due to non-availability of trucks and a host of other reasons. Hence, the need for a centralized monitoring system was felt for understanding the various causes leading to non-performance of HEMM units. Between the late 1970's and the present, the concept of fleet management has developed gradually from a simple monitoring system to the state-of-the-art HEMM management and control system in real time. The various eras in the evolution of fleet management are:

Era 1 : The mine operations management performed a fixed allocation of trucks to respective shovels at the start of the shift. Dispatchers, equipped with mobile radios, were assigned to various points in the mine to report on anomalies in the operating status. This practice could shore up the performance of HEMM to reach a plateau at approximately 12% of the existing norms. This process was cumbersome in mines where large fleet was in deployment and night operations reporting were incorrect. Incorrect reporting resulted in falsified documentation of production figures.

Era 2 : Keeping in mind the shortcomings inherent in the system in Era 1, the mine management decided to invest in providing mobile radios to all the operators on board shovels and trucks to report their operating statistics to a central control station. Dispatchers at the central control would update their database based on RF inputs from the operators in the field. This era did not yield any significant performance improvements owing to errors in reporting/incorrect reporting from the operators or non-reporting of trips etc.

Era 3 : With a view to further upgrade the performance of HEMM in deployment, a couple of top mining corporations set up engineering units in early 1980's to develop a software based system for monitoring the performance of the HEMM units in real time. The operators were required to send in the inputs through mobile radios and the dispatchers were required to update the system in real time. In this system, while the profitability of error or falsified reporting continued, the scope of not reporting was eliminated. The system would prompt the dispatcher at the control room regarding the units that have not transmitted their reports for more than a stipulated time period (since last transmission). This resulted in a pseudo real time monitoring system. The performance improvement registered an increase of approximately 8% over the prevailing performance statistics.

Era 4 : Sustained developments in this arena resulted in the evolution of a real time computerized monitoring and detection system with mandatory partial operator interaction. The concept of availability and utilization became a Key Performance Indicator (KPI). The Operator cabins were mounted with an interface data terminal with keys for sending messages and an electro-luminescent display for reading incoming messages. The keys were clearly marked as LOAD START, LOAD END, WAITING AT SHOVEL, WAITING AT DUMP YARD, DUMP START, DUMP END, IN TRANSIT, FUELLING, BREAKDOWN etc. The operators were required to manually activate the relevant keys for enabling the system to capture the data. This could have been a true real time system provided the operators would comply with the standards and norms for them. Production reporting suffered owing to operators not activating the right keys at the right time, operators activating the right keys at scheduled intervals without actually hauling any load. In effect, despite an upgradation in technology, the increase in utilization or productivity was marginal. However, this practice provided for a new development route in fleet management systems.

Era 5: Encouraged by the "performance improvement through innovation" syndrome as also precipitated by the line managers complaining of high degree of supervision required for making the above system work, the research groups concentrated on formulating a fully real time monitoring system with specialized algorithms for automatic reassignment of trucks to different shovels depending upon the mine dynamics. Strategic decision points in a mine were identified and beacons emitting radio waves were installed. These beacons were also installed onboard the shovels, at the crusher and at the waste dump end for the purpose of sensing the arrival of trucks. The receivers for these radio signals were mounted on board the trucks and the internal truck data radio tuned such that reception of any of the en-route radio signals could then be validated by transmitting a second signal to the host computer at the dispatcher terminal, thereby recording its position in the mine in real time. The system also ensured optimized reallocation of the trucks to different shovels upon validation at each of the decision points. This required operator intervention only for status change from EMPTY to LOADED or to report a BREAKDOWN status. Implementation of these systems resulted in a net productivity improvement of 32% over current operating statistics of the HEMM. The concern areas in these systems were:

1. Inability to track the trucks in between two beacons.
2. Operators would often forget to change status from EMPTY to LOADED.
3. The base monitoring and dispatching systems had problems in a situation where shovels were in close proximity and Radio Beacons on board both shovels sensed an

approaching truck. This was subsequently solved at the application software level through masking techniques.

4. The software was unable to determine when trucks carrying waste was asked to offload the material at locations where mine roadway preparation was in progress.

5. The beacons were prone to tampering by unauthorized personnel.

Era 6 : GPS based systems, VIMS, PMS, Grade Control, Blending, and Maintenance Time Management

Towards the end of 1991, a major innovative breakthrough was announced with the incorporation of Global Positioning Systems for HEMM and material movement tracking. NAVSTARS provided accurate tracking techniques using differential correction parameters. Use of GPS technology in metalliferrous mining segment provided tremendous advantage in facilitating operations pertaining to ore quality and blending especially in instances where slight differential in ROM quality may impact organizational financial bottom line. The concept of Dual GPS Zonation (Ref. Fig. 53.1) ensured redundancy of the dump switches and status change from LOADED to EMPTY or vice versa was automatic. Incorporation of GPS technology in dispatch management systems ensured reduction in cost, accuracy of data collection, reduced field hardware interference (by unauthorized personnel) and low response time with decrease in field hardware components like beacons and infra red close proximity detectors. Productivity registered an increase of approximately 28-33%.

With HEMM manufacturers incorporating advanced microprocessor based vital signs management interfaces, truck dispatch systems also extended their horizons to incorporate the monitoring systems in their software for management of machine critical information with adequate mechanism to trigger alarms.

Most mines across the world complained of non-availability of information on "What happened to HEMM resources after they entered the maintenance shed ?". This question came to the forefront after it was realized that availability statistics were vital in defining the operational viability of the machinery. An integrated extension of the current day truck dispatch system ensures that each activity within the maintenance shed is timed and apportioned to various support departments for tracking the performance of the concerned departments in ensuring availability of the HEMM. Trackings of such functions provide the necessary bandwidth to effect control over both - the operations personnel (responsible for utilization) as well as support departments (responsible for ensuring availability).

Hence, it has been demonstrated in mines across the world that the current day truck dispatching systems are extending beyond mining boundaries for effecting enterprise level bottom line.

2.1 FLEET MANAGEMENT—DEVELOPMENT TRENDS UPTO 2020

It is expected that by the year 2020 the concept of fleet management would become deeply entrenched and not be limited only to large surface mines but would also extend to smaller mines. Fleet Management systems would populate approximately 90% of large surface mines and approximately 60% of the smaller mines. It needs to be understood that there are quite a few issues that need to be addressed in a real time fleet management system for countering the business concerns of the new millenium.

Advancement in fleet management systems would encompass not only the basic aspects of automatic data collection, automatic dispatching of the hauling units to loading and unloading points but also to ensure integration with Mine Planning and Enterprise Decision Support Systems, it is envisaged that major mining corporations would move out of traditional mining mould, thereby, overcoming the barrier which has eternally segregated mine field operations from the corporate level operations. Discrete beta testing of the above concept at various coal and iron ore mines in India and abroad, through advancements in communication services, IT and HEMM management enables the management of large mining corporations with an opportunity to develop a Digital Mine linking operations related systems to integrated enterprise wide business systems.

The Digital Mine Concept, if conceived, developed and implemented successfully, is likely to provide following benefits :

- Effect optimum control over operations and production planning through integration of real time data from mining processes to the central enterprise information management hub.
- Quick transfer of information to the right destination at the right time for timely decision making, thereby, overcoming the traditional interdepartmental firewalls.
- Automatic system generated optimized mine operational plans, including HEMM deployment plans obtained after defining the operational constraints. Using the best fit plan, the mine management may optimally control operations from the control desk with LIVE figures of material movement plan v/s actual, material grade plan v/s actual, HEMM utilization plan v/s actual etc.
- Executive Information Systems to understand and isolate stray incidences impacting upon the profitability and strategize on action plan in isolation.

The concept of the Digital Mine (Ref. Fig. 53.2) is also likely to provide the gateway for implementing Total Mine Management Systems (TMMS), phase wise from the time of exploration to formulation of the HEMM deployment plan and further graduating into deposit modeling, preparation of production plans (both short term and long term), execution of optimized plans and finally to provide the top management with tools for facilitating detailed insight into core mining operations. The proposed TMMS is also likely to segregate information pertaining to financial viability of individual mines and provide information technology applications (Fleet Management Systems, Mine Planning & Geological Modeling Systems, Geographic Information Systems, evolution of Parallel Database Management Systems, Enterprise Asset Management Systems & Enterprise Resource Planning Systems), Global Positioning Systems (GPS) through use of NAVSTARS, bi-directional mobile communications network with adequate responsiveness and bandwidth, has ensured the feasibility of Intelligent Mining with upgraded reliability and precision in decision making.

2.2 Challenges to Fleet Management of the 21st Century—An Overview

In the current day scenario, Fleet Management Systems operate in isolation in most mines. Hence, the performance of surface mining systems would tend to plateau after a modest to sizeable increase. With Fleet Management systems operating in isolation, the profitability of the mine cannot register any further increase primarily owing to non availability of support functionality – maintenance, materials and finance. In isolation, core truck dispatching systems have to depend upon manual data feed on short term and long term mine planning. Inaccurate data feed creates a situation where the planning and simulation modules of dispatching systems are unable to devise the "should-be" operating plans for HEMM deployment. Typical examples are inaccurate information on deposit model or grade block parameter, which may jeopardize the actual ROM grade of material (ore) at the blend point, thereby, affecting the operational profitability directly.

With increasing focus on improving the tonnage kilometers (TKPH) of material moved by individual units, foolproof methods aimed at eliminating under-loading of high capacity trucks must be devised. Current methods using infra-red cone beacons and on-board weighing systems on board trucks often provide inaccurate tonnage readings due to inconsistent and irregular maintenance and calibration. Methods using infra-red cone beacons have not been successful since the concept did not perceive incidences of unproductive boom swings by the shovel as also incidences of tampering by unauthorized personnel. On-board weighing systems, comprising of load cells on the suspensions over the rear axles on board trucks, provide inaccurate tonnage readings due to the transducers getting damaged owing to ambient levels of dust and humidity.

The entire mineral extraction process would commence as a supply chain from the face to the crusher/blend point and, thereafter, to storage bays for onward (long distance haul) transportation to customer site. 21st century practices would demand a coherence between ore/mineral extraction rate (SUPPLY) and net dispatch rate (DEMAND) from storage bays to end customer premises. This practice is likely to come into vogue primarily because :

- Non-availability of ore at the right time would result in payment of penalty to end customer. This is as a result of stringent global business practices.
- Excess ore would lead to stockpiling (increase in inventory levels) and in the case of sensitive, low incubation period minerals like Coal, may result in coal fires.

3.0 RE-ENGINEERING FLEET MANAGEMENT IN A COAL MINE – A CASE STUDY

A typical case study was conducted at one of the large coal mine in north central India. This mine has a mineable coal reserve of 492 MTe and production capacity is expected to reach 10 MTY by Year 2003-2004. The average stripping ratio is 3.73 cubic meters per tonne of coal produced with multi seam operations. The mine is expected to operate in two sections. However, as on date, the eastern section is only being operated. The distribution of key HEMM is provided Table 53.1.

Table 53.1 : HEMM population at the coal mine discussed.

Sl. No.	*Description of HEMM*	*On Roll for achieving 4.8 MTY*	*Proposed for achieving 10 MTY*
1.	Walking Dragline (24/96)	02	04
2.	Overburden Electric Rope Shovel (12.5 cum.)	05	10
3.	Coal Electric Rope Shovel (10 cum.)	03	08
4.	Coal Electric Rope Shovel (8 cum.)	02	02
5.	Truck 120T (Overburden)	12	48
6.	Truck 85T (Overburden)	51	51
7.	Truck 85T (Coal)	X	10

The scope of the study took into consideration the dynamic situation comprising only of the shovels and trucks. The productivity/performance of the dragline has been intentionally excluded from the scope of the study primarily to focus on the extent of improvement that can be realized through truck dispatch system implementation from the shovel/truck combination only.

As is evident, the control group mine has high capital investment in terms of HEMM fleet deployed/planned for deployment. With a view to demonstrate the effects of implementing a truck dispatch system in a high capacity mine of this stature, it is mandatory to understand the business process inefficiencies, which impact the bottomline of the mine as well as the company. Table 53.2 highlights the areas of process improvement and the effect of dispatching systems on the areas of process change.

Table 53.2: Proposed benefits to a coal mine in Central India by implementation of a Truck Dispatch System – Real Life Case Study.

Sl. No.	*Process Improvement Mandate*	*Expected Improvement through implementation of truck dispatching systems.*
1	2	3
1.	Current Utilization Factors • 120T Trucks : 53% • 85T Trucks : 38%	Integration of coal off-take rate by clients with truck dispatch systems coupled with real time monitoring can improve the utilization factors by 13.5% for coal trucks and 9% for overburden trucks.
2.	Late Start and Early Finish account for 25% of shift time.	Computerized truck dispatch systems take into account the concept of "Hot-Seat Exchange". Based on prevailing mine culture, this concept will ensure reduction in Late Start and Early Finish hours by a minimum of 30% - 50%.
3.	Fueling Practices : Since all trucks queue for fueling at the end of the shift, time lag between first truck and last truck amount to an average of 135 minutes.	Automatic dispatching algorithms in truck dispatch systems provide for allocation of the trucks to the fueling bay based on the fuel status of the trucks. This automatically increases the utilization factor since the trucks will not be required to queue at the fueling bay in such a situation. It is expected that the average queuing time for fueling would reduce to 9 – 11 minutes.
4.	Availability of Truck Fleet is at 33% (21 operating units out of 63). The mine incurs an additional cost of operating excess trucks.	Advanced scheduling and planning sub-systems inherent in truck dispatch systems provide constraint based computation and analysis on deployment of HEMM units. Excess trucks in a shovel/dumper haulage system are identified and segregated with the aim of saving on operating costs. This is likely to improve non spare parts related availability by 21 – 27%.

1	2	3
5.	Shovel Performance at 80-85 trips per shift against industry average of 140 trips per shift under same techno-mining conditions.	Real time HEMM performance monitoring coupled with the imperative for high speed communications and business object triggers is likely to ensure an increase in shovel operational performance improvement by 14-18%.
6.	It is foreseen that with the current automation levels, blending between Grade B and Grade F coal seams are not possible.	Blending is an important feature of modern day truck dispatch systems. In a multi seam operation, the coal source parameters (i.e. quality) are fed into the system either manually or captured through computerized seam models. The system creates an allocation plan based on sequencing the trips between either seam to achieve the budgeted figure. This practice ensures conservation of the richer seam and maintaining the requisite quality demanded by the end user. (Ref. Fig. 53.3)
7.	High Idle time at Workshop : 60% of the time being due to non availability of spare parts.	Fleet Management solution provider Wenco International of Canada has formulated an integrated management support tool which can effectively capture and track all activities performed on an HEMM unit, the reasons for having stopped work, non availability of spare parts, integration with Materials and Maintenance subsystems.
8.	Expenditure on Tires is high.	Integrated monitoring systems for tracking the history of the usage of individual units of tires, failure statistics, tonnage kilometers per hour provide the management with figures on tire utilization, analysis of failure causes and provides the basis for formulating cost effective methods of using the truck tires. It is estimated to increase the life of tires by a quantum of 28%.
9.	Fuel/POL Costs amount to INR 29.68 per tonne of Coal produced.	Fuel Management systems form an integral part of a truck dispatch system. Global Positioning Systems provides information on the terrain and the duration the HEMM unit has traveled in that terrain (incline, decline, level) and whether FULL or EMPTY. Through real time tracking and fuel content discounting methodology, it is expected to reduce fuel costs to INR 27.50 per tonne of Coal produced.

Given the cost of production of coal (at 85% efficiency) be INR 442.50 and the average selling price be INR 534.73, with a minimum of 8 additional trips of coal (using 85T coal body trucks) per shift using a Truck Dispatch System (as computed by the Wenco PMCS 3000's Scheduler module), the system can provide for a profitability increase of INR 43 Million per year (assuming 300 working days per year and fill factor for 85T Coal body trucks to be at 65T). Hence, deployment of such a system will pay for itself in a time span of 40 months further assuming that the mine is not plagued by land acquisition problems, problems of industrial relations, sustained non availability of working face and other operational issues. However, other benefits accruing from the implementation of such a system will ensure that the payback period further reduces to 8 months.

4.0 BUSINESS CONCERN ANALYSIS

Computerized fleet management systems with a population base of approximately 120 across the globe has provided major impetus to mining corporations employing shovel/ dumper fleets of high capacity and numbers. However, it is interesting to note that in over 60% of the installations, the mine management, primarily in light of the following has questioned the usage statistics of such systems :

- High capital investment on truck dispatching systems procured.
- Incorporation of new technology without understanding the realistic bottlenecks in implementing the new technology.
- Payback period had not been well defined.
- Systems did not have productivity/potential savings computational algorithms.
- Had been primarily considered as real time data managers leading to questions like "What do I do with the data that is made available to me ?"
- Forward planners and simulators had not been there.
- Work of line managers for control of ROM grade ore continued to be cumbersome.
- Optimum usage of the system could not happen due to apprehensive nature of the users leading to "natural death" of the system.

The new generation systems came into existence when mining majors realized their inability to formulate process driven methodology for achieving the preset productivity norms. A host of associated problems came into the forefront, which precipitated a chain reaction among HEMM manufacturers, mining software system consultants as well as IT giants.

5.0 FLEET MANAGEMENT SYSTEMS : A FUTURISTIC SOLUTION

In a surface mining setup, the HEMM fleet comprises of more than 60% of the capital investment. Management of the fleet, therefore, assumes utmost importance in ensuring profitable operation of the mine. Advances in mechanization have added different dimensions to more accurately manage the performance of the fleet. The adoption of Total Quality Management by a number of mining corporations during this decade has resulted in significant technological advance in monitoring, planning, implementing and decision support systems for ensuring efficiency and economics of operation of the fleet units.

Wenco International Mining Systems Limited, Canada is a solution provider for Fleet Management and Performance Management systems the world over – the primary objective being to provide state-of-the-art technology evenly balanced with the realistic feasibility of technology deployment in real life mining scenario over various mining conditions. For example, incorporation of high precision GPS in coal mines is not feasible owing to the high price to performance improvement ratio as also incorporation of Touch Screen Graphical Terminals on board mobile units of HEMM operators in developing countries, owing to high level of vibration, high levels of dust and moisture rendering operator problems with the touch screen after a period of use, operator eye strain due to screen reflection during night operations, high levels of undesired operator interference rendering the overall system unusable, high initial cost and cost of maintaining such systems. Having been founded in the year 1983, Wenco's technical expertise is well proven by the following major milestones:

1. First to incorporate GPS technology in truck dispatching for the first time in the year 1992 at Kalgoorlie Consolidated Gold Mines, Australia. Fig. 53.4 provides an overview of a moving mine map through real time GPS layer and other related layers defining the mine map.

2. First to incorporate the Concept of Mobile Supervisor Terminal (MST) – Touch Screen based as well as Notebook computer based for effecting prompt, real time, on-the-spot decisions.

3. First to promulgate the concept of Open Architecture based system with PMCS 3000 Truck Dispatch Systems running on Intel based systems running on Windows NT and Oracle RDBMS, aimed for user friendliness.

4. First to use of the concept of "Dual GPS Zonation" at software level aimed at eliminating the need for physical devices for data capture of status change from EMPTY to LOADED.

Realizing the needs of the mining industry of the new millennium, new features aimed at effecting a transformation to the Intelligent Mine, has been developed by software developers and are currently undergoing extensive tests to evaluate the feasibility and extent of performance improvement by deploying such additional modules. This section deals with the new series of ongoing developments aimed at aiding the mining industry to reach higher levels of automation through optimum use of information technology in fleet management processes.

5.1 Autonomous Dispatching

Mining corporates across the world aim to improve productivity and profitability by introducing systems to take care of repetitive and duplicative processes. With advancements in system technology, the role of the dispatcher is gradually taking a back seat. Autonomous dispatching senses the change in dynamics of the mine and based on historical information of similar circumstances and responses, generates an automatic instruction set routed directly to the operator. Unique and specialized anomalies are referred to the mine supervisor for action. This eliminates duplicated supervisory action.

5.2 Supervisory Control Systems

Mine supervisors and foremen are highly mobile. Hence, they require the services of a complete real time unit for alerting them on the ongoing mine dynamics and to generate required instructions. The real time unit comprises of a mobile touch screen graphics monitor with a CPU as well as a dedicated radio link for real time information availability, reporting, decision making, scheduling and issuing instruction sets. Windows environment and distributed computing architecture provides for the feasibility of this option.

5.3 Enterprise Level Integration

The critical success factors associated with the success of a large surface mining project depends, to a large extent, upon the performance of individual units of HEMM in operation. The profitability of a holding corporation comprising of a group of mines is derived from the performance of the individual mining units. Hence, the operational level solutions so associated with each of the mines must have an integration with the enterprise wide MIS so that the real time data may be compiled at the corporate level for decision making purposes. Information on availability and procurement plan of spare parts for effective operation of the HEMM in the digital mine, real time profit and loss statement of individual mines, corporate level cost planning and computation of business strategy, all form a part of a proposal for creating a neural network for information flow from the mining face to the desk of the top management. Symmetric drill down facilities and simulation tools enable the management to have a better understanding of existing business processes as well as a foresight into various futuristic scenarios of business operations.

With the evolution of the "Digital Mine", relevant information from the pit would automatically transmit to the various support departments through a corporate wide Microsoft Exchange or a Lotus Notes intranet and work flow system. Each level of work flow would be secured through access control mechanisms as well as time cycle alerts for effecting paperless response at the right time. All systems within the enterprise would no longer work in isolation.

5.4 Scheduling and Costing

Current day dispatching and fleet management systems are equipped with operational costing engines and advanced scheduling engines operating in isolation. Research and Developments by industry specialists indicate that the mine management must be privy to information which would attribute a cost parameter to various feasible options in operation arrived at through a scheduler. The integration of the scheduling and the costing engines provide a balance between the best possible plan and the optimum cost operations and optimized HEMM deployment plan.

The cost parameters include cost of maintenance, lead times, idle times, fuel and POL, spares, depreciation, interest costs, overheads besides others. The scheduling algorithm bases its computation on quality constraining parameters, lead distances, number of hauling and loading units, end user quantity and quality requirements, production required, maintenance schedules, fueling hours and a host of other parameters. The prime objective being to schedule production in such a way so as to increase production in a cost effective manner without sacrificing the essential requirements of maintenance, health of HEMM and other forms of operational breaks.

5.5 Integrated Maintenance Time Manager

With a view to extend the boundaries of dispatching beyond the pit, it was realized that the mission critical HEMM units be first linked with the maintenance department for ensuring system availability at the lowest possible time. Immediately after receipt of a maintenance request from the core dispatching system, the system would allocate the maintenance crew and check upon the availability of the necessary resources. Every process associated with the maintenance request is mapped and the return time to Operations is communicated over the system intranet facility. The time count would commence immediately after the HEMM units enters the maintenance assembly line. Each of the processes for the particular HEMM unit is time tracked in accordance with the preset time interval. Accommodative deviations to actual are incorporated in the delivery schedule and communicated to Operations.

Through this process, an audit trail can be established on errant departments and process deviations, if any. For example, materials not having been ordered for requisite spares or maintenance not having been indented for the spares.

5.6 Software based Payload monitoring

HEMM manufacturers and third party vendors provide on board weighing systems, which after a sustained period of time do not work owing to due to high dust and moisture levels and vibration levels affecting the electronics of the system. Joint research and development programs between HEMM manufacturers and information technology solution providers have identified the process for evolution of a foolproof system. Truck dispatch system software integration with the hydraulics of the dump cylinder can be used as a reference vis-à-vis the differential between full body load dumping and partial body load dumping. Differential in hydraulic pressure (from standards) in the dumping process can be used to compute the actual load carried by the truck from the shovel to the dump point. This has been the process tested for achieving the optimal performance.

5.7 Mine operations Playback for effecting Audit Trail

Real time monitoring has its own benefits but equally important during the course of management reviews is the process of conducting a postmortem exercise of the performance of critical shifts. The purpose of this module is to record the time stamped events from the start of the shift to the end of the shift. This is likely to provide a better understanding of the critical success factors associated with the performance of the shift. The module will also enable the management to track individuals responsible for a set of success or failure thereby serving to bolster the audit trailing mechanism.

5.8 Pseudo Tele-Mining

Industry analysts feel that the concept of "Tele-Mining" and "operator less trucks" are seemingly technology savvy but, in effect, have inherent shortcomings in view of extremely frequent changes in mine dynamics and safety parameters. Hence, the evolution of the term "Pseudo Tele-Mining" where systems governing the "Intelligent Mine" are in place and the HEMM operator is expected to follow the instructions in toto. With a view to effect automation in pre defined processes, it is mandatory that external interference from operators and unwanted elements be minimized. Hence, the sophisticated systems so implemented must be robust and must have the lowest form of system/operator interface. The operator will be expected only to follow the instruction sets issued by the system instructions either through screen based text format or through system generated audio format. Such a level of automation integrates fleet operations with all support functions – mine planning, safety, surveying, maintenance, materials/logistics and finance. Data on anomalies and pre-set course are immediately captured, processed, analyzed, responded to and, if required, relayed to respective support departments for further action. Proposed improvements in vital information management systems and their interfaces are expected to ensure that real time information on the type of breakdown would be communicated to maintenance without operator interaction.

5.9 Web Enabling Technology

With a view to adopt remote monitoring and effect remote decision support mechanism, dispatching system servers can be integrated with web servers for remote login by the top management for viewing specific mine/set of mines. Real time dispatch screens can be uploaded to the web and viewed through TCP/IP or leased line connections from the corporate office, thereby simulating the "As-Is" scenario at the corporate office.

5.10 Standardization of Products

Wenco uses industry standard products for the PMCS truck dispatch system. namely, GPS processors and receivers from Trimble, Ashtec and Rockwell, Microsoft Windows NT as operating system, Oracle as the database, Motorola/TAIT as radio communication sets which provide extremely open architecture systems for smooth user accessibility, future customization, addition of specialized site specific applications, as are likely to be required over a period of time by the mine management.

6.0 CONCLUSIONS

Technological innovations in fleet management are gradually paving the way for evolution of an "Intelligent Mine". However, technological overtures must not dilute the basic need for deployment of dispatching and fleet management systems for improvement of productivity of shovel dumper systems. The paper indicates a paradigm shift in the scope and gamut of fleet management systems from erstwhile operations level to enterprise wide level. The case study provided serves to allay the fears, uncertainties and doubts which plague the industry top management and prevent them from taking decisions aimed at implementing truck dispatch systems. There is no doubt whatsoever that the new millennium will witness major applications of the truck dispatching systems in the surface mining arena, both large and small, and will be increasingly integrated with mine planning and resource planning software.

REFERENCES

1. Amitabh C, Duce Rod, Ghose AK, Upgrading Surface Mining Performance in the Indian Mineral Industry – Managerial and Technological Options, Proc. International Seminar on Opencast Mining, July 1998,Mining, Geological & Metallurgical Institute of India, Calcutta, pp. 339-356.

2. Amitabh C, Chattopadhaya T, Computerized Management of Fleet for Iron Ore Mines, proc. National Seminar on Iron Ore Mining and Beneficiation, Puri January-February 1997, pp. 16.1-16.6.

3. Amitabh C, Ghose AK, Richard PL, Quality Control in Surface Mining Systems using Information Technology – Challenges in the next millennium, Proc. National Seminar on Information Technology in the Mineral Industry, Mining, Geological & Metallurgical Institute of India, Calcutta, July 1997, pp. 87-106.

4. Amitabh C, Ghose AK, Wan RN, Management of HEMM through focussed application of Information Technology, proc. International Seminar on Management of Mining Machinery, Mining, Geological & Metallurgical Institute of India, Calcutta, July 1999, pp. 179-193.

5. Amitabh C, Ghose AK, and Walshe PW, Information Technology for shoring up the performance of the Mineral Industry sector in the 21st. Century, Proc. International Seminar on Mining Technology, Confederation Indian Industries, New Delhi, October 1998, pp. 7-9.

6. Chadwick J, Surface Loading and Hauling, Mining Magazine, May 1996, pp. 296-302.

7. Czaplicki, JM, Productivity of the Shovel-Truck System based on an Analysis of different Models, Journal of Mineral Resources Engineering, Volume 6, Number 3, July-September 1997, pp. 117-127.

8. Ghose AK, Productivity Enhancement in Surface Coal mining section in India through new technology, Journal of Mines, Metals and Fuels, September 1996, pp. 273-275.

9. Lema NM, Price, ADF, Benchmarking : Performance improvement towards competitive advantage, Journal of Management in Engineering, January-February 1995, pp. 28-37.

10. Rainbird, Mark W, Whole of Enterprise Solutions for Mining, December 1997.

11. Scoble, M, Strategic Planning for Advanced Mining Technology: A Canadian Perspective, Journal of Mineral Resources Engineering, Volume 6, Number 3, July-September 1997, pp. 97-116.

12. Wusaty, E, Paulus, L, Mining's new frontier – The Fording Coal GPS Experience, proc. International Conference on Mine Planning and Equipment Selection (ed. Singhal), Balkema 1995, pp. 303-311.

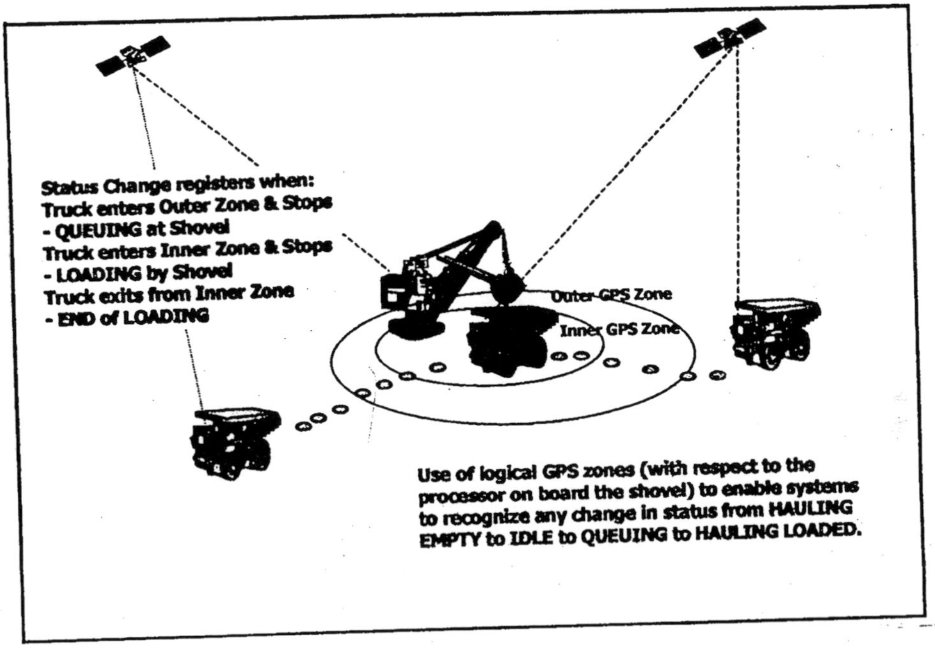

Fig. 53.1 : Innovative use of GPS Technology

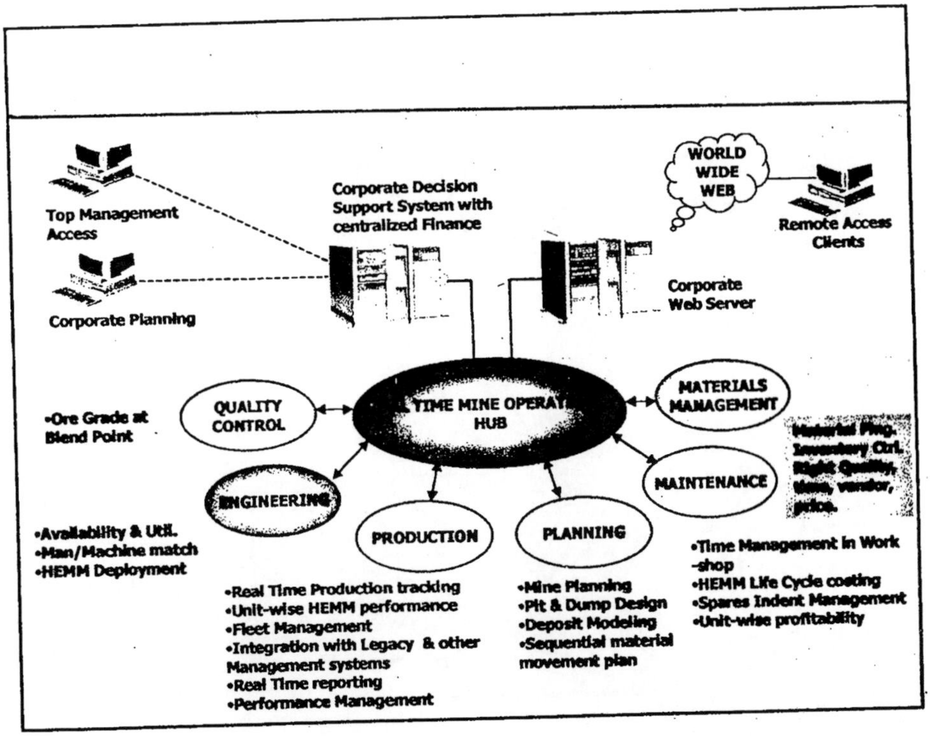

Fig. 53.2 : Proposed Architecture of a Digital Mine

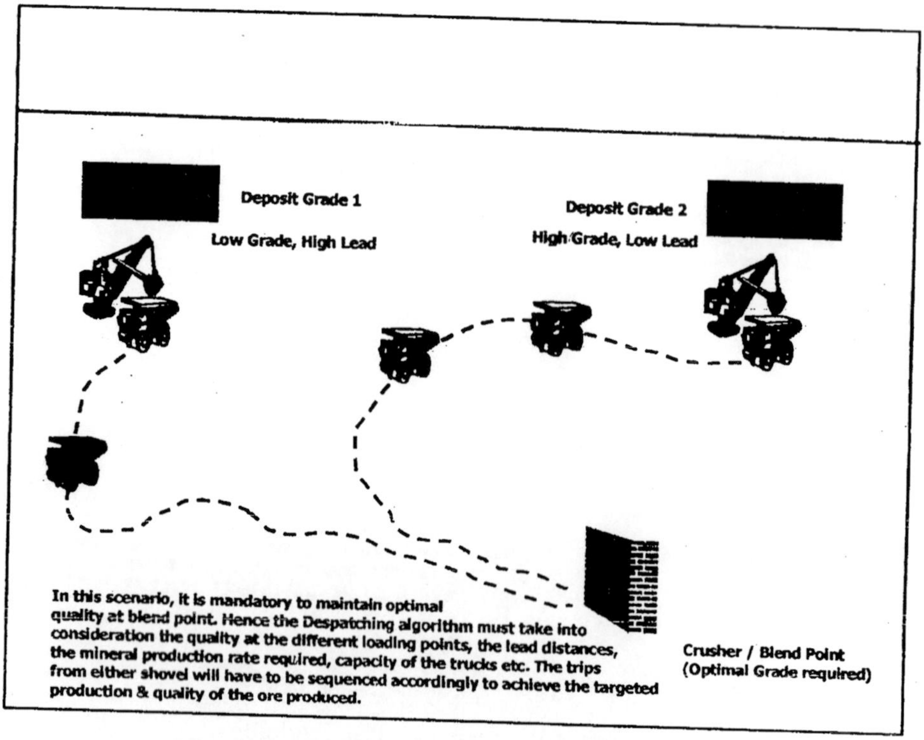

Fig. 53.3 : Sequential Dumping & Blending

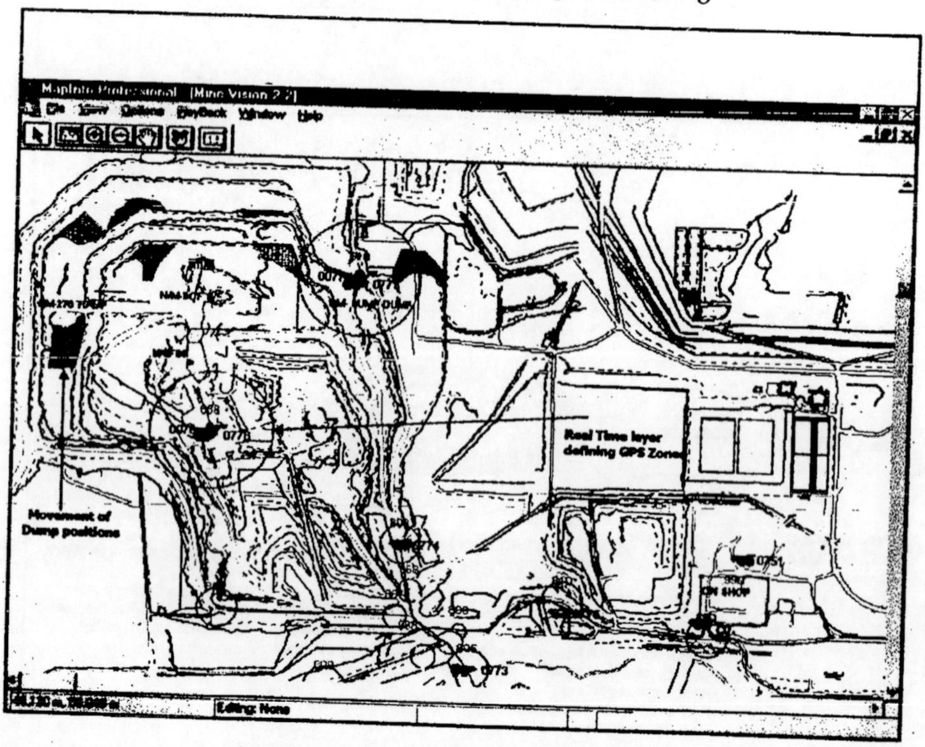

Fig. 53.4 : Real Time GPS Mine Map

CHAPTER 54

The Impact of Information Superhighway on Rock Mechanics Instrumentation

Tadisetty Srinivasulu and R.N. Gupta**

ABSTRACT

The new paradigm exaggerated by the notion that the information superhighway for rock mechanics applications. This new information technology can be effectively utilised by embedding with rock mechanics instrumentation for solving many strata control problems online. This is like a highway where all the expertise is readily available for solving many typical strata control problems quickly. The expert neither to leave his place nor his presence required at instrumentation site. However, his expertise is available at instrumentation site. Moreover, it connects any expert quickly from any corner of the world. Therefore, quick decisions are possible, which are essential for dynámic strata behaviour particularly during depillaring operations. This paper investigates the role of information superhighway in rock mechanics instrumentation.

1. INTRODUCTION

The information superhighway (ISH) will transform one's own mind as well as the way one thinks of problems, the way they synthesise information, and the way they anticipate the consequences of their own actions (Editorial, 1999). This new technical environment created new opportunities in every discipline. It provides many facilities like firstly makes people very close, not just one, but also person to person even though they are thousands of kilometers away. Second, anything created on the highway computer like text, data, graphics, digitised audio or video or software could be transmitted along the PC drive highway. Third, you may receive the imaged information more often than the digitised. Fourth, all the above works can be completed without wasting one's valuable time (Qing

*National Institute of Rock Mechanics, Champion Reefs – 563 117, Kolar Gold Fields, India
tadisetty_s@hotmail.com and *gupta_rn57@yahoo.com*

Zhang, 1995). Therefore, ISH can be effectively utilised for creating rock mass database for wide accesses to experts working in the field, from any corner of the world. However, standardisation of database is necessary for general use. Since distances become so small due to ISH, users are more interested in consulting with experts from around the world then the local ones. This will accelerate the process of knowledge specialisation and implementation.

2. ROCK MECHANICS INSTRUMENTATION VS INFORMATION SUPER HIGHWAY

Generally, mining industry is facing various strata control problems like roof fall in working panel, air blast in side the goaf, side fall, over-riding of the pillars, etc. particularly during depillaring operations. Thus most of the roof falls prediction and forecasting mainly depends on the following points.

- Instrumentation system development and commissioning in the field
- Continuous monitoring of a strata behaviour
- Online evaluation of strata behaviour
- Quick decision by field experts
- Quick implementation of expert suggestions
- Forewarning of the impending roof falls
- Data quality and regular maintenance of the system

For executing the above tasks, services of various experts are necessary. All the experts may not be at one place at one time. For example, design experts will take instrumentation system development and commissioning at the field. Operators monitor and maintain the system. Field experts handle data interpretation and decision making. Finally, the decisions have to be implemented quickly by field engineers. The ISH will interconnect all the experts and systems via computer network. So the interactions with each other could be quickened and delay in implementation avoided. The ISH could be effectively utilised for solving various dynamic mining problems by interconnecting a number of experts.

3. ROCK MECHANICS INSTRUMENTATION

Remote and manual types of instrumentation are shown in Figs 54.1 and 54.2 respectively. In either case, instrumentation system cannot take decisions by itself. However, it can log and store the various parameters for investigation particularly in the case of Fig. 54.1. In the second case the operator has to go near to the instrument to record observations (Fig 54.2). Manual observations may not be continuous due to regular progress of mining operations. Hence, sometimes the manual instruments may be in inaccessible place or in hazardous area. Moreover, generally the person who records the observations or who operates the system need not be an expert. So consulting experts for data interpretation and conclusions is inevitable. Conventional way of solving the problems is tedious and takes more time. None of the instrument is capable of taking decisions on its own. Nevertheless, due to revolution in instrumentation knowledge-based instruments are now available. The experts' decision is final and knowledge-based systems implement experts instructions quickly. Moreover, ISH can reduce the process time drastically and provide environment for quick decision making. Therefore, ISH is indispensable for effective safety and production.

Field Surface Laboratory

Fig. 54.1 : Remote online PC-based instrumentation system for Continuous monitoring of BG panel convergence

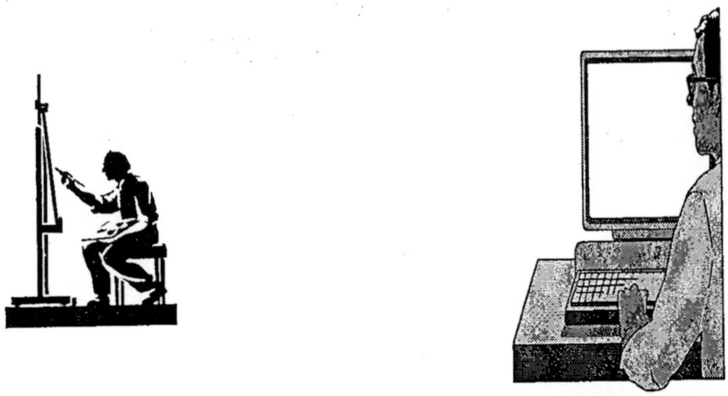

Recording the observations manually Off-line analysis

Fig. 54.2 Instruments for BG panel convergence recording

4. THE NEED FOR ISH

Frequent roof falls are common in underground coal mines. Most of the investigations are off-line and empirical in nature. None of them is implemented effectively due to insufficient time. Moreover, most of the roof falls are preceded by exponential rise in roof closure within a short time. Therefore, the Information superhighway embedded with online PC-based system could solve this problem and allow the experts to interact with the system easily. Thus, ISH helps experts to observe roof fall and forecast.

5. IMPACT OF ISH

The ISH connects the system, operators, experts, field engineers and forecasting modules together as shown in Fig. 54.3. The system logs various mining parameters and stores in memory. Then it would contact experts for data interpretation and conclusions. Always the system would contact odd number of experts for finalising majority suggestion for quick implementation as shown in Fig. 54.3. Thus, the instrumentation system will implement quickly and forewarn the impending roof falls for necessary precautionary actions.

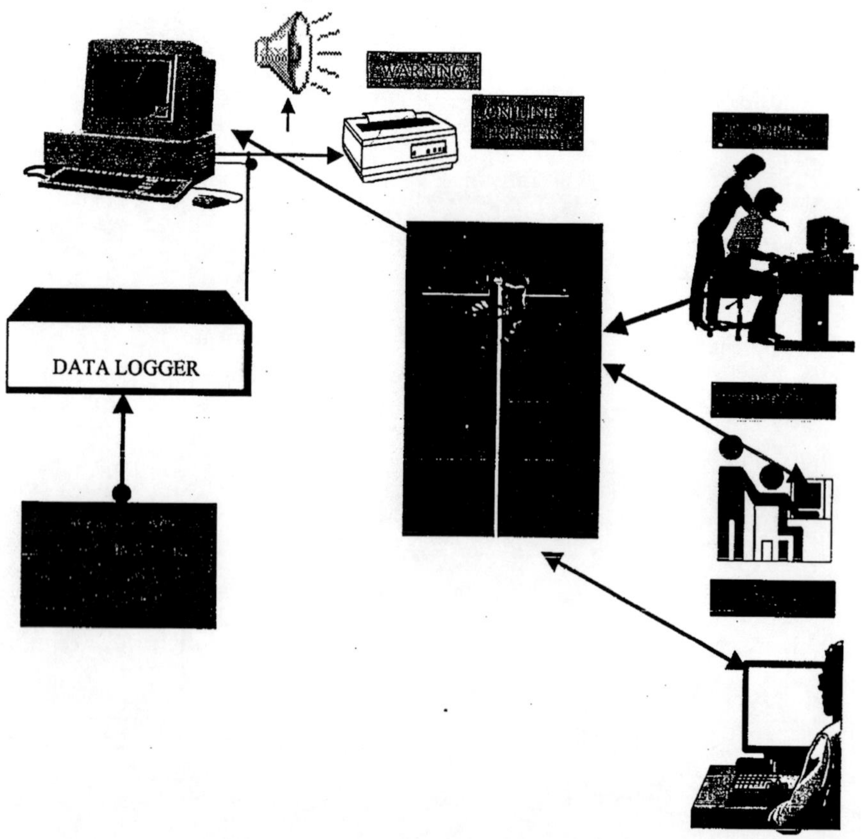

Fig. 54.3 : Schematic diagram of the remote data logging system embedded with Information superhighway

CONCLUSIONS

Information super highway is extremely useful for rock mechanics instrumentation. Most of the problems relating to strata control can be solved by embedding the information superhighway with online instrumentation systems. This is a cost-effective solution to improve safety and production.

ACKNOWLEDGEMENTS

The authors acknowledge Profs.R.V.Raja Kumar and S.Maka of IIT Kharagpur for fruitful discussions.

REFERENCES

Editorial Gaining ground with information technology, Jl of Tunnels and Tunnelling International, Apr.1999. PP1-2

Qing Zhang, The information super highway-what can it do for rock mechanics? Jl of International Society of Rock Mechanics, Vol.3 No.2 Sept.'95 PP 27-29

CHAPTER 55

Decision Support System — A Tool for Assessment of Risk in Mining Investment

S. Nakra & R. Arora***

ABSTRACT

In the present global scenario and more so in the inherently uncertain and volatile economic environment surrounding any Mining Project the need for risk analysis has attained increased importance. **"Decision Support System – A Tool for Assessment of Risk in Mining Investment"** provides a tool with the help of which uncertainties and risk, that can affect project-estimates, schedule and production can be quantified and the combined impact of these uncertainties on project economics can in turn, be examined.

In recent years several methods have been developed which specifically deal with the uncertainties surrounding future cash flows. Research by Hertz find and point out that Computer Simulation for assessing the effect of risk on investment policies provides a means towards effective investment for growth.

In carrying out Computer Simulation for risk analysis several parameters can be treated probabilistically so as to produce a distribution of a profitability measure. The key input factors that any management uses for this purpose are: Grade of Coal, Mineable Reserve, Production Tonnage, Fixed cost, Useful life of project.

Analysis is based upon sample from each distribution for a value of each parameter. The execution of a cash flow analysis is based upon this set of variables. This process is repeated a number of times, the result of each cash flow analysis is stored and ultimately processed into a probability distribution for the measure of profitability.

*Dy. General Manager (Mines), GIPCL-SLPP, Surat, India.

**Under Manager (Mines), GIPCL-SLPP, Surat, India.

To illustrate the technique a hypothetical case of typical Lignite Mine in Gujarat, India has been taken and result of computer simulation has been demonstrated.

INTRODUCTION

Risk can be defined as "exposure to the possibility of economic or financial loss or gain, physical damage or injury, or delay, as a consequence of the uncertainty associated with pursuing a particular course of action". Risk Management/Analysis involves a number of approaches in dealing with the problems created by uncertainty, including the identification, evaluation and control.

A business executive is usually baffled by a situation where the basic assumptions, on which financial viability of a project are based, it self goes wrong. The assumptions going wrong and the consequent impact on the viability of a Mining venture is devastating. The difficulty therefore, lies not only in the assumptions but also in its impact.

In the inherent uncertain and volatile economic environment each assumption presents its own degree (often a high degree) of uncertainty; and taken together, the combined effect of the uncertainties of all the assumptions can multiply to create total uncertainty often assuming critical proportions. This is where the element of risk enters. The element of risk increases manifold in Mining because the characteristics of a mineral deposit lying below the surface of the earth cannot be known with certainty.

"Decision Support System – a tool for assessment of risk in mining investment" is a tool by which uncertainties and risk, that can impact on Mining Project estimates, schedules and production are quantified and the combined effect of these uncertainties on project economics can be assessed.

Effective and efficient performance of risk management involves a number of contributing elements. These include models, methods and computer software as well as tangible skill related elements such as methodology design, specialist expertise and study team management.

PROCEDURE

The steps involved in computer simulation analysis are as follow:

Step I : *Model the project*: The model of the project shows how the net present value (or any other criteria of merit) is related to the parameters and the exogenous variables. (Parameters are input variable, which are constant over all simulation runs. Exogenous variables are input variables, which are stochastic in nature and outside the control of the decision-maker.)

Step 2 : Specify the value of parameters and the probability distribution of the exogenous variables.

Step 3 : Select a value at random from the probability distribution of each of the exogenous variables.

Step 4 : Determine the net present value or any other criterion of merit corresponding to the randomly generated values of exogenous variables and pre-specified parameters value.

Step 5 : Repeat step (3) and (4) a number of times and by computer simulation run to get a large number of simulated net present value or any other criterion of merit.

Step 6: Plot the frequency distribution of the net present value or any other criterion of merit.

The above procedure is shown in **Fig. 55.1.**

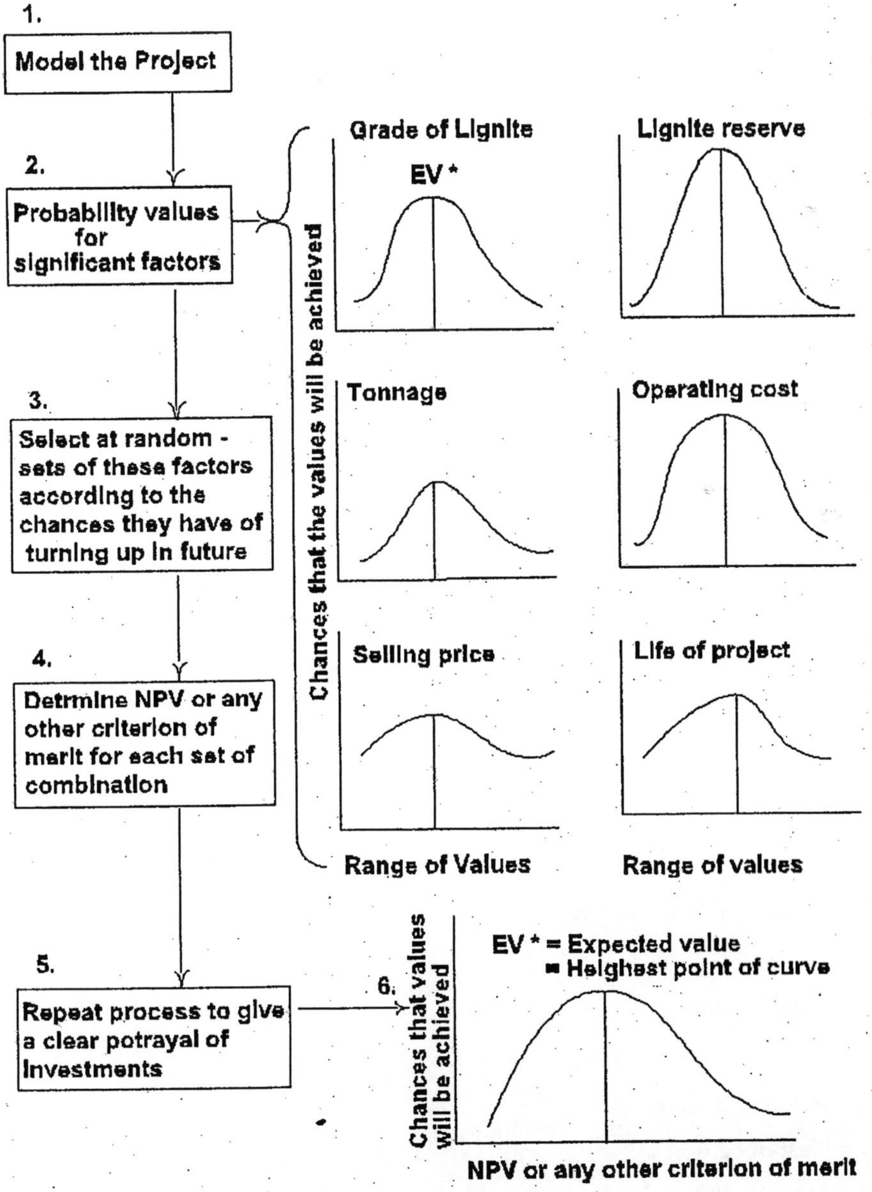

Fig. 55.1 : Simulation for risk analysis and investment planning

SUMMARIZED INFORMATION OF A TYPICAL LIGNITE MINE

1.	Target Production of Lignite	:	1.2 Mc Tonne per Annual
2.	Mine Life	:	35 Years
3.	Initial Capital Investment	:	Rs. 265.90 Lac.
4.	Debt - Equity Ratio Loan	:	Rs. 14667 Lac
	Equity	:	Rs. 11923 Lac
5.	Selling Price at PitHead	:	Rs 999 / Tonne
6.	Total Fixed Cost excluding depreciation	:	Rs. 414 Lac
7.	Variable Cost including Royalty (Royalty @ Rs. 2.50 per Tonne)	:	Rs. 325.34 Lac
8.	Unit Cost of Production	:	Rs.739.35 / Tonne
9.	Cost of Capital	:	14 %.
10.	Depreciation Charge	:	Rs. 1731.27 Lac/ Year
11.	Net Salvage Value	:	Rs. 0.00 Lac
12.	Tax Rate	:	30%

CASTE STYDY

To illustrate the technique a computer simulation run is performed for mining investment of hypothetical case of typical lignite mines in Gujarat.

Step 1: Model for the mine investment is as follow:

$$NPV = \frac{[Q(P-V)-F-D](1-T)+D}{(1+k)^t} + \frac{S}{(1+K)^n} - I \qquad \textbf{(i)}$$

Where, NPV = Net present value of the mine investment

Q = Annual production

P = Selling price Rs. per Tonne

V = Variable cost Rs. per Tonne

F = Fixed cost in Rs.

D = Annual depreciation charge

T = Income tax rate

k = Cost of capital

n = Mine life in years

S = Net salvage value

I = Initial investment

Step 2: The key input factor that management has to decide and use are:

(i) Selling price of lignite(P),

(ii) Variable cost (V), and

(iii) Life of mine (n).

In the NPV model embodied in equation (I), the parameters are Q, F, D, T, k, S, and I whereas P, V, and n are exogenous variables. The probability profile/distribution of exogenous variables are shown in **Table** (55.1) and **Fig.** 55.2, **Fig.** 55.3, and **Fig.** 55.4, where as parameters value are as follow :

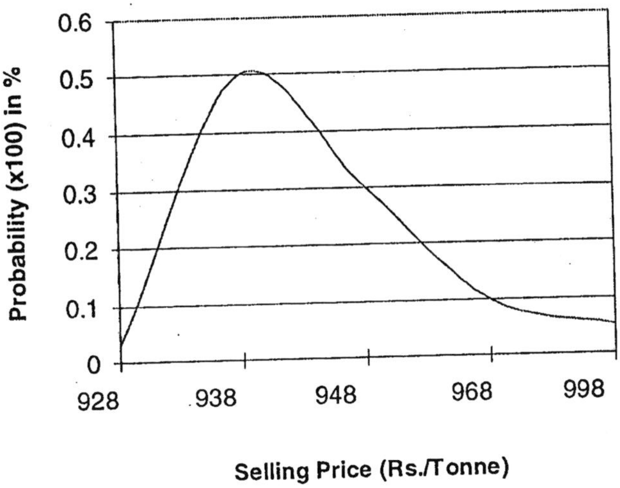

Fig. 55.2 : Selling Price V/s Probability

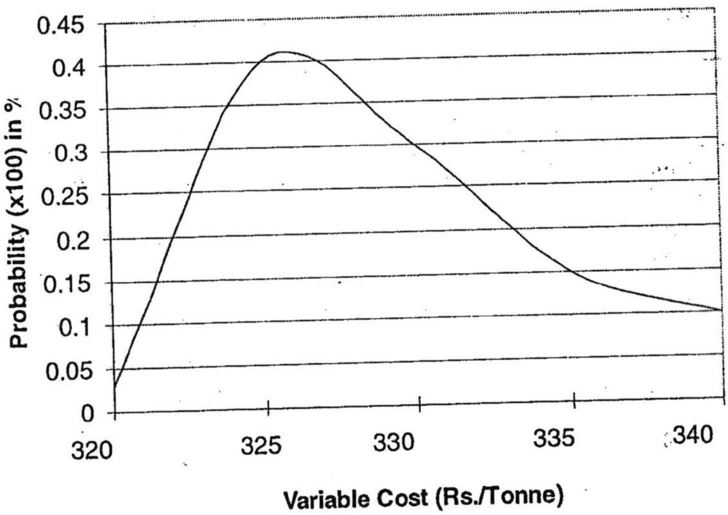

Fig. 55.3 : Variable Cost V/s Probability

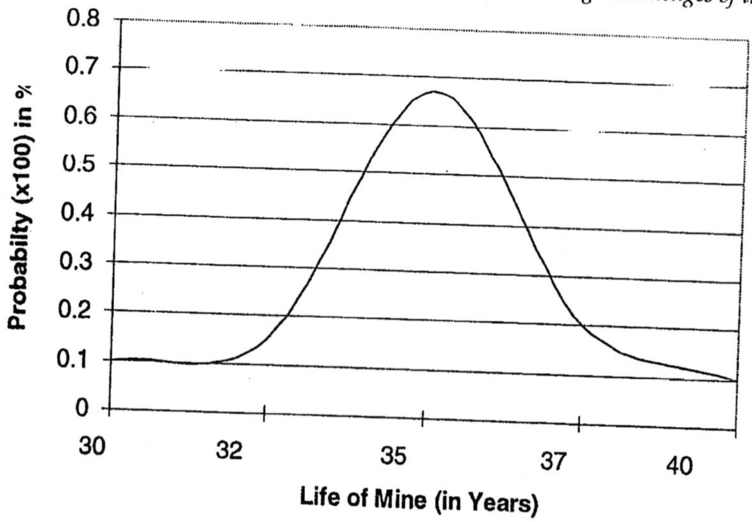

Fig. 55.4 : Line of Mine V/s Probability

Parameters value:

Q: 12 Lakh Tonne

F: Rs.414.0 Lakh

D: Rs.1731.27 Lakh

T: 30%

k: 14%

S: 0.0

I: Rs.265, 90 Lakh

Table 55.1 : Probability distribution of exogenous variables:

Selling Price(P) (Rs./Tonne)			*Variable Cost(V) (Rs./Tonne)*			*Life of Mine(n) (in Years)*		
Value	*Prob.*	*Cum. Prob.*	*Value*	*Prob.*	*Cum. Prob.*	*Value*	*Prob.*	*Cum. Prob.*
918	0.02	0.02	310	0.02	0.02	28	0.05	0.05
928	0.03	0.05	320	0.03	0.05	30	0.10	0.15
938	0.50	0.55	325	0.40	0.45	32	0.15	0.30
948	0.30	0.85	330	0.30	0.75	35	0.67	0.97
968	0.10	0.97	335	0.15	0.90	37	0.20	0.99
998	0.05	1.00	340	0.10	1.00	40	0.10	1.00

Step 3: The correspondence between these exogenous variable and random numbers is as shown in Table 55.2.

Table 55.2 : Correspondence between these exogenous variable and random numbers:

Selling Price(P) (Rs./Tonne)				*Variable Cost(V) (Rs./Tonne)*				*Life of Mine(n) (in Years)*			
Value	*Prob.*	*Cumm. Prob.*	*Random No.*	*Value*	*Prob.*	*Cumm. Prob.*	*Random No.*	*Value*	*Prob.*	*Cumm. Prob.*	*Random No.*
918	0.02	0.02	00-01	310	0.02	0.02	00-01	28	0.05	0.05	00-04
928	0.03	0.05	02-04	320	0.03	0.05	02-04	30	0.10	0.15	05-14
938	0.50	0.55	05-54	325	0.40	0.45	05-44	32	0.15	0.30	15-29
948	0.30	0.85	55-84	330	0.30	0.75	45-74	35	0.67	0.97	30-96
968	0.10	0.97	85-94	335	0.15	0.90	75-89	37	0.02	0.99	97-98
998	0.05	1.00	95-99	340	0.10	1.00	90-99	40	0.01	1.00	98-99

Step 4: Computer simulates an actual situation by selecting one value from the profile of each factor according to the chances values have of turning up.

Step 5: Computer combines values selected to determine a NPV for this individual situation.

Step 6: Computer repeats this process of selection and NPV calculation several numbers of times. (By iteration)

After 100 runs from computer the NPV as obtained in simulation results drawn on a probability profile of NPV which is shown in Fig. 55.5.

CONCLUSIONS DRAWN

- Computer Simulation offers management a tool that will enable it to examine and quantify the risk consequences of investment decision. Computers Simulation techniques are useful and understable which give clear picture by way of probability profile.
- The result of 100 simulation runs shows that there is 30 % chances that the Net Present Value will Rs.-7.5X107 and 10% chances that the Net Present Value will be Rs. +27.5X107.
- In short, the skewness and kurtosis of Fig. 55.5. Probability profile of NPV from Simulation is towards (-ve) side that indicates the investment is risky.
- In today's operating climate there are lots of uncertainties due to external environment, which influence the price and cost of lignite. Taking into account these considerations the management should focus on minimizing the cost of production.
- Management should pay more attention to minimize the risk in the following manner:
  1. More scientific working should be practiced so as to reduce overburden: Lignite (Stripping) ratios.
  2. Detailed geological exploration should be carried out so as to enhance the confidence levels.

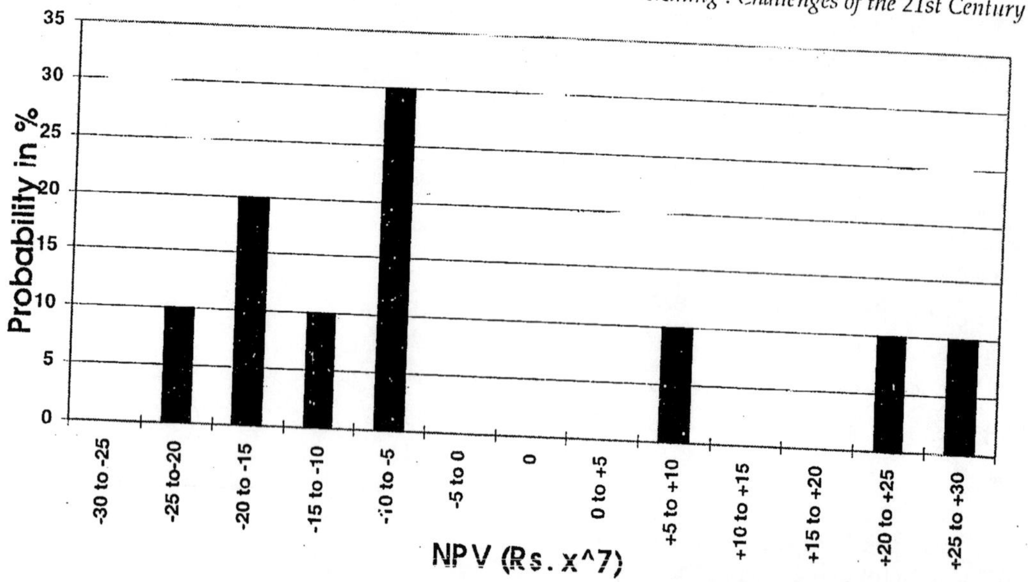

Fig. 55.5 : Probability Profile of NPV from Simulation

3. More efforts towards reduction of operation cost/cost of production by minimizing overhead expenditure.

4. Look for alternate Lignite bearing areas with more favorable stripping ratios.

- The success of project is critically dependent on the Mining Investment and acceptable norms of the risk involved in it. In this regard knowing the quantum of risk is important as it provides the necessary yardstick required by a corporation/company for allocation of resources efficiently.

ADVANTAGES

A method that appears promising and is becoming a popular tool of risk management/analysis, offers definite advantages:

1. It's principal strength lies in its versatility. It can handle problems characterized by (1) numerous exogenous variables following any kind of distribution, and (2) complex interrelationship among parameters and exogenous variables.
2. It compels the decision-maker to explicitly consider the interdependencies and uncertainties characterizing the project.
3. The simulation approach has the inherent advantage of simplicity. It requires only an extension of the input estimates in terms of probabilities.

LIMITATION

Simulation, however, is a controversial tool, which suffers from certain shortcomings:

1. It is difficult to model the project and specify the probability distribution of exogenous variables
2. Simulation is inherently imprecise. It provides a rough approximation of the probability distribution of net present value or any other criterion of merit.

3. A realistic simulation model, likely to be complex, would most likely to be constructed by management scientist, and not the decision-maker. The decision-maker, lacking in understanding of the model, may not use it.

4. In the case study the uncertainty profiles of key variables should be drawn with the help of experienced managerial person because result of the simulation depends on the accuracy of these risk profile.

5. To reduce risk critical variables should be properly analyzed and more precise data should be collected.

REFERENCES

1. Arora, R., "Mining Investment Appraisal: A Case Study of Captive Limestone Mines of Binani Cement", MBF Project Thesis, Indian Institute of Finance, Delhi, April-1997.
2. Arora, R., "Mining Project Appraisal-Wealth Growth Rate Approach", 5th National Seminar on Surface Mining, IIT Kharagpur, Dec. 4th-5th, 1995
3. Arora, R., "Mining Project Investment Appraisal Technique and Practice, Mineral Wealth, Directorate Mines & Geology, Ahmedabad, Jan-June, 1996.
4. Harris D., "Risk Analysis in Mineral Investment Decision", Society of Mining Engineer, AIME Transaction, Vol. 247.
5. Hertz, D.B., "Investment Policies that pay off", Harvard Business Review, Jan-Feb., 1968.
6. Hertz, D.B., "Risk analysis in capital investment", Harvard Business Review, Vol. 42, Jan-Feb., 1964.
7. O' Hara, T.A., Analysis of Risk in Mining Project, CIM Bulletin, Vol. 75, No. 843.

CHAPTER 56

Emerging Role of IT in the Management of Mining Enterprises

*Chinmoy Sarkar**

ABSTRACT

With the declining ore reserves and rising production costs, surface mining have been looking for means to automate certain processes to obtain timely and quantitative feedback on how well each component in the mine is operating. Proper and timely use of this information would permit more dynamic planning with fluctuating market conditions, aging equipment fleets and varying mineral grade.

Business survival in the 21st century is dependent on the ability of mining managers to understand the impact of any changes in the external environment to his business enterprise. Information is an indispensable feature in meeting these challenges of competitive world. Information Technology (IT) is harnessing the decision-making issues for the managers. From mine planning to equipment positioning and control system, IT plays a tremendous important role in today's mining world towards achieving enhanced productivity level. The mine-wide integration of information flow is providing the managers the access to refer to the database for instant decision making in the functioning of each mining subsystems.

This paper looks into present market scenario for the implementation of information technology in Indian mining sector. The article also focuses on the areas of operations where IT can be implemented in the survival and growth of mining enterprises in competitive business environment. The problems and issues in aligning IT in the functioning of organisation have also been dicussed.

INTRODUCTION

Since last few years, the industrial policy in India is under revolutionary change. The notable change is free and open economy as compared to protected socialistic economy

*Tata Consultancy Services, ITC Centre, 10th Floor, 4, Russel Street, Calcutta - 700 071,
Email: chinmoy@tcscal.co.in Phone: (33) 2260707 Fax: (33) 2265530.

of earlier years. The industries like mining, steel, petroleum which were mostly under the control of public sector undertakings are now encountering a new challenge from deregulation and privatisation. The message is loud and clear that the government expects the mining business of the next millenium to check deteriorating efficiency and productivity to be global and competitive. Apart from that, mining business has a set of characteristics that make its operations capital intensive. The government is thus worried about the productivity of capital in mechanized mining units. The changes in the policy of the government in mining sector aim at creating a favourable outlook towards the utilisation of the acquired system for higher production and improved productivity. To achieve the goal, information is vital at every step of decision making in mining operational environment where numerous technical parameters and business processes need to be synthesized and evaluated. The modernization and expansion programmes of mining units thus, need to be integrated with a separate policy of implementation of information technology (IT).

EVOLUTION OF IT IN MINING

The new technologies for IT in mining are introduced gradually over a period of last 40 years to assist the evolution in mine management process. In retrospect, we can identify four fairly distinct phases.

Electronic Data Processing (EDP)

During 1960s, a specialized IT staff function was developed which worked in isolation from the business. The economics of computer use during this era pushed firms toward automating large-scale clerical activities-hence the label "data processing." Payroll and accounting applications were natural targets of opportunity.

Management Information System (MIS)

During 1970s, the focus of attention in Information Systems was shifted from basic clerical and accounting processes to designing and building reporting systems to meet managers' information needs. The rationale of this era was to organize and present this data to managers, assuming almost by definition that the more data they had the better.

Information Integration

The MIS era marked the extreme of centralization in computing. During 1980s, the early use of personal computers and then in the transformation of both the technology and its uses, new applications were developed with the help of shared data resources through local area networks. The business processes embraced IT as a way to integrate information from executive information system, mine planning system to truck dispatch system.

Business Integration

During 1990s, change has accelerated in just about every area of both business and technology. Business is dominated by radical not incremental shifts. Most large companies are working actively to free their employees from the constraints, waste, and inflexibility of the hierarchical structures With the development of IT, new classifications are continually emerging for the application systems. The trend is gradually moving towards the use of an application system to drive and measure their business activities. Mining enterprise has a huge physical asset base in the form of heavy earth moving machineries involved in its production process. It uses complex integrated production systems, where the production and maintenance planning are very sensitive to equipment availability. Mining production throughout is dependent on the maximum up-time of the machineries and mining managers have to devote considerably more management effort to increase the availability of the machineries. It is from this difference that the philosophy of mining enterprise solution derives its benefits over the regular enterprise resource planning concept.

INDIAN MINING IT SCENARIO

Indian mining sector has a monopoly structure of operations by government undertakings. Though Indian public sector organisations have always recognised the significant role of information technology as a potential tool to contribute to the development of the organisations, they are perceived as having relatively low priority in introducing IT for decision making processes at various levels of the organisations. Mining industry has mostly been restricted to transaction processing (e. g. payroll, accounts payable), operational control (e. g. maintenance, inventory, production) for executive information system with very limited applications in mining planning and geographic information system.

PUBLIC SECTOR ATTITUDE

IT contributes mainly towards overcoming two perceived problems that act as barriers for the improvement of productivity in public sector organisation. First, centralisation of decision making, which makes decision makers too remote from the locus of decision information and action and second, unaccountability, which makes decision makers too remote from those outside the organisation who are affected by their decisions. There is no magic recipe for ensuring widespread use of the integrated approach to IT without very strong attempt to build the hybrid managers – spanning managerial, IT skills. Some managers in mining organisations may have computing skills but they do not gain information systems knowledge to use the existing system to its optimal level.

COMPUTERISATION ISSUES

Despite the potential of business reform in every sector of economy in India, there remain substantial problems for the Indian mining sector to enter the reform era due to socio-political reasons. The sector has traditionally fought automation due to pro-labor policies followed by governments. Embracing reform in this sector was very slow in many organisations. They invested in huge capital for increasing production tonnage, but they concentrated little for the improvement of productivity. The same is a reality in respect of the infrastructure created to make use of information technology in day to day decision making. The use of computers was restricted to word processing than anything else. They could not hire computer professionals due to employment freezes. Apart from all these, the organizations tend to use computers for a longer life period due to the low cost of maintenance and high cost of new hardware.

INFORM AGE REFORM

In spite of discouraging scenario in the past, the crucial advantage for India to go in for information age era will be the ability to leapfrog over many intermediate technologies and use the latest. Adaptability is an essential part of Indian culture. Hence there should not be any cultural factors that could inhibit the growth of information technology. The present trend of integrated approach to relate IT and business administration also places information in the driving seat in any organisation, relegating technology to an important but enabling role. This means that one can therefore see a trend of 'information age reform' that combines the existing reform agenda of mining industry with the promise of the information age.

MINING IT REFORM PROCESS

The rationale of mining information age reform process is to address the following in the organisations with a more overt role for information and with greater use of information technology. If it functions as anticipated; the computerisation in mining sector has greater ability to manage organisational resources and their performance. Within these systems, and by comparison with manual systems, creation of a computerised working environment can produce a number of organisational benefits. These include faster decision making and control through provision of timely information and better decision making and control through provision of relevant information. The direct benefits of mining reform with IT perspective are as follows:

Increasing efficiency: Improvement of the input to output ratio of expenditure and inefficiency of many of its processes

Decentralisation: Transfer of decision making to lower, more localised levels of the organisation. This reduces the costs of centralised decision making and creates more flexible and responsive decision making.

Increased accountability: Making staff more accountable for their decisions and actions. This increases the pressure on staff to perform well, to make them more responsive to recipient groups, and to reduce inefficient or corrupt practices.

Improved resource management: Effective use of human, capital and other resources. It includes a refocusing of the way the performance of these resources is planned, measured and managed.

Marketisation: Uuse of market forces to cover relationships within the consumers. This addresses the market relation issues that drive costs down and increase efficiency and/or effectiveness of service delivery.

Barriers in the implementation process

The mining information technology marketplace is different to many others. It has a set of characteristics that combine to make it unique. Application is dispersed geographically and operations are diversified with involvement of heavy mining machineries at various stages of development. The sector has to address the following issues before they invest on IT.

Telecommunication: India's biggest weakness is in the area of telecommunication infrastructure which was in Government monopoly until recently. Now, private companies are participating in telecom sector making it more effective and efficient.

Labour Unions: Indian politicians were hesitant to antagonize the powerful mining labor unions. There were attempts to introduce IT in mining organsiations in the past but the employees promptly went on strike as a protest against this policy. The scenario is gradually changing as their leaders understands the potential of computerisation in their day to day life.

Bureaucratic controls: The government has taken a bold step in allowing private competition in the area of mining. In order for India to become a strong player in international mining market, the government has to dismantle bureaucratic controls which discourage foreign firms from investing in India.

Lack of infrastructure: Organisations are forced to invest on high cost items like satellite links and high speed lines due to lack of infrastructural support. Not all firms would be able to afford such capital costs both in terms of money and time.

Government as Customer: The present monopoly of Government in mining sector restricts the introduction of IT in mining. The mining undertakings have to become major customers for the IT consultants and suppliers to encourage the growth of IT in mining sector. At present, there is a favourable outlook of mining companies in using the services of private consultancy firms in the implementation of their IT projects.

Essential features for Success

The success of mining sector information age reform depends on the following:

People: Employees should have proper motivation to act accountably. This, in turn, depends on many other factors such as current rewards within the job for performing well; the likelihood of detection and punishment for wrong-doing; and the organisational culture's viewpoint on accountability.

Management: Lack of accountability in running the system that helps the decision makers shedding their responsibilities has to be looked into and a formal organisational information structure need to in place. In case of failure, the responsible person should be made accountable. Information flows also enable accountability.

Standards: Some type of performance indicators must be set that can be compared against standard to determine what is not satisfactory.

Processes: The processes must exist to gather accountability information and deliver it to relevant recipients.

Inertia: The costs of IT implementation is sometimes, the deterrent factor in keeping the project viable for a longer period. Also the mining managers who are directly responsible for the production of the unit, are not used to making strategic decisions within the timeframe. They may not wish to give up their information, which are the inputs to the information systems and become more accountable in the process. Their support is thus, essential at every stage of operations.

UNDERSTANDING MINING ENTERPRISE MODEL

The importance of understanding the framework of information technology in the organization has become more apparent as computers move from their traditional role in the "back-office" to supporting the day-to-day activities of production managers. Beyond being a production man in the mine site, a thorough understanding of mining production and maintenance processes are essential for all mining professionals in order to appreciate what can be achieved in mining through the use of IT. The benefits of IT applications in mining depend on what stage of development cycle, the mine is in.

Exploration: to reduce the cost and uncertainty associated with identifying and characterizing mineral resources and with developing efficient mining plans.

Extraction: to improve resource efficiency in mineral extraction activities and to reduce the disturbance of the earth associated with them.

Processing: to reduce processing costs and to improve resource efficiencies.

Utilization: to promote continued and new uses of all mining products while increasing productivity, decreasing costs and improving the efficiency of product utilization.

Environment, health and safety: to minimize air and water emissions, solid waste production and worker exposure to hazards in all segments of the mineral product cycle (production, processing, utilization, recycling, disposal and the associated transportation systems).

MINING ENTERPRISE VISION

The key to meeting IT challenges in mining enterprise has been the "Enterprise Backbone" developed through the communications infrastructure with which the system would be deployed in the organization. As pressure increases to reduce costs, and maintain profitability in the minesite activity, mine management must be able to evaluate the mining processes influenced by the outside world like external competition, government regulation and environmental restrictions. Central to the continuous improvement cycle are the difficulties associated with the gathering and analysis of mine site activities and production information. The proposed framework under which mining organization would use information in new mining business era with an aim to integrate enterprise wide mining solutions is presented in Fig 56.1.

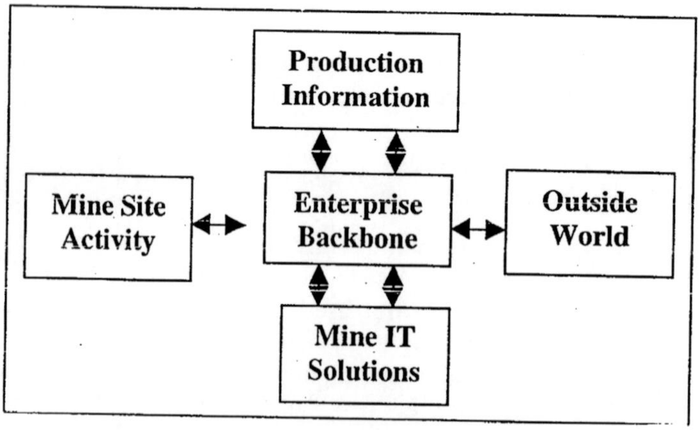

Fig. 56.1 : Mining Enterprise Vision

ENTERPRISE MISSION

There is always a mission for any business enterprise; for mining, it is the production of right quality and grade of coal/ore (after removal of overburden for opencast mine) keeping due importance of safety, conservation and environment. The framework of mining enterprise mission is presented in Fig 56.2. The core element of mining enterprise is production or stockpile inventory. The input required for the production system is mainly mother earth, the raw material for mining operation. Apart from the raw material, production assets like heavy earth moving machineries (HEMM) are also very critical inputs to the system. The demand side of the supply chain is outbound logistics or stockpile inventory that is responsible for the supply of right grade and size of material for customer's satisfaction. The mission of mining enterprise is supported by enterprise asset management that controls the use and allocation of capital resources for the production unit.

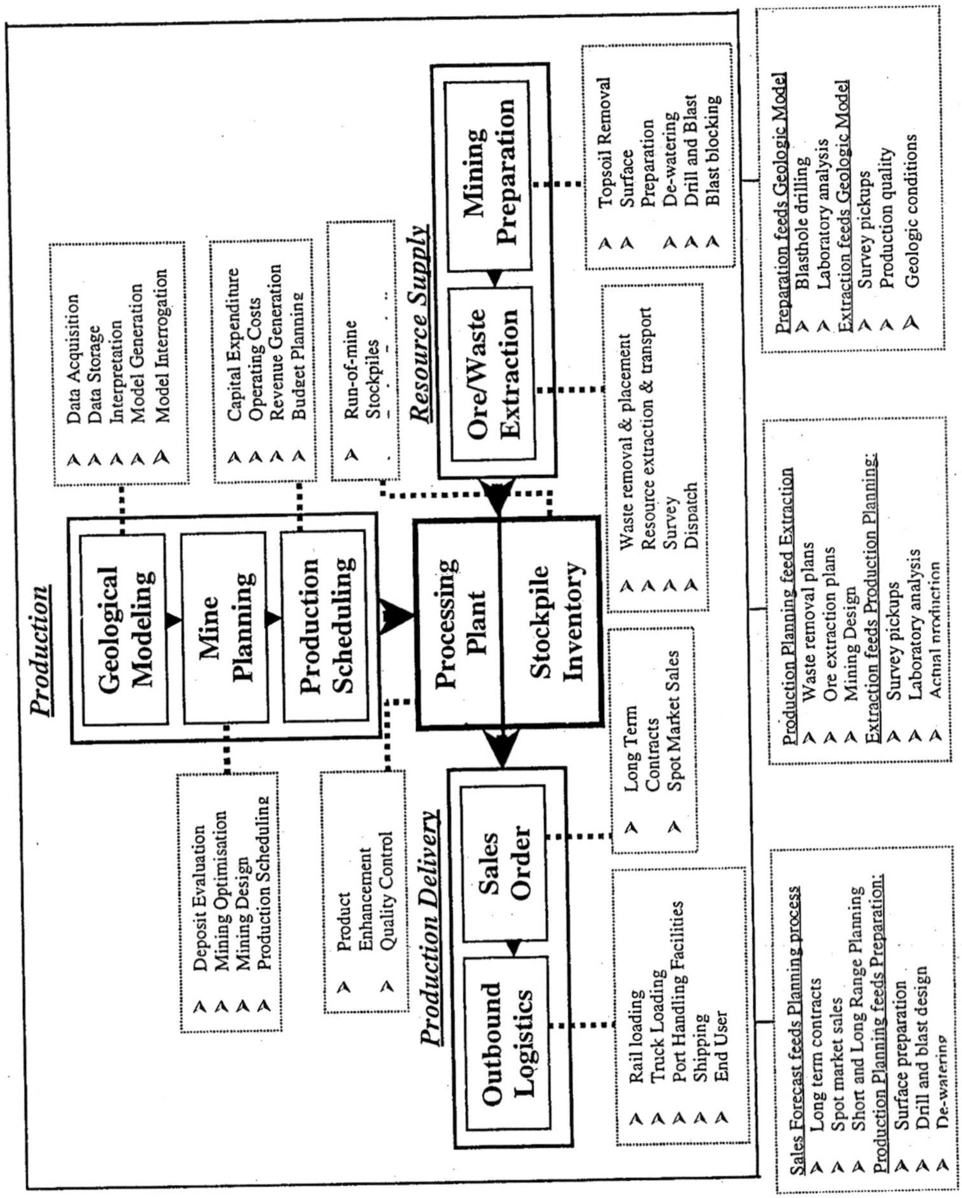

Fig. 56.2 : Mining Enterprise Mission

ENTERPRISE ASSET MANAGEMENT

Enterprise Asset Management (EAM) evolved from needs of maintenance departments to schedule planned maintenance and breakdown repairs for the fixed and mobile plant required to run the enterprise. Variously known as MRO (Maintenance, Repair and Operations) and also CMMS (Computerised Maintenance Management Systems), this concept has been expanded to include the efficient usage of all assets, including human resources and inventory. Using the same process to analyse asset management rather than production, it can be shown that the three main sub-systems of an EAM system are Maintenance, Purchasing and Human Resources. These systems can also be further broken down into detailed components to show operational activities. EAM systems are typically used in enterprises where there is only one product such as coal, iron ore. The defining characteristic of these organisations, which at first glance have little in common, is that they rely on the availability of their capital equipment for their success much more than their ability to respond to changing demand. Mining industry has a large amount of equipment that must be kept 'available' while containing costs for the maintenance and the holding of spare parts. In these organisations the goal is not to repsond to market change, but to provide maximum 'up time'. The typical user of EAM systems would devote considerably more management effort to the control and prediction of spare parts purchasing than to the purchase of raw materials. The core elements of enterprise asset management presented in Fig. 56.3.

MINING ENTERPRISE SOLUTIONS

Mining operations encounters difficulties in consolidating and using information from non-integrated software systems which complicate the information flow and causes the inevitable re-keying or manual transfer of data. The need to connect various mining activities drives the development of enterprise wide solution where managers can share the same set of data for individual decision making. The breakthrough approach lowers costs through better use of all information generated and enabling decisions to be made in real time. The solution controls day-to-day mining operations, including all aspects of financial, materials, maintenance and human resources management. It also includes technical mine planning package that allows mine managers to use all the information collected during a mining operation to develop the most effective extraction and production schedules. The production information system adds tremendous value to mining operations by enabling managers to use information to make real-time production decisions. It allows mining activities to be viewed in 3D and interrogated with information flowing from the field to the office and vice versa. The system provides sophisticated control by allowing vehicles to be automatically tracked throughout the mine operation and to be automatically shut down based on critical condition alarms. Automatic dispatch will form a future stage of development. Condition monitoring information such as oil temperature, oil pressure, hydraulic pressure, fluid levels, etc. can be monitored and used to initiate control actions. This information can also be archived and trended for analytical purposes and integrated with maintenance management systems. This will also allow maximum utilisation of equipment and can be used to better plan and manage maintenance and day-to-day operations. The core of implementation mainly relies on the acquisition and synthesis of data to produce meaningful and timely information that can be used to make strategic and tactical decisions. The proposed framework is presented in Fig. 56.4.

HIGHLIGHTS

- Activities across an entire mining operation to be inter-connected.

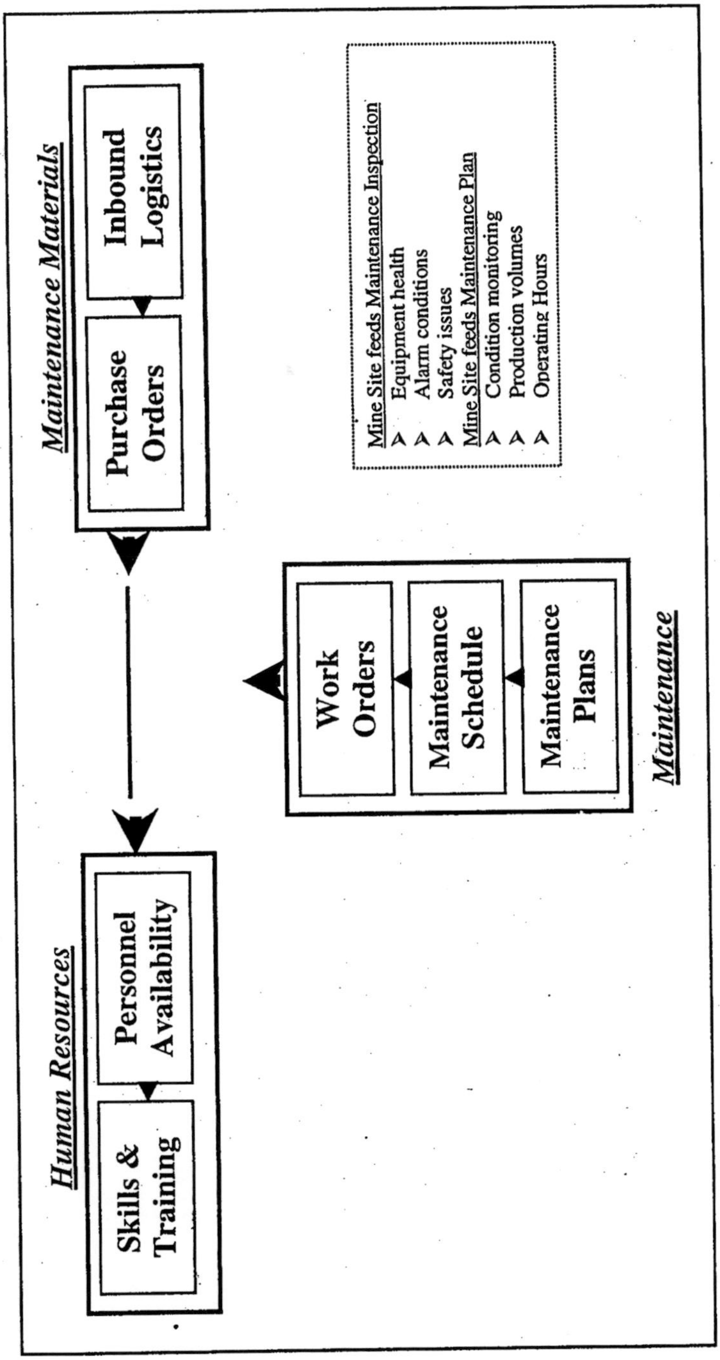

Fig. 56.3 : Mining Enterprise Asset Management

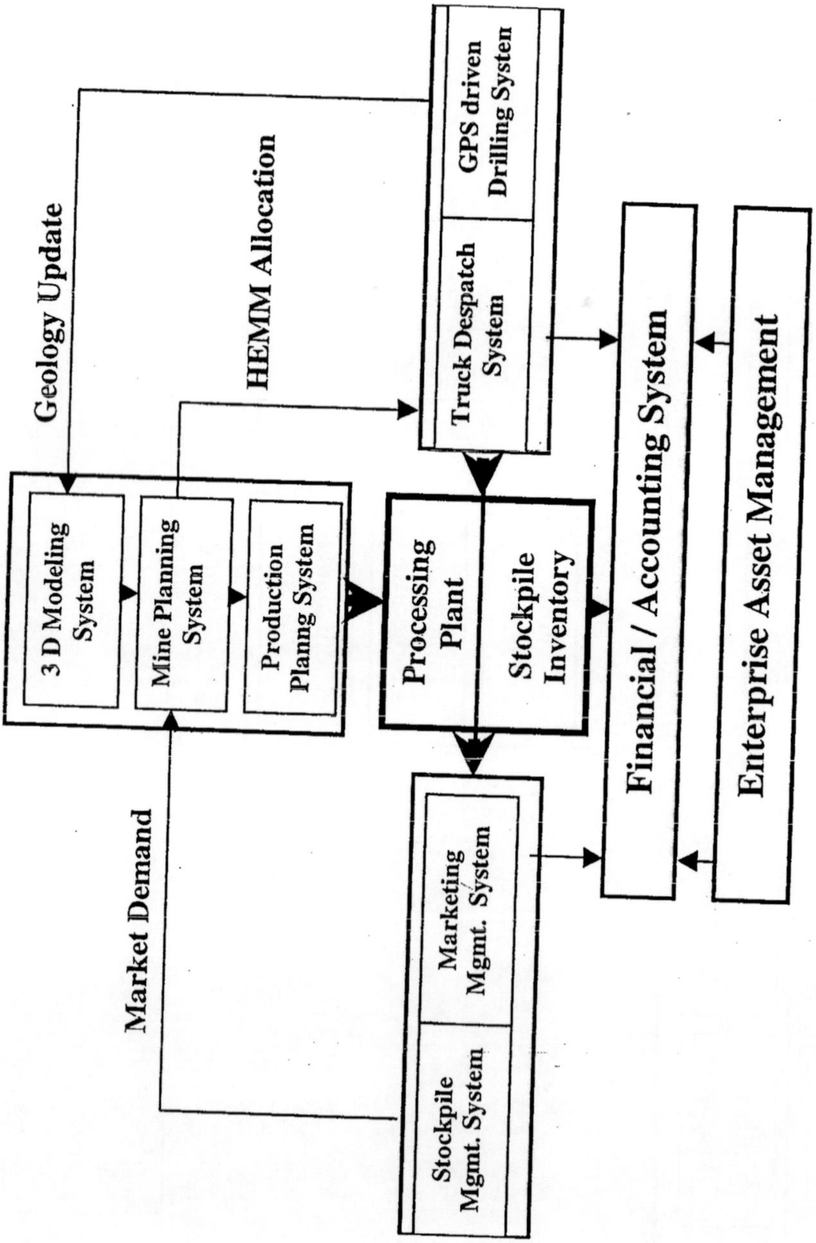

Fig. 56.4 : Mining Enterprise Asset Solutions

- Lowering costs through better use of all information generated
- Real time mining and connection to production and business processes
- Integration of both technical and business systems.
- Ability to integrate other home-made applications in the system
- Optimise future extraction and planning
- Models and plans that parallel the actual mining operation
- Basis of all reconciliation & comparison between planned and actual events/activities
- Basis for all costing, budgeting and forecasting
- Visual application allows mining activities to be visualised & interrogated
- Mine planning is a core offering of the mining solution

MODULES

- Materials and Maintenance management
- Financial and Human resources management
- Production and Despatch System
- Geological Modeling and Mine Planning & Scheduling System
- Stockpile Management System

BENEFITS

- Assist day-to-day operations of asset-intensive organisations
- Real integration rather than simply interfaces
- Delivering an open, flexible, customised business solution
- Increasing Profits by reducing costs and streamlining management systems
- Higher Equipment availability
- Reduced cost of maintenance
- Longer Equipment Productive life
- Reduction in Inventory value (without loss of service level)
- Reduction in Procurement Costs
- Increased Labour Productivity
- Improved Safety
- Improved inventory service levels at reduced cost
- More effective capital / project management

CONCLUSION

It is now possible to create a mining enterprise model in reality as the environment within organization has been favourable in recent times for the leader and the management team to bring transformation for such a change. It is accepted that the desired change brought about by information technology hinges on cultural commitment and network of the organisation. The benefits of mining enterprise solutions, though tangible, can only be defined in generic terms and guarantees are difficult to come by. Diversion of manpower from routine activities to more profitable ones can be tangibly projected, though assessment of the gains derived therefrom would still be subjective.

CHAPTER 57

Information Technology in Mining - CMRI Contributions for the Next Millennium

A.K. Chakraborty, M. Ramulu*, A.K. Raina*, P.B. Choudhury, A. Dhudhankar* and J.L. Jethwa**

ABSTRACT

Information Technology (IT) is today's buzzword of any organisation. IT is such a powerful tool that it can completely control the next millennium. In order to take the advantage of "Computer Revolution' to the mining industry, CMRI Regional Centre Nagpur developed two software packages in Rock Blasting and a few are in development stage. The software are FRAGALYST and TUNNEL BLAST. FRAGALYST is used for grain size measurement of the blast fragments by digital imaging. The software was designed to form a handy tool in order to find optimum fragment size in the open-pit blasting and optimise the blast design. TUNNEL BLAST is used for optimum blast design for tunnelling keeping in view the drive rate and economics. Major output parameters of the software are discussed in the paper.

INTRODUCTION

Blasting forms an important component in mining and civil (tunnelling) projects. The focus of such projects is to achieve maximum economic benefits with less input and simultaneously achieve better performance of the manpower and machinery, involved. A well-known fact is that reducing or increasing explosive consumption about an **"OPTIMUM"** has adverse effect on the performance of the system. The blasting officer is often faced with the problem, as how to decide upon the **"OPTIMUM"** because of inhomogeneous nature of rock, changing working conditions, and absence of tools to quantify the blast performance. This paves way for ingress of softwares that can store information and side by side perform analysis of the results. The results of the analysis are

* Central Mining Research Institute, Regional Centre, 54-B Shankar Nagar, Nagpur-440 010, India.

again utilised for deciding upon the optimum blast design, optimum fragment size for a machine and so on.

FRAGALYST and **TUNNEL BLAST** are steps in the same direction. The former is a fragmentation size measuring system and later generates a blast design for various types of tunnels. These in a way not only help in decision making but save considerable time of the engineer. The paper discusses the capabilities of the software and usage with possible benefits. The two packages are discussed below separately.

FRAGALYST IMAGE ANALYSIS SOFTWARE

Digital image processing technique uses sophisticated software to quantify the geometric aspects of images in two dimensions such as area, number, perimeter, shape, size and orientation. The method includes image capturing of muck pile, scaling the image, filtering the image, segmentation of image, binary image manipulation, measurement and stereometric interpretation. The method is quick and reliable.

The technique involves importing video/ photographic pictures to a computer by a frame grabber card. The movie or still images can be processed with the help of specific software. The final images are transported to the analytical software, which after certain calculations based on the grey scale of the photo/ image provide grain size or information about the joints.

ADVANTAGES OF DIGITAL IMAGE ANALYSIS OVER VISUAL ANALYSIS

1. The qualitative visual analysis method is subjective assessment. Changes in any joint pattern, and size of particles can only be distinguished when the differences are great. The technique is not very precise and is only good for initial assessment so that the specialists can later make a complete study. Noren & Porter(1974) and Reid (1979) have used the method extensively.
2. The digital images can be interfaced with a computer and lot of analysis can be performed depending upon specific requirements.
3. The method is accurate.
4. Can handle large areas at a time and analysis can be done simultaneously.
5. Grain size analysis of large fragments with large amount of material is not possible with screening by sieving method.
6. Several clips of a face/ rock exposure can be stored in a camera and imported to computer for averaging joint properties.

FEATURES OF THE SOFTWARE AVAILABLE

Indigenous software in this field is "FRAGALYST" developed by Central Mining Research Institute (Nagpur) and Wavelet Group (Pune). The software and total system is about 300% cheaper than imported versions. The software provides following information about grain sizes:

1. Grain size in terms of area.
2. Grain shape.
3. Spherocity.

4. Cumulative percentages of the grains in different groups of size.

5. Perimeter of the grains.

Information about the joints can be found in the following manner :

1. Number of joints in an area scanned.

2. Distance between the joints.

3. Lengths of short and long joint in the area.

4. Qualitative information about joints.

Not only does analytical part help in quantifying the rock mass, the added advantage of a record is that it allows to recapitulate the rock exposure or any static or dynamic phenomenon for analysis and decision making. Fig. 57.1 shows typical output forms of the software.

PROBLEMS IN DIGITAL PHOTO ANALYSIS APPROACH

Many problems are encountered when trying to estimate fragmentation using photographic images. The individual rock fragments in the image must be delineated, either by hand or by using an image processing technique. Problems with non-uniform lighting, shadows, noise and the large range of fragment sizes make delineation very difficult using standard edge-detection routines. Other problem is correct extraction of three dimensional information from the two-dimensional images for which assumption and site specific calibration for obtaining the third dimension is to be made from the two dimensional images. In addition, a correction is to be made for overlapping of fragments. Finally, to include small and large size fractions from a site, images at different scales must be taken and appropriately processed together to obtain a final size distribution curve.

"In their opinion, the largest block should occupy about 10-20 per cent of the width of the image but no single block should occupy more than 20 per cent of the width of the image. They also suggested measures to avoid wide angle close up photography and oblique shots which distort the scale. Palangio et al. (1995) concluded that the photograph of each muck pile should contain the image of at least 400 fragments. They also suggested that uniform and diffused lighting should be provided without excessively sharp or one-sided shadows."

DIGITAL IMAGE ANALYSIS TECHNIQUES - GLOBAL REVIEW

Different image analysis methods developed and used at various corners of the world for blast fragmentation analysis are detailed below:

1. Sweden

In 1982 Lulea University of Technology, sponsored by the Swedish Mining Research Foundation, developed an automated fragmentation measuring system using a TV camera, microcomputer and printer. The system was tested in a research mine in 1983. In the automatic mode, pictures were taken at constant intervals while a loader removed materials from the rock pile. Scale was determined by placing an object of known dimensions on the rock pile. Areas were converted to volumetric size distributions based on "stereometric considerations". The operating resolution ranged from about 0.2 to 1.7 mm/pixel. The results from several images were averaged before printing.

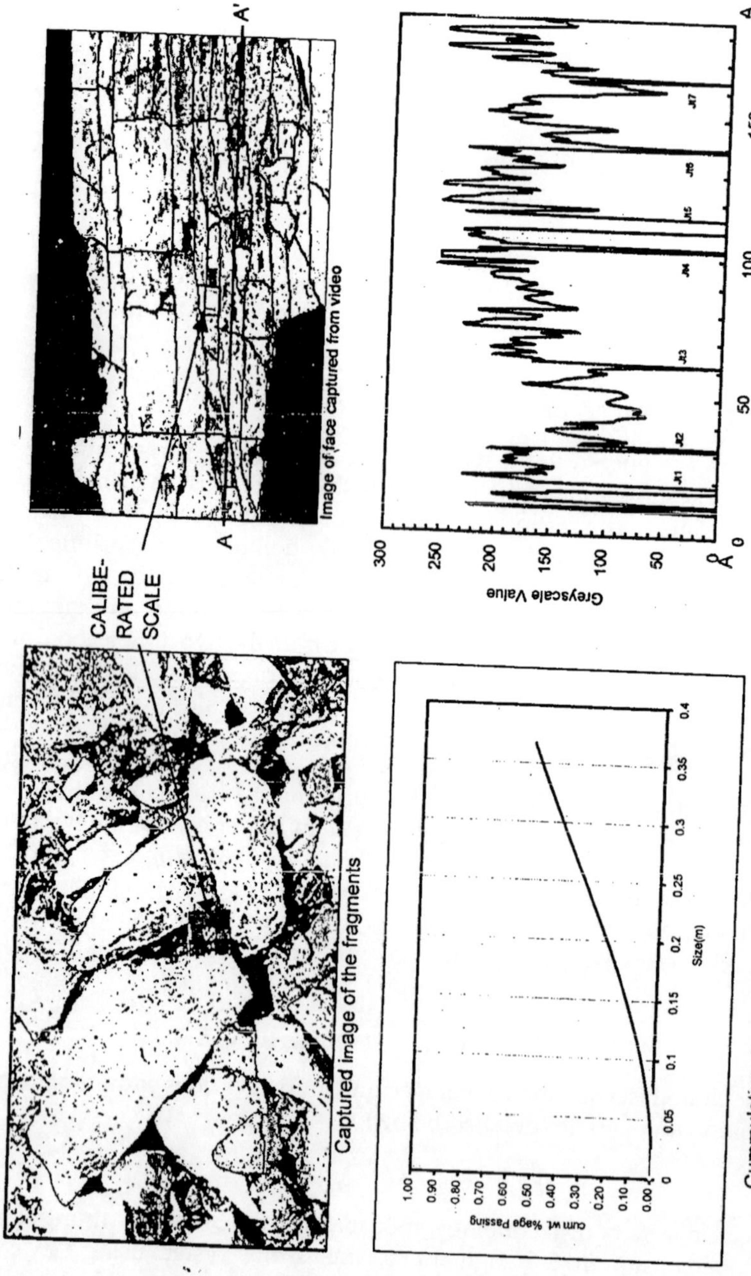

Fig. 57.1 : Output as obtained from the analysis of fragments and blast face

The method was tested against sieved samples and by comparing images of separated and overlapped fragments. In nearly every case, the fine fraction was overestimated, which as attributed to an overlapping effect.

The development of image analysis has been continued by a company called IPACS, a collaboration between the ' Swedish Board for Technical development, the Chinese Government and Lulea University of Technology. The new system includes enhanced subroutines and is being evaluated by the Swedish Road Administration for testing ballast material in the road prior to asphalting. A video camera is mounted on a car and ballast is imaged by the video camera as the car is driven along. Other applications developed at IPACS included an on-line system for combustion control as well as fragmentation on moving conveyor belts and after blasting.

2. France

The system "FRAGSCAN" commercialized by Paris School of Mines and Nitrobickford, has been used to take images of rock in a quarry dump truck just before the primary crusher. A black and white CCD camera was installed on an 8 m high mast and the dumper position was linked to a 486 microcomputer with an image acquisition card, User-friendly software managed the different operations and triggered image processing. FRAGSCAN did not use edge detection to find the particles but used a series of 10 opening operations simulated sieving through ten different mesh sieves. A threshold was obtained by successive filters and automatic threshold selection, the image was converted to binary, then a series of 10 opening operations were completed. An X-size erosion eliminated all fragments smaller than X and when followed by dilation, the remaining fragments were rebuilt to the previous size. The proportion of "lost" fragments smaller than the given value was computed. Because of time limitations, this treatment was completed only for ten sizes. Surface areas were converted to volumes or weights, using experiments on small-scale rock piles and assuming a spherical model. The cumulative distribution curves were approximated by the Rosin-Rammler model. The program has been adapted to compute size distribution of material carried in haul trucks to optimize blasting parameters for crushing operations in quarry.

3. Germany

Vogt and Abbrock (1993) from the Technical University of Clausthal used fast automatic image processing to measure production blast. Photographs were digitized by TV camera. Edge detection took into account roundness or convexity of the fragment to reduce the fusing of small fragments into large ones. The fragment net was shown as an overlay on the image. Fragments smaller than 50 pixels were not shown on screen, but were counted in the particle size distributions. The program name is TUCIPS

4. United States

(a) University of Arizona

The University of Arizona method uses a high resolution video camera or CCD camera, an image capture board and a Macintosh computer. The image analysis software is built on the public domain software NIH Image by the U.S. National Institute of Health (NIH). This supports live image capture and editing and has many built in functions for grayscale manipulations and shape analysis. It includes a built-in macro programming language and also allows source code modifications to be made.

The Arizona researchers developed procedures for assigning a probability for a particle falling into any given size class, and for correcting for overlap and fragment shape. A percentage of the non-particle area is assumed to be fines and is added to the histogram of delineated fragment sizes. An equation was also developed for zoom-merging, i.e. combining images at different scales of magnification.

(b) U.S. Bureau of Mines, Minneapolis

Grannes (1986) introduced an image analysis method to investigate the size distribution of taconite ore. In a test in a quarry in Minnesota, 424 tons of rock were screened using techniques similar to those described by Stagg (1987). Comparison with sieving showed that the photographic technique underestimated the amount of fines.

The fragmentation distribution determined photographically was adjusted to include fines predicted by an empirical equation based on explosive type, diameter, decoupling ratio, breakout angle and length of charge . Use of this equation "avoids the need for close-up photograph". After correcting for missing fines, the adjusted curves agreed closely with sieve data.

The method was developed by the U.S. Bureau of Mines in collaboration with the Natural Resources Institute of Minnesota Department of natural Resources and U.S.X. Corporation.

(c) NORAMCO, Minnesota

In 1992 NORAMCO Engineering Corporation of Hibbing, Minnesota developed a Computer Image Analysis System (CIAS), to be released in October 1994.

The system was used for measurement of blast fragmentation of magnetite ore, and agglomeration of taconite concentrate, pellet sizing and the sizing of bubbles in flotation froth.

The overall results were described as "promising". The acceptable range of smallest to largest particle size was about 1:10. The analysis of one picture frame typically took less than 30 seconds. The software was made using C software subroutines, with a command sequence that included setting up the board, digitizing the video image, edge detection, measurements, and display.

4. Canada

(a) CANMET, in their experimental mine in Val d' or, Quebec, used a video camera, a 386 micro computer with an additional monitor for image display and image analysis by software obtained from Universal Imaging Corporation. Colour photographs were preferred for ease of processing. The images included a rectangular scale marker resting on the rock pile surface. The photo print was traced with a fine pen. Small particles were grouped and traced on a separate transparency to give a separate measure of "fines".

The traced fragment net was viewed over white paper. After correction for tilt, the area was measured as the number pixels multiplied by the area per pixel (the scaling factor). The traced edge was excluded because only white pixels were counted. Other measurements available included perimeters, chord lengths, diameters and a shape factor.

(b) McGill University, Montreal

The McGill method was based on an early version of WIEP (Waterloo Image Enhancement Program) extended and improved by the McGill researchers for speed precision and systematic fragmentation records. The photographs typically required initial manual tracing. The percentage of fines was determined separately and the size distribution corrected accordingly.

Lizotte et al. (1993) described an application to LHD and FEL performance assessment at the Brunswick Mining and Smelting Corporation No. 12 Mine. The Kuz-Ram fragmentation prediction model was calibrated for blast hole sloping. The McGill version of WIEP was used with four photos per shift per work place, sufficient to characterize fragmentation at the given production rate.

(c) ICI Canada

The ICI fragmentation-measuring method was developed for use with the ICI SABREX computer blast model. SABREX predicted fragmentation from data of the rock, the explosives and the blast geometry and could be worked backward using fragmentation measurements to estimate rock quality.

(d) University of Waterloo, Canada

The University of Waterloo started the use of digital photoanalysis system (WIEP) for measurements of jointing, including orientation, spacing and roughness as well as blast overbreak in fragmentation. At that time stockpiles of broken rock were digitized vectorially with a digitizing board.

By early 1987, an unfolding function had been derived theoretically and then calibrated to convert the visible areas in a 2-D image of overlapped fragments into a 3-D volume or eight distribution. The method was tried at a rock quarry in Stafford, Virginia, using photographs of rock in the backs of trucks. A x-y digitizing board continued to be used, and the particles were delineated by hand.

After substantial developments of the edge detection algorithms for underground applications and research funded by INCO, the WIEP system was since been commercialized as the WipFrag system by Maerz, Franklin, and Palangio of Wipware, Inc. In addition to sizing rock fragments, this program has been applied to ammonium nitrate prills, glass beads and zinc concentrates. WipFrag also uses a zoom-merge technique for combining images of different scales.

5. India

CMRI Regional Centre, Nagpur and M/S Wavelet Group of Pune have jointly developed an image analysis system for this first time in India. The system consists of capturing video photographs of the muck pile, downloading the photographs to the computer, converting the image/s to greyscale, image enhancement, calibration and blob analysis. With the aid of menu-driven software developed by them, it is possible to determine the area, size, shape, and spherocity of the fragments in a muck pile on the basis of greyscale difference.

Pre-blast block size and joint orientation details can be obtained from photographs of opencast benches. The information is significant for obtaining RQD and joint properties, which help in blast design. Details of overlapping fragments can be also determined. The output consists of x-y curves and histograms. Moreover, output data of each fragment are available on the screen itself. Regarding processing of the images, the software has a facility of cropping of a particular portion of the image for analysis and accurately deciding the threshold parameters for precise analysis. The threshold values have been categorised as global and local types. The global one ranges from 0 to 255 and the local one ranges from 1 to 10. The zoom merge technique is also advised here for analysis when the fragment size varies to a wide scale. The 2-D information available from the software can further be processed for stereological analysis for 3-D information. Knowledge on the principles of geometric probability and stereology are required to construct or unfold the 3-D information. The scaling accuracy in case of wide angle zooming can be reduced by averaging the calibration constant obtained for near-field and far-field fragments. The system can be used also for optical surveying and slope monitoring. The use of the system for explosive energy and excavation system optimisation is a matter of in-depth blasting research. The software can be used for joint mapping also. Row or column scanning is done along a particular line after orientation of the image to make it vertical. The peaks showing the least gray levels are identified as the joints. The distance between the joints are provided in pixels number which can be converted in desired scale after proper calibration. This facility helps in estimation of joint frequency and pre-blast block size.

TUNNEL BLAST

Till recent times, the task of optimum blast design in a given geo-mining condition would be dependent on individual experience and could be completed by several trials. Blasting research in late sixties has developed some in-depth understanding of rock-explosive interaction. Determination of blast parameters for different combinations of explosives and site parameters can now be precisely made using such knowledge. The site engineers are sometimes unaware of such developments and require the help of some experts to optimise the pattern. With the advent of computer, it has been possible to develop a blast design software, named TUNNEL BLAST, with the help of already developed know-how. This allows the site engineers to carry out the task of optimisation without updating their level of knowledge. The software accounts for change in the rock mass condition, tunnel design and explosives properties. Further, the use of computer can make the optimisation process very fast and store large amount of information for a long time. TUNNEL BLAST has been developed in Autolisp language. The language, which runs through AutoCAD, is a very powerful tool and is used for computer aided designing. It has wide range of facilities and is interactive. Further development in the software becomes very easy in Autolisp programming.

D-shaped tunnels are most common in civil engineering projects and rectangular openings are most common in mining industries. The first version of the software is made for D-shaped excavations only. Rectangular openings will be included in the next version of the software. The software has been designed for V-cut and four-section parallel cut as these types of cut are commonly used in the civil and mining projects in India. In view of the success of smooth blasting over the conventional blasting due to the reasons of improved productivity, better safety and reduced support cost, the software is designed for smooth blasting only.

Various design parameters considered in the TUNNEL BLAST software are:

(a) Rock Properties: Compressive strength, density, joint frequency

(b) Tunnel design : Size of the opening: Width and vertical wall height

(c) Blast design: Type of cut (V/ Parallel cut), blast hole and relief hole size and blast hole length, explosive strength, cartridge specifications.

The output of the software consists of:

(a) Size of the tunnel.

(b) Probable deviation of blast holes.

(c) Optimum depth of the round.

(d) Look out angle for peripheral holes.

(e) Burden, spacing and charge of holes in cut area, floor, periphery and in the middle of tunnel section (sloping holes).

(f) The front and sectional views of the blast pattern on the computer screen.

The views can be printed and then projected on the tunnel section to maintain accuracy in the burden and spacing. The hole to hole distance and direction of each hole can be computed on the screen itself. Any portion of the views can be zoomed fpr better appreciation.

(g) The pull, powder factor and total number of holes per round.

Different blast design parameters and graphical output obtained using TUNNEL BLAST for an assumed set ofgeo-mining conditions are given below (Table 57.1):

INPUT PARAMETERS FOR TUNNEL BLAST

Width of tunnel (m), Height of the vertical wall (m), V cut/Parallel cut, Blast hole diameter (mm) Maximum blast hole length (m), Rock mass density (T/cu.m), Uniaxial compressive strength (MPa), Average spacing of discontinuity (m), Weight strength-of explosive (%), Length of cartridge (mm), Weight of a cartridge (g)

OUTPUT OF TUNNEL BLAST

A. Blast design information

Cross-sectional area (m^2), Optimum advance per round (m), Look out angle (degree), Deviation of hole (m)

B. Detailed blast pattern informations:

Burden (m), Spacing (m), Charge/ round, No. of holes, Total charge(kg), of Cut holes (baby,main), Easer holes, Floor holes, Sloping holes, Contour holes (Alternately charged)

C. Blast performance:

(i) Expected pull

(ii) Powder factor: 1.41 kg/m3

D. Graphical output of blast patterns

The blast results of an inclined tunnel (site A) has been compared below with the estimated obtained from TUNNEL BLAST. It can be seen that the software TUNNEL BLAST is quite reliable:

(A) Geo-mining conditions and design parameters of site A

* Size: 15 m^2
* Shape: D-type
* Formation: Sandstone
* Uniaxial compressive strength: 20-40 MPa
* Specific gravity: 2.32
* Joint spacing: 0.3-0.4m
* Blast hole depth: 2.3m (drill rods are 2.4m long)
* Blast hole diameter: 38mm
* Type of cut: Four-section parallel cut
* Relief hole diameter: 100mm
* Type of explosive: Ajax-G, Make: ICI, India

(B) Comparative results obtained in field and from TUNNEL BLAST

Parameter	*Results*	
	TUNNEL BLAST	*Field*
Total charge/round	63.68	65
Pull (m)	1.96	1.8
Powder factor (Kg/m^3)	2.13	2.3

The following advantages of the said software TUNNEL BLAST should prove it to be a useful tool for the tunnel engineers to achieve faster drivage and to economise the operation:

(i) It eliminates the necessity of a blasting expert for tunnel blast design.

(ii) It provides quick information and, thus useful, in decision making. The use of the software is specially appreciated in changing geo-mining condition. The blast design can be optimised each time, in such condition, without much of trials.

(iii) It has potential of application in wide range of geo-mining conditions.

(iv) It provides estimation very close to actual field results.

(v) It is simple to use, menu driven and user-interactive.

(vi) The blast pattern viewed on the computer screen can be printed and projected on the tunnel face for accurate drilling.

(vi) The configurations like angle, depth, burden, spacing etc. of each hole can be easily determined on the screen itself. This was possible due to the use of Autolisp software.

ACKNOWLEDGEMENTS

The authors are thankful to Director, Central Mining Research Institute for permission to publish this paper.

REFERENCES

Adhikari G.R. and Gupta R.N. (1988). Influence of Natural Blocks on Rock Fragmentation by Blasting. Int. Jr. Mining and Geological Engg., Chapman & Hall, U.K., 1988.

Chakraborty A.K., Jethwa J.L. and Paithankar A.G. (1994). Effects of Joint Orientation and Rock Mass Quality on Tunnel Blasting Engineering Geology, No. 37, pp. 247-262.

Cheimanoff N.M., Charez R. and Schleifer J. (1993). Fragscan : A Scanning Tool for Fragmentation after Blasting. Proc. Fourth International Symposium on Rock Fragmentation by Blasting - Fragblast'4, Vienna, Austria, July 5-8. pp. 325-329.

Chiapetta R.F. and Borg D.G. (1983). Increasing Productivity through Field Control and High-speed Photography. First Symp. on Rock Fragmentation by Blasting, Lulea, Sweden, 1983.

Chung S. H. and Noy M. J. (1996). Experience in fragmentation control, Measurement of blast fragmentation, Balkema, Netherlands, pp. 247-252.

Chung S. H., Burchell S. -(1991), Computer based 3-D blasting analysis system, Proc. 7th. Annual Symp. Expl. and Blasting Res., Int. Soc. of Explosives Engrs., Las Vegas, Neveda, pp. 141-149.

Chakraborty A. K., Jhanwar J. C. and Jethwa J. L. (1998), Evolution of computer aided image analysis technique for fragmentation assessment - A global overview, Proc. Nat. Seminar on Indian Mining Industry: Problems & Prospects, MINPROPS'98, Feb 27-28, Chandrapur Engg. College, Nagpur.

Da Gamma C.D. and Jimeno C.L. (1993). Rock Fragmentation Control for Blasting Cost Minimization and Environmental Impact Abatement. Prod. Rock Fragmentation by Blasting, Balkema, pp. 273-279.

Dhar, B.B., Pal Roy, P and Singh R.B. (1993) Optimum Blasting for Indian Geomining Conditions - Suggestions Standards and Guidelines, Published by CMRI, Dhanbad.

Farmer I.W., Kemeny J.M. and McDoniel C., (1991). Analysis of Rock Fragmentation in Bench Blasting Using Digital Image Processing. Proc. of. International Congress on Rock Mechanics, Aachen, Germany, pp. 1037-1042.

Franklin J. A., Kemeny J. M. and Girdner K. K. (1996). Evolution of measuring systems: A review, Measurement of blast fragmentation, Balkema, Netherlands, pp. 47-52.

Franklin J.A., Marez N.H., Bennett C.P. (1988). Rock Mass Characterisation using Photoanalysis, Int. J. Mining Geology on Engineering, Vol. 6, pp. 97-112.

Gardner K. K., Kemeny J. M., SrikantA. and McGill R. (1996), The split system for analyzing the size distribution of fragmented rock, Measurement of blast fragmentation, Balkema, Netherlands, pp. 101-108.

Ghosh A., Daemen J.J.K. and Vanzy! D. (1990). Fractal based Approach to Determine the Effect of Discontinuities on Blast Fragmentation. Proc. of the 31st U.S. Symp. on Rock Mechanics, Golden Co.

Granes S. G. (1986). Determining size distribution of moving pallets by computer image processing, In: Applications of computer and operational research in the mineral industry, ed. R. V. Ramani, Soc. of Mining Engrs., pp. 545-552.

Gupta R.N. (1989). Some Aspect of the Optimum Blast Design. Lecture delivered at CMRS on 14.9.84 in the course 'Extraction of Minerals by Open Pit Mining, CMRS, Dhanbad.

Hunter, G.C. et al.(1990), Review of image analysis techniques for measuring blast fragmentation, Mining Science and Technology, Vol. 11, No. 1, pp. 19-36.

Jimeno L. et al. (1995). Evaluation of Blast Results, Drilling and Blasting of Rocks, Balkema, Rotterdam, pp. 290-311.

Kemeny J. (1994). Practical Technique for Determining the Size Distribution of Blasted Benches, Waste Dumps and heap leach sites, Mining Engineering, Nov. pp. 1281-1284.

LizotteY.C. and Scoble J.M. (1994). Geological Control over Blast Fragmentation, CIM Bulletin, September, pp. 57-20.

Lizotte Y., Scoble M. J., Sing A. and Mohanty B. (1993). Prediction and assessment of fragmentation in underground mine, Proc. 4th. Int. Symp. on Rock Fragmentation by Blasting, Vienna, Balkema, pp. 361-368.

Mckenzie A.S. (1966). Cost of Explosives - Do you Evaluate it Properly ? Mining Congress Journal, May. pp. 32-41. Motajbai N. Cetintas A., Farmer I.W. and Savely J.P. (1989). Jn-place and Excavated Block Size Distributions, Proc. Rock Mechanics as a Guide for Efficient Utilisation of Natural Resources, Balkema, Rotterdam, pp. 537-543.

Nielson K. (1983). Optimisation of Open Pit Bench Blasting, Proc. of 1st International Symposium on Rock Fragmentation by Blasting, Lulea, Sweden, pp. 653-654.

Pal Roy P. & Dhar B. B. (1996). Fragmentation analyzing scale - A new tool for breakage assessment, Proc. 5th. Int. Symp. on Rock Fragmentation by blasting - FRAGBLAST 5, Montreal, 25-29 Aug., pp. 448.

Pal Roy P. & Dhar B. B. (1993). Comparative assessment of ground vibration and fragmentation in three iron ore deposits in India, Proc. 4th. Int. Symp. on Rock Frag. by Blasting, Vienna, July 5-8, Balkema, pp. 287-293.

Rhzevsky V.V. (1985): Opencast Mining Unit Operations, Mir Publishers, Moscow

Rholl S. A. (1993). Digital image based fragmentation assessment. Proc. 66th. Mtg. Minnesota Subsection SME.

Seam M.M. (1993). Optimal Fragmentation ; Main Objective of Drilling and Blasting. Proc. National Symposium on Advances in Drilling & Blasting, KIOCL, Kudremukh, 15-16 Oct., pp. 139-146.

Singh B. and Pal Roy P. (1991). Blasting in Ground Excavations and Mines, Oxford & IBH Publishing Co. Pvt. Ltd., New Delhi, pp. 114-131.

Stagg M. S. and Rholl S. A. (1987). Effects of accurate delays on fragmentation for single row blasting in a 6.6m bench, Proc. 2nd Int. Symp. Rock Fragmentation, Keystone, Colorado.

VogtW. and Abbrock 0. (1993). Digital Image Processing as an Instrument to Evaluate Rock Fragmentation by Blasting in Open Pit Mines. Proc. Fourth International Symposium on Rock Fragmentation by Blasting - Fragblast'4, Vienna, Austria, July 5-8. pp. 325-330.

Wu X. and Kemeny J. M. (1992), A segmentation method for multiconnected particle delineation, Proc. IEEE Computer Society Press, Loss Alamitos, CA, pp. 240-247.

Zeggeren F. Van and Chung S.H. (1993). A Model for the Prediction of Fragmentation Patterns and Cost in Rock Blasting, Rock Mechanics, pp. 557-567.

POSTER SESSION PAPERS

CHAPTER 58

Challenges of Mineral Development in India

D.K. Sahni, S.V. Ali** and P.K. Jha***

INTRODUCTION

From the days of pre-independence era to the threshold of the next millennium, a period of five decades has witnessed appreciable progress in certain segments of the mining and mineral sector in India. The value of mineral production on the eve of independence was merely Rs. 58 crore which took rapid strides to the level of Rs. 40,000 crore as at present. The sector thus contributes around 3 per cent to the GDP and accounts for as much as 20 per cent of the aggregate export business. The mineral production comes from about 3,600 mines producing 84 minerals (comprising 4 fuel, 11 metallic, 20 minor and 49 non-metallic industrial minerals) spread over 10,000 mining leases and 700,000 hects which is as follows in respect of principal minerals / ores:

Coal	300	million tonnes
Limestone	100	"
Iron ore	67	"
Petroleum crude	34	"
Bauxite	6	"
Copper ore	4	"

The mineral production is close to 500 mil ton aggregate. In the year 1947, the resource details were available only in respect of coal, iron ore, chromite, bauxite, manganese ore and magnesite. Presently, the inventory figures are known for 61 minerals spread over 13,000 locations and are updated every five years.

The public sector which was dormant before independence assumed dominant role thereafter so much so that 97 per cent of the total mineral quantity and 95 per cent of the

*CMD, MOIL, Nagpur.

**MGMI, Nagpur chapter, Nagpur

total production value are contributed by the public sector. As much as 100 per cent of copper, silver, zinc, lead and diamond, 98 per cent of coal, 60 per cent of iron ore and 49 percent of manganese ore come from this sector.

In producing several minerals, India is very close to the top as is evident from the following–

Coal and lignite	3rd in the world
Chromite	3rd
Barytes	3rd
Iron ore	4th
Talc/steatite	5th
Bauxite	6th
Manganese ore	6th
Magnesite	9th
Aluminium	10th

Undoubtedly, the Indian mining industry is sizeable. A sector with such stupendous dimensions as reflected above, has to be naturally beset with a myriad of problems which can also be perceived as challenges.

ENVIRONMENT

Meeting the challenge of preservation of environment lies in the thoughtful use of natural resources such that the maximum sustainable benefit is made available to the current generation while the potential is maintained for the needs and aspirations of the coming generations. It is some thing akin to the fact that the capital should never be eroded as we must know how to live off the income provided as interest from that capital.

India is a growing economy with sizeable industrialization and mining sector. It has already been stated that about 500 ml t aggregate minerals are produced in the country annually with 80 per cent of coal and almost the entire quantity of industrial minerals coming from opencast mining. Our population is at the threshold of one billion mark which is as much as 16 per cent of the world population residing on only fortieth part of the global area. There is no doubt that the environment is of greatest relevance to India. As much as 500 hect. of land is degraded every year by opencast mining alone.

In India, the required forest cover is 33 per cent of its area whereas the actual cover at present is only 14 per cent. Per capita forest cover comes to 0.07 hect. which is as low as 9 per cent of the world's average. As much as 66 per cent forests are in 133 tribal districts, close to 100 of which have only 19 per cent coverage. The forests are shrinking at the rate of 5,500 sq kms. annually in spite of all the recent awareness. Naturally, obtaining mining leases have become increasingly difficult. Example of Gandhamardan can be cited in this case. Delay in renewal of mining lease in respect of Kudremukh iron ore mine was apprehended as the mine is located in the proximity of an environment park. Acacia trees were planted as compensatory forest while rose wood and sandal wood trees were cut away, pleading that it takes about forty years for the rose wood trees to grow. It was completely ignored that better variety of trees were replaced by an inferior one which can also not withstand winds upto 120 km / hour which is quite common in the area. Only recently lease extension for one year has been granted.

Mining damages environment by causing fissures, subsidence, land slides etc. It also results in barren lands and generates dust, smoke and vibrations. There is no doubt that a few organizations mustered appreciable resources and stood admirably well for the cause of improved environment. In this connection, it is relevant to mention the names of NLC, BALCO, TISCO, BGML, etc. for the commendable work carried out by them. ISO 14001 has been obtained by the mines of NALCO and INDAL Bureau of Research for Geology and Mining, France has been associated with IBM to work on environmental parameters in Goa iron ore and Sukinda nickel. Tata Energy Research Institute has also prepared a report on Goa environmental quality management for its minings belts. Socio-economic baseline for 57 villages have also been prepared.

Some progress has been made on mineral waste utilization. Progress on commercial utilization of flyash and remud has been somewhat tardy so far despite the fact that some good work has been done on pilot scale. KIOCL has taken steps to formulate a bold project at the cost of Rs. 26 crore to reclaim 117 mil t of tailings lying in Lakhya dam to recover 21 mil t concentrates. Some useful clue can be had from the proposed recovery of ferrochrome from slag (60 ml t) at Zimasco in South Africa, worth about one billion dollars with possible recovery of 96 per cent.

More recycling of used metals such as steel, aluminium, lead, zinc and copper will not only help in conservation of their ores but also will promote the cause of environment. However, it seems that a lot of work is required to be addressed with regard to systematic collection of scrap and energy inputs and at appropriate rates.

The concept of Environmental Audit introduced in our country a couple of years back, is expected to yield envisaged results as in case of a number of countries where it has also been in force.

PRODUCTIVITY

One may not feel pride for the productivity levels achieved in the Indian mining and mineral sector as it is much less compared to other countries even under comparable mining conditions and technology. The rate of shaft sinking in India at 15 m per month may be quite a standard progress whereas at Hertebeestfontain in South Africa this was as high as 350 m per month a few decades earlier. The fastest driving rate in our country is 100 m per month in a cross section of 2.5x2.5 m2 whereas in Granduc in Australia, the progress was as much as 700 m per month in a tunnel of 4.5x4.5 m^2. The rate of root bolting at bolts/man/shift may not be regarded as poor in our mines whereas in Australia this could be ten times as much. Stope productivity also has similar comparison figures with regard to Canada, Australia, Sweden and the USA.

Low level of productivity is also the experience in mineral consuming industry. In India, the productivity in steel plants is 60 t/man/year whereas the figure for Japan is 200 and for Germany 150. In aluminium smelters, the productivity in India is 30 t/man/year whereas in the USA it is 200 and in the UK it is 125. Productivity in Indian cement plants is 200 t / man / year. The figure in the UK is 5.5 times as much.

As the productivity directly affects the economic health of a running venture, its improvement is a matter of constant endeavour. The reasons for low productivity and the remedy to improve the same are very well known. Some more important reasons include substandard management, lack of proper working atmosphere, substantial gaps between consecutive shifts, poor communication facilities underground, improper work sequence, poor ventilation, erratic power supply, ill maintained equipment deployed, lack of genuine spare parts, too high and abrupt absenteeism, etc.

Productivity in private sector may not always be higher than the one attained in the public sector under all the circumstances and at all times. As for example, the productivity in TISCO and SAIL were respectively 46 t and 83 t per man per year. Share of the private sector in steel manufacture has gone up from 51 per cent in the year 1992-93 to 64 per cent in 1997-98. However, its impact on the productivity is not very evident. SAIL has 175,000 workers on its role of which 75,000 are stated to be surplus. These figures are 63,000 and 3,000 respectively for TISCO. VRS was introduced in SAIL on 1.3.1998 for workers of 50 years of age and 20 years of length of service. This has now been further liberalized to 40 and 15 years respectively since June 1999.

Post nationalization scenario in coal mining appears to be over stressing on opencast mining as this variant has the advantage of large production, small construction period and amenability to large scale mechanisation. Whereas in the year 1974-75, only 20 ml of coal were produced, the figure was 225 mil t for the year 1997-98. However, in our zeal to produce more, all of us ignored the fact that the proportion of production by opencast mining is far in excess than the reserves amenable for the purpose. Therefore, a time will come once again when stress will be tilted to underground mining, and if this is not done timely, production collapse vis-a-vis requirement of coal quantity is inevitable. As easy mining by opencast is the order of the day, hardly any shaft is being sunk for going deeper to exploit coal. The problem of fire in Jharia coalfield is well known and solution is needed. Almost one lakh workers are redundant. About 50 uneconomic mines of CIL are required to be closed. In deeper seams longwall mining will have to be adopted for which technology will have to be imported and indigenised.

As far as mechanisation in India is concerned, it is almost on par with the advanced countries in opencast mining. 20 cu m shovels, 39 cu m bucket / 120 m boom draglines and 200 t dumpers are in use in the country in junction with drills of 250 / 311 mm diameter. This is not in any way inferior to the mechanisation elsewhere. Even in the USA, the 'Big Muskie' (168 cu m bucket) deployed in the late sixties, continues to be the largest in the world for over three decades. Large underground metalliferous mines are fairly mechanised in India. However, in underground coal mining, decrease in 'head-load-mining' is still a matter of discussion, while experimentation on longwall mining continues. Continuous miners with shuttle car combination is under consideration for room and pillar mining whereas in other countries the emerging innovations include shearer initiated electrohydraulic controls for advancing powered supports.

CIL has planned to introduce the most up-to-date technology in its underground mines. Global tenders have already been issued.

SIZE OF OPERATIONS

The Indian mining sector is regarded to be comprising of small mines on account of their significant contribution in terms of 50 per cent of the total production value and 95 per cent number of mines.

The above also means that 5 per cent of the mines contribute half of the mineral production in value terms. There are a number of mines which produce 10 mil t of coal at 4 to 5 stripping ratio. In such operations, the total volume of coal plus overburden may exceed 4o to 50 mil t per annum.

With the passage of time larger number of mines in the above size range will be designed in India. Stripping ratio may be 1:6 and the mines may go upto a depth of 300 m. Formulation, planning, construction and operation of such large projects may pose challenges with which we may not be familiar. Quality control and transportation of such large quantities may also warrant careful planning. A couple of large opencast projects are likely to be of bauxite in Gujarat and Orissa. NALCO also has a programme of expansion from 2.4 mil t per annum to 4.5 mil to for its Panchpatmalli mine.

Another challenge that the mining operations are facing is on account of the deteriorating mining conditions which makes the operations the ceaseless saga of struggle against the imponderables. In this context, citing the example of Kolar is relevant where due to hostile mining conditions, fast depleting reserves, dwindling grades, frequent rock bursts and burning temperature mining could not be continued. A mining operation with such hostilities coupled with poor grade of 3 gm per tonne of ore down from 40 gm obtainable at the beginning of production of this mine, can not be economic.

However, the example of Neyveli Lignite is also relevant, but on the other extreme of the scale, as a successful story of the endeavour to subdue the hostile conditions such as excess water, clayey top and hard lignite.

PRICES

In the recent past, several metals have fallen from favour in the international market with regard to their prices on account of breaking of the Soviet Union into CIS countries and collapse of the fast rising economies in South East Asia. Metals which suffered serious setback in their prices, include copper, lead, zinc, nickel, steel, gold, etc. The decline in prices have thrown up challenge to the very existence of some of the running ventures. Aluminium is fast moving to the state of glut in the market and copper is already having more than 200,000 t of quantity accumulated in the international market. The price of copper came down to $ 1,500 per t in 1998 which was equal to the price prevailing in 1935 in the real terms. Only three years back in 1995, copper prices were double at $ 3,000 per t. No doubt an organization such as HCL could not cope up with the challenge of declining prices and incurred heavy losses to the extent of Rs. 130 crore in 1996-97, Rs. 170 crore in 1997-98 and Rs. 200 crore (e) in 1998-99. The reduction in duty on imported copper compounded the problem further as the imported copper became much cheaper in the Indian market than the indigenously produced one. All this made HCL financially crippled as they needed Rs. 600 crore for revamping, Rs. 150 crore to pay suppliers and Rs. 120 crore for employees who opted for VRS.

During the period from 1992-93 to 1996-97, the total expenses on raw materials for steel production went up by 51% in case of SAIL and 61% for TISCO. The increase per tonne of steel produced comes to 34.3% and 28.2% respectively. The increase in the cost of raw material coupled with other factors caused serious erosion in the profitability. As for example, during the year 1996-97, against a sale of Rs. 13,000 crore, SAIL made a profit of Rs. 515 crore which came down to Rs. 104 crore in 1997-98. During the next year, SAIL incurred a loss of Rs. 1,200 crore in spite of cost saving efforts bearing fruits to the extent of Rs. 900 crore.

The combined export target for 1997-98 for pig iron and steel was merely one million tonnes compared to the actual export performance of one million tonnes for pig iron and 1.4 mil to for steel as far back as 1976-77.

The floor price for steel was fixed at $ 302 per t but the business was mostly carried out at $ 245. However, the imported steel products were available in the Indian market at $ 165 during 1998-99. Vizag had to suspend its production of pig iron for exports as the prices came crashing down to $ 77 per t from $ 116 per t between June and October 1998.

No wonder the steel producers witness their steel inventories piling up. Several producers have cut their production activities as much as 30 per cent. Export targets were set higher than the production targets to cut on inventories without perceptible results. The year 1999 may be even more difficult though anti-dumping duty was introduced for steel in November 1998, and the special import duty of 4% on certain steel raw materials have already been waived. A five year tax holiday for steel industry has also been proposed. Under such circumstances the Expert Committee Report of the Steel Ministry could not find favour with some of the producers as the domestic consumption estimated as 32.5 mil t by the year 2001-02 appears to be unrealistic since the growth rate of 12 per cent witnessed in 1993-94 has come down to 1.49 per cent in 1997-98. Domestic demand thus hovers around 24 mil t. In fact, the financial institutions had cleared an investment of Rs. 22,000 crore in the private sector comprising 17 mills of 11 mil t annual capacity of which 7 are ready for commissioning. Construction of the remaining 10 mills has been asked to go slow.

The steel producers are also withdrawing from non- core activities. TISCO has sold away its cement plant to Lafarge for a consideration of Rs. 550 crore. They have also withdrawn from one million tonne capacity greenfield Utkal alumina project (Rs. 4,000 crore) in which they were to give 20 per cent of the Rs. 1,600 crore equity.

Crash in gold prices created panic in a country such as South Africa whose 53% exports are due to this commodity. In Western Australia and Brazil, iron ore projects had to be slowed down. The fall in prices could not be absorbed by application of new technology, increase in size of equipment, better transport system, improved management practices, etc.

Net profit of NALCO also came down from Rs. 547 crore in 1997-98 to Rs. 248 crore against a turnover of Rs. 1,854 crore and Rs. 1,507 crore respectively on account of 25% cut in production and closure of one of the pot lines.

Modernisation may be helpful in cutting down costs appreciably. Bhilai steel plant is currently undergoing modernisation so also South Eastern Coalfields at Chirmiri, Kapildhara and Pinora. Eastern Coalifields Ltd. has been referred to BIFR as its accumulated losses exceeded Rs. 1,000 crore. The financial problem in respect of this PSU is primarily on account of the high class rate by West Bengal government at 47 per cent ad valorem as against Rs. 50 to 135 per tonne in other States. This makes coal costlier by 30 per cent in West Bengal. Steel market is trying to divert itself to construction industry. In India, only 15 per cent steel goes for construction whereas the world average is 45%. New markets in Gulf and Middle East are also being explored.

The country is doing very well as far as exports of cut and polished diamonds are concerned. It is likely to exceed $ 5.2 billion during this year which is about 50% of the total world trade. Even against a global drop of 3 per cent, the Indian exports are going up by 12 per cent.

POWER

As such there is overall power shortage in the country. Leakage, theft, etc. has made the power availability much less than the quantum generated. This has led to a situation in which efficiency is rather poor which in turn results in higher power rates. High energy cost adversely affects the production cost in mines, processing plants, smelters and other power consuming units. In such circumstances, value addition may not be rewarding. The economy of steel production has been adversely affected by higher power cost in the country. In India, the cost of iron ore per tonne of saleable steel was merely $ 20 whereas the energy cost was $ 136. When compared with other countries, the following picture emerges:

($ per t)

Item	*India*	*USA*	*UK*	*Japan*
Iron ore	20	61	70	50
Energy cost	136	98	76	88

Ferro alloy producers in Orissa such as Factor, Nav Bharat, TISCO and Ispat Alloys have therefore, requested for reduction in power charges in order to remain in the export market. Ferro alloy plants in MP and Maharashtra are closed due to high power cost.

Power generating organisations are also going through financial problems as they are not getting their dues from consumers, and are not therefore in a position to make timely payments to coal producers. Inter departmental loans have mounted to thousands of crores of rupees.

LPG

Liberalization, privatization and globalisation have found favour in a number of countries specially in the mineral and mining sector. Several countries have already modified their statue to tailor to the requirement while the others are in the process of doing so.

The entire steel industry in Latin America has been privatized in 1997. Poscod of South Korea, one of the largest steel producers in the world, is now fully privatized Poland has programme to restructure its steel industry completely by the year 2010. Rumania is currently upgrading its outdated facilities and is planning to cut its labour force by 50%. Bulgaria is mobilising resources for massive investments. Guinea has undertaken provatization of its bauxite and alumina industry. Mynmar has awarded modernization of its mines to Australia. Indonesia is awarding contracts to foreign countries to develop its silver, nickel, gold and diamond deposits. Ashton, Australia has undertaken development of diamond deposits in Angola. Jordan is inviting foreign companies to develop copper, gold, oil shales, kaolin, etc. China has proposed joint venture with ZCCM to operate copper mines in Zambia. ALCOA and China are joining hands for mining its non-ferrous mineral deposits. Tanzania and Botswana have modified their legislation to suit the policy of liberalization.

In India, the liberalization as reflected in the National Mineral Policy, 1993 is an outcome of the policy initiatives taken in the Industrial Policy, 1991. This has provided a perceptible tilt from the public sector towards the private sector, and facilitated mobilization of capital from resources in India and abroad Provision of 50 per cent equity holding in

Indian mining sector through automatic route except in case of gold, diamond, etc. (the figure is 74 per cent for service sector) served the desired policy for private investment mobilization even from external sources. Equity participation upto 100 per cent is also possible with the approval of the Foreign Investment Promotion Board. The States such as Orissa, Karnataka, Andhra Pradesh, Madhya Pradesh, etc. have taken bold initiatives for private participation, and invitation to foreign companies have been extended to explore and develop deposits of diamond, bauxite, granite, etc. Setting up of a number of large greenfield alumina plants is on the anvil in Gujarat and Orissa. Jhamarkotra rockphosphate mine is taking the assistance from SNC and Geostat International for modernization. India and South Africa have signed protocol to co-operate in geology, mining technology and mapping for diamond. Jindal Vijaynagar Steel Ltd. is to set up a colour coating mill under joint venture with British Steel. SAIL was reported to be negotiating with Mittals for taking over its sick units as joint venture with majority stake and management control.

Disinvestment has also been proposed in several organizations. A decision has been taken to privatize 51 per cent stake in KIOCL. 50 per cent stake of BALCO is to be awarded to a strategic partner through global tender which will have management control. Further shedding of stake will be made in the domestic market and the government will retain only 26 per cent shares. NALCO, HZL and HCL have also been identified to disinvestment. Acquisitions and mergers are also being witnessed galore. Low metal prices and share values make it cheaper to buy operating companies. This helps in restructuring and better utilization of available infrastructure. During the period January 1995 and April 1998, merger and acquisition worth about $ 10 billion took place in the USA, Canada, Australia and South Africa and $ 6.6 billion in Latin America Figure for gold was $ 20 billion and for aluminium, lead, zinc and nickel, it was $ 7 billion. In the USA, Peabody have combined operations of North Antelope and Rochelle mines making it the largest coal mine in the country which produced 59 mil t during 1998. Jamalco and Alpout have merged their operations in Jamaica.

Sterlite Industries (India is to acquire Mt Lyell copper mine in Tasmania which will assure supply of 100,000 t of copper concentrates annually to Tuticorin smelter of the Company. Binani is considering investment in Zambian copper.

Mining legislation is further being modified in India to attract foreign capital. Foreign investor should know the procedures which will now be a single package. As recommended by the Tandon Committee, more powers with regard to renewal of mining leases and approval of mining plans will be entrusted to State Governments. Unrestricted privatization of coal which was hitherto for only captive consumption in power, steel and cement, is likely to materialise soon.

CONSERVATION AND SCIENTIFIC DEVELOPMENT

Mineral endowment is an enntity which we think that we have inherited from out forefathers whereas we have actually borrowed it from our next generation. Once minerals are extracted and consumed, they can not be put back in the earth.

Conservation of minerals is therefore, of prime importance. Industrialists would desire that the burgeoning mineral needs should be responded to while the conservationists would want them to remain in the mother earth. The balance between the two extremities can be struck by utilization of lower grade minerals through mineral beneficiation, developing mining and processing technology enabling mining of poor grades, development of technology for use of cheaper minerals in place of costly ones, recovery of by-products in the interest of improving economy, development of new uses for utilization of lower grade deposits, technology development for reduction in norms of consumption, utilization of complex ores, non-mineral substitutes and recycling of metal scrap.

Jharia Coalfield have posed the greatest challenge with regard to conservation. Here, as much as 2,500 mil to of coking coal can not be recovered as 750 mil t are locked in pillars, 550 mil t lie below townships and 1,200 mil t are below railway sidings. Coal is also under spontaneous combustion in 100 odd locations and for over eight decades.

R & D work is required in the areas of conservation, environment, exploration, beneficiation, mining methods, high speed shaft sinking, driving and raising, productivity improvement, energy saving, by-product recovery, etc. With increasing operational depth and size of operations ; rock mechanics, strata control and slope stability have gained significance. Beneficiation of base metals, polymetallic ores, ocean nodules, tin, tungsten, diamond, phosphate, kyanite, magnesite, china clay, etc. and *in situ* leaching of copper and heap leaching for gold are also some areas for R & D. Utilization of flyash and remud has not picked up very much on commercial scale though a lot of work has already been done on pilot scale. Removal of phosphorus from manganese ore, use of blast mats for controlling flyrock, obtaining methane gas from methane beds and sea-water magnesia are some of the area gaining momentum.

Several organizations such as IBM, CIL, NALCO, BALCO, HZL, HCL NMDC, BGML, NLC, etc. have useful R&D units. Work done by CMRI and ISM is no less significant. SAIL has programmed to increase its R&D earnings through RDCIS, Ranchi by extending services even to its competitors in steel production.

SAFETY

The challenge of safety appears to have been effectively tackled in Indian mines in the recent past. On average, about 800 accidents take place in the mining industry of the country causing 170 fatalities. This compares very well with several other countries. Over the last decade, the fatality rate per million tonnes of coal output and injury rate per 1,000 persons employed have both reduced by 33 per cent.

In the recent past, 50 miners met their grave due to mud slides in Tipuani gold operations in Bolivia, 60 died due to methane explosion in Ural mine in Ukraine and 56 miners drowned on account of inundation in a tanzanite mine in Tanzania. In South African gold mines, 267 fatalities and 5,449 serious injuries occurred in 1997. In India, 55 deaths took place in New Kenda colliery in 1994 and 64 in Ghaslit and next year.

Recently, the Ministry of Labour has asked the DGMS to chalk out a special drive to conduct thorough inspection of all mines in the country on time bound basis. They have also asked the DGMS to identify accident prone mines.

ZAP (Zero Accident Potential) does no more seem to be a mere hollow slogan or a distant dream. It will be a reality soon with dedication and endeavour in the right direction.

INDUSTRIAL RELATIONS

Industrial relations and trade unionism by far appear to be peaceful. However, sometimes simmering dis-satisfaction manifests itself in several ways. This aspects comes to the fore when any concept for higher production and productivity such as automation, modernisation, computerisation, diversification, restructuring, rehabilitation, etc. is sought to be introduced. Over manpower is a propblem in several organisations but cutting down flab to the desired extent has not been easy. VRS though introduced in a number of organisations in mining and mineral sector, did not have a very smooth sailing.

Fighting informal practices is also important. The International Finance Corporation (World Bank) which looks after economic development in member countries of the World Bank through investment in private sector conducted a survey by preparing a list of 15 obstacles to investment. One of the findings of its report is that the corruption is amongst three top obstacles in developing countries. In this context, it is pertinent to make mention of the efforts by the Ministry of Mines of Rajasthan Government which has brought itself on website in order to enhance transparency in mining leases.

CONCLUDING REMARKS

Mineral and mining industries should stick only to their core business and should do away with their non core activities. Here the example of TISCO selling away their cement plant is relevant However, it will be useful if within the core function, some diversification is undertaken so that if the market for one business is not doing well, advantage can be taken of the availability of the business opportunities in the other. MOIL has undertaken the production of electrolytic manganese dioxide and ferro alloy along with manganese ore. Here, it may be a useful suggestion if the coal producing companies go in for power generation also. This will reduce at least the landed cost of coal at the power plants which will result in cutting the cost of power generation as well.

It appears necessary to go in for judicious and appropriate technology otherwise the profitability of a running venture will continue to be dictated by the prices prevailing in the international market. If the prices are down, the producing company finds itself in financial problems. In case the prices improves, the financial performance of the same company improves. Thus the management or technology do not seem to play a significant role to accept the challenge of price fluctuation.

Efforts needs to be made to cultivate new markets for mineral products. SAIL has taken significant steps in this direction in seeking export market in the Middle East and Gulf. Market for new uses should also be nurtured. Again, the example of steel sector is relevant which is making effort in building up market in construction industry.

Surplus manpower is posing a problem which has to be reduced to the extent required. VRS has been introduced in several organisations which may not fetch desired result. Schemes for retraining, rehabilitation and deployment of such workers need to be taken up. China has a work-force of 3.3 million in steel sector which has been planned to be reduced to 2.6 million in steel sector which has been planned to be reduced to 2.6 million in over a couple of years. As many as 110,000 workers were shifted to tourism in 1998 alone.

From the above, it would be seen that the Indian mineral sector is faced with a myriad of challenges, some of which are presently being tackled. However, the agenda is not yet finished, and the challenges will no doubt percolate deep into the next century. It will require considerable endeavour and persistent efforts to effectively overcome the same.

CHAPTER 59

Life Cycle Assessment Analysis For Indian Mining Industry - A Clean Environment Approach

B.K.Prusty, B.K.Tewary*, A.K. Dube**

ABSTRACT

As environmental issues have gained greater public recognition, there is an increasing awareness that the consumption of natural resources and production and consumption of manufactured products have an impact on the quality of our lives. Due to increased public concerns, government and industry have both intensified the development and application of methods to identify the adverse environmental effects of all activities so that appropriate actions can be taken. Life Cycle Assessment(LCA) is a set of tools that is used to evaluate the environmental consequences of either a product or a set of activities. The assessments can be used to evaluate the opportunities for reducing the environmental effects associated with the production and use of products or production processes or to assess the impact of options. LCA is an important tool used in pollution prevention and green design efforts. Selection of product design, materials, processes, reuse or recycle strategies and final disposal options requires careful examination of energy and resources consumption as well as environmental discharges associated with each prevention or design alternative. In this regard a number of models have been developed ; but the SETAC-EPA Model is the most widely accepted model for life cycle assessment. Three main stages of LCA in this model are inventory, impact assessment and improvement. It focuses first on manufacturing process, estimating fuels consumed, other resources used and the amount of each waste discharged into the environment. The procedure then estimates the resources consumed and environmental discharges produced by the most important upstream suppliers and downstream activities. LCA analysis will be very helpful in designing greener and cleaner mining activities. It will help in selecting the methodology of mining so that the environmental discharge and the cost will be minimised. The concept of LCA analysis is

*Central Mining Research Institute, Barwa Road, Dhanbad - 826 001, INDIA

new in India but is widely followed in the developed countries. Such study must be undertaken for the Indian Mining Industry, as mining is generally perceived to cause great damage to the ecology and environment.

INTRODUCTION

There is an increasing awareness that the consumption of natural resources and production and consumption of manufactured products have an impact on the quality of our lives. The effects occur at all stages of the life cycle of an industrial activity. As public concerns have increased government and industries have both intensified and the development and application of methods to identify the adverse environmental effects of all activities so that appropriate actions can be taken up. Of late in India, mining activities have come under the vigil eyes of public, environmentalist and green pressure groups. Mining activities have raised the eye brows of the environmentalists in different parts of India such as the coastal fragile areas of Goa, the high altitude and young mountain range of Doon valley region, and biologically rich Gandhamardan region in Orissa. The increased environmental awareness of the people, government and NGOs require the mining activities to be undertaken in more environmentally sound manner. The statutory requirement of EIA study and the EMP are proper steps in this direction. But new knowledge and techniques have come in recent times to plan for any industrial activity for better econmoy, resource conservation and least environmental discharge. Life Cycle Analysis(LCA) is a new concept, now used in the developed countries like USA, UK and other European countries. This concept of LCA can be applied to Indian mining industry in the planning stage to select the optimum mining plan, methodology, environmental management, recycling strategies and post-mining land use and reclamation planning.

What is LCA ?

Life Cycle Assessment(LCA) is a set of tools that is used to evaluate the environmental consequences of either a product or a set of activities. The assessments can be used to evaluate the opportunities for reducing the environmental effécts associated with the production and use of products or production processes or to assess the impact of options. LCA is an important tool used in pollution prevention and green design efforts. Selection of product design, materials, processes, reuse or recycle strategies and final disposal options requires careful examination of energy and resources consumption as well as environmental discharges associated with each prevention or design alternative. At a basic level, LCA involves looking at what goes into a process, e.g. raw materials, energy and water and what comes out: products, co products, by-products, emissions, effluents and wastes. **Fig. 59.1.** illustrates the most basic components of a system and highlights the critical elements and the system boundary. Beyond the system boundary the 'environment' acts as a source of both raw materials and energy and it is the ultimate sink or receptacle for emission and waste.

Different models of LCA

There are two approaches or models of LCA analysis.

1. Process Model (SETAC-EPA LCA approach)
2, Economic input-output (EIO) model

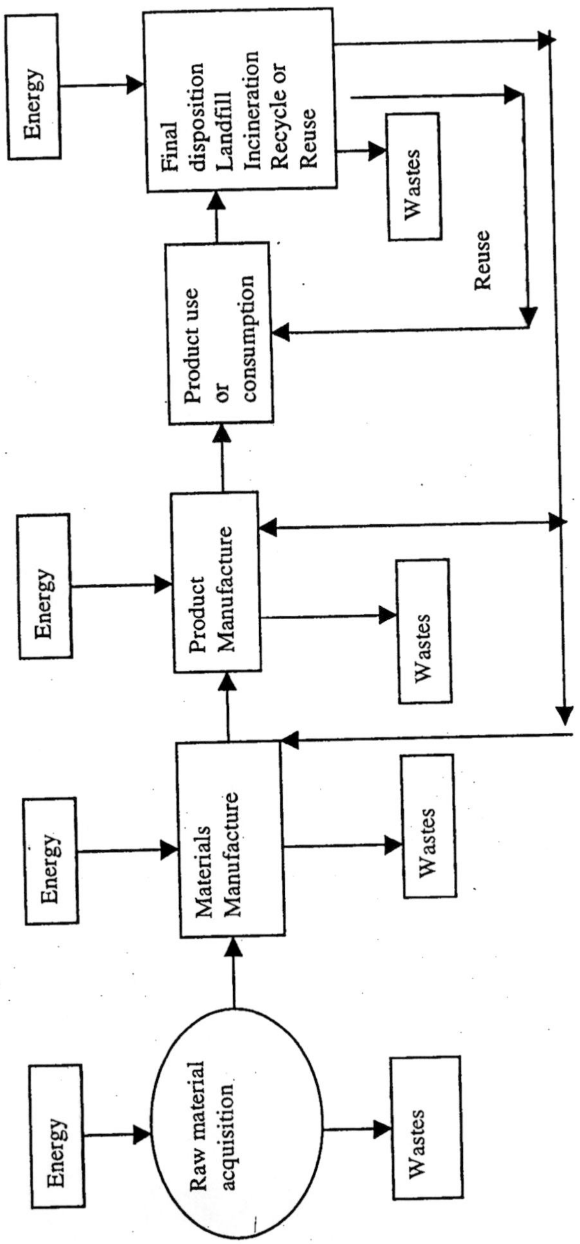

Fig. 59.1 : Overview of a framework for life cycle assessment

SETAC-EPA Model

The most widely accepted model for life cycle assessment has been established by the Society of Environmental Toxicology and Chemistry(SETAC). Three main stages of LCA have been identified.

- Inventory
- Impact assessment
- Improvement

The initiation phase defines the problem and establishes the objectives of the study. The inventory phase provides a detailed picture of the raw materials and energy input used by the system and the solid, liquid and gaseous wastes produced as outputs. The impact assessment phase links inventory to inputs and outputs to real world environmental problems. The improvement phase focuses on changes to the base system so as to improve its overall environmental performances.

SETAC-LCA Approach

It focuses first on manufacturing process, estimating fuels consumed, other resources used and the amount of each waste discharged into the environment. The procedure then estimates the resources consumed and environmental discharges produced by the most important upstream suppliers and downstream activities. Because of the difficulty in estimating resources consumption and environmental discharges produced by process associated with the life cycle of a product, LCA analysis is simplified by drawing an adhoc system boundary that excludes all but a few upstream and downstream processes. **Fig. 59.2.** shows a generic, seven-stage life-cycle assessment product chain.

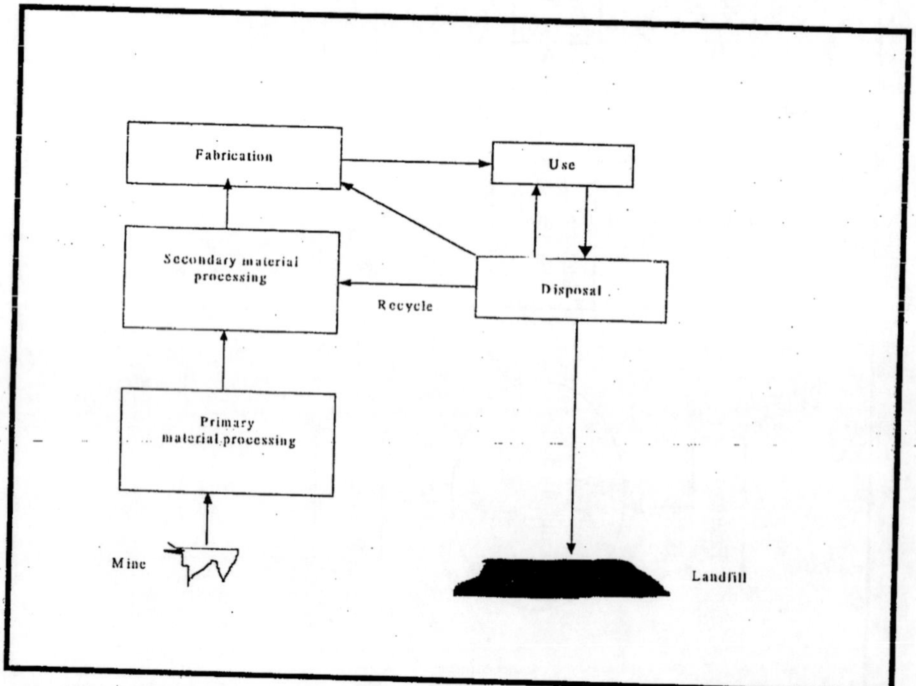

Fig. 59.2 : Generic seven stage life-cycle assessment product chain.

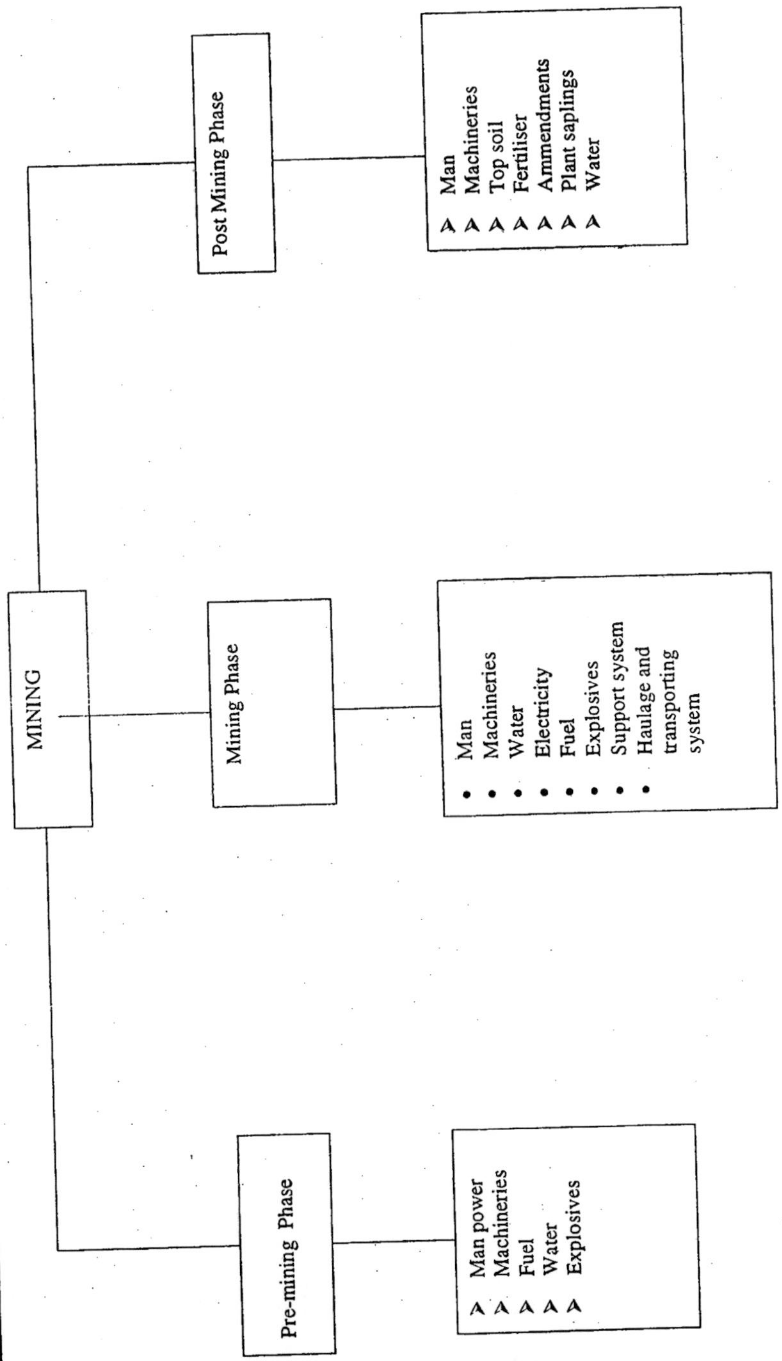

Fig. 59.3 : Resource input in different stages of mining activity

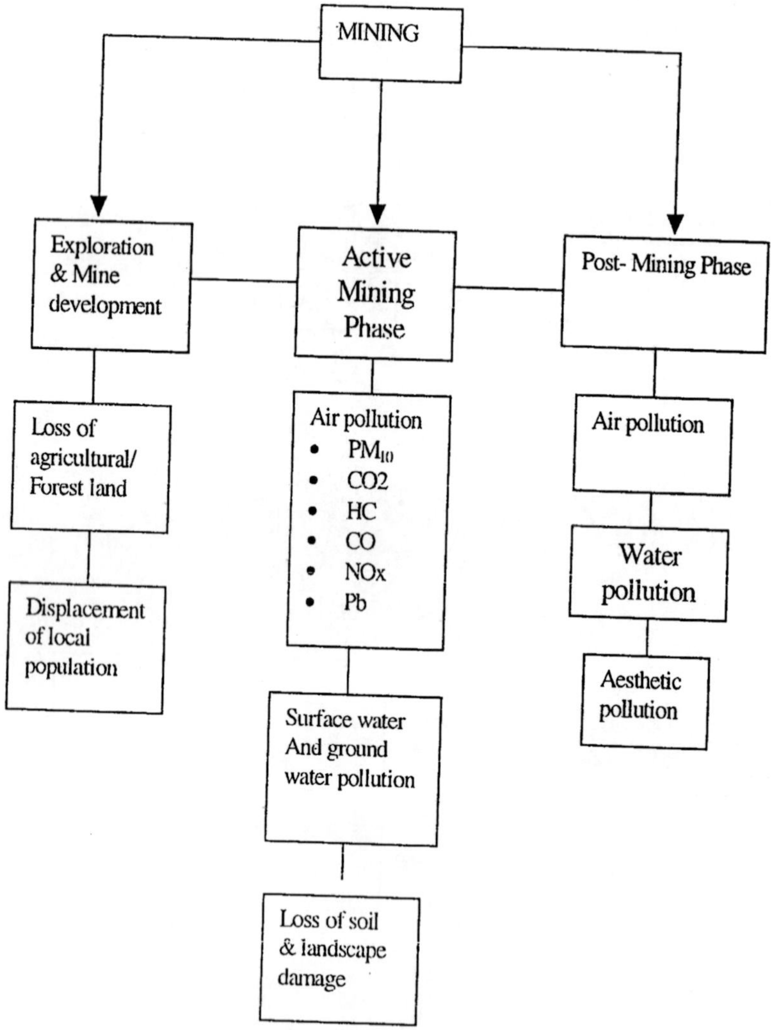

Fig. 59.4 : Schematic Environmental discharge model of mining projects

Alternatively, the LCA could be represented as **nxn** matrix of direct process stage interactions, where 'n' is the number of individual process associated in the life cycle of the product. Data entries in the matrix can be expressed in monetary value (i.e. Rs., $, etc.), or some other relevant units, i.e. raw materials required can be expressed in tones or energy inputs in KAmp. Valuation in monetary units provides consistency and conversion to mass or energy can be done using average prices.

The direct supplier process inputs vector $X_{direct\ suppliers}$ can be obtained by multiplying the matrix (I + D) by the vector of required outputs.

$$X_{\text{direct suppliers}} = (I + D)\,F \qquad \text{.........................(1)}$$

The elements of $X_{\text{direct suppliers}}$ are the direct supplier inputs to produce a product.

I - An identity matrix accounting for the product output.

D - Direct requirement matrix (matrix of different individual processes)

F - A vector of desired final output.

In equation (1) only direct supplier inputs are considered (i.e. the supplier chain is one level deep). But more generally, an LCA should take into account the entire supply chain, including indirect suppliers to a supplier as in the case of multilevel deep supply chain. For a multi-level deep supply chain product, the equation (1) can be rewritten as

$$X = (I + D + DD + DDD + \ldots\ldots\ldots)\,F \qquad \text{................(2)}$$

X now takes into account all suppliers input levels. The supplier requirements services (I + D + DD + DDD +...) is equal to $(I - D)^{-1}$. So that total input

$$X = (I - D)^{-1} F \qquad \text{......................................(3)}$$

Environmental Output Calculation

Once the economic output for each process stage is calculated, a vector of environmental outputs can be obtained by multiplying the economic output at each stage by the environmental impact per dollar of output.

$$B_i = R_i\, X = R_i\, (I - D)^{-1} F$$

B_i = Vector of environmental outputs (subscript i denotes the type of environmental discharge like solid waste, emission, effluent etc.)

R_i = Matrix that has diagonal elements representing the environmental impact per dollar of output for each process.

EIO Model for LCA

For modeling the interdependencies in producing goods and services in the economy, Leontief(3) proposed a general equilibrium model that requires specifying the inputs that any sector of the economy needs from all other sectors to produce a unit of output. In this LCA model, the system boundary includes the entire economy. The economy is divided into number of small sectors. For producing any goods and services, in any sector of the economy, it requires inputs from other sectors. EIO model requires all these inputs to be specified. This model is based on a simplifying assumption that increasing the output of goods and services from any sector requires a proportional increase in each input received from all other sectors. Here also, the LCA can be represented as a transaction matrix. The 1987 EIO matrices for the USA divide the US economy into 519 commodity sectors. Elements in rows, the direct coefficients (X_{ij}) are normalized by the total economic output (X_j) of the sector. Each coefficient represents the dollar input from the sector of the economy named at the left of the row for each $1 of output from the industry named at the head of the column. The general format of the transaction matrix for the basic EIO model is shown in Table 59.1.

Table : 59.1

Output from sectors	*Input sectors* 1	2	3	n	*Intermediate output O*	*Final demand F*	*Total output X*
1	X_{11}	X_{12}	X_{13}	X_{1n}	O_1	F_1	X_1
2	X_{21}	X_{22}	X_{23}	X_{2n}	O_2	F_2	X_2
3	X_{31}	X_{32}	X_{33}	X_{3n}	O_3	F_3	X_3
n	X_{n1}	X_{n2}	X_{n3}	X_{2n}	O_n	F_n	X_n
Intermediate input I	I_1	I_2	I_3	I_4			
Value added V	V_1	V_2	V_3	V_4		GDP	
Total input X	X_1	X_2	X_3	X_4			

Total input (Xj) is the sum of the intermediate input (I) and the value added (Vj). Total output (Xj) is the sum of the intermediate output (Oi) and the final demand (Fi). The gross domestic product (GDP) is the sum of all the value added (ΣVj) and final demand terms (ΣFi).

Analogus to process model, EIO model can be represented as a chain. However, each box in the chain would represent an economic sector rather than specific process. Because this EIO model is comprehensive for the entire economy, the process chain diagram would include connections among all sectors. Indirect interactions between suppliers can be important. Although indirect demands may be small individually, taken as a whole they can be significant because they are likely to be numerous. Moreover, economic sectors also purchase some of their own individual products. With a matrix representation of a process supply chain, indirect economic impacts can be included by applying equation (3). EIO tables greatly facilitates a first order approximation of the economy-wide environmental impacts of a production decision and in the economy wide treatment the extent of hazardous waste and emission generations is revealed more fully. Indirect environmental effects associated with these sectors can be estimated by applying equation (4).

LCA of Mining Projects

For the life cycle analysis study of mining projects the mining life can be divided into three stages such as

1. Pre-mining Phase
2. Active mining phase
3. Post-mining phase

The resource use and the environmental discharge for each phase are studied separately. The schematic model of resource use in mining activity is shown in the Fig. 59.3. The environmental discharge model for mining is shown in the Fig. 59.4. Pre-mining phase includes the activity from exploration upto the development of the mine. The energy and material input at this stage are labour, machinery, servicing, fuel and electricity, explosives and water. The environmental discharge is not much in this stage. But there may be loss of forest in case of forestland, loss of agricultural land, loss of topsoil and displacement of local population. The resource input during the active mining phase are man, machineries, explosives, blasting accessories, fuel, electricity, water and mineral processing machineries. The resource use for different component activities of mining can be listed separately and so also the environmental discharge. The environmental discharge is maximum in the active mining phase. There are gaseous discharges like CO_2, CO, NO_x, SO_x, Pb due to the movement of diesel driven heavy vehicles. The air borne dust is a major problem causing health problems for man, animals and plants. The liquid effluent from the mines causes pollution to the surface water and the surrounding soil. Solid wastes like dumps are long-term source of pollution to surface water, soil and ground water. In the post-mining phase the resource inputs are manpower, topsoil, fertiliser, soil amendment, earth moving machineries, plant saplings and water. The environmental discharge in this phase continues for a long time. For the above three phase of mining, both resource use and environmental discharge should be computed in quantity and then they should be converted to monetary values .The comparison of the monetary values will be an indicator regarding the long term profitability of the project . This study conducted at the planning level will help in choosing the suitable method of mining, environmental protection scheme, resource conservation, recycling strategy, environmental management, reclamation scheme and post-mining land use. The LCA study should be undertaken during the planning stage of the mining projects. It should be integrated with the mine plan.

CONCLUSION

The study of life cycle analysis is a new concept in India. But it may prove to be very helpful for the Indian Mining industry, which is now under the vigil eye of the environmental activist and green pressure groups, funding agencies like the World Bank. Since the discipline is new there is no proper expertise available in India. The academic institutes and research organisation has a big role in developing and refining the discipline in the country. To formulate comprehensive procedure, guidelines for the LCA study they should collaborate with the mining industry, Ministry of Mines, Ministry of Coal and Ministry of Environment and Forests .

REFERENCES

Bisio, A. & Boots, S. *The Wiley Encyclopodia of energy and the Environment- Vol. 2.*, pp.1032.

Hendrickson, C. , Horvath, A., Joshi, S. & Lave, L. 1998. Economic Input-Output Models for Environmental Life-cycle Assessment., *Environmental Science & Technology*, April 1, pp.184A.

Leontief, W.1986. *Input-Output Economics*, 2nd ed., Oxford University Press: New York

Maclean, H.L.& Lave, L.B.1998. A Life-Cycle Model of an Automobile. *Environmental Science & Technology*, July 1, pp.322A

Powell, J.C., Craighill, A.L., Parfitt, J.P. & Turner, R.K. 1996. A Lifecycle assessment and economic valuation of recycling, *Journal of Environmental Planning and Management*, 39(1), pp.97-112.

Steele, N.L.C. & Allen, D.T.1998. An Abridged Life-Cycle Assessment of Electric Vehicle Batteries, *Environmental Science & Technology*, Jan 1, pp.40A.

CHAPTER 60

Waste Problems in Some Metallurgical Industries- Their Abatement and Control

A. K. Basu & M. K. Singh**

SYNOPSIS

Our abode–The mother earth–is being constantly subjected to the vagaries of vilification of her environmental day by day through various nefarious agencies. These agencies–better known as pollutants–have become objects of great concern to the health and hygiene of the inhabitants of the world. With the rapid pace of industrialisation throughout the globe the problem of pollutants of the environment is becoming more and more awful. The effluents from various industries–toxic gases, waste products/by products, radio–active fallouts etc. are endangering the environment to an alarming proposition.

In this article, the factors leading to the pollution of the environment by waste products of some metallurgical industries viz. coal carbonization plants and blast furnaces of integrated steel plants, metal finishing industries like electroplating plants, disposal problem of *Red - Mud* in the extraction of Aluminium have been discussed along with suggestion for appropriate corrective action/remedial measures in respective cases.

1. INTRODUCTION

A lot of resources in the form of raw material, fuel, water etc. are invested in an industry to achieve a target product. But in return, it does not get 100% of what it has invested. This is analogous to the example of any machine whose output is always less than its input and whose efficiency can never be 100%. A considerable portion of the valuable resources is lost as waste in industries. Treatment and disposal of this huge quantity of industrial waste is a major problem to which the scientific world is aware of. Effluents from some industries are greater contributors to environmental pollution. It is surprising

*Scientists, Central Mining Research Institute, Barwa Road, Dhanbad - 826001, Bihar India. Fax - 0091-326-202429-202093 E-mail : director@cscmri.ren.in/root@cscmri,ren.nic.in

to note that a about one–third of the phosphate mined in Florida, one-half of the tin mined in Bolivia and one fifth of the world tungsten are lost as slimes. A lot of work is being done throughout the world to tackle this problem. The increased use of mining and metallurgical wastes in construction work will decrease the demand on the available natural resources and help to solve many perplexing disposal problems in mining and industrial areas.

2. RECOVERY FROM WASTE

Waste generated by mining, mineral processing and metallurgical industries is estimated to be 2-3 billion tons annually. The waste include gases, dusts, solution and a variety of massive solid minerals such as tailing and slags, all of which now must be disposed of with minimum environmental degradation and at an acceptable cost. Some metal/mineral constituents have potential economic value. The resource potential of waste generated by iron and steel, aluminium, clay, phosphate, copper, coal mining and processing, non-metallic and secondary metal industries should be examined.

For instance, about seven million tons of *'Red Mud'* (dry basis) containing substantial amount of valuable mineral residues and dissolved process salts are generated annually in the USA during production of aluminium by the Bayer process. It is estimated that these wastes contain caustic soda, alumina, iron and titania. In addition to loss of these mineral values, impoundment of red mud poses a serious environmental problem.

Slags from ferrous and non–ferrous smelting operation are being utilised to some extent but still there is scope for wider use.

3. WASTE PROBLE IN SOME METALLURGICAL INDUSTRIES

Waste problem with the following metallurgical industries are discussed below:

(i) Problem of coal carbonization in integrated steel plants.

(ii) Waste from steel mills and blast furnace.

(iii) Metal finishing industries.

(iv) Disposal of Red Mud.

3.1 Coal carbonization in steel plants

The production of coke by high temp. Carbonization is an integral part of steel plants. Water is used in different steps ranging from quenching of coke, cooling, washing and scrubbing of coke oven gas for different bye-product recovery processes.

The effluents discharged from bye-product coke plant consists of ammonia, phenol, cyanides, H_2S, pyridines and oils. The effluents are both of inorganic and organic type, some of them being toxic while others are non-toxic. Steel industries adopt abatement measures of waste pollution by recovering useful bye-products and reusing water for cooling and other processes, such as wet scrubbing of flue gases in blast furnaces. But still various natural water sources downstream these industries are heavily polluted by coke even effluents.

The Table 60.1 given below shows the sources and volume of waste water from a steel plant.

Table : 60.1

Sl. No.	Unit Operation	Quantity (m^3/hr) Max	Min	Composition
1.	Coke Quenching	-	small	Tar, ammonia, phenol cyanides and H_2S
2.	Cooling, scrubbing & Refining operation			
	(a) Flushing liquor	4	1	Tar & condensate ammonia, phenol & oils
	(b) Primary coolers	2	1	Tar & ammonia
	(c) Condensate pits	10	1	Ammonia liquor
	(d) Machine House	10	2	Contaminated condensate
	(e) Dephenolization plant	10	2	Ammonia
	(f) Potash plant	5	2	Ferrocyanide carbonate, cyanides & oils
	(g) Final cooler	1	-	Phenols & light oils
3.	Ammonia steel	60	30	Fixed ammonia
4.	(a) Benzol plant	6	2	Phenol, wash oil and benzol
	(b) Tar distillation	30	-	light oil, tar, cresote, oil, phenols & condensate.

The total volume of process effluents from a bye-product coke plant of one million tonne capacity is about 80 m^3/hr (0.42 mgd).

Treatment of coke oven effluents

1. Coke of suitable quality is heated, in absence of air, to 1000°C for 12-18 hrs and then quenched in retorts. The gases from the retorts are condensed. Tar and Ammonia liquor at 90°C is sent to bye product plant by suction pump and recirculated in coke ovens. Gas is further cooled in primary coolers where tar, ammoniacal liquor and naphalene separated out. Complete removal of tar from gas is done in an electrostatic separator.
2. Absorption of ammonia with refrigerated water or by scrubbing with dil H_2So_4 (sulphuric acid).
3. After final cooling of the gas with high velocity water spray in a cooler, crude napthelene separated out.
4. Absorption of benzol is done by scrubbing with wash oil.
5. Removal of H_2S from the gas by scrubbing with a dilute solution of NaoH.
6. Finally the gas is compressed and stored in gas stacks under pressure.

Some *auxiliary processes* are carried out simultaneously in bye-product coke oven plants

(a) *Fractional distillation* of crude benzol and refining of benzene, toluene and xylene.

(b) *Tar distillation* to yield ammonia liquor, light oils, carbolic oil and napthelene oil.

(c) *Dephenolization* : By kopper's recirculation process, ammonia liquor stripped by steam and the blow down is absorbed in NaoH soln. to form sodium phenolate. Steam is recycled and ammonia liquor is sent still for final processing.

(d) *Ammonia still* : Ammonia liquor obtained from various stages of scrubbing operations is treated with steam. NH_3 gas is absorbed in H_2So_4. Sizeable quantity of hot effluent containing ammonia, phenol, cyanogen compounds is discharged to waste. 1 ml

4. STEEL MILL WASTE

The effluents of a steel mill are :

1. *Wet scrubber effluents* : Originate from flue gases of blast furnaces. The effluents contain iron oxide, alumina (Al_2O_3), silica, carbon, lime and magnesia. However, the concentration of each constituents varies with the type of ore used, furnace lining, quality of coke, amount of air used in the blast etc.
2. *Waste picking liquor* : Originating from cold rolling mills. It is highly acidic and contain 1-2% H_2So_4 and $FeSo_4$
3. *Waste from Hot Rolling Mill* : It contains, mostly, of the suspended matter derived from the descaling of ferric oxide from various stages of operation.
4. *Steel melting shop (SMS) Waste* : It consists of water used for cooling purposes.
5. *Miscellaneous effluents* :

   (a) Thermal power station : Waste water containing flyash, effluents from demineralization plant containing acid or alkali or both and other chemicals for conditioning of water.

   (b) Sintering process and other miscellaneous operations.

Chemical characteristics of steel mill effluents are shown Table 60.2.

Table : 60.2

Sl. No.	*Characteristics*	*Blast furnace*	*Cold Rolling mill*	*Thermal power station*	*Sinter ring plant*
1	2	3	4	5	6
1.	Flow (m^3/hr.)	2	70	6	2
2.	PH	8.4-7.0	2.0-1.1	7.5-2.8	7.6-7.3
3.	Dissolved solid (mg/l)	425-168	18500-2240	920-200	150-75

1	2	3	4	5	6
4.	Suspended (solid mg/l)	3660-920	39750-90	1175-35	80-1-
5.	Acidity (mg/l)	-	12700-1600	400-40	-
6.	CoD. (mg/l)	1075-192	560-40	400-40	200-20

4.1 Methods of Treatment and Disposal

1. *Blast Furnace effluent* : Removal of flue dust by plain sedimentation or conjunction with chemical coagulation and recirculation of the clarified effluent back into the system. The sludge is disposed of by dumping.

2. *Effluents from cold rolling mills* : Treatment with lime for neutralisation. This is a problem of great magnitude. The only useful bye-product from the liquor is $FeSo_4$. The lime sludge is voluminous and is a slow settling type.

3. Effluents from the various operations in steel making are not of a polluting type and their treatment and disposal do not pose serious difficulty.

4. Bio-degradation of the toxic chemicals in effluents from bye-product coke oven plants may be efficiently carried out in continuous fermentation process [8]. Suitable organisms for this process may be isolated from the soil.

EFFLUENTS FROM METAL FINISHING INDUSTRIES

With the rapid growth of industries, the need for applying attractive finishes has become a necessity as well as a reality. The phenomenal development of the metal finishing industry has brought in its wake the problem of waste water disposal and consequent pollution of the water resources. Electroplating industry uses large quantites of water. The contaminants in the effluents are from ringe water drag outs and occasionally from spent baths. Electroplating effluents are primarily of the following types :

(1) Effluents containing acids or alkalis/metallic salts or mixture of metal/acid or metal and alkalis from pickling soln., chemical electro polishing soln., plating baths.

(2) Effluents containing impurities which are immiscible with water such as grease, oils or their solvents such as benzene, kerosene etc.

(3) Effluents containing cyanides from plating baths for Cu, Zn, Brass, etc.

(4) Effluents carrying chromates from chromebaths, etching baths, electro polishing, passivating baths and chromating solns. Most of the constituents in the effluents are highly toxic and toxicity persists even at low concentration.

The metal finishing industry has to use a considerable quantity of water but no part of water is held in the final product. All the water consumed by the plating establishment must be disposed of as waste. If the cost of treatment outweighs the cost of water there is no point in reusing the water except reducing the toxicity to the permissible limits.

The concentration of the toxic substance in the final effluents are :

(1) highly acidic effluents to be mixed with alkaline (excluding cyanide) effluents.

(2) In effluents containing chromic acid, the highly toxic compound is converted into less poisonous trivalent chromium by reduction which is then removed by $FeSo_4$. Subsequent addition of alkali will precipitate the chromium.

(3) Effluents containing cyanide should be treated with bleaching power or hypochlorite solution at shop end so as to destroy the toxic constituents in the solution itself.

(4) Any oily matter is to be skimmed off.

Table : 60.3

Sl. No.	*Constituents*	*Before treatment*	*After treatment*
1.	P	1.3-2.2	9.0-9.5
2.	Total Acidity, mg/l	1000-2000	-
3.	Minral Acidity, mg/l	900-1400	-
4.	Sulphate, mg/l	1470-2000	129
5.	Total dissolved solids, mg/l	850-1600	1484
6.	Suspended solids	Nil	Nil
7.	Nickel, mg/l	8-30	Nil
8.	Copper, mg/l	25-100	1.2
9.	Iron Oxide, mg/l	60-100	Nil
10.	Silicon, mg/l	14-40	Nil
11.	Chloride, mg/l	200-400	Nil
12.	Cyanide, mg/l	Nil	Nil
13.	Chromate, mg/l	Nil-40	Nil

Rate of flow = 18000 Lit/Hr.

5.1 Treatment of effluents

There are two types of process:

(a) Batch process for a smaller volume of liquid (Greater than 5000 lit./day)

(b) Continuous process for larger vol. (5000 lit/day).

In the batch process, the effluent is collected in a large tank, the treatment reaction is carried out, liquid allowed to settle and the clear, harmless effluent is discharged. This type of batch process is applicable to cyanide, chromic acid or acidic/alkaline type of effluents.

In continuous process the effluents are collected in small volume and treated continuously.

Disposal of Red Mud

Red mud, the causic insoluble residue from bauxite processing is the largest volume solid waste from non-ferrous industries. Its disposal poses increasing problems of land costs, storage and environmental pollution.[9]

Red mud is a complex material containing of Fe_2o_3, AI_2O_3, TiO_2, Sio_2, Na_2O and CaO as major constituents besides other mineral phases. It is very fine sized with poor setting properties, thixotropic and cracks on drying. It is harmful for the soil. However, P.M. Prasad et al improved its soil mechanical properties by diluting with fly ash or lime. Red mud can be commercially exploited for making bricks, construction of dams, pavements etc.

It has been found in some studies that bricks made only Red mud deteriorates rapidly with time. But when diluted with 10-15% ash, good quality bricks can be produced from Red Mud.

Red mud can be disposed of commercially as an additive to cement. Studies have shown that crushing strength of cement and mortars can be increased upto 10 wt % dilution. It has sown that it is possible to consume at least 10% of cement by its dilution with red mud. This application alone has the potential to use all the red mud.

CONCLUSION

Like other industries–some metallurgical industries too–have their well established roles in polluting the environment, besides waste problem. The problem of colossal waste of resources in metal and mineral industries, treatment and disposal of effluents from pollution have proved to be of great concern to the scientists, Engineers and technologists. It took about 700 generations to evolve from stone age to the space age of 1960's. But it has taken about one generation to reach the age of scarcities which the 1960's and 1970's may well be called. Supply of waste is plentiful. Only what is needed is the formulation of economic and technical studies/feasibilities of recovering these resources and converting them into useful products.

Availability of minerals and metals are depleting while the demand is increasing day by day. This has compelled our planners to look for better utilisation. Waste recovery and resources management and extend life of deposits/resources for future mankind.

ACKNOWLEDGEMENT

The authors express their sincere thanks to Dr. A. K. Dube, Director, CMRI for his kind consent in allowing the paper to be published. The views expressed in this paper are of the authors and not by the institute they belong to.

REFERENCES

1. Waste Recovery–A potential source of minerals/metals : IBM Role–G.M. Rao-Indian Mining and Engineering Journal ; Vol 33, No. 2, February 1994.
2. Industrial waste problem in the coal carbonization and steel industries in India–S. Raja Gopala and S. R. Aligraswamy–DoC : S-4/9, 12th Indian Standards Convention, Bhubneshwar, 14-21 Dec '68.
3. Rudolf(W). Industrial waste–their disposal and treatment, 1953. Reinhold Publishing Corporation, New York.
4. Ludzack (FJ) and Etlinger (MB)–Journal of Water Pollution Control Federation 32, 1960; 1173.
5. Control of effluents from the metal finishing industry–S. Krishnamurti (I.T.I.), Banglore)–DoC : S-4/11, 12th Indian Standards Convention, Bhubneshwar, 14-21 Dec '68.
6. Gurnham CF–Principles of Industrial Waste Treatment,/56 John Willey & Sons, New York.

7. Weiner R.–Effluents treatment in the metal finishing industry 1963–Robert Draper Ltd. Teddington, U.K.

8. Lovette (M)–chem. & ind. 70; 1951; 813.

9. P.M. Prasad–supply and Solid waste–G. D. Birla Medal lecture '90 at 28th National Metallurgical 'Day and 44th ATM, (IIM)–Tiruchirapally, 1990.

CHAPTER 61

Failure of Wire Ropes — A few Case Studies

M. K. Singh, A.K. Basu* & S.P. Choudhary**

SYNOPSIS

In the mining industry, wire ropes are widely used for winding/haulage purpose and are considered as safety equipment as these are related to the safety of the working personnel and damage to structure. Because of the complex geometry of the mining operations, wire ropes are subjected to the misuse of bending/crushing compressive/tensile, fatigue and dynamic load cycles. These hostile load condition damage the rope surface seriously whereby there occurs localised stress concentration at the defective sites. Fatigue strength is reduced to a considerable extent leading to a premature failure. Failure of a wire rope depends upon several factors like design, manufacturing defects, service defects, operational lapses and maintenance, environmental effects, etc.

In this paper the cause of premature failures of different types of wire ropes used in Indian Mines and oil drilling rigs have been analysed and preventive measures discussed in each case in this article.

INTRODUCTION

Wire ropes are extensively used in collieries for winding and hauling of men and material as well as for transmission of power. These are available in different forms such as winding ropes, haulage ropes, guide ropes, rubbing ropes etc. and in different shapes such as round ropes, locked coil ropes, flattened ropes, galvanised round/locked coil ropes etc. of various sizes and constructions. Wire ropes are considered as safety equipment in the mining industry because the safety of the personnel underground is related to their safe behaviour in operating conditions. A number of precautionary measures, therefore have been prescribed by the competent authority to ensure safety to the working personnel and reliability in operation.

---

*Scientists, Central Mining Research Institute, Barwa Road, Dhanbad - 826001, Bihar, India. Fax - 0091-326-202429/202093, E-mail: director@cscmri,ren.in/root@cscmri.ren.nic.in

The precautionary measures include:

(1) fixing of a high value of factor of safety (FOS) ranging from 6-10, depending upon pit depth, to account for operating static and dynamic loads.

(2) application of rigid quality control measures during manufacture of a rope.

(3) Inspection of the surface quality of the rope and assessment of the extent of surface deterioration after every six months of its use in a mine.

(4) *In-situ* inspection of the rope by NDT methods to detect flaws, if any, in the rope.

(5) Rope life is fixed by statutory rules for each colliery after considering the working environment, at the end of which the rope is thoroughly examined in an approved establishment before further use.

In spite of so many precautionary measures being adopted by the manufacturers and the users, there have been cases of catastrophic failure of wire ropes in many collieries. In fact, a wire rope is a precision machine of many moving parts carefully designed and fabricated to withstand a multiplicity of stress like bending, crushing, tensile, fatigue, bending fatigue in actual use in a hostile working environment. But if it is not given due care, consideration and lubrication for maintaining its health, the life of a rope gets shortened and its behaviour becomes unpredictable.

There are two types of defects which cause failure of a wire rope. They are :

(a) manufacturing defects, and

(b) service defects.

Manufacturing defects in a wire originate in the manufacturing stage. Green rod fore wire drawing may contain defects coming from the rolled ingot. Defects may also come during heat treatment or wire drawing stage through the die. The principal manufacturing defects include incipient crack, lap, micro-pipe, decarburisation, partial decarburisation and inadequate micro structure during patenting. The chemical composition of the steel also plays a vital role in deciding the quality of the rope wires.

The common service defects include abrasive/plastic wear, pitting corrosion, fatigue corrosion, stress corrosion cracking, martensitic structure, fatigue and crack in zinc coating.

However, failure of any machine component or material is due to the cumulative effect of a number of cause working simultaneously to generate a nucleus of fracture. From the failure data collected, it is seen that the cases of rope failure due to service defects far out number those due to manufacturing defects.

A new interesting cases of rope failure investigate in the laboratory have been cited and are being discussed hereunder to analyses the causes of failure in each case.

DISCUSSION

The analysis on data of physico–mechanical values as well as micro-structural analysis of the failed rope wires throw light to identify the possible causes of failure of the rope in each case.

Table 61.1

	Case I Winding Rope	Case II Haulage Rope	
I. History of Failure	Two broken strands found 11 ft. from drum when the cage was at pit bottom and withdrawn from service. i. SWL = 5 Tonnes ii. Drum/sheave Dia = 94" cage driven by steam engine.	(a) Failed while two loaded tubs were hauled upto the incline in the colliery. (b) Details of haulage Installation : i. HP of haulage = 120 HP ii. Drum Dia = 1250 mm iii. Length of haulage plane from surface to level = 417 mm. iv. Normal load attached to haulage rope = 20 Tonnes v. Gradient of haulage plane = 29°	
II. Particulars of Rope		Sample I	Sample II
(a) Dia of Rope	(a) 25 mm	(a) 19 mm ϕ	19.40 mm ϕ
(b) B.L. specification	(b) –	(b) 20 Tons (calculated)	
(c) Construction of rope	(c) 6x7 (6/1) FMC	(c) 6x6(5/1) FMC	6x6(5/1) FMC
(d) Type of lay	(d) RHL	(d) RHL	RHL
(e) condition of pit	(e) Not known	(e) Not known	Not known
(f) Frequency of operation	(f) 350 Cycles/day	(f) Maximum output 15 Tonnes/hauling	
(g) Pit depth	(g) 230 ft.	(g) –	–
(h) Date Snapping of installation	(h) & (i) Two and a half months after recapping	(h) & (i) Rope snapped after three months of installation	
III. Mechanical Test Results		Sample I	Sample II
(a) Breaking load of rope	(a) 40 Tonnes	(a) 13 Tonnes	14.5 Tonnes
(b) Location of fracture	(b) 3 strands snapped at centre (cup & cone fracture)	(b) 4 strands broken at max load 50 cm from left left hand grip	4 strands broken at max load fracture at center
(c) Range of T.S. of individual wires (Kg/sq. mm)	(c) (i) 170 - 180 Away from fracture end (ii) 176 (around fracture end (iii) 174 (core	(c)(i) 69.31-87.62 137. 50-200.40 core	42. 79-79.47 137.5-176.8 core

contd......

		Case I Winding Rope	Case II Haulage Rope Sample I	Sample II
(d)	Reduction in area	(d) 35.90	(d) –	–
(e)	Elongation	(e) 4.33 %	(e) –	–
(f)	Range of torsion	(f) (i) 23-26 (Away from fracture end	(f) i. –	–
		(ii) 30 (Around fracture end	ii. –	–
		(iii) 33 (core)	iii. –	–
(g)	R.B. alue on 7.5 mm radius or 1800	(g) (i) 10-13 (Away from fracture end)	(g) i. 5-11	5-9
		(ii) 8-11 (Around fracture end)	ii. –	–
		(iii) 10 (core)	iii. –	–
(h)	hardness value, VHN at 1 kg load	(h) (i) 407 - 426 (Away from fracture end)	(h) i. –	–
		(ii) 370-384 (Around fracture end)	ii. –	–
		(iii) –	iii. –	–

Table : 61.2

1. Physical Examination of Wire Rope

	Parameters of Physical Examination	*Case-I Winding Rope*	*Case-II Haulage Rope* Sample I	Sample II
(a)	Lubrication of rope	Satisfactorily Rope	Poor lubrication, fibre Core almost dry & was damaged	
(b)	Wear / corrosion of individual wires of the strand	Average reduction in dia ~ 3.57 (light wear)	40-76% for broken end 40-60% away from broken end	40-75% for broken end 30-75% away from broken end
(c)	External damage of the rope	–		
(d)	Characteristics of Fracture	Square fracture surface. 50% of the failed wires had radial crack upto 1/3 of dia & then bend. 25% fracture of ends had smooth surfaces. Rest of the wires had equal proportion of fine and fibruous structure.	Kinked	Kinked
(e)	Condition of the strands in the rope	Wires found infact in the strands. Two strands in the rope were found broken.	Fibre core was found between two strands	

Table : 61.2A

2. Micro-Examination of the wires

Location of micro sample	*Case I* *Winding Rope*	*Case II* *Haulage Rope*
(a) Away from facture end	(I) Polished transverse section revealed (i) Intermittent corrosion pittings in the periphery (ii) No harmful inclusion	(I) Broken end sample : (a) Transverse section revealed (i) uniform sorbitic Pearlite in the matrix (ii) Corrosion pittings at the periphery (b) Good drawn structure
	(II) Ethced transverse section (i) Uniform distribution of ferrite and pearlite grains. (ii) Intermittent pitting corrosion.	(II) Away from borken end : Structrue as above (a)
	(III) Polished longtudinal section (i) No harmful inclusion, no crack	(III) Broken end structure : (a) Transverse section : Same as I(a) above (b) Longitudinal section : (i) Good drawn structure (ii) Surface stress corrosion cracks visible.
	(IV) Etched longitudinal section : (i) Good cold-drawn structure	
(b) Good portion near fracture end © Fracture end	Transverse and longitudinal section of the sample revealed same structure as above (I) Polished transverse section revealed: (i) Intermittent corrosion pittings in the periphery (ii) No harmful inclusion (II) Etched transverse section (i) Uniform sorbitic pearlite structure (ii) Intermittent peripheral corrosion pits (HV1=373) (III) Polished longitudinal section (i) No harmful inclusion (ii) Multiple cracks along axial and radial direction of fractured surface (IV) Etched longitudinal section (i) Good cold-drawn structure Raidal cracks from surface to centre propagated axially (HV1=384)	

Table 61.3A : Details of Rope, history of failure and mechanical values of cases III, IV & V

Details of Rope		Case III Winding Rope		Case IV Oil drilling Rig Rope	Case V Suspension Rope
(a)	History of rope failure	Snapped 84' away from cappel when a loaded tub was raised 5' from ground.		Not provided	Not provided
(b)	Particulars of Rope				
(i)	Dia	25 mm		25.6 mm	22.5
(ii)	Breaking load	—		—	—
(iii)	Construction	6×7 (6/1) FMC		6 × 19 (9/6/1) FMC	6 × 19 (12/6+6F/1) IWRC
(iv)	Lay	RHL		RHL	6 × 7 (6/1) RHOL
(v)	Pit condition	wet (humid)		—	—
(vi)	Frequency of operation	3 shifts and 200 winding/shift		3.37 Round trip operation with 27/8 inch 1-55 grade tubing, 6 R.T. operations with D-grade drill pipes (RH)	—
(vii)	Pit depth	—		1900 m deep	—
(viii)	Date of installation & snapping	Ropes failed after 1 yr. & 9 months of installation (Statutory life is equal to 2 yrs.)		Snapped after 4 months use	
c)	Mechanical Test Results :	Broken end	Drum end		
(i)	Breaking load of rope	15 Ton 22.7 mm ϕ	37 Ton 25.7 mm ϕ	37.00 Tonnes, Dia - 25.6 mm Two strands broken at max. load, 20 cm from R.H. Grip.	Calculated B.L. 17.44 Tonnes
(ii)	Range of T.S. of individual wires (Kg/sq.mm)	81.8 - 126.4	170.4 81.3	184.5 - 212.3 (Broken end 1st layer 150.3-185.7 (Broken end II layer 141.5-194.2 (failed end)	64.9 - 84.4 (Ist sample) 64.9 - 142 (IInd sample)
(iii)	Reduction in area	0 - 1	4 - 21	19 - 32 (Good portion of broken end) 0 - 15 (Near failed end)	6 - 26 (Ist sample) 1 - 11 (IInd Sample)
(iv)	% Elongation	—	—	—	—
(v)	Range of torsion	1-12	12-16	—	6 - 12 (Ist sample) 3 - 10 (IInd sample)

Table 61.3B : Physical Examination of Wire/Rope and Micro-examination of wires

	Case III Winding Rope	*Case IV Oil drilling Rig Rope*	*Case V Suspension Rope*
I) PHYSICAL EDUCATION			
(a) Visual examination of Rope/Wires	Outer wires of rope severely corroded and worn out wires near fibre core more affected by corrosion.	—	Thimble not found in wire slings.
(b) Lubrication of rope	Rope devoid of lubricant (broken end portion), core dry and brittle, rest of rope properly lubricated.	Devoid of lubricant and dried	Completely dry and devoid of lubricant
(c) Extent of wear/ corrosion of individual wires	Avg. Reduction in dia of snapped rope = 11.32% Reduction in dia of individual wires = 24.5% - 51%	(i) 4.87-26.82% (failed end, 1st layer) (ii) 20.83-33.33% (failed end, IInd layer) (iii) 4.87-12.19% (Away failed end, Ist layer) (iv) 0-25% (Away from failed end, IInd layer)	6.66 - 50% in different strands 6.67-5.5-11.10% for I.W.R.C.
(d) Characteristics of fracture/Appearance of fracture surface	Wires of broken end (Heavy corrosion & pittings). Tensile fracture (abundant) Corrosion fatigue (less)	Mainly tensile fracture wires fractured at a particular place of rope	Tensile fracture with plastic wear, abrasive wear & corrosion pittings
(e) Condition of strand/ fibre core of rope	Fibre core missing (failed rope) strands opened out	Fibre core—dry, no corrosion pittings in adjacent wires	One strand kinked, others loosened, IWRC not in position & exposed at some place of rope
(f) Nature lubricant	Acid value high 1.75 mg KoH/gm of lubricant (Max permitted value = 1 mg KoH/gm lubricant)		
II) MICRO-EXAMINATION			
(a) Location of micro sample	(i) Cappel end (ii) Drum end		
(b) Polished sample	No harmful inclusion, intermittent corrosion pittings in sample (i) & (ii)	No harmful inclusion found	
(c) Etched sample		(sample from broken end)	(Broken end sample)
(i) Transverse section	Uniform sorbitic pearlite, corrosion pittings in periphery	Same as in CASE III also revealed martensite at periphery & crack	No structural discrepancy
(ii) Longitudinal section	drawn structure stress corrosion cracks visible in sampel (i) & (ii)	Drawn structure some fibres were deformed at heavy plastic nicking zone	cold-drawn structure micro-structural change in the zone of heavy plastic deformation.

CASE-I

In the first case cited above, the mechanical properties of the individual wires as well as that of the rope were within the specified standard limits. But the fracture surface appearance revealed bending fatigue type of failure of the rope.

Fatigue might have been generated into the rope by

(a) Kinking during/prior to installation.

(b) Looseness of the strands.

(c) Flaws in tensioning.

(d) Fatigue resistance decline sharply due to surface defects on the rope.

CASE-II

(a) The individual wires of the strands were subjected to severe wear and reduction in area ranging from 40 - 76%.

(b) Excessive loss of material from the wire surface caused loosening of strands leading ultimately to kinking of the rope.

(c) Low value of torsion obtained from the tested indicate notch brittleness on account of surface corrosion pittings.

(d) Endurance limit of the wire material as well as its loading bearing capacity were reduced abruptly due to excessive wear on the wire surface. Thus the rope failed in service under operation cyclic and dynamic stresses.

(e) No micro-structural change was observed in the failed wire sample.

CASE-III

(a) The outer wires of the rope were severely corroded and worn, wires near the fibre core being more affected.

(b) The rope was devoid of lubricant, the core being completely dry.

(c) Poor torsional value of the wires were due to corrosion pittings on the wire surface.

(d) The acid vale of the lubricant was too high (1.75 mg KOH/gm lubricant) and beyond permitted value of the relevant standard.

(e) Adverse pit condition (i.e. wet condition) was extremely harmful and favourable for initiation of corrosion pittings.

(f) Due to the effect of high reduction in area and corrosion pittings on the surface of the wires, the factor of safety was reduced to 3 which made the rope incapable of withstanding operational stress.

CASE-IV

(a) The fibre main core (FMC) of the rope was dry and corrosion pittings were noticed on the wire surface.

(b) Presence of surface defects like nicks, grooves etc. revealed that there was heavy friction and hindrance in the relative movement of the wires and strands. This happened due to lack of lubrication.

(c) Heavy friction and subsequent critical cooling led to the formation of brittle matensite phase in the micro-structure of the wire surface.

(d) Under operating stress the martensite phase cracked and propagated subsequently.

(e) Evidence of heavy localised deformation was revealed in the micro-structure. This led to notch effect on the wire surface which was responsible for low value of torsion of the individual wires.

CASE-V

(a) Poor torsion and R.B. value on wires in the good portion of the rope suggested deficiency in toughness or residual ductility.

(b) Change in micro-structure was due to heat generated from friction of moving parts of the rope and subsequent cooling in the ambient atmosphere.

(c) There was heavy loss of material from the wires due to wear.

(d) Due to notch effect presence of the brittle micro constituent and wearing, there was stress concentration in the wire material. Fracture initiated from "Notch" site and propagated due to high operating stress.

CONCLUSION AND REMEDIAL MEASURES (CASEWISE)

Different factors responsible for premature failure and its remedial measures are shown in Table 61.4.

Table : 61.4

Case No.	*Predominant Cause (s) of Failure*	*Remedial Measures*
I	Kinking, looseness of the strands, Nicking & grooving, and loss of fatigue resistance	Proper handling of the rope during/ prior to installation.
II	Inadequate lubrication, Excessive wear, Kinking, looseness of the strands, Obstruction in the movement of rope/wires.	Proper Maintenance by lubrication, handling and design of installation
III	Heavy wear of wire surface, severe pittings corrosion, high acid value of lubricant, adverse pit condition	Use of suitable lubrication and proper Maintenance schedule
IV	Formation of nicks and grooves on wire surface total absence of lubricant in the rope core.	Proper lubrication & Maintenance schedule
V	Dry fibre core & heavy reduction in cross sectional area of the wires.	Maintenance of proper lubrication schedule.

In conclusion, it can be said that the cause of premature failure of wire ropes, in a majority of cases was due to improper maintenance/lubrication and wrong handling of the rope.

ACKNOWLEDGEMENT

The authors express their sincere thanks to Dr. A. K. Dube, Director, CMRI for his kind consent in allowing the paper to be published. The views expressed in this paper are of the authors and not by the institute they belong to.

REFERENCES

1. Metal Hand book, vol. 10, failure Analysis and prevention, American Society of Metals, ASM, Metal Park Ohio.
2. Mechanical Metallurgy–G. E. Dieter, Mc Graw Hill–Kogakusha Limited, International Studies Edition, PP 280-284.
3. R.D. Barer and B. F. Peters " Why metals fail"–Gordon and Breach Science publisher, 1991 edition
4. A. K. Basu, K. K. Ganguli, J. C. Ghosh, R P Chakraborty & M. N. Tarafdar– "Standardisation and safety in Mining Industry"–I M & EJ. August 1991.
5. Wire Rope Hand Book–Edited by Allan B Dove, Vol. II, PP. 10-17, The Wire Association, INC, Branford, Conn.
6. Role of Surface Quality on Fracture Susceptibility of Cold drawn high carbon steel used in Mines–D. Mukherjee, R.P. Chakraborty & P. R. Roy–New Sketch, Annual Number, 1979
7. Performance study of Indigenous Mine winding Ropes–R. R. No. 48. CMRS E6/48, Feb'68, T. K. Sengupta & P. R. Ro, PP. 20-21.
8. (International publication) Rope Man's Hand book Published by N.C.B., U.K.
9. The Wire rope–an incredible Machine–R.K. Kejriwal–Paper presented in Colloquium on Wire Ropes at CMRS Auditorium on 10th August, 1980.
10. A study to correlate behavioural relationship of torsion to tensile strength in rope wires under different states of deterioration in Indian mines–A.K. Basu & S. C. Goswami, Journal of mines, Metal and Fuels, February, 1983.

CHAPTER 62

Work Place Dust : A Growing Challenge for Mining Industry

*J.K.Pandey**

ABSTRACT

Mechanisation is critical to meet the growing demand of coal and minerals for national development. Some of the mining machines are known to produce more dust in a minute than what three men can do in a day. There exists stringent environmental regulatory burden related to atmospheric SPM load. However, work place dust so critical to miners' health and productivity is neglected, and if not taken care of, it will take up disastrous proportions. Present paper gives an assessment of harmful effects of work place dust, magnitude of the problem and future risk associated with it. Besides, it also deals with different methods of controlling dust hazard in mines with particular reference to CMRI contributions in the area.

INTRODUCTION

Today, perhaps no industry is free from fine dust in their working environment; mining is no exception. Any form of dust, be it mineral dust, coal dust or radioactive particle entering into the lung through respiratory tract can cause havoc to the workers receiving these dust. Although air borne dust at work place has many other harmful effects viz. increasing suspended particulate matter(SPM) load of general atmosphere, affecting health of machinery and increasing maintenance cost, and adding to nuisance value affecting workers efficiency but its human health aspect started gaining prominence mainly because mining is a labour intensive. Sometimes the cost of labour goes to as high as 80% of cost of coal production.

Recent trend of mechanisation is probably the only answer to meet the constantly growing demand of coal and mineral which is very critical for the developing country like India. There has always been a chase between mechanisation and dust generation as one gives rise to the other. Dust related health hazard has an incubation period of about 5-15

*Scientist , Central Mining Research Institute, Barwa Road, Dhanbad-826001.

years. Therefore, if not checked, dust related hazards could grow into disastrous proportion and could lead to labour unrest and production losses.

AIRBORNE DUST HAZARDS

Air borne dust affects the workers environment in more than one way. Considerable energy is wasted for producing the unwanted fines during mining operations. Cost in fact goes in geometric proportion with increasing degree of fines. The fines in turn adversely affects productivity and health of workers; thus adding to the overall cost. Continued inhalation of air borne dust of certain size range and concentration damages the lung of the miner and impairs their normal function. Percentage of miners suffering from this dust disease which is collectively known as pneumoconiosis, depends upon nature and composition of dust, the duration of exposure, the peak values of respirable air borne dust concentration, exposure to peak concentration and place of work.

Dust is also great menace to the workers as it impairs the visibility at the working place while handling the coal, it irritates eyes, nose and skin. It even affects the nervous system, while the dust on the floor is a great nuisance during working. The dust affects the machines and warrants more frequent servicing and maintenance of the machines thus becoming a cause for higher maintenance cost, production loss due to breakdown and increased depreciation cost. The fine coal dust is explosive when it is in right quantity and size distribution in mine atmosphere. One could never over emphasis the ill effects of a coal dust explosion in a mine. Finally, this dust contributes to the suspended particulate matter (SPM) load of general environment and become a concern for every living being.

Of these, pneumoconiosis hazard in workers probably over shadows all others. This disease is due to prolonged inhalation and retention of dust in the lung. It is seen that particle greater than 7.1 micron U.D.S. are totally cut off by the upper respiratory tract allowing particle below this size to find its way into lung. Along with the decrease in the size the penetration increases, being maximum (50%) at 1 to 2 micron and then decreases

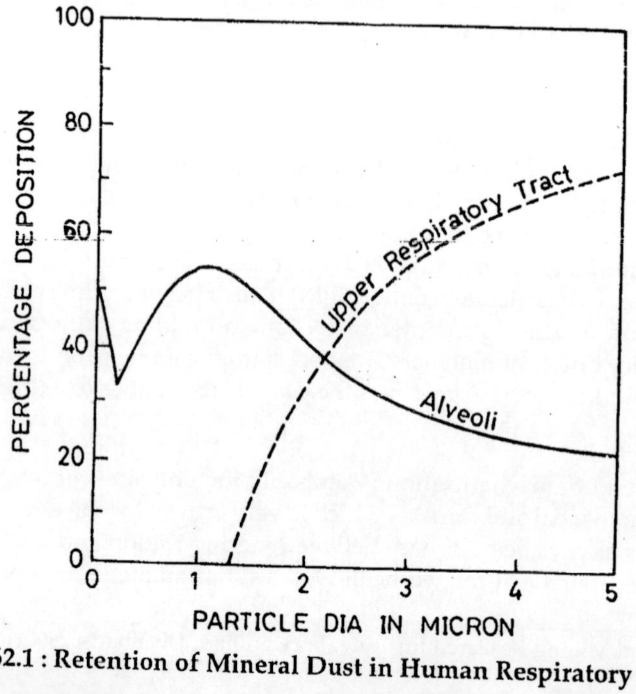

Fig. 62.1 : Retention of Mineral Dust in Human Respiratory System

up to the size of 0.3 micron but shoots up again at the lowest end of the size spectrum. Fig. 62.1 gives the penetration and retention of mineral dust in human respiratory system. All these sizes together contribute to the total dust deposited into the alveolar sacs. Study (Nair and Sinha,1988) shows that upto 100 grams of dust below 5 micron size is produced per ton of coal and 1-10 % of this is dispersed as air borne dust. Thus, workers in work places inhale a large of quantity of dust every day. But our lungs have unique self-clearing mechanism which clears most of the dust leaving only 1-2% (Hatch and Paul, 1964) to remain there. Investigations carried out by CMRI shows that 75% of dust found in lungs are under 1 micron size (Table 62.1). When this accumulation exceeds a certain minimum limit, which varies from person to person and nature of dust, manifestation of disease occur.

Most severe form of pneumoconiosis has been found to be silicosis. As its name implies it is caused by mineral dust containing free silica. The industrial research on dust hazards has now clearly established that dust in coal mining too are dangerous to the health of the workers which causes Coal Workers Pneumoconosis(CWP) and Anthrasilicosis. Incidences of CWP are higher in anthrasite miners than bituminous mining. Workers in coal mines are not only exposed to coal dust but also to stone dust while working in stone drifts or dirt band in the seam containing silica. This causes an early damage of the lung. Other dust like asbestos causes asbestosis, iron oxide causes siderosis, titanium oxide causes titaniosis. Some dust like asbestos is carcinogenic and its long exposure among workers causes many other complicated disorders responsible for up to 8000 deaths per year (Rahman, 1995). Similarly poisonous nature of chromium dust affects human health in a more damaging way. In fact, it has not been worked out as to what factors cause this different type of pneumoconiosis. It is still a controversial matter but one thing is very clear that any dust if inhaled for a long time will cause lung diseases, though the extent of the hazard will mostly depend upon the nature of the dust.

Table 62.1 : Size distribution of dust collected from pneumoconiotic lung of Indian coal miners (CMRS,1986)

Mean size micron	*Cumulative % of particle*	*Cumulative % of surface area*
0.2	48.1	-
0.28	60.1	1.8
0.39	65.8	-
0.55	71.5	4.6
1.10	85.5	-
1.55	91.5	21.40
2.20	97.2	-
3.10	100*	100

* Only two particles greater than 3 micron size were found.

DUST HAZARD CONTROL

Best method of avoiding any such dust hazard is to do not let it happen. Which is possible only by maintaining dust exposure level of the workers below threshold limit value (TLV). It is determined on the assumption that worker's lung will not accumulate enough dust to cause pneumoconiosis while working for his entire service duration. For coal containing not more than 5% free silica this limit has been fixed as 3 mg/m^3 in India. To achieve the TLV value at work places it is often required to go for some dust prevention measures, dust suppression or dust consolidation, personal protection from dust or at last, job rotation.

Of these, last two are of less importance. Sometimes, dust generation is very significant but the workers exposed to them are only few and management does not find it economical to go for any large-scale dust prevention and suppression measures. Management then prefers to provide personal protective devices to workers exposed to fine dust. When miners find it inconvenient to bear while working, management is left with the option of job rotation at suitable time interval so that permanent damage to miner's health does not take place. Doyle (1970) has suggested the concept of critical doze for dust which is related to product of dust concentration and the number of working days during the entire period of under ground employment. The critical doze for coalmine was established to be 20 gram. This is useful to calculate the permissible duration of employment before which a coal worker could be shifted from his dusty occupation to safer places. But in either case, contribution to the environmental damage is not checked. More over in the case of job rotation, even miner's health is not given the proper justice. The preventive and suppression measures, which are vital for dust control, are being given in detail.

(a) Preventive Measures

Prevention strategy for dust generation involves either optimising the moisture content of coal seam or physical parameters of the cutter. It has been found that a coal seam is relatively free from air borne dust problems when its inherent moisture was 7 to 5 percent (Sinha, 1987). On the other hand coal seams containing less than 1.5% moisture have been found to be generally more dusty. This indicates that if the inherent moisture of coal seam is increased to optimum figure, the dustiness of the working face would be reduced. This is normally achieved by wetting the coal seam by high-pressure water infusion technique. This involves water infusion into coal at high pressure till the water start bleeding from the coal seam. This method is suitable where the coal seam is not much broken as otherwise the water will just short circuit and leak out from the face and no pressure will develop. On a suitable coal face, the water consumption is about 8-10 litre of water per minute at a pressure of 2 to 15 kg/sq.cm which vary considerably depending upon the seam condition. Water infusion in coal seam works in three ways; (a) Wetting the coal and the dust which may already present in the coal (b) Loosening the coal which will ease the operation by the machine e.g. cutter-loader and will reduce the consumption of explosive (c) give better fragmentation with less fines.

Coal seams condition in India varies widely and the infusibility of the various coal seams have not yet been carried out. In certain development heading, water pressure up to 900 psi have been obtained but in longwall and depillaring workings, it can be expected to be much lower.

As an alternative to this, pulse infusion shot firing technique primarily involves firing of explosive charges in a bore hole filled with water under pressure. This suppresses the dust and also prevents ignition. In India preliminary trial were started as early as 1960, in which holes up to 5 ft deep were fired in uncut coal.

Better roof support is conducive to better dust control by preventing coal seam crushing in-situ, avoiding roof falls. Also, it has been found that sharper picks produce less quantity of fines compared to blunt picks while cutting or drilling. Conical picks produces less dust than a normal picks. Speed of cutting and drilling has also a direct relationship to the production of fines which decreases with the rate of speed of cutting of coals or drilling. Therefore an optimum speed of cutting should be achieved. In developed countries optimum speed of coal cutting has been arrived but India has never attempted any such study.

(b) Dust Suppression

Water has, by and large, been used as a main element in reducing the dust, which is already formed. At times wetting agents have also been used with water to increase the wetting property of coal. To get better result water is sprayed in atomised mist form. In other countries, water sprays have been fitted to the coal cutting/loading machines and jet of water is directed to the picks, which are cutting the coal. Water directed through picks improves dust control than water jet directed at the picks. Experiments have also carried out in Indian coal mines to find out the dustiness and to reduce the dust concentration at different operation, such as undercutting of coal, drilling of shot holes for blasting, pneumatic stowing in board and pillar working, belt conveyor transfer points etc. Contribution of CMRI in dust suppression is remarkably significant with respect to following three major dust producing operation.

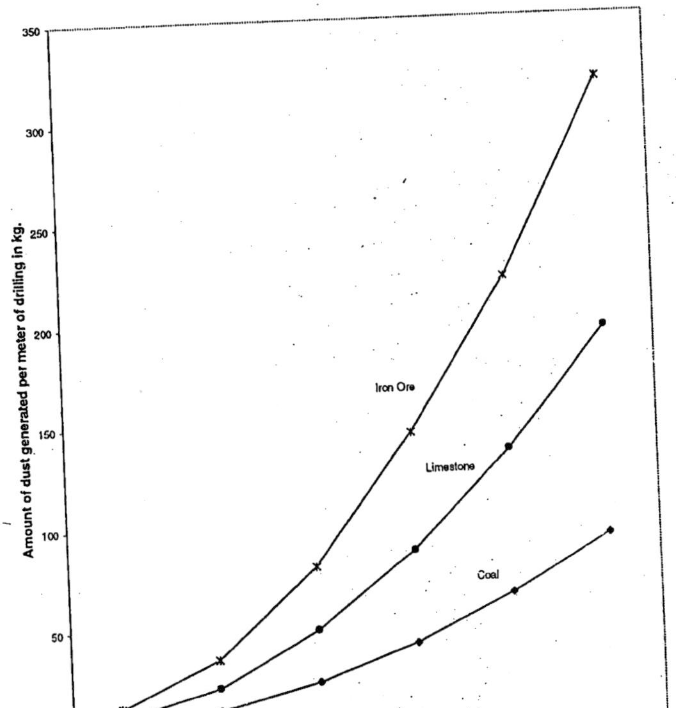

Fig. 62.2 : Quantity of total dust generated from drilling for coal, limestone and iron ore for different drill hole diamster

(i) Dust Control During Drilling

Drilling is an essential process in mining. It produces largest quantity of respirable dust per unit weight during the shortest time. On an average, total dust generated per meter of drilling for 45 mm diameter in coal is 1.87 kg (Chaira et al, 1983). But this increases with increase in drill diameter and hardness of rock (Fig. 62.2). 1-2 % of this dust lies in respirable range. Study shows that a 43 mm diameter electric rotary drilling generates up to 28 grams of respirable dust per meter of drilling whereas, larger diameter drill of 160 mm and 250 mm diameters produce 680 gms and 1460 gms of respirable dust per meter of drilling (Nair et al, 1999). Dilution cost of the dust to health compliance level of 2 to 3 mg/$m^3$by ventilation will be very high. Dust suppression by water spraying is not very effective because of very poor water wettability of coal. In confined places of underground mines it also contribute to humidity problem. CMRI has developed dust extractors (Nair and Sinha, 1984; Sinha and Nair,1982) for dust control in all the four major categories of drilling used in mines viz.

- Electric rotary drilling for shot hole in u/g mines.
- Jack hammer drilling for secondary blasting.
- Electric rotary drilling in mine roof for roof bolting,
- Large diameter deep hole drilling in open cast mines for bench blastng.

These dust collects are portable , easy to use and do not use any source of energy for their operation, and yet collects almost total respirable dust (except in electric rotary drilling in mine roof) and keeps atmosphere clean (Table 62.2). Fig. 62.4 shows the performance of dust arrester with time for two cases of 150 mm drill master drilling machine in iron ore mine. The average dust collection for both cases is 99.8% and 98.1% respectively.

Table 62.2 : Performance of CMRI dust collector for different drillling

Type of drilling	*Dust collection effeciency by Conventional method*	*Dust collection effeciency using CMRI dust collectors*
1. Electric rotary drilling for shot holes	40%	95-100%
2. Electric rotary drilling in mine roof	-	85-90 %
3. Jack hammer drilling	40-60 %	85-90%
4. Large diameter deep hole drilling	20-40 %	95-100%

(ii) Dust Control in Coal Handling Plants

Coal handling plants (CHP) account for a significant amount of total dust produced in mining and allied process. Main dust generating operation in CHP are unloading at bunker, crushing, screening and multistage conveyor transfer points. Conventionally, dust control at these points involves either water spraying or dry dust collection by hood and suction method. Water is poor dust suppresant due to its poor coal wettability. Dry dust collection using hood and suction system is costly and less efficient. Moreover, it throws away dust in to the atmosphere itself. CMRI has recently proposed (Nair and Pandey,1999) a new canopy curtain method for absolute dust control at these dust generating points in CHPs by assimilating the advantageous features of both dry and wet system. The method utilises, energy of the material flow system and atmospheric disturbance for controlling air borne dust. Hence, running cost of the system is negligible, yet it gives almost total dust control in respirable range.

(iii) Dust Control on haul Road

Dust control on haul road is effected with water, oil emulsions or by encrustation. None of this method has been found to be very useful, having some major disadvantage. A recent wet encrustation process introduced by CMRI (Nair and Singh, 1990) has remarkable potential to dust suppression and it is cost effective too. It takes advantage of properties of a super absorbent chemical for dust consolidation on mine haul roads which is very effective and economic. In a case study reported by Pandey, Nair and Gupta (1999), application of the chemical for the dust suppression on a three km. dusty haul road provided an overall

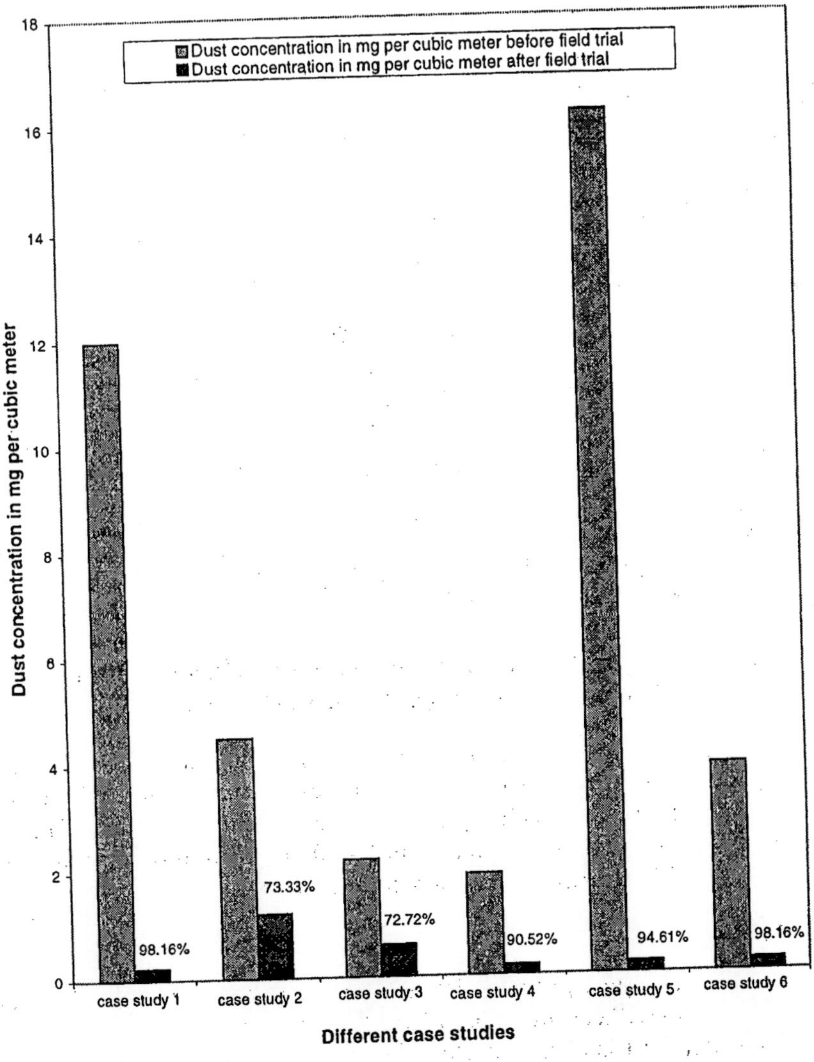

Fig. 62.3 : Performance of haul road dust consolidation using CMRI technique

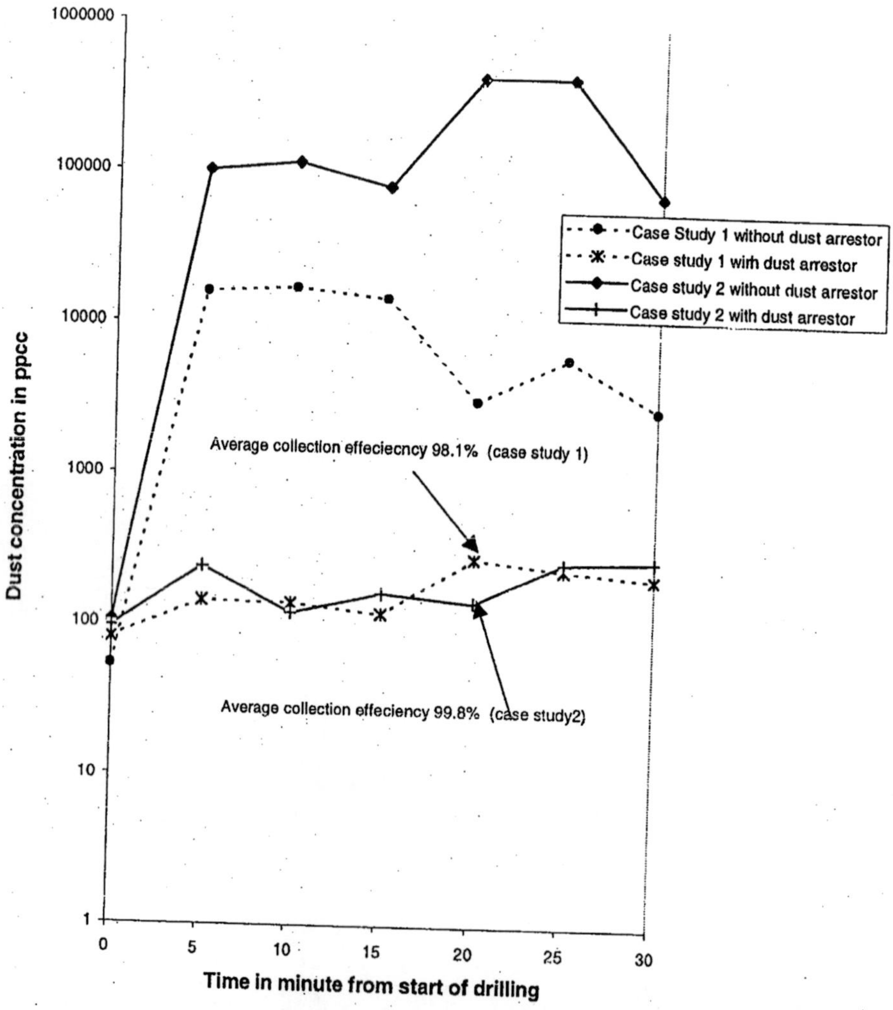

Fig. 62.4 : Air borne respirable dust concentration measured during drilling in iron ore with and without dust arrestor using 150 mm drill master drilling machine

cost saving of Rs. 93,000 per months as compared water spraying. The effectiveness of the method is being depicted in the Fig. 62.3 for six case studies conducted by CMRI. It can be observed from the figure that CMRI technique of dust consolidation is more effective in very dusty haul roads. Its reduced performance in less dusty or bouldered road can be improved effectively by adding additional dust before treatment.

FUTURE SCENARIO

Coal is very critical energy input for a developing country like India. Its dominance on national energy scenario will continue for at least few decades. Time interval required to double the present demand is decreasing day by day and now it is not more than 8 years. Only a technological break through can provide answer to widening gap of demand

and supply. Even today, coal demand is increasing at the rate of 4.28 mt/year whereas we are able to maintain coal production growth rate as 3.42 mt/year. An unfortunate side effect of this solution is its potential for dust generation. Some of the mining machines are known to produce more dust in a minute than what three men can do in a day. Thus, dust generation will be increased not only due to increase in total coal production but also due to increase in dust generation per tone. This will result in many fold increases in contribution of air borne dust to atmosphere. Today, there may be only few places where work place dust concentration exceeds the permissible limit. The situation is changing very rapidly. Moreover, there has been rather rapid increase in coal production since nationalisation but dust control measures was always neglected. As the pneumoconiosis has a incubation period of about 5-15 year, we may have to start realising its economical aspect sooner than later. USA had to bear an annual compensation of $2billion in as early as 1981 at the rate of 0.5 million dollar per case (Sinha,1987). CMRI survey (1969) reveals that 11.63% (1969) of coal miners were suffering from coal workers pneumoconiosis while subsequent survey have shown alarming progression of the disease among workers. Indian fleet of miners is one of the largest in the world and if it wont go for timely preventive and protective measures in a comprehensive and systematic way, one can easily visualise the impact of the blow due to huge amount of compensation likely to visit our mining companies. Our metalliferrous sector is not behind; a study in Mica-pencil factory reveals up to 54.6% incidences of pneumoconiosis. There is another equally important issue hidden in it. The disease acts slowly on workers and before making totally incapable to work, it allows these workers to work in mines with continuously decreasing work efficiency and productivity for few years. Though it is not possible to calculate these cost but one can simply imagine how enormous it would be.

CONCLUSIONS

Coal production in India is taking rapid strides. Mechanisation of coal extraction technique and coal transportation has accelerated a race between production and pollution due to coal mine dust. Inhalation of this dust affects the health of worker and reduces useful period of activity in long run. Therefore, there is an urgent need to contain dust induced diseases before it takes disastrous proportion. In contrast to this, there has been comparative slackness on the part of industry in implementing various measures for dust control and hazard minimisation. And regulatory bodies too do not foresee the likely adverse impact of work place dust on workers. CMRI initiatives in developing dust control devices for drilling and haul road transportation have proved to be economical and effective.

In view of this, it is emphasised that CMRI, regulatory bodies like Directorate General of Mines Safety (DGMS), and implementing industries should look for working with greater co-operation before the situation goes out of control.

ACKNOWLEDGEMENT

Author is grateful to Director, Central Mining Research Institute, Dhanbad for his permission to present the paper. He is also thankful to Sri P.K.Nair, Sri P.K.Gupta Sri M.L.Gupta and Sri I. Sinha, Scientist CMRI for their contribution in preparation of the paper.

REFERENCES

Chaira B.B.,Bose J.M.,Nair P.K.,Das B.N.,Rashidi N.A.,Verma B.K. and Sinha J.K.,1983, Assessment of Particulate Pollution of Mine Atmosphere due to Drilling in Coal Mines, Journal of Mines Metals and Fuels, Apil 1983 pp.122-124.

CMRS,1986, Status of Research Report on Mine Environment, Central Mining Research Station, Dhanbad.

Doyle H.N.,1970, A model for evaluating coal mine dust standard, Journal of Occupational Medicine, Vol.12, pp.370-371.

Hatch,T. and Paul Gross,1964, Pulmonary Deposition and Retention of Inhaled Aerosoles, Academic Press, London, pp.3-4.

Nair P.K. and Sinha J.K ,1988, Dust, Coal Workers' Pneumoconiosis and Some Control Measures, Indian J Environmental of Environmental Protection, Vol. 8, No.4, pp.279 - 286.

Nair P.K. and Sinha J.K ,1984, Dust Extractor- A safety Device, Indian J. Environmental Protection, Vol.4,No.3, pp.128 –132.

Nair P.K. and Singh B.,1990, Haul Road Dust Consolidation in Open Cast Mines - A New Approach, Indian Journal of Environmental Protection, Vol.10, No.1, p. 62.

Nair P.K. Pandey J.K. and Gupta M.L., 1999, Control of Dust in Drillling – Some Recent CMRI Initiatives, Proc. International Symposium on Clean Coal Initiatives, New Delhi, 22.23 January, 1999, pp. 627-632.

Nair P.K. & Pandey J.K, 1999, Canopy Curtain Technique for Dust Suppression in CHPs, Discussion paper for S&T funding – SSRC, Ministry of Coal, Govt. of India.

Pandey J.K., Nair P.K. and Gupta M.L.,1999, A New Technique for Haul Road Dust Consolidation, Proc. International Symposium on Clean Coal Initiatives, New Delhi, 22.23 January, 1999, p. 403.

Rahman Qamar,1995, Asbestos: An Occupational and Environmental Carcinogen, Proc. First World Mining Environment Congress (WOMEC-95), New Delhi,11-14 December 1995, pp. 549-564.

Sinha J.K., Nair P.K, 1981, Dust Arrestor: A Novel Safety Device, Proc. All India Seminar on Air Pollution, Indore 19-21 April 1982, pp. 294-314.

Sinha J.K,1987 , Pneumoconiosis – Its Effects and Control, Invited Lecture, 27 February 1987, Indian Toxicological Research Institute, Lucknow.

CHAPTER 63

Slurry Transportation—Technology Option for Next Century : A Case Study

D.K. Ghosh & Y.G. Kale***

ABSTRACT

Mineral industry is subjected to various constraints from all the possible angles, such as financial, technical, environmental, social and marketing to name few. With the gradual improvement in the quality of living standards of mankind, the demand for mineral commodities are expected to grow. On the other hand the creamy layers of our mineral resources are depleting fast and therefore, the net effect of all these factors results in increased cost of production. Slurry transportation is one of the options to reduce the cost of production in general and material handling cost in particular. Although, the concept of slurry transportation is not new, its application in Indian mining industry so far, has been to a limited extent. One such application is successfully implemented at Kudremukh iron ore mine in South India where iron ore concentrate is being transported over 67 km long slurry pipe line. Economics of this project have been radically altered by the use of slurry transportation, which were otherwise not favourable for such a low-grade ore. The present paper deals with the concept of slurry transportation with illustrations of Kudremukh case study.

1.0 INTRODUCTION

Growing population and continuous improvement in the quality of living standards of mankind gave rise to the industrialisation. Mining, being the basic industry which caters to the entire requirement of raw materials, also marked rapid growth, which is evident from the increasing per capita consumption of mineral resources throughout world in general and developing countries in particular. The proportion of world production

*Ag. Chief Controller of Mines.

**Asstt. Controller of Mines , Indian Bureau of Mines, Indira Bhavan, Civil Lines, NAGPUR – 440 001

accounted by mining, which is at present 7% is likely to increase in next 20 years.[1] Therefore, the mining industry has to gear up to meet the present and future demands of raw material resources. At the same time, it has to maintain balance between the quality output and the cost of production. Material handling cost constitutes one of the major share in overall cost of mining and slurry transportation because of its unique characteristics can prove technological as well as cost effective break-through where conventional modes of transport are not feasible.

2.0 PRINCIPLES OF SLURRY TRANSPORTATION

When a certain mass of solid material in very fine form is mixed with water and the slurry is maintained at a velocity above the Limit Deposit Velocity (L. D.V.) for that material at that concentration, the whole slurry behaves and moves ahead more or less like a homogeneous fluid through the pipe line. It is important that the desired velocity of the fluid is very carefully maintained because velocity lower than the LDV will result in deposition of the solid particles whereas unnecessary high velocity will lead to undue loss of energy due to pipe friction. The relationship between fluid velocity and pressure drop due to friction of the pipe line walls is indicated in the Fig. 63.1. The curve 1-2-3-4 relates to water which is a homogeneous fluid at all velocities. When, however, fine mineral ore is mixed with this water, the curve 1-2-3-4 behaves differently except at higher velocities when the slurry is also nearly as homogenous as water (curve 1-2). At lower velocities, the slurry no longer remains homogeneous as the solid concentration increases near the pipe invert. This brings in more mineral solids in contact with the pipe wall and results in greater pressure drop due to increasing wall friction than would be expected in case of homogeneous fluid like water at the same velocity. The form of flow in this region is known as 'Saltation'.

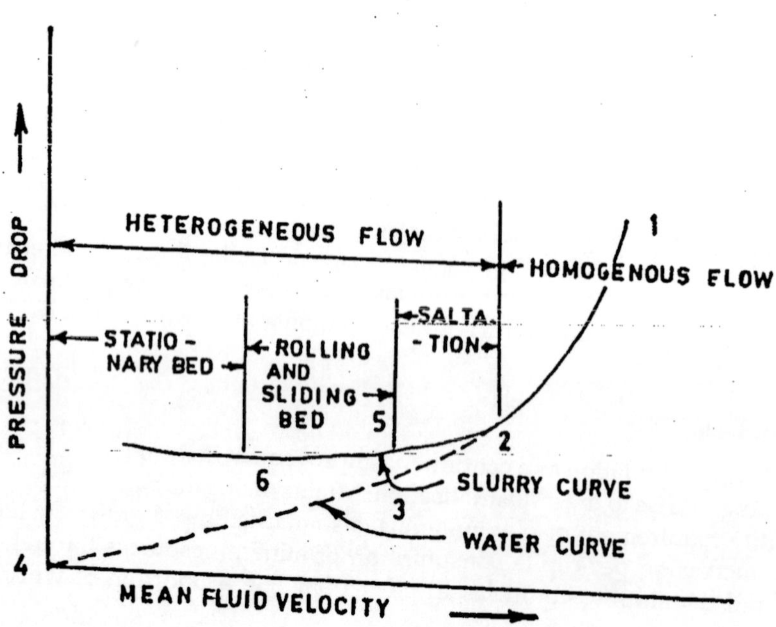

Fig. 63.1 : Relationship between fluid velocity and pressure drop

A further reduction in velocity results in some particles starting to roll and slide along the pipe invert (curve 5-6) and the rate of decrease of pressure loss with decrease in velocity is reduced at this zone. Finally, a point 6 is reached beyond which the mineral bed becomes stationary. It is, therefore, apparent that in any hydraulic transportation system, the slurry must be maintained at a velocity exceeding that corresponding to point '6' to prevent formation of a stationary bed. At the same time, the velocity should not be too high which will otherwise result in frictional loss of energy.

2.1 Merit of slurry transportation

The notable merits of slurry transportation over that of conventional system of transport can be enumerated below:

(i) It offers significant saving in the transportation cost.

(ii) It could cope with difficult terrain which is plagued by frequent landslides. Long distance transportation of mineral is possible without much infrastructure.

(iii) It is best suitable for transportation of minerals in ecologically fragile areas. No dust, fumes are produced and no hindrance to traffic and to prevailing land use pattern. Therefore, it is more eco-friendly.

(iv) Although, the initial investment is high, the operating cost is very less.

(v) It reduces the risk of accidents, extremely safe and easy for operations.

(vi) It has high degree of efficiency, reliability and round the year availability.

(vii) Direct loading of minerals into the ships in deep sea is possible.

3.0 KUDREMUKH SLURRY TRANSPORTATION

3.1 Kudremukh Iron Ore Mine

Kudremukh iron ore mine, India's largest opencast mine in metalliferous sector, is designed to produce 22.6 million tonnes of ROM ore to generate 7.5 million tonnes of iron ore concentrate per annum. It has been also proposed to expand its production capacity to 30 million tonnes of ROM ore and to 10 million tonnes of iron ore concentrate per annum. The mine is nestled high in the picturesque mountains of the Western Ghats in Karnataka. The region is subjected to heavy rainfall of 7000 mm per annum and is a source of several perennial rivers. The unique features of the Kudremukh mine are, it is the only mine in India where low grade magnetite ore is mined. It is the first long distance slurry pumping operation in India and one of the few firsts in the world. The method adopted for mining is opencast with fully mechanised means.

3.2 Material Handling

The material handling at the mine is being carried out in three stages. The first stage involved in transportation of blasted ROM ore from 14m bench face to the crushers located at different suitably placed sites with the help of 10.6 m^3 capacity shovels and 108 tonnes capacity dumpers.

In the second stage the crushed ore is transported on specially designed conveyor 1800 mm wide capable of transporting 4000 to 6000 tonnes per hour to autogenous mills in the concentrator plant.

The third stage involving hydraulic transportation of iron ore concentrate from slurry storage tank to port handling plant, comprises of 67 km long slurry pipe line.

3.3 Pipeline Design

The slurry pipeline is burried 2 meters below ground along most parts of the routes and passes through a tunnel of 1.6 km long to avoid pumping of slurry over the hill. The pipeline is burried, double coated and wrapped except for 1.6 km length in tunnel. Special cathodic protection is installed to avoid pipeline to soil polarisation. From the slurry storage tanks, a bank of five slurry pumps push the ore slurry with 65% solids into a pipe to transport the ore concentrate to the port of Mangalore. The pipe has diameter of 457 mm over initial 13 km followed 406 mm diameter over the length of the remaining 54 km. The pipeline with a wall thickness of 10.3 to 18.3 mm has been designed for the entire life of the project. The slurry pipeline has been laid aesthetically to follow the ground contour with slope not exceeding 10%. The tunnel section is 2.25 m wide and 2.25 m high with semi-circular ceiling. The tunnel rising at 1% upto half the way from east end and then falling at 7.25% towards west. The slurry is pumped from an elevation of 800 mRL to the midpoint in the tunnel at 847 mRL from where it flow down by gravity to the port of Mangalore. The slurry pipeline route map is given in Fig. 63.2.

3.4 Parameters of Slurry Flow

Slurry is stored in two receiving tanks at each end of the pipeline, each having capacity of 4500 m^3. The contents of these tanks are continuously agitated to maintain the fine particles in suspension and homogenity of the fluid. The pump house at concentrator plant consists of a series of five pumps. Slurry with 60 to 70 weight percentages of solids is drawn from the storage tank into the pump house and out along a total length of 67 kms pipeline. This bank of five pumps is supported by a stand by bank of five centrifugal pumps to cut down the idle time. Power rating for each pump is 400 KW and the maximum expected pump discharge pressure is 450 PSI. Before and after pumping slurry, the pipeline is flushed with water with the help of flush vertical pump. Planned operating hours per annum of the system are 5760 hours. The planned Limit Deposit Velocity is 2.8 meter per second at 65 percent solid. Stream rate of the slurry is 1500 tonnes per hour or 1115 m^3 per hour. Minimum and maximum iron ore concentrate size is 325 mesh and 100 mesh respectively. The slurry takes about five hours to reach the port. At the port, slurry is discharged into 2 storage tanks from where it is taken out into filtration plant for the purpose of dewatering the slurry.

3.5 Maintenance

Although, slurry pipeline is designed for the entire life of the mine, proper maintenance is necessary for the smooth flow of slurry. The following steps are being taken for the maintenance purpose.

(i) Concentrate is screened and ensured by time to time sampling to keep it free from the material coarser than 100 mesh.

(ii) Flow velocity is maintained above minimum of 1.8 m/sec by ensuring that slurry pumps are in good working order and that the variable speed drive is properly set. Velocity maximum upto 2.8 m/sec. is being controlled.

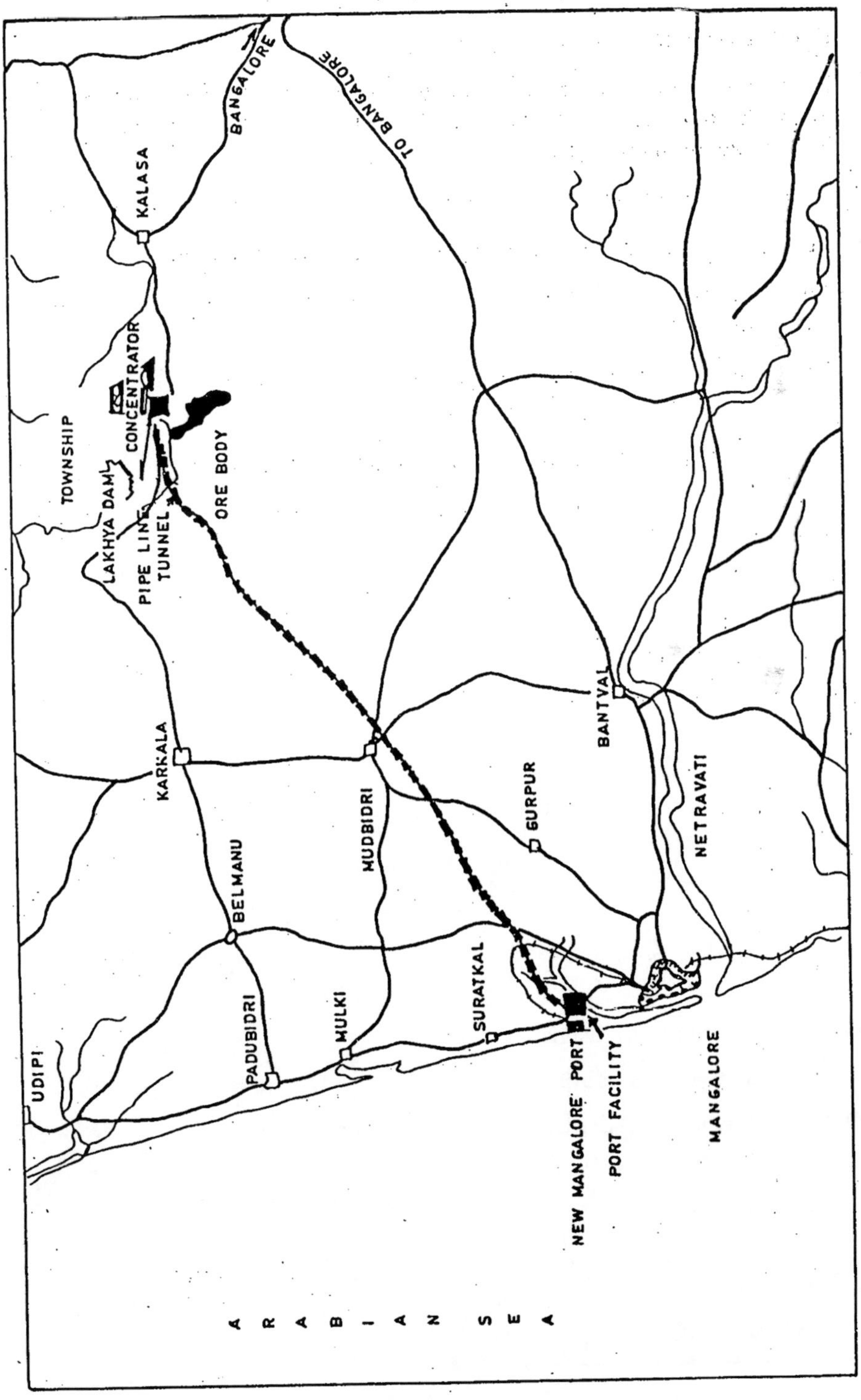

Fig. 63.2 : Slurry Pipe Line Route Map

(iii) Percentage of solids in slurry is controlled by adding water, if the slurry density is too high, the slurry is re-circulated to the concentrate thickener, if the density is too low. A density between 1.80 to 2.20 is being maintained.

(iv) Pipeline is being cleaned using pig.

(v) Internal corrosion is controlled using lime solution. Oxygen scavengers are also added to remove free oxygen. A pH of 9 to 10.5 and dissolved oxygen content of 0 to 2 mg/litre are being maintained.

(vi) For prevention of external corrosion cathodic protection system adopted.

(vii) Aerial as well as ground inspections are regularly carried out.

(viii) An emergency power (diesel generator) takes over in case of power failure during pumping.

The whole system from pumping head to dewatering head is highly automatic and the features are summarised in Table 63.1.

Table 63.1 : Features of Slurry Transportation at Kudremukh

Pipeline route length	67.0 Kms
Pipeline Diameter	457 mm for 13.0 Kms 406 mm for 54.0
Kms Concentrate Tonnage	7.5 Million Tonnes
Planned operating hours per annum	5760 hours
Feasible range of Slurry Cocentration (Wt.% solids)	60-70
Design velocity (m/sec @ 65 % solids)	1.8 Minimum 2.8 Maximum
Design Stream rate	1500 Tonnes/hour or 1115m^3/hour @ 65% solids
Iron ore Concentrate size	Min.325 mesh Max.100 mesh
Maximum expected pump discharge pressure	450 PSI
Number of Slurry pumps (Horizontal)	5 Operating
Number of Slurry pumps (Centrifugal)	5 Standby
Flush Vertical Centrifugal pump	1
Power rating for each pump	400 KW
Number of Slurry storage tanks	At Concentrator-2 At Port –2
Capacity of each tank	4500 m^3

3.6 Economic Parameters

The capital expenditure of laying and operating the slurry pipeline at Kudremukh as installed in the year 1980 was about Rs. 1.0 crore per km, which amounts to say Rs.70.0 crores for the entire system. The present operation and maintenance cost per tonne is about Rs. 15–20, which amounts to Rs.11.25 crores per annum considering the 7.5 million tonnes of concentrate per year. The equivalent approximate freight of bulk transportation of dry material per km is around Rs. 1.50 - 2.00 per tonne which amounts to about Rs. 123.75 crores per annum keeping in view of the road distance of 110 km from Kudremukh mines to Mangalore port. This means there is a net saving of Rs. 112.50 crores every year in the material handling cost. The slurry transportation process consumes only 1.75 to 2 units of electricity per tonne of dry iron ore concentrate, which is the bare minimum compared to any other form of surface transportation(2). Therefore, the economnics of this project has been radically altered by the use of slurry transportation and the proximity of the port is an added advantage for the mining of low grade iron ore (38% Fe), which would otherwise become not possible.

SCOPE FOR WIDER APPLICATIONS

The concept of slurry transportation is not new. In 1750, the Hungarian Engineer Hell invented a hydraulic column engine by means of which, salt, the first mineral in the form of brine was lifted to a cooling tower for evaporation of a part of the water in air.(3) However, the hydraulic transport pipeline of any major significance was first installed by Consolidated Coal Company of USA in 1957 for transporting 1.5 million tonnes of coal per year in slurry form over a distance of 108 miles.(4) Since then this system of transport has been gaining popularity. However, its application in Indian mining industry so far, has been to a limited extent. Since the conventional modes of transport causes environmental pollution depending on magnitude of operation and in future environmental control regulations are likely to become more and more stringent, the slurry transportation of minerals can pave option for the sustainable mineral development.

Table 63.2 shows some of the examples where slurry transportation has been successfully employed. Considering the merits of the slurry transportation, in India also its application can be used in more mining areas. If implemented successfully, it can prove a technology option for coming century. Table 3 shows some of the potential areas for slurry transportation in Indian mining fields.

Table 63.2 : Some of the examples of Slurry Transportation of Minerals

Sl.No	*Mineral*	*Length (In Km)*	*Capacity (Million tonnes /year)*	*Country*
1	2	3	4	5
1.	Iron ore	85	2 - 3	Tasmania
2.	Iron ore	6	1.0	New Zealand
3.	Iron ore	3	1.0	New Zealand
4.	Iron ore	40	1.8	Mexico
5.	Iron ore	27	1.5	Mexico

1	2	3	4	5
6.	Iron ore	32	2.0	Agentinar
7.	Iron ore	395	12.0	Brazil
8.	Iron ore	67	7.5	Kudremukh, India
9.	Sulphide	27	0.7	Bougain Valley
10.	Sulphide	111	0.3	West Uran
11.	Sulphide	61	0.9	Turkey
12.	Sulphide	18	0.4	Arizona, USA
13.	Limestone	27	0.4	Columbia
14.	Limestone	9	0.6	Columbia
15.	Limestone	4	0.1	Columbia
16.	Limestone	10	0.6	Trinidad
17.	Limestone	92	1.6	Rugby, England
18.	Limestone	27	1.8	California, USA
19.	Iron ore	225	2.3	Kama Bay, Canada
20.	Iron ore	402	4.0	Minas Gerais, Brazil
21.	Gold Slimes	9.60	0.75	Brazil
22.	Gold Slimes	10.4	0.54	Brazil
23.	Nickel & Copper	19.3	3.6	Onrario, Canada
24.	Limestone	24.3	1.5	Calaveras, USA
25.	Limestone	70.8	0.9	Australia
26.	Limestone	51.0	1.15	Washington, USA
27.	Limestone	24.0	1.8	Gladstone, Australia

Source : Ingle et al 1996, Das 1981 & Basu et al 1984.

5.0 CONCLUSIONS

Slurry transportation of minerals is an established and proven technology. In spite of this, its application has so far, remained limited. Since, slurry transportation of minerals offers various advantages over conventional modes of transport, specially in cost minimisation and environment protection, it can be a technology option for transportation of minerals. Kudremukh slurry transportation is an established and successful example in India. Therefore, it is necessary to explore the possibility of the transportation of minerals through slurry in the existing mines as well as in new upcoming projects in India.

ACKNOWLEDGEMENT

The authors would like to thank Shri A. N. Bose, Controller General, Indian Bureau of Mines for his kind permission to present this paper. The authors are also thankful to

Table 63.3 : Potential areas for Slurry Transportation in Indian Mining Industry

Sl.No.	Mineral	Location From	Location To	Pipe length (in km)	Pipe diameter (in cm)	Pipeline Capacity (in million tonnes/year)	Company
1	Iron ore	Bailadila	Vizag	277	40.6/50.8	5/10	NMDC
2	Iron ore	Bababudan	Mangalore	140	45.72	7.5	NMDC
3	Bauxite	Dhangarwadi	Ratnagiri	73	20.32	0.3/0.6	BALCO
4	Rock Phosphate	Durmala	Harrawala	40	10.16/15.24	0.3/0.6	PPCL
5	Rock Phosphate	Jhamarkotra	Umra Rly. Station	12	15.24	0.6	RSMML
6	Rock Phosphate	Maton	Debari Zinc Smelter	10	10.16	0.1	HZL
7	Sand	Damodar River	Jharia Coal Fields	40	20.32/30.48	1.0/3.0	BCCL
8	Limestone	Choli	Kandroori	78	25/20	2.0	CCIL

Source : Das 1981 & Basu et al 1984.

M/s. Kudremukh Iron Ore Co. Ltd. for providing necessary technical details about the project. The views expressed in this paper are those of authors own and not necessarily of Indian Bureau of Mines.

REFERENCES

1. Information brochure on MINETIME 99 : Challenges and Opportunities for the 21st Century. Journal of Mines, Metal and Fuels, Aug – Sept. 1998, p. 281.
2. News letter of Science and Technology Advisory Committees (STACs), Vol. 4, No. 1 Jan. 1996, p. 4.
3. Hydraulic transport of solid materials – A Technical alternative. Engineering and Product News. The Indian Mining and Engineering Journal. July 1981 p. 26.
4. J. Das, 1981, Hydraulic transportation of minerals. The Indian Mining and Engineering Journal. July 1981, p. 5.
5. Ingle K. G., Makode M. N. & Raju K. S. 1996, Mineral Beneficiation : A tool to overcome certain constraints of mineral/mining industry. Seminar on constraints of mineral/mining industry in Central India and its future outlook, Nagpur. 27.2.1996 Technical papers volume, p. 33.
6. Basu S. R. & Saxena A. K. 1984, Techno-economic feasibility of transporting limestone in slurry form to feed a cement plant at the foothills of the Himalayas. Bulk Solids Handling Vol.4, No. 3, Sept. 1984. Trans Tech. Publications, p. 584, 587.

CHAPTER 64

Studies on Ground Vibration Due to Eco-Friendly Blasting Operations—A Case Study

V.P. Chatterjee, B.S. Roy* & S.J. Raina**

ABSTRACT

The blasting operations in open cast mines generates certain adverse effects namely structural damage due to vibrations, problems of safety due to flyrock and damage and annoyance potential of air blast and noise. However, on the brighter side, the process enables securing proper size of fragmented rock conducive to optimal utilization of commercial explosives. Ground vibration studies were conducted by NCB in one of the limestone mines in western part of the country keeping in view the above problems. The main objective of the study was to protect the nearby Jamwanti limestone cave of pre-historic times, a PWD road and the scanty habitation in the close proximity of the mine. The present paper mainly deals with the determination of safe charge and safe distances of the blastings, blast geometry and configuration, powder factor and the maximum charge per delay to be used for ecofriendly blasting operations. The peak particle velocity (ppv) measured during blasting operations ranges from 0.03 to 4.11 mm/sec and the air blast recorded ranges from 98 to 125 dB. The results obtained are well within the safe permissible limits as for the pre-historic Jamwanti limestone cave is concerned. This could be achieved because optimised blasting parameters suggested by the NCB team were adopted to conduct the blasting operations.

1.0 INTRODUCTION

The blasting operations in open cast mines generates certain adverse effects namely structural damage due to vibrations, problems of safety due to flyrock, and damage and potential of air blast and noise. However, on the brighter side, the process enables securing

* National Council for Cement and Building Materials, New Delhi

proper size of fragmented rock conducive to optimal utilization of commercial explosives. Environmental problems due to mine blasting revolve around four important aspects :

1. Characteristics of blast – induced ground vibrations in the vicinity of various structures and chances of their damage;
2. Impact of flyrock ejected from the blast area;
3. Human problems associated with the effects of ground vibrations and air blast over pressure; and
4. Damage potential of airblast.

NCB had taken up the ground vibration studies due to blasting operations in a captive limestone mine of Gujarat in respect of :

1. Control ground vibration with a view to protect the Jamwanti cave and temple of historical importance and a PWD road nearby the mine,
2. Estimate noise level and suggest measures to keep it below permissible limit,
3. Suggest measure to control flyrock, and
4. Suggest measures to improve fragmentation where oversize boulders are formed.

2.0 BLASTING PARAMETERS

The basic functions of explosives in the geomining operations are :

— To displace the limestone and marl (wherever required) from the bed rock strata.

— Fragmentation of limestone to suitable size that can be handled and loaded into dumpers by hydraulic excavators/loaders.

To achieve these objectives, the blasting parameters are so designed that the burden, spacing, angle of holes etc. with corresponding depth of holes would result in the required output and maximum economy.

2.1 Techno-economy and energy conservation of explosive

In the present study the mining authorities of the mine were suggested to drill the blast holes at an angle of 8°-10° from vertical in order to reduce adverse impact of flyrocks on the immediate surroundings and to achieve maximum efficiency of explosive used.

Explosive

The common type of explosive and accessories readily available and used in this limestone mine are :

— Ammonium Nitrate (ANFO) – Diesel Oil Mixture

— Powder Gel C

— D-Cord-II

— Cord relay

— Excel Shock tube (200 mm)

— Excel Connector (42m)

— Ordinary detonator

— Blasting safety fuse

ANFO being a low density explosive and cheaper in cost is largely used. Slurry explosive is also used to a limited extent in more dense and harder limestone of top bench.

3.0 GROUND VIBRATION

The limestone mine under study has a total area of 643 hectares under mining lease. It produces about 10 lakh tonnes limestone per annum to manufacture 1.1 million tonnes of cement clinker per annum.

3.1 Drill and Blast Methods

The drill machine is of DTH type and can drill 150 mm diameter shot holes. D-cord-II, Cord-Relay, Excel shock tube, excel detonator and blasting safety fuse are used. Delay interval are used for sequential firing of the shot holes.

3.2 Site conditions vis-à-vis Ground Vibrations

The thickness of the deposit is about 12 to 30 meters. The mine has only one pit and there are two main working benches, third and fourth benches are partially developed and used. The cement plant is located almost on the east of the mine. Structure in the plant are either made of RCC or steel fabrications and are of extremely good quality. Jamwanti limesteone cave is also located almost on the east of the mine, which is more than 500m away from the mining lease area. On all other sides, small agricultural fields are situated. Temporary settlement of stone houses are present in the northern side of the mine and a PWD road passes about 500 m away from the mining lease area.

3.3 Monitoring of Ground Vibrations

Eight test blasts were monitored using Blast mate Model-DS-477 manufacture by Instantel Inc., Canada for the present study. Locations showing the instrument stations and blasting sites with distances are shown in Figure 64.1. Data of each blast was stored in the inbuilt micoprocessor of the seismograph. Further analyses of data was done on the computer for establishing relations between peak particle velocity and scaled distance factor. Blast wise description including RL, bench number and exact position of instrument is given in Table 64.1. Log-Log graph is plotted between scaled distance factor and peak particle velocity is shown in Figure 64.2.

The equation showing relationship between peak particle velocity (ppv) and scaled distance factor (SD) is given below :

$$ppv = 2339.7\ (SD)^{-2.361} \qquad \ldots\ \ldots (1)$$

Considering the location of Jamwanti limestone cave and PWD road in and around the ML area, number of holes per blast, charge per delay, hole depths and blast configuration maximum value of 12.5 mm/sec for ppv is considered as a safe value to calculate the maximum safe values of charge per delay for various scaled distances. Details of recorded ppv, distance between the blast point and the instrument, charge/delay and scaled distance are given in Table 64.2.

Table 64.1 : Details of Blasting Measurements

Sl. No.	*Location of Blasting Block*	*Instrument Location*	*Distance (in metres)*
1.	Bench-II RL-33	On the same bench and level below the first bench	250
2.	Bench-I RL-41	50 m away from Jamwanti cave	800
3.	Bench-I RL-41	Towards Jamwanti cave	850
4.	Bench-II RL-33	On the same bench and level towards Jamwanti cave	300
5.	Bench-I RL-41	Private Agricultural Land towards Tata Chemical Road	400
6.	Bench-I RL-41	25m away from Jamwanti cave	300
7.	Bench-II RL-33	On the same bench and same level	250
8.	Bench-II RL-33	On the same bench and same level	250

Table 64.2 : Results of Blasting Operations

Sl.No.	*Trial Blast No.*	*Distance (m)*	*Charge/ delay (kg)*	*Scaled Distance (m)*	*ppv mm/sec*	*Bench No. & RL*	*Noise Level (in dB)*
1.	I	250	110.41	29.1	1.21	II 33	120
2.	II	800	143.18	66.9	1.16	I 41	119
3.	III	850	117.76	68.3	2.13	I 41	123
4.	IV	300	73.6	34.2	0.25	II 33	114
5.	V	400	125	35.8	3.62	I 41	125
6.	VI	300	95.2	25.8	4.11	I 41	112
7.	VII	250	54.16	28.5	0.25	II 33	115
8.	VIII	250	45.45	29.4	0.03	II 33	98

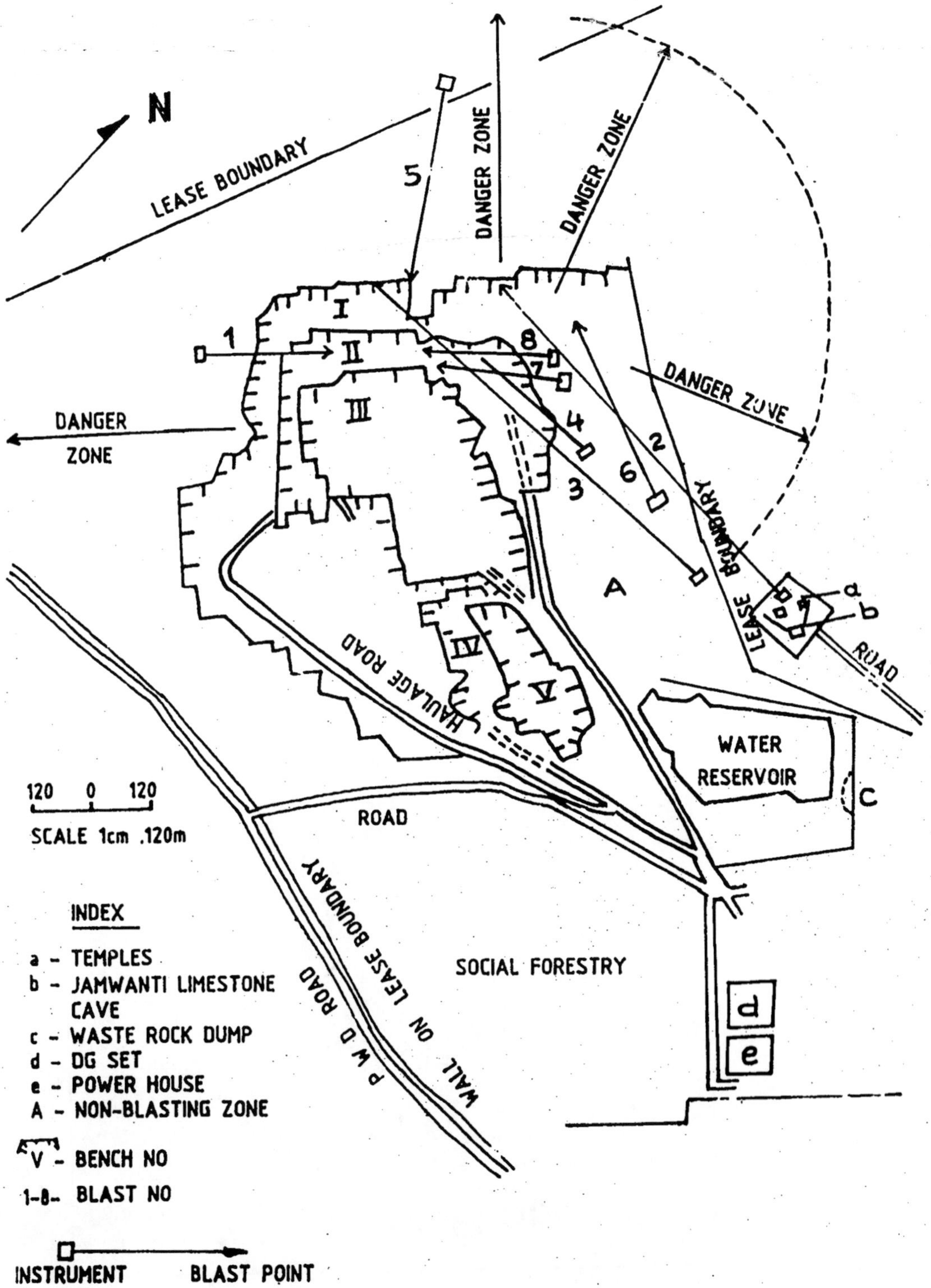

Fig. 64.1 : Surface plan of the mine showing blasting stations

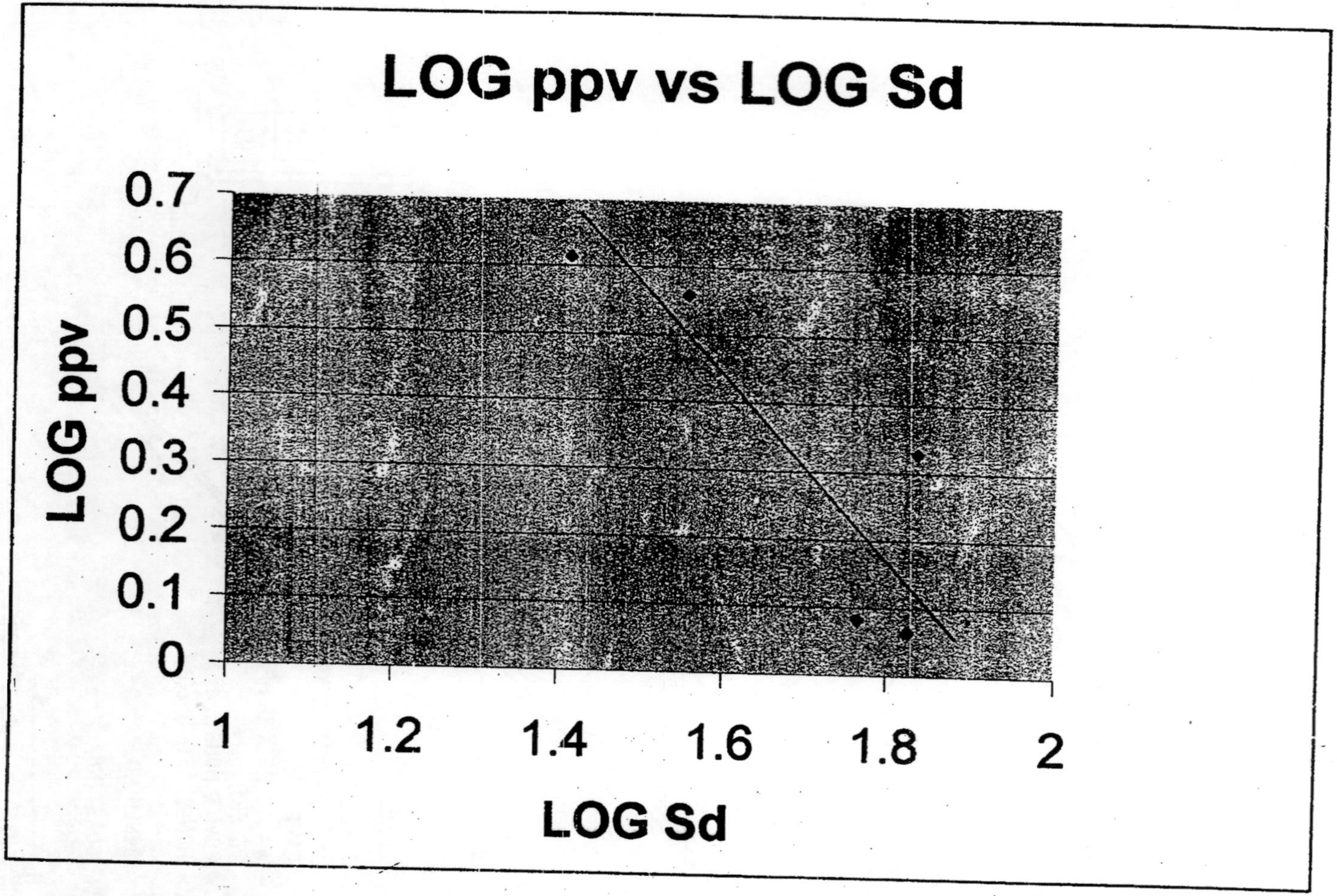

Fig. 64.2 : Log-Log Graph showing relationship between scaled distance (Sd) Factor and Particle Velocity (ppv)

3.3.1 *Discussions on results*

The powder factor for the eight trial blasts conducted during the study varies from 2.5 to 3.76 T/kg. The ppv values ranges from 0.03 to 4.11 mm/sec and the air blast values varies from 98 to 125 dB. Fly rock was restricted to a radius of about 8 to 10 meters towards open face. All these values could be arrived at because optimised blast designs suggested by NCB were followed.

3.3.2 *Assessment*

Bench-I limestone are comparatively hard and compact with maximum 4 vertical and 3 horizontal joint sets. Most of the joint sets are filled either with clay or black cotton soil. Bench-II limestone are soft, biscuity, thinly laminated and porous. This bench limestone occasionally contains laterite filled cracks. In one of the blasts 3-4 meters back bench cracks were observed. Due to presence of joint sets few blasts resulted into poor fragmentation (8-10% boulders). This even brought down the powder factor. However, most of the blasts were effective and generated homogeneous fragmentation. Moreover, for improved fragmentation of the compact and strong first bench limestone, drilling of satellite holes of smaller diameter in between the main holes were recommended.

4.0 RECOMMENDATIONS

1. The peak particle velocity (ppv) measured on I and II benches during blasting operations ranges from 0.03 to 4.11 suggests that the ground vibration is well within the permissible limits. (12.5 mm/sec. for houses in poor repair and 5.0 mm/sec. for objects of historical importance and sensitive structures, CMRI standard).

2. The air blast recorded ranges from 98 to 125 dB are also safe. There are no settlement of human population surrounding the mining lease area or intense agricultural activity is carried. Hence the noise level is also within the permissible limits. (128-136dB, DGMS, 1974).

3. Maximum charge per delay interval used for blasting operations is 143 kg. It was recommended that the maximum charge per delay could safely be used up to 160 kg. in this captive limestone mine. Jamwanti Limestone Cave is the only historical structure present in the vicinity of the mine, which will not be affected by such low intensity of blasting.

4. In general back deformation on the same stratum of the blasting source is almost nil in both the benches. However, in one instance the cracks developed due to wave propagation during blasting on I-bench was traced up to 4 meters from the last row of the holes. These fractures were developed due to presence of weak geological structure within the limestone. Proper site condition investigations are suggested prior to take up the drilling operations.

5. Fly rock is restricted to about 8 to 10 meters within the active mining area and towards open face site. hence there is no danger due to this. However, proper precautions should be adopted if danger arises due to this cause at any time in future.

6. Oversize boulders are produced only on the I-bench. In case of II-bench blastings, the fragmentation is homogeneous. Even the lower portion limestone of I-bench are quite porous and soft, which generates uniform fragmentation. In one of the blasts, the oversize boulders were 8 to 10 per cent. However, the same size boulders are

directly fed to the crusher without any further fragmentation. The generation of boulders can be further controlled by putting satellite holes in between the main holes depending on the site conditions. But in case of II and III benches, no such methods have to be adopted.

5.0 CONCLUSION

Eco-friendly optimised blastings should be conducted in this limestone mine to protect the historically important Jamwanti limestone cave. Prior to blasting operations pre-requisite ground condition investigations are always better. Based on that blast design should be made with proper blast geometry and configuration to achieve higher powder factor, better fragmentation and less ground vibration. Limestone of I-bench is hard and compact, hence to conduct blasting on this bench stemming length, bench spacing, sub drilling should be up to the required level. Based on ppv (0.03 – 4.11 mm/sec) and air blast (98-125 dB) results obtained in this study it is recommended that maximum charge per delay should be restricted to 200 kg.

ACKNOWLEDGEMENT

The authors have freely drawn upon completed sponsored project report of NCB. This paper is being published with the permission of the Director General of NCB.

CHAPTER 65

Quality Evaluation of Mining & Engineering Components by Non-Destructive Evaluation

V.N. Pathak, M.K. Singh* & S.P. Choudhary**

Non-destructive evaluation is a basic engineering method to ensure safety, quality, reliability and durability with destruction of the component. It is an important tool which can prevent disaster in the industry. By definition, NDE differs from all the other test which damages the serviceability of the tested components. By NDE techniques the material condition could be known to the designer so that higher factor could be maintained without the components become heavier is desirable now-a-days.

Secondly, to reduce the manufacturing cost, NDT is required for row material of the first stage because the finished product of defective ingot may not be free from flows. Naturally, the finished components will be rejected which will effect the cost of the components. This technique provide an assurance about reliability and durability of the components. In brief the utility of NDT technique in engineering field or material science are as follows:

(i) To aid better design;

(ii) To ensure product reliability and durability;

(iii) To maintain a uniform quality level;

(iv) To control manufacturing cost;

(v) To prevent accident and save human life.

Mining industry is one of the vital sector where NDT is extensively utilised right from the row material selection to the finish product to assure product reliability and further for service inspection for operational reliability. The application of Non-destructive

*Scientists, CMRI, Dhanbad.

evaluation in mining industry has an important role for the assurance of safety and economical installation as because of system has been introduced in shaft mines for safe and easy transportation of man and material through cages. Between the cage and wire rope some critical and vulnerable components like friction wedge type rope cappel, safety hook, distribution plate and chains are incorporated called safety components or parts of cage suspension gear. Since all the safety components have been manufactured by small scale industries, which undergo various metallurgical and mechanical process. Therefore, it is imperative to examine the components carefully by non-destructive techniques before putting the items into the service for maintaining the safety and reliability during the service period.

The purpose of NDE is also to determine whether a material or a parts will satisfactorily performs its intended function upto optimum period. To be more sure about the safety and reliability the parts like winder engine drum shafts, tie rod, head gear pulley shaft, ventilation fan shaft are also essential to evaluate by NDE for their quality and mechanical integrity before and after these are installed into the service. These components are being evaluated periodically by non-destruction evaluation techniques to avoid the premature failure.

To assess the quality and soundness of the components by NDE, the following techniques are used as follows in CMRI:

(i) Visual examination;

(ii) Magnetic particle crack detection;

(iii) Ultrasonic flaw detection;

(iv) Dye penetrant technique.

(i) Visual Inspection

After applying the proof load (three times of safe working load) on the part of cage suspension gear, they are dismantled and cleaned to remove the dirt, grease and scaling from the surface of the component. The parts are examined visually with magnifying glass to reveal surface imperfection, appreciable deformation or loss of material in the body, thermal crack and the surface condition of the component. The harmful surface imperfection and appreciable indentation is recorded to maintained in the report. Moreover, the visual inspection is not sufficient to detect fine flaws because ensitivity of inspection depends upon the size and width of the defect and degree of light reflection from the flow as well as surface smoothness, so other techniques are also required.

(ii) Magnetic particle crack detection

Since the mining components are of ferrous material of ferromagnetic nature, the magnetic particle crack detection is the best method to reveal the fine surface and subsurface flaws. When the ferrous material is kept in magnetic field, it is magnetised. The magnetic line of force starts to flow in proper direction through the material. The magnetic permeability decreases and creating two mini opposites poles where surface or subsurface discontinuities exist and causing leakage of magnetic flux. In this condition, when any magnetic particle (ferroux oxide) will be spread over the surface will be attracted by leakage of flux and forming the indication of defects.

The formation and appearance of magnetic particle depends upon the direction, strength of the magnetic field, method of magnetisation, characteristics of the magnetic particle, size, shape and orientation of the discontinuities. The appearance of magnetic particle is maximum when the plan of the magnetic flux and defects are perpendicular to each other but it is minimum when it is parallel.

Method of Magnetisation

(1) Longitudinal magnetisation

It is performed to magnetised the component in its axial direction when the material is kept between the two poles of permanent magnet or the pieces coiled.

With a conductor mainly with electrical cable called soil magnetisation. The magnetic field is produced surrounding the test piece after passing the current through conducter. This magnetic field magnetised the test piece longitudinally to the extent depending upon the amount of the current passing through the coil and the number of turns in the coil. The effect of this magnetic field so produced remain effective upto distance of 15 cm to 20cm from the either side of the turn in the coil. The minimum diameter of the test piece should be 1/6th and 1/8th of the diameter of the coil. The cross sectional area of the part to be magnetised is greater then 1/10th of the cross sectional area of the coil. The amount of the current required to test a specimen should be calculated from the given relation :

$$\text{Amp turns} = \frac{45000}{L/D}$$

L = Length of the specimen; D = Dia of the sample.

L/D should not be less than 2 and more than 15. Number of turn is known by the supplier of the instrument.

(2) Circular magnetisation may be produced by passing the current through the specimen and secondly by passing the current through the conductor which is inserted into the hollow section of a test piece. The amount of current required in this technique should be in the range of 250 Amp to 400 Amp per centimeter of the diameter or thickness of the sample. Circular magnetic field may be produced by means of a prod also. In this method, two prods are placed at the two points on a test sample. There by passing the current through the sample, the circular magnetic field is produced whose direction can be determined using right hand rule.

The effective distance between in two prods is 15 to 20 cm.

The defect indication is governed by a magnetic developer which is deposited during the magnetic field is produced. The magnetic developer or fluid are of three types :

(i) Dry magnetic particle;

(ii) Wet magnetic particle; and

(iii) Fluorescent magnetic particle.

The second and third are prepared by mixing the particle should be 2 to 10 micron and the concentration should be 2% by volume. The fluorescent magnetic particle are coated with a phosphor. This is a type of material, gives off light when exposed to ultraviolet light (black light). Here, the black light is absorbed by the fluorescent material and re-emitted as light of longer wave length in the visible range. The advantage of the material is a brilliant glow under black light and it is faster more reliable and sensitive to the very fine crack.

Properties of the magnetic particle

1. It should have soft iron which has maximum permeability, minimum coercively and low retentivity.
2. It should have proper size (1/8 of the micron) and shape (sphericalor elongated).
3. Viscosity of the fluid should be suitable for magnetic particles.
4. Gravitational attraction should be minimum.
5. Mobility of the particle should be free to the medium.
6. The colour of the particle and fluid should be quite satisfactory to give a better contrast.
7. Setting time of the particle should be 30 minutes for the separation in two layer. The dry magnetic particle are more sensitive to reveal subsurface defects, wet magnetic particles are suitable for fine cracks and fluorescent particle is the most suitable for hair crack.

The accumulation or deposition of magnetic particle depends upon.

(i) The amount of flux leakage at the place of the defect on the specimen.

(ii) Proper strength and concentration of the magnetic fluid.

(iii) The proper setting of the component in the magnetic field.

(iv) The proper time of magnetisation.

(v) The surrounding condition.

Acceptance

In no any case transverse surface imperfection or crack is allowed in the components of any size. Only longitudinal surface imperfection of certain length in stress and non stress zone is allowed. Indian standards (IS: 3703-1980) and NCB norms being considered for the acceptance and the rejection of the components.

Limitation

(i) This is only for ferrous material

(ii) The test does not reveal the defect which is below the 5mm depth.

(iii) Ultrasonic Flow Detection

The ultrasonic flow detection technique is used to reveal the location, orientation and shape and size of internal flaws like forging cracks, non-metallic inclusion, cavity, porosity, flakes and others which might have developed during the manufacturing. The results of ultrasonic flaw detection depends upon the condition and smoothness of the surface, acoustic couping used to remove the air gap between the probe and specimen, geometry of the components and the characteristics of the transducer and constriction of the probe used for scanning. The surface roughness should not be less than 0.1mm because polished flat surface are often awkward to test. The probe stick due to suction and therefore doesn't side along easily. But due to very rough surface the ultrasonic beam scattered in all directions. In this case the surface roughness should not exceed 6.3 micro meter.

To ensure adequate transmission of ultrasonic energy a suitable couplant having good wetting property is used. Thick oil, grease, cellulose paste are used as a coupling medium. For forged components the selection of frequency and diameter of the probe should be 2-5 MHz and 20-25 mm respectively. Smaller diameter probe may be used for small diameter forging components. Selection of testing procedure is made after giving due consideration of manufacturing process and shape and size of the objects. But in all cases maximum possible area should be scanned.

Acceptance

In any case rectification is not allowed by welding in the forge components. Forgings are classified in three categories depending on acceptable flaw size. In Category 1 the flaw size should not be greater than 2 mm equivalent size. The machinery and its accessories where human life is evolved come in the 1st category. In second the flaw size should not be more than 5mm equivalent size. And in this group medium stressed components of various equipment come. In the third category the flaw size is allowed to 10mm this category covers general forged components like line shaft, floating shaft of cranes and various other machines parts. Crack, flake, seam and laps are not acceptable of all categories of forging. In any area where one or more discontinuities produces a continuous loss of back reflection is unacceptable. In stress zone area where 40-50% loss of back echo produces due to flaws is unacceptable. If the flaw area is greater than the half of the plate thickness is not acceptable.

Indian Standard, British Standard, ASTM and DIN are also looked into for the acceptance and rejection of materials.

The authors are grateful to the Director CMRI, Dhanbad for his kind permission to present the paper in the International Conference.

CHAPTER 66

Protection of Land-Environment in Coal Mining Areas in India in 21st Century

Abha Rani & Rekha Ghosh***

ABSTRACT

It is a known fact that coal mining damages land-use and environment of the concerned area. The matter had been assessed over thirty coal mining areas in India.

It was noted that forest, one of the important components of environment has been depleted seriously. Damages to forest invites land degradation and hence alters land-use pattern over the region.

Realisation for protection of environment was generated more than twenty-seven years ago and discussed in the Stockholm conference on 5th June 1972. Since then, attempts are going on to protect different aspects of environment. While on 5th June 1999 the Honourable Prime Minister of India has stressed on the requirement of 33% forest cover over the country, India is far behind the target. Though huge plantation activities are going on even then it was observed by the present study that these areas are gradually getting the forest cover depleted.

The above fact indicates that there must be some unidentified lacuna which is not allowing the desired increase of forest-cover. It is realised that some natural factors must have some control over the change in land-use pattern being caused. Some of these may be geological and demographic characteristics of the area, climate etc. Further, as mining changes land-use pattern in an area while planning for developing any new (post-mining) land-use through an appropriate land-use plan it is imperative that the need of the growing population should be supported by the newly developed land-use plan and should be such that the land (with all its geological and other earth-science characteristics) should be capable to support it.

*Department of Applied Mathematics, Indian School of Mines, Dhanbad, 826004, India.

**Centre of Mining Environment, Indian School of Mines, Dhanbad, 826004, India.

In the above background this paper attempts to analyse statistically how the above referred natural factors are reflecting over development of a post-mining land-use pattern in such areas. This information will help in planning for development of land-use pattern in coal mining areas of India in 21st Century.

PROLOGUE

In the present century India has seen development in many aspects and it is expected that the 21st century will put the country more towards the climax. While development includes industrial development, so also the development of mining industry, the fact is, any interaction with land damages land; it is more so for the very important developmental activity, mining.

Land forms an important component of environment, and the need for environmental protection does not require any explanation in today's world situation. This land is a finite natural resource to be used by the human beings for almost all purposes. To meet this, in the present days world situation, the per-capita land availability is even less than 0.3 hectare. The situation is more servere in India as this country holds 16% of global population in only about 2.3% of usable land area of the globe.

Mining forms the stepping stone for development of those regions which hold mineable wealths and hence is an unavoidable very important developmental activity. This results damage to land, a very scarce and important natural resource, in the name of development, which is not acceptable under any condition.

Need of environmental protection was realised in the second half of 20th century, as was officially recorded in the Stockholm conference on 5th June, 1972, the first "World Environment Day", but even at this fag end of 20th century the situation has not improved much. It has been a matter of concern because if the activities run uncontrolled as in the present century, the 21st century my encounter with an unhabitable earth. Hence there is a very serious requirement to have a compromise between damage to land environment (i.e. land and land-use pattern) by mining and protection of land.

In this background the research was planned to have an attempt to generate a strategy for protection of land-environment in coal mining areas in India in 21st century.

ANALYSIS OF THE PROBLEM

Impact of mining on land-environment gets reflected in land-use pattern of the respective area because the more the land gets exposed to erosion by loosing its green cover or by getting disturbed otherwise due to mining (excavation, overburden dumping etc.) and related activities, its water resources get damaged, the more the damages go on proceeding in accelerated rates and the cumulative effects push the land towards degradation. The process works through a cycle termed as land deterioration cycle (Fig. 66.1), (Ghosh & Ghosh, 1993).

The extent of damage to land-use pattern or water resource by mining has been analysed in parts of many coalfields in India as well as in some metal mining areas, namely Raniganj, Singrauli, Jharia and Makum coalfields (Chatterjee & Ghosh 1994, Sekhar 1996, Ghosh et. al 1997 and Dutta 1997) and also in Mosaboni (Ghosh & Das 1998). In each of the cases damage has been noted to be reflected on decrease in green cover or water resource or both.

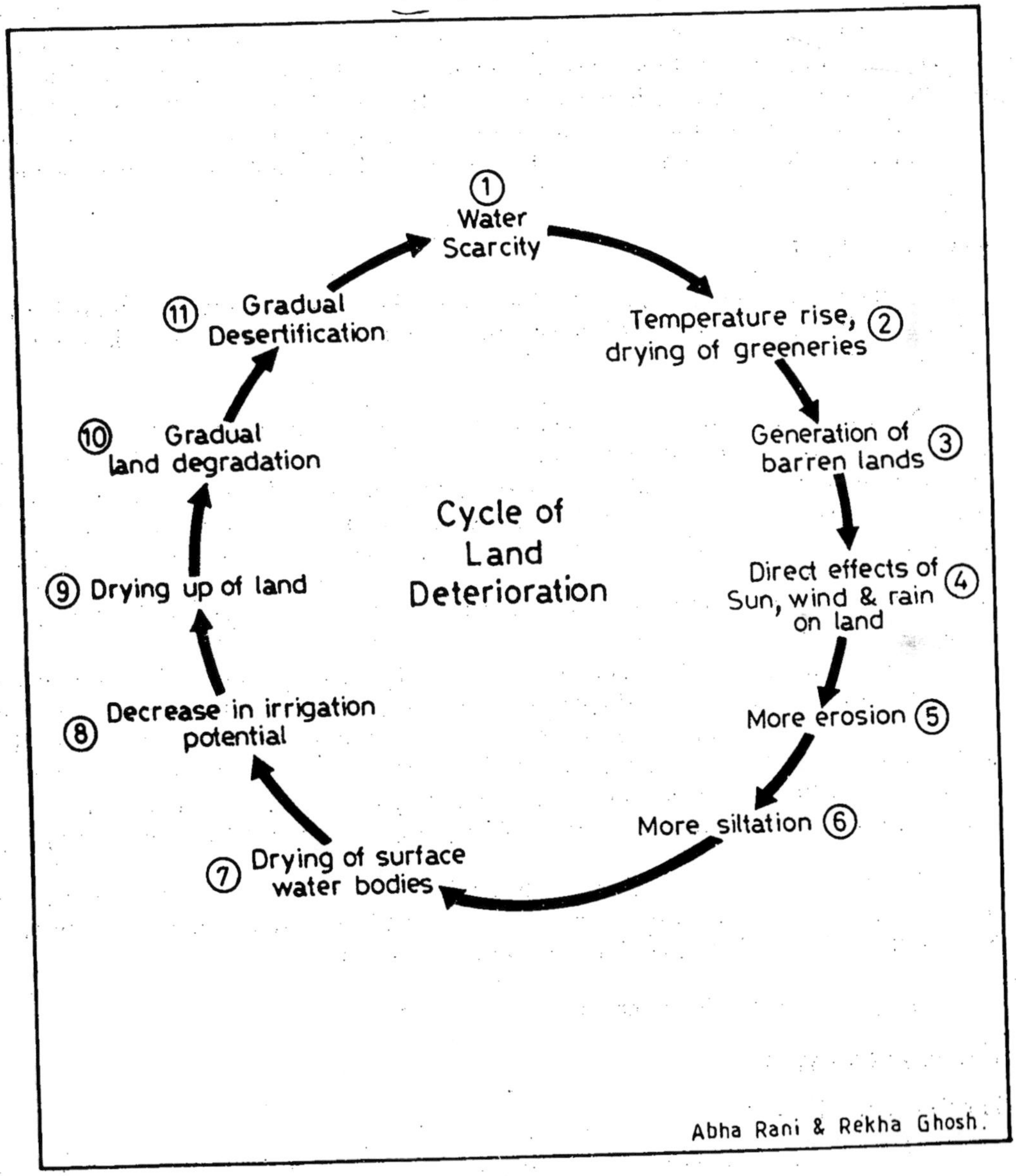

Fig. 66.1 : Land Deterioration Cycle

Further, greenery bears a pivotal role in protecting the quality of all aspects of environment, (Fig. 66.2) (Ghosh & Ghosh, op. cit.), also in many cases it is considered as an indicator or protector of environmental quality in the region. Regarding desired land-use pattern some notification has been made in the National Forest Policy, the notified level of required national average forest cover is 33%. The Honourable Prime Minister of India has stressed upon this fact in the "World environment day", 1999 (Ministry of Environment and Forests, Government of India notification made on 5th June, 1999). In spite of the fact that this realisation generated long back the country is still far behind the desired level in this matter. This fact gives an indication that there is some drawback in the land-use management system. This has been tried to identify in the present research, with reference to the coalfields in India.

PREVIOUS WORK

In the present research project informations were collected on land-use pattern of different coalfields of India at two different time levels, i.e. premining (as far back as possible) and 1994.

Next, an attempt was made to measure the change in land-use pattern through time, keeping in mind the fact that a perfectly eco-friendly activity should not result in any change in (nature induced) land-use pattern i.e. the amount of land under different uses. Thus if **m** represents the average rate of change in amount of land under any category of use over the total area, **m** should equal to "zero" in the most eco-friendly activity condition.

While the rates of change experienced by different components of land-use were being studied, special stress was given on change in "forest-cover" as it is the chief component of land-use with respect to protection of land-environment. It was observed that the rate of decrease of forest cover in coal mining areas in India is significant. This points towards some unidentified lacuna in the presently followed coal mining system with reference to activities like reclamation, reclamation planning or land-use planning (Ghosh & Rani 1999). Some such problems of mining and reclamation plans followed in Jharia coalfield was indicated by findings of some previous study also (Ghosh 1988).

As a next step of activity changes (through time) in different aspects of land-use in coal mining areas in the country were analysed statistically and an attempt was made to find out the trend of the change (Rani & Ghosh, communicated).

The present paper documents the follow-up action.

THE PRESENT WORK

The Concept followed

The land-uses/land-covers reported in the coalfields of India were grouped into six major land-use/land-cover types, i.e. forest land, agricultural land, built-up areas, mining areas water bodies and others (barren land and any other type, whichever is present). It was gathered from the analysis of data received and field conditions that, while the major developmental activities in mining areas are all related to mining, reclamation of all such lands degraded (or partially degraded) by mining and related activities was not practically feasible. The actual attempt was made to reclaim only those lands which were visibly identifiable as severely disturbed lands e.g. fire areas, quarry areas etc., that too upto a very limited extent. Thus, for the major parts of the mining areas, the newly developed

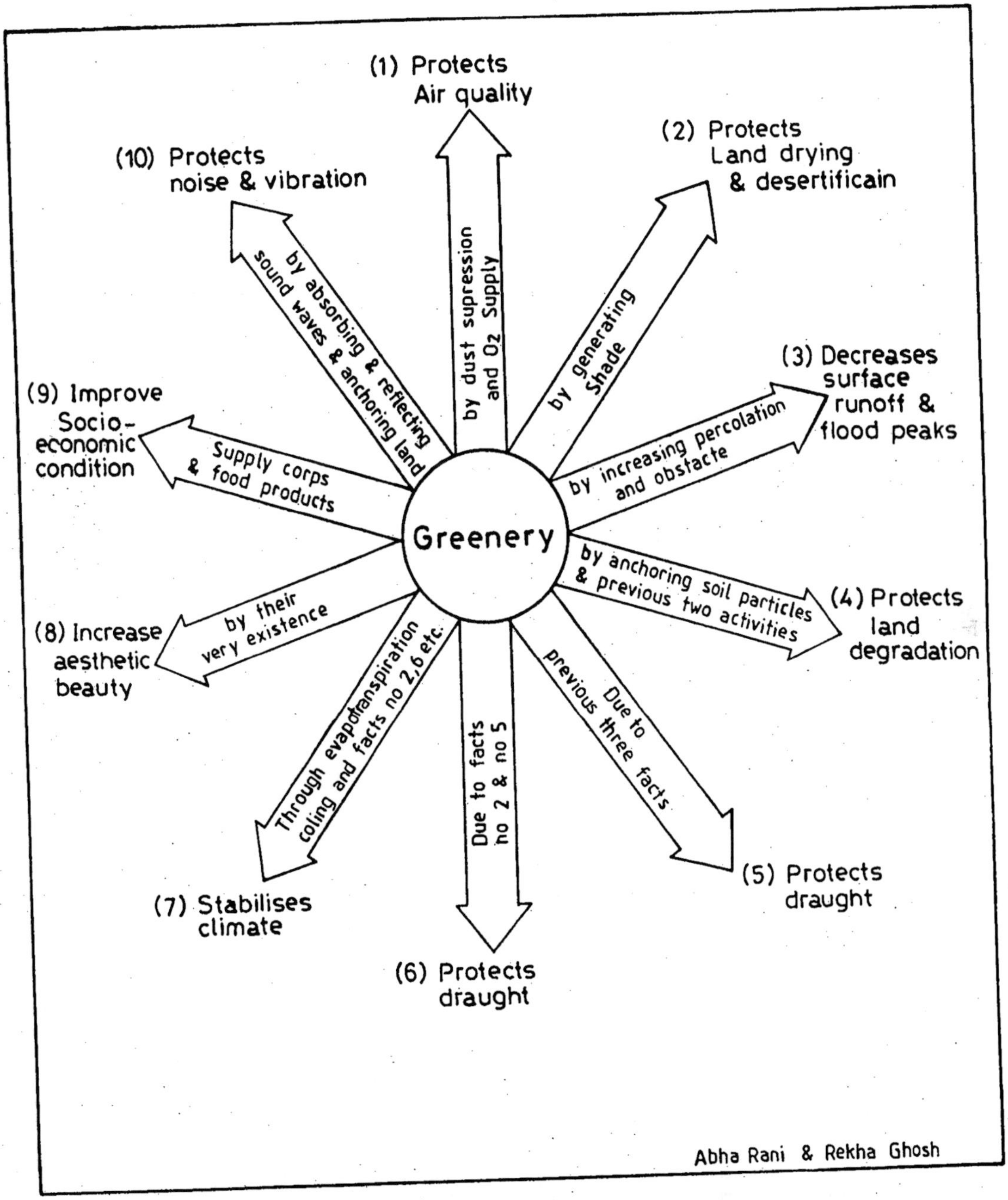

Fig. 66.2 : Pivothal role of greenery in protecting environment

land-use patterns were generated mainly by nature's care. This nature however includes climatic, physiographic (including drainage), geological and soil conditions (nature, thickness etc.) and even demographic conditions (population density, income status etc) all affected by the mining activity. Naturally mining criteria like excavation details (depth, thickness etc), life of the mine and area of mining, etc. all have some control on impact of mining on land-use of the area. Thus, the final land-use pattern generated is an expression of interaction of all these natural and human induced factors.

Henceforth the change in land-use pattern was considered as an impact of mining on land, controlled by the above referred natural and nature controlled factors and mining factors.The basic concept behind this is, in many cases such natural factors provide limitations and potentials of land for particular uses (Spangle et. al., 1976). Further, coal mining in any area is basically dependent upon occurrence of mineable coal seam(s), which is also a geologic fact, i.e. a natural factor.

Data Treatment and Analysis

When the informations on land-use change in coalfields of India were collected, parallelly informations were gathered on the respective controlling factors, as discussed above. For treatment of the data the following activities were conducted successively **with respect to each of these controlling factors individually.**

However, regarding location it was noted that India being a tropical country all the coalfields are falling in tropical zone. So no further analysis was done with latitude and longitude. Moreover rainfall has been dealt separately as a controlling factor.

(i) Conditions existing in different coalfields were noted.

(ii) The total range of the conditions existing in all the coalfields was noted.

(iii) The total range was devided into five class intervals numbered as 1, 2, 3, 4, 5 so that number 1 refers to condition best suitable for growing of greenery, and gradually through higher numbers, number 5 refers to the toughest condition for survival and growth of greenery.

(iv) The exact field conditions (sample numbers) falling in each class intervals were listed.

(v) Class 3 was considered as the "moderate condition", hence detailed analysis was made for samples falling in class interval number three, to analyse statistically the change in land-use pattern the areas have suffered with respect to each land-use components, under moderate natural conditions.

(vi) At the time of analysing the results of above study it was considered that whatever might be the change in land-use pattern by mining under any natural condition, the total amount of land remained same. Therefore, under any condition the decreased amount of land under any definite type of use must had gone under some other type of land-use. Hence while m with different suffixes represent change in amount of land under different uses respectively, the positive and negative values of **m** should sum-up to "zero". Therefore, any slightest discrepancy observed on this point was readjusted at $\mathbf{m}_{(O)}$ i.e. change in land-use types other than the five different uses listed as $\mathbf{m}_{(F)}$, $\mathbf{m}_{(C)}$, $\mathbf{m}_{(B)}$, $\mathbf{m}_{(W)}$ and $\mathbf{m}_{(M)}$ representing change in amount of forest land, agricultural lands, built-up areas, water bodies and mining areas respectively.

For each of the controlling factors a linear relation was obtained as impact of unit mining on land-use pattern considering all those samples which fall in the moderate range of class intervals of each controlling factors.

In the linear relations obtained "m" was omitted and the magnitude of "m" has been written as prefix of the respective land-uses i.e. (M), (F), (C), (B), (W) and (O).

RESULTS & DISCUSSION

Analysis of data on land use pattern in coal mining areas in India as it existed in 1994 and in the premining days was represented in terms of rate of change with respect to time. (i.e. slope of line representing change). The mean of the slope of data was calculated to represent the overall change with respect to time for each of the components of land-use.

These means were now used to represent the effect of mining on the land-use pattern. The trend had been calculated which came out to be linear in nature. As it has been said earlier also that the trend somehow is definitely effected by the natural factors, analysis was carried out taking each of these factors (for which data could be made available) into account one by one.

The following thirteen (13) equations could be obtained thus, which represent impact of unit mining on land-use under moderate natural conditions as listed against respective equations:

1. Surface water availability: $1.0\,(M) - 1.02(F) - 2.65(C) + 0.22(B) + 0.35(W) + 2.10(O) = 0$ (1)

2. Topographical set-up: $1.0(M) - 1.38(F) - 2.38(C) + 1.88(B) - 0.25(W) + 1.13(O) = 0$ (2)

3. Rainfall: $1.0(M) - 0.79(F) - 5.88(C) + 0.38(B) + 0.12(W) + 5.17(O) = 0$ (3)

4. Excavation depth: $1.0\,M - 0.79(F) - 0.37(C) + 1.67(B) + 0.0(W) - 1.51(O) = 0$ (4)

5. Life of mine: $1.0\,M - 0.29(F) - 2.18(C) + 0.19(B) - 0.34(W) + 1.62(O) = 0$ (5)

6. Project area: $1.0\,M - 0.88(F) - 18.0(C) + 0.79(B) + 0.05(W) + 17.04(O)$ (6)

7. Excavation thickness: $1.0\,(M) - 1.36(F) - 1.64(C) + 1.64(B) + 0.0(W) + 0.36(O) = 0$ (7)

8. Habitated by people of moderate income: $1.0\,(M) - 0.67(F) - 5.28(C) + 0.51(B) + 2.16(W) + 2.28(O) = 0$ (8)

9. Moderate population density: $1.0\,(M) - 0.96(F) - 14.86(C) + 1.26(B) + 0.17(W) + 13.39(O) = 0$ (9)

10. Moderate soil cover: $1.0(M) - 0.89(F) - 27.56(C) + 1.56(B) + 0.0(W) + 25.89(O) = 0$ (10)

11. Geological conditions with respect to rock type: $1.0 (M) - 0.59(F) - 3.63(C) + 0.10(B) + 0.09(W) + 3.03(O) = 0$ (11)

12. Faulting effect: $1.0 (M) - 0.58(F) - 1.49(C) + 0.91(B) - 0.09(W) + 0.25(O) = 0$ (12)

13. Strata dip: $1.0 (M) - 0.31(F) - 0.91(C) + 0.15(B) + 0.0(W) + 0.07(O) = 0$ (13)

Next an attempt was made to identify the relative importance of the natural factors in terms of their responsibility for creating change in land-environment. The basic concept followed in the analysis was, since the "Forest" is the chief determinant of the quality of land-environment, **m** for forest should be "zero", i.e., there should be no change in forest cover of the region (under consideration) in an ideal condition.

As in most of the cases studied presently, the fact was not as desired, a hierarchi was searched out, from the previously given thirteen linear equations, to identify the natural condition that causes maximum damage to forest cover and that causes the minimum damage to this indicator of the quality of land-environment and also the intermediate ones, in order, among all natural factors, considered in the study.

The order of the natural factors in terms of magnitude of their capability to support damage to forest cover caused by mining, thus established, is as given below:

Topography, Excavation thickness, River, Population density, Soil cover, Area of mining, Rainfall, Excavation depth, Income (per head), Rock type, Fault, Dip (of strata), and lastly Life of the (mining) project.

CONCLUSIONS AND RECOMMENDATIONS

When land-use pattern changes are caused by coal mining in India, in most of the cases the quality and quantity of changes are controlled effectively by some natural factors more than human care.

The relative control of such natural factors on change in forest cover has been analysed. Such relative control of all effective natural factors on change in other land-use components should be studied.

Results of such analysis should receive due importance while formulating land-use management plans for coal-mining areas in India so that the plan can be implemented more effectively, successfully and in an echo-friendly manner to produce a habitable environment in 21st century.

ACKNOWLEDGEMENT

The authors are thankful to the Director, Indian School of Mines (ISM), Dhanbad for providing institutional facilities for the work and kindly permitting the presentation of this paper. The authors are grateful to Head, Centre of Mining Environment (CME) and Head, Department of Applied Mathematics, ISM, Dhanbad for providing facilities to work for this project. The second author gratefully acknowledges the financial help provided by the Ministry of Environment & Forests, Govt. of India by way of supporting the CME to which she belongs. Special grateful thanks are being extended towards Prof. N.C. Saxena, the Head, CME for his all encompassing help and encouragement for the work.

REFERENCES

Chatterjee, D.S. and R.Ghosh. 1994. Human induced land-use change around Raniganj, Raniganj Coalfield, Eastern India, In: Indian Journal of Earth Sciences, Vol. 21, No.1, pp 44-60.

Dutta, S.K. 1997. Analysis of changes in land use and land cover due to mining and other human activities in and around Makum Coalfield, Assam - a remote sensig approach. Unpublished thesis ISM, p. 88.

Ghosh, R. 1988. Environmental impact of Coal Mining in Jharia Coalfield, Eastern India, its past, present and future, In: Transaction, Mining Geological & Metallurgical Institute of India, Vol.85, No. 2, pp. 90-98.

Ghosh, R and S.K. Das. 1998. Strategies for protection of water resources in a part of Subarnrekha river basin, Bihar. In: Indian Journal of Environment and Ecoplan, Vol.1&2, pp. 27-32.

Ghosh, R & D.N. Ghosh. 1993. Hydrology & environment the environmental deterioration cycle, In: Quaterly Journal of The Geological Association and Research Centre, Vol. 1, No. 2, pp. 1-9.

Ghosh, R., Y. Kumar & N.C. Saxena. 1997. Impact of industrialisation on land use and employment, a case study of Jharia Coalfield, In: Minetech, Vol. 18, No. 1&2, pp. 49-58.

Ghosh, R. & A. Rani. 1999. A step toward eco-friendly coal mining in India, In: Proceedings of the International symposium on Clean Coal Initiatives, N. Delhi, pp. 587-592.

Rani A. & R. Ghosh. 1999. Land use pattern change in Coal Mining areas in India - a trend finding exercise. Communicated to the International Journal of Surface Mining and Reclamation, Calgary, Alberata, Canada..

Sekhar, T.S. 1996. Impact of mining on land-use/land cover of Rihand reservoir & environs using remote sensing techniques. Unpublished thesis, ISM, p. 65.

William Spangale and Associates, F. Beach Leighton &Associates, and Boxter, McDonald & Company. 1976. Earth-Science Information in Land-Use Planning - Guidelines for Earth Scientists and Planners. In: United States Geological Society Circular No. 721.

CHAPTER 67

Water quality simulation of Damodar River

M.K. Chakraborty*, R.S. Singh*, S.K. Chaulya* & M. Ahmad

ABSTRACT

Damodar river is one of the largest river of eastern India and serves as source of drinking water and also fulfill the water requirement of irrigation and industries of this region. The basin of the river is one of the most mineralized zones of the country and has witnessed prolific industrial growth leading to serious environmental degradation of the river.

Therefore, this paper deals with the characterization of various effluents generated and water quality simulation model of the Damodar river in the selected critical stretch. Results indicated that Total Suspended Solids (TSS), Total Dissolved Solids (TDS), Biological Oxygen Demand (BOD) and Dissolved Oxygen (DO) were the major pollutant contributors in the Damodar river. Water quality simulation for the values of DO and BOD parameters have been carried out employing QUAL2 E model. The model shows that BOD concentration can not be brought down to the tolerance limit, however the critical DO concentration will remain above tolerance limit, prescribed by Indian Pollution Control Board, unless some more additional preventing measures are implemented for the sustainable development of the region.

Keyword : Water quality; Modeling; Simulation; Assimilative capacity

1.0 INTRODUCTION

Presently there are several environmental issues that have grown in size and complexity threatening the survival of people on the earth day by day. Rapid increase in world population, intensive industrial growth and substantial increase in per capita utilization of natural resources have tremendously affected our environment. Industrialization has further added another new dimension to the environment by

*Scientist, Central Mining Research Institute, Dhanbad - 826 001, India

introducing unwanted substances into the air, water and land mass that was not present during pre industrialization period. Many environmental incidents had occurred in the sixties, like the outbreak of Itai-Itai diseases in Japan (Tucher, 1972) followed by the consumption of rice contaminated with high concentration of Cadmium and Mina mata disease (Grant, 1971, and Fujki, 1972) due the consumption of fish contaminated with methyl Mercury, out break of malaria and filaria (Mudgal, 1985), have drawn attention of researchers towards the study of water pollution.

Coal mining and it's allied industries constitute one of the major man made pollution sources in and around mining areas. Opencast coal mining results in serious disruption of surface topography and water course, whereas underground mining is associated with lowering of ground water table. Acid mine drainage is also a very common affair in underground operation. In addition to these, coal utilization industries create more nuisance to the existing aquatic flora and fauna. Coal washeries, Cokeoven plants and thermal power plants are the major coal utilization industries which create water pollution and ultimately affect the receiving water bodies. (Singh, 1995, Intraratria et al., 1995). Number of studies have been carried out by various researcher (Dhaneshwar, 1972, Tripathi et al., 1980, Singh, 1985). But there is no comprehensive study for the area was adopted to find out river assimilation capacity.

The basic phenomenon governing the assimilative capacity of surface water source includes self purification capacity and dilution potential. The assimilative capacity for conservation substance can be estimated through mass balance involving determination of streams flow during critical season, where as no straight forward methodology exists for non-conservative substance. (Dorth et.al., 1988, Mccutcheon, 1983, 1985, 1986.) Therefore, in the present study stream flow method was employed.

The river not only serves as a source of drinking water but also fulfills the water requirement of irrigation and industries of the region. The basin being one of the most mineralized zones of the country has witnessed prolific industrial growth leading to serious environmental degradation. Industrial activities in this region are concentrated along the river and, which led to migration of population into some of the pockets, straining the already water resources in terms of quality and quantity, ultimately threatening Damodar's self purification capacity in its various stretches.

The paper deals with characterization of various effluent generated and water quality simulation model of Damodar river in the most critical stretch under the two scenario that is if industries continued to discharge effluent under the business as usual scenario or comply the prescribed regulatory standards.

2.0 STUDY SITE

2.1 Damodar river basin

The Damodar River Basin area extends from 22°45'N to 24° 30'N and 84° 45'E to 88° 00'E (Fig. 67.1). It is one of the largest rivers of eastern India and the course of its length, about 542 km, flows through the cities of Ramgarh, Bokaro, Dhanbad, Asansol, Durgapur, Burdwan and Howrah and joins the lower Ganga (Hooghly estuary) at Shayampur, 55 km downstream of Howrah. The river is fed by a number of tributaries at different reaches, the major ones being the Bokaro, Konar, Jamunia and Barakar. The total catchment area of the basin is about 23170 km; of this about three quarters of the total basin area lies in Bihar and

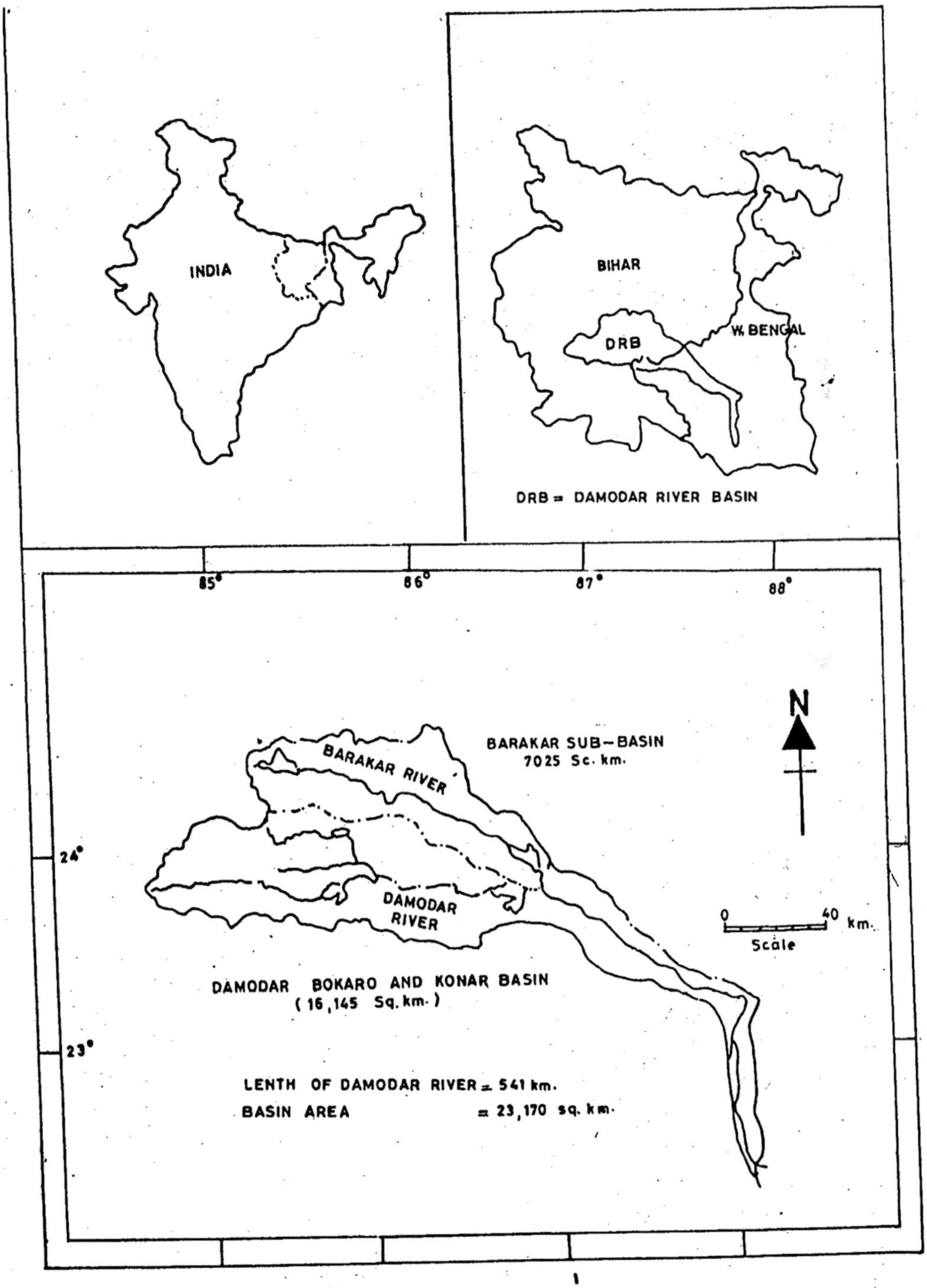

Fig. 67.1 : Location of Damodar River Basin

one quarter in West Bengal. The basin is characterized by a tropical climate with an annual rainfall of about 1200 mm. The major part of the rainfall (82%) occurs in the monsoon season (June-October) with a few sporadic rain in winter.

The Damodar River flows through a vast track of eastern India with variable topography and geology. The upland catchment area in Bihar comprises highly dissected paneplained upland with deep ravines and bare hills and the lower catchment area is within the delta plains of West Bengal. The basin geology is characterized by rocks consisting of granites granitic gneisses of Archean age, sandstones and shales of the Gondwana age and the Recent alluvial. In the lower part of basin rocks of Tertiary age are covered by a thick veneer of alluvium. The upland area of the basin is occupied by rocks of the Archean crystalline complex and basic intrusive. The bulk of the crystalline complex consists of Late-Precambrian gneisses and granites of batholithic dimension. The gneisses are non-uniform in character and mostly granitic in composition. The texture of the granitic rocks varies from fine to coarse-grained and porphyritic, massive to highly gneissic type. Quartizites, quartz schists and quartz-mica schists are other metamorphics in the basin. Field and petrographic data suggest that the gneisses are older than the associated granites that show an intrusive relationship with country rocks (Ghose 1983). The crystalline complex is divided into a northern and southern belt by the east-west extending outcrops of the coal-bearing Gondwana formations of Permo-Carboniferous age. Though not rich in metallic minerals, the Damodar basin has rich deposits of coal. Other than coal, fire day, bauxite, mica and limestone are associated with the geological formations of the basin.

Fig. 67.2 shows the location of the river basin along with study site. The master stream of the region, not only serves as sources of drinking water, but also fulfill the water requirement of irrigation and industries of the region. The basin being one of the most mineralized zones of the country has witnessed prolific industrial growth leading to serious environmental degradation, and shortfall in basic natural resources. Industrial activities in this region are concentrated around the river, and as a result the waste/effluent generated by these industrial activities has led to migration of population into some of the pockets, straining the already week water resources in terms of quality and quantity. Besides this it comprises of some major coalfields like Jharia, Raniganj, Ramgarh, East & West Bokaro and South & North Karanpura (Fig. 67.2). These activities ultimately, threatening Damodar's self purification capacity in various stretches. For the present study, stretch between Tenughat dam to Upstream of Panchet Reservoir is selected for detailed water quality analysis vis a vis simulation of water quality.

3.0 METHODOLOGY

Reconnaissance survey was carried out to find out different types of point area and line sources. Samples were collected from these sites along with flow rate in the critical seasons.

3.1 Sampling

Three seasons water samplings were carried out for river system. Whereas for effluent samplings were considered only in critical season of 1995-96. Samples were taken from the depth of 1.0 m in boro-sillicate glass bottle along with rate of discharge. Narrow mouth ground stopped glass bottle of capacity 300 ml was used for dissolved oxygen. Standard physical and chemical procedure were employed to preserve the samples for at least 24 -36 h.

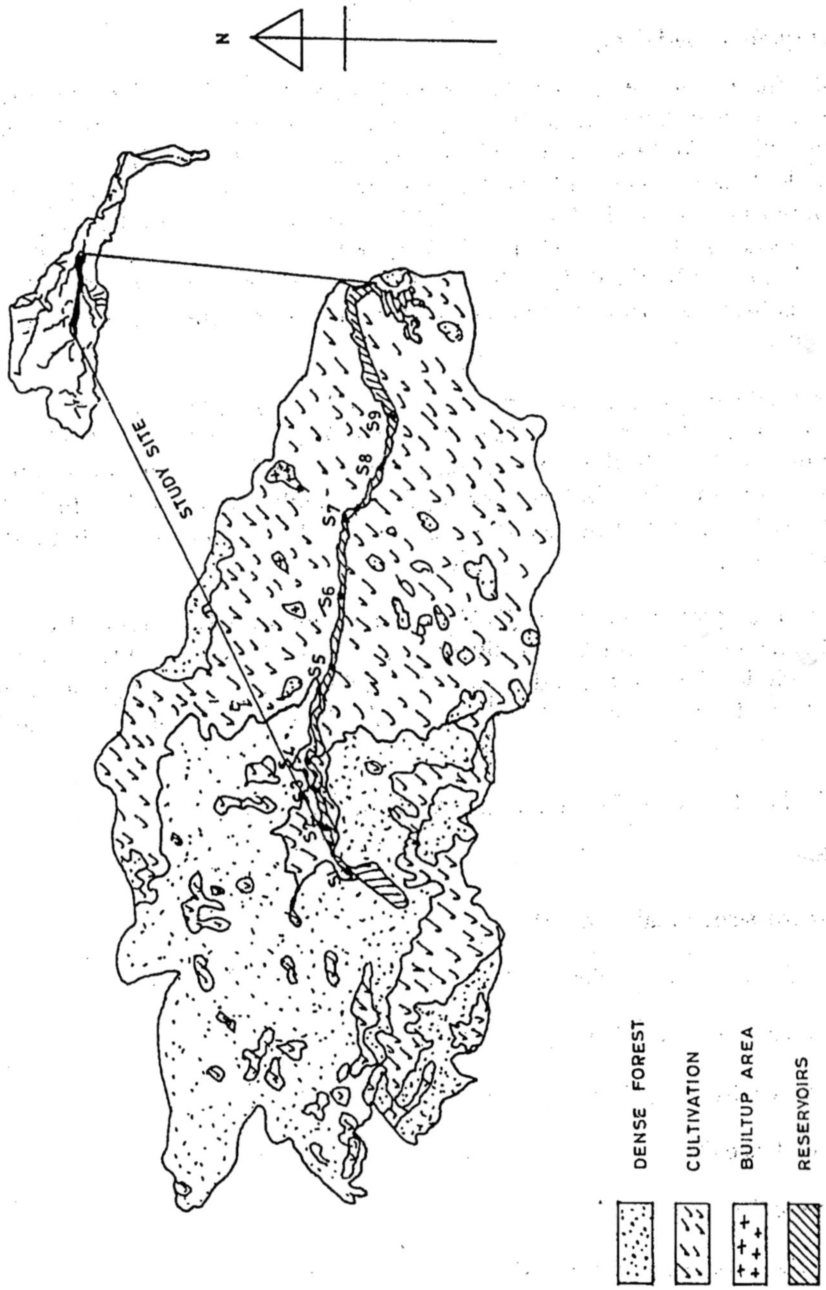

Fig. 67.2 : Location of Sampling Points

3.2 Chemical analysis

The standard method were followed for analysis of water and waste water following: AWWA & APHA (1992). All the reagents and colorimatric solutions were prepared using analytical reagent grade chemicals and standard glass apparatus.

3.3 Water quality modeling

Literature survey was carried out to find out best suited model for the river system to understand assimilative capacity. QUAL2E model as was found suitable under given stream condition. (Ambrose et al. 1982, JRB associates, 1983). QUAL2E is comprehensive and versatile stream water quality model. It can simulate up to 15 water quality constituents in any combination desired by the user. Constituents which can be simulated are a Dissolved Oxygen, Biochemical Oxygen Demand, Temperature, Algae as Chlorophyll a, Organic Nitrogen, Ammonia N, Nitrite N, Nitrate N, Organic Phosphorus P, Dissolved Phosphorus P, Coliforms bacteria, Arbitrary Non-conservative Constituent and three Conservative Constituents.

The model is applicable to dendritic streams that are well mixed. It assumes that the major transport mechanisms, advection and dispersion are significant only along the main direction of flow (longitudinal axis of the stream or canal). It allows for multiple waste discharges, withdrawals, tributary flows and incremental inflow and outflow. It also has the capability to compute required dilution flows for flow augmentation to meet any pre-specified DO level.

The basic equation solved by QUAL2E is one dimensional advection-dispersion mass transport equation which is numerically integrated for each water quality constituent. This equation includes the effects of advection, dispersion, dilution, constituent reactions and interactions, and sources and sinks. Under steady state mode for any constituent, C, this equation can be written as :

$$d[Ax\ DL\ dc/dx/Axdx - d\ (Ax\ uc)/Ax\ dx\ dc/dt + s/v \quad ...(1)$$

Where,

A_x = cross-sectional area (L^2)

D_L = dispersion coefficient (L^2T^{-1})

u = mean velocity (LT^{-1})

c = concentration (ML^{-3})

x = distance (M)

t = time (T)

s = external source or sinks (MT^{-1}) and

v = volume (L^3)

The above equation represent the effect of, dispersion, advection, constituent changes, external sources/sinks, and dilution respectively.

In the present application, the water quality simulations have been carried out in terms of DO and BOD. The differential equations for river DO and BOD simulation as used in QUAL2E are as follows :

$$dL/dt = -K_d L \qquad ...(2)$$

$$do/dt = K_a(C_s\text{-}O)\text{-}K_dL - \sigma 5_1 N_1 - \sigma 6_2 N_2 \qquad ...(3)$$

Where,

L = concentration of ultimate carbonaceous BOD (mg/l)

D = concentration of DO (mg/l)

Cs= saturation concentration of DO at local temperature and pressure (mg/l)

K_d= de-oxygenation rate coefficient (day^{-1})

K_a= re-aeration rate co-efficient(day^{-1})

$\sigma 5_1$ = rate of oxygen uptake per unit of NH_3-N oxidation (mg-O/mg-N)

B_1 = ammonia oxidation rate co-efficient, (day^{-1})

N_1 = ammonia nitrogen concentration (mg-N/l)

$\sigma 6_2$ = rate of oxygen uptake per unit of NO_2-N oxidation (mg-O/mg-N)

B_2 = nitrite oxidation rate coefficient (day^{-1})

N_2 = nitrite nitrogen concentration (mg-N/l)

For the conversion of 5-day BOD to ultimate BOD following equation was used :

$$BOD_5 = BOD_u (1.0 - \exp(-5 * KBOD)) \qquad ...(4)$$

Where,

BOD_5 = 5-day BOD (mg/l)

BOD_u = ultimate BOD (mg/l)

KBOD = BOD conversion rate coefficient (day^{-1})

Following differential equations governing transformations of organic nitrogen to ammonia, to nitrite and finally to nitrate are used in QUAL2E.

$$DN_4 /dt = -B_3N_4 - \qquad ...(5)$$

N_4 = concentration of organic nitrogen (mg-N/l)

B_3 = rate constant for hydrolysis of organic nitrogen to ammonia nitrogen (day^{-1})

$$DN_1 /dt = - B_3N_4 - B_1N_1 \qquad ...(6)$$

N_1 = conc. of ammonia nitrogen (mg-N/l)

B_1 = rate constant for oxidation of ammonia nitrogen (day^{-1})

$$DN_2/dt = B_1N_1 - B_2N_2 \qquad ...(7)$$

N_2 = concentration of nitrite nitrogen (mg-N/l)

B_2 = rate constant for oxidation of nitrite nitrogen (day^{-1})

$$dN_3/dt = B_3N_2 \qquad ...(8)$$

In order to calculate river flow velocity, when the cross-sectional properties of the stream segments are available as a function of the depth d, the wetted area A_x is calculated as a function of discharge by solving Mannings Equation using Newton-Raphson method:

$$Q = 1/_n A_x R_x^{2/3} S^{1/2} \qquad ...(9)$$

Where,

R_x = mean effective hydraulic radius (m)

n = manning roughness factor

s = slope of energy grade line

q = discharge (m^3/s)

The average flow velocity is then calculated by the following equation :

$$u = Q/A_x \qquad ...(10)$$

3.4 Estimation of assimilative capacity

The assimilative capacity of Damodar river water environment in Jharia coalfield (D/s of Tenughat to U/s of Panchet) is assessed keeping in view the level of contaminants reaching the surface water body and quality of river water during most critical summer season. Most appropriate geographical entity for estimating the assimilative capacity is the watershed and designing the monitoring water quality, quantity and other information (Edward, et.al, 1997). To estimate the assimilative capacity, assessment was carried out for the determination of critical parameters like Biological Oxygen Demand (BOD), Chemical Oxygen Demand (COD), Dissolve Oxygen (DO), Total Dissolve Solids (TDS) and Total Suspended Solids in the above mentioned stretch. Schematic diagram of river stretch along with location of discharge/withdrawal/sampling point is shown in Fig. 67.3 and described in Table 67.1 and 67.2.

3.5 Segmentation of Damodar River

The river stretch for modeling was divided into different riches taking into consideration the location of waste water outfall, withdrawal points, tributaries and mining areas which were assumed as non point source of pollutant. The each riches was further divided in small elements of 5 km length to study the stretch for improvement of resolution and minimum averaging effect, as the same would help for simulation by studying affect of all discharges independently within each elements shown in Fig. 67.3.

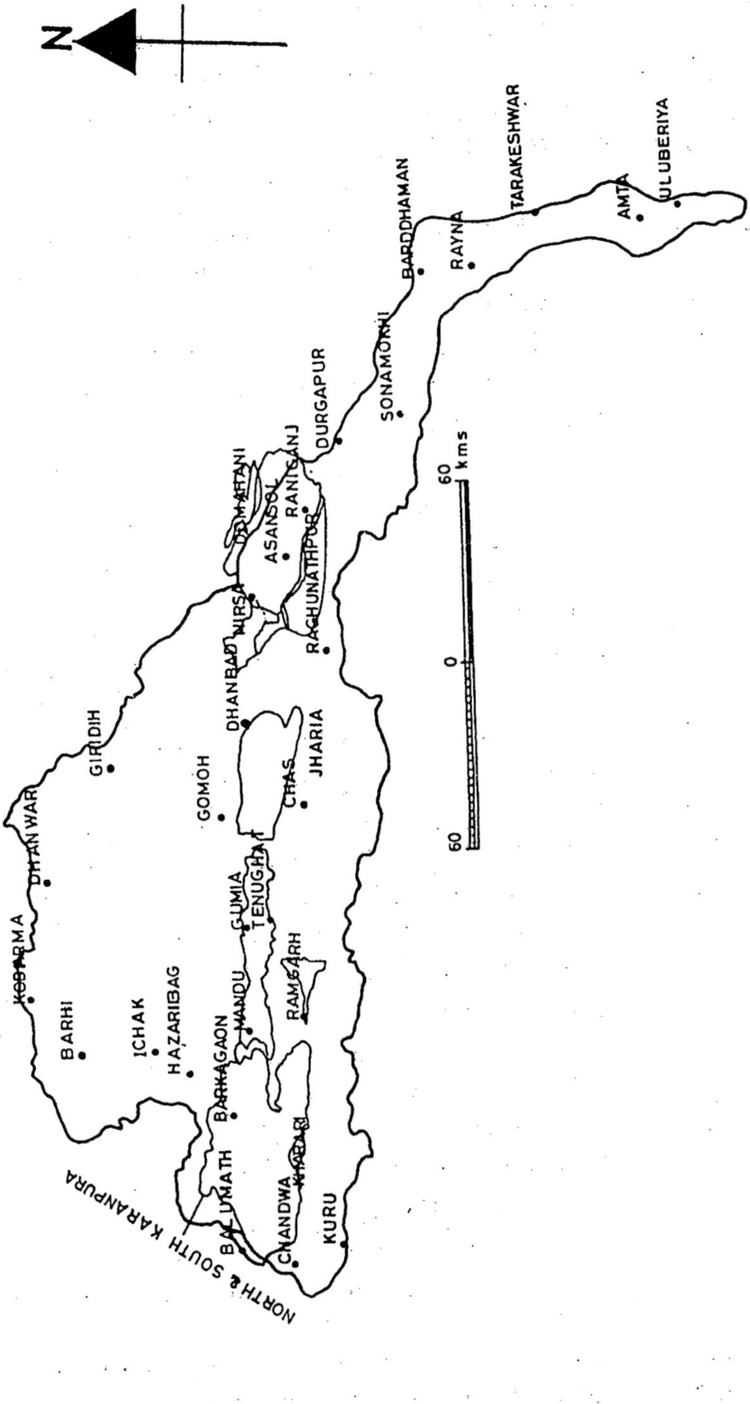

Fig. 67.3 : Location of Major Coalfield Areas of Damodar Basin

Table 67.1 : Sampling Locations For Water Quality Survey

Sampling Point No.	*Identification*	*Description*
1	S1	Damodar river water d/s of Tenughat Dam
2	S2	Damodar river water u/s of confluence with Konar river
	T1	Konar river water before confluence with Damodar river
3	S3	Damodar river water u/s of Kargali Coal Washery
4	E1	Kargali washery effluents
5	S4	Damodar river water d/s of Kargali Coal Washery
6	E2	Effluent from Chandrapura Thermal Power Station
7	S5	Damodar river u/s of Bokaro Steel Ltd.
8	E3	Effluent from Bokaro Steel Plant outfall-I
9	E4	Bokaro Steel Plant effluent outfall-II
10	E5	Effluent from Bokaro Steel Plant outfall-III
11	S6	Damodar river d/s of Bokaro Steel Ltd. near Dugdha
12	E6	Effluent from Kathara Coal Washery
13	S7	Damodar river water u/s of Telmachu bridge
14	T2	Jamunia river water u/s of confluence with Damodar river
15	T3	Garga nullah
	T4	Jharia Nullah u/s of confluence with Damodar river
16	S8	Damodar river water at confluence of Damodar & Katri
17	E7	Effluent from Loyabad coke oven plant
18	E8	Effluent from Jamadoba Washery (TISCO)
19	E9	Effluent from Jamadoba Power House No. 1 (TISCO)
20	E10	Effluent from Jamadoba Power House No. 3 (TISCO)
21	E11	Pathardih Coal Washery effluents
22	E12	Effluent from Sudamdih Coal Washery
23	E13	Effluent from Chasnallah Coal Washery
24	E14	Effluent from FCI, Sindri
25	E15	Effluent from PND, Sindri and Township
26	E16	Effluent from Santhaldih T.P.S. and Bhojudih Washery
27	S9	Downstream STPS and Bhojudih
28	S10	Damodar River Water U/S of Panchet Dam
29	E17	Effluent from FCI Sindri
30	E18	Effluent from PND, Sindri and Township
31	E19	Effluent from Santhaldih Thermal Power Station and Bhojudih Washery
32	E20	Effluent from Dishergarh Thermal Power Station
33	E21	Effluent from Carew and Phipson Co. Ltd., Asansol

Table 67.2 : Damodar River Stretch - Tenughat to Panchet

Identification	Flow (m^3/hr)	BOD (mg/L)	DO (mg/L)	TDS (mg/L)	TSS (mg/L)	Temp. (°C)
1	2	3	4	5	6	7
S1	-	5.50	6.80	124.0	42.0	33.5
S2	-	4.00	6.90	173.0	30.0	33.0
S3	-	9.50	5.40	277.0	1170.0	32.3
S4	-	8.00	6.60	325.0	404.0	32.0
S5	-	9.00	6.30	195.0	1234.0	31.5
S6	-	10.50	5.60	263.0	114.0	31.5
S7	-	6.00	4.30	248.0	58.0	32.0
S8	-	7.30	-	179.0	76.0	31.0
S9	-	2.90	5.10	129.0	25.0	31.0
S10	-	11.0	7.40	198.0	94.0	32.0
T1	6500.0	4.00	5.10	276.0	204.2	33.0
T2	6750.0	19.00	6.60	232.0	179.0	31.0
T3	141.0	15.00	6.10	282.0	399.0	31.0
T4	132.0	6.70	6.40	672.0	84.0	32.0
E1	116.7	20.00	-	-	40.0	29.0
E2	500.0	4.70	4.00	500.0	205.0	32.0
E3	19.0	10.00	7.10	368.0	100.0	34.5
E4	719.0	7.00	5.40	729.0	22310.0	32.0
E5	50.0	12.00	5.50	250.0	700.0	30.0
E6	170.0	10.00	6.00	250.0	700.0	30.0
E7	203.0	8.00	5.80	264.0	1541.0	30.0
E8	1987.0	23.00	7.30	327.0	36.0	31.0
E9	970.0	20.50	7.30	250.0	12.0	31.9
E10	298.0	16.00	7.30	250.0	17.0	32.0
E11	33.0	2.60	6.40	571.0	90.0	23.0
E12	200.0	10.00	5.50	250.0	700.0	30.5
E13	147.6	51.80	0.85	525.0	556.0	32.0
E14	26.1	16.20	0.85	-		-
E15	200.0	10.00	5.50	250.0	700.0	30.5
E16	350.0	25.00	2.00	418.0	669.0	28.0
E17	26.1	16.20	0.80	-	-	25.5
E18	200.0	10.00	5.50	250.0	700.0	30.5
E19	350.0	25.00	2.00	418.0	669.0	28.0
E20	1387.0	20.00	2.10	471.0	1792.0	31.0

1	2	3	4	5	6	7
E21	562.0	7.20	3.77	564.0	2842.0	30.5
C1	278.0	200.00	0	500.0	400.0	30.0
C2	129.0	200.00	0	500.0	400.0	30.0
C3	124.0	200.00	0	500.0	400.0	30.0
C4	181.0	200.00	0	500.0	400.0	30.0
C5	459.0	200.00	0	500.0	400.0	30.0
C6	2.0	200.00	0		500.0	400.0
C7	48.0	200.00	0	500.0	400.0	30.0
C8	432.0	200.00	0	500.0	400.0	30.0
C9	2189.0	200.00	0	500.0	400.0	30.0
C10	152.0	200.00	0	500.0	400.0	30.0
C11	450.0	200.00	0	500.0	400.0	30.0

The simulated water quality profile for a given stretch, in addition to head water quality and quantity and forcing functions such as wastewater flows is greatly influenced by the selection of values for model parameters.

Hydrodynamic data

The river water flow is mainly governed by Damodar valley corporation (DVC). They had installed a few flow monitoring stations in the Patratu to Durgapur stretch of the river. Field data available on cross sectional profiles of Damodar and its tributaries at some locations and flow data reported have also been taken into consideration. QUAL2E uses Mannin's equation for estimation of stream velocity. The bed slope obtained by analyzing the maps prepared for GIS from survey of India was used during modeling exercise. Dispersion coefficients were estimated by making use of the appropriate formulae given in relevant literature(Demuren, et.al., 1983, Denhil and Gulliver, 1988)

Estimation of coefficient

In any study connected with dissolved oxygen (DO) balance of a water body, sinks and sources of DO must be considered (Martin, 1986). In this study, the sources of oxygen considered are atmospheric re-aeration, evolution of oxygen by phytoplankton during growth, and evolution of oxygen during denitrification. The sinks of DO considered are algal respiration, oxidation of detrital carbon and carbonaceous material from untreated effluents and non-point discharges and nitrification.

The model uses flow induced re-aeration coefficient expressed as a function of velocity and depth. The formulae employed for estimation of re-aeration coefficients for these simulations are given below :

(a) $K2o = (Dm\ u/d^{1.5})^{.5}$ O'Connor and Dobbins (1985)

Where,

Dm = Molecular diffusion coefficient (L2T-1)

u = Mean velocity (LT-1)

d = Stream depth (L)

(b) K2o = 3.3 u/d1.33 x 2.31 Longbien and Durum

Segment temperature was used to adjust Kaj by using the standard formula :

Kaj (t) = Kaj (20oC) (O) (T - 20)

Where,

T = Segment temp, °C

Kaj = Re-aeration rate coefficient of segment at ambient segment temperature (day^{-1})

O = Temperature coefficient (Model default value 1.028)

The dissolved oxygen saturation concentration is determined as a function of temperature and expressed as :

DO = 14.6532 - 0.41022T - 0.007991T2 -0.000077T3

Due to the non-availability of site specific reaction coefficients and rate constants for Damodar river the field estimated reaction rate coefficients for Kharkai and Subernarekha rivers of Jamshedpur region (NEERI, 1995) were used as input to the model, as these rivers are falling in the same region and subjected to similar industrial pollution. Other values were taken from literature (Demuren, 1983, Bowie, et.al., 1985, Denhil and Rodi, 1988)

4.0 RESULT AND DISCUSSION

Sixteen effluent samples nine river samples and four tributaries samples were collected from different locations, and all the critical source values were considered for the study. Results indicated that TSS, TDS, BOD and DO were the major pollutant contributors in the Damodar river. Water quality simulations for the values of DO and BOD parameters have been carried out employing QUAL2E model. The objective of this exercise was to find out the levels of treatment required for waste water discharge into Damodar river and to maintain the desired maximum DO and minimum BOD levels in the river. Since the river water in this stretch is used for supplying water to the adjacent major industries and townships in this coalfield region, it is felt that the river water quality should meet the requirement of class C level (Table 67.3) to start with the minimum DO and maximum BOD levels that need to be maintained for class C waters are 4 and 3 mg/l, respectively as per Indian Central Pollution Control Board norms.

Our study has considered two scenarios based on the available data.

(i) Discharge of effluents from various domestic, industrial and mining area sources into Damodar river as it is a common practice at present time.

(ii) Discharge of treated effluents to meet the prescribed regulatory standards

After first run the model parameters were readjusted to match the model DO and BOD values in the segments with the observed DO and BOD values. The simulated baseline water quality profiles with respect to DO and BOD parameters for the Damodar river stretch extending from Tenughat to Panchet. Second simulation run has been obtained using the calibrated model but with waste loads confirming to prescribed regulatory standards. The predicted water quality profiles with respect to DO and BOD parameters for the same stretch. This has been presented in Fig. 67.4 and 67.5. It is observed that in the predicted water quality profile BOD concentration can be brought down to 6 mg/l, however the critical DO concentration remains above 6 mg/l. This indicated that even under the scenario of industrial effluent meeting the prescribed regulatory standards it is not possible to bring down the BOD concentration in Damodar river unless some more additional preventing measure are implemented. Thus it can be concluded that assimilative capacity of Damodar river in this stretch (with respect to designated usage) is under continuous threat. Therefore, additional and scientific management should be initiated at earliest to meet out this environmental threats. Due to industrialization and Urbanization in this coal

Table 3 : Classification of Inland Surface Waters (CPCB Standards)

Sl.	*Characteristics*	*A*	*B*	*C*	*D*	*E*
1.	Dissolved oxygen, (mg/l), min.	6	5	4	4	-
2.	BOD (5 days at 20°C) (mg/l), min.	2	3	3	-	-
3.	Total Coliforms (MPN/100 ml), max.	50	500	5000	-	-
4.	Total Dissolved Solids (mg/l), max.	500	-	1500	-	2100
5.	Chloride as Cl (mg/l), max.	250	-	600	-	600
6.	Colour, Hazen units, max.	10	300	300	-	-
7.	Sodium absorption ratio, max.	-	-	-	-	26
8.	Boron as B (mg/l), max.	-	-	-	-	2
9.	Sulphates as SO_4 (mg/l), max.	400	-	400	-	1000
10.	Nitrates as NO_3 (mg/l), max.	20	-	50	-	-
11.	Free Ammonia as N (mg/l), max.	-	-	12	-	
12.	Conductivity at 25°C (cm), max.	-	-	-	1.0	2.25
13.	pH	6.5	6.5	6.5	6.5	6.0
14.	Arsenic as As (mg/l), max.	0.05	0.2	0.2	-	-
15.	Iron as Fe (mg/l), max.	0.3	-	50	-	-
16.	Fluorides as F (mg/l), max.	1.5	1.5	1	-	-
17.	Lead as Pb (mg/l), max.	0.1	-	0.1	-	-
18.	Copper as Cu (mg/l), max.	1.5	-	1.5	-	-
19.	Zinc as Zn (mg/l), max.	15	-	15	-	-

* If the coliform count is found to be more than the prescribed tolerance limits, the criteria for Coliforms shall be satisfied if not more than 20 percent of samples show more than the tolerance limits specified, and not more than 5 percent of samples show values more than 4 times the tolerance limits. Further, the faecal coliform should not be more than 20 percent of the coliform.

*Source : Indian Standard IS : 2296-1982

#A - Drinking water source without conventional treatment but after disinfection

B - Outdoor bathing (organized)

C - Drinking water source with conventional treatment followed by disinfection

D - Propagation of wildlife, fisheries

E - Irrigation, industrial cooling, controlled waste disposal

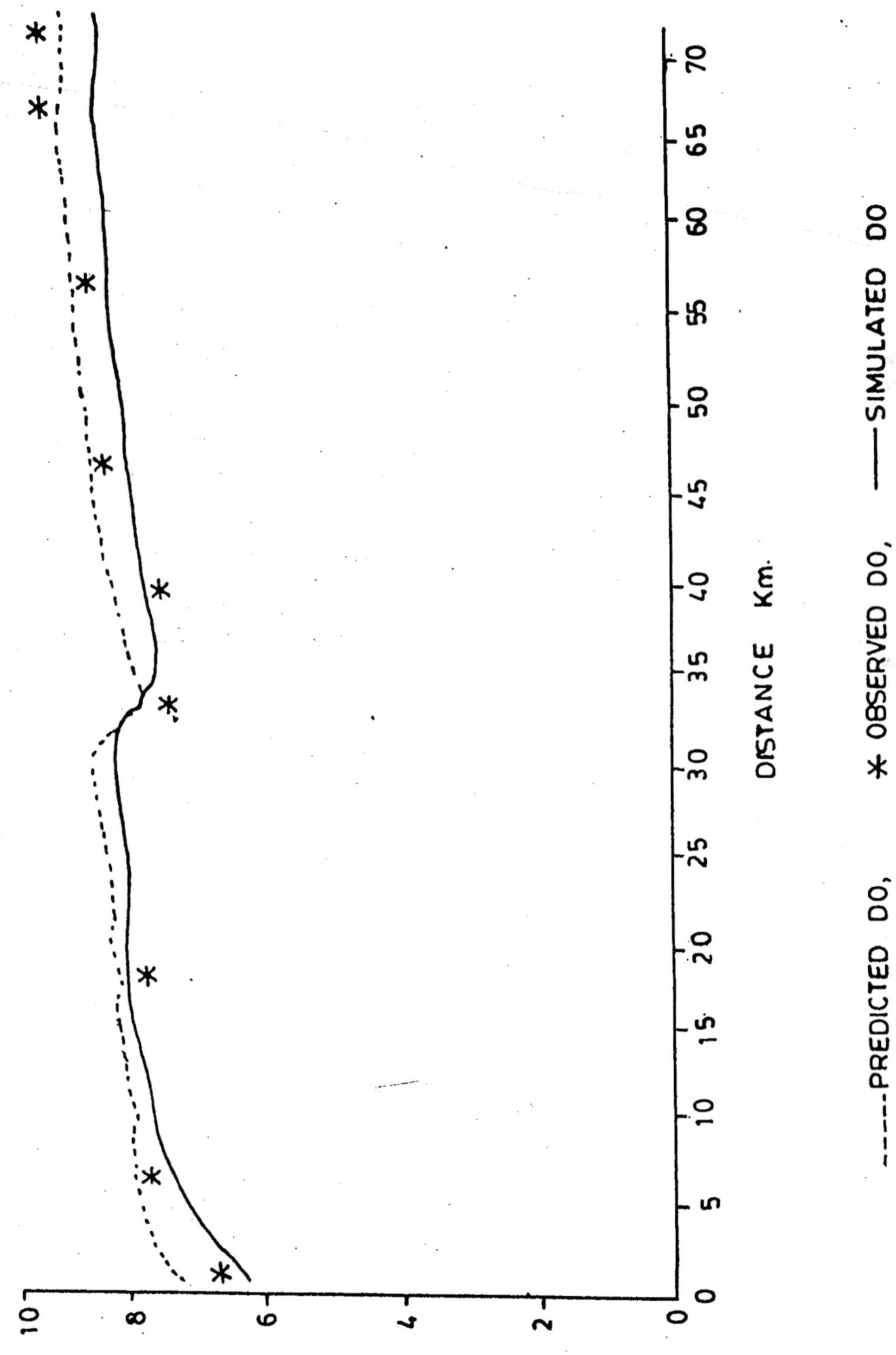

Fig. 67.4 : Predicted Water Quality with respect to DO

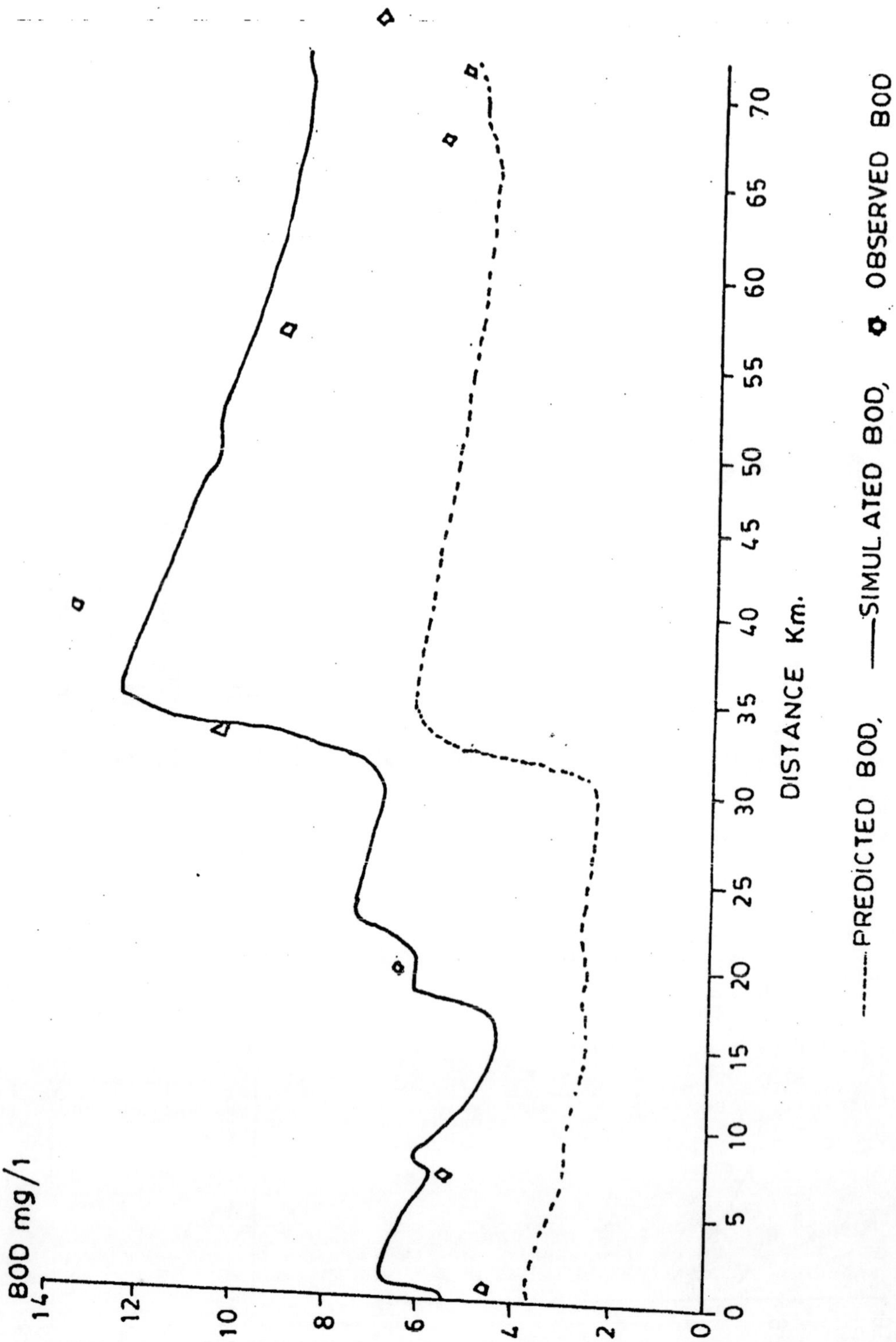

Fig. 67.5 : Predicted Water Quality with respect to BOD

belt areas.

However, cleaner technologies of industrial production, alternative industrial mix and typology, technological and policy interventions in addition to advanced EOP (End Of Pipe) treatment technologies adoption need to be considered under long term environmentally compatible economic alternative developmental scenarios of Damodar Basin.

In order to arrest further deterioration in all the quality of Damodar river water under existing and short term scenario (before the long term strategies start yielding results), it is desirable to take immediate short term measures which mainly include the following. Implication for management of Damodar River Water.

(i) Provision of effluent treatment facilities in all the industries wherever they are not present

(ii) Provision of additional treatment facilities in all the industries wherever they are found to be inadequate. Document published by CPCB may also be consulted to identify the industries under this category.

(iii) Provision of advanced treatment units in all the industries in addition to inplant and water conservation measures to facilitate recycle and reuse of wastewater.

(iv) Wastewater minimization opportunities identification in each industry.

(v) Provision of sewerage to facilitate collection of wastewater from all the towns.

(vi) Provision of adequate domestic waste treatment facilities in each town and these could possibly include the following

  (a) Conventional activated sludge

  (b) Aerated lagoon

  (c) Extended aeration

  (d) Oxidation ponds

  (e) UASBR (Upflow Anaerobic Sludge Blanket Reactor)

(vii) Mining wastewater treatment to facilitate recycle and reuse

ACKNOWLEDGMENTS

The authors are grateful to Dr. S.C. Mudgal, Sr Advisor, of MOEF Government of India for sponsoring the project study and constant encouragement. Prof. B.B. Dhar, Principal investigator of the project is gratefully acknowledged for providing necessary facility constant encouragement and suggestion during the study. At last sincere thanks to all the scientist of CMRI , who directly or indirectly helped during the study.

REFERENCES

American Public Health Association, Inc., (1985) (16th Edition). Standard Method for the examination of water and waste water.

American Water Works Association (AWWA), (1992) Standard method for the examination of water & waste water, Greenberg, Arnold E., Lenore, Clesceri S., and Eatten, Andrew D., (Ed)., Washington DC 20005.

Bowie, G.L., Mills, W.B., Procella, D.B., Camphell, C.L., Pagen Kopf, J.R., Rupp, G.L. Johnson, K.M., Chan, P.W.H and Gherini, S.A. (1985) Rates Constant and Kinetics formulation in surface water quality modeling, 2nd edition, EPA/600/3-85/040, U.S. Environmental protection agency, Athens, GA.

Demuren, A.O. and Rodi, W. (1983) Side Discharge into Open Channels : Mathematical Model, J. Hydraul, Eng., 109(12), 707

Deniil, E.I. and Gulliver, J.S. (1988) Temperature dependence of the liquid film coefficient for gas transfer, J. Environ, Eng., American Society of Civil Engineers, 114(5), 1224

Dhaneshwar, R.S. (1972) Proc. of All India Conf. on Abatement of Env. Poll. CMWSA and Jadavpur University, Calcutta 14 - 16 Dec., pp. 25-40.

Edward Jones E.S. The river valley project : A participatory approach to integrated catchment planning and management in Scotland, Journal of Environmental Planning and Management, 40(I), 125-141, 1997

Fujiki, M. (1972) Proc. 6th int. Conf. Water poll, Research paper No.12

Grant, N. (1971) Environment, Vol.13, No.4 pp 2.

Indratna, B., Sullivian, J., Methery, A. (1995) "Effect of ground water table on the formation of acid sulphate soils, Mine Water and the Environment Vol.14, 1995, pp - 77-84.

JRB Associates, (1983) User manual for Vermont QUAL-II, model, prepared for U.S. Environmental Protection Agency, Washington, D C

Martin, J.L. (1986) Simplified, Steady-State Temperature and Dissolved Oxygen Model : User's Guide, Instruction Report E-86-4, U.S. Army Corps. of Engineers Waterways Experiment Station, Vicksburg, MS

Mudgal, S.C. (1985) Jan, Guideline for Environmental Impact Assessment of river valley project Department of Environment

Mudgal, S.C. (1985) Environmental Aspect of Water resource development project, Indo-Soviet workshop 17-18 Sept.

O'Connor, D.J. and Dobbins, W.E. (1985). "Mechanism of re-aeration in natural stream" Trans, ASCE, Vol.123 pp. 641-684.

Singh, R.N., Sivakumar, M. and Atkin, A.S. (1995) "Application of Executive informative system to mine site water pollution control, mine water and environment, Vol.14, Annual issue, 1995, pp. 95-106.

Singh, S.K. (1985). The Civil Affairs, Vol.30, pp. 23-29.

Tripathi, P.S.M., Singh, J., Jha, S.K. and Hag, R. (1980) Proc. on Water Pollution and Land reclamation, I.S.M. Dhanbad.

Tucher, A. (1972) The Toxic Metals, Earth Island Ltd., London

CHAPTER 68

Information Management Need for Eco-Friendly Mining in Next Millenium

B.K. Tewary, R.K.Tiwary*, B. Nityanand*, P.K. Singh* and A.K. Dube**

ABSTRACT

The economically strong and stable country requires the large scale exploitation of minerals with minimum damage to environmental system. This necessitates large scale data generation and handling of environmental parameters. The decision on the data required and its collection will be influenced by a range of factors including the existing data, local capabilities in sampling and analysis; the existing information infrastructure; the projected life of monitoring system; and of course the costs of establishing and maintaining the data collection system.

The determination of institutional responsibilities for management of information is a difficult issue. A pragmatic approach is that the organization which needs the data should initially collect and process it in a manner as simple as possible. The networked information not only provides speedy and timely information but also broadening the universe of possible questions on conservation, sustainable development etc.

INTRODUCTION

Throughout history, the mining industry has been greatly influenced by technological advancements. As with any information based systems a high capacity communication network is the backbone of the system. Once communication is established, the positioning and navigation system must be developed followed by the control systems and mining equipment that could perform the required functions. Underground communication technology has progressed rapidly in recent years. Currently most mining operations are using or installing systems for voice communication, computer control and video surveillance and online monitoring of environmental parameters. The implementation of robotics mining technology will accelerate the need for faster and higher capacity communication networks.

---

*Central Mining Research Institute, Dhanbad.

The impacts of mining operations commence with exploration activities, extend through extraction and processing of minerals and continue up to post closure of operation. The potential, social and environmental issues associated with mining and mineral processing operations are both highly significant and complex to manage. The fixed location of mineralized zone of interest imposes constraints on all aspects of mining developments which in turn profoundly influences the environmental impacts of mining developments.

Most mining operations share a number of common activities, each of which have potential impacts on natural environment. These fall into:

- Exploration of the mineral.
- Disposal of overburden and waste rock.
- Treatment of disposal.
- Ore processing.

The potential environmental implications of exploration depend on number of factors viz.

- Proximity of surface water to drill sites.
- Ecological significance of affected habitat.

The key issues to consider in determining the magnitude and significance of environmental impacts include:

- The aerial extent and depth of mineralized zone.
- Effects on dump location.
- Inherent toxicity of wastes.
- Management of slurries and liquid effluents.

EXPLORATION WITH MODERN TECHNOLOGY

Various software program exist that allow the miner to take the raw exploration data and investigate the design of mine. By creating a virtual mine, the entire mine life cycle can be viewed based on exploration data. The type of software and databases useful to anyone involved in mineral exploration include:

- Data resources such as satellite imagery, topography, land grids etc.
- Geologic modelling and visualization software.
- Economic analysis.

Field Geologist may run mapping software program and drill hole log plots to make sure that they have enough data. They save the detail work on modelling for the office workstation and more sophisticated 3D modelling software. An engineer can change the cut off grade or increase the equipment operating costs 2% and view the feasibility of ore body all before a single access road is constructed to the site. A full exploration package includes features for:

- Design and definition of database for spatial data including point samples, trenches, traverses, underground samples and drill holes.
- Surface modelling.
- Drill hole data analysis.
- Reserve estimate.
- 2D /3D graphical interpretation.

Some of the tools addresses the areas of functionality for resource development from exploration to mining and rehabilitation including:

- Resource modelling with solids and block modelling.
- Openpit and Underground mine design.
- Tailing modelling.
- Ground water management.
- Blast design.

One area of special significance is the close integration of the solids and block modelling tools together with the central real time graphics module designed for work in areas such as:

- Reserve calculation during evaluation.
- Powerful modelling of underground stopes and workings.

Software programs in conjuction with databases have given the computer power necessary to run its complex mining options.

The synoptic view of terrain obtained through remotely sensed data helps in easy and reasonably accurate delineation of extent quarries, mining waste dumps, stockyards, land routes etc. It offer to locate the subsidence areas caused by underground mine roof collapse.

The thermal imagery is a very useful tool which could help to detect the underground coal fires. Aerial and satellite images are very effective tools for mapping and evaluation of the vegetation and water resources.

Demarcation of unorganised mining areas in mountainous regions and related impacts in downstream regions is time consuming by conventional survey techniques. Remotely sensed data provides a clear view of the terrain for assessment and detailed analysis. This could be used in planning, execution and restoration of mine activities in addition to monitoring.

Role of IT in embedded systems i.e. UV-Visible Spectrometer (UV Visble), Atomic Absorption Spectrophotometer, Inductively Coupled Plasma Atomic Emission Spectroscopy (ICP-AES) is very important. In meteorological studies many advanced models of weather monitoring systems have been developed. They can store the data on regular interval of time which can be retrieved easily. The data can also be vigorously analyzed as per the requirement through available software.

SOFTWARE – A NECESSITY FOR 2000 AND BEYOND

PROGRAMS	APPLICATION
GEMCON	ORE BODY MODELS
	RESOURCE ESTIMATION
SURPAC	RESOURCE MODELLING
	MINE DESIGN
	TAILING MANAGEMENT
	SURVEYING ETC.
ENTEC/EASI/PACE/SPANS/ARC/INFO	ENVIRONMENTAL MODELLING

INFORMATION MANAGEMENT AND THE MINING INDUSTRY

EXPLORATION

↓

Powerful new exploration technologies

Implication for the industry

Allows real time exploration

Improves accuracy

Permits easy integration of database

Allows sharing of knowledge

MODERN CADASTRE AND TITLE MANAGEMENT SYSTEMS

↓

increases speed and security

integrates with land management

Data base

PRODUCTIVITY INCREASES

↓

Better Mine Planning

GPS and Grade Control

Sensors and control room displays

Improved safety

ENVIRONMENT

↓

through out project cycle

impact statements

environment protection design

continuous monitoring

CONCLUSION

As the Next Millennium approaches we are facing paradigm changes brought on by I.T. Effective Software Management System and appropriate networking between institutions to facilitate dissemination and sharing of information should not be underestimated. A coordinated efforts must be undertaken in order to make sense of all the Environmental Management System for undertaking large scale mining activities with minimum damage to environment.

CHAPTER 69

Computer Aided Allied Services Management System for Mining Industry

*Ajit Jere**

Information Technology in mining industry is still in its infancy stage. Though all other sectors of the economy have been influenced by this, unfortunately mining industry has still not caught up the wave of computerization. As one understands the IT opens up greater opportunities for the exploitation of available resources and allows more flexible utilization of labor and equipment. It also gives the organization a competitive advantage of internationalization of mining activities. Many processes in mining industry have been automated and the industry has been transformed by computer aided design (CAD) and geographical Information system (GIS), but still lot of activities have to be computerized to keep in pace with the mining activities in other countries.

Mining is a continuous process which some time requires mission critical on-line decisions. The organizations involved in mining activities are realizing the importance of computer tools, which will help them in automating the decision making process. For the successful implementation of any mining related tasks the concerned authority needs the authenticated and near accurate data and powerful computer tools which make decision making easy and a way of life. Geographical Information System (GIS) are such tools, which help the authority to accurately create and manage their data.

1.0 INTRODUCTION

Geographical Information Systems are computer aided decision support and planning tools which integrate data from maps (spatial data) and other auxiliary data (attribute data) for a geographical area of interest. They can be used to create and maintain geographic databases and are eminently suited for **what-if analysis** in any planning related activity.

*Autodesk India Pvt. Ltd., Bangalore – 25

Once the GIS package integrates the spatial and attributes data, the planner has a powerful tool, which can be used for information dissemination or analysis. By querying the geographic database in several ways the planner can present the available information in a variety of formats: as printed tabular reports, graphically as map display and map outputs on paper.

The analysis capabilities of a GIS package allow the planner to address what-if questions and work out a variety of action plans in a scientific manner. A number of problems can be solved by geographic analysis. The planner can conduct analysis on finding out the best location for a particular economic activity or finding out the best possible route for the vehicles in case of traffic congestion on major roads.

2.0 APPLICATION AREA

In mining industry, information technology has been actively used at different stages of a project. The software tools, which are used, from exploration to production give competitive advantage to the organization and give useful utility to market mining services. In the recent times, with advent of high-speed hardware components and user friendly software products have given rise to wider acceptability of computers in mining sector. Though high-end computer tools are available to carry out precise mining activity less attention has been paid to the allied services which are equally important. The activities such as township development, site location and facility management and distribution management for mining sector are important areas where computers can be effectively used. The paper gives an overview of possible allied activities where computer tools can be used for effective management of mining activities.

Following application areas have good scope for employing distributed computer aided tools to better manage their resources.

- Township development
- Vendor Information System
- Ancillary Unit Information System
- Vehicle Management System
- Demand Forecasting Model

1. Township Development

One of the key areas for computerization in mining sector is to provide a decent and cost effective accommodation to the large labor force. Mining is a environmentally sensitive segment and it becomes prudent to provide good and clean living environment to its work force. The townships thus designated not only should be closer to the work place but also should be clean from all the possible pollution generated by the mining activities.

GIS tools such as Land Development Desktop allows the user to capture data right from the surveying and interpret to generate a terrain on which one can plan the township with all the possible amenities.

2. Vendor Information System

One of the key application areas for mining industry is to manage the service provider network. The service provider network constitutes a group of companies who own the responsibility of delivering different task at different stages of mining. Hence managing the service provider network becomes extremely important not only for the logistic part of the business but also for the smooth business transaction.

Presently, these operations are managed manually which take considerable amount of time, creating many bottlenecks at critical times. The mining company will not be able to trace the exact problem area and provide the much-needed support.

By employing the computer aided management system, the mining company can have the on-line information on the status of various transaction taking place – either on a day-to-day basis or on a stipulated time frame.

The solution could be on an Internet based tool called Autodesk Map Guide which keeps track of graphical component of the vendor location with complete plant layout and also the real time attribute data such as the inventory details, exploration details, infrastructure details etc. The entire network is then brought into the digital format of the application software to carry out on-line query on various logistic management aspects.

Just to quote an example of similar implementation, M/s. Bonzar Iron and Steel Company in Brazil have captured their entire dealer channel in digital format as illustrated in the Figure 69.1. The company has captured both graphics and real time data using Autodesk MapGuide and queries on information such as total steel sheets moved out on a particular day. These informations are then used to plan out future manufacturing process by making assessment on the consumption pattern of a particular item.

M/s. Bonzar also uses Autodesk MapGuide for infrastructure management of its dealer to optimally utilise the available resources such as dealer storage space, vehicles available for transportation of ordered goods, etc.

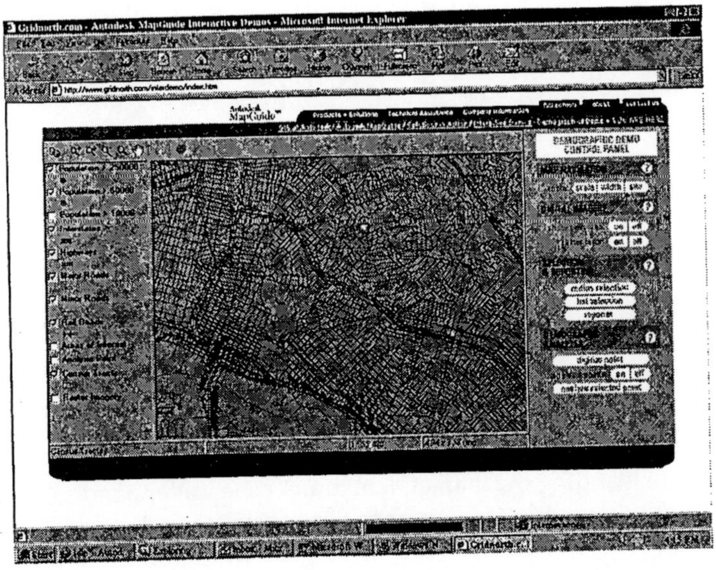

Figure : 69.1

3. Ancillary Unit Information Module

The success of any mining company depends on the way the company manages its ancillary units. As understood, nearly 80% of the mining activities are managed by the ancillary companies, it becomes prudent to manage the flow of components from all ancillary units. The requirement becomes still critical when the operations are intra-dependant. As observed in many component assemblies - 'B' can not start functioning unless and until 'A' is finished its activites. In such dependant activities computer aided tools, which are designed to take care of on-line, dynamic data, comes as handy and provide much needed solutions.

All the ancillary units can be geographically captured with all the details of plant layout and activity flow. Then either Internet or Intranet can link these units with client server architecture. The ancillary units can update status of component design and feed the results to the server. This will enable the mining company to decide and manage its activity accordingly.

4. Vehicle Tracking System

As understood, a road network is a complex system that requires constant monitoring of vehicles that are operating on the system. For various logistic and economic reasons, it is important to keep track of the geographical locations of each vehicle carrying mining ore. With the use of computer-aided tools and global positioning equipment, the system operator can remotely monitor vehicles from a central location.

The advantages of using such automated tools are cost-effectiveness, easy to implement; faster deployment of resources; many end users can use the service at the same time; can use real-time data, etc.

As a methodology, each truck on the road is identified with a unique identification number. All the vehicles on the road can be tracked and important information such as the date, time, latitude, longitude, speed, and heading can be obtained through the automated process in the computer tools function. For example, to generate a report for truck with identification number, say, 3602, one can select 3602 from the list then click on the "Report" button provided in the system software.

The way application works is very simple. A network connects each vehicle with GPS fixed in the vehicle. As soon as a vehicle reaches a destination or before start of a journey, the GPS records the location details and is sent across to the main server. Any one accessing the server will have the latest information of the select vehicle. The software can also compute the average speed of a vehicle and identify the location anywhere between a start and end location.

5. Demand Forecasting Model

This is another area of application where the mining industry can employ computer-aided tools for understanding the market demand. As one understands, the demand for a particular commodity depends on a number of physical as well as natural parameters. These parameters vary from socio-economics of a city to more specific individual requirement. But it is possible to bring these parameters in a statistical model which then can be combined with geographical inputs to find out the demand for a particular mining ore in the study area.

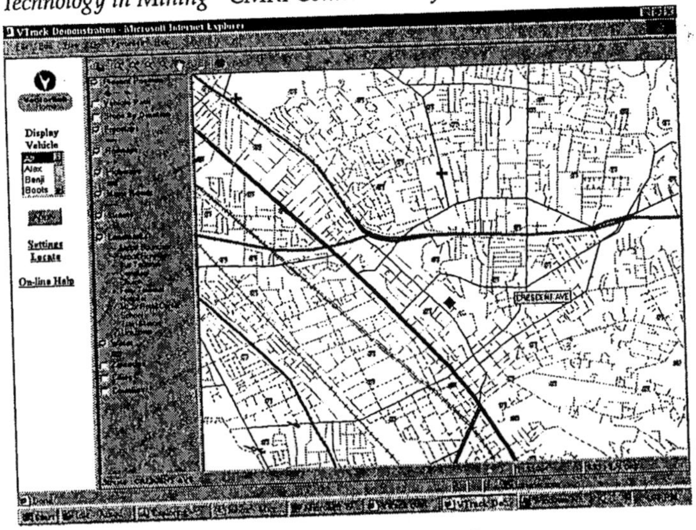

Figure : 69.2

All the above applications can be addressed by a computer aided tool called Autodesk MapGuide, which is essentially a Network-Centric Management tool.

4.0 APPROACH

Network-centric management tool provides several distinct advantages to individuals and groups wishing to develop a map-enabled application for deployment within or outside an organization.

First, network-centric architectures use Internet standards as a foundation; thus the developer builds an application within an open standard–based, widely adopted networking environment. Typically, the developer will already have access to an internal network and will be able to use this backbone along with the existing server hardware and software.

The Autodesk MapGuide plug-in provides a mapping/graphical component for client applications that are called to execute full size or within an HTML frame-set. When a browser opens an HTML document that has a line referencing a map/graphical window file (MWF), parameters that follow the filename alter the plug-ins Default State.

Using the Autodesk MapGuide Author, the developer has the flexibility to configure a pop-up menu system that adds new features to the client plug-in. The pop-up menu accesses server-based processes and configures client-based actions to interact with spatial objects, communicate with the server, navigate the map/graphical window, and perform other application-specific functions.

At the client end, the Autodesk MapGuide plug-in and the emerging JavaScript standard are tightly integrated. Using JavaScript and Java applets embedded within HTML documents, the developer can enable for two-way communication between the end user's interaction with the HTML source document and the Autodesk MapGuide software's internal functionality. This gives the developer several options for providing custom features to the client.

Autodesk MapGuide provides a core-mapping/drawing component at client and server that permits fast development and deployment. It has graphical tools for configuration, and it integrates with open and evolving standards, ensuring that investments in this technology will continue to provide returns on investment for a long time.

Because Autodesk MapGuide was developed as a network centric technology, it is positioned to maximize the benefits of a new age in computing architecture. And as the first management system platform of its kind, it sets the standard for a world of distributed map/drawing related decision support tools.

6.0 CONCLUSION

Autodesk MapGuide which is the first network centric management tool provides useful functionality to manage inter-dependant activities in a corporate industry (involving graphical data). The tool provides dynamic, on-line logistic utilities that can be used for distributed graphical data for access and management.

In a competitive environment like mining industry, wherein, every one is trying to out-beat the other, computer application tool such as Autodesk MapGuide which are on-line and dynamic will help in managing the logistic part of the business. With the help of these tools companies can automate many of their prototype activities to get the best results.

The applications, which are discussed above, are just for reference only. Mining companies can think of these kinds of solutions wherever dynamic data is involved and on-line information or status is required.

CHAPTER 70

Thé Indian Coal Market in The New Millennium

A.K.Tooley & Ashok Verma***

1.0 CHANGING ECONOMIC SCENARIO

Competitive market economy

1.1 Since the beginning of this decade, dramatic political and economic changes are taking place in India. The economic reforms initiated by the Government in 1991 mark a watershed in India's economic thinking and aim at lowering protectionism and integrating the country's economy with global market. Barriers to import and foreign capital are being lowered. Changes in industrial, foreign trade, exchanges rate, and tax policies have led to a paradigm shift towards market forces and competition. New concepts, procedures, treaties and legislative measures are replacing the earlier concepts and processes in businesses and business practices. Due to this globalisation and internationalisation of businesses and business practices, we as a developing nation, have to prepare ourselves to meet the dramatic change in the business and economic scenario in our country.

1.2 India is a sub - continent with an area of 3.3 million-sq. km. and a population of over 900 million. India is endowed with rich natural resources spread over the diverse geographic and climatic conditions. Its growing domestic market promises a vast potential and is now considered one of the locomotives of world economy. It is the third largest economic country in Asia and the eleventh largest in the world. India is the largest emerging market in the world next only to China. Being the second largest stable and mature democracy India is a vibrant nation poised for rapid economic growth.

1.3 As we stand today, at the threshold of a Global Market, there are many challenges looming in the horizon for Indian business and corporate management. Management

*Resident Director, Bengal EMTA Coal Mines Ltd.

**Dy. Chief of Geology, CIL, New Delhi

at all levels has to react to these structural changes in the Indian economy and demonstrate the ability to make tangible contribution towards effectiveness and improving the performance of the organisation in new and different ways to suit the changing scene.

1.4 With development proceeding uninterruptedly, with more than 97% of the primary energy needs being met by fossil fuels and with large reserves of coal available in the country compared to other fossil fuel, the role of coal in the country's economy will continue to be vital. Today as the India economy prepares for momentous changes in its bid to join global mainstream, every sector of the economy, particularly the Indian coal industry has also geared itself up to meet these challenges.

2.0 ENERGY USE IN INDIA

The primary sources of commercial energy in India are Coal, Crude oil, Natural gas, Hydro-electricity and nuclear power. While Coal remains the predominant source of commercial energy in India, Petroleum and Natural Gas are the other major energy contributors to the Indian economy. Natural gas is a relatively new entrant in India's energy sector and has potential to make significant contribution. The share of hydel energy has diminished over the years whereas no major breakthroughs have been achieved in the nuclear energy sector. Coal's potential to contribute to the security of the energy supply is demonstrated in the figure below. The consumption pattern of primary commercial energy in India at present is as follows:

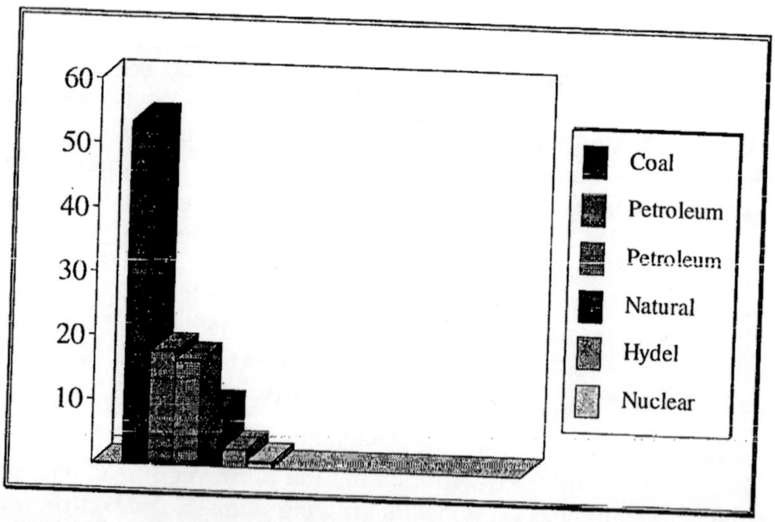

2.1 The development of commercial energy has been taking place at a rapid pace in India since the beginning of the planning era in 1950's. Over the years there has been steady change in pattern of energy supply and the share of commercial fuel has increased from a level of 25% in 1950-51 to about 60% at present. Energy use in India grew at 6.0 % per annum during the last decade. In spite of this growth rate the per capita commercial energy consumption is as low as 9 GJ as against the world average of 60 GJ. Availability of adequate energy and Infrastructure facilities has been recognised to be the two most vital requirements to sustain a GDP growth of 7% and more during the next decade. Interest in India's energy sector continues to grow with the continuing expansion of the Indian Economy.

2.2 For an expected GDP growth of 7% and above for the next decade, the energy requirement of the country would need to rise by 7 - 8%. In order to meet the increasing demand for energy now being anticipated in the next 15 years it will require large investible resources. The Indian energy sectors, which have been largely in the public domain, will witness increasingly the participation of private sector to supplement the efforts of the Govt. to achieve the planned economic growth.

3.0 COAL IN INDIA

3.1 Coal one of the primary sources of energy, as compared to Crude oil and Natural Gas has been the choice of Fuel Policy Planners, as the mainstay of commercial energy in the country. Primacy of coal in the energy policy planning of the country lies in its relatively comfortable resource position vis-a-vis that of the hydrocarbons. The present available coal resources going by the reserve to production ratio is expected to last for another 247 years as compared to 26 years for oil and 40 years for Natural Gas.The present proven recoverable resources of fossil fuels are:

Coal	70 billion tonnes
Crude oil	0.8 billion tonnes
Natural Gas	707 billion Cu. meter.

(*Source*: TERI Energy Data Directory and Yearbook)

Coal reserve

3.2 Geological coal resources as per the recent assessment made by the Geological Survey of India (GSI) is around 209 billion tonnes in seams of 0.9 m and above in thickness and upto 1200 m depth. Coal resources in India account for 0.8 % of the global coal resources. However, proved reserves are 5.7% of the global proved coal reserves.

3.3 Coal deposits in India are confined to the Eastern, South Eastern and Central parts of India consisting of 27 major coalfields. Jharia Coalfield of Bihar is the main source of prime coking coal. Superior grade non-coking coal is available in Raniganj Coalfield of West Bengal, Central India Coalfield of Madhya Pradesh and partly in Talcher Coalfield of Orissa. The shares of overall coal resources of different States are:

Bihar	32 %
Orissa	23 %
Madhya Pradesh	20 %
West Bengal	13 %
Andhra Pradesh	6 %
Maharashtra	3 %

(*Source*: Geological Survey of India)

Balance 3% of coal resources is distributed over Uttar Pradesh, Arunachal Pradesh, Assam, Meghalaya and Nagaland. Quality wise resource position is as under:

Coking / Metallurgical grade coal	15 %
Non coking / Steaming coal	85 %

(*Source*: Geological Survey of India)

Out of this 85 % non coking coal superior grade (A, B, and C having ash content of 24 % or less) is 20 % while the remaining 80 % is of inferior grade having an ash content of 24 - 45 %.

3.4 The ash (mineral matter) of Indian coal is of inherent nature and has high presence of near gravity mineral (NGM). This makes washing of Indian coal very difficult. Indian coal has, however, some very positive features in that it contains low sulphur and low toxic elements. Also the ash is of high fusion temperature.

4.0 PROSPECTS & CHALLENGES - IN THE COAL SECTOR

Pressure on performance

4.1 The environment for the coal industry in India is changing rapidly. For creating a competitive coal market in the country and making the coal mining Industry in India commercially and economically viable, its Budgetary support has been gradually phased out, tariffs on import of coal has been substantially reduced, pricing and distribution of coal is being gradually deregulated. The frame work for future reforms provides for full deregulation of remaining regulated coal prices by January 2000, and opening up non-captive undeveloped mining blocks to competitive bidding and private ownership.

4.2 These substantial reforms, are a major departure from the traditional regime of control and protectionism exercised by Government on the coal sector. Public ownership and state monopolies are giving way to private enterprise and competitive market. The government is content to restrict its role to defining the necessary regulatory frameworks. Lowering of barriers to import and foreign capital and changes in industrial, foreign trade, and tax policies is leading to a paradigm shift towards market forces and competition.

4.3 The emerging business scenario indicates that the Indian coal industry has to prepare itself for these momentous changes and gear itself up to join the global mainstream in the new millennium. There is an urgent need for the Indian coal industry to become proactive, improve its overall performance and understanding of market and customer's need. In order to achieve this, the coal sector has to go through a painful transition. There has to be an increase in the awareness and concern for cost effective coal production, improving quality, upgrading technology, efficiency improvement, downsizing manpower, and increasing productivity and profitability. Coal India has to develop a marketing strategy which uses its own competitive advantages to meet the challenges of growing competition from imported coal and the new entrants in the Indian coal market.

4.4 With more than 60% of the primary energy needs being met by coal and with large reserves of coal available in the country compared to other fossil fuel, the role of coal in the country's economy will continue to be vital. The liberal economic policies are encouraging the rapid expansion of Industry in India, which is in turn creating unprecedented demand for coal to fuel power stations, steel mills, cement and various other industries.

4.5 Given the above scenario some of the considerations for the long term strategic objectives for the Indian coal industry has to be to become the cost leader in terms of being the lowest cost provider of a given value. It should be noted that liberalisation is removing most of the regulatory entry barriers for competition and hence there is a strong need for superior positioning in terms of efficiency. The coal producers are in a position to enter into agreement for supply of coal directly with the consumers as well as charge the prices, which the market can bear. Marketing for coal has to be perceived in terms of pricing for a link in the value chain of consumers rather than in terms of pricing for a commodity. Marketing provides incentive to managers to be efficient with respect to incurrance of costs. This would ensure the strategic objective of remaining the dominant player in the industry.

5.0 DEVELOPMENT OF COAL INDUSTRY - EMERGING ISSUES

5.1 The coal development plan of India has to be seen in the context of the global developments, which are also reflected in the recent changes in the country's development policies. In order to sustain the trend and pace of development of the economy, there may be a need for reconsideration of the policy related to coal development plan keeping in mind that in the past two decades, Coal sector development was centralised with the nationalisation of the coal mining industry in 1973.

5.2 A look at the development of the coal industry from the embryonic stage to now reveals that the successive thrusts on the development of coal sector have been dictated by the needs of the consumer. Initially it was the railways, which were the main consumer, and all the coal, whether superior or inferior in quality was mainly used in the steam locomotives. Subsequently, as the needs of the steel industry grew in importance, the development of the coal industry was moulded accordingly, with a new thrust on exploitation of the coking coal reserves in Jharia field. The power sector has now emerged as the major consumer of coal with coal for metallurgical sector coming next; coal consumption in the railways has become insignificant due to rapid electrification and diselisation. In recent plans, coal based generation has shown the most rapid increase and today power sector (utilities) alone accounts for nearly 66% of the coal demand in the country. Analysis of the trend of coal consumption in different sectors show that the growth in coal demand is now mostly oriented towards thermal power generation programme.

Emerging Issues

5.3 With growing demand in power sector, a number of issues have come up. Critical issues are:

- Adequate availability of coal in terms of quantity, supply of coal as per boiler designs, delivered cost of coal at power stations and movement of coal.
- A long term policy for development of coal resources for meeting the coal requirement of power generation and other sectors of quantity, quality and transportation infrastructure.
- The possible technologies for improving the quality of the coal that is supplied and for more efficient utilisation of coal as also the economics of the same are becoming increasingly relevant.

- An optimal exploitation of opencast and underground coal resources and environment, issues also need to be looked into concurrently for ensuring optimum exploitation of domestic coal resources.

- Next to the power sector, there is a growing need of coal of appropriate quality for steel (hot metal production) cement and fertiliser industry. The import of coking coal is on the rise, as there have been both quantity and quality constraints in respect of domestic coking coal. The Import of coking coal, which is around 10.65 million tonnes is likely to increase further and threatening to become a permanent feature of the economy. The possible strategies that can be followed in future with regard to domestic coking coal supplies including the technological aspects need to be looked into. As regards imports of non-coking coal the comparison of costs of imported and domestic non-coking coal reveals that the cost of the imported coal on calorific value basis may become cheaper in the coastal areas.

- Coal production in the coming years would need to be nearly double the present level and hence large investment are required for development of coal supplies. The policy package necessary to facilitate private participation keeping in view, at the same time, the need for ensuring consumer satisfaction and coal supplies at the least possible cost has assumed particular significance in the changing context.

- The problems of the coal industry, which was nationalised in two spells in 1972 and 1973, have steadily accumulated over the years. Although there has been an impressive increase in production, this achievement has not been accompanied by a satisfactory increase in productivity levels and is still far below international standards. One of the major constraints on the profitability of the coal sector is low productivity level in underground mines. This has stagnated at an OMS (output per man-shift) of 0.55 tonnes for the last two decades, despite massive investments made in the mechanisation of underground mines. The underground mines employ 80% of manpower, but contribute to only 24% of the total output.

- One of the reasons for this state of affairs is the excessively large labour force employed in the industry we employ over 600,000 men in CIL to produce about 251 mt of coal, while USA manages to produce over 900 mt. with around 120000 workers.

- Nationalisation has been the turning point in the history of the coal industry in India. It was possible to structure the coal development programme to cater to the new and emerging demands. To have fulfilled and even surpassed the heavy target of production set before it is certainly a matter of pride for Coal India considering the odds that many of its mines had to face in the wake of nationalisation. It confirms the strength of the organisation. An incremental coal production of 300 million tonnes would be required in the course of the next fifteen years to meet the future coal demand, Meeting the challenges posed by such a rapid rate of growth in the requirement of coal imposes an enormous responsibility on the Indian coal Industry.

- In this context of the emerging issues the various options pertaining to Demand and supply of coal, Technology development required in the area of coal production and utilisation, infrastructure facilities, Coal benificiation, Encouraging Private sector participation in coal mining require immediate attention.

6.0 COAL DEMAND SCENARIO

6.1 Consumption Pattern of Coal in India

The emerging scenario indicates that the coal sector has to strive for a higher growth rate in future years. Long Term Coal demand is projected to have an average annual growth of 6.85 % in the Ninth Plan against about 5.3 % in the Eighth Plan. The level of demand is likely to be 412 million tonne by 2001-02.

Coal is an intermediate and its demand is linked to the consumption pattern in the Indian Economy. While Power sector is the most important consumer, the other important sectors are cement, steel and fertilisers. The balance is being consumed by thousands of other industries like textiles, fertilisers, refractories, brick-kiln, glass, chemical and railways etc. The current consumption level of **296** million tonnes is shared by consuming sectors in the following pattern:

Power Utility & captive	64 %
Steel sector	12 %
Cement	5 %
Other Industries including Fertilisers, Textiles, Brick Klins & Railways	19 %

(*Source*: Planning Commission, Government of India)

6.2 Projected Growth in Coal Demand

According to the recent projections by the Working Group on Coal and Lignite the coal consumption profile is expected to be as follows:

(Million Tonnes)

Sector	*1996 -97*	*2001-02*	*2006-07*
Steel	41.4	49.60	64
Electricity	205.90	262.00	447
Cement	18.2	21.40	30
Others	57.88	79.2	112
TOTAL	323.38	412.20	563

(*Source*: Planning Commission, Government of India)

7.0 COAL AVAILABILITY SCENARIO

7.1 Coal Sector in India is under State Control since the early seventies. One of the objectives of Nationalisation of Coal Industry was rapid increase in exploitation of indigenous Coal resources on a planned basis to meet the growing demand. Massive development programme was undertaken since then, in opening up of new projects, re-organisation of existing mines and development of related infrastructure. As a result, Coal production rose from 78 million tonnes in 1974-75 to 296 million tonnes by 1997-98. Today India is the third largest coal producing country in the world. The impressive growth in coal production is depicted in the table below:

Year	*Production (MT)*
1973-74	78.20
1983-84	138.24
1990-91	211.74
1991-92	229.28
1992-93	238.14
1993-94	246.14
1994-95	257.56
1995-96	283.00
1996-97	285.60
1997-98	295.80

(*Source*: Ministry of Coal, Government of India)

7.2 Projected growth in Indigenous Coal Availability

Based on availability of financial resources indigenous coal availability is expected to be as follows:

(Million tonnes)

Company	*1997-98*	*2001-02*	*2006-07*
Coal India Ltd.	260.50	314.00	372.00
Singareni Collieries Co. Ltd	28.80	36.00	39.00
Tisco/Isco/DVC	6.50	7.60	7.60
Private Coal Companies		13.00	25.00
Total	295.80	370.60	443.00

(*Source*: Ministry of Coal, Government of India)

8.0 DEMAND SUPPLY MANAGEMENT

8.1 When compared with the domestic availability there lies a large gap in meeting the projected demand of coal in the year 2001-02. Against the assessed demand of 412.20 mt the indigenous availability is 370.60 mt only leaving a gap of 41.60 mt. The steel sector has proposed to import 19 mt of coking coal and two thermal power stations have proposed to import 2.3 mt of non-coking coal in 2001-02. This still leaves a gap of 20.3 mt in 2001-02.

8.2 In view of this large demand-supply mismatch there is a need for policy changes to augment coal production in the coming years by measures like facilitating private sector participation in a big way, including the commercial mining by rationalisation of coal prices. In addition to these measures, proper demand management for coal by adopting suitable coal utilisation technologies by the end-use sectors, locating more and more pithead power stations to overcome the rail movement constraints, necessary amendments to the various Acts/Regulations, competitive bidding of coal blocks, restructuring of coal industry etc. are some of the areas which need immediate attention,

9.0 POLICY INITIATIVES TO MEET THE CHALLENGES

9.1 Keeping in view the widening gap between demand and domestic supply, steeply increasing energy demand, shortage of financial resources with public sector coal companies and considering the important position of coal and lignite in India's energy strategy and the urgent need for rapid development of these resources, a Committee was constituted on 'Integrated Coal Policy'. The Committee has made a number of recommendations for augmentation of coal and lignite production and private sector participation in the coal sector. The major recommendations include

- Coal conservation through reduction in specific coal consumption levels for power generation and adoption of improved coal utilisation technologies for power generation and steel making,
- Augmentation of domestic coal production by involvement of private sector, offering of coal blocks both for exploration as well as mining on competitive bidding basis.
- Integrating the exploratory efforts for coal and lignite, creation of a Special Fund and an independent body for coal and lignite exploration,
- Deregulation of coal prices, creation of a Regulatory body for looking into price disputes etc.,
- Setting up of more pithead coal/lignite-based thermal power stations,
- Augmentation of railways and port infrastructure facilities,
- Amendments/modifications in various Acts viz. Coal Mines (Nationalisation) Act 1973,
- Colliery Control Order, 1945, Coal Bearing Areas (Acquisition & Development) Act, 1957,
- Mines and Minerals (Regulation & Development) Act, 1970,
- Forest Conservation Act, 1980 etc. for facilitating private sector participation and expediting project clearance procedures.

9.2 Following recommendations of the Committee have been accepted by the Government:

(i) Deregulation of the prices and distribution of the D Grade of non-coking coal, hard coke and soft coke.

(ii) Allowing CIL and SCCL to fix the prices of E, F & G Grades of non-coking coal till 1st January, 2000, once in every six months by updating the cost indices as per the escalation formula contained in the 1987 Report of the Bureau of Industrial Costs and Prices (BICP).

(iii) Allowing CIL and SCCL to fix the prices of E, F and G Grades of non-coking coal in relation with the market prices and to distribute these grades of coal after 1st January, 2000.

(iv) Setting up of an Independent Body to monitor the detailed and regional exploration of coal and lignite resources in the country.

(v) Allocation of new blocks on the basis of a competitive bidding process in which Indian companies including National Coal Companies (NCCs) may participate.

(vi) Establishment of a Regulatory Body which would perform an appellate function to resolve any price disputes between the producers and consumers.

(vii) Subject to legislative approval, permission to any Indian company to mine coal without the restriction of captive consumption. Implementation of some of these recommendations would need amendments to various statutes governing coal mining in the country.

10.0 THE PRIORITY AREAS

10.1 The priority areas identified for the coal sector in the coming years are:

- Measures for augmenting domestic coal production to reduce the huge gap between demand and supply since large-scale imports are not feasible due to port and foreign exchange constraints.
- Improve productivity of men and machines, maximise capacity utilisation, improvement in technology, improved project implementation etc.
- Improve coal quality as also share of UG mines in total coal production.
- Improve customer service, customer satisfaction and totally computerise the sales operationsand make them totally transparent.
- Encouraging private sector participation to increase domestic coal production.
- Extension of infrastructure benefits to coal and lignite sector.
- National Coal Companies to enter into joint ventures.
- Beneficiation of non-coking coal supplied to consumers beyond 700 kms to be made mandatory with utilisation of rejects for power generation through FBC route at pitheads.
- Proper commercial mechanism for collection of outstanding coal and power sale dues.
- More and more power plants at pitheads to ease movement constraints.
- Streamlining of forestry and environmental clearance procedures.
- Development of clean coal technologies.

- Utilisation of Coal Bed Methane as a source of commercial energy .
- Improvement of environmental management in coal mining areas.
- Restructuring of coal industry.
- Proper coal pricing policy.
- Proper development of rail and port infrastructure facilities.
- Improved safety and welfare measures for coal mine workers.

11.0 INCREASING THE SCOPE OF INVESTMENT IN COAL SECTOR

11.1 To meet the future coal demand, an incremental coal production of 160 million tonnes would be required in the course of the next fifteen years. This will mean an investment of approximately Rs. 30000 crores (US $ 10 billion) at the current price level in coal projects and an additional Rs 3500 crores (US $ 3 billion) in creation of related infrastructures including setting up of coal beneficiation plants. An investment to the tune of Rs. 21000 crores (US $ 6 billion) mainly through government budgetary support, has been made till now in existing mines, new projects and infrastructure. This has now come down sharply and at present there is practically no budgetary support. Public sector coal companies have, therefore, to rely solely on their internal resources and borrowings for sustaining their development programmes. For raising such large resources, multilateral and bilateral funding as well as participation of private/joint ventures on large scale has become inevitable.

11.2 Encourage Private participation in Coal Mining

Efforts are continuing to expand the scope of private entreprenureship in Coal sector. Private Power Generating Cos. setting up new coal based thermal power stations can establish Fuel linkage sufficiently in advance Various options now open to Independent Power Producers (IPPs) are :

- IPPs can enter into a legally enforceable contract with existing public sector coal companies for supply of coal.
- IPPs may Import coal .
- Coal Companies may undertake mining on behalf of IPPs or may decide on join tventures.
- Develop Captive coal mines

A number of Power and Steel Companies have shown interest in development of captive coal mines. Fourteen (14) coal blocks having a geological reserves of 3553 million tonnes and a potential of 37 Mt. production have been allotted for power generation and four (4) blocks having a geological reserve of 2150 million tonnes resources have been allotted to the Steel Authority of India Ltd. SAIL) for mining of Coking Coal.

11.3 Establishing New Washeries

Beneficiation of high ash Indian Coal has been identified as a thrust area in Indian Coal Mining Industry to improve plant performance, reduce strain on transportation system as well as to mitigate environmental impact of coal combustion. The coking coal supplied to Steel mills is washed. In case of non coking coals considering the high ash content, it has been decided that Coal supply to bulk consumers located 750 km away or more from supplying sources should invariably be washed. CIL companies have initiated action for setting up of washeries on Build-Own-Operate (BOO) basis.It has been estimated that about 150 million tonnes per year of non coking coal will be required to be washed during the next 15 years for supplying to bulk consumers located away from the pitheads.

12.0 DEREGULATION OF DISTRIBUTION AND PRICES

Coal distribution and prices till recently were totally administered by the Government. With effect from March 1996 however, Coking coal and superior grade non-coking coal (Gr. A,B,C & D) have now been taken out of the ambit of 'Administered Prices' and deregulated. Ministry of Coal recently (March,1997) deregulated the prices and distribution of D grade Non-coking coal, hard coke, soft coke and empowered the Coal Companies to fix up the coal prices biannually for non-coking coal grades E, F & G on the basis of escalation formula recommended by BICP. Till 1st January 2000 prices of E, F & G grade of Non-coking coal can be revised on every six months basis by updating the cost of indices as per the escalation formula contained in the report submitted by BICP on 1997. After that date these Companies may fix the price of coal in relation with the market price. Distribution of these coal will also be deregulated with effect from 1.1.2000. This is a positive step allowing the market forces to decide the price of the commodities and removing all restrictions on distribution and pricing of coal by the government.

13.0 ENCOURAGING INVESTMENT

The Government is encouraging the International Coal mining and Power generating companies to invest in opening new mines, coal washeries and equipment manufacturing plants in India on turnkey basis. Coal India has also invited entrepreneurs in India and abroad to come and participate on a "Build-Own-Operate" scheme for coking coal as well as non-coking washing. The custom duty on import of steam coal and coke has also been drastically reduced by the Indian Government. This has been done to assist the private power project, which are counting on import as their supply source. The coke levy has been cut to give steel producers the option of importing coke instead of coking coal.

14.0 INTRODUCTION OF MODERN TECHNOLOGY

Under the economic liberalisation policy the coal sector has been opened up to the private sector for investment through joint ventures and strategic alliance and the policy offers competitive terms for the Worlds Coal Industry to participate in the development of Indian Coal Sector. Indian Coal Industry is seeking private capital/Joint venture participation in coal projects for augmenting coal production. Global tenders have been invited for optimisation of coal production from some selected underground mines on gain/.risk sharing basis. A number of Indian and Foreign Companies have shown interest in development of these mines.

Indian Coal Industry has been seeking foreign technical and financial collaboration with various coal producing countries like former USSR (CIS), France, UK, Canada, Australia, Germany, Poland and China. World Bank has also financed a number of open cast projects under 'Coal Sector Rehabilitation Project'. In the area of underground mining technology import of modern roof bolting technology, introduction of high production mining technology, continuous miners etc. are being considered. Joint Working Group on Coal has been set up with UK, France, Germany, Russia, Canada, Australia and China to identify projects for bilateral co-operation.

15.0 SUMMING UP - CREATING COMPETITIVENESS

The coal-mining sector in Indian is now poised for a major leap forward. With the changes in the policy, new and unlimited opportunities have opened up for the first time for investments in this sector for the Indian private sector companies. They can invest on their own in exploration and mining and induct foreign capital and technology through joint ventures or otherwise. The policy initiatives taken by Government of India, we hope will go a long way in increasing the flow of private investment and new technologies in future years.

It has got to be appreciated that for the type of growth that the country needs in this sector, the public sector alone would not be able to deliver the goods. The gap has got to be filled up by private initiatives both Indian and abroad. The growth in coal mining Industry would lead to multiplier effect on the economy through generation of Power and growth of manufacturing and Industrial sector .The participation of private sector in coal mining will improve the competitiveness of the industry and puts pressure upon the efficiency of existing companies. The ultimate policy objective is to provide coal at minimum cost, at internationally competitive prices. The new policy has paved the way for the existence of Public and Private sectors in coal mining and emergence of market forces in this vital energy segment. The initiatives are directed towards:

- Creating a competitive market which will induce and increase efficiency through compulsion of market forces.
- Encouraging all modes of investment - Public and Private, Domestic and Foreign.
- Mitigate the difficulties experienced by potential investors and developers.

CHAPTER 71

Explosives for Servicing Mining Beyond 2000

T. K. Varadarajan and R. Vijaykumar***

ABSTRACT

The industrial explosives sector is intimately associated with the mining industry. Competitive efficiencies that are affecting primary mining companies and secondary users will intensify as the economy restructures and opens out. These effects appear likely to determine the survival of explosive manufacturers beyond the year 2000 by modulating the nature and pattern of growth in demand.

The Indian explosives industry is now critically placed and directions for change need to be properly assessed by its main actors. Because coal absorbs 80 percent of explosives consumed in mining, the modernization emphasis of Coal India Limited is giving a healthy push towards products that are safer, eco-friendly and cost-effective. The obverse face of CIL is its monopsonistic slicing of margins that reduce the ability of the industry to modernize and quickly respond to precisely articulated demand of users. An explosion in supply and large underutilized capacities has further reduced incentives for technological upgradation.

In the restructuring process that is underway, some firms are already developing products by local innovation and through technical collaborations. The industry needs clear guides for the future from the mining sector by demand assessment, careful articulation of demand to achieve sustainable use of resources, precision in establishing the standards for performance and supply comparison and a cooperative approach that rewards technological quality and performance.

INTRODUCTION

Industrial explosives are crucial to the development of modern industrial infrastructure. India has made rapid progress in achieving a great degree of self-reliance

*Assistant Manager (R&D), Tamil Nadu Industrial Explosives Limited, Vellore - 59.

**Chairman and Managing Director, Tamil Nadu Industrial Explosives Limited.

in creating the capacity for producing a wide range of explosives and accessories. Global developments continue to progress rapidly and it is necessary for us to keep pace and to develop technologies and innovate to meet requirements unique to our country.

NG-based explosives have met the needs of the country for more than four decades. Slurry explosives were introduced later and have well accepted. Emulsion explosives are a new generation of explosives and are gradually finding acceptance with consumers in view of their special features and specific applications.

In this paper, we first set out the demand for explosives by the mining industry and how that demand is articulated by the coal sector in terms of technology and services. We identify the factors of Liberalisation, Efficiency, Safety, Quality and Environmental Sustainability that have actuated the nature of demand. This is described in the context of underground and opencast mines. We then describe the response of explosives manufacturers to the demand and their constraints in adequately satisfying requirements. We conclude with suggestions for institutional changes, training needs, attitudinal alterations and procedural improvements.

DEMAND FOR EXPLOSIVES

In this paper, we examine the need for explosives in the mining industry in India. As Table 71.1 shows, the mining sector supports 84 percent of the explosives currently consumed in the country. The non-mining sector is dominated by a miscellaneous category of Construction, Quarrying and other uses that also include important mining sectors - the so-called minor minerals, that are significant for their revenue generation, trade and employment creation. A quick survey would reveal that they have created demand for innovative and cost-effective products from the explosives' sector. For this paper, we will, focus on the mining sector because of its importance in India's ongoing transformation.

Table 71.1 : Demand of Explosives

Mineral	*Explosive Demand (in Tesla)*			
	1995	*2000*	*2005*	*2010*
1	2	3	4	5
MINING				
Coal	1,64,722	2,44,400	2,94,708	3,36,666
Lignite	3,789	5,378	12,467	13,567
Iron/Manganese	14,000	17,170	19,000	20,429
Limestone/Dolomite	16,667	23,333	33,333	46,667
Copper/Zinc	5,650	6,000	6,200	6,300
Others	3,400	5,300	5,800	6,300
NON – MINING				
Seismic	900	1,000	1,200	1,500
Roads & Railways	1,500	1,500	1,750	2,000

1	2	3	4	5
Hydel	6,900	9,950	10,600	11,700
Construction, Quarrying & others	29,900	31,00	33,250	34,400
TOTAL	2,47,428	3,17,131	4,18,308	4,79,529

Source: Demand perspective for explosives & accessories by 2010 AD

The Table 72.1 also includes exports made of explosives, largely to the Middle East, Africa and to some countries in the Far East. Although the industry has the required capacity and technological capability to service this demand, a number of factors inhibit growth of exports. Briefly, these arise from the high cost structure of the economy that raises production costs, poor infrastructure - especially the fact that shipping facilities for IMCO cargo are available only at Bombay and are neither adequate nor cost-effective. In consequence and also because of unfair competition, our subsidised competitors from Chinaare able to undercut prices by 60 percent for customers in the Far East.

Within the mining sector, coal and lignite contribute to 68 percent of explosives' demand. Most of the extracted coal supports the power sector which absorbed in 1996, nearly 77 percent of coal produced by underground and opencast mines. As Table 71.2 indicates, share of the power sector will rise to 83 percent by the year 2010.

Table 71.2 : Demand of Coal by Various Sectors

Sector	*1996*	*2000*	*2005*	*2010*
Power	209	280	447	559
Steel Sector	41.0	51.6	64.0	78.0
Cement	17.5	21.4	30.0	45.0
TOTAL	270.5	353.0	541.0	682.0

Source: Report of the IX Five-Year Plan Working Group on Coal & Lignite.

These trends are supported by the projected increase in energy demand over the next decade. Table 71.3 shows that energy demand will accelerate in the coming decade and a significant share of this growth is likely to rest on the coal sector.

Table 71.3 : Energy Demand

S. No.	*FUEL (MT)*	*1996*	*1999*	*2001*	*2006*	*2011*
1.	Coal	300	360	386	494	628
2.	Petroleum Products	26	30	36	43	61
3.	Crop Waste	110	115	121	129	145
4.	Fire Wood	270	280	295	315	344

Source: Tata Energy Research Institute, New Delhi, 1997

1	2	3	4	5	6	7

The coal production program that has been drawn up by the Working Group on Coal and Lignite commissioned by the Government of India projects an addition of 20.5 percent in coal production in five years from 2001 to 2006 and another 37.1 percent in five years from 2006 to 2011. We present these figures in summary form in Table 71.4.

Table 71.4 : Coal Production Program

Companies	*1996*	*2001*	*2006*	*2010*
CIL	223.2	303.1	370.8	572.0
SCCL	25.7	36.0	39.0	42.5
TISCO	4.6	5.3	5.3	5.3
OTHERS	1.8	2.1	2.3	2.3
TOTAL	255.3	346.5	417.4	572.1

Source: Report of the Working Group on Coal & Lignite

Although these production guesstimates appear excessively ambitious, they indicate two features of significance for the explosives industry and concurrently for all consumers of explosives. The first feature is the increasing emphasis on coal consumption. As coal requirements for power escalate by 59.6 percent in the first half of the coming decade and decrease to 25.4 percent for the rest of the decade, the need for higher quantities of good quality coal will multiply. The second feature is the projected increase of production in Coal India by 22.3 percent in the first half of the decade and by 40 percent in the later half of the decade. Therefore, while other coal producers stagnate, Coal India will increase its share in coal production to nearly 91 percent. Therefore, the systems and procedures of Coal India Limited will continue to dominate the industry and pattern the response of the explosives' industry.

ISSUES IN DEMAND

The coal production program drawn up by the Working Group and which is currently available as the only available perspective on the future of the coal industry does appear to be excessively ambitious. Current trends do not support any hope that the targets set out will be achieved even with changes in the time frame. However, it may become possible to expect improvements if coal production becomes more efficient and the quality of coal improves significantly. In such a scenario, we might expect that import substitution may occur and more financial resources can become available for existing and new coal mines.

In the following paragraphs, we will discuss a number of factors that impact on coal production prospects and thereby having a bearing on the demand for explosives. These factors are interlinked in their operation. For simplicity in analysis, we will here discuss only the effects of Liberalisation, Efficiency, Safety, Quality and the emphasis on Environmental Sustainability. We will initially make brief comments on the global situation before reverting to the Indian situation.

Global Scenario

Globally, the preference for open cast mining appears likely to decrease because of the increasing emphasis on environmental conservation. There is increasing concern about global warming from the rising atmospheric content of green house gases and the emission of gases that affect the ozone hole. There is great concern about acid rain from sulphur dioxide (SO_x) and other emissions that are closely associated with coal use, especially with inadequately beneficiated coal. However, in the absence of cost-effective substitutes, coal is a known, cheap alternative for energy that comes from oil resources.

Remediation costs for most open cast mines, especially those located near points with high population density can be reasonably expected to be high, even as much as 50 percent of mining production costs. In consequence, there may be a return towards underground mining. As coal resources become scarcer or production costs rise, efficiencies in extraction can be legitimately expected to increase with little wasteful mining of coal beyond the year 2000. Coalmines will generally supply multi-products for different consumers and there may be less dumping of dull coal or discard material containing carbon and heat value.

This scenario will, however, be substantially affected by policies and performance in India and China, the largest countries by population and because they are expected to achieve high growth rates well into the next century. Energy consumption for direct consumption and in materials manufacture will soar in these countries and their current dependence on coal as a readily available, local resource seems likely to persist and perhaps intensify. Global trends towards efficiency and environmental sustainability will certainly influence the industry in these two countries.

Indian Scenario

It appears likely that open cast mining operations will continue to dominate the Indian mining scene for the first half of 21st century. However, with the exhaustion of surface-extractable coal seams, underground mining operations will dominate in the second half of the century. Increasing environmental constraints resulting in the higher open cast mining costs will also gradually encourage underground coal mining.

Thermal power plants in the country are already insisting on reasonably low ash coal necessitating the installation of more washers near the pitheads. In comparison to China where value addition is over 24 percent in raw coal that is washed, value addition by beneficiation of coal is currently very low in India. Poor quality of coal poses major problems for steel and power industries because of the need to dispose watery rejects and fly ash, with attendant environmental problems. It is also inefficient to transport non-fuel waste along with coal. From this viewpoint, it appears that underground excavation of coal that yields better quality will be preferred. The solid effluents of washeries will necessarily have to be used for remediation and restoration in surface mines and will hike up costs substantially.

We may collaterally expect increasing use of computer-based systems in planning and design for mining activity. We expect that greater emphasis will be given to reliability, utilization and availability of different mining equipment. There will therefore be greater pressure on explosives manufacturers to meet these priorities and working constraints of mine managers in the future.

If we consider the alternative of easier imports as an important feature of liberalisation, secondary users appear to be increasingly looking at this aspect for achieving greater efficiencies in production and in input costs. This has a direct bearing on the demand for explosives.

Many of the factors that we have discussed in the previous paragraphs are already perceived in India. The liberalised regime, stress on efficiency in price and output of coalmines and the need for safe mining operations has, over the past few. years given a thrust towards open cast mining. This will continue for some time till the constraints attendant to this method make it less feasible. Considering the high level of environmental interest in the country, we may safely expect that time is rapidly running out for this alternative. The Table 71.5 depicts the rapid growth of open cast mining in the country over the last few years and its continuance into the next decade.

Table 71.5 : Trends in Open Cast and Underground Mining

Year	*Coal India*		*Singareni*		*Total*
	UG	*OC*	*UG*	*OC*	
1995-96	54.83	182.44	13.10	13.60	264.04
1996-97	55.18	195.47	13.34	15.00	278.99
1997-98	57.39	213.62	14.51	15.53	301.22
1998-99	59.17	221.50	15.62	16.30	312.89
1999-00	62.00	235.00	16.00	17.00	330.00
2004-05	68.00	305.00	18.00	18.00	409.00
2009-10	75.00	360.00	20.00	19.00	473.00

Source: TEL

All these aspects can be expected to influence the policies of Coal India Limited. As we observed earlier in this paper, Coal India will increase its dominance of the industry in the coming years. Coal India itself comprises a number of subsidiaries. Growth in their mutual competitiveness can reasonably be expected to lead to the promotion of cost-effective and modern explosives and accessories.

The explosives industry in India has been keeping pace with developments taking place in the Indian mining scene. Today nearly all the modern explosives technologies of the world are available in our country. The salient developments in each of the technologies that have taken place to suit the specific requirements of mining industry are discussed below in the context of specific mining methods that are being adopted.

Underground Mining

The underground coal and metal mining industry in India is an agglomeration of relatively small mines and may be distinguished from patterns available in North America. These mines have since been taken over in the structure that we see now in the form of Coal India Limited and its eight large subsidiaries. The size imposes constraints on the transfer of technologies developed under different structural conditions. Moreover, the nature and volume of investment in Indian mines are significantly different. Therefore, technologies and methodologies applied there may not yield similar cost economies in the Indian situation.

Technical innovations within the explosives industry and the heightened awareness of cost and productivity have, however, promoted some changes and innovations in blasting practices. However, the underground coal-mining sector needs greater emphasis and encouragement for technological change oriented towards improved productivity. Under Indian conditions as they exist today, the aim of high production and low investment has supported two popular methods of blasting: blasting gallery method of mining and blasting with auger drills. We, however, expect that long wall mining will gain ground because it can raise extraction rates with improved safety. Automation possibilities also exist. The older board and pillar method may persist into the future.

Explosive manufacturers have been keeping pace with the need of underground mines. The traditional NG explosives have been supplemented by permitted slurries but its lower performance features limit the share of slurries. However, neither NG nor Slurry explosives are able to meet the challenges of mining industry in the areas of safety, environmental hazards and cost-efficiency. Research and development efforts worldwide have yielded superior and novel emulsion explosives.

In emulsion explosives, the intimate interfacial contact between oxidizer and fuel results in a high degree of sensitivity and high reaction rate. The very high velocity of detonation that yields a high detonation pressure and shattering effects improves fragmentation. Because the fuel phase is water immiscible, emulsion explosives are highly waterproof. They have made a major impact in the Indian market as may be seen from the Table 71.6.

Table 71.6 : Consumption of Different Types of Permitted Explosives

Explosive	*1996 – 97*		*1997 – 98*		*1998 – 99*		*1999 – 2000* *	
	CIL	*SCCL*	*CIL*	*SCCL*	*CIL*	*SCCL*	*CIL*	*SCCL*
Quantity								
NG	8199	6890	7705	6740	7816	5880	22920	5350
Slurry	7112	340	7312	310	6193	560	21500	450
Emulsion	7073	578	7246	770	6572	810	20000	840
TOTAL	22384	7808	22263	7820	20581	7250	54420	6640
	% Share							
NG	36%	88%	34%	86%	38%	81%	35%	76%
Slurry	31%	4%	33%	4%	30%	8%	33%	6%
Emulsion	33%	8%	33%	10%	32%	11%	32%	18%

Source: TEL

* Figures for 1999-2000 are order quantities and are typically thrice actual supplies.

The Table 71.6 makes it clear that the demand for permitted explosives in India is fairly stagnant. Forecasts suggest that future growth will be limited to not more than 1-2 percent. Because emulsion explosives have many attractive features, it appears likely that they may entirely phase out NG explosives in the future. The Government of India has already issued a notification denying expansion or new licensing for NG based products. The emulsion explosives manufacturers have also been lobbying the Government to speed up the phasing out of NG products. However, Table 71.6 also makes it apparent that after making an initial impact, emulsion explosives have not succeeded in improving their market position. Closer study, however, reveals that the performance of permitted emulsion explosives is far from adequate. Their inconsistent initiation sensitivities compel use of accessories that compromise safety and attenuate a chief feature of emulsions.

Among accessories used for mining, explosive manufacturers have indigenously developed or entered into technical collaborations to respond to the needs of the industry. These are (NONEL) non-electric initiation systems. NONELS have a small diameter plastic tube internally coated with an explosive powder. A propagating shock wave initiated using a detonator or detonating cord travels down the plastic tube to initiate the delay element of a detonator that is used as a primer charge for the explosive. NONELs may be used in places where electrostatic hazards are likely and more particularly in underground metalliferous mines where a large number of holes have to be blasted simultaneously.

Opencast Mining

A large number of open cast mines are currently operated. As we have commented earlier, secular growth and expansion of open cast mining appears likely to be limited, although short-term trends show substantial growth as evident from Table 71.5. Changes in mining techniques, mainly due to the decline of mineral prices has, however, created demand for more advanced, cost effective and reliable explosives. Emphasis is building up to optimise blasting results and fragmentation, maximise safety, and minimise misfire rates while avoiding cut-off.

The open cast situation permits economies in the delivery systems for explosives and greater safety in transport and handling. Against the earlier formulations that were packaged, bulk explosives have become popular. Dramatic changes have therefore taken place in developing explosive formulations and delivery systems for bulk explosives. Among the greater technical and economic advantages of bulk explosives over packaged explosives are the lower capital costs, adaptability to larger scales in operations, shorter gestation periods, lower manpower needs and higher rates of production.

Bulk explosives are handled as 'bulk' and are brought to the point of use and adapted to the specific needs. They may be broadly classified into the three categories of Plant mixed systems, Site mixed systems and Repumpables. Each of these systems has its specific advantages and can be used to suit the site design and the objectives established by the mining engineer.

Current consumption of bulk explosive systems is shown in Table 71.7. The mining industry has been quick to realize the benefits of bulk explosives and can be expected to continue to support this sector.

Table 71.7 : Consumption of Bulk Explosives

Explosives	*Consumption(metric tonnes)*		
	1997-98	*1998-99*	*1999 – 2000*
Slurry	47,331	39,081	35,200
Emulsion	11,436	13,472	25,000
Doped emulsion	38,506	64,855	71,450
Heavy ANFO	15,247	13,540	6,500
TOTAL	1,12,520	1,30,948	1,38,150

Source: TEL

The primary beneficiaries of expanded use of bulk explosives are the manufacturers of emulsion explosives. However, there has been substantial increase in the use of large diameter slurry explosives that provide higher levels of sensitivity and power required in some applications. The consumption of non-permitted small diameter slurry explosives for coal and in rock excavation, estimated to be about one-fifth, is expected to stagnate at this level.

Although the manufacture of NG explosives involves greater costs and restrictions compelled by considerations of safety, non-permitted small diameter NG explosives continue to be used. This is primarily because of problems in inconsistent initiation sensitivities of slurries and emulsions and because they are susceptible to pressure desensitization.

Explosives manufacturers have enthusiastically responded to these requirements and are currently providing the full range of explosives. They have established plant capacities and have invested in temporary systems near the user to service needs of mine managers. We will discuss in greater detail the explosion in licensed capacities in the next section. Technical collaborations have been an important route to this situation. They are also trying to adapt to local needs and the perceived shortcomings that we have discussed above by innovations and by the development of appropriate products and new types of services. These include training programs for mining engineers and forward integration by taking up contracts for blasting using their own explosives.

CAPACITY LICENSING

The present available licensed and installed capacity for explosives and accessories is shown in Table 71.8. In a situation where substantial unutilised capacities exist, there has been a simultaneous expansion of licenses issued for potential manufacturers. In about one year from April 1998 to May 1999, the Government of India has issued Letters of Intent for creating additional capacities of 34,810 metric tonnes of Explosives, 50 million numbers of Detonators and 112 million metres of detonating fuses.

These new entrants will enlarge the pressure on explosives manufacturers who have invested considerably in modern equipment. Moreover, licensing does not ensure that manufacturers establish quality control systems that enable them to assure quality in supply and ability to develop new products or innovate. If anything, ad-hoc licensing only inhibits investment in these areas and in modern equipment by existing manufacturers. Although this comment appears to act against the theme of liberalisation that is currently popular, a similar situation does not appear to prevail in the developed countries. There appears to be a need to examine the institutions and processes set in place abroad and develop systems suited to India that achieve specific goals. To reiterate, the explosives industry is a vital associate of the mining industry. We consider that inadequate policy attention to this sector will definitely have negative consequences for the mining sector.

Table 71.8 : Licensing of Explosives and Accessories

Product	*Capacity*		*Projected Demand*		
	Licensed	*Installed*	*2000*	*2005*	*2010*
Explosives ('000 tonnes)	502.00	414.00	350.00	425.00	515.00
Detonators (M nos.)	648.50	386.00	360.00	273.00	290.00
Det. Fuses (m metres)	211.10	141.00	71.00	91.00	113.00

Source: Chief Controller of Explosives, Nagpur.

With demand far outstripping supply of coal, Coal India Limited, as a holding company of several subsidiary coal mining companies, dominates in the supply of coal and therefore in the consumption of explosives. CIL in under tremendous pressure to improve productivity. The explosive industry is therefore also under considerable pressure to improve performance. The institutional systems of CIL and its processes for purchase and use of explosives are critical to the future of the explosive industry in India. This critical feature of CIL has been growing because capacity utilization in dropping fast in the explosives industry.

We are of the view that CIL has a primary duty to create the systems and procedures that can assure health in the explosive industry and thereby improve its own productivity. We acknowledge that CIL has already taken many steps in this direction by the promotion of new types of explosives. Recently, CIL also organised a team to inspect suppliers to grade their quality assurance systems. Obviously, much more needs to be done in the methods of purchase and evaluation of products; in promoting innovation in the mines and among suppliers; and in rewarding those explosive manufacturers who commit long-term investments and are in the business of supplying quality goods over the long-term. In the following section, we discuss in some greater depth, the systems that are currently followed for the evaluation of explosives and the purchase systems in place.

PERFORMANCE EVALUATION OF EXPLOSIVES

The availability of many different kinds of explosives systems for use in the same location and the large number of manufacturers and suppliers poses many problems in the systems and procedures adopted for comparing performance.

Some of the most important parameters that are examined by testing laboratories and explosive suppliers are Energy, Strength, Impulse, Relative Weight Strength and Bubble Energy. From the point of view of the user, the performance of an explosive must be related to rock and blast conditions. Because variable rock conditions are encountered in mining operations, the mining engineer needs to access more than one type of explosive and vary compositions to suit his purpose. The parameters that buyers use to compare suppliers and evaluate offers are therefore Price per tonne, Reliability, Quality of Supply, Scope and Products Range of the manufacturer, Technical Support capability and Past Performance.

Currently, two systems are followed for rating suppliers of explosives. One of the major mining companies selects the type of explosives based on the lowest quotation while the other major mining company adopts the lowest rate arrived for the other company and adds the requirement of a "Guaranteed Powder Factor." The choice of type and quantity of explosives is left to the manufacturer. Both these systems have a serious drawback because undue emphasis is laid on cost. The need for optimum blasting is not sufficiently reflected in the purchase decision. Moreover, the concept of a guaranteed powder factor as a sole criterion is technically inappropriate and controversial in application in an open tender format as practised. Its application is also problematic because it does not appear to support the important long-term objective of demand articulation that will help the explosives industry support increased productivity in the coal sector.

By and large, the comparison depends on the quoted price per tonne which is by no means a valid measure. Explosive performance must relate in the final analysis to the comprehensive objective of minimizing total cost of both blasting and excavation.

Because explosives with similar characteristics may perform differently, the choice of and preference between explosives have necessarily to be made based on their performance in the field at the time of testing and under actual operational conditions. The most suitable explosive for a particular situation may be selected by an analysis of the powder factor, yield and power consumption. The appropriate explosive is the one that is characterized by low powder factor, high yield and low power consumption. If two or more explosive systems fall within the same category, costs may then be compared to arrive at the best offer.

We are of the view that the systems that exist in the country in the coal sector are highly centralised and rigid. Perhaps a more federal set-up within the main coal companies with clearly established systems for decentralisation of authority and structures for accountability may yield better results. The objective of any such reform should be the development of new products through collaboration and innovations as appropriate to the unique conditions of Indian coal mines.

We have observed a long-standing problem of attitudes. Mining companies tend to look upon explosive manufacturers as suppliers of a low-cost commodity. One of chief reasons for this communication gap is the lack of a procedural or institutional arrangement for ensuring that explosive manufacturers and mining companies work in cooperation to achieve mining objectives. As a result of this lack of exchange, explosives manufacturers currently concentrate on importing or developing technology to produce better, low-cost commodities. If we wish to obtain more efficient recovery of minerals and achieve international standards of performance in the mining industry, all players and empowered agencies in the areas of mine planning, mine engineering, equipment manufacture, explosives manufacture, testing and certification have to learn to work together as a team. For these to be effective, effective mid-course training programmes need to be organised for mine planners and engineers. These training programs should be in addition to the programmes arranged by explosive manufacturers to demonstrate new technologies and to advertise their products.

Centralised purchase procedures for explosives are unavoidable but also necessary to ensure fair and transparent treatment of suppliers. As we mentioned in the previous section, CIL has already taken steps to assess the capability of explosives suppliers and ensure that better supplies are made over the long-term. Tender procedures, however, may need to look at a longer time-frame in order to support the proliferation of bulk explosives and thereby achieve cost-economies. Tender procedures promote fierce competition, which is healthy. However, the slicing of margins becomes counter-productive. If cartels among the explosive manufacturers fail, as they have always done in the past, then disincentives begin to operate in technological development and encourage unprofessional methods of getting over the losses that are the consequence of unbridled competition.

CONCLUSIONS

We have attempted to present in this brief paper, an overview of the demand for explosives by the mining industry. We have also tried to describe how this demand is articulated to the explosive industry in terms of the technologies and competitive efficiencies desired. A large number of factors have come together to initiate this process and these include liberalisation, safety, environment, etc. As a result, the industry can access nearly all types of explosive technologies available the world over. Much more effort is still needed from all parties to this mission that provides valuable infrastructural support to India's growing economy.

We have also attempted to identify issues that need to be addressed for the joint efforts of explosive manufacturers and mining professionals to yield the required results. We reiterate the main features of these aspects below.

Information sharing is not adequate on the need and use for explosives; and on their actual performance features as determined in the field. Although privacy is necessary, such information is critical for explosives suppliers to respond in an informed manner with appropriate technology.

We have highlighted the need for institutional development in the mining industry. In addition, we have highlighted the need for institutional and procedural changes in the interaction between the mining sector and explosives manufacturers. We have particularly underlined the need for training and updating skills of field mining engineers. We feel that it is the responsibility of mining institutions to sharpen the skills of mining professionals through inputs that are overtly unaffected by explosives' interests. However, in practice explosive companies are providing a variety of technical inputs while introducing their products.

We have emphasised the variety and depth of the response of explosives manufacturers to the demand of the mining industry. Perhaps better quality of research and development may become possible in the explosives sector through a dedicated science and technology development center that could be funded by the government, explosive manufacturers and users.

We are confident that this Mining Conference will invite attention to all these factors and assist in creating an atmosphere where all players in the mining sector and the explosive industry work together.

CHAPTER 72

Management of Safety in Indian Mines — Need for a Fresh Approach

*Vinay Mahajan**

ABSTRACT

India, being well endowed with mineral reserves, has taken giant strides in the production of minerals in last 50 years. In 1997-98, some 85 minerals with an aggregate value of app Rs 40,000 crores constituting about 4.5 % of the national GDP were produced from the mines. On an average app 1 million persons are employed in the mineral industry every day. In exploitation and production of minerals such as Coal, Iron Ore, Limestone, Bauxite, Bayrites, Chromite, Manganese etc India holds an eminent position in the world.

Though the output of various minerals may have increased greatly with the introduction of high capacity machines and new technologies but a commensurate reduction in the fatal and serious accidents in the mines have not been achieved so far. The fatality or serious accident rates, indicative of the hazard potential in the mines, continue to be distressingly high. Occurrences of major accidents in Indian mines at frequent intervals of time continue to haunt the Industry.

Safety in mines is a barometer of excellence of organizations operating the mines, regulatory legislative measures and the employees or their Unions. There cannot be an excellent industry where a lot of accidents, inflicting injuries on persons working therein, occur regularly. Barring isolated ups and downs, the accident curves for accidents in mines are stubbornly maintaining a flat profile for the past two decades or so. While on one hand the Safety and Occupational Health problems are likely to get compounded in the coming years, modern society on the other hand is demanding a more safe and decent place of work than ever before. It must be understood at this point that our ability to improve safety standards in mines is directly related to our ability to change according to requirements of the industry- a process, which is evolutionary in nature.

*Former Director General of Mines Safety, Ministry of Labour, Govt. of India.

The task may be daunting but it is necessary now to ponder upon the need for a fresh approach towards achieving higher and better safety standards in Indian mines. The approach should be such that measurable results, over a period of time, can be achieved for the benefit of all associated with the industry. Higher and better safety standards are not only necessary to enable the Indian Mining Engineers and the Nation to hold the heads high in the globally competitive environment but also to attract intelligent, diligent and dedicated manpower into the sunrise industry like mineral industry in India.

A. PREAMBLE

There are numerous references of Gold, Silver, Iron, Gems etc. in ancient and medieval Sanskrit literature. These metals and gems played an important role in the Aryan religious and secular life. There are numerous references to metals and gems in the hymns of Vedas etc. Even in the post Vedic Sanskrit literature numerous references can be found e.g. Manu's Law of Rajdharma etc. There is ample evidence that these metals, gems etc. were produced from indigenously mined minerals. Even rules regulating the lease and licence of miner, for conducting mining operations and the qualifications of mine Superintendent etc. can be found in Kautilya's Arthashastra. The presence of metallic coins and other artifacts found in archeological diggings of Harrapan civilization are material evidence of ancient mining, smelting and forging/stamping techniques etc. as practised then. There are many evidences that exploitation of minerals like Coal, Iron Ore, Copper, Lead and Zinc has been going on in the country from the time immemorial.

Minerals constitute one of the main pillars of economic growth of any country and India has been eminently endowed with this gift of nature. Exploitation of minerals to obtain critical metals and minerals for achieving rapid growth of industrialization of the country with resultant growth of the national economy was considered desirable and essential. After India became independent, the growth of mining industry under the impact of successive five-year Plans has been very rapid.

B. GROWTH OF THE INDUSTRY

The post independence era witnessed a massive expansion of exploration activities, which led to the augmentation of the mineral inventory as well as addition of mineral reserves to the existing ones. The search was not only confined to the Indian landmass but was carried out in off shore areas too. These efforts yielded the presence of large reserves of petroleum in off shore areas as well as large reserves of poly- metallic nodules bearing metals such as Copper, Cobalt, Nickel, Manganese etc. lying on the ocean floor.

Ambitious programs were launched to increase the production of minerals to meet the ever-growing demands of the core sector industries like Steel, Non- Ferrous metals, Power, fertilizers etc. besides exporting some of the minerals to earn much needed foreign exchange. Many public sector organisations were set up to take up exploration for and exploitation of minerals. The State thus assumed direct responsibility for developing mines of important minerals and establishing mineral based industries. This not only resulted in increases in the production of metals but also items like Thermal Energy, Cement, Chemicals, Fertilizers and other mineral based products. The trend in growth of some of the important minerals in the country is shown in the Table 72.1.

Table : 72.1

Year	*Coal*	*Copper Ore*	*Lead and Zinc ore*	*Iron Ore*	*Limestone*	*Bauxite*
1951	34.98	0.37	0.01	3.71	2.96	0.06
1961	55.71	0.42	0.15	12.26	15.73	0.48
1971	75.64	0.68	0.30	32.97	25.26	1.45
1981	127.32	2.01	0.96	42.78	32.56	1.75
1991	237.76	5.05	1.82	60.03	75.02	3.86
1997	316.68	4.76	2.01	68.69	156.02	5.04

**All figures in million tonnes

The strategies adopted for rapid expansion of production capacities included increased mechanization, adoption of new technologies and their adaptation under our geo-mining conditions and assimilation of latest scientific innovations in the concerned areas. However, this pursuit for higher production has also brought in its wake-increased hazard potential in the mines posing new problems of safety management.

C. MINERALS AND ECONOMY

Some 85 minerals are presently being mined in our country. The gross value of the total mineral production in 1995 is estimated to be app Rs 40,500 crores i.e. app 4.5 % of the GNP. The contribution of mineral production has steadily increased from 1.02% in 1960-61 to 1.54% in 1981 to 4.5% in 1997. Minerals continue to play a vital role in India's overseas trade too. The total value of the minerals produced in the country has been growing at a rate of 4-5 % annually during the last 2 decades.

D. EMPLOYMENT

With the growth in production of the minerals, the employment in the mineral industry has also grown substantially. The Table 72.2 indicates the trends in employment over the years: -

Table 72.2 : Average Daily employment in Mines (in '000)

Year	*Coal*	*Oil*	*Copper ore*	*Gold ore*	*Iron ore*	*Lime-stone*	*Mang ore*	*Mica*	*Stone*	*Others*	*Total*
1951	351.9	N.A	3.7	21.7	20.2	16.0	55.5	52.1	5.1	22.7	548.9
1961	411.2	N.A	4.2	21.7	54.5	54.7	47	29.6	8.5	44.9	676.3
1971	382.3	13.6	7.6	12.4	52.8	53.2	30.4	12.2	8.8	57.5	630.8
1981	513.4	14.5	13.4	12.3	44.9	49.8	26.5	6.7	7.7	60.6	749.8
1991	554.1	35.5	12.8	9.3	40.0	43.5	17.9	2.2	11.2	63.3	789.8
1997	503.5	28.6	10.3	6.4	37.6	31.1	14.8	1.2	3.9	48.4	685.8

* Figures for 1997 are provisional

There has been a gross reduction in the employment in mines producing Gold ore, Mica and Manganese ore. This reduction has come about either due to the poor quality of the ores concerned (thus making the exploitation uneconomically viable) or upgradation in technology thereby reducing the requirement of the mineral in various processes or development of substitute products at cheaper prices.

Out of app. 6000 metalliferous mines coming under the purview of Mines Act, 1952 only about 2000 mines are submitting the returns. Further a large number of mines engaged in the exploitation of minor minerals (not falling under the purview of the Mines Act, 1952) are not submitting any returns to DGMS and hence the employment potential therein can not be known exactly but may be only guessed. As such it may be safely assumed that about one million persons are engaged every day in the mines.

E. NATIONAL POLICY

National Labour Policy formulated during the Sixth Plan *inter alia* attached emphasis on

> "......Working conditions include not only wage structures, fixing a minimum wage and protection of income, but also fixing of working hours, periods of rest, paid holidays, provision of canteen facilities and provision of crèches for children. Today, safety includes not only protection of workers against accidents at work but also against occupational diseases. Indeed with the growth and diversification of industry and agriculture, the safety and health of workers go beyond mere measures of prevention. It is equally important to improve the environment since safe and healthy working conditions are the best protection for the worker and best guarantee for increased production......."

F. EXPLOITATION OF MINERAL RESOURCES: PROBLEMS FACED

The new liberalized Industrial Policy may lead to high and accelerated growth in mineral industry to complement and supplement the revival and rapid growth of national economy. However 'Growth at all costs" cannot be allowed to become the order of the day. Ill conceived quest or 'Mad Rush" to produce more may lead to unsustainable developments. The task of providing the nation with much needed minerals in larger quantities has got enmeshed with many problems carried over from the past, which may have come in the form of tradition or legacy. All these need to be addressed to sincerely at this point of time.

As the mine operators begin to look forward for ways and means to improve productivity and efficiency to cut costs, safety considerations certainly assume more and more importance and emerge as a significant factor even purely on economic considerations.

Side by side, the modern society is also awakening more and more to demand a safe and decent work environment as a social need. Hence it must be understood and accepted by all that production of minerals at a high cost especially in terms of human lives or human suffering is not acceptable any more no matter how laudable or pressing the need be.

G. MINING: A HAZARDOUS PROFESSION

It is accepted all over the world that mining is a hazardous profession but there shouldn't be any complacency in our thinking that the hazards cannot be eliminated or reduced and the workers in the industry must suffer. As seen in other hazardous industries/ professions (viz. space industry, aviation industry, petroleum refining, transport industry engaged in conveyance of petroleum products or other hazardous chemicals etc.), it is possible to bring down the hazards by available means to an acceptable level.

As in any industrial accident, unsafe act and unsafe conditions of work lead to an accident in a mine i.e. it is a combination of working environment, the equipment and the worker. Most accidents are preventable- they don't just happen but are caused. Ultimately most of the accidents are either directly or indirectly attributable to human failings.

H. ACCIDENT EXPERIENCE

Incidence of accidents in mines is an important indicator of overall status of safety in mines. Every accident leading to a bodily injury causes suffering to the victim and his/her family. In case of a fatal injury or permanent disablement, the effect is catastrophic on family life. We are still paying heavily for accidents both in terms of human suffering and economic waste. Safety at work is still a serious problem.

The incident of accidents with resultant bodily injuries is shown in Table 72.3.

H.1 Trends in rates of Fatal and Serious Injuries in Mines.

H.2 Coal Mines

Table 72.3 : Death Rate per 1000 persons employed in coal mines

Year	*Number of accidents*	*Number of persons killed*	*Death rate per 1000 persons Belowground*	*Death rate per 1000 persons Opencast*	*Death rate per 1000 persons Aboveground*	*Death rate per 1000 persons Overall*
1951	81	319				0.91
1961	222	268				0.65
1971	199	231				0.60
1981	165	184				0.36
1982	158	185				0.35
1983	156	191				0.36
1984	160	176				0.32
1985	176	204				0.36
1986	180	214				0.39
1987	162	176	0.39	0.44	0.15	0.32
1988	159	175	0.36	0.45	0.20	0.32
1989	156	177	0.39	0.42	0.17	0.32
1990	151	166	0.33	0.42	0.20	0.30
1991	138	143	0.27	0.39	0.19	0.26
992	165	183	0.39	0.52	0.16	0.33
1993	156	176	0.39	0.37	0.19	0.32
1994	156	241	0.59	0.64	0.15	0.46
1995	137	219	0.60	0.38	0.13	0.43
1996	131	146	0.31	0.42	0.19	0.29
1997	143	165	0.41	0.44	0.13	0.33

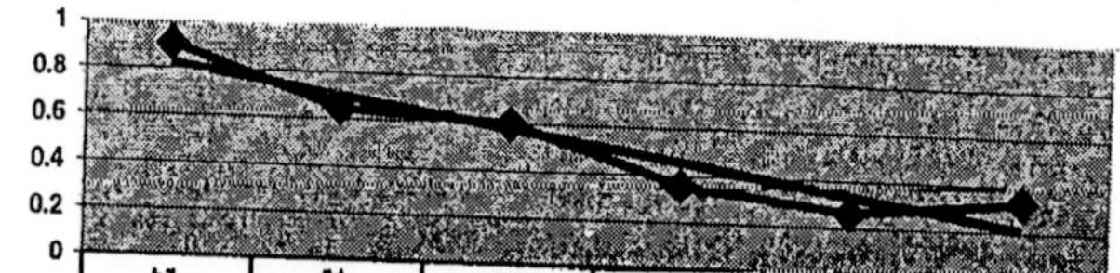

	51	61	71	81	91	97
Death Rate-Overall	0.91	0.65	0.6	0.36	0.26	0.33

Death Rate-Overall — Linear (Death Rate-Overall)

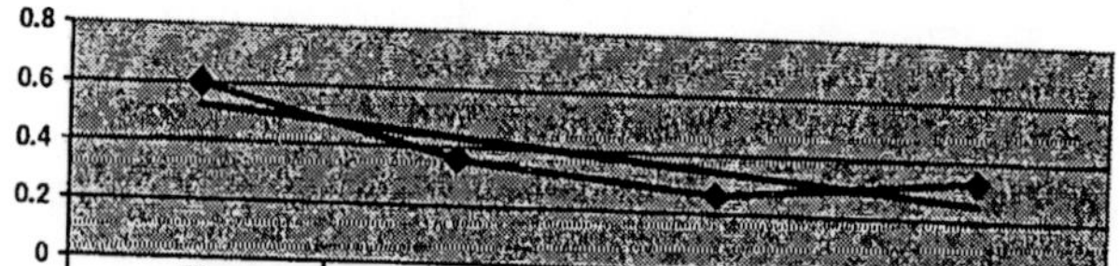

	71	81	91	97
Death Rate-Overall	0.6	0.36	0.26	0.33

Death Rate-Overall — Linear (Death Rate-Overall)

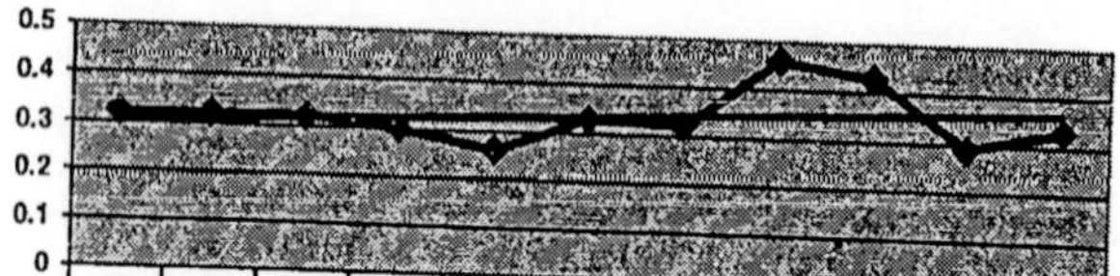

	87	88	89	90	91	92	93	94	95	96	97
Death Rate -Overall	0.32	0.32	0.32	0.3	0.26	0.33	0.32	0.46	0.43	0.29	0.33

Death Rate -Overall — Linear (Death Rate -Overall)

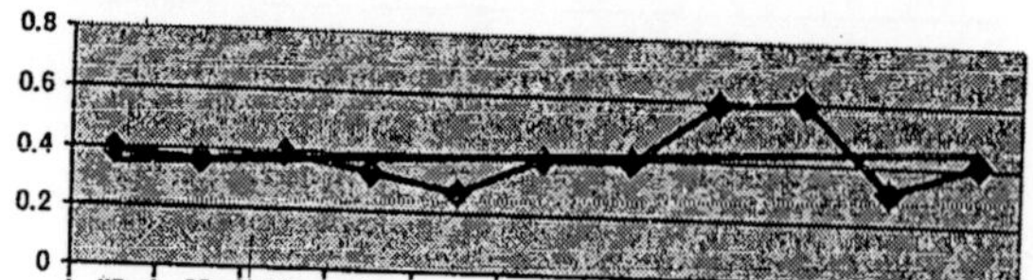

	87	88	89	90	91	92	93	94	95	96	97
Death Rate_Belowground	0.39	0.36	0.39	0.33	0.27	0.39	0.39	0.59	0.6	0.31	0.41

Death Rate_Belowground — Linear (Death Rate_Belowground)

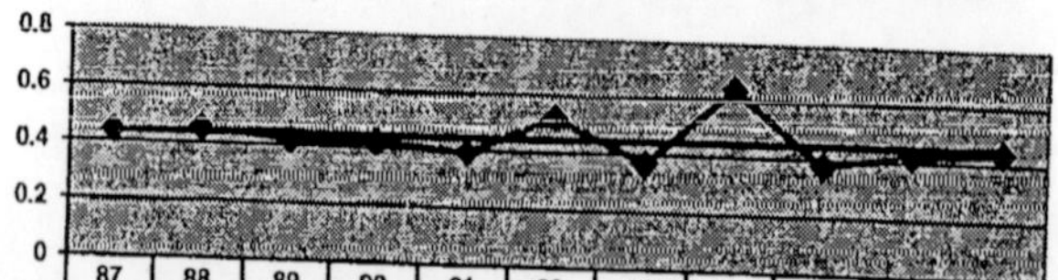

	87	88	89	90	91	92	93	94	95	96	97
Death Rate_Opencast	0.44	0.45	0.42	0.42	0.39	0.52	0.37	0.64	0.38	0.42	0.44

Death Rate_Opencast — Linear (Death Rate_Opencast)

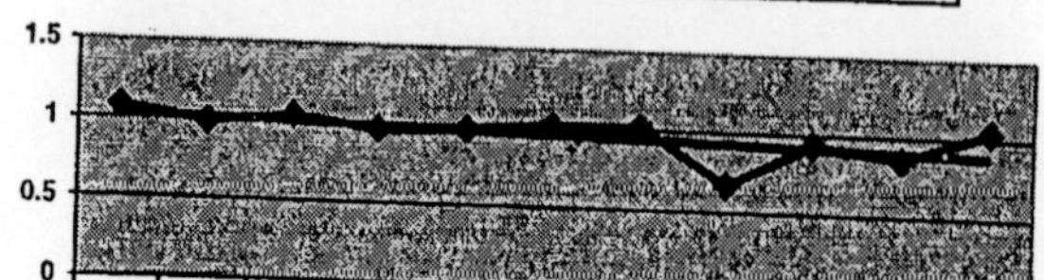

	87	88	89	90	91	92	93	94	95	96	97
Death-Rate_Aboveground	1.08	0.99	1.04	0.96	0.98	1.01	1.01	0.67	0.93	0.86	1.04

Death-Rate_Aboveground — Linear (Death-Rate_Aboveground)

H.3 Serious Injuries

Table 72.4 : Serious Injury Rate per 1000 persons employed in coal mines.

Year	*Number of accidents*	*Number of persons injured*	*SI rate per 1000 persons Belowground*	*SI rate per 1000 persons Opencast*	*SI rate per 1000 persons Aboveground*	*SI rate per 1000 persons Overall*
1951	1893	1931				5.69
1961	3515	3569				8.77
1971	1460	1542				4.03
1981	1151	1213				2.36
1982	1135	1184				2.23
1983	1169	1219				2.28
1984	1196	1236				2.24
1985	1007	1060				1.92
1986	1167	1199				2.20
1987	910	946	2.29	0.59	1.08	1.72
1988	771	815	1.97	0.58	0.99	1.51
1989	899	946	2.21	1.11	1.04	1.73
1990	893	983	2.38	1.16	0.96	1.79
1991	803	854	1.96	1.02	0.98	1.54
1992	810	894	2.14	0.79	1.01	1.62
1993	854	903	2.19	0.83	1.01	1.65
1994	717	775	2.08	0.86	0.67	1.48
1995	757	813	2.07	1.05	0.93	1.58
1996	677	723	1.83	1.10	0.86	1.43
1997	678	726	1.70	1.26	1.04	1.43

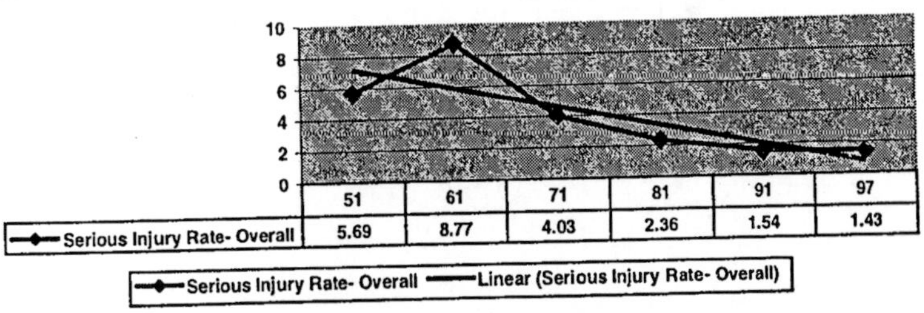

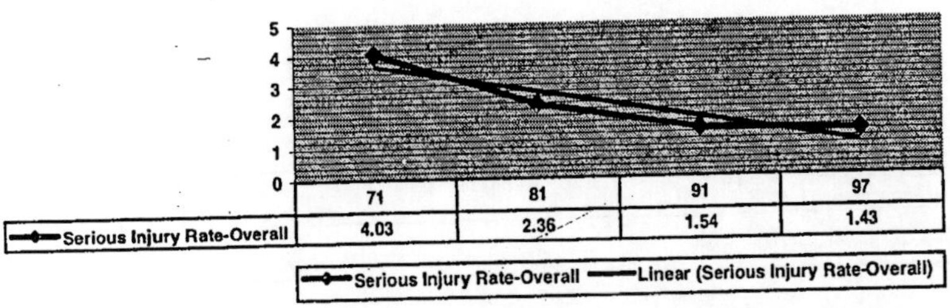

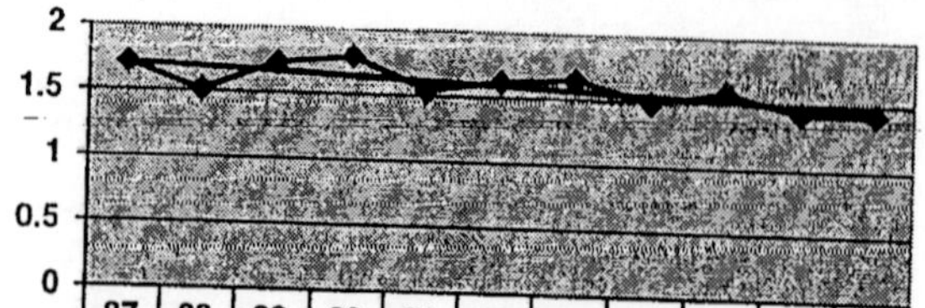

	87	88	89	90	91	92	93	94	95	96	97
Serious Injury Rate-Overall	1.72	1.51	1.73	1.79	1.54	1.62	1.65	1.48	1.58	1.43	1.43

Serious Injury Rate-Overall —— Linear (Serious Injury Rate-Overall)

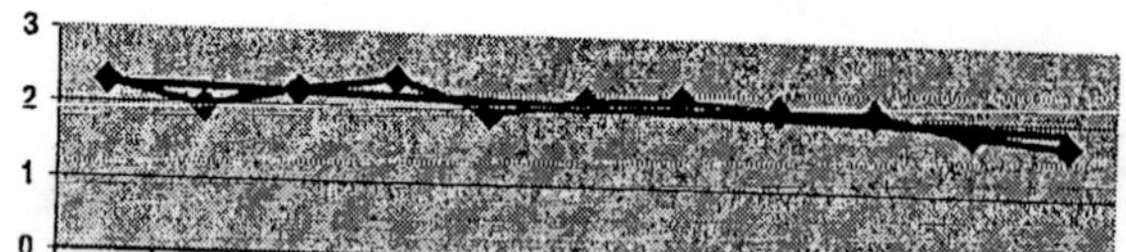

	87	88	89	90	91	92	93	94	95	96	97
Serious Injury Rate-Belowground	2.29	1.97	2.21	2.38	1.96	2.14	2.19	2.08	2.07	1.83	1.7

Serious Injury Rate- Belowground —— Linear (Serious Injury Rate- Belowground)

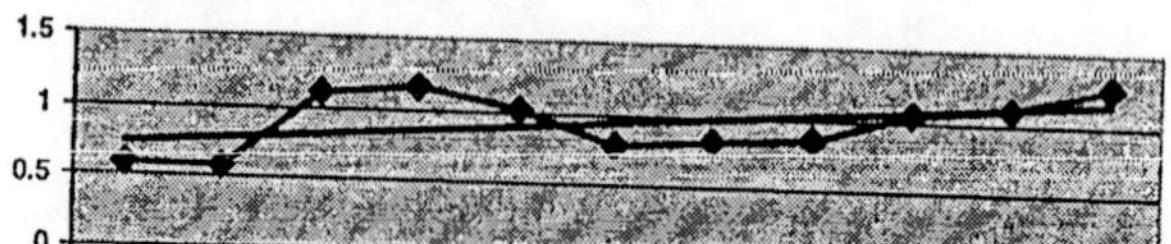

	87	88	89	90	91	92	93	94	95	96	97
Serious Injury Rate-Opencast	0.59	0.58	1.11	1.16	1.02	0.79	0.83	0.86	1.05	1.1	1.26

Serious Injury Rate-Opencast —— Linear (Serious Injury Rate-Opencast)

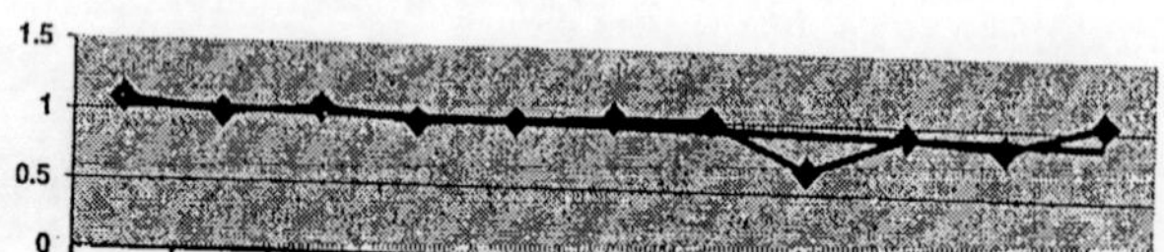

	87	88	89	90	91	92	93	94	95	96	97
Serious Injury Rate_Aboveground	1.08	0.99	1.04	0.96	0.98	1.01	1.01	0.67	0.93	0.86	1.04

Serious Injury Rate_Aboveground —— Linear (Serious Injury Rate_Aboveground)

H.4 Metalliferous Mines

Table 72.5 : Death Rate per 1000 persons employed in Metalliferous mines.

Year	Number of accidents	Number of persons killed	Death rate per 1000 persons Belowground	Death rate per 1000 persons Opencast	Death rate per 1000 persons Aboveground	Death rate per 1000 persons Overall
1951	81	106				0.53
1961	67	76				0.29
1971	65	73				0.29
1981	63	70				0.3
1982	71	78				0.31
1983	64	75				0.31
1984	62	75				0.32
1985	67	69				0.29
1986	70	86				0.35
1987	68	74	0.54	0.28	0.16	0.29
1988	67	75	0.52	0.35	0.19	0.32
1989	51	55	0.61	0.20	0.19	0.23
1990	63	76	0.56	0.41	0.25	0.35
1991	84	102	0.42	0.45	0.4	0.44
1992	68	78	0.52	0.39	0.25	0.33
1993	58	73	0.44	0.37	0.22	0.31
1994	61	86	1.4	0.32	0.23	0.38
1995	69	82	0.35	0.39	0.27	0.38
1996	63	73	0.42	0.54	0.19	0.41
1997	69	75	0.42	0.45	0.39	0.43

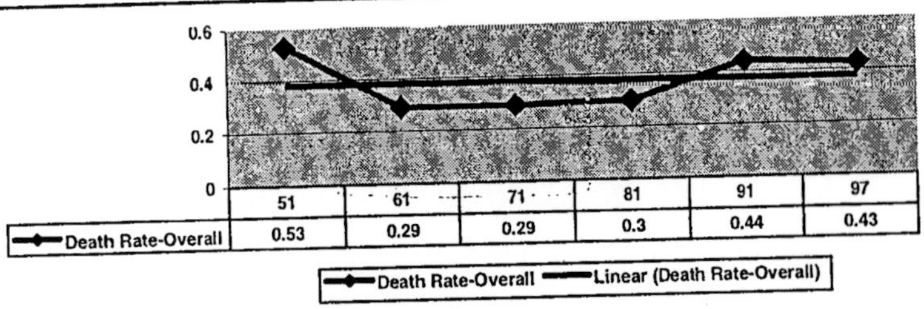

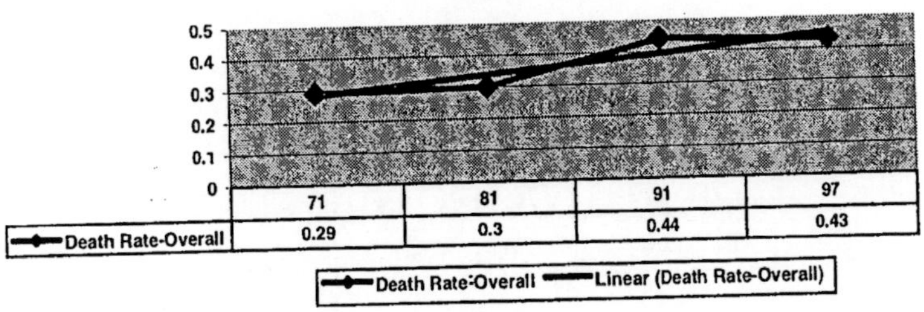

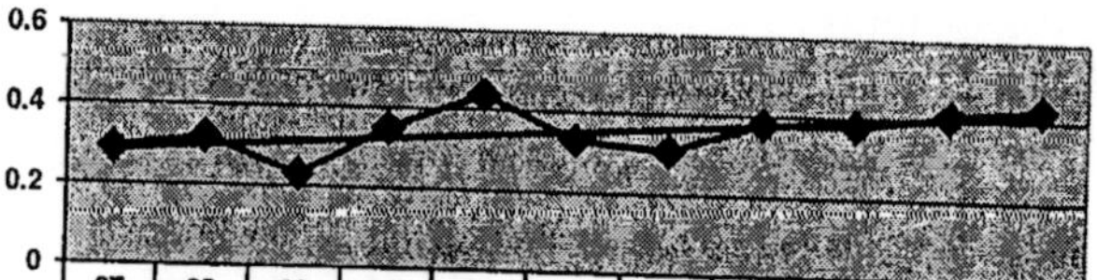

	87	88	89	90	01	92	93	94	95	96	97
Death Rate-Overall	0.29	0.32	0.23	0.35	0.44	0.33	0.31	0.38	0.38	0.41	0.43

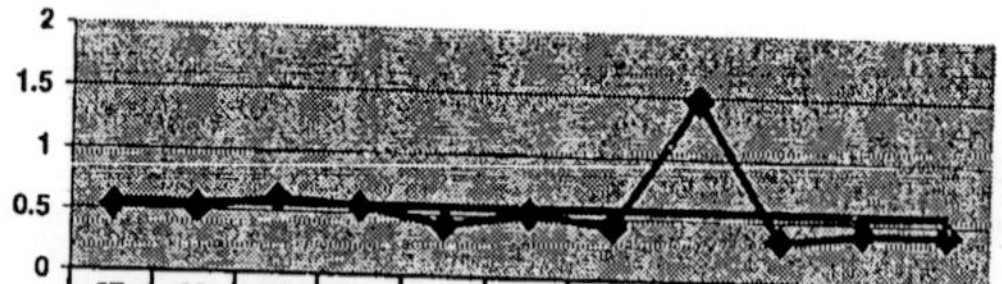

	87	88	89	90	91	92	93	94	95	96	97
Death Rate-Belowground	0.54	0.52	0.61	0.56	0.42	0.52	0.44	1.46	0.35	0.42	0.42

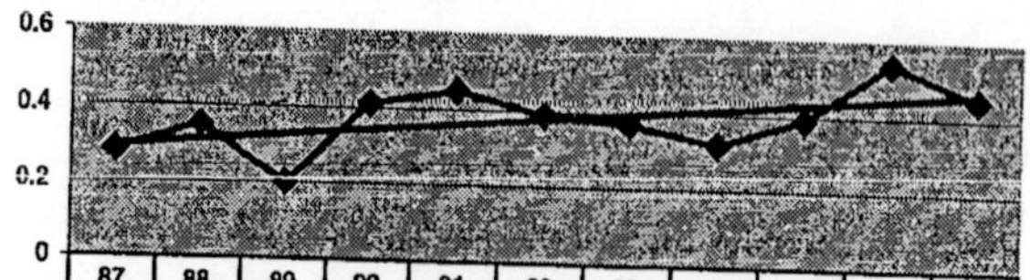

	87	88	89	90	91	92	93	94	95	96	97
Death Rate_Opencast	0.28	0.35	0.2	0.41	0.45	0.39	0.37	0.32	0.39	0.54	0.45

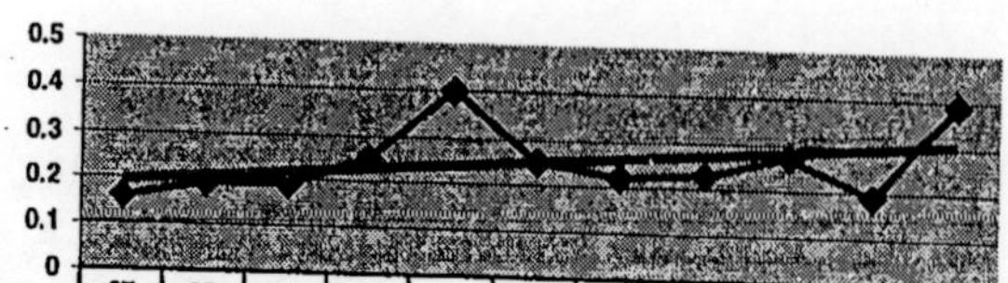

	87	88	89	90	91	92	93	94	95	96	97
Death Rate-Aboveground	0.16	0.19	0.19	0.25	0.4	0.25	0.22	0.23	0.27	0.19	0.39

H.4.2 Serious injuries

Table 72.6 : Serious Injury Rate per 1000 persons employed in metalliferous mines

Year	*Number of accidents*	*Number of persons injured*	*SI rate per 1000 persons Belowground*	*SI rate per 1000 persons Opencast*	*SI rate per 1000 persons Aboveground*	*SI rate per 1000 persons Overall*
1951	596	669				3.23
1961	1523	1618				6.04
1971	784	804				3.24
1981	808	833				3.53
1986	391	409	6.24	0.50	2.4	1.66
1987	402	414	5.88	0.55	2.24	6.24
1988	296	305	4.67	0.33	2.12	5.88
1989	249	255	4.86	0.39	1.37	4.67
1990	313	333	5.13	0.48	2.53	4.86
1991	269	290	5.71	0.40	1.79	5.13
1992	259	282	4.98	0.49	1.79	5.71
1993	286	301	6.06	0.49	1.79	4.98
1994	225	239	5.46	0.39	1.38	6.06
1995	250	260	3.93	0.6	1.96	5.46
1996	235	255	4.85	0.58	1.86	3.93
1997	223	239	5.06	0.34	1.94	4.85

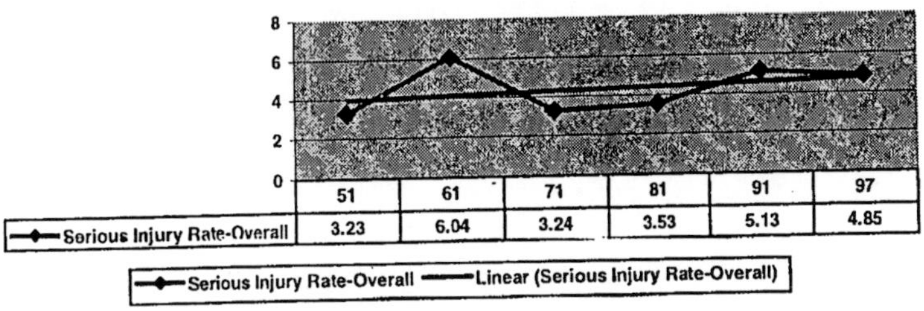

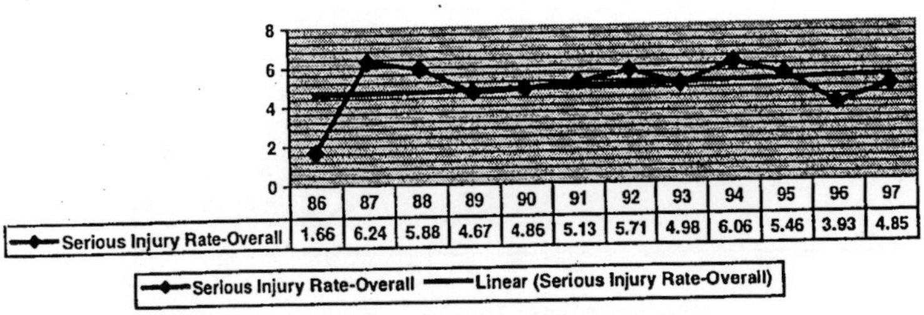

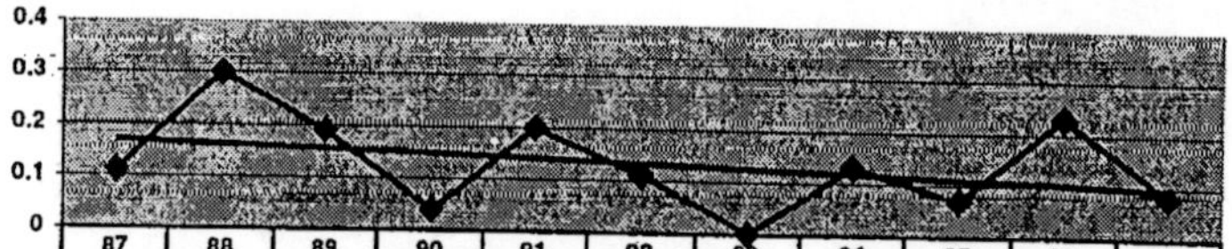

	87	88	89	90	91	92	93	94	95	96	97
France	0.11	0.3	0.19	0.04	0.2	0.11	0	0.13	0.07	0.23	0.08

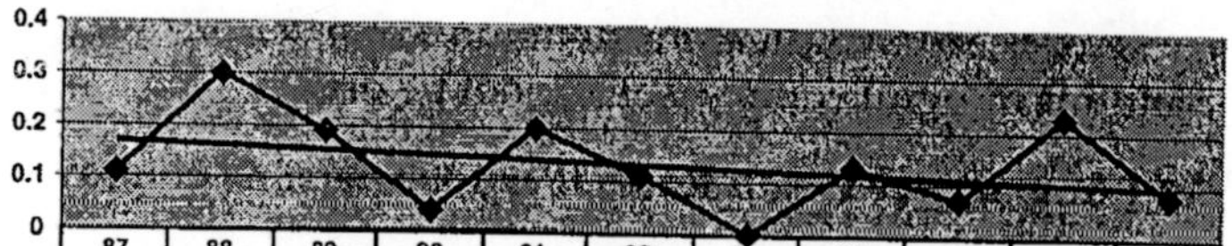

	87	88	89	90	91	92	93	94	95	96	97
France	0.11	0.3	0.19	0.04	0.2	0.11	0	0.13	0.07	0.23	0.08

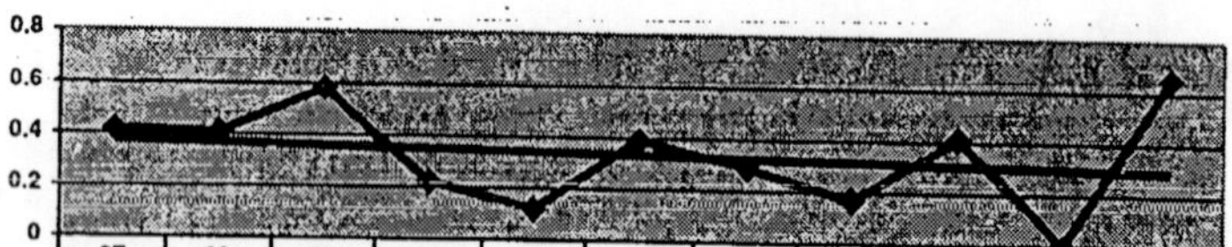

	87	88	89	90	91	92	93	94	95	96	97
Japan	0.42	0.41	0.58	0.22	0.12	0.39	0.29	0.17	0.41	0	0.65

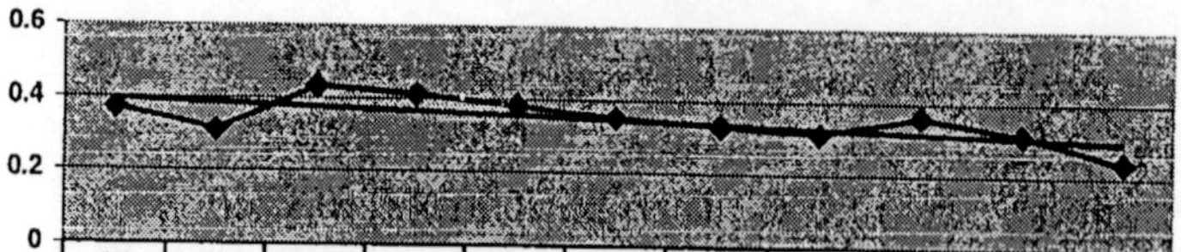

	87	88	89	90	91	92	93	94	95	96	97
USA	0.37	0.31	0.43	0.41	0.38	0.35	0.33	0.31	0.36	0.31	0.24

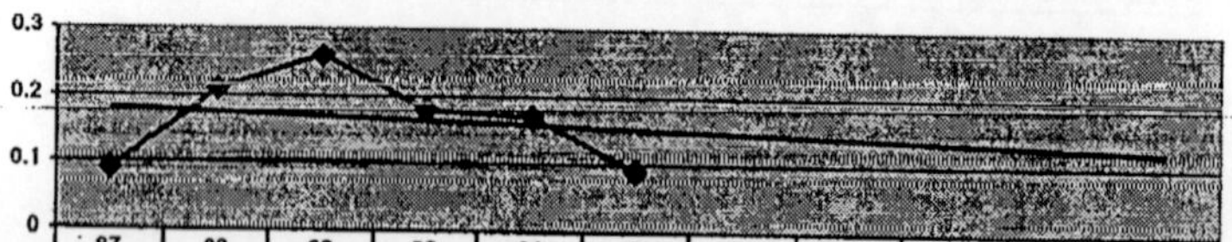

	87	88	89	90	91	92	93	94	95	96	97
UK	0.09	0.21	0.26	0.18	0.17	0.09					

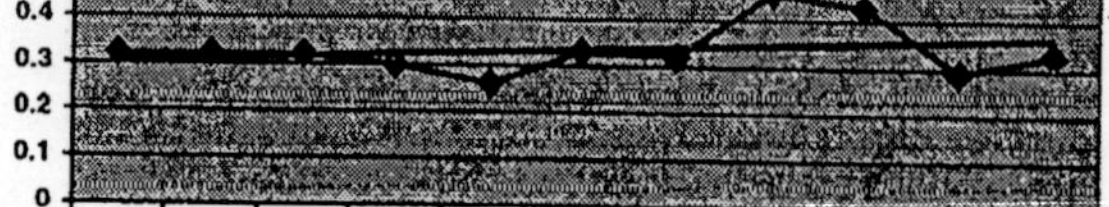

	87	88	89	90	91	92	93	94	95	96	97
Death Rate-Overall	0.32	0.32	0.32	0.3	0.26	0.33	0.32	0.46	0.43	0.29	0.33

I.2 Conclusions : International Comparison

World over efforts are being made to bring down the "Death Rate" in mines to a socially acceptable level. During the period 1987-97, **the comparison reveals that, the** "Death Rate" (per 1000 persons employed in the coal mines for a year) **show a declining trend in USA, France, UK, Germany, Japan, Czechoslovakia.** "Death Rate" in **Indian coal mines during the same period however shows a rising trend.**

I.3 Conclusions : Coal Industry

- Death rate/Serious Injury Rate per 1000 persons employed for a year in coal mines (proportionately equivalent to 3,00,000 manshifts worked/one million hours worked) has been considered as more true/representative figure for hazard potential in mines than the Death Rate/Serious Injury Rate per million tonnes of output of coal (which is more reflective of the production and productivity). The study and analysis of the figures reveal that
- Death Rate for the periods 1951-97 and 1971-97 show a declining trend. **However the Overall Death Rate for the period 1987-97 shows a rising trend.** As per the trends, Death Rate for 97 should have been 0.20 (1951-97) and 0.24 (1971-97) against the actual figure of 0.33 indicating thereby that the status of safety in coal mines has suffered and slipped during the period.
- During the period 1987-97, the Death Rates in the Belowground as well as Opencast workings have remained consistently higher than the Overall Death rate every year. **Further during this period both Death Rates also are showing a rising trend,** indicating thereby a highly disturbing and alarming situation.
- Serious Injury Rates for the periods 1951-97, 1971-97 and 1987-97 all show declining trends. Based on the trends, the rate of Serious Injuries in 1997 should have been 0.4 (1951-97) and 1(1971-97) against the actual rate of 1.43 **indicative of a positive slow down.**
- **Serious Injury Rate** for the period 1987-97 in respect of **Opencast Workings shows a rising trend.**
- During the period 1987-97, **in Opencast Workings**-Death Rate was consistently higher than the Overall Rate every year coupled with a rising trend in the Serious Injury Rate **points to inherent shortcomings and contradictions in the operations therein.**

1.4 Conclusions : Non- Coal Industry

Death rate/Serious Injury Rate per 1000 persons employed for a year in coal mines (proportionately equivalent to 3,00,000 manshifts worked/one million hours worked) has been considered as more true/representative figure for hazard potential in mines than the Death Rate/Serious Injury Rate per million tonnes of output of coal (which is more reflective of the production and productivity). The study and analysis of the figures reveal that :

- Whereas the trend of Death Rate for the period 1951-97 is nearly horizontal, the trends of Death Rates for the periods 1971-97 and 1987-97 are rising indicating thereby that the **Status of Safety in Metalliferous Mines has deteriorated in the past 3 decades- a truly alarming situation.**

- During the period 1987-97, **the trend of Death Rate in Opencast Workings and Aboveground are seen rising indicating inherent shortcomings in the practices, procedures etc in operations thereat.**
- Whereas the trend of Serious Injury Rate in Underground workings for the period 1986-97 is seen as declining gently, but the same is very high as compared to Serious Injury Rate either in opencast or aboveground.
- For the periods 1951-97, 1971-97 and 1987-97, **the trend of Serious Injury Rate is seen rising – a truly disturbing and alarming picture.**

J. CHALLENGES TO BE FACED BY MINERAL INDUSTRY AND NEED FOR A FRESH APPROACH.

J.1 "Philosophy"' or "Paradigm" of Safety in Indian Mines.

Thomas Kuhn, father of the term "Paradigm" stated that the scientific paradigm are "accepted examples of actual scientific practice, examples which include law, theory, application and instrumentation together- that apply provide models from which spring coherent traditions of scientific research". Joel Baker has asserted that "A paradigm is a set of rules and regulations (written or not) that: (1) establishes or defines boundaries of how people believe: and (2) tells how people how to behave inside boundaries in order to be successful. He further believed that the concept of paradigms could be applied outside the "hard science" too.

Mine Safety Management in Indian mines has its own "Paradigm" which had its birth/origin in the first Mines Act that came into force in the year 1902. Based on the experience gained as well as the desire of the State to protect the workers in the mines. The Mines Act was enlarged and reenacted in 1923 and 1952 without altering the basic concepts. The twin concepts were that the operations in the mines need to be controlled and directed by the State to achieve the best possible results.

The control on the conduct of operations in mines was exercised by

- prescribing that the operations in the mines were supervised by competent persons duly certified by the competent authority i.e. Chief Inspector of Miners/Director General of Mines Safety. To achieve these objectives the scale of employment and duties and responsibilities of nearly all supervisory persons likely to be employed in the mines were laid down and prescribed,
- detailing out the hazards likely to be met in the mining operations and the remedial actions to be taken to eliminate or to contain them.
- grading the contraventions (and their likely impact on the safety, health and welfare of the work persons employed in the mine). The procedure for enforcing and prescribing the penalties for the various types of contraventions were also laid down.

The directions to the mining operations in the mines were provided by

- laying down/prescribing the minimum limits in a normal and common mining operations: and
- requiring the management of the mine to obtain prior approval/permission/ relaxation/exemption from the Chief Inspector of Mines/Director General of Mines

Safety before commencement of certain operations or use of material/plant/machinery/equipment etc in the underground coal mine.

This approach led to the formulation of detailed and prescriptive rules, regulations etc to be enforced in the mines. This approach has inherent shortcomings and may be good only in static environment- a situation totally different from one which is likely to emerge in a rapidly growing industry with fresh induction of higher technology regularly to attain bigger/higher production/productivity levels than ever before. This approach suffers from the following defects:

- It is not always possible to develop prescriptive requirements to cover all hazards at all mines in all circumstances. Prescriptive legislation lead parties conducting and engaged in mine operations to assume mistakenly that "all risk will be adequately covered **by general** prescriptive legislative requirements."
- It contributes to apathy on the part of the mine workers and mine management. Apathy helps to perpetuate dangerous attitude to safety on the part of the managers (compliance with the provisions of the Mines Act and the subordinate legislation enacted under it, is **all that is needed**) and employees (**safety is a management issue)**
- Breaches of safety rules by mine worker (s) may reflect attempts to assert an alternative code of safety, a code arising out of the his/their own experiences.
- The workers **maynot "own" and implement safe operating procedures** unless they were involved in their design. The prescriptiveness of coal industry legislation limits the scope for workers and managers to design procedures suited to their circumstances, and in this way may contribute to apathy and the violation of safety rules.
- It does not permit the mine management **to import successful** safety practices from other parts of the world **in toto.** While hazards vary between mines, the approaches to safety can be common (i.e. delineating responsibilities, involving work place parties and integrating safety into the way work is done, ability to cross-pollinate good management practices from a range of disciplines).
- Approval mechanism in respect of materials/plants/machinery/equipment etc intended/required to be used in the coal mine can also can raise capital costs besides causing delay in it's deployment.
- **Statutory positions also perpetuate the view that safety is the responsibility of particular individuals, rather than developing a culture in which safety is everyone's responsibility.**
- **The involvement of DGMS (the Inspectorate) in mine safety decisions diminishes** the responsibilities of work persons and supervisors engaged at work sites/local management for safety. While the inspectorate's involvement is beneficial if it leads to an extra level of scrutiny, it is not conducive to safety if work place parties/local management come to rely on the assessment of inspectorate rather than rigorously assessing hazards themselves. Firstly an Inspector of Mines may only be able to inspect a mine a few times in a year. Secondly it is unrealistic to expect an Inspector to be familiar with all the aspects of a mine (such as skills of the work force and the

way the work is done at that mine) that bear on how work should be performed safely at that mine.

- Permissions/exemptions/relaxations/approvals by DGMS (the inspectorate) may also perpetuate a static approach of the management to safety by ensuring that **arrangements are safe when scrutinized/approved by the inspectorate** rather than a dynamic approach where by the arrangements are reviewed regularly to take account of factors that are changing at mines.
- Legislative provisions prescribing supervision by persons holding specific competency certificates and appointment of other persons in traditional cadres/ capacities/grades inhibited the introduction of team work, perpetuated traditional hierarchical models of work, impeded the flexible deployment of labour, and perpetuated demarcation of work.
- The responsibility of any contravention of the provision(s) of the Mines Act, rules, regulations etc. framed thereunder or an accident/dangerous occurrence/disaster, at present, rests on the Owner, Agent and the Manager of the mine. The only exception to this general rule occurs when the management is able to prove the negligence of such a worker and get him prosecuted within 3 months under the provisions of Sec 77 of Mines Act, 1952- an impossibility under our prevalent conditions.

J.2 Inquiries into accidents/disasters

A Court of Inquiry appointed to inquire into the circumstances attending to an accident/disaster holds a public inquiry and often make recommendations as well as assign responsibility for the said accident. While action is normally generated on the recommendations but such an **Inquiry rarely reach closure as the prosecutions under the provisions of Mines Act seldom come to a conclusion.**

The overall Fatality rate (based on employment) in coal industry shows a rising trend for the period 1987-97. For the period 1981-97, the trend is flat/horizontal with the value of 0.34. **Fatality rates in mechanized opencast mines have been consistently equal to or higher than the fatality rates in underground mine in the last 12-13 years. This trend is in contrast to the trends witnessed in other advanced countries where the fatality rates are a fraction of the fatality rates in underground mines.**

Some of the reasons for the present state of affairs are: -

- An Inspector of Mines though empowered to hold an Inquiry into an accident/ dangerous occurrence under Mines Act **doesn't have powers to hold a public inquiry.** At times he is not is not able to secure the presence of key persons or relevant documents. The inquiry as such maynot be complete and conclusive. Further as and when a prosecution is launched in a Court, the Inspector of Mines is required to prove his case before the learned court whereas the delinquents enjoy all the benefits of doubts/errors of omissions etc. The inbuilt delays in the deliverance of judgments in multi-tiered judicial system in such cases compounds the problem still further.
- Penalties likely to be imposed on successful conviction in the Court of law are insignificant in today's context. Low fines ranging from 0 to Rs. 5000/- and jail terms ranging from 0 to 2 years are just too small and meagre looking at the economic profits/loss sustained as result of contravention. **These penalties don't act as a**

deterrent against the commission of a contravention nor do they act as an incentive for any person or management not to infringe the provision of law. Much higher penalties have been prescribed and are in force under Factories Act, 1948 and Environment Protection Act, 1970.

The report of inquiry by an Inspector of Mines into any accident/dangerous occurrence becomes a public document only when submitted in a Court of law.

- As and when a prosecution is launched against any person of the management, the management directly and readily bears all expenses for the defence of the delinquent person. **Morally speaking, does it not amount to the acceptance of the fact that the said delinquent had contravened the law on behalf of and for the management?**
- **Any shortcoming or deficiency in the operations/procedure/process can be very easily identified provided the true cause of the accident is known.** Identification of the cause and development of its remedial measures are the first and most important findings of any inquiry into an accident/incident/disaster. **Fixation of responsibility can only be the spillover effect of such an inquiry.** However if the fixation of the responsibility assumes a role other/higher than the identification of the cause, it is likely to give rise to an impression of **witch- hunting. Under these conditions the evidence shall be invariably tempered with and moulded so as to avoid any personal harm to a living person.**
- At present there is no mechanism in place **for in-depth analysis of accidents (both at macro level and micro level) to define the occurrence of the accidents cause wise and duly related to the mine/area or a company.** All such analysis would bear fruits provided necessary solutions are developed, discussed and adopted at an appropriate tripartite forum at a fixed interval of time for implementation in the concerned mines. **Further there appears to be no mechanism for prompt sharing of the knowledge of all causes of all accidents occurring in the mines on industry basis so that similar accidents could be prevented elsewhere.**

J.3 Responsibility for accidents in Mines

Under the provision of prevalent Mines Act, **the Owner, Agent and the Manager of a mine all the responsibility for all contraventions of any provisions of the Act, rules, regulations etc. made there under.** The only exception to this general rule is when they are able to cite a worker specifically for the contravention and ensure his appearance in the Court within 3 months as provided under Sec. 77 of the Mines Act. This has created an unbridgeable divide between the Management and supervisors on one hand and the workers on the other.

J.4 National and Corporate Safety Policy

Worldwide, mining employees represent 1% of the work force but 8% of the fatal accidents happen to them and according to International Labour Office (ILO), although a reduction in mining accidents and disease has been recorded over the recent years, there is a large gap between the safest and least safe mines. **The ILO has also pointed out to evidence to support the premise that safe mines are more productive.**

No mission statement specifying the level of Safety to be achieved for any year (for that matter a period of time) has ever been issued at the National level or by any corporate body.

"Zero Accident Potential" is very often cited as the National or Corporate Safety Policy. ZAP is in fact an **"UTOPIAN"** goal that we wish to achieve. ZAP cannot be achieved in a single day in a single stroke. In order to reach this goal, one would have to traverse the road of gradual reduction in accidents (with consequent reduction in bodily injuries).

Targets for production, productivity, profitability etc (**except for desired Status of Safety in a mine/company**) are set and announced for every year. Achievement of such targets is not only commendable but is also highly acclaimed by all.

J.5 Mine Safety Legislation and Self-Regulation

It is difficult to illustrate an "Ideal Mines Act". This cannot be static document.

Mines Act should always remain a dynamic document and must keep on improving as the distilled and pure knowledge, gained out of inquiries into accidents in mines, pitfalls observed in the various mining procedures being conducted in the mines, technical/ administrative and management problems surfacing in the mine operations due to induction of/upgradation of technology in use in mines etc, becomes available.

J Davitt McAteer, Assistant Secretary for Mine Safety and Health, Government of United States stated

"... Success has not been easily achieved. The evolution of federal mine safety legislation is a story of series of laws passed in response to the tragic loss of life in mine disasters during the century. "

As materials are extracted, a mine that is safe on one day can be hazardous the next. Some dangerous conditions/situations may be seen and foreseen too. However with the induction of newer technology new hazards may be introduced in a mine. **Under these circumstances a universally valid prescription for best safety practice is all but impossible to draw up.**

Certain cornerstones are needed to underpin a safe environment. The fundamental cornerstones in protection are

- Systematic inspections and enforcement
- Education of mine workers
- Regular miner training and
- Compliance assistance

As a step for **"Self-Regulation" and minimum necessary adherence to the provisions of the Mines Act etc. Internal Safety organisation (ISO)** has been created and is functioning in each production company (large organised sector). It was also anticipated that the said organisation would be able to develop and enforce improved codes of practices in mines thereby improving the Status of Safety therein. ISO has not been able to deliver the expected results for the simple reason that the said organisation (including the officials employed therein) is **neither insulated nor protected from the main line functions of the Company i.e. the production process.**

Self–Regulation can be very good in case of large operations/big mines/large organised sectors where it is possible to take a long-term view and necessary resources are also available. **However, small mines have historically, presented higher risks of injury and death than large operations and, in this area, the State's role may need to go further than framing and enforcing safety legislation.**

J.6 Workers participation in Management of Safety in mines.

Institutions of Workmen's Inspector and Safety Committees have been created and are in place at most of the mines in the organised sector. However their role is mostly advisory in nature and the local mine management has been given more than enough accommodation to either act on the advice or ignore it completely. It may be borne in mind that invariably (in majority of cases) it is the workers who get directly affected due to the lack of safety at and in the work place. "Workers Representatives" AND "Labour Unions" have a positive role to play in it.

K. FRESH APPROACH

It will not be appropriate on my part to suggest a blue print for the "Fresh Approach". However it stands to coherent reason and the demands of the society that the issues raised above are debated openly amongst all concerned i.e. the Government Institutions involved with the Mineral Industry, managements of the mines, Worker's representative, Trade Unions, experts, knowledgeable persons etc to develop, recommend, adopt and implement the same to achieve acceptable level of safety in mines.

REFERENCES

1. Dasgupta G., et al : Report on "Status of Safety in Mines"- Report of the Sub-Committee of the Consultative Committee of Parliament attached to Ministry of Labour.
2. DGMS : Standard note on DGMS as on 1.1.99- an internal publication
3. World Bank document : Review of Legal and Regulatory framework in Coal Industry.
4. V. Mahajan : "Enforcement of OSH Legislation in Mines- Recent Experiences and Initiatives. - A paper presented at 11th APOSHO Conference, New Delhi, 16-18 Nov 1995.
5. Simon, Rosa A and Steven I : "Redesigning the Safety, Health and Environmental functions for the year 2000
6. Petersen D : "The Role of Safety in 1990s- Our Choice"
7. Kuhn TS : "Profile: Reluctant Revolutionary". Scientific American. May 91. The Structure of Scientific Revolution. University of Chicago Press.
8. V Mahajan, et al : Management of Safety in Indian Mines- Need for a Paradigm Shift- A paper presented at 27th Conference of Safety in Mines Research Institutes, New Delhi
9. V Mahajan, et al : Strategies for developing Safety Awareness in Indian Mining Sector- A paper presented at 27th Conference of Safety in Mines Research Institutes, New Delhi
10. V Mahajan, et al : Indian Mine Safety- Problems and Challenges. - A paper presented at 27th Conference of safety in Mines Research Institutes, New Delhi
11. Protect and Survive : Article in "International Mining" Quarterly Review 3/99.